In Silico Toxicology
Principles and Applications

Issues in Toxicology

Series Editors:
Professor Diana Anderson, *University of Bradford, UK*
Dr Michael D Waters, *Integrated Laboratory Systems, Inc, N Carolina, USA*
Dr Timothy C Marrs, *Edentox Associates, Kent, UK*

Titles in the Series:
1: Hair in Toxicology: An Important Bio-Monitor
2: Male-mediated Developmental Toxicity
3: Cytochrome P450: Role in the Metabolism and Toxicity of Drugs and other Xenobiotics
4: Bile Acids: Toxicology and Bioactivity
5: The Comet Assay in Toxicology
6: Silver in Healthcare
7: *In Silico* Toxicology: Principles and Applications

How to obtain future titles on publication:
A standing order plan is available for this series. A standing order will bring delivery of each new volume immediately on publication.

For further information please contact:
Book Sales Department, Royal Society of Chemistry,
Thomas Graham House, Science Park, Milton Road, Cambridge,
CB4 0WF, UK
Telephone: +44 (0)1223 420066, Fax: +44 (0)1223 420247, Email: books@rsc.org
Visit our website at http://www.rsc.org/Shop/Books/

In Silico Toxicology
Principles and Applications

Edited by

Mark T. D. Cronin and Judith C. Madden
Liverpool John Moores University, Liverpool, UK

RSCPublishing

Issues in Toxicology No.7

ISBN: 978-1-84973-004-4
ISSN: 1757-7179

A catalogue record for this book is available from the British Library

Published by The Royal Society of Chemistry,
Thomas Graham House, Science Park, Milton Road,
Cambridge CB4 0WF, UK

Registered Charity Number 207890

For further information see our web site at www.rsc.org

From JCM
To JGA and MMA

Preface

Over recent years, there have been great advances in the use of *in silico* tools to predict toxicity. Significant drivers for this include legislative demands for more information on chemicals to ensure their safe use and minimise toxicity to both human health and the environment, combined with an emphasis on the necessity to reduce and replace animal testing wherever possible.

The aim of this book was to provide a single 'from start to finish' guide to cover all aspects of developing and using *in silico* models in predictive toxicology. Hence the book covers all aspects of modelling from data collation and assessment of quality, to generating and interpreting physico-chemical descriptors, selecting the most appropriate statistical methods for analysis, consideration of applicability domain of the models and validation. Factors that may modulate toxicity such as external and internal exposure are presented and the use of expert systems, grouping and read-across approaches is discussed. *In silico* toxicology is becoming less of an academic exercise and more of a practical approach to filling the gaps in our knowledge concerning the toxicity of chemicals in use. The practical applications of these techniques in risk assessment are also considered in terms of their use in overall weight-of-evidence approaches and integrated testing strategies.

This book is intended to be a useful resource for those with experience in modelling who require more detail of the individual steps within the process, but should also provide the fundamental background information for those less experienced in *in silico* toxicology who wish to investigate the use of the techniques. Each chapter provides the reader with useful information on the individual aspects of the modelling approaches and relevant literature references are provided for those seeking more detail.

During the time of writing this book there are growing signs of change in toxicology and attitudes towards alternatives. This has been stimulated as much by the need to change in response to legislation as the vision to

Issues in Toxicology No.7
In Silico Toxicology: Principles and Applications
Edited by Mark T. D. Cronin and Judith C. Madden
© The Royal Society of Chemistry 2010
Published by the Royal Society of Chemistry, www.rsc.org

incorporate the new technologies emerging in molecular biology, high throughput screening and toxicological pathways. The National Academy of Science's report, *Toxicity Testing in the 21st Century*, puts that vision into words, encapsulating the deficiencies of the current paradigm and the possibilities for a future without animal testing, but with real relevance for the effects of long term and low exposure of chemicals to humans. *In silico* techniques are a part of this process, whether it is in providing direct predictions of toxicity, rationalising the results, grouping chemicals with relevance to pathways or assessing the exposure to a chemical through toxicokinetic considerations. This book represents the state-of-the-art of *in silico* toxicology. We hope it is obvious to the reader that we have as many (if not more) tools and techniques as the toxicological data will support. What is needed in the near future is an illustration of how to use these *in silico* models. This includes integrating results and methods together to obtain greater certainty. Methods need to be developed to demonstrate the power of computational workflows and how they can provide estimates of the confidence in a prediction. *In silico* toxicology has an important role to play in toxicity testing in the 21st century; it must stand up and show the way to predicting effects.

Mark Cronin and Judith Madden

Contents

Issues in Toxicology No.7
In Silico Toxicology: Principles and Applications
Edited by Mark T. D. Cronin and Judith C. Madden
© The Royal Society of Chemistry 2010
Published by the Royal Society of Chemistry, www.rsc.org

CHAPTER 1

In Silico *Toxicology—*
An Introduction

M. T. D. CRONIN AND J. C. MADDEN

School of Pharmacy and Chemistry, Liverpool John Moores University, Byrom Street, Liverpool L3 3AF, UK

1.1 Introduction

Chemistry is a vital part of everyday life. In order for our interactions with chemicals to be safe, we must understand their properties. Traditional methods to determine the safety of chemicals are centred around toxicological assessment and testing, often using animals. There is, however, great interest and a need to develop alternatives to the traditional testing regime. Given the breadth and complexity of toxicological endpoints it is likely that, to ensure the safety of all chemicals, a variety of techniques will be required. This will require a paradigm shift in thinking, both in terms of acceptance of alternatives and the recognition that these alternatives will seldom be 'one for one' replacements.

In silico toxicology is viewed as one of the alternatives to animal testing. It is a broad term that is taken, in this book, to indicate a variety of computational techniques which relate the structure of a chemical to its toxicity or fate. The purpose of *in silico* toxicology is to provide techniques to retrieve relevant data and/or make predictions regarding the fate and effects of chemicals. In this sense the term '*in silico*' is used in the same manner as *in vitro* and *in vivo*, with '*silico*' relating to the computational nature of the work. There are, obviously, many advantages to *in silico* techniques, including their cost-effectiveness, speed compared with traditional testing, and reduction in animal use.

Issues in Toxicology No.7
In Silico Toxicology: Principles and Applications
Edited by Mark T. D. Cronin and Judith C. Madden
© The Royal Society of Chemistry 2010
Published by the Royal Society of Chemistry, www.rsc.org

The science of *in silico* toxicology encompasses many techniques. These include:

- Use of existing data. If suitable data exist for a compound, there should be no requirement to initiate a new test or make a new prediction (unless prediction is for the purposes of model validation). If data are lacking for the chemical of interest, then other data can be used to develop (and subsequently evaluate) a new predictive model. Data sources include the ever increasing number of available databases as well as the open scientific literature. In addition, those working in industry may be able to utilise their own in-house data. More details on the retrieval and use of existing data are given in Chapter 3.
- Structure–activity relationships (SARs) are qualitative and can be used to demonstrate that a fragment of a molecule or a sub-structural feature is associated with a particular event. SARs become particularly powerful if they are formalised into structural alerts. A structural alert can be used to associate a particular toxicity endpoint with a specific molecular fragment such that, if the fragment is present in a new molecule, that molecule may elicit the same toxicity. The use of SARs and structural fragments is discussed in more detail in Chapters 8, 13, 16 and 19.
- There is a strong theme in this book towards forming groups of similar molecules. These groupings are also termed chemical categories. There are a number of approaches to 'categorise' a molecule including mechanistic profilers (structural alerts) and chemical similarity. Once a robust group of structures has been formed, it can be populated with toxicity data for those members of the group where experimental measurements are available. This allows for a read-across approach to be used to predict the toxicity of those members of the group for which no data are available. Various strategies for category formation and read across are discussed in Chapters 13–17.
- Quantitative structure–activity relationships (QSARs) provide a statistical relationship between the effects (toxicity and fate) of a chemical and its physico-chemical properties and structural characteristics. Linear regression analysis is often used but a variety of other multivariate statistical techniques are also used. The generation and use of QSARs are discussed in many chapters in this book.
- Expert systems (in the sense of *in silico* toxicology used in this book) are formalised and computerised software packages intended to make the use of SARs and QSARs easier. They usually provide an interface to enter a molecular structure and a suitable means of displaying the prediction (*i.e.* the result), in certain cases other supporting information is also given. Expert systems may be distributed on a commercial basis, although some are freely available. Increasingly they are integratable with other software. These types of software are described in more detail in Chapters 16, 17 and 19.
- Some *in silico* models can be derived to extrapolate the toxic effect measured in one species to predict toxicity in another species, thus reducing the requirement for further testing. This is discussed in more detail in Chapter 18.

- Models for other effects are increasingly becoming included under the remit of *in silico* toxicology. For example, a number of pieces of software can be used to estimate the likely exposure of an organism to a toxicant. Models exist for both external exposure (*i.e.* determining the amount present in the environment) and internal exposure (*i.e.* the amount taken up and distributed within an organism). These will not, themselves predict toxicity, but provide useful supplementary information for the overall risk assessment process. External and internal exposure scenarios are described in Chapters 20 and 21 respectively.

A variety of other tools and software also included within *in silico* toxicology are considered within this book. These include numerous applications for modelling the properties of chemicals (Chapters 5–8) defining the applicability domains of models (Chapter 12) and assisting with weight-of-evidence predictions (Chapter 22).

A note on terminology: it is increasingly common to see the use of the abbreviation (Q)SAR, indicating both SAR and QSAR. Where possible, the term (Q)SAR has been applied in this book and is intended to apply to both. A broad description of the use of alternatives to animal testing is provided by Hester and Harrison.[1] Alternatives can be generalised into *in vitro* tests and *in silico* approaches. In addition to activities such as optimising testing and reducing harm, these approaches form a framework within the 3Rs philosophy (Replacement, Refinement and Reduction) to replace animal tests. *In vitro* toxicology includes the use of cellular systems, -omics, *etc.* and while it will support the *in silico* approaches discussed in this book and is often referred to, it is not the main focus here.

1.2 Factors that Have Impacted on *In Silico* Toxicology: the Current State-of-the-Art

Much has been written on the history of QSAR and related techniques, and it is not the purpose to review much of it in this chapter.[2,3] However, it is worth considering some of the key factors to reach the current state-of-the-art. The initial factors listed below (Sections 1.2.1–1.2.4) are the drivers for the search of *in silico* toxicology; the remaining factors (Sections 1.2.5–1.2.7) have assisted in progress to the current state-of-the-art.

1.2.1 Environmental and Human Health

There is a need to ensure that any species exposed to a chemical is at minimal or no risk. The chemical cocktail to which man and environmental species is exposed potentially comprises a vast number of different substances, with more being added to that list annually. Whilst it is never the intention to allow exposure to a hazardous chemical at a concentration that may harm (with the exception of pesticides and pharmaceuticals *etc.*), for the vast majority of chemicals there is little knowledge regarding their effects. Traditional testing of chemicals to assess toxicological properties has been heavily reliant on animal testing. Such tests require specialist facilities, are time-consuming and costly— even before animal welfare considerations are taken into account. It is widely acknowledged that to assess all the chemicals that are commonly used, animal testing will not solve the problem of ensuring that harmful chemicals are identified. Therefore, *in silico* alternatives that can make predictions from chemical structure alone have the potential to be very powerful tools.

1.2.2   Legislation

The manufacture and release of chemicals is carefully regulated across the world. The aim is to ensure safety through legislation. In addition, companies consider it a corporate responsibility to ensure the safety of their workers and consumers, and are highly aware of the possibility of litigation if they fail to do so. Each country and geographical region has a raft of legislation allowing the risk assessment and risk management of chemicals; it is beyond the scope of this volume to discuss this further and readers are referred to the excellent 'standard text' from van Leeuwen and Vermeire.[4]

No one single piece of legislation has promoted the use and development of *in silico* approaches. However, many have included it implicitly or explicitly. At the time of writing much work is being performed as a result of the European Union's Registration, Evaluation, Authorisation and restriction of Chemicals (REACH) Regulation, not to mention the Cosmetics Regulation. Elsewhere globally there has been similar legislation such as the Domestic Substances List in Canada and the Chemical Substances Control List in Japan.[5] Within all of these pieces of legislation there is the expectation that *in silico* toxicology will be applied. In each case new tools, methods and techniques have been developed as a direct response to the legislative requirements.

1.2.3   Commercial—Product Development

Linked to the needs to comply with chemicals legislation, businesses have long recognised the need to predict toxicological properties from structure. This provides many competitive advantages including possibilities to identify toxic compounds early on in the development pipeline, designing out toxicity in new molecules, registration of products with the use of fewer animals and hence at lower cost—in addition to the rationalisation of testing procedures. *In silico* approaches are broadly applied across many industries with a particular emphasis on the development of pharmaceuticals.[6] Therefore, there is a commercial need for reliable tools and approaches.

1.2.4   Societal

Not only does society desire safe chemicals, but it would prefer that animals were not used in the assessment of the properties of molecules. Therefore, there is commercial and consumer pressure to find and use alternatives. 'Computer models' are often cited as a method to replace animal tests. There is an opportunity here to gain public support of this area of science. In so doing, this may improve the perception of what science can provide to society. As often happens, public opinion moves ahead of the science and what might be realistically achievable. Everyone reading this book and considering using these methods should be encouraged to promote their use and realistic expectations of what they can provide.

1.2.5 Commercial—Software

Sections 1.2.1–1.2.3 reveal a potentially huge marketplace for *in silico* approaches to predict toxicity. This has been realised by a number of companies which have developed commercial software systems; an overview of many of these is provided in Chapter 19 and elsewhere in this book. These companies have helped to raise the profile of *in silico* toxicology and make it more than an academic exercise. Many of the software products are now considered 'standard' and, whilst they cannot be considered the 'finished product' (the models will always need updating and refining), the commercial impact on *in silico* toxicology should never be underestimated. A competitive marketplace is being developed; a number of commercial companies are offering free 'taster' products with the assumption that the user may want more from the company. While often incomplete, these free products can provide invaluable training and educational tools, and allow the novice to familiarise themselves with the concepts and practice of *in silico* toxicology.

1.2.6 Computational—Hardware and Software

Anyone reading this book will appreciate the exceptional advances in computational power, software and networking capabilities in their lifetime. Both authors fondly remember performing early computational chemistry calculations by building the classic 'straw' model of a molecule which was manipulated manually to obtain a 'visual' optimum geometry before guessing at reasonable bond lengths and angles. The computational input file for the three-dimensional (3-D) structure was then written by hand allowing for optimisation of a single bond that was often an overnight calculation. Happy days indeed! The rapid progression of technology has, however, meant that calculations can be performed at previously unthought of rates and for vast inventories—millions of compounds very rapidly. *In silico* methods for computational chemistry and toxicology have been quick to embrace the new, affordable computational power and also use the internet to compile, organise and distribute information. Many of the tools described in this book would not have been possible ten or even five years ago.

1.2.7 New and Better Solutions to Complex (Toxicological) Problems

It is true to say that *in silico* models may be better able to predict certain endpoints than others. This is a result of a number of factors, in particular, the

Table 1.1 Summary of the different types of *in silico* model and software for the prediction of chemistry and toxicology.

In silico tools and resources	Information retrieved or predicted	Description	Chapter in this book for further information
Databases	Records of toxicological data and information (existing data rather than predictions)	Usually searchable by chemical identifier, substructure or similarity	3, 4
Calculation of physico-chemical properties (descriptors) to be used in models—may provide implicit information	Various physico-chemical properties (descriptors)	Fundamental properties such as log P, solubility, pK_a, etc.	5
Calculation of chemical structure-based properties—descriptors to be used in model	2-D properties	Various software calculates properties from 2-D structure, e.g. molecular connectivities	6
	Molecular orbital properties	Quantum chemical calculations requiring a 3-D optimised structure	7
Calculation of toxicological effects—direct prediction of toxicity	Structural alert based expert systems	Fragment systems relating a particular sub-structure in a molecule to a toxic effect	16, 19
	Multivariate and/or quantitative expert systems	Systems automating the QSAR approach to allow for seamless prediction of toxicity	19
	Grouping or category approaches	Formation of groups of molecules on the basis of rationally defined similarity	14, 15, 16, 17

Estimates of external exposure	Complex models for various exposure scenarios	May calculate potential for exposure from a variety of routes, *e.g.* inhalation, dermal, *etc.*	20
Predictions for internal exposure	Absorption, distribution, metabolism and elimination characteristics of chemicals	Assessment of factors that may modulate overall toxic potential by considering extent of internal exposure at site of action	21
Validation of models	Applicability domain definition	Various statistical methods for defining the domain of a (Q)SAR	12
	QSAR model reporting format	Automated methods to compile the details of a (Q)SAR suitable for validation	11
Integration of models	Pipelines for integration of models	Automated methods to link together algorithms into a predictive workflow	24

data available for modelling, the extent of knowledge concerning the mechanisms involved and the complexity of the endpoint. The current real challenge to find alternatives in toxicology is for the chronic, low-dose, long-term effects to mammals (and understanding the effects on man in particular).

The most difficult endpoints to address with alternatives include developmental toxicity and repeated dose toxicity. For these endpoints it is necessary to move away from the direct replacement dogma that has driven *in vitro* toxicology. There are likely to be many alternatives proposed, but one way forward is capturing the chemistry (*i.e.* structural attributes) of compounds associated with particular pathways which lead to toxicological events; more discussion is given in Chapter 14. In a related manner, this is (partially) the vision of the report, *Toxicity Testing in the 21st Century: A Vision and A Strategy*, published by the US National Research Council of the National Academies,[7] which has subsequently spawned the Tox21 collaboration between the US National Institute of Health (NIH) institutes and the US Environmental Protection Agency (EPA).[8]

1.3   Types of *In Silico* Models

There are many different types of models for computational chemistry and *in silico* toxicology. Table 1.1 shows a broad distinction between the types of models referred to in this book. As illustrated in Table 1.1, only some of the available software can be used to predict toxicity directly. Other models provide the building blocks to (Q)SAR formation and their application. How these may all fit together is described in more detail in Chapter 2.

1.4   Uses of *In Silico* Models

As suggested in Section 1.2 and Table 1.1, there are likely to be many potential uses of *in silico* models and related software. For the purposes of this book, these uses can be summarised in terms of which aspects of the life of the chemical or product they relate to. These are summarised in Table 1.2—but remember that this is only a small proportion of the possibilities for use.

With regards to the assessment of the toxicological properties of molecules, there will be much greater emphasis in the future on so-called Integrated Testing Strategies (ITS).[9] These are initially being used in response to legislative requirements, but future use will hopefully extend their application to provide realistic frameworks to replace animal tests. These are described in more detail in Chapter 23 and require a variety of *in silico* building blocks, *e.g.* use of existing databases, (Q)SAR predictions, *etc.* A part of ITS will undoubtedly be the combination of predictions from different methods.[10] This will be an important process, and whilst models are usually considered in isolation, it must be remembered that greater confidence will be obtained if all possible information is combined together.

Table 1.2 Uses of *in silico* methods to predict toxicity and properties through the various stages of chemical development.

Chemical life stage	Use of in silico methods
Development of substance, *i.e.* pre-patent or registration	*In silico* screening to eliminate potentially toxic compounds prior to synthesis
	Designing out of harmful features during development
	Prediction of properties to assist in formulation, *e.g.* solubility, melting point
New chemical, *i.e.* at registration	Assessment of toxicological profile or confirmation of animal tests
	Prediction of properties, *e.g.* log *P* required for registration
Existing chemical	Databases may be used to retrieve information on existing chemicals
	Prediction of toxic effects to assist in prioritisation of existing chemicals
	Grouping of compounds within inventories to assist in read-across

1.5 How to Use this Book

This brief chapter is not intended to do anything more than set the scene for the reader. It is anticipated that readers will be users and/or developers of models. Users of models will find the book a useful place to find basic definitions, obtain in-depth details of models and assess the different approaches for the *in silico* prediction of toxicity. Both novice and experienced modellers will find Chapter 2 the ideal starting place for tackling the problems of creating a meaningful workflow for *in silico* toxicology development. It is intended that this book will lead developers through the process of identifying relevant data, characterisation of molecules, development of significant (statistical) relationships, the interpretation and documentation for regulatory purposes—and allow them to be able to place the models in the context of current knowledge.

1.6 Acknowledgements

Funding from the European Union 6th Framework Programme OSIRIS Integrated Project (GOCE-037017-OSIRIS) and the ReProTect Integrated Project (LSHB-CT-2004-503257) is gratefully acknowledged.

References

1. R. E. Hester and R. M. Harrison, *Alternatives to Animal Testing,* Royal Society of Chemistry, Cambridge, 2006.
2. M. T. D. Cronin, in *Predicting Chemical Toxicity and Fate*, ed. M. T. D. Cronin and D. J. Livingstone, CRC Press, Boca Raton, FL, 2004, pp. 3–13.

3. M. T. D. Cronin, in *Recent Advances in QSAR Studies: Methods and Applications*, ed. T. Puzyn, J. Leszczynski and M. T. D. Cronin, Springer, Dordrecht, The Netherlands, 2010, pp. 3–11.

4. C. J. van Leeuwen and T. G. Vermeire, *Risk Assessment of Chemicals: An Introduction,* Springer, Dordrecht, The Netherlands, 2007.

5. A. P. Worth, in *Recent Advances in QSAR Studies: Methods and Applications*, ed. T. Puzyn, J. Leszczynski and M. T. D. Cronin, Springer, Dordrecht, The Netherlands, 2010, pp. 367–382.

6. S. Boyer, *Altern. Lab. Anim.*, 2009, **37**, 467.

7. National Research Council of the National Academies (NRC), *Toxicity Testing in the 21st Century: A Vision and a Strategy*, National Academies Press, Washington DC, 2007.

8. F. S. Collins, G. M. Gray and J. R. Bucher, *Science*, 2008, **319**, 907.

9. C. Grindon, R. Combes, M. T. D. Cronin, D. W. Roberts and J. F. Garrod, *Altern. Lab. Anim.*, 2008, **36**(Suppl 1), 7.

10. C. M. Ellison, J. C. Madden, P. Judson and M. T. D. Cronin, *Mol. Inform.*, 2010, **29**, 97.

CHAPTER 2

Introduction to QSAR and Other In Silico *Methods to Predict Toxicity*

J. C. MADDEN

School of Pharmacy and Chemistry, Liverpool John Moores University, Byrom Street, Liverpool L3 3AF, UK

2.1 Introduction

Scientific analysis of the world around us is based on collecting information on what is known, or can be measured, and structuring the information to enable investigation of why systems behave the way in which they do. The information must be organised into a framework from which the relationships between the different aspects of complex systems can be determined and how they interact to produce the effects observed. Scientifically, this current knowledge is used to create theories or models from which predictions of unknown phenomena can be made. As scientific knowledge advances and more information becomes available, this can be used to test the theories or models that have been built.

In silico prediction of toxicity is based on such scientific principles. Initially, information is gathered from previous observations such as collation of measured toxicities of a group of chemicals. The properties of these chemicals are investigated to establish which features are responsible for their toxic activity, *i.e.* to determine the relationship between the specific molecular properties of the compound and its associated toxicity. This information can then be used to build models that can explain why a given compound does (or does not) elicit a

Issues in Toxicology No.7
In Silico Toxicology: Principles and Applications
Edited by Mark T. D. Cronin and Judith C. Madden
© The Royal Society of Chemistry 2010
Published by the Royal Society of Chemistry, www.rsc.org

particular effect and to predict the effects likely to be elicited by compounds for which the measured data are not available. As more data become available, the validity of the models can be tested and adjustments made as necessary. This is an iterative process allowing for continual refinement of predictive models.

In silico models use computational methods to predict activity of compounds based on knowledge of their chemical structure and selected properties. The properties themselves (*e.g.* physico-chemical or structural properties) may be computationally calculated using a range of software, or determined experimentally. *In silico* methods include (quantitative) structure–activity relationships [(Q)SARs], expert systems, grouping and read-across techniques. Each of these techniques is introduced below and expanded upon in subsequent chapters.

The basic tenet of quantitative structure–activity relationship (QSAR) analysis is that a given biological activity can be correlated with the physico-chemical properties of a compound using a quantitative mathematical relationship. QSAR (and other *in silico* methods to predict toxicity) are powerful and highly attractive tools in science. In part this is due to the diversity of the knowledge base the techniques employ, drawing together information from several disciplines. There are the chemical and physical properties of compounds to consider in addition to knowledge of the chemical, physiological or toxicological mechanisms that give rise to the effects.

There is an ever increasing array of techniques available from which to build models employing the advances in mathematical and computational sciences. Overall, the number of models available is rapidly expanding. All models are, by definition, surrogates for real systems. The usefulness of models is determined by the extent to which they offer a greater understanding of the system and enable predictions beyond current knowledge.

In silico methods in toxicology provide a framework for combining knowledge from several disciplines, offer mechanistic insight into chemical and biological processes, help to identify anomalous observations and promote savings in time, money and animal use where estimations of toxicity are required.

The philosophy of this book is to guide readers through the processes involved in generating and using *in silico* techniques to make predictions for toxicity. The aim of this chapter is to provide an overview of how the different sections of the book link together to enable such predictions to be made. This chapter serves as an overall introduction to QSAR and *in silico* techniques, outlining how to go about generating and using the models. This general overview is supplemented by subsequent chapters which provide a more detailed analysis of each individual step in the model building process. This chapter focuses on how to develop a QSAR for a toxicological endpoint. However, the methods described are equally applicable to developing QSARs for other endpoints such as predicting drug activity or pharmacokinetic/toxicokinetic properties. The use of other *in silico* techniques (*e.g.* category formation and read-across) are also introduced in this chapter.

2.1.1 Fundamentals of QSAR

The origins of QSAR date back to the 19th century when researchers including Cros,[1] Crum Brown and Fraser,[2] and Richardson[3] all identified a relationship between the activity of a compound and its chemical properties. However, it was the pioneering work of Hansch *et al.*[4] which is most often quoted as the beginning of modern QSAR. Equation 2.1[4] encapsulates the philosophy of QSAR, *i.e.* that a given biological activity can be correlated with the physico-chemical properties of a compound using a quantitative mathematical relationship.

$$\log 1/C = 4.08\pi - 2.14\pi^2 + 2.78\sigma + 3.36 \tag{2.1}$$

[statistics not given]
where:

C is the concentration to produce a herbicidal effect
π is an indicator of hydrophobicity
σ a measure of electronic effects within the molecule (discussed below).

Many elegant descriptions concerning the chronological development of this field have been published in the literature.[5,6] Readers are referred to these articles for a comprehensive review of QSAR from a historical perspective. Here the emphasis is on the current state-of-the-art, with comment on the future of QSAR and its potential utility in predictive toxicology.

QSAR and other *in silico* techniques have been widely used by the drug industry for many years. However, new European legislation such as the REACH Regulation[7] and the Cosmetics Directive[8] have led to increased interest in these methods. This is because the legislation promotes the use of alternatives to using laboratory animals to provide estimates of toxicity for risk assessment purposes. QSAR enables the relationship between activity (toxicity) and physico-chemical properties of molecules to be determined (Figure 2.1).

Although QSAR can be applied to diverse areas of science covering a range of endpoints (drug activity, pharmacokinetics/toxicokinetics, pesticide toxicity,

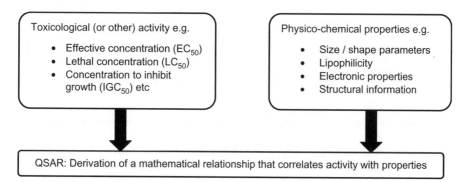

Figure 2.1 Overview of QSAR.

fragrance, *etc.*), the fundamental approach to developing, validating and using QSARs is relatively consistent. A flow diagram indicating the steps involved in this process is shown in Figure 2.2.

While this chapter is mostly concerned with the steps involved in generating a new *in silico* model, existing models such as published (Q)SARs or expert

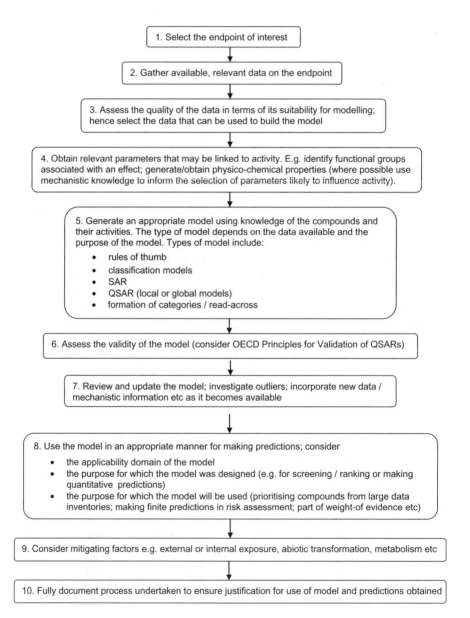

Figure 2.2   Steps involved in generating and using *in silico* predictions.

systems (see Chapter 19) can also provide useful information in terms of predicting activity, elucidating mechanisms or confirming the validity of newer models.

2.2 Building an *In Silico* Model

2.2.1 Step 1: Selecting the Endpoint to Model

Selecting the endpoint for which a model is to be built determines the nature of the data required to build the model. Although choosing the endpoint may appear trivial at first, it may not be a trivial issue in practice. In some cases the endpoint may be obvious, *e.g.* a mouse oral LD_{50} measured at a specific time point. However, consider building a model for toxicity to fish, here a 96-hour LD_{50} value measured in guppy is a different endpoint to a 96-hour LD_{50} in fathead minnow. The question is this: is the endpoint of interest (and hence the model to be built) toxicity to all fish species, an individual species or the most sensitive species? For *Daphnia* toxicity, should a model to be generated for LD_{50} at 24 hours or 48 hours, or can data from either endpoint be combined into a single endpoint for LD_{50}? In Ames testing for mutagenic effects, is the endpoint mutagenic activity in the absence or presence of metabolising enzymes? If enzymes are present, is this in the form of microsomes or the S9 fraction? Also consider the case of the human health endpoint skin sensitisation: to build a model for skin sensitisation there are several endpoints on which data may be gathered including the guinea pig maximisation test, the occluded patch (Beuhler test), Freund's complete adjuvant test, mouse ear swelling test, local lymph node assay, human repeat insult patch test, *etc.* The modeller needs to determine which is the most relevant endpoint for their purpose. In many cases, availability of data will influence the endpoint selected; however, it is important to have a clear idea from the outset which is the specific endpoint to be modelled and hence which data are relevant.

2.2.2 Step 2: Gathering the Data

Once the relevant endpoint has been identified, the next step is to gather available data for that endpoint. There are many ways in which data can be gathered for modelling purposes. Information can be retrieved based on searches for specific compounds of interest or for endpoints of interest. Sources of data include both in-house and publicly available resources. In-house data resources should be considered where available as these have several advantages:

- details of specific experimental protocols and associated metadata can be obtained and verified;
- the causes of anomalies in results are easier to investigate;
- the chemical space of in-house data sets is more likely to be representative of the chemical space of the compounds for which a prediction is sought.

Publicly available sources of data include:

- primary literature (for individual compounds and data sets);
- internet–based resources: ChemSpider;[9] ChemIDplus Advanced;[10] and the Cheminformatics and QSAR Society's web pages;[11]
- completed and ongoing efforts within EU projects such as CAESAR[12] and OSIRIS,[13] which are releasing data sets that have been highly curated;
- global (meta) portals such as ACToR[14] and the Organisation for Economic Co-operation and Development (OECD) eChem portal.[15]

These are only a few examples of the extensive resources available; further information on data resources is given in Chapter 3.

Once an endpoint has been selected and the search for data has begun, availability of data should guide the refinement of the search criteria. It may be pragmatic to expand the search criteria where data are scarce (*e.g.* to include more species within the same taxa, additional time points or related effects) or to limit the search criteria where data are plentiful (*e.g.* to a single species, study duration or effect). The more stringent the criteria for including data in the model, the more reliable the model should be as sources of variability are reduced.

As discussed in Chapter 3, there is a fine balance between making search criteria so broad as to introduce too much variability and too narrow to obtain a reasonably sized data set. Chapter 3 provides a detailed review of available resources, along with recommendations for using these and example case studies in obtaining data for different purposes. When starting to search for toxicity (or other) data, readers are referred to Chapter 3 for guidance in how to go about finding relevant data. Note, however, this is a rapidly evolving area and new sources of data with user-friendly interfaces are continually being developed.

When acquiring data it is essential to capture all the relevant information and ensure that the data storage system is reliable and portable. In many instances, a simple spreadsheet (e.g. Microsoft® Excel or Accelrys® Accord for Excel) will suffice and allows the flexibility to store required information that can be read directly into a range of statistical analysis or model development software packages.

It is essential to ensure that all transcriptions of data are correct as any errors introduced in the data gathering stage will later translate into inaccurate models being developed. It is crucial to ensure that the structure itself is correct and that the data correlate to that specific compound. The validity and accuracy of the SMILES string/CAS number should be checked rigorously before storing any associated data. Chapter 4 discusses, in more detail, issues to ensure accuracy in data collection.

Gathering data can potentially be a very long process. A pragmatic approach must therefore be taken to ensure a reasonable number of compounds and their associated data are collected for analysis without spending an excessive amount of time tracking down every last compound, as is often the case the law of diminishing returns applies!

2.2.3 Step 3: Assessing the Quality of the Data

Once the data have been acquired, the quality of the data and their suitability for the purpose of modelling needs to be assessed before their inclusion in the model building process can be justified.

Important issues in assessing the quality and/or suitability of data include:

- Can the data be unequivocally associated with the given compound? Potential problems here include:
  - incorrect nomenclature, CAS numbers or chemical structures;
  - the use of salts in place of parent compounds;
  - tautomeric and isomeric forms of the compound; and
  - questions concerning purity of the tested compound.
- Has the stated quantity or concentration of the compound been verified? Issues here include limited solubility in the test system or volatility of the compound resulting in the actual amount present being lower than the nominal (stated) concentration. Passive dosing systems or monitoring of concentration throughout the time course of the experiment may provide greater confidence in results.
- Are the data from a reliable source? For example was the result produced in a laboratory with acknowledged expertise in the area by appropriately trained staff performing work to standards commensurate with Good Laboratory Practice?

Formal scoring methods (*e.g.* Klimisch criteria)[16] for assessing data quality are available and serve as a useful guide as to whether or not the data are ideal, reasonable or should only be used with caution.

Chapter 4 deals with these issues and other aspects of assessing data quality in much greater detail. Only data whose inclusion can be justified should be incorporated into building or testing of models.

2.2.4 Step 4: Obtaining Parameters Potentially Related to the Activity

The purpose of building a model for use in *in silico* prediction of activity (toxicity) is to derive a relationship between the properties of the compounds and their biological effects. Thus, if the relevant properties of a new compound can be determined its biological effect can be predicted from these. (Note that the 'properties' of molecules may be variously referred to as 'parameters', 'variables' or 'descriptors'.)

Key to building a useful model is determining which properties of the compounds are linked causally to their effects. While it is possible to obtain properties from the literature, computationally generate or measure *de novo* thousands of properties of molecules, the majority will not have a causal relationship with effect. In some cases no correlation between the property and biological effect will be demonstrated; in other cases, a superficial relationship may be determined but this apparent relationship will not withstand rigorous statistical investigation. Statistical analysis and the problems of spurious correlations are discussed further in Chapters 9 and 10.

In building a model it is more useful to incorporate only those properties of the compounds for which a rationale for the relationship with effect can be justified. Notwithstanding, in certain cases, so little is understood about the mechanisms involved in the processes that it may be necessary to resort to the inclusion of less readily defensible molecular properties. There are a myriad of one-, two- and three-dimensional properties that can be obtained for individual chemicals. These encompass steric, electronic, hydrophobic and topological properties in addition to composite parameters produced from combinations of such properties, e.g. the electrotopological indices which unite electronic and topological effects.[17] Properties may relate to the whole molecule or, where structural analogues are investigated, substituent parameters may be employed. Table 2.1 provides an overview of some of the key molecular properties of compounds encountered in *in silico* models for predicting activity and a rationale for their inclusion in models.

Chemical structure encodes all possible information of the nature of a chemical and how it will interact from whether it will be a solid, liquid or gas at room temperature and pressure to how it may interact with biologically relevant molecules such as enzymes or DNA. Although all of this information exists within the chemical structure, our understanding of these systems and ability to interpret the information is less than perfect.

In certain cases it is the presence of a particular structural feature or functional group that influences or determines activity; hence identifying these features may serve as useful 'parameters' in model building. Simple physico-chemical properties such as the logarithm of the octanol/water partition coefficient (log P) or logarithm of the aqueous solubility (log Saq), *etc.* can be obtained from the literature, measured directly or estimated using a range of software. The ability of a molecule to elicit a biological effect may be correlated simply to such a property; for example, the narcotic effect of alcohols can be related to their hydrophobicity as it involves a non-specific interaction perturbing the cell membrane. Chapter 5 provides more detail on simple physico-chemical properties and their measurement.

Two-dimensional (2-D) properties relating to the chemical's structure (*i.e.* those determined from a graphical representation of the connectivity of atoms within molecules) are discussed in Chapter 6. These allow for shape characteristics, in terms of molecular topology, to be considered which can influence the ability to reach a target site or to fit to a receptor.

Chapter 7 provides an in-depth analysis of electronic properties derived from molecular orbital calculations. These are of particular significance for more reactive compounds as they provide a quantitative indication of the reactivity of chemicals and hence determine the likelihood and degree of interaction with biologically important macromolecules such as DNA and proteins. The ability of a molecule to elicit a biological effect may be related to relatively simple physico-chemical properties. However, many biological effects are the result of specific binding to individual receptors or biological targets (*e.g.* the binding of

Table 2.1   Example descriptors used in (Q)SAR and the rationale for their inclusion in models.

Descriptor	Definition of descriptor and rationale for inclusion
Functional groups/ structural alerts	Certain toxicities may be associated with specific structural features, *e.g.* presence of nitroso groups associated with carcinogenic potential or Michael type acceptors associated with skin sensitisation.
Similarity indices or similarity scores	Based on the concept that 'similar' molecules will produce 'similar' effects. There are many ways in which 'similarity' may be defined (*e.g.* similarity of size, shape, spatial distribution of key atoms/functional groups, reactive potential, *etc.*). What makes one molecule 'similar' to another is a debatable issue.
Indicator variables	These indicate the presence or absence (usually denoted by 1 or 0 respectively) of specific structural features (*e.g.* hydrogen bond donating/accepting groups, presence of particular functional group, *etc.*).
Hydrophobic/hydrophilic descriptors	Indicate solubility in aqueous and/or organic medium and relative partitioning between phases. Descriptors may correlate with the ability of compounds to cross biological membranes, accumulate within biophases or relate to specific hydrophobic interactions within receptor sites.
Log P	Logarithm of the partition coefficient, *i.e.* the ratio of concentrations of a compound between an organic and an aqueous phase (usually 1-octanol/water).
Log D	Logarithm of the distribution coefficient (or apparent partition coefficient). Derived from log P but taking into account the partitioning of ionised or associated species.
Log k'	Logarithm of the high performance liquid chromatography (HPLC) capacity factor. Retention on an HPLC column is correlated with log P. It can be determined more rapidly and can be applied to compounds for which standard methods of measuring log P are not appropriate.
Log S_{aq}	Logarithm of the aqueous solubility.
Lipole	Distribution of lipophilicity within a substituent or whole molecule.
Molecular lipophilicity potential (MLP)	Geometric distribution of lipophilicity within a molecule.
π	Substituent constant indicating the influence of individual substituents on overall partitioning behaviour: $\pi = \log P_{\text{(substituted derivative)}} - \log P_{\text{(parent)}}$.
Electronic descriptors	These represent diverse properties which are associated with many effects. Examples include the ability to cross biological membranes or bind to macromolecules (correlated with hydrogen bond donating/accepting ability or dipole interactions) and chemical reactivity associated with (covalent) binding that may elicit DNA damage or immune

Table 2.1   (*Continued*)

Descriptor	Definition of descriptor and rationale for inclusion
	responses (correlated with electronegativity, E_{HOMO}, E_{LUMO}, etc.).
HD/HA	Hydrogen bond donating/accepting ability. This may be represented as an indicator variable for ability or lack of ability to hydrogen bond or be quantified as to strength of bonding ability.
E_{HOMO}	Energy of the highest occupied molecular orbital (negative of the ionisation potential).
E_{LUMO}	Energy of the lowest unoccupied molecular orbital, (a measure of the ability to accept electrons).
Electronegativity (x)	Ability of an atom (or group) to attract electrons, associated with reactivity.
Electrophilicity (ω)	Models the ability of a molecule to accept electron density, associated with reactivity.
Superdelocalisability	A measure of reactivity determined from: $\sum$(charges of atoms in molecular oribtial) $\div$ (energy of each molecular orbital).
Atomic charge (q_n)	Charge associated with atom 'n'.
Dipole moment (δ)	Distribution of charge within a molecule.
σ	The Hammett substituent constant, indicating the electron directing effects of aromatic substituents (positive for electron attracting and negative for electron donating groups).
Steric descriptors	Associated with ability to reach target site (*e.g.* be absorbed across relevant biological membranes) and fit within specific receptors.
Molecular weight	Relative molecular mass indicating general size of the molecule.
Molecular volume	This may be calculated using the sum of van der Waals atomic volumes; indicates general size.
Molecular surface area/solvent accessible surface area	Computationally, a probe molecule can be 'rolled' over the surface of a molecule to determine the area that is accessible (to solvents or interacting molecules such as receptors).
K	The kappa index is a shape parameter based on the degree of branching of the molecular graph.
Sterimol (L_1, B_1-B_5)	Shape descriptors that indicate the length (L) of a substituent and its widths in different directions (B_1-B_5).
Es	The Taft steric constant indicates the size contribution of substituents on a parent molecule.
Topological descriptors	These are based on graph theory and relate to overall topology, dictated by the way in which atoms are connected to each other. Whilst many studies relate topology to molecular properties, their use remains controversial due to the difficulty in interpreting some of these parameters.
$^{n}\chi$	N^{th} order connectivity index. Zero order connectivity is obtained by counting the number of non-hydrogen links to each atom and taking the reciprocal square

Table 2.1 (*Continued*)

Descriptor	Definition of descriptor and rationale for inclusion
	root. Higher order is obtained by summing across different numbers of bonds.
$^{n}\chi^{v}$	Valence corrected connectivity indices are used to distinguish between heteroatoms.
3-D descriptors	Three-dimensional representation of molecules provides a more accurate description of molecular dimensionality. Such parameters are important in terms of receptor fit and binding.
Composite parameters	These represent combination effects and can provide additional information reflective of more than one feature.
Polar surface area; hydrophobic surface area	Dividing the surface area of a molecule into regions of polarity or hydrophobicity can provide useful information for example in terms of specific receptor binding interactions.
Electrotopological state indices (*S*)	A combination of electronic features and topological environment for given atoms.

drugs to enzymes inhibiting their activity or the binding of an oestrogen-mimic to the oestrogen receptor, associated with endocrine disrupting effects). These processes require specific interaction between a chemical and its target macromolecule. The correct orientation of the atoms and the distribution of their electronic and hydrophobic features in space can only be accurately determined by consideration of three-dimensional (3-D) descriptors. Three-dimensional interactions between chemicals and biological targets are considered further in Chapter 8.

Clearly, it is beyond the scope of a single chapter to describe all potential molecular properties of interest. Thus for further information on the use and generation of one-, two- and three-dimensional properties as well as comparisons of the software that may be employed to generate such descriptors, readers are referred to Chapters 5–8.

2.2.5 Step 5: Generating the Model

As Figure 2.2 shows, once the data on the endpoint have been gathered and assessed and the appropriate parameters generated, the next stage is to generate the model itself. This is the key step in the process and careful consideration needs to be given as to how to approach model generation.

There are many different modelling approaches and statistical methods of analysis from which to choose. Selection of the most appropriate method must take into account:

- the nature of the endpoint to be modelled and the accuracy of the data for the endpoint (*i.e.* accurately measured continuous data or categorical classifications based on positive/negative indicators or rank ordering);

- the amount of data available (small or large data sets);
- the information available on the compounds (*i.e.* which parameters can be reasonably obtained and what they can tell you about the molecule); and
- the purpose for which the model is being built (*i.e.* for general screening/ rank ordering purposes or for quantitative prediction).

The simplest models are those based on 'rules of thumb', which are broadly applicable and useful for general screening purposes. A notable example of this is Lipinski's 'rule-of-fives'[18] which has gained broad acceptance in drug development because of its simplicity and interpretability. The rule states that for any given compound: log $P > 5$; molecular weight > 500; number of hydrogen bond donors > 5; number of hydrogen bond acceptors > 10, are all factors associated with poor intestinal absorption. Similarly, cut-off values for certain properties can be useful for classifying compounds into broad categories. For example in toxicological assessment, compounds with log P values > 4.5 have been classified as having the potential to bioaccumulate.[19]

Where specific structural features can be identified as being associated with a particular activity, structure–activity relationship (SAR) models may be generated. For example the presence of nitroso groups has been associated with carcinogenicity,[20] glycol ethers have been associated with developmental toxicity,[21] and Ashby and Tenant indicated a number of structural alerts associated with carcinogenicity and mutagenicity endpoints.[22] The identification of structural alerts such as these can be formalised into knowledge-based prediction methods; these form the basis (or rule-base) of many 'expert systems' to predict toxicity. The methods employed by these systems, their use and application, are discussed further in Chapter 19.

Moving beyond these simple and intuitive relationships, more mathematical models can be derived providing quantitative estimates of potency, *i.e.* quantitative structure–activity relationships (QSARs). These models may be global (*i.e.* covering large data sets of diverse molecules) or may be more local models applicable to fewer compounds representing a narrower area of chemical space. Global models have the advantage of being generally applicable (although detailed information on interactions may be lost), whereas local models may provide more insight into the mechanisms involved in the process (at the cost of being applicable to fewer chemicals).

A QSAR may be developed using simple linear regression, *i.e.* where a single parameter is correlated with biological activity. More commonly, more than one parameter determines activity; hence multiple linear regression is necessary to generate the model.

There are many statistical methods that can be used to generate QSAR models; the most appropriate depends on the nature of the data set, the descriptors available and the ratio between the numbers of compounds and descriptors. Chapters 9 and 10 discuss in detail the most appropriate choice of statistical methods for analysing continuous and categorical data, respectively, and how to interpret the statistics generated for the models.

Recently there has been a great deal of interest in category formation and read-across methods as tools for *in silico* prediction of toxicity. These are based on the premise that 'similar' molecules will possess 'similar' activity. Similarity itself can be defined in many ways. For example compounds may be similar in terms of parameter values (*e.g.* log P, log S_{aq}), size, molecular shape, reactivity, presence of structural features, *etc.*

Similarity in one respect does not mean that they will be similar in others (*e.g.* compounds of the same molecular weight may have vastly different log P values). Hence it is important to determine which property is related to the activity and look for molecules that are similar in respect of that property. This allows for 'groups' of chemicals to be formed; the common feature of the group should be associated with the activity. If the activity of one (or preferably more) of the members of the group is known, then the activity of other members of the same group can be predicted. This can be done in either a qualitative or quantitative manner.

One useful way to group chemicals together is by mechanism of action. For example, chemicals acting as Michael acceptors are known to react with skin proteins resulting in skin sensitisation. This means that a category can be built for chemicals that may act as Michael acceptors. If several members of the category are known to be skin sensitisers, then the prediction could be made that other category members are also likely to be skin sensitisers (note this is a simplified example, further details of this example can be found in Enoch *et al.*[23]).

Chapter 13 discusses the role of the underlying mechanisms in toxicity, information which could be useful in forming toxicologically meaningful categories; this topic is developed further in Chapter 14.

An increasing number of tools are becoming available for category formation and read across. Chapter 15 provides a brief introduction to the principles of read-across and the theme is developed further in Chapter 16, which discusses the use of the OECD (Q)SAR Application Toolbox for making predictions using read-across. Chapter 17 considers other tools that may be used to investigate similarity of molecules which can be used to form categories.

Whilst there is a vast array of techniques available for generating models, the data available and the purpose for which the model is being developed should provide guidance on which is the most appropriate method to select for a given query.

2.3 Assessing the Validity of the Model

Given sufficient data and a high number of descriptors, it is possible to generate a large number of models; however, in order to be useful, the model needs to be valid. Whilst there has been a great deal of interest lately in what constitutes a 'valid' model, key principles of developing a good model were formally proposed more than 35 years ago. Unger and Hansch[24] proposed five criteria for selecting the 'best equation' to correlate activity, these were:

(i) selection of independent variables, *i.e.* those that are not inter-correlated with other variables;

(ii) justification of the choice of independent variables, *i.e.* these should be validated by an appropriate statistical procedure;

(iii) principle of parsimony, *i.e.* to use the simplest model;

(iv) number of terms, *i.e.* to have sufficient data points available per descriptor to avoid chance correlations. (The Topliss and Costello[25] rule recommends at least five data points to be incorporated per descriptor added.);

(v) qualitative model, *i.e.* one which is consistent with the known physical-organic and biomedicinal chemistry of the process involved (now generally referred to as 'mechanistically interpretable').

Moving forward to 2004, the OECD Principles for the Validation for Regulatory Purposes of (Q)SARs (which can be used also to guide assessment of the validity of other *in silico* models) were proposed.[26] These state that the model should be associated with:

(i) a defined endpoint;

(ii) an unambiguous algorithm;

(iii) a defined domain of applicability;

(iv) appropriate measures of goodness-of-fit, robustness and predictivity;

(v) a mechanistic interpretation, if possible.

Sections 2.2.1 and 2.2.5 have dealt with selecting an appropriate endpoint and method to generate the model (the algorithm). In terms of selecting an unambiguous algorithm, more transparent models such as multiple linear regression (MLR) are preferred to non-transparent models such as certain neural network methods.

The domain of applicability is used to determine the chemical space for which the model is applicable. This may be defined in terms of structural features of the compounds, physico-chemical properties or mechanism of action (or a combination of these). For example, if a model relating log P to toxicity was built entirely using compounds with log P values between 2 and 5, then it would be questionable to use the same model to predict the toxicity of a compound with a log P value of 8, as this is beyond the scope of the parameters used to generate the model (*i.e.* the compound falls outside of the applicability domain of the model). There are many ways in which the applicability domain of the model can be defined and these are presented in detail in Chapter 12.

In terms of assessing 'appropriate measures of goodness-of-fit, robustness and predictivity', the method by which the performance of a model is judged is dependent upon the type of model generated. For a classification model, a measure of concordance (*i.e.* the percentage of compounds placed into the correct category) gives a useful indication of overall model performance. However, in terms of toxicity prediction, sensitivity and specificity may be more important. Sensitivity is the proportion of compounds correctly classified as 'toxic' (or active) relative to the total actual number of toxic (active) compounds; specificity is the proportion of compounds correctly classified as 'non-toxic' (or inactive) compared to the total number of actual non-toxic (inactive) compounds. These are important in toxicity as potentially the consequences of a false negative prediction (*i.e.* predicting a toxic compound to be non-toxic) are much more serious than a false positive prediction (*i.e.* predicting a non-toxic compound to be toxic). In the case of the former, inadequate measures may be put in place to protect humans and the environment resulting in harm; in the latter case, excessive protective measures may be put in place which could be costly to industry.

For linear and multi-linear regression based models, the proportion of variability in the data accounted for by the model (r^2 values) are often used to determine model performance. Statistical measures should also include an assessment of the predictivity of the model, *i.e.* how well it performs in predicting compounds that were not present as part of the training set. Appropriate statistical measures on which to judge model performance are given in Chapters 9 and 10.

Considering point (v) of the Principles above, the model should be interpretable, *i.e.* the parameters included in the model should 'make sense' in terms of what is known about the process.

If the process is well understood, then it is generally easier to rationalise the presence of readily interpretable parameters; where little is known of the mechanisms of the process this may not be readily achievable. Chapter 11 discusses the validation of *in silico* models in more detail.

2.4   Reviewing and Updating the Model

Once the model has been generated and the validity assessed, this should not be considered as the end of the process in terms of using *in silico* models for toxicity prediction. The presence of outliers (*i.e.* compounds for which the model poorly predicts toxicity/activity) can provide useful information which can be used to revise the current model or generate new models. In some cases the outlier may have the potential to act *via* a different mechanism to other compounds used to generate the model. This can provide useful insight into mechanisms of action involved in toxicity.

An outlier may be due to an erroneous experimental measurement (in which case it should be omitted) or may provide information on the limitations of the experimental procedure. For example, highly volatile compounds may be lost from the experimental media before a true measure of their toxicity (at a given concentration) could be measured. Similarly a compound may be metabolised or abiotically transformed to another compound before it reaches its site of action. Such compounds may also appear as outliers as the parent compound is no longer present (or is present in much lower concentration) and the measured effect may in fact be due to metabolite or other transformation product.

Acquisition of new information of this type or of further data for a given endpoint should be used to re-develop models in an iterative manner. In this way models are continually updated and improved. Within a given industry, the chemical space of compounds of interest may gradually evolve over time; hence the applicability of a model developed in the 'old' chemical space may not be predictive for compounds in the 'new' chemical space. Model development should be a dynamic, ongoing concern utilising the most up-to-date information and techniques.

2.5   Using the Model

Many *in silico* models already exist and there is the potential to generate infinitely more. When electing to use a model whether obtained from the literature, formalised in an expert system or generated in-house, very careful consideration must be given to its appropriate use. Models developed have often been highly criticised because they fail to perform well when tested with a given set of compounds. However, such apparent 'failures' of models can often be traced back to inappropriate use.[27] To make a prediction for any compound it must be determined as to whether or not it falls within the applicability domain. Compounds that do not fall within the domain may be poorly predicted.

Models may be developed for very general purposes and be useful for these. For example, models to predict intestinal absorption of drug candidates can be used to screen large in-house virtual libraries in drug companies; however, these may not be appropriate for predicting absorption of pollutants that enter into the food chain. In this case more specific models giving a more accurate prediction for compounds of a different chemistry are needed.

The confidence needed for a prediction also depends on the use to which it will be put. A global model may be useful for prioritising testing of compounds from a large inventory, where the *in silico* model is used to select which of the compounds are more likely to be associated with a toxic effect. In Integrated Testing Strategies (ITS), which are discussed further in Chapter 23, *in silico* models may be used to select which of a group of compounds should be selected, based on their predicted activities, for further testing. For example, category formation can be used to assign chemicals to appropriate groups, one or more members from each group can then be selected for testing to help fill data gaps. Another use is in weight-of-evidence approaches in cases where several models exist for the same endpoint. Whilst use of each individual model may not result in a high confidence estimate, the use of several models in combination may increase confidence, particularly where results from all models are concordant. Hewitt *et al.* demonstrated an example of this for reproductive toxicity endpoints.[28] The weight-of-evidence concept is discussed further in Chapter 22.

2.6   Consideration of Mitigating Factors

Although a good *in silico* model may be able to provide an accurate prediction of inherent toxicity of a given compound, the ability of the compound to actually elicit such an effect can be significantly modulated by other factors.

Chapter 20 deals with external exposure potential. Compounds may not be released in sufficient concentration to elicit an effect, or they may be so reactive that they do not persist for long enough. Others may only be released in industrial settings where containment measures mean they do not come into contact with humans or the environment. In such cases the risk of a potential toxicant actually eliciting a toxic effect is negated.

Internal exposure considerations are of equal importance and are dealt with in Chapter 21. If a toxic compound is present in the environment, food chain or water supply, there is still the requirement for it to reach its biological site of action in sufficient concentration to elicit an effect. This is determined by the absorption, distribution, metabolism and excretion (ADME) properties of the compound (*i.e.* the factors that dictate its toxicokinetic profile). Many models exist to predict oral, dermal or inhalational uptake of compounds to determine whether or not they are likely to enter the body; there are also a small number of models to predict distribution.

Metabolism is the most studied of all ADME endpoints as it plays a central role in determining biological effects. Metabolism may render a potentially toxic parent compound non-toxic or may convert a non-toxic parent into a highly reactive or toxic metabolite, the toxicity of which will need additional consideration. Many programmes have been developed to predict metabolism and these are also presented in Chapter 21.

Excretion of the compound (and its metabolites) will ultimately remove it from the system and so end its potential to elicit toxicity directly. However, the potential for bioaccumulation must also be considered, as a weakly toxic compound that persists in the body or in the environment may build up to toxic levels.

Such modulating factors need to be taken into consideration in predicting the true toxic potential of any compound.

2.7 Documenting the Process

In the interests of transparency, a complete record should be kept of the entire model generation process. To aid the formalisation of this process, the Computational Toxicology Group at the European Commission's Joint Research Centre, Institute for Health and Consumer Protection[29] has developed a QSAR Model Reporting Format (QMRF) which is a harmonised template for summarising and reporting key information on (Q)SAR models. The information required reflects the OECD Validation Principles for (Q)SARs[26] and relates to:

(i) information on the endpoint: units of measurement, species used, age, test duration, experimental protocol, assessment of quality. *etc.*;

(ii) a description of the algorithm: type of model used (*e.g.* SAR, QSAR, expert system, neural network, *etc.*); the descriptors used in the model, how these were generated (name and version of software, literature reference *etc.*); how the relevant descriptors were selected;

(iii) a description of the applicability domain of the model: structural and physico-chemical space of the model;

(iv) a report on the statistics associated with the model (for training and validation sets): information on the training and test set compounds;

(v) information relating to the mechanistic interpretation of the model.

If outliers have been removed from the model, a justification for this should be included.

Analogous to QMRF, a QSAR Prediction Reporting Format (QPRF) is also available. This is a harmonised template for summarising and reporting substance-specific predictions generated by (Q)SAR models. These reporting formats are discussed in more detail in Chapter 11.

Accurate, harmonised documentation for QSARs, ensuring the transparency and repeatability of the methodology, should increase the uptake in use of high quality *in silico* models and greater acceptance of predictions, generated from such models, by regulatory authorities.

2.8 Pitfalls in Generating and Using QSAR Models

Previous sections have provided guidance on the methodology behind producing good quality *in silico* models and using these appropriately to generate predictions. However, there are many potential pitfalls in using the techniques. Many potential problems have been implicitly addressed in the guidance above such as:

- ensuring adequate good quality data (in the correct units) are used;
- the model has a defined applicability domain;
- the model uses interpretable descriptors; and
- the model is statistically sound.

However, there are many other potential pitfalls that (in)experienced modellers should be aware of. Readers are referred to recent literature which provides further information on how to ensure that *in silico* models are good quality and that common errors in modelling are avoided.[30–33]

2.9 Conclusions

The building of *in silico* models is an exciting field as it draws on so many areas of science, is dynamic and continually evolving. Looking to the future there are many opportunities for the further development of this field. This will include the use of more powerful computational techniques to gain insights into fundamental molecular and chemical processes, as well as novel methods for analysing and combining information from different sources to enable weight-of-evidence decisions and support ITS.

Following a few simple guidelines it is possible to organise the information on known compounds into a structured framework and use this to generate robust reliable models to predict properties of unknowns. In this way *in silico* models can be used to solve many problems in predictive toxicology, furthering knowledge and reducing the need to resort to animal testing.

This chapter provides a framework for the generation, validation and use of *in silico* models. It serves as an introduction to subsequent chapters which provide more detail on each of the individual steps.

2.10   Acknowledgements

Funding from the EU FP6 OSIRIS project (GOCE-CT-2007-037017) is gratefully acknowledged.

References

1. A. F. A. Cros, Ph.D. thesis, University of Strasbourg, 1863.
2. A. Crum Brown and T. R. Fraser, *Trans. Roy. Soc. Edinburgh*, 1868–1869, **25**, 151.
3. B. J. Richardson, *Med. Times Gaz.*, 1869, **2**, 703.
4. C. Hansch, P. P. Maloney, T. Fujita and R. M. Muir, *Nature*, 1962, **194**, 178.
5. T. W. Schultz, M. T. D. Cronin, J. D. Walker and A. O. Aptula, *J. Mol. Struct. (Theochem.)*, 2003, **622**, 1.
6. C. D. Selassie, S. B. Mekapati and R. P. Verma, *Curr. Topics. Med. Chem.*, 2002, **2**, 1357.
7. Regulation (EC) No. 1907/2006 of the European Parliament and of the Council of 18 December 2006 concerning the Registration, Evaluation, Authorisation and Restriction of Chemicals (REACH), establishing a European Chemicals Agency, amending Directive 1999/45/EC and repealing Council Regulation (EEC) No. 793/93 and Commission Regulation (EC) No. 1488/94 as well as Council Directive 76/769/EEC and Commission Directives 91/155/EEC, 93/67/EEC, 93/105/EC and 2000/21/EC, *Off. J. Eur. Union*, 2006, **L396**, 1–849.
8. Directive 2003/15/EC of the European Parliament and of the Council of 27 February 2003 amending Council Directive 76/768/EEC on the approximation of the laws of the Member States relating to cosmetic products, *Off. J. Eur. Union*, 2003, **L66**, 26–35.
9. ChemSpider, www.chemspider.com [accessed March 2010].
10. ChemIDplus Advanced, http://chem.sis.nlm.nih.gov/chemidplus/ [accessed March 2010].
11. The Cheminformatics and QSAR Society data sets, www.qsar.org/resource/datasets.htm [accessed March 2010].
12. Computer Assisted Evaluation of industrial chemical Substances According to Regulations (CAESAR) Project, www.caesar-project.eu [accessed March 2010].
13. Optimized Strategies for Risk Assessment of Industrial Chemicals through Integration of Non-Test and Test Information (OSIRIS) Project, www.osiris.ufz.de [accessed March 2010].

14. ACToR: Aggregated Computational Toxicology Resource, http://actor. epa.gov/actor/faces/ACToRHome.jsp [accessed March 2010].

15. ChemPortal, http://webnet3.oecd.org/echemportal/ [accessed March 2010].

16. H. J. Klimisch, E. Andreae and U. Tillmann, *Regul. Toxicol. Pharmacol.*, 1997, **25**, 1.

17. L. B. Kier and L. H. Hall, *J. Math. Chem.*, 1991, **7**, 229.

18. C. A. Lipinski, F. Lombardo, B. W. Dominy and P. J. Feeney, *Adv. Drug Deliv. Rev.*, 1997, **23**, 3.

19. T. Feijtel, P. Kloepper-Sams, K. den Haan, R. van Egmond, M. Comber, R. Heusel, P. Wierich, W. Ten Berge, A. Gard, W. de Wolf and H. Niessen, *Chemosphere*, 1997, **34**, 2337.

20. A. M. Helguera, M. N. D. S. Cordeiro, M. A. C. Perez, R. D. Combes and M. P. Gonzalez, *Toxicol. Appl. Pharmacol.*, 2008, **231**, 197.

21. E. de Jong, J. Louisse, M. Verwei, B. J. Blaauboer, J. J. M. van de Sandt, R. A. Woutersen, I. M. C. M. Rietjens and A. H. Piersma, *Toxicol. Sci.*, 2009, **110**, 117.

22. J. Ashby and R. W. Tennant, *Mutat. Res.*, 1988, **204**, 17.

23. S. J. Enoch, J. C. Madden and M. T. D. Cronin, *SAR QSAR Environ. Res.*, 2008, **19**, 555.

24. S. H. Unger and C. Hansch, *J. Med. Chem.*, 1973, **16**, 745.

25. J. G. Topliss and R. J. Costello, *J. Med Chem.*, 1972, **15**, 1068.

26. Organisation for Economic Co-operation and Development, *The Report from the Expert Group on (Quantitative) Structure–Activity Relationship ([Q]SARs) on the Principles for the Validation of (Q)SARs*, OECD, Paris, 2004, OECD Environment Health and Safety Publications Series on Testing and Assessment No. 49, ENV/JM/MONO(2004)24, www.olis. oecd.org/olis/2004doc.nsf/LinkTo/NT00009192/$FILE/JT00176183.pdf [accessed March 2010].

27. T. R. Stouch, J. R. Kenyon, S. R. Johnson, X. Q. Chen, A. Doweyko and Y. Li, *J. Comput. Aided Mol. Des.*, 2003, **17**, 83.

28. M. Hewitt, C. M. Ellison, S. J. Enoch, J. C. Madden and M. T. D. Cronin, *Reprod. Toxicol.*, 2010, **29**, in press.

29. European Commission Joint Research Centre, Institute for Health and Consumer Protection, Computational Toxicology Group, http://ecb.jrc. ec.europa.eu/qsar/ [accessed March 2010].

30. M. T. D. Cronin and T. W. Schultz, *J. Mol. Struct. (Theochem.)*, 2003, **622**, 1.

31. J. C. Dearden, M. T. D. Cronin and K. L. E. Kaiser, *SAR QSAR Environ. Res.*, 2009, **20**, 241.

32. E. Zvinavashe, J. M. Albertinka and I. M. C. M. Rietjens, *Chem. Res. Toxicol.*, 2008, **21**, 2229.

33. T. Scior, J. L. Medina-Franco, Q. T. Do, K. Martinez-Mayorga, J. A. Y. Rojas and P. Bernard, *Curr. Med. Chem.*, 2009, **16**, 4297.

CHAPTER 3

Finding the Data to Develop and Evaluate (Q)SARs and Populate Categories for Toxicity Prediction

M. T. D. CRONIN

School of Pharmacy and Chemistry, Liverpool John Moores University, Byrom Street, Liverpool L3 3AF, UK

3.1 Introduction

Data are at the heart of any modelling approach to make rational predictions. In the context of *in silico* approaches to predict toxicity, data are required to develop and evaluate models. The purpose of this chapter is to inform the reader how to compile toxicological data and information for the development of *in silico* models for toxicity prediction.

3.2 Which Data Can be Used for *In Silico* Modelling?

Two types of data are usually required to develop an *in silico* model for the effect of a chemical:

- Data relating to the description of chemical structure and/or properties. These are usually what the model itself is based on. They are not discussed *per se* in this chapter but are addressed fully in Chapters 5–7.

Issues in Toxicology No.7
In Silico Toxicology: Principles and Applications
Edited by Mark T. D. Cronin and Judith C. Madden
© The Royal Society of Chemistry 2010
Published by the Royal Society of Chemistry, www.rsc.org

- The effect or property of a molecule that the user of the model may wish to predict. This may include biological activities and also the types of properties which could be used to develop models (thus there are very well-established predictive approaches for octanol/water partition coefficients, aqueous solubility, melting point, *etc.*).

As a generalisation, *in silico* models can be developed to predict nearly any type endpoint based on reliable and consistent data. Since the focus of this book is the prediction of toxicity, other biological activities (*i.e.* pharmacology) will not be considered explicitly, although the techniques are equally applicable to these other areas. Thus the emphasis here is on prediction of toxicity endpoints and effects, as well as fate, distribution and physico-chemical properties.

Toxicity data can usually be thought of as some kind of measurable effect from a test. This may include observations (*e.g.* presence or absence of irritation after the application of a chemical to the skin) or a potency (*i.e.* concentration required to bring about an effect such as 50% lethality in a (sub-)population). It is always worthwhile for modellers to first consider what they are attempting to model in consultation with an expert toxicologist. Thus, a modeller may wish to predict carcinogenic potency but will need to consider in more detail what this means (*i.e.* which species, which specific endpoint, *etc.*)

The results of the tests, and data to be modelled, will dictate the modelling approaches that can be undertaken. From a statistical perspective, continuous data (*e.g.* potency or a quantitative physico-chemical effect) will usually require different modelling approaches from categorical data (*e.g.* the presence or absence of an effect). Applicable statistical approaches are described in Chapters 9 and 10.

3.3   Uses of Data

This chapter is intended to help readers find toxicological and other data. There are a number of uses for these data and this type of information. The discussion below does not cover every use or application but focuses on how they can be applied for *in silico* modelling.

It is essential that, before they start, data hunters and gatherers know how they want to use the data and have some form of expectation for their application—however (un)likely it may be to meet that expectation.

There are a variety of uses of toxicological data. Note that each use may possibly have different data requirements. Some of these uses are described below, with the emphasis on *in silico* toxicology.

  i) **To develop a SAR.** Structural alerts are a powerful and popular method to develop *in silico* toxicology models and have been applied in expert systems such as Derek Nexus (see Chapter 19) and have been developed extensively for certain endpoints such as mutagenicity.[1] This approach involves the identification of the structural basis (usually a structural fragment) of a toxicological event, or assignment to a mode or mechanism of action, from a small number of data, or even a single datum point. It is ideally supported by further toxicological information. In certain circumstances, this can also form the basis of a category—see (iii) below.

ii) **To develop a QSAR model.** So far the most common application in *in silico* modelling has been the development of QSAR models. Depending on the approach applied, this may potentially require data for large numbers of compounds which are consistent and of high quality.

iii) **To populate a category.** Data can also be used to populate a category or grouping of similar compounds (see Chapter 15). It is important to note that, whilst there are a number of tools to develop a structural or toxicological category (*e.g.* OECD (Q)SAR Application Toolbox, Leadscope, Toxmatch, *etc.*) (see Chapters 16 and 17), it does not guarantee that the category will be populated with data. As described below, and in Chapter 15, even a small number of significant data supported with a sound toxicological basis can make the formation of a category very powerful.

iv) **The start of an Integrated Testing Strategy (ITS).** *In silico* toxicology is the 'usual' starting place for an ITS. Here the toxicity data sources may assist in the filling of the step often labelled 'existing data'. Much guidance is becoming available on how this may be achieved, for example, from the European Chemicals Agency (ECHA).[2] The use of data for ITS, and possible waiving of subsequent testing, will rely greatly on assurance of data quality (see Chapter 4) and the acceptance of non-standard data. Discussion of ITS and the possible implications of using non-standard data is given in Chapter 23 and elsewhere.[2]

v) **To evaluate and, if possible, validate a (Q)SAR.** Since the acceptance and practical application of the OECD Principles for the Validation of (Q)SARs (see Chapter 11), there has been a greater emphasis on the assessment of the performance of (Q)SARs through external validation. This requires data that have not been included in the original model.

vi) **Individual risk assessment of a chemical.** This is clearly not an *in silico* modelling approach, but many of the data sources listed in this chapter may provide information to assist in risk assessment. For more information on how to perform risk assessment and the potential data requirements, see for example van Leeuwen and Vermeire.[3]

Whatever the use of the toxicological data and ultimately the model, the user is required to understand the meaning and significance of the data and, where possible, check them for accuracy and suitability. When using the data for whatever purpose, it is the user's responsibility to assess data quality (see Chapter 4) and, where possible, go back to the original source of the data to confirm that no transcription error or omissions in methodology have occurred.

3.4   How Many Data are Required?

Before the search for data begins, a frequent question from both inexperienced and experienced modellers alike is 'how many data are required to develop a (Q)SAR?' There is no straightforward answer to this question. However the general characteristics of a data set include the following:

- There should be sufficient data to ensure the statistical significance of a quantitative model. Estimates of what this means vary; for a QSAR, the often quoted Topliss and Costello convention suggests that this should be a minimum of five compounds to every single descriptor used in the model.[4] Other estimates have suggested a higher figure, *e.g.* ten compounds to every descriptor.[5] My recommendation is as 'many data' as reasonably possible and realistically a minimum of ten compounds per descriptor.
- For a quantitative model, a good spread of activity/potency values is required. It is possible that a relatively small number of data spanning a large range of potencies may provide a more adequate model than a larger number of data in a narrower range of activity.
- More data are required if there are a greater number of mechanisms of action.
- There are different issues for developing a qualitative model as opposed to a quantitative model. The developer must ensure that, in a qualitative model, the numbers of active and inactive chemicals are reasonably balanced. One could imagine, for instance, a large dataset of pre-dominantly either active or inactive compounds which may be very restrictive for modelling.

Possibly a better question is to ask 'what can be achieved with the available data?'

- Even a single datum point may be useful within a given context. Should the information be significant and of high quality, it may assist in the population of a category or may be helpful in other modelling approaches. This has been the case, somewhat controversially, in the US Environmental Protection Agency's ECOSAR expert system (for environmental effects) where some QSARs have used data for only one compound to adjust a regression line for a set of compounds.[6] In addition, a limited amount of toxicological information may be sufficient to develop a SAR or structural alert, particularly if supported by mechanistic knowledge. Obviously, using only a single datum point to develop a predictive approach must be treated extremely cautiously.
- A small number of toxicity data may be sufficient to develop a simple *in silico* model. For instance, a QSAR can be developed using simple linear techniques such as a regression technique using a single or small number of descriptors. Categories, SARs and descriptions of structural alerts can be populated with even a small number of data.

- A large number of data is, at first appearance, the ideal situation for modelling. It allows for more complex approaches to be applied. For QSARs, non-linear models may be developed using many descriptors. SARs and categories will be well-supported and populated, and may even allow for local models and read-across to be performed. There is, however, greater effort to organise, check and comprehend the significance of a larger data set. This means, at some point, the law of 'diminishing returns' comes into play, *i.e.* the inclusion of more data into a model may not be worth the effort involved.

There may, thus, be instances where adding data for further compounds does not improve a (Q)SAR. Is it possible to have too many compounds for analysis? The answer is probably not, providing a sufficiently thorough analysis is performed. Should one be in the fortunate position of having a large database to analyse, the same thorough principles of model development should be applied. One particular issue is that poor quality or erroneous data can be hidden in a large and diverse data set; therefore the same care is required to evaluate and check data. As ever, this requires the effort of the developer to curate and organise the data.

At the end of the day, a great deal of pragmatism is required. For instance, in the discussion above, care has been taken not to define what is meant by a 'small' and 'large' number of data. Such decisions come with experience and are different in every case. The modelling approach should be dictated by the nature of the data, endpoint being modelled, mechanisms, potency range, and the problem to be solved and addressed.

There are many instances in the literature where workers have developed their own suite of software, which may be a set of descriptors and/or statistical method and then move through all the established QSAR data sets available. Such an approach does increase the number of models, but it does not necessarily produce usable, useful or repeatable models. Instead, the data should drive the modelling approach (*i.e.* modelling should be appropriate for the data) rather than data being modelled without thought or consideration.

3.5 Sources of Data

There are a number of potential sources of data for *in silico* modelling. The modeller may, in some circumstances, be able to direct testing (particularly of *in vitro* tests). Otherwise, the modeller will have to use existing data sources.

3.5.1 Creation of New Data (Measurement)

The most reliable method of obtaining toxicological information is through the direct measurement of the effect. Measured toxicity data are still required in some instances (*e.g.* for regulatory submission of new chemicals, pharmaceuticals *etc.*). Whilst it is an obvious statement to make, should the toxicity of a compound be measured, it negates the requirement for a prediction. Despite this, in the event that testing is required, it may be advantageous to make a prediction alongside the measurement to enable the veracity of the model (and indeed the measured value) to be determined. Assessment of the accuracy of a predicted value will highlight strengths and weaknesses of the *in silico* approach, as well as provide further evidence of the applicability domain of a model. Clearly the disadvantages of measuring toxicological data are the cost, animal use and time that may be required.

Another aspect of measuring toxicity data will be more apparent as integrated testing strategies (ITS) are more widely applied. Whilst probably not requiring full toxicological assessment, limited *in vitro* testing may assist or give greater confidence to an *in silico* prediction, *e.g.* confirming the narcotic nature of an environmental pollutant or receptor binding of a mammalian toxicant (see Chapter 13). The issues of using alternative or supplementary data in an integrated testing strategy are discussed in Chapter 23.

3.5.2   Existing Data

There are a number of potential sources from which existing data may be retrieved. These are described briefly below. An excellent overview of data resources has been published recently by Valerio[7] and readers are also referred to that information.

3.5.2.1   In-house

The most significant untapped sources of toxicological data are those held by private companies themselves. A (Q)SAR developer or user with this resource is strongly encouraged to investigate possible internal sources of data at the start of a study. The advantages of using internal or in-house data include:

- the full study result may be available, thus giving an insight into the quality of the data as well as potential problems such as impurities and solubility;
- detailed methodology will also provide important meta information on the protocol;
- commercial advantages may be gained from developing models on data that are not publicly available as well as capturing the knowledge from within an organisation; and
- further commercial advantages may be derived from internal evaluation and or validation of models.

Unfortunately, there are a number of possible disadvantages of using internal corporate data:

- Often, internal toxicity data will be treated as confidential business information (CBI), meaning that the results from studies cannot be published. Whilst this may provide a commercial advantage, it may also provide a block to the regulatory acceptance of a prediction as the *in silico* model may not be fully describable.

- Internal information is often not stored or overseen sufficiently well, and thus effort (time and human resource) may be required to compile the data. Whilst there may be an initial outlay, the benefit in the long term should far outweigh the initial expense, particularly if this can be combined with a data storage and retrieval system.

3.5.2.2   Publicly Available Data Sources

Most (Q)SAR developers and/ or users will at some time have to resort to publicly available toxicity data. There are a variety of sources of toxicological data and information available for use.

There are a number of advantages of using publicly available toxicity data. These include:

- large numbers of data and toxicological information may be available effectively free of charge (or at trivial cost);
- the data and information may be integratable with, or even form the basis of, internal data compilations; and
- certain datasets (*e.g.* DSSTox) are well curated and contain potentially high quality data.[8]

As will be illustrated in the more detailed discussion of toxicity data, there are a number of disadvantages in using publicly available toxicity data and information.

- For non-curated data sources, it is well established that errors and inconsistencies exist and are propagated within databases.
- The user will often have to consider and assign a value for the quality of data.
- The format of the database or retrieved information may not be compatible with the user's requirements.

The remainder of this chapter attempts to direct readers on how to obtain and use publicly available data.

3.6   Retrieving Publicly Available Toxicity Data and Information for *In Silico* Modelling

There are a number of approaches to obtaining data for *in silico* modelling. It should be remembered that, with regard to toxicological information, there are few sources of information in the scientific literature that have not already been mined. In no particular order, the approaches to retrieve data are described below. It must be stressed that the information is for illustration only; collation

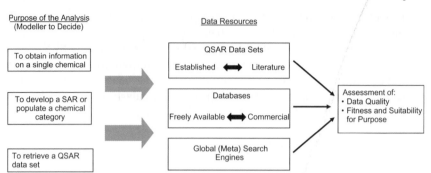

Figure 3.1   Workflow for data retrieval for *in silico* toxicology.

of toxicity data is a fast moving area and the determined data miner will find more data sets and databases than listed here.

Readers should also appreciate that there is considerable overlap between data sets and databases. So the same toxicity data may appear often in the sections below—the databases for cancer and mutagenicity are a good example. National Toxicology Program (NTP) data have been used in many QSAR models, as well being recorded in the ISSCAN database which is freely available on a number of websites and in the OECD (Q)SAR Application Toolbox. These data will also be retrieved by 'meta' search engines such as TOXNET and eChemPortal as described below. A workflow for using these resources is provided in Figure 3.1.

3.6.1   Using Well Established QSAR and Toxicity Data Sets

For a scientist interested in developing (Q)SARs who wishes to begin model-ling, there are a number of places to retrieve toxicity data. It might be wrong to describe these as 'classic' or 'traditional' QSAR datasets, but some have been (too) heavily modelled in the literature. The advantage of going to these data sets is that much effort has been made for several of them into ensuring the accuracy of structures and data. In addition, should (Q)SAR developers wish to learn more about the science, then they can attempt to repeat existing models or develop their own. A note of caution—it may be difficult to repeat some QSAR studies due to errors or inconsistencies in the analysis.[9]

In the first instance, some non-commercial websites list a wide variety of donated 'QSAR-ready' data sets which have been subject to a number of analyses (usually published). A good place to start is with the data sets on the website of the Chemoinformatics and QSAR Society (www.qsar.org/resource/datasets.htm). These data sets tend to emphasise drug design end-points. More data sets are available on the 'QSAR World' website (www.qsarworld.com). Again, there is a concentration on drug design, but this website also includes a number of data sets for ADME and limited human health effects.

A further web resource is www.cheminformatics.org which brings together a number of well-known toxicity databases. The style of information varies, *e.g.* two-dimensional (2-D) or three-dimensional (3-D) structures which may or may not incorporate isomerism, and vary in the degree of detail regarding the biological activity. Therefore, there may still be some effort required by modellers to reproduce models or create their own.

In addition to general web resources, developers of toxicological (Q)SARs should acquaint themselves with the US Environmental Protection Agency's Distributed Structure-Searchable Toxicity (DSSTox) Database Network.[8] This is a project to build a public data resource and its website provides a public forum for publishing downloadable, structure-searchable, standardised chemical structure files associated with toxicity data. Many of the datasets in DSSTox have been highly curated and cover a wide range of human health and environmental endpoints.

In an attempt to provide information on particular endpoints, some notable (Q)SAR datasets are listed in Table 3.1 and the references it contains.[10–17] Table 3.1 is by no means an exhaustive list—neither in the coverage of toxicological endpoints nor the data sets, but will give modellers an opportunity to make a start in developing (Q)SARs.

There are a number of advantages to the use of 'well-established' (Q)SAR data sets for toxicity. These include:

- These data sets are often well curated and may have undergone a thorough quality control process.
- The data are available free of charge.
- Some data sets represent the 'gold standard' of publicly available toxicity data.
- Many data sets are well established and respected by modellers, toxicologists, *etc.*
- Toxicity data may be associated with defined mechanisms or modes of action.
- The data sets maybe easily retrieved and downloaded in a usable format.

There are some disadvantages to the use of these well-established (Q)SAR data sets. The main one is that many of the data sets have been subjected to many modelling approaches and thus there may be limited possibilities to develop new models.

3.6.2 Obtaining Toxicity Data from the Open Scientific Literature

Although it may appear almost too trivial to mention, toxicological data for (Q)SAR studies may be obtained from the published literature. As all who have training in scientific research will know, it is now an easy task to search for journal articles. There are many computerised and internet-based tools for

Table 3.1 A non-exhaustive listing of publicly available toxicological data sets and databases that may provide a starting point for a (Q)SAR modeller. Readers are recommended to retrieve the relevant data sets and databases to determine their size and suitability.

Endpoint(s)	*Database*	*Reference or Source*
Human health		
Cancer/mutagenicity	ISSCAN	DSSTox[a]
		OECD Toolbox[b]
	Carcinogenic Potency Database (CPDB) Summary	DSSTox
	OASIS Genetox	OECD Toolbox
Skin sensitisation	Local Lymph Node Assay (LLNA)	Kern *et al.*[10]
		Roberts *et al.*[11]
	Guinea Pig Maximisation Test (GPMT)	Cronin and Basketter[12]
	BfR	Schlede *et al.*[13]
	ECETOC skin sensitisation	OECD Toolbox
	OASIS skin sensitisation	OECD Toolbox
Mammalian single dose toxicity studies	Japan EXCHEM	OECD Toolbox
Mammalian repeated dose toxicity studies	RepDose	Bitsch *et al.*[14] www.fraun-hofer-repdose.de
Multiple endpoint	US Food and Drug Administration (FDA) Genetic Toxicity, Reproductive and Development Toxicity, and Carcinogenicity Database	www.fda.gov/AboutFDA/CentersOffices/CDER/ucm092217.htm Yang *et al.*[15]
Eye irritation	ECETOC eye irritation	OECD Toolbox
Skin irritation	RIVM skin irritation	OECD Toolbox
Skin penetration	Human *in vitro* permeability coefficients	Moss and Cronin[16]
	Human *in vitro* permeability coefficients	Wilschut *et al.*[17]
	EDETOX	www.ncl.ac.uk/edetox/
Environmental toxicity		
Bioaccumulation	CANADA bioaccumulation	OECD Toolbox
	CEFIC – LRI BCF	OECD Toolbox
	OASIS bioaccumulation	OECD Toolbox
Acute aquatic toxicity	ECETOC aquatic toxicity	OECD Toolbox
	EPA fathead minnow acute toxicity	DSSTox
	Japan aquatic	OECD Toolbox
	OASIS aquatic	OECD Toolbox
	US EPA ECOTOX	OECD Toolbox

Table 3.1 (*Continued*)

Endpoint(s)	Database	Reference or Source
Oestrogen receptor binding	OASIS ERBA	OECD Toolbox
	EPA oestrogen receptor K_i binding study	DSSTox
	FDA National Center for Toxicological Research (NCTR)—oestrogen receptor binding	DSSTox
Miscellaneous endpoints	US EPA ToxCast	DSSTox

[a]DSSTox is available from the U.S. EPA web-site.[8]
[b]OECD (Q)SAR Application Toolbox, see Chapter 16.

performing a literature search. Some of these are also able to retrieve freely available (open source) articles and may link to any institution or corporate licensing agreements for electronic access. However, many of the better search tools can be costly to purchase and maintain; a list of literature searching tools is not provided here so not to prejudice opinion. Should readers not be aware of the availability of tools, they are recommended to contact their institutional or corporate library and/or information services.

In addition to commercial literature searching, freely available search engines are increasingly providing literature searching tools. For instance, Google Scholar (accessed *via* www.google.com) is becoming more sophisticated. The advantage of using an Internet search engine is that it may be possible to combine the literature search with a search of the Internet.

As with all types of electronic searching, the key is the correct and efficient use of the keywords to perform the search. Generally two types of searches could be considered. The first is to find published (Q)SAR analyses for a particular endpoint; the second is to search for publications on a particular toxicological effect or endpoint (which has not been previously been incorporated into an *in silico* study). Once the appropriate literature has been established, the chemical structures and biological activity can be extracted and used to create a database. Whilst this may be a manual operation requiring the effort of the researcher, many journals now publish (and allow free distribution) of supplementary information on their website which may assist in the data collation process.

There are a number of advantages of going to the primary literature to obtain toxicity data:

- A comprehensive search should provide a thorough assessment of the current/past literature and be able to retrieve the majority of publicly available data.
- A literature search incorporating (Q)SAR studies will give a good appreciation of the state-of-the-art in that area of *in silico* toxicology.
- Retrieving the primary literature or reference source will allow the modeller to determine experimental protocols as well as ensuring data have been transcribed accurately.

There are also a number of disadvantages:

- A thorough search of the literature may be time-consuming.
- Electronic recording of structures may not be available, so the transfer of structure and activity may be a source of error.
- There is much overlap between (Q)SAR data sets and also databases in the literature, so the modeller must be careful not to have duplicate data.
- It may be difficult and/or costly to obtain historical literature articles.
- There may be difficulty in using correct search criteria, *e.g.* correct description of an endpoint, inconsistency in English and technical terms such as UK English *vs*. US English. In addition, the spelling of compound names often varies and may be inconsistent.

As a final, personal, point on using literature searches, it is easy for native English speakers to ignore publications in languages other than English. In particular, resources from the former Eastern European countries and the USSR may provide high quality data. A recent study has brought together scientific literature from Russia.[18] Other sources such as from China should also not be overlooked.

3.6.3   Other Databases and Toxicity Data Sources

As well as the databases mentioned above, there are some other notable sources of toxicity information (please remember that there will be overlap between those noted above and these).

The previously mentioned data sources are free of charge (with the potential exception of charges for literature searching). There are also a number of commercial databases (or at least with some type of nominal charge). These tend to be larger and may allow for easy extraction of data to enable *in silico* model development. A selection of these databases is listed in Table 3.2 use of the term 'commercial' in this chapter is relative; some activities listed in Table 3.2 do little more than make a nominal charge to cover the costs associated with the upkeep of the database and its distribution, rather than using it as a profit-making venture.

There are a number of advantages to the use of these commercial toxicity databases. These include:

- They often incorporate large numbers of toxicity data, many of which may not be available elsewhere.
- They are highly curated and quality assured.
- There may be a direct relationship with chemical structure; such a facility may allow for the easy processing of *in silico* models.
- Being commercial ventures means there may be resources to develop the database technology and search for new data.
- Some are active in sharing and promoting the use of regulatory data (*e.g.* Leadscope and Lhasa Ltd have developed databases from the US Food and Drug Administration) or corporate data (*e.g.* the VITIC project from Lhasa Ltd).

Table 3.2 Commercially compiled databases of toxicological information (available often at nominal cost).

Name and Reference or source	Supplier	Endpoints
Leadscope database and FDA databases www.leadscope.com	Leadscope	Carcinogenicity Genetic toxicity Chronic and sub-chronic toxicity Acute toxicity Reproductive and developmental toxicity
VITIC Nexus www.lhasalimited.org	Lhasa Ltd	Carcinogenicity Mutagenicity Genetic toxicity HERG Hepatoxicity Skin sensitisation and skin irritation
TerraTox www.terrabase-inc.com	TerraBase Inc	Acute toxicity to a large number of environmental and mammalians (*in vitro* and *in vivo*) endpoints
Toxicity Database—includes the Registry of Toxic Effects of Chemical Substances (RTECS) database www.symyx.com	MDL	Acute toxicity Mutagenicity Skin/eye irritation Tumorigenicity and carcinogenicity Reproductive effects Multiple-dose effects

The main disadvantage is that, as commercial ventures, the database may be costly to obtain, although it is recognised that these costs are often kept to a minimum.

A further set of databases are worth considering as sources of toxicological information. These are a relatively recent phenomenon on the Internet and bring together regulatory submission information, hazard ratings, classifications and warnings as well as compilations of other information (*e.g.* literature data, physico-chemical properties, use, *etc.*). A non-exhaustive selection of such database resources is given in Table 3.3.

One key issue here is that the databases listed in Table 3.3 are generally designed with a specific purpose in mind and that is not normally *in silico* modelling. Thus some databases may be developed for regulatory (industrial and government agency) use; others may be made available to inform the educated lay person. All databases are likely to provide a wealth of information, but only when prompted to do so for a single chemical substance. Thus these resources may be particularly useful with regard to risk assessment. They are seldom searchable; for instance, if one wished to develop a model for a particular toxicity endpoint using data from a single protocol.

Table 3.3 Selection of freely available Internet-based resources providing information in a regulatory context relating to human health and/ or environmental endpoints.

Name and Reference or source	Description
Agency for Toxic Substances and Disease Registry (ATSDR) toxicological profiles www.atsdr.cdc.gov	Information on selected toxic substances and how they may affect human health including their characteristics, exposure risks, associated health effects and related Centers for Disease Control and Prevention (CDC) and ATSDR health studies and assessment
Hazardous Substances Data Bank (HSDB) www.toxnet.nlm.nih.gov	Database with comprehensive, peer-reviewed (eco)toxicological data and physicochemical data
European chemical Substances Information System (ESIS) http://ecb.jrc.ec.europa.eu/esis/	Database with (eco)toxicity data on industrial chemicals including information from: • European Inventory of Existing Commercial chemical Substances (EINECS); • European List of Notified Chemical Substances (ELINCS); • No-Longer Polymers (NLP); • Biocidal Products Directive (BPD); • Persistent, Bioaccumulative, and Toxic (PBT) or very Persistent and very Bioaccumulative (vPvB) chemicals; • Classification and Labelling (C&L), Export and Import of Dangerous Chemicals; • High Production Volume Chemicals (HPVCs) and Low Production Volume Chemicals (LPVCs); • IUCLID Chemical Data Sheets; • IUCLID Export Files; • OECD-IUCLID Export Files; • EUSES Export Files; • Priority Lists.
OECD Screening Information Data Sets (SIDS) www.inchem.org/pages/sids.html	Comprehensive review of all (eco)toxicity endpoints
US EPA Integrated Risk Information System (IRIS) www.epa.gov/iris/	A compilation of electronic reports on specific substances found in the environment and their potential to cause human health effects.

The advantages of the types of data sources listed in Table 3.3 are:

• These are mainly freely available.
• There are often large numbers of data for many different types of chemicals.

- The incorporation of other types of chemical identifiers, *e.g.* SMILES strings may be related electronically to structure.
- They may allow for the population of a chemical category where there is less emphasis on obtaining a data set with consistent toxicity values (*i.e.* all the same protocol).

There are a number of disadvantages:

- Most of these databases are not designed with (Q)SAR in mind and therefore may not allow for download of a complete 'modellable' dataset.
- By their nature, they comprise very variable data quality and users have to determine for themselves the suitability of the data for modelling – see Chapter 4 for guidance.
- The data and information (and even chemical structures) may be prone to error. It is the user who must do the checking.

3.6.4 Global (Meta) Data Search Engines

In order to combine a number of the databases and resources together, several very useful tools have recently become available. These are most likely to be useful for obtaining information about a single compound. Of these, the most significant are TOXNET, the OECD's eChemPortal and the US EPA ACToR resources. More information regarding these three global resources is shown in Table 3.4. These are remarkable search facilities—not least as they are freely available over the Internet and easy to use, bringing together many disparate data resources.

Other internet sites that may provide useful information, particularly on structure, identification and toxicity are ChemSpider (www.chemspider.com) and PubChem (http://pubchem.ncbi.nlm.nih.gov). ChemSpider may also search material safety data sheets (MSDS) as well as providing some calculated estimates of physico-chemical properties.

These impressive global and meta- resources do have a number of advantages (many are in common with the data sources noted in Table 3.3):

- They allow for searching of many sources of information for large numbers of compounds.
- These Internet resources are freely available.
- The information maybe retrievable by, and related to, the 2-D or 3-D structure.
- The information from the searches crosses endpoints and effects.
- The information maybe suitable for populating categories.
- Information on exposure, use, calculated and measured physico-chemical properties, *etc.* is compiled in one place.

Table 3.4 Further information on global (meta) portals and toxicity data resources..

Name and Reference or source	Description	Data resources incorporated
Aggregated Computational Toxicology Resource (ACToR) http://actor.epa.gov/actor/faces/ACToRHome.jsp	Collection of databases collated or developed by the US EPA NCCT for environmental chemicals.	Over 200 databases including: • EPA databases (DSSTox); • PubChem; • other NIH and FDA databases; • state and other national sources; • data resources from academic groups; • data collection ToxRefDB.
eChemPortal http://webnet3.oecd.org/echemportal/	Offers free public access to information on properties of chemicals. Allows for simultaneous search of multiple databases. As such it will provide information on physico-chemical properties, environmental fate and behaviour; ecotoxicity and toxicity.	• Canada's Existing Substances Assessment Repository (CESAR) • Information on Biodegradation and Bioconcentration of the Existing Chemical Substances in the Chemical Risk information platform (CHRIP) • Data Bank of Environmental Properties of Chemicals (EnviChem) • ESIS • US EPA High Production Volume Information System (HPVIS) • HSDB • New Zealand Hazardous Substances and New Organisms Chemical Classification Information Database (HSNO CCID) • Chemical Safety Information from Intergovernmental Organizations (INCHEM) • Japan Existing Chemical Data Base (JECDB) • Australian National Industrial Chemicals Notification and Assessment Scheme (NICNAS) • Priority Existing Chemical (PEC) Assessment Reports

TOXNET (TOXicology Data NETwork) www.toxnet.nlm.nih.gov	A cluster of databases covering toxicology, hazardous chemicals, environmental health and related areas. Managed by the Toxicology and Environmental Health Information Program (TEHIP) in the Division of Specialized Information Services (SIS) of the National Library of Medicine (NLM) in the USA. Provides free access to and easy searching of a significant number of databases.	• OECD High Production Volume Database • Screening Information Data Sets (SIDS) export files for HPV Chemicals in International Uniform Chemical Identification Database (IUCLID) format as maintained by the OECD • OECD Initial Assessment Reports for HPV chemicals including Screening Information Data Sets (SIDS) as maintained by United Nations Environment Programme (UNEP) Chemicals • US Environmental Protection Agency Substance Registry Services (US EPA SRS) • HSDB • US EPA IRIS • International Toxicity Estimates for Risk (ITER) • Chemical Carcinogenesis Research Information System (CCRIS) • GENE-TOX (Genetic Toxicology) • Tox Town • Household Products Database • Haz-Map • TOXMAP • LactMed (Drugs and Lactation) • Carcinogenic Potency Database (CPDB) • TOXLINE • Development and Reproductive Toxicology/Environmental Teratology Information Center (DART/ETIC) • Toxics Release Inventory (TRI) • ChemIDplus

As with the sources noted in Table 3.3, there are also a number of disadvantages:

- The data retrieved will be very complex in structure and difficult to decipher.
- The outputs from the searches are not designed with QSAR in mind (with the exception of ACToR) and they may not allow for download of a complete 'modellable' dataset.
- The toxicity results may be very variable in terms of data quality.
- The results, structures, *etc.* may be prone to errors.

3.7 Ongoing Data Compilation Activities

In addition to the approaches and databases mentioned above, there are a number of other projects that will, in the future, result in significant resources in terms of databases and toxicological information.

A number of European Union funded projects are attempting to compile data. This is by no means an exhaustive list, but the following may provide some further sources of publicly available data. It is hoped that these databases will be a highly quality assured data resource. It is not the intention that these data resources should exist in isolation, so it is likely that the data retrieved will be incorporated elsewhere, *e.g.* DSSTox and the OECD Toolbox.

EU projects that may be of interest include:

- CAESAR (www.caesar-project.eu);
- OSIRIS (www.osiris.ufz.de);
- CADASTER (www.cadaster.eu);
- ACuteTox (www.acutetox.org);
- Sens-it-iv (www.sens-it-iv.eu) and
- OpenTox (www.opentox.org).

Although these projects may have been completed, the results in terms of new toxicological information, *etc.* should be a legacy that will go on into the future. There are also likely to be many other national projects and readers are recommended not to ignore these initiatives as potential sources of information.

As well as publicly funded resources, there have been a number of industry and other initiatives to collate toxicity data. For instance, the European Centre for Ecotoxicology and Toxicology of Chemicals (ECETOC) has compiled a number of databases, some of which are in the OECD Toolbox (see Table 3.2). Many of these compilations have been stimulated through funding from the European Chemical Industry Council Long-range Research Initiative (CEFIC LRI). Other compilations are being developed by organisations such as the International Life Sciences Institute's Health and Environmental Sciences Institute (ILSI HESI).

3.7.1 ACToR and ToxCast

Two further activities are ongoing at the time of writing of this chapter. Both are publicly funded and progressing in the USA, including funding from the US EPA. The two activities are:[19]

- Aggregated Computational Toxicology Resource (ACToR)—a data management system; and
- ToxCast—a comprehensive testing programme.

These projects are very different in their scope, and what they will achieve and provide to the *in silico* modeller of toxicity. They are described briefly below.

ACToR is a collection of databases collated or developed by the US EPA National Center for Computational Toxicology (NCCT). The resource is freely available *via* the Internet (see Table 3.4). More than 200 sources of publicly available human health and environmental effect data for chemicals have been brought together and made searchable by chemical identifiers and structure. Data include chemical structure, physico-chemical values, *in vitro* assay data and *in vivo* toxicology data. Chemicals include, but are not limited to, high and medium production volume industrial chemicals, pesticides (active and inert ingredients) and potential groundwater and drinking water contaminants. Judson *et al.* have described the following key uses for ACToR:[20]

- derivation of training and validation data sets for *in silico* modelling;
- a resource for the development of models linking chemical structure with *in vitro* and *in vivo* assays; and
- a valuable resource for regulatory agency reviewers who are examining new chemicals submitted for marketing approval.

One important application of ACToR has been a survey of the toxicity data available on key environmental chemicals. In a recently published paper,[21] Judson *et al.* analysed the extent of publicly available toxicity information on approximately 10,000 substances including industrial chemicals, pesticide ingredients, and air and water pollutants. Key findings are that, while acute hazard data are available for 59% of the surveyed chemicals, detailed testing information is much more limited for carcinogenicity (26%), developmental toxicity (29%) and reproductive toxicity (11%). The EPA ToxCast screening and prioritisation program is designed to address this toxicity data gap.

ToxCast is a testing programme (and hence not initially a database resource) which was initiated by the US EPA's National Center for Computational Toxicology (NCCT). The purpose was to develop a cost-effective approach for prioritising the toxicity testing of large numbers of chemicals in a short period of time. It is intended to use data from high throughput screening (HTS) bioassays to build computational models to forecast the potential human toxicity of chemicals. This should lead to more efficient use of animal testing.

In its first phase, ToxCast is profiling over 300 well-characterised chemicals (primarily pesticides) in over 400 HTS endpoints. These endpoints include:

- biochemical assays of protein function;
- cell-based transcriptional reporter assays;
- multi-cell interaction assays;
- transcriptomics on primary cell cultures; and
- developmental assays in zebrafish embryos.

Almost all the compounds being examined in Phase 1 of ToxCast have been tested in traditional toxicology tests, including developmental toxicity, multi-generation studies, and sub-chronic and chronic rodent bioassays. ToxRefDB is a relational database being created to house this information and this will be integrated into the ACToR database. The second phase of ToxCast will screen additional compounds from broader chemical structure and use classes in order to evaluate the predictive bioactivity signatures developed in Phase I.[21]

Readers are encouraged to keep aware of the progress, results and findings of these two initiatives which will have very important implications for the future of (*in silico*) toxicology.

3.8 Ensuring Success in Using Toxicity Data in *In Silico* Models: Hierarchy of Data Consistency and Quality

The majority of this chapter provides readers with sources of information to build *in silico* models for toxicity. In order to be successful in using these data, modellers (whatever their background) need to appreciate the following issues:

- All biological data, whatever the method used or how standardised they may appear, are variable.
- Many databases and data sets contain errors in the chemistry (*e.g.* description of the chemical) and biology (*e.g.* incorrect recording of results), making thorough checking of information vital.
- The modeller should obtain, even at a basic level, some appreciation of what the toxicity is that is being modelled, how this was obtained and how the experimental protocols and methods impact on the test result (*e.g.* solubility and use of solvents, route of exposure, *etc.*).
- As such, the modeller should address the issue of data quality. Criteria to assign data quality are given in Chapter 4. It is very important to appreciate that, within a data set, the assignment of data quality should be on an individual basis for each datum point, rather than a blanket assignment to a data set. This is because, even within what may be

Table 3.5 Minimum requirements in terms of data set consistency and individual data quality for different aspects of *in silico* modelling.

Minimum requirement for:	Data set consistency	Data quality
High quality QSAR	High	High
Low quality QSAR or high quality SAR	Moderate	Moderate
Low quality SAR, populating a chemical category	Low	Low

considered a data set of good quality, there may be values that are incorrect (*e.g.* for non-soluble chemicals).

Thus, one may consider a hierarchy of toxicity data set consistency.[22] To do this one must consider as many as possible of the following factors.

The consistency of a data set will be considered to be high if:

- The chemicals are characterised properly and structures are unambiguous.
- The same experimental protocol (preferably a standardised guideline method such as OECD) has been used.
- Data are measured in the same laboratory and even, preferably, by the same technicians.
- The same endpoint is considered.
- The compounds tested are within the experimental limits of the test system, *e.g.* are within solubility limits, volatile chemicals are not lost from an open system, *etc.*

A brief consideration of the minimum requirements for different types of *in silico* models is given in Table 3.5. While the ideal for an accurate (Q)SAR is a consistent data set of high quality data, in terms of success of *in silico* modelling even low quality data or inconsistent methodologies may be useful for populating a chemical category provided the restrictions and implications are properly documented and understood.

3.9 Case Studies

To illustrate how the varied resources could be utilised, three case studies are provided which indicate (very briefly) how an interested scientist may be able to obtain information. The first case study illustrates how information can be obtained for various toxicity endpoints for a single chemical; this will not provide sufficient information for *in silico* modelling (other than populating a chemical category), but may be the starting point of an integrated testing strategy or risk assessment. The remaining two case studies show how data sets could be identified for modelling a human health and an environmental endpoint.

Before readers continue this section, they must be aware that this information is provided in good faith and is believed to be accurate at the time of preparation. Should readers attempt any of these data retrieval processes, they should review the availability of data at that time and independently confirm data quality. It should also be noted that these case studies are for illustration and are non-exhaustive; there are likely to many other sources of information and it is the effort and imagination of the scientist that will obtain them!

3.9.1 Case Study 1: Data Retrieval for a Single Chemical

This case study provides a simple illustration of how toxicological, physico-chemical and other data and information can be retrieved for a single substance. To obtain these data, global (meta) search engines were utilised. These produce a bewildering quantity of information for this compound (a common industrial chemical). To obtain the most relevant information from these sources, it is recommended that the information is interpreted by someone with some toxicological or risk assessment experience.

The chemical chosen for this illustrative case study is 1-butanol (SMILES: CCCCO), which was chosen at random as a common industrial chemical for which there should be a reasonable amount of toxicity data. It is hoped that it is a relatively 'uncontroversial' chemical in that it is not associated with high profile and popularised human health effects (e.g. the phthalates). 1-Butanol is also a High Production Volume (HPV) chemical according to *The 2004 OECD List of High Production Volume Chemicals.*[23]

Following a very brief Internet based search of freely available resources, the following information can be established regarding 1-butanol:

- As a HPV chemical it has a published OECD Screening Information Data Set (SIDS) dossier (available at www.inchem.org/documents/sids/sids/71363.pdf). This dossier includes information on use, exposure, physicochemical properties and human health and environmental effects. Much of the information in this dossier will be common to, or repeated in, the following sources.
- ChemSpider and PubChem can be used to determine the structure (this would be important for chemicals for which there is no SIDS dossier, *i.e.* the majority of industrial chemicals). In addition, ChemSpider and PubChem will provide physico-chemical information as well as toxicological information relevant for human health and environmental effects. Links are provided to a large number of other databases.
- TOXNET provides approximately 2500 literature references for 1-butanol to toxicity information (but not the original reference which the user would have to obtain separately). In addition, over 100 literature references to developmental toxicity information are provided. There is a listing of over 250 entries in the Hazardous Substances Data Bank

(HSDB) and much more toxicological information such as genotoxicity results.

- eChemPortal confirms the above information in addition to regulatory information and classifications from, amongst others, ESIS, OECD, US EPA, *etc.*

Thus, using a small number of freely available web resources, a wealth of physico-chemical, use, exposure, regulatory, hazard and toxicological information is available in minutes. It would take considerably longer to sift through this information and make relevant decisions for risk assessment. It must, of course, be noted that 1-butanol is one of the more common (and simpler) HPV chemicals and therefore there is likely to be considerable information. There will be considerably fewer data and less information for a less common chemical.

3.9.2 Case Study 2: Obtaining Data for *In Silico* Modelling—Human Health

To illustrate the process of how a modeller may attempt to collate data for a human health endpoint for *in silico* modelling, skin sensitisation was chosen as the endpoint. This was chosen as the author is familiar with *in silico* modelling for this endpoint and aware, as illustrated below, that there are some easily accessible data sets.

To retrieve data for skin sensitisation, the modeller could start either with a literature search or go directly to the established data sets (or compilations such as in the OECD Toolbox). For the purposes of this illustration it will be assumed that a literature search has first been performed—the literature search and use of established databases will result in obtaining the same information.

A literature search for *in silico* models for predicting skin sensitisation illustrates a number of modelling approaches in the last three decades (see also Chapter 13). It should be noted that skin sensitisation is an endpoint where there are variations in spelling (*i.e.* sensitisation or sensitization) which may result in some citations being missed. We are fortunate in having a number of good reviews on the use of QSAR to predict skin sensitisation[24,25] and the further citations contained therein, which highlight a number of key data sets:

- A historical data compilation of Guinea Pig Maximisation Test (GPMT) and Buehler tests from Cronin and Basketter.[12]
- A data compilation from the Local Lymph Node Assay (LLNA) from Kern *et al.*[10]
- BfR [German Federal Institute for Risk Assessment] data from Schlede *et al.*[13]

Further data sets can be added to the above three main data sets from the literature.[11]

Consideration of the data from the OECD (Q)SAR Toolbox shows that the three main data sets have been combined into a database and supplemented by the ECETOC skin sensitisation database. It should, however, be noted that this data compilation includes data from at least four different sources (GPMT, Buehler, LLNA, human patch test) and these may be of varying quality. Modellers should consider carefully whether they wish to combine together the values and if this is appropriate, *i.e.* is it appropriate to consider a compound to be sensitiser from any assay or are some assays more sensitive? In addition, there are further interesting aspects to the data; some can be considered to be in a qualitative format and some semi-quantitative (*e.g.* GPMT) or quantitative (*e.g.* LLNA). Modellers therefore need to decide both on data quality and which data and techniques to use. It is exactly these types of questions that require expert toxicological advice.

In the case of skin sensitisation, it is unlikely that further extensive literature searching will add significant numbers of data from the standard assays such as LLNA. However, further information may be obtainable from clinical data or adverse reaction data. Should these data be available, and they can be used very carefully in an appropriate context, they may help in populating categories or evaluating/validating models. Similarly, the global (meta) databases are unlikely to provide any significant further data. However, there may be safety data for HPV chemicals, pharmaceuticals and pesticides that may provide further information, *e.g.* for categories or validation, *etc.* There does not seem to be any further data that could be obtained from commercial data sources

3.9.3 Case Study 3: Obtaining Data for *In Silico* Modelling—Environmental Toxicity

Acute aquatic toxicity has been chosen to illustrate the retrieval of environmental toxicity data and a quantitative potency value. This is actually a very broad endpoint and could include information on a number of trophic levels (*e.g.* microbial, invertebrate and vertebrate) and different species within those trophic levels. In addition, there are further complications regarding which species are favoured for testing in different geographical regions (*e.g.* different fish species are used in Europe, North America, Japan, *etc.*). Therefore, only a small number of key data sets are described here. This is also a good example of where modellers need to decide what exactly they wish to predict before they put too much effort into *in silico* modeling.

Much effort since the early 1980s has been placed on the prediction, through (Q)SAR, of acute aquatic toxicity. Therefore, it is good idea to start with the existing literature to guide data retrieval. A good recent review is provided by Netzeva *et al.*[26] As well as an appreciation of the state-of-the-art of QSARs, Netzeva *et al.* also give an overview of suitable aquatic databases.[26] From their listing and the existing scientific literature, the following specific data sets may be of use:

- 96-hour LC_{50} to the fathead minnow (*Pimephales promelas*) of approximately 550 diverse chemicals (available electronically *via* DSSTox);[27]
- 96-hour LC_{50} to the rainbow trout (*Oncorhynchus mykiss*) of approximately 270 chemicals;[28,29]
- various LC_{50} values to *Daphnia* sp. for 370 organic chemicals[30] and 262 pesticide;[31]
- 40-hour IGC_{50} values to *Tetrahymena pyriformis* for over 1500 chemicals.[32–34]
- 5, 15 and 30 minute toxicity to *Vibrio fischeri* for over 1000 organic chemicals.[35]

Many of these data sets—especially the toxicity values to *P. promelas* and *T. pyriformis*—are well-established and 'classic' data sets that have been subjected to numerous QSAR analyses.[26] They should, therefore, give some confidence to the QSAR modeller.

These data sets can be supplemented from the US EPA's ECOTOX database (available electronically or through the OECD (Q)SAR Toolbox), ESIS, *etc.* Others are available through the OECD (Q)SAR Toolbox (*e.g.* Japanese aquatic toxicity) and the ECETOC database. There seems little added value in continuing with an extensive literature or database search to determine environmental toxicity values since the US EPA's ECOTOX database is so comprehensive. Of the other databases, the commercial TerraTox data resources (Table 3.2) may provide supplementary information, but obviously there will be cost implications.

Acute aquatic toxicity is therefore an area where many data are available for a wide range of species. To obtain the data and data sets to develop high quality (Q)SAR models, modellers must decide which species to collate data for. It will also be essential to assess the quality of the data.

3.10 Conclusions and Recommendations

Data are a prerequisite for the creation of *in silico* models for toxicology. As well as for their development, they are required for their evaluation and possible validation, and for populating chemical categories. Significant toxicity data are available for these purposes for some endpoints, but are currently lacking for others. It should also be borne in mind that the available data sets for modelling do not cover the complete chemical universe and that further effort is required to collate data for better coverage.

In the last decade, the availability of data and information in most aspects of life has been revolutionised by the Internet. It is gratifying to see that a number of workers (particularly in the US EPA, FDA, OECD, *etc.*) have grasped this opportunity to increase data availability. This has resulted in increasingly easy and, mostly free, access to toxicity data.

For *in silico* modelling, there are particular requirements of the data and how it is structured. Currently, especially when the global (meta) search engines are used, there is the impression of a toxicity data swamp in which large amounts of individual unsophisticated information float; some of this information has been co-ordinated into more complex and sophisticated structures which have different functions. Therefore, further efforts are required to structure and formalise the information for the various purposes. This is at least partly the responsibility of *in silico* modellers to put some structure behind the complex data matrices to release the knowledge and formalise predictive methodologies.

From the information currently available, the followings conclusions can be drawn:

- The model developer or user must decide what types of data are required and even which endpoints and protocols.
- There are a variety of sources of information to develop and use in *in silico* models for toxicology.
- There are different approaches to collating data.
- The model developer should decide on the use of the data being sought, *e.g.* if data are required to populate a category, then different criteria for acceptability may be appropriate than for a 'high quality' (Q)SAR.

This chapter is intended to provide the modeller with data sources. The next step is for the modeller to assess data quality—a process which is addressed in Chapter 4.

3.11   Acknowledgements

This project was sponsored by the UK Department for Environment, Food and Rural Affairs (Defra) through the Sustainable Arable Link Programme. This work also was supported by the EU 6th Framework Integrated Project OSIRIS (contract no. GOCE-ET-2007-037017). I am also indebted to Dr Ann Richard of the US Environmental Protection Agency for helpful and stimulating discussions on many aspects of this chapter.

References

1. R. Benigni and C. Bossa, *Mutat. Res.*, 2008, **659**, 248.
2. European Chemicals Agency, Guidance on Information Requirements and Chemical Safety Assessment, EChA, 2008, http://guidance.echa.europa.eu/guidance_en.htm#GD_METH [accessed March 2010].
3. C. J. van Leeuwen and T. G. Vermeire, *Risk Assessment of Chemicals. An Introduction,* Springer, Dordrecht, The Netherlands, 2007.
4. J. G. Topliss and R. J. Costello, *J. Med. Chem.*, 1972, **15**, 1066.
5. T. W. Schultz, T. I. Netzeva and M. T. D. Cronin, *SAR QSAR Environ. Res.*, 2003, **14**, 59.
6. K. L. E. Kaiser, J. C. Dearden, W. Klein and T. W. Schultz, *Water Qual. Res. J. Can.*, 1999, **34**, 179.
7. L. G. Valerio Jr, *Toxicol. Appl. Pharmacol.*, 2009, **241**, 356.

8. C. L. R. Williams-DeVane, M. A. Wolf and A. M. Richard, *Bioinformatics*, 2009, **25**, 692.

9. J. C. Dearden, M. T. D. Cronin and K. L. E. Kaiser, *SAR QSAR Environ. Res.*, 2009, **20**, 241.

10. P. S. Kern, G. F. Gerberick, C. A. Ryan, I. Kimber, A. Aptula and D. A. Basketter, *Dermatitis*, 2010, **21**, 8.

11. D. W. Roberts, G. Patlewicz, S. D. Dimitrov, L. K. Low, A. O. Aptula, P. S. Kern, G. D. Dimitrova, M. I. H. Comber, R. D. Phillips, J. Niemelä, C. Madsen, E. B. Wedebye, P. T. Bailey and O. G. Mekenyan, *Chem. Res. Toxicol.*, 2007, **20**, 1321.

12. M. T. D. Cronin and D. A. Basketter, *SAR QSAR Environ. Res.*, 1994, **2**, 159.

13. E. Schlede, W. Aberer, T. Fuchs, I. Gerner, H. Lessmann, T. Maurer, R. Rossbacher, G. Stropp, E. Wagner and D. Kayser, *Toxicology*, 2003, **193**, 219.

14. A. Bitsch, S. Jacobi, C. Melber, U. Wahnschaffe, N. Simetska and I. Mangelsdorf, *Reg. Toxicol. Pharmacol.*, 2006, **46**, 202.

15. C. Yang, L. G. Valerio and K. B. Anderson, *Altern. Lab. Anim.*, 2009, **37**, 523.

16. G. P. Moss and M. T. D. Cronin, *Int. J. Pharmacol.*, 2002, **238**, 105.

17. A. Wilschut, W. F. ten Berge, P. J. Robinson and T. E. McKone, *Chemosphere*, 1995, **30**, 1275.

18. M. Sihtmäe, H. C. Dubourguier and A. Kahru, *Toxicology*, 2009, **262**, 27.

19. A. M. Richard, C. Yang and R. S. Judson, *Toxicol. Mech. Methods*, 2008, **18**, 103.

20. R. Judson, A. Richard, D. Dix, K. Houck, F. Elloumi, M. Martin, T. Cathey, T. R. Transue, R. Spencer and M. Wolf, *Toxicol. Appl. Pharmacol.*, 2008, **233**, 7–13.

21. R. Judson, A. Richard, D. J. Dix, K. Houck, M. Martin, R. Kavlock, V. Dellarco, T. Henry, T. Holderman, P. Sayre, S. Tan, T. Carpenter and E. Smith, *Environ. Health Perspect.*, 2009, **117**, 685.

22. M. T. D. Cronin and T. W. Schultz, *J. Mol. Struct. Theochem.*, 2003, **622**, 39.

23. OECD Environment Directorate, The 2004 OECD List of High Production Volume Chemicals, OECD Environment Directorate, Paris, 2004, www.oecd.org/dataoecd/55/38/33883530.pdf [accessed March 2010].

24. G. Patlewicz, A. O. Aptula, D. W. Roberts and E. Uriarte, *QSAR Comb. Sci.*, 2008, **27**, 60.

25. M. T. D. Cronin, in *Recent Advances in QSAR Studies. Methods and Applications*, ed. T. Puzyn, J. Leszczynski and M. T. D. Cronin, Springer, Dordrecht, The Netherlands, 2010, pp. 305–325.

26. T. I. Netzeva, M. Pavan and A. P. Worth, *QSAR Comb. Sci.*, 2008, **27**, 77.

27. C. L. Russom, S. P. Bradbury, S. J. Broderius, D. E. Hammermeister and R. A. Drummond, *Environ. Toxicol. Chem.*, 1997, **16**, 948.

28. P. Mazzatorta, M. Smiesko, E. Lo Piparo and E. Benfenati, *J. Chem. Inf. Model.*, 2005, **45**, 1767.

29. M. Casalegno, G. Sello and E. Benfenati, *Chem. Res. Toxicol.*, 2006, **19**, 1533.
30. P. C. von der Ohe, R. Kühneo, R.-U. Ebert, R. Altenburger, M. Liess and G. Schüürmann, *Chem. Res. Toxicol.*, 2005, **18**, 536.
31. A. A. Toropov and E. Benfenati, *Bioorg. Med. Chem.*, 2006, **14**, 2779.
32. T. W. Schultz, M. T. D. Cronin, T. I. Netzeva and A. O. Aptula, *Chem. Res. Toxicol.*, 2002, **15**, 1602.
33. T. W. Schultz and T. I. Netzeva, in *Predicting Chemical Toxicity and Fate*, ed. M. T. D. Cronin and D. J. Livingstone, CRC Press, Boca Raton, FL, 2004, pp. 265–284.
34. T. W. Schultz, T. I. Netzeva, D. W. Roberts and M. T. D. Cronin, *Chem. Res. Toxicol.*, 2005, **18**, 330.
35. K. L. E. Kaiser and V. S. Palabrica, *Water Pollut. Res. J. Canada*, 1991, **26**, 361.

CHAPTER 4

Data Quality Assessment for In Silico Methods: A Survey of Approaches and Needs

M. NENDZA,[a] T. ALDENBERG,[b] E. BENFENATI,[c]
R. BENIGNI,[d] M.T.D. CRONIN,[e] S. ESCHER,[e]
A. FERNANDEZ,[f] S. GABBERT,[g] F. GIRALT,[h]
M. HEWITT,[e] M. HROVAT,[i] S. JERAM,[i] D. KROESE,[j]
J. C. MADDEN,[e] I. MANGELSDORF,[f] R. RALLO,[h]
A. RONCAGLIONI,[c] E. RORIJE,[b] H. SEGNER,[k]
B. SIMON-HETTICH[l] AND T. VERMEIRE[b]

[a] Analytisches Laboratorium, Luhnstedt, Germany; [b] RIVM, Bilthoven, The Netherlands; [c] IRFMN, Milan, Italy; [d] Environment and Health Department, Istituto Superiore di Sanita, Rome, Italy; [e] School of Pharmacy and Chemistry, Liverpool John Moores University, Liverpool, UK; [f] FhG-ITEM, Hannover, Germany; [g] WUR, Wageningen, The Netherlands; [h] URV, Tarragona, Spain; [i] Institute of Public Health of the Republic of Slovenia; [j] TNO, Zeist, The Netherlands; [k] University of Bern, Switzerland; [l] Merck, Darmstadt, Germany

4.1 Introduction

The REACH regulation aims to establish an effective risk management system of chemicals in the European Union. As laid out in Article 1 and 2,[1] REACH is

Issues in Toxicology No.7
In Silico Toxicology: Principles and Applications
Edited by Mark T. D. Cronin and Judith C. Madden
© The Royal Society of Chemistry 2010
Published by the Royal Society of Chemistry, www.rsc.org

targeted towards:

- ensuring a high level of protection of human health and the environment;
- maintaining good function of the European market of chemical substances; and
- supporting the development and implementation of alternative testing methods.

The techniques currently available for chemical hazard and risk assessment address these concepts only in part. For example, typical animal models for toxic endpoints are a simplification of the actual targets. The challenge of REACH is to:

- develop new strategies that cope more efficiently with the intended objectives; and
- improve the existing tools for chemical hazard and risk assessment with minimal animal experiments.

Since REACH explicitly calls for innovation, it would clearly be inappropriate to focus chemical assessments on established existing methodology.

Integrated Testing Strategies (ITS) [*i.e.* sequential combinations of existing (eco)toxicological and physico-chemical data (from testing as well as non-testing sources) along with exposure information] have been considered to allow 'speeding up' of the hazard and risk assessment of chemicals whilst reducing testing costs and animal use,[2,3] (ITS are discussed further in Chapter 23). Thus, ITS aim to use and combine existing data optimally for human and environmental risk assessment purposes, while minimising the need for new testing.

REACH has advocated a weight-of-evidence (WoE) approach to decide whether information is adequate to reach a conclusion, *e.g.* on the toxicological properties of a substance (see Chapter 22). In order to determine how much a piece of information should contribute to the overall conclusion, the validity of the methods used needs to be assessed as well as the reliability and relevance (fitness-for-purpose) of this information. Many aspects influence the quality of individual results and may feed as weight factors into WoE-based ITS. Hence, the evaluation of data quality is an essential step in the overall process of developing ITS for different endpoints.

4.2 Principles of Data Quality Assessment

Data quality is not an absolute feature but depends on data suitability. For example, the grouped or ranked results of a particular study may be unsuitable ('poor quality') for modelling purposes though they have been perfectly adequate ('high quality') for the purposes of that study. This chapter discusses relevant data quality issues for *in silico* methods—an important component of ITS. However, the quality of data is fundamental for risk assessment regardless of the method used to generate the data.

The reliability and relevance of *in silico* models such as (quantitative) structure–activity relationships [(Q)SARs], read-across and interspecies correlations depend greatly on quality assurance throughout the model development and prediction stages. Only if highly reliable and relevant data are used as a starting point (*i.e.* the experimental input data) will there be a high probability that non-test methods can produce useful results and that the outputs from the model will be acceptable to end-users, *e.g.* industry and regulators. For further quality criteria in the derivation and application of QSARs, readers are referred to the OECD guidance on the validation of (Q)SAR models.[4]

The assessment of data quality in ITS frameworks under REACH relates to (eco)toxicological and physico-chemical data and calculated descriptors in QSARs. It also concerns information needs in data integration and cost-effectiveness analyses of test methods and test results. There are also many issues applicable to exposure data (see Chapter 20), but these are not explicitly considered in this chapter.

Data quality assessment needs to address multiple issues at several levels:

- **Individual data quality.** The inherent variability in pieces of information depends on any confounding factors in the (experimental) procedure used to generate the data. *In silico* predictions must include an assessment, at the very least, of the error range of the experimental data used to derive the model. In the case of toxic effect data, variability can be due to either technical (*e.g.* identity of test substance, deviations of test protocols, differences in exposure conditions) or inherent biological (*e.g.* species, strain, age and sex of test animals, seasonal influence) factors.
- **Combined data quality.** Less reliable data can still be adequate for risk assessment in combination with other evidence. The pooling of several studies, one or more of which may be inadequate by itself, may collectively satisfy the overall requirement for valid data.
- **Context-dependent data quality.** Different levels of data quality are required for different purposes. For example, read-across requires a very high confidence in each of the few data points it uses; QSARs demand increasing confidence in the experimental input data with decreasing number of substances in the training/test set; and evidence-based toxicology (EbT) may cope with mixed variabilities.

Variability is an obvious first measure of data quality, but what is actually required is an understanding of the degree of (un)certainty. The ultimate objective of data quality assessment is to contribute to identification of uncertainty, reduction of uncertainty and communication of uncertainty in decisions based on data; this is consistent with current practice in the regulatory assessment of substances.

The acceptable uncertainty of data (measured or generated *in silico*) in the regulatory context depends on the outcome of the hazard identification and risk assessment, and can also be weighted for the 'cost' of errors. For example, low confidence in a positive prediction may pose less risk if it means that precautionary measures are adopted (which in the case of a false positive may not have been necessary). Within the same scheme, however, low confidence in a negative prediction would not be acceptable as a false negative could result in severe problems if adequate precaution was then not taken.

Data quality assessment is a complex and time-consuming task. It is, however, exceptionally important as models derived from poor quality data will only deliver poor predictions (corresponding to the oft-quoted phrase from computer science: garbage in → garbage out).

This chapter aims to support and encourage the crucial process of data quality assessment with focus on specific needs in *in silico* toxicology by providing:

- common terms and definitions;
- background information on formal data quality scoring schemes;
- descriptions of chemical and biological factors affecting data variability;
- checklist approaches to data quality assessment;
- data quality considerations for physico-chemical data, calculated QSAR descriptors as well as environmental and human toxicity; and
- data quality needs in data integration and socio-economic evaluations of ITS by means of cost-effectiveness analysis.

This chapter contributes to the discussion on data quality from different (inter)disciplinary perspectives. It is not the intention to detail any methodology (*e.g.* bioaccumulation assessment or information fusion), but to display data quality considerations in different facets of the chemical hazard and risk assessment.

4.2.1 Terms and Definitions

For proper evaluation of available data (*in vivo*, *in vitro*, *in silico*, human data), the definitions used in the (overall) assessments should be clear cut and their interrelationships well characterised.

The following terms and definitions are only partly agreed upon and some are still under development. While several data quality parameters are objective and defined by mathematical equations, others need more subjective evaluations and may vary depending on the manner or context in which they are assessed. As a result, it may be that very different quality ratings are obtained by different panels (see also Section 4.4.2).

Furthermore, some definitions involve the use of other parameters which may be included in more than one term. As a result, the same parameter may be used more than once in the overall assessment. A general point is that some of these definitions are tailored depending on their purpose. The choice of the most suitable definition should take this into account.

4.2.1.1 Validity—relates to the method itself

Validity is defined as:

Evaluating the method used for the generation of data for a specific endpoint relative to accepted guidelines.

or

The extent to which the methods used for the generation of data result in finding the truth as a result of the investigator actually measuring what the study intended to measure.[5]

4.2.1.2 Reliability—relates to the result produced with a (valid) method

Reliability is defined as:

A measure of the extent that a test method can be performed reproducibly within and between laboratories over time when using the same protocol. It is assessed by calculating intra- and inter-laboratory reproducibility and intra-laboratory repeatability.[6] [While repeatability relates to measurements performed under exactly the same conditions (same operator, laboratory, *etc.*), reproducibility describes the closeness of agreement between independent results obtained with the same method on identical test material, but under different conditions (*e.g.* operators, laboratory, *etc.*). Reproducibility is a key parameter in the standardisation of methods.]

or

Evaluating an individual result with regard to the inherent quality of a test report or publication relating to a, preferably standardised, methodology and the way that the experimental procedure and results are described to give evidence of the clarity and plausibility of the findings.[7]

or

The extent to which results are consistent over time and are an accurate representation of the total population under study. Of central importance is that the results of a study must be replicable under a similar methodology to be considered to be reliable.[5]

4.2.1.3 Variability

Variability is defined as:

The confidence interval associated with a test result (*in vivo* or *in vitro*) or an estimate (*in silico*) from a particular method. Statistically it can be assessed with a series of parameters such as standard deviation, range, *etc.* It is recognised that for most methods there is no true appreciation of variability, in particular for *in vivo* data.

4.2.1.4 *Uncertainty*

Uncertainty is defined as:

The estimated amount or percentage (margin of error) by which an observed or calculated value obtained by a method may differ from the true value.

Uncertainty depends on both accuracy and precision, and is caused by either variability or lack of knowledge.

4.2.1.5 *Accuracy*

Accuracy is defined as:

The closeness of agreement between test method results and accepted reference values.

It is a measure of test method performance and one aspect of relevance. The term is often used interchangeably with 'concordance' to denote the proportion of correct outcomes of a test method.[6]
In case of an endpoint that is a binary classification, accuracy is measured as the proportion of substances correctly classified as either positive or negative by the test relative to the total number of substances as given by equation 4.1:

$$\text{Accuracy} = \frac{(TP + TN)}{Tot} \times 100 \tag{4.1}$$

where:

$$TP = \text{number of true positives}$$
$$TN = \text{number of true negatives}$$
$$Tot = \text{total number of items.}$$

In the case of continuous values, accuracy can be measured as the difference between the mean of a set of results and the value which is accepted as correct.

4.2.1.6 *Sensitivity*

Sensitivity is defined as:

The proportion of substances correctly classified as positive/active by the test relative to the total number of positive/active substances.

It is a measure of accuracy for a test method that produces categorical results and is an important consideration in assessing the relevance of a test method.[6]

$$\text{Sensitivity} = \frac{TP}{TP + FN} \times 100 \qquad (4.2)$$

where:

$$TP = \text{number of true positives}$$
$$FN = \text{number of false negatives.}$$

4.2.1.7   Specificity

Specificity is defined as:

The proportion of substances correctly classified as negative/inactive by the test relative to the total number of negative/inactive substances.

It is a measure of accuracy for a test method that produces categorical results and is an important consideration in assessing the relevance of a test method.[6]

$$\text{Specificity} = \frac{TN}{TN + FP} \times 100 \qquad (4.3)$$

where:

$$TN = \text{number of true negatives}$$
$$FP = \text{number of false positives.}$$

4.2.1.8   Relevance

Relevance is defined as:

Description of the relationship of the test to the effects of interest and whether it is meaningful and useful for a particular purpose—this property is clearly context dependent. It is the extent to which the test correctly measures or predicts the biological effect of interest. Relevance incorporates consideration of the accuracy (concordance) and of the scientific appropriateness of a test method.[5,6]

or

Covering the extent to which data and/or tests are appropriate for the intended purpose (*i.e.* safety and risk assessment of chemicals) with regard to, for example, representativeness of species, route of exposure, effect of concern for target population *etc.*[7,8]

4.2.1.9   *Adequacy*

Adequacy is defined as:

The usefulness of data for risk assessment purposes related to the reliability and relevance of the available information and supported by weight-of-evidence. If there is more than one set of data for each effect, the greatest weight is attached to the most reliable and relevant.[7,8]

or

The usefulness of data for risk assessment purposes, *i.e.* whether the available information allows clear decision making by the registrant [from REACH Implementation Project 3.2.2]

or

Weight factor composed of the elements reliability (data quality) and relevance (accuracy, sensitivity, specificity of a method).

4.2.1.10   *Good Laboratory Practice (GLP)*

The principles of Good Laboratory Practice (GLP) intend to ensure the generation of high quality and reliable test data related to the safety of industrial chemical substances and preparations in the framework of harmonising testing procedures for the mutual acceptance of data.

4.2.1.11   *Gold Standard (no official definition)*

This is the recommended test by REACH for each (part of an) endpoint (usually an OECD-guideline test).

4.2.1.12   *Weight (no official definition)*

This is a factor between 0 and 1 applied to test results to rate their impact on risk assessment outcome relative to some target weight (e.g. 1 for the 'gold standard'). The term 'weight' is closely related to the term 'adequacy'.

4.2.1.13   *Coverage (no official definition)*

This is the extent to which an alternative or non-guideline method provides results comparable to an OECD guideline test with respect to reliability and regulatory need (*i.e.* the 'gold standard'). *In vitro* tests that are scientifically validated according to internationally agreed principles (OECD GD 34) may fully or partly replace an *in vivo* test depending on the purpose for which the test method was validated or adopted. 'Old' animal studies, which have not been conducted according to an OECD guideline, may replace a guideline study when for example the missing information can be classified as non-relevant for Derived No Effect Level Concentration (DNEL).

4.2.2 Formal Data Quality Scoring Schemes

A theoretical approach to rank scientific evidence to support a hypothesis[9,10] should consider the theoretical basis and the scientific method used including auditability, calibration and validation as well as possible bias. This is summarised in Table 4.1.

With regard to the quality scoring of scientific methods, it needs to be stressed that, particularly in the case of animal work, the paradigm that large sample size equates with better quality is invalid. High quality is obtained from an adequate sample size that depends on the endpoint. For example, for a very obvious effect only a few animals are required; for a subtle effect much larger numbers may be necessary—but both are of equal quality. Experimental design is recommended to optimise the power (*i.e.* ability to detect the response) of experiments ensuring the minimum number of animals is used whilst remaining statistically valid.[11] With REACH promoting a reduction in animal use, it is important that no more tests are performed than are absolutely necessary.

The objectivity factor in Table 4.1 relates to where the data have been generated. Laboratories that are well-established and have staff with considerable expertise would be expected to produce higher quality data. It is also important to ensure that there could be no bias in the generation of the data, *i.e.* that the laboratory is independent and the study has been well-designed so as not to introduce systematic bias, including the use of positive and negative controls as well as adequate calibration and validation of the scientific method to ensure that the experiment is running correctly and no external factor can be responsible for the results.[5,12]

A variety of formal systems have been established for rating the quality of experimental data in terms of accuracy and reliability,[13,14] mostly based on the scoring system of Klimisch *et al.*[7] as outlined in Table 4.2. Different classification systems have been compared by Lepper[15] as shown in Table 4.3.

Good Laboratory Practice (GLP) principles are supposed to ensure that test data produced in GLP compliant laboratories meet certain quality criteria. However, GLP simply indicates that the laboratory technicians/scientists performing experiments follow highly detailed requirements for record keeping, including details of the conduct of the experiment and archiving relevant biological and chemical materials:[5] GLP specifies nothing about the quality of the research design, the sensitivity of the assays, or whether the methods employed are current or out-of-date. Furthermore, data generated under GLP can also have considerable variability. In other words, flawed studies can be GLP compliant if documented!

Table 4.1 Indicators of evidence quality (adapted from Bowden[9] and Pollard *et al.*[10]).

Quality rank	Theoretical basis	Scientific method	Auditability	Calibration	Validation	Objectivity
Very high	Well established theory	Best available practice; adequate sample size; direct measure	Well documented trace to data	Exact fit to data	Independent measurement of same variable	No discernable bias
High	Accepted theory; high degree of consensus	Accepted reliable method; small sample size; direct measure	Poorly documented but traceable to data	Good fit to data	Independent measurement of high correlation variable	Weak bias
Moderate	Accepted theory; low consensus	Accepted method; derived or surrogate data; limanalogue; limited reliability	Traceable to data with difficulty	Moderately well correlated with data	Validation measure not truly independent	Moderate bias
Low	Preliminary theory	Preliminary method of unknown reliability	Weak and obscure link to data	Weak correlation to data	Weak indirect validation	Strong bias
Very low	Crude speculation	No discernable rigour	No link back to data	No apparent correlation with data	No validation presented	Obvious bias

Table 4.2 Scoring system to categorise the reliability of a study (according to Klimisch *et al.*[7]).

Score	Description
1	**Reliable without restrictions:** Studies or data . . . . generated according to generally valid and/or internationally accepted testing guidelines (preferably performed according to GLP) or in which the test parameters documented are based on a specific (national) testing guideline ... or in which all parameters described are closely related/comparable to a guideline method.
2	**Reliable with restrictions:** Studies or data . . . . (mostly not performed according to GLP) in which the test parameters documented do not totally comply with the specific testing guideline, but are sufficient to accept the data or in which investigations are described which cannot be subsumed under a testing guideline, but which are nevertheless well documented and scientifically acceptable.
3	**Not reliable:** Studies or data . . . . in which there were interferences between the measuring system and the test substance or in which organisms/test systems were used which are not relevant in relation to the exposure (*e.g.* non-physiological pathways of application) or which were carried out or generated according to a method which is not acceptable, the documentation of which is not sufficient for assessment and which is not convincing for an expert judgement.
4	**Not assignable:** Studies or data .... which do not give sufficient experimental details and which are only listed in short abstracts or secondary literature (books, reviews, *etc.*).

Table 4.3 Comparison of different classification systems according to Lepper.[15]

Category	TGD Reliability Index (RI)	US EPA	IUCLID
I	I (highly reliable)	high confidence	valid without restrictions
II	II (reliable)	moderate confidence	valid with restrictions
III	III (not reliable)	low confidence	invalid
IV	IV (unknown reliability)	unknown confidence	not assignable

There are several reasons why existing study data may be of variable quality, *e.g.*:

- the use of different test guidelines (compared with today's standards) or of different procedures within the same test guidelines;
- the use of different methods for data evaluation (statistical methods);
- the inability to characterise the test substance properly (*e.g.* purity, physical characteristics, use of salts where only parent compound name is reported, *etc.*);
- the use of crude techniques/procedures which have since become refined;

- certain information may have not been recorded (or possibly even measured), but that has since been recognised as being important;
- variability existing even within standardised test procedures performed according to GLP principles; and
- inherent biological variation of the test organisms.

Furthermore, it is important to consider what the aims of the original study were. Care is required when using data from one study to inform another study as the data may or may not be suitable for another purpose.

Study results may be considered valid and robust, with or without GLP (*e.g.* from well-described scientific publications that have been peer-reviewed) if they satisfy a number of criteria addressing the overall scientific integrity and validity of the information in a study (*i.e.* reliability) with particular emphasis on:

- a description of the test substance (identity, purity and stability);
- a description of the test procedure including exposure period and control of exposure concentration;
- data on the test species including the number of individuals tested and information on biology/physiology of test animals (sex, age, feeding, disease control, strains, *etc.*);
- demonstration of sufficient sensitivity (positive control);
- demonstration that no extraneous (*e.g.* environmental effects) were apparent (*i.e.* the application of a negative control);
- a description of measured parameters, observations, endpoints;
- a description of the methods used for data evaluation [*e.g.* method for determination of LC_{50}, no observed effect concentration (NOEC) or statistical analysis]; and
- independent replication of results.

Any data based on a test not providing complete information would be considered as less reliable, but can still be adequate for risk assessment in combination with other evidence. Three similar results of less reliable tests are more convincing than one poor result alone.

Irrespective of whether or not data meet the full set of quality criteria, consideration should be given as to whether the data:

- are outliers in a large data set for a particular substance; and
- fit with what is known about the effects of other related substances.

Uniform scoring systems (as indicated in Table 4.3) provide information on the reliability of data, but they do not address their relevance and adequacy. These aspects of data quality are context-dependent and require expert judgement with detailed consideration of how representative, and how readily generalised, the results are. Strategies need to be adjusted to avoid the ranking of data as less reliable because they do not cover a specific (regulatory) endpoint well enough, but provide reliable information that is useful for hazard and risk characterisation.

Since WoE-based ITS aim to make best use of available data, measures for both reliability and relevance need to be considered. Thought has to be given to the extent to which reliability scores (*i.e.* validity of results of individual hazard tests, based on a limited number of study parameters) truly capture the extent to which the results contribute in an overall assessment. Relying only on simple indicators of individual study quality may detract from a broader strategy based on the assimilation of what is known about each chemical in the context of others and may impact negatively on the potential to minimise required testing through maximal utilisation of available data.

4.2.3 Chemical Factors in Data Variability

Several factors may contribute to the substantial variability observed in measured data for toxicity, ecotoxicity and physico-chemical properties of chemicals. Depending on chemical class, testing protocol and regime, either higher or lower values may occur for the same compound. One example is the variable bioconcentration factors (BCFs) for pentachlorobenzene for which values between 900 and 250,000 have been reported.[16] Even data controlled by GLP measures may range over several orders of magnitude.[17]

Sources of data variability are manifold; apart from deficiencies in testing practices, different factors may affect test results in multiple ways. In particular, the quality of biological data for chemicals which possess properties that make experimentation difficult (e.g. high lipophilicity, low water solubility, high vapour pressure) is frequently questionable.

Confidence in data starts with the unambiguous characterisation of the tested substance(s) and includes confirmation of values in data compilations within primary literature sources, since it is known that there are error rates of up to 10% in (non)commercial databases.[18]

The following considerations regarding the identity and characterisation of test substances apply equally to biological data, physico-chemical data, calculated descriptors and model predictions—though to different extents and with different main emphasis.

4.2.3.1 Identity and Characterisation of Test Substances

The correct communication of the substance(s) actually tested in a study is a key issue. This includes the following information:

- chemical identity [two-dimensional (2-D) and three-dimensional (3-D) structure(s)];
- purity; and
- amount and type of impurities.

All *in silico* techniques in toxicology have in common the assumption that the underlying structures used to estimate or predict a toxicological behaviour are correct. Chemical structures for these efforts are often gathered from a variety of public and private databases. These databases contain records for thousands of chemicals with information about a variety of chemical, physical and toxicological properties specified for each chemical. It is unrealistic to assume that every record in every database has been entered or translated correctly. A recent paper[19] reported that incorrect chemical structures in six public and private databases ranged from 0.1% to 3.4%. It was found that slight errors in chemical structures (*e.g.* incorrect location of a chlorine atom) can cause significant differences in the predictive accuracy of QSARs.

Two European Union (EU) funded projects on the development of QSAR models for regulatory purposes, DEMETRA and CAESAR, devoted many months of work to simply checking data for pesticides and industrial chemicals, respectively. It was found that data reported in the US EPA DSSTox database[20] have a very low rate of error (<0.1%), while data presented in scientific papers, even recent ones, may have inconsistent values for up to 10% of the chemical structures.[18]

In addition to improper translations (systematic errors) or incorrect entries (random errors) of chemical structures into databases, several issues affect unambiguous characterisation of test substances. Most substance identifiers are not unique and disguise differences in 2D and 3D structures. In particular for QSAR modelling, it is of utmost importance to ensure that the toxicity data and the (calculated) descriptor values are obtained for the same well-defined structure.

Chemical Identifier

CAS number: Many substances have more than one CAS registry number that only partly differentiates between pure substances and isomeric mixtures, *e.g.* ChemIndustry.com (www.chemindustry.com/apps/chemicals) lists three CAS numbers for acrolein and five for cyanoguanidine. Major problems are associated with incorrect or ambiguous CAS numbers in different databases (even those maintained by competent authorities), particularly when automated retrieval and combination of information is performed for hundreds of compounds. One example of such a mistake is that the CAS number of a pure metal is frequently listed when actually a metal salt has been tested.

Names: Most compounds have many names (trivial, trade, IUPAC) that can be written in different ways or even be misspelled (or with different spelling in different languages). For example, ChemIndustry.com (www.chemindustry.com/apps/chemicals) contains more than 20 names each for acrolein and tetracycline.

SMILES: Multiple correct SMILES (Simplified Molecular Input Line Entry System) notations are possible for most chemicals depending on the starting point in, and path through, the molecule. It is, however, possible to write unique SMILES strings (www.daylight.com/dayhtml/doc/theory/theory.smiles.html).

InChI™: IUPAC's International Chemical Identifier (www.iupac.org/inchi) provides unique structural codes, but a lack of consolidated databases and automated conversion routines limit its ease of use at the current time (for more information and useful links the reader is referred to www.qsarworld.com/INCHI1.php).

Mixtures

Many chemical entries in databases represent mixtures, *i.e.* the test substances contain impurities, are formulated products or (stereo)isomeric mixtures. According to MolGen (www.molgen.de/?src = documents/molgenonline), examples of isomeric mixtures include chlorophenol (three isomers), dichloro-naphthol (42 isomers) and nonylphenol (633 isomers with multiple substitution and branching patterns).

Attention also has to be paid to mixtures of enantiomers and diastereomers (a minimum of two optical centres, different physico-chemical properties) with identical 2-D, but different 3-D, structures. In the case of possible tautomerism, it is necessary to check the likely (pH dependent) form(s) under experimental conditions against the database entry.

Further (unintentional) mixtures may result from (a)biotic degradation of test substances during exposure (*i.e.* parent compounds and metabolites) such as benzaldehyde → benzoic acid. Because the extent to which a component of a mixture contributes to the observed response is not clear, mixture data should generally not be used for *in silico* toxicity modelling. They may, however, provide useful information about synergistic or antagonistic effects of the compounds when present together if, for example, data are available for individual compounds as well as the mixture.

Some databases contain the toxicity of a salt along with the structure of its corresponding neutral chemical, primarily because the chemical structures were automatically generated from their SMILES codes. Only verification with primary literature sources can resolve such pitfalls.

Dissociation of organic acids or bases needs further attention as fractions of neutral and charged species vary with ambient pH conditions; as well as possible zwitterionic compounds (*e.g.* amino acids). Mostly, descriptors for QSAR models are created on the basis of the neutral form of molecules, regardless of dissociation under experimental conditions of the target endpoint. Failure in predictive power is then inevitable.

2-D Structures versus 3-D Structures

Two-dimensional structural representations of molecules ignore different 3-D structures (*e.g.* cis/trans and stereoisomers); because the 2-D notations are input to many parameter calculations, this will incorrectly result in identical descriptor values for different compounds.

When using 3-D descriptor calculations, consideration has to be given to the fact that most programmes optimise chemical geometry for the gas phase. This conformation is unlikely to be the same as the mostly aqueous target site of biological action. Furthermore, even 3-D based descriptors typically do not distinguish between enantiomers, but only diastereomers.

Hydrogen Suppression

Starting from hydrogen suppressed SMILES codes or chemical graphs, some parameter calculation programmes are prone to erroneous allocation of hydrogen atoms (*e.g.* on an aromatic nitrogen) or even may omit some or all hydrogen atoms such that incorrect descriptor values are delivered. Good indicators of such problems are the output for molecular weight and the empirical formula.

The final comment on this topic can be taken from Young et al.[19] 'With the growing importance of using computational techniques for regulatory purposes, it is of utmost importance that the chemical structure information being used is accurate.'

4.3   Biological Data Variability

Variability in biological data may be context-dependent and sometimes the borderline between technical and inherent biological variability is somewhat vague:

1. Acute toxicity test results for a chemical at the limit of solubility will be more variable than those obtained at lower concentrations. This type of variability is technical, whereas a difference due to the use of different species is inherent biological variability.
2. If two laboratories perform the same test at different pH, the consequences of the respective pH value on speciation/charge of the test compound and its toxicity would be a technical source of variability.
3. If, however, the altered pH from example 2 (above) also influences the physiology of the test organism, effects on test outcome may be attributed to biological variability.

To make things not too complex, it is suggested that variability related to test protocol and administration of test compounds is termed technical variability (which could be overcome by more stringent control of test conditions), while biological variability is inherent to the organisms and therefore cannot be controlled *via* test conditions.

With regard to biological factors potentially causing variability of test results, inherent (genotypically fixed) sources of variability such as species differences may be masked by variability caused by deviations in test conditions and their impact on the phenotypic performance of the test organisms. For instance, environmental temperature modulates the physiology of ectothermic species, and this in turn, influences their vulnerability. With respect to aquatic tests, water temperature is the prominent example.[21] In effect, elevated temperature and the presence of a toxicant may both increase the (adaptive) metabolic oxygen demand of organisms and the two stressors may even potentiate each other. Another interaction may arise from decreasing levels of dissolved oxygen in the water with increasing temperature. Changing environmental conditions may influence the test outcome not only by modulating the animal's physiology, but also by their effective on the test chemicals. The pH conditions in water influence not only the physiological status of the organisms, but also the speciation and charge of organic chemicals with impact on their bioavailability and uptake by the animals.[21]

Importantly, it is not only the environmental conditions during testing that have an effect on the toxic responses of test organisms, but also their history. For instance, the nutritional status prior to testing can significantly modify the apparent toxicity of chemicals to fish, as shown by Lanno *et al.*[22] and Braunbeck and Segner.[23]

Another confounding factor may be 'hidden', non-obvious diseases in a test animal stock. A prominent example is zebrafish (*Danio rerio*), a frequently used test species in environmental toxicology;[24] a bacterial disease, mycobacteriosis, is a tenacious problem at culture facilities of this species. Infected fish do not necessarily exhibit overt signs of disease, but their response to toxicants and, thus, test outcome can be confounded by the infection.[25] Disease-free stocks are available for mammalian toxicology, but for non-mammalian vertebrates and invertebrates used in ecotoxicological testing, disease-free stocks are usually not available.

Variability between species in their response to toxicants may contain relevant information about possible ecological hazards of compounds. However, regulatory risk assessment often does not take advantage of this information (except for approaches such as species sensitivity distributions). An example of how informative a careful and systematic consideration of species differences in toxicity values can be is provided by the study by Kwok *et al.*[26] who investigated differences in acute toxicities between animal species from tropical and temperate zones. Toxicity can also vary within a given species, particularly with life stage and gender.[27,28]

Reliance on test data generated according to harmonised test guidelines and under GLP principles may reduce test data variability to some extent, but it will not eliminate it.

4.4 Checklist Approach

The assessment of inherent data quality (reliability) is frequently guided by checklists,[8,29,30] addressing criteria with regard to:

- test substance identification and purity;
- test organism/test system characterisation;
- study design description;
- study results' documentation;

Table 4.4   Checklist for data reliability assessment (extended from Zweers and Vermeire[8]).

Are the identity, purity and source of the test substance fully known?	✔
Is sufficient information on solubility, volatility, impurities, pH shifts, osmolality, *etc.* available (possibly indicating (i) problems in test agent handling, (ii) secondary effects)?	✔
Has the study been well-designed with adequate sample size to detect the effect under study and due consideration of avoidance of bias, *etc.*?	✔
Is a complete description of the test system available, including environmental factors (*e.g.* housing, feeding, bedding, light–dark cycles, *etc.*)?	✔
Is the test conducted in accordance with a well-established guideline and do those conducting the test have appropriate expertise?	✔
Are deviations from the guideline sufficiently supported?	✔
Is the test species properly identified and described (*e.g.* age, size, sex, lipid content, genetic strain, health status or specific disease state induced)?	✔
Are positive and negative controls reported and do they substantiate sufficient sensitivity of the test?	✔
Is the exposure fully characterized (*e.g.* measured or nominal exposure concentrations for aquatic systems?)	✔
Are the methods of analysis performed correctly and described in sufficient detail (including analytical quality assurance)?	✔
Is the test conducted in compliance with GLP principles?	✔
Is a complete test report available?	✔
Are the test method and results reported in a peer-reviewed journal?	✔
Is a statistical analysis performed and is it done correctly?	✔
Is independent (external) replication of the study results available?	✔
Are other tests on structural analogues available confirming the results obtained?	✔

- plausibility of study design and data, including relevance and suitability of the study for the given endpoint; and
- methods for data evaluation (statistical analysis).

Table 4.4 provides examples of relevant questions that were derived for individual data in the regulatory context.

It is likely that many studies do not comply with all issues raised. Perhaps (not) surprisingly, within an ECVAM project, experiments using scientists to rate a questionnaire-based quality assessment tool for toxicological data[29] revealed major differences in the evaluation of the results of the same study, despite the notion that uniform criteria should deliver similar results. The diverging interpretations partly reflect problems of transfer of qualitative aspects to a binary (yes/no) system. Further variability may be attributed to the different experience and background of the scientists undertaking the rating. In general, old studies are assessed more critically than newer ones, and the further away a study is from a guideline study, the more strict is the judgement.

Most scoring systems reward guideline studies of high quality with category I (Table 4.3: highly reliable, high confidence, valid without restrictions), while comparable non-guideline studies mostly fall in category II (Table 4.3: reliable, moderate confidence, valid with restrictions). Whether this differentiation is relevant for *in silico* modelling purposes needs to be decided on a case-by-case basis depending on the particular endpoint.

To provide *in silico* models with ample databases, category I and II data are usually taken as suitable input. Invalid data [*e.g.* measured above water solubility (category III)] are clearly not appropriate. Data of unknown reliability (category IV) can be useful under certain circumstances because these data may be ranked as less reliable for not sufficiently covering a specific (regulatory) endpoint; regardless, they may still provide valuable information for ITS building blocks.

High reliability scores (category I or II) for individual data are necessary, but not sufficient quality criteria, for compiling training and test datasets for QSAR modelling. Considerations of data consistency gain major weight in model building with regard to having a comparable experimental procedure (requirement for a defined endpoint→OECD Principles on validation of (Q)SAR models);[4] see Chapter 11).

Ideally, the data in training and test sets for QSAR modelling should be obtained by the same procedure and from the same laboratory so as to minimise unsystematic variability. Table 4.5 provides a checklist of additional data quality requirements for datasets for QSAR modelling; for more details on proper QSAR modelling and application see, for example, Nendza,[16] OECD,[31] and Dearden *et al.*[32]

The aspect of data relevance applies to QSARs to the extent that individual data comply with the model endpoint. Entering QSAR estimates and other pieces of information into ITS requires further evaluation of their relevance. It addresses the extent to which data are useful for a particular purpose in a context-dependent manner. The evaluation requires expert judgement with regard to how representative are the species, the route of exposure and the effects of concern for the target population, *etc.* These factors are considered in Table 4.6.

4.5 Endpoint-specific Considerations of Inherent Data Variability

In this section, the theoretical issues presented in the previous sections are illustrated with practical examples from human (mammalian) and environmental toxicology, physico-chemical data and calculated QSAR descriptors. Particular attention is paid to

- **Reliability:**
  - **data generation:** standardised endpoint, defined procedure (reliable, validated?);
  - **applicability domain:** chemical domain, mechanistic domain;
  - **variability and uncertainty;**
  - **validation of the overall performance/theory.**

Table 4.5   Checklist of additional data quality requirements for datasets for QSAR modelling.

Activity data

Are the data 'high quality' (*e.g.* according to the checklist Table 4.4)?	✔
Do the data cover a broad activity range (relative to the inherent data variability)?	✔
Are the data evenly distributed over the activity range?	✔
Do the chemicals act by the same mode of action (*e.g.* parallel dose response curves)?	✔
Is the activity expressed as a function of applied concentration (*e.g.* EC_{50})?	✔
Is the activity expressed in molar concentrations? (Applied doses on a weight/weight basis are not comparable for different compounds.)	✔
Are percentage effects data transformed accordingly?	✔

Descriptor data

Are the descriptor data obtained for the same (3D) structures as used to measure the activity data?	✔
Are the descriptors comprehensible and relevant to the activity data?	✔
Are the descriptor data calculated correctly?	✔
Has collinearity of descriptors been properly considered in variable selection?	✔

Statistics

Is the training set properly designed?	✔
Are adequate statistics applied properly?	✔
Is the number of descriptors reasonably small (relative to the number of compounds in the data set)?	✔
Is the algorithm reproducible?	✔
Is the QSAR properly validated?	✔
Are outliers explained (not just omitted)?	✔
Is the applicability domain defined in terms of structural, physico-chemical and functional similarity?	✔

Table 4.6   Checklist for data relevance assessment (adapted from Zweers and Vermeire[8]).

Is the species relevant with regard to the target trophic levels in the population, community and ecosystem under consideration?	✔
Is the route of exposure relevant to the target trophic levels in the population, community, ecosystem and exposure scenario under consideration?	✔
Is the exposure regime relevant for the target trophic levels in the population, community and ecosystem under consideration?	✔
Is the effect (in terms of test endpoint and life stage) relevant and/or representative for growth and survival of the target communities?	✔

- **Relevance:**
  - **mechanism/mode of interaction**: well/less well defined;
  - **fitness-for-purpose:** transparency, acceptance by regulators, effects on predictive models (QSAR, read-across, ITS);
  - **(eco)toxicological target:** species or processes for which the data should be used.

4.5.1 Physico-Chemical Data and Calculated Descriptors

A standardised endpoint and defined procedure (gold standard) is available and recommended for all REACH-relevant physico-chemical properties of chemicals. Inherently, physico-chemical as well as calculated stereoelectronic and thermodynamic properties have a clear mechanistic meaning, but care has to be taken if they can be expressed in different units (*e.g.* Henry's Law constants). The mechanistic meaning is less evident or even lacking for many empirical descriptors that have been derived more recently (*e.g.* in the past two or three decades). Their indirect interpretation *via* correlation with physico-chemical properties provokes the question of why the latter is not used directly.

4.5.1.1 Lipophilicity (log P*)*

Lipophilicity in terms of 1-octanol/water partition coefficient (log *P*) is discussed here as an example of both a physico-chemical property as well as its use as a calculated descriptor in many QSAR studies. Log *P* is the surrogate parameter for diverse partitioning processes, related to a major aspect of bioavailability. Confounding factors concern differences in lipophilic phase properties between 1-octanol and cellular lipids and mitigating transport processes (*e.g.* size-limited diffusion rates).

Experimental determination of log *P* by the standard shake-flask technique[33] usually provides reliable results, but problems (*i.e.* variability of experimental data) may arise from:

- compound impurities;
- solvent impurities;
- glass/surface adsorption effects;
- formation of stable emulsions;
- concentration dependency;
- non-attainment of equilibrium;
- ion pair formation; and
- extractable ions in buffered aqueous phases

Alternative experimental methods, especially for highly lipophilic compounds (log *P* > 5), comprise several modifications of the shake-flask procedure such as, for example, the slow-stirring method[34,35] and various chromatographic techniques.[36,37]

There are many commercial and freely available software packages to calculate log *P* such as those discussed by Cronin and Livingstone[38] and Mannhold *et al.*[39] These differ in their computational method, *e.g.* group contribution, fragmental, atomic values, linear solvation energy relationships (LSER) and topological descriptors.

Most computational methods for log *P* are so-called global models for inert organic compounds, but they frequently fail for:

- inorganic compounds;
- surface-active compounds;
- chelating compounds;
- organometallic compounds;
- partly/fully dissociated compounds;
- compounds of extremely high or low lipophilicity;
- mixtures; and
- impure compounds.

The currently available programmes may provide inaccurate estimates for a variety of compounds to different extents. This is of particular concern for the very high (log $P > 5$) and very low (log $P < 0$) ranges of lipophilicity and/or if intramolecular interactions are not adequately accounted for by the respective algorithms. Examples where variability exceeds two log units relative to experimental data[40,41] are presented in Table 4.7. It is evident that none of the methods perform in a superior manner throughout. The most reliable estimates of log P are consensus log P^{39} (*i.e.* the geometric mean of multiple log P values, calculated independently by different methods, bears the best approximation to the true value).

With regard to so-called superlipophilic substances, where the calculated data frequently deviate from experimental values, the latter are also subject to major uncertainties (often ± 2 log P units and greater). The calculated values that were apparently too high were, in several cases, later confirmed by refined experimental procedures. For practical purposes, a computed log $P > 5$ should be read: This compound is lipophilic with a log P of $\pm 20\%$ about the

Table 4.7 Comparison of different log P calculations with experimental data for different example compounds.

CAS	Name	Experimental log P^a	ACD v10.0	KOW-WIN v1.66	Dragon v5.4 Moriguchi log P
81-11-8	4,4′-diamino-2,2′-stilbene disulfonic acid	**−1.7**	−1.72	−1.42	0.93
88-74-4	2-Nitroaniline	**1.85**	1.83	2.02	1.35
87-65-0	2,6-Dichlorophenol	**2.75**	2.61	2.80	2.73
119-12-0	Pyridaphenthion	**3.2**	3.08	3.66	2.88
67-72-1	Hexachloroethane	**4.14**	4.47	4.03	3.29
101-02-0	Phosphorous acid, triphenyl ester	**6.62**	7.43	6.62	4.45
70-30-4	Bis(2,3,5-trichloro-6-hydroxyphenyl) methane	**7.54**	7.20	6.92	4.80
78-42-2	Tris(2-ethylhexyl) phosphate	**4.23**	10.1	9.49	6.08

[a]Experimental data extracted from ref. 40 or from EPI Suite experimental log P database.[41]

estimated value, so a calculated log *P* of 7.5 can be taken as evidence that the actual value may be between 6 and 9.

The transferability and reproducibility of *in silico* results requires explicit specification of the software used, including the version number. The transparency of many computational methods for log *P* is given through the publication of parameters and coefficients (*i.e.* the algorithm), such that individual data can be confirmed manually. However, not all factors are scientifically derived (*e.g.* Rekker's empirical 'magic constant') to account for systematic deviations.[42]

The log *P* descriptor data have to be consistent (obtained by the same method) within a QSAR model and attention has to be paid to ensure that log *P* is calculated for exactly the same structure as has been present in the experiment to determine activity, accounting for isomerism, dissociation, *etc.* The aforementioned considerations evidently apply not only to log *P*, but to any calculated descriptors!

Furthermore, it has to be considered that variability (error range) in log *P* (and any other) descriptor data is propagated through *in silico* model derivation and contributes to uncertainty in predictions in addition to the variability in the activity input data. A striking example of the consequences of even minor variability in descriptors can be illustrated with equation 74 of the EU Technical Guidance Document on Risk Assessment (TGD), Part II[43] for log BCF estimation: If a compound has log *P* 4.9 ± 0.2, it is predicted *B* (bioaccumulative), but if it has log *P* 4.9 ± 0.3, it may be either non-*B* (at log *P* 4.6), *B* (at log *P* 4.7-5.1), or even very (v)B (at log *P* 5.2). While errors in log *P* of ± 0.2 or ± 0.3 are rather optimistic, the example even leaves out the impact of any further (most likely even larger) uncertainty in BCF input data and from statistical modelling.

4.5.2 Bioaccumulation

The bioaccumulation potential of chemicals is frequently assessed from bioconcentration, conventionally measured according to OECD 305 (Bioaccumulation: Flow-through Fish Test).[44] These studies deliberately reduce the manifold uptake and elimination mechanisms in aquatic organisms to respiratory absorption *via* gills and diffusion through the skin.[45,46] Despite the simplifications, testing costs are high and a minimum of 108 fish are used in each OECD 305 bioconcentration guideline study.[47]

A number of factors contribute to the experimental variability of measured BCF data:[40,48]

- **Methodological factors:**
- – test method (*e.g.* OECD 305);
- – duration of uptake and depuration phase;
- – exposure typology (*e.g.* flow-through);
- – tissue analysis (*e.g.* total body, lipid content, specific tissue);
- – detection method (*e.g.* radiolabel, analytical);
- – water conditions (*e.g.* temperature, pH, hardness, total/dissolved organic carbon content);

 – light conditions (intensity, spectral quality).
- **Biological factors:**
  – fish species, age, life stage, gender, size and physiological condition (*e.g.* lipid content, test organism health), respiration rate and growth rate.
- **Chemical factors:**
  – purity of the test chemical, its log P, water solubility and toxicity.

Harmonised test protocols and detailed documentation may reduce uncertainty due to methodological and chemical factors. For example, differences between wet weight and lipid-based BCF values [*e.g.* for acenaphthylene (CAS 208-96-8) in *Cyprinus carpio*: 271 L kg^{-1} (whole body wet weight) *versus* 11,631 L kg^{-1} (lipid)] are easily explained.

Three experimental BCF data sources (whole fish, wet weight) from the public domain have been comparatively analysed, after pruning the databases for compounds with ambiguous chemical identity. The data sources considered were:

- **EURAS database:**[17] 'validated' BCF data for 465 compounds (multiple experimental values with Klimisch reliability scores);
- **Arnot data set:**[40] multiple BCF and BAF (bioaccumulation factor) values for 403 compounds in diverse aquatic species [reliability scores from 1 (best) to 3 (worst) based on six evaluation criteria: water analysis, radiolabel, water solubility, exposure (time), tissue analysed and other factors]; and
- **Dimitrov data set:**[49] selected BCF data for 511 compounds (one experimental value for each chemical plus log P) for QSAR modelling.

Table 4.8 shows the variability in the BCF values from these data sources.

The two data sources with multiple entries per chemical reveal a similar pattern of variability. Only 10% of the chemicals show major divergences of greater than one log unit. About 65% of the chemicals show differences between minimum and maximum BCF values of less than half a log unit. This deviation is reasonable in view of experimental procedures, but very

Table 4.8 Variability in experimental BCF values within databases.

Difference max-min (log units)	EURAS database[17]		Arnot dataset[40]	
	Number of compounds	%	Number of compounds	%
≥ 2	1	0.2	0	–
1–2	52	12	46	11
0.5–1	88	21	100	25
<0.5	284	67	257	64
total	*465*	*100*	*403*	*100*
<0.4	255	60	233	58

Table 4.9   Variability in experimental BCF values between three databases.[17,40,49]

Difference max-min (log units)	*Number of compounds*	*%*
≥ 2	3	1
1–2	55	18
0.5–1	80	27
<0.5	162	54
total	*300*	*100*
<0.4	137	46

problematic when considering the distance of only 0.4 log units between the *B* and *vB* criteria of REACH.

Perhaps not surprisingly, variability increases between data sources: of the 300 compounds that are common to all three databases[17,40,49] only 54% have multiple BCF values within 0.5 log units as shown in Table 4.9.

Table 4.10 details minimum, average and maximum values of experimental log BCF values for selected chemicals. For instance, for DDT the Arnot[40] and Dimitrov[49] data sources are concordant, while the EURAS database[17] has a larger variability. The ten BCF values for DDT reported in the EURAS database,[17] all with a reliability score of 2, range between a minimum of $24\,L\,kg^{-1}$ and a maximum of $89,100\,L\,kg^{-1}$ (difference of 3.57 log units), with a standard deviation of 1.21. In this case, documentation and data quality rating does not eliminate variability. In two other examples (CAS 117-81-7 and CAS 84-51-5), the variability within the same data source is limited while it is larger between different databases. Only for some compounds (*e.g.* CAS 57-15-8) are the experimental BCF data in agreement in all three data sources, partly due to using the same original data source. Figure 4.1 displays the variability of BCF data for 1,3-dichloro-2-propanolphosphate with mean log values of 1.5 in Arnot *et al.*[40], 1.0 in the EURAS dataset[17] and 0.13 in the Dimitrov dataset.[49]

The comparative analysis of experimental BCF in fish from three data sources in the public domain identified three reasons for major data variability:[40]

- exposure time insufficient to reach 80% of steady state;
- incorrect use of radiolabelled compounds; and
- different or insufficient analytical methods.

These examples highlight the fact that large variation in BCF occurs despite formal data quality rating. The inherent variability of mostly <0.5 log units reasonably reflects experimental complexity, but may have major impact on the identification of persistent, bioaccumulative, toxic (PBT) and very persistent, very bioaccumulative (vPvB) compounds under REACH. The relevant thresholds of concern are log BCF 3.3 and 3.7, respectively, and particular accuracy is required for BCF data close to this narrow range of 0.4 log units.

Table 4.10 Minimum, average and maximum values of experimental log BCF data in three databases[17,40,49] and differences between total maximum and minimum values (log units).

CAS	Name	Arnot exp logBCF			EURAS exp logBCF			Dimitrov exp logBCF	Δ_{max} log BCF
		min	average	max	min	average	max		
50-29-3	DDT	4.17	4.19	4.20	1.38	3.36	4.95	4.57	3.57
117-81-7	1,2-Benzene dicarboxylic acid, bis (2-ethylhexyl) ester	2.43	2.69	2.80	0.53	1.00	1.47	1.19	2.27
84-51-5	9,10-Anthracene dione, 2-ethyl-	0.98	0.99	1.00	3.09	3.11	3.13	2.83	2.15
57-15-8	2-Propanol, 1,1,1-trichloro-2-methyl-	0.23	0.26	0.29	0.23	0.31	0.38	0.23	0.15

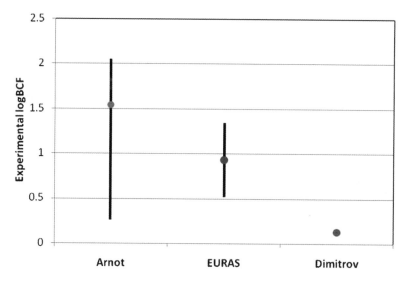

Figure 4.1 Variability of experimental BCF data for 1,3-dichloro-2-propanol-phosphate.[17,40,49]

Further refinement is needed for clear decision-making based on experimental BCF values as well as alternative methods and QSAR predictions.

4.5.3 Aquatic Toxicity

4.5.3.1. Fish Toxicity

Acute aquatic toxicity testing with invertebrates, algae and fish is required for the classification and labelling of substances, and may be also used for environmental risk assessment. Compared to the high variety of species and climatic variability in (aquatic) environments, the established testing protocols are an extreme simplification and reduction of complexity. Only for the evaluation of specific ecosystems is the selection of diverse test species reasonable, but data needed in the framework of GHS (global harmonised system) or REACH must match consolidated standards.

Internationally accepted guidelines exist for algae, daphnia and fish tests (OECD 201, OECD 202 and OECD 203). These guidelines allow considerable freedom in the selection of test species, test conditions and methods for the evaluation of test results. For example, two methods are available for the evaluation of test results in the acute toxicity test for algae:

- comparison of growth rate; and
- comparison of the area under the growth curves.

In fish acute toxicity, seven test species are recommended for use leading to major variability between (GLP compliant) test results performed according to the same standard procedure. Therefore, the standard test guidelines should be optimised for regulatory purposes, *e.g.* by selecting rainbow trout (*Oncorhynchus mykiss*) as the first option for species selection. Rainbow trout is already frequently used in acute toxicity testing and it is known to be among the most sensitive fish species.[50–52] This decision would further reduce the variability in test conditions (*e.g.* temperature, pH, water quality) required for the different fish species from different environments (*e.g.* sea, lakes, rivers).

A current investigation[53] explored the variability in multiple fish acute toxicity test results [96h LC_{50} (mg L^{-1})] in the US EPA ECOTOX database (http://cfpub.epa.gov/ecotox/). Fish species and fish life stages were identified as important factors affecting 96h LC_{50} data as well as test conditions (temperature, pH, water hardness) and purity of the test substances as shown in Figure 4.2. For the 44 compounds with at least ten data entries, a total of 4,654 test records were evaluated based on reported test conditions. The first observation concerns lack of reporting of relevant information, *e.g.* fish life stage (66.5%), water temperature (19.6%), hardness (48.2%) and pH (41.2%). The most influential factor affecting variability by orders of magnitude in 96 h LC_{50} test results was found to be test substance purity, as shown in Table 4.11.

4.5.3.2   *Toxicity to* Tetrahymena Pyriformis

There are many aquatic toxicity endpoints that may provide useful information relating to the harmful effects of compounds in the aquatic environment. These range across the trophic levels from photosynthetic algae, other microorganisms, invertebrates (including *Daphnia magna*) to fish and mammalian species. The effects measured range from acute lethality to chronic effects including reproduction.

In terms of QSAR modelling of the acute effects of organic compounds to aquatic species, there are at least two 'classic' datasets that have often been used—and many other smaller data sets and database compilations. The two most commonly modelled databases are those for the 96-hour acute lethality to the fathead minnow (*Pimephales promelas*) from the US EPA and the 40-hour inhibition of growth of ciliated protozoan *Tetrahymena pyriformis*.

Tetrahymena sp. are freshwater ciliated protozoans which feed/predate on organic material and bacteria. Two species in particular have been used for the assessment of the effects of chemicals to aquatic organisms, namely *T. thermophila* and, more commonly, *T. pyriformis*. In addition, there have been a number of variations in the assay, varying in factors including the time and volume of inoculum. The discussion here is based on the use of data from a 40-hour assay that has been used and optimised for over three decades in the laboratory of Professor Terry Schultz at the College of Veterinary Medicine, University of Tennessee, Knoxville TN, USA. This is discussed in detail as the toxicological effects of many compounds have been assessed in this assay and these data have been applied in many QSAR analyses by Professor Schultz. The

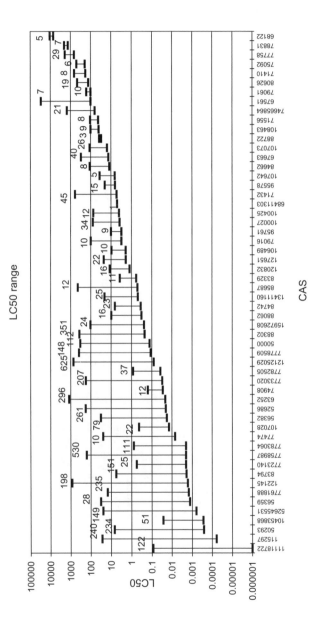

Figure 4.2 Variability in acute fish test results (96h LC$_{50}$ [mgL^{-1}]) for 56 selected substances in the US EPA ECOTOX database. The numbers in the graph indicate the number of tests per substance.

Table 4.11 Comparison of variability between all 96h LC_{50} test results for four example compounds and the respective test results with defined chemical purity (98–100%) from the US EPA ECOTOX database.

	CAS	Chemical name	No. of results	LC_{50} min ($\mu g\ L^{-1}$)	LC_{50} max ($\mu g\ L^{-1}$)	$Log\ LC_{50}$ min	$Log\ LC_{50}$ max	$\Delta\ Log\ LC_{50}$
All 96h LC_{50} test results	63-25-2	1-Naphthalenol methylcarbamate	328	21.05	1,200,000	1.32	6.08	4.76
	67-66-3	Trichloromethane	42	13,300	300,000	4.12	5.48	1.35
	7758-98-7	Sulfuric acid, Copper(2+) salt (1 : 1)	587	0.057	2,500,000	−1.24	6.40	7.64
	52-68-6	2,2,2-Trichloro-1-hydroxyethyl phosphonic acid, dimethyl ester	273	20	180,000	1.30	5.26	3.95
96h LC_{50} test results with defined chemical purity (98–100%)	63-25-2	1-Naphthalenol methylcarbamate	10	140	2,800	2.15	3.45	1.30
	67-66-3	Trichloromethane	4	43,800	115,000	4.64	5.06	0.42
	7758-98-7	Sulfuric acid, Copper(2+) salt (1 : 1)	7	135	7,340	2.13	3.87	1.74
	52-68-6	2,2,2-Trichloro-1-hydroxyethyl phosphonic acid, dimethyl ester	192	72	50,000	1.86	4.70	2.84

number of publications based on these data has exploded as other QSAR modellers have analysed these data.

The *T. pyriformis* assay is a short-term, static protocol which measures the 50% impairment of growth concentration (IGC_{50}) after 40 hours as the recorded endpoint. Cultures are reared in 50 ml of a semi-defined medium in 250 ml Erlenmeyer flasks. During the assay, a range-finding assay followed by three replicate definitive tests is performed on each test material. Definitive test replicates consist of a minimum of five different concentrations of each test material with duplicate flasks of each concentration. Thus, a minimum of 30 data points comprise each analysis. Duplicate controls, which have no test material but are inoculated with *T. pyriformis*, and a 'blank' are used to provide a measure of the acceptability of the test by indicating the suitability of the medium and test conditions, and provide a basis for interpreting data from other treatments. The viability (population density) of the test organism is measured spectrophotometrically against a control. The IGC_{50} (mg L^{-1}) and the 95% fiducial interval are determined for each test compound. The IGC_{50} is calculated by probit analysis using the percentage control-normalised absorbance as the dependent variable and the toxicant concentration in mg L^{-1} as the independent variable. The information on the assay has been summarised from the University of Tennessee website (www.vet.utk.edu/research/bioactivity.php) with additional details taken as required from the protocol described by Schultz.[54]

In terms of data quality, a number of (unpublished) analyses and replicate tests over time have shown considerable reproducibility and accuracy. Reproducibility has been shown to be consistent within and across recognised mechanisms of action.[55] This conclusion is supported, to some extent, by the development of high quality QSARs (*e.g.* Ellison *et al.*[56]) and good correlation with other nominally high quality toxicity data.[57]

The tests performed at the University of Tennessee form a highly consistent database for modelling and other purposes. They have been performed within the same laboratory, led by the same person and performed by a small number of highly trained personnel, with the results undergoing rigorous internal scrutiny. It should be noted, however, that the *Tetrahymena pyriformis* inhibition of growth assay is not a regulatory endpoint. Neither have the tests been performed according to GLP principles.

A further aspect of the testing is slight alterations in test protocols (particularly in historical assays of more than 15 years of age). While subtle differences in the protocol (duration, temperature, medium composition, *etc.*) were used over the course of more than 30 years of testing, each was used with a test regime to allow for 8–9 cell cycles in controls. Duplicate flasks are inoculated with an initial density of approximately 2,500 cells ml^{-1} with log-growth-phase ciliates. Following approximately 40 hours of incubation at 27 ± 1 °C, population density is measured spectrophotometrically and 50% effect levels are determined.

The limits of the *T. pyriformis* are well understood. Some chemicals (*e.g.* neutral organics with 1-octanol/water partition coefficients greater than 5.0) are not toxic at saturation; others do not attain the measured 50% effect endpoint at saturation, and still other chemicals (*e.g.* highly bioreactive toxicants) have a very narrow concentration-response range that precludes proper statistical analyses.

In terms of ITS, information from the *T. pyriformis* assay (and indeed any evaluated cytotoxicity assay) can be utilised in a number of ways. First, within carefully defined mechanistic categories (most notably compounds known to be narcotic), there could be a direct (inter-species) correlation between toxicity to *T. pyriformis* and fish. Care must be taken in defining the domains of these mechanistic groupings since the protozoan lacks some of the activating enzymes found in fish. Should a compound be shown to exhibit 'excess toxicity' in *T. pyriformis* (*i.e.* toxic potency above that associated with non-polar narcosis), then that could be used as an indication of a specific mechanism (*e.g.* bioreactivity) in the fish. This might indicate the requirement for prioritisation for testing, or more elaborate *in silico* evaluation.

4.5.4   Mutagenicity and Carcinogenicity

For mutagens and carcinogens, the main existing databases are:

- the Carcinogenic Potency DataBase (CPDB)—now also available from DSSTox;
- ISSCAN; and
- a group of around 4,000 mutagens/non-mutagens referred to as the Bursi or Kazius–Bursi database.

The Kazius–Bursi database can be obtained by downloading data from the TOXNET Chemical Carcinogenesis Research Information System (CCRIS) database (http://toxnet.nlm.nih.gov/cgi-bin/sis/htmlgen?CCRIS).

CPDB and TOXNET are significant collections of data that list, for every chemical, all the references available and a brief summary of each result. They therefore form a precious source of basic information. However, they contain data of very different quality. For example, a number of chemicals in the CPDB are characterised only by old carcinogenicity experiments performed with a very small number of animals. In addition, for chemicals tested by different experimenters with contradictory results, there is no final 'summary call' that takes into account the number and, most importantly, the quality of the different results. To resolve this ambiguity, in their modelling work the Kazius–Bursi group calculated a 'summary call' which was based exclusively on statistical criteria. For these reasons, the Kazius–Bursi elaboration of TOXNET is a good basis for large-scale analyses (*e.g.* to derive 'average' overall characteristics of the data), but has to be used with caution for fine-tuned analyses (*e.g.* mutagenic properties of individual chemicals, selection of subsets of chemicals for local QSARs): here experts have to intervene using their critical judgment on a one-by-one chemical basis. The same applies to the CPDB database on carcinogens.

ISSCAN has different characteristics, since it was planned for use in QSAR studies. The carcinogenicity and mutagenicity calls are summaries that are derived from a critical review of the available data and are coded in such a way that no further transformation is necessary.

The different scopes determine the different characteristics of the databases. In ISSCAN the data are filtered for a very specific aim (mainly QSAR) and do not need further pre-treatment. CPDB and TOXNET are more broadly aimed, but because of this have to be pre-treated for the development of QSARs. The Kazius–Bursi version of TOXNET is the result of a statistical pre-treatment; thus it is good for some purposes but needs extensive further filtering for more fine-tuned aims.

Consideration of different database concepts and objectives is important and may have consequences on the final results of the applicative work. The respective issues are generally valid also for databases for endpoints other than mutagenicity and carcinogenicity. Thus, it may be useful to characterise the databases from this point of view as well.[58]

4.5.5 Skin Sensitisation

Skin sensitisation can be assessed by a number of methods; a good overview of the state-of-the-art and historical basis for testing (also including skin irritation) is provided by Basketter *et al.*[59]

Only two tests are now globally accepted by regulatory authorities, *i.e.* the Buehler occluded patch test and the guinea pig maximisation test (GPMT). These two tests are described formally in OECD Guideline No. 406. In the last decade, a murine test, the Local Lymph Node Assay (LLNA), has been formally validated as a complete replacement for the guinea pig test and has its own test guideline (OECD Guideline No. 429).

From these two protocols, two main databases are available for modelling purposes. These are a compilation of GPMT and Buehler data published by Cronin and Basketter,[60] and those for the LLNA published by Gerberick *et al.*[61] which have since been supplemented by data published by Roberts *et al.*[62]

In brief, the local lymph node assay protocol used is typically as follows. Groups of four CBA/Ca female mice (7–12 weeks of age) are treated topically on the dorsum of both ears with 25 µl of test material, or with an equal volume of the vehicle alone (4 : 1 acetone : olive oil, v/v). Treatment is performed once daily for three consecutive days. Five days following the initiation of exposure, all mice are injected *via* the tail vein with 250 µl of phosphate-buffered saline containing 20 µCi (740 kBq) of tritiated thymidine. The mice are sacrificed five hours later and the draining lymph nodes excised and pooled for each experimental group. The lymph node cell suspension is washed twice in an excess of phosphate-buffered saline and then precipitated with 5% trichloroacetic acid at 4 °C for 18 hours. Pellets are resuspended in trichloroacetic acid and the incorporation of tritiated thymidine is measured by liquid scintillation counting.

The LLNA has been shown to be repeatable and reproducible across different laboratories.[59] The database compiled by Gerberick *et al.*[61] represents an historical compilation from various sources (including publications) as well as corporate information and internal studies. Therefore experiments have not been performed in the same laboratory or by the same technicians. As some of the data are up to 15 years old they cannot be said to have been performed exactly according to the OECD guidelines. In addition, not all studies may comply with GLP. Regardless of this, these data have been analysed and rationalised extensively (Roberts *et al.*)[63] and they can be considered to be accurate.

One possible source of variability within the database has been the use of different vehicles to apply the compounds; this has been shown to have a significant, although not extreme contribution to skin sensitisation potency in the LLNA.[64] Another issue that should be borne in mind in utilising skin sensitisation data is the occlusion of the application area. The current 'clean' version of the database, with a small number of errors having been rectified (*e.g.* a duplicate chemical, incorrect structures) is available on the website page of the European Union CAESAR project (www.caesar-project.eu).

In the context of ITS for skin sensitisation, data from LLNA are generally accepted to form the gold standard. However, GPMT and Buehler data are also very valuable. It is usually acknowledged that the LLNA is not as sensitive an assay as the GPMT and Buehler assays. In terms of variability, LLNA test results are probably best utilised to provide a (semi-)qualitative assessment of skin sensitisation (*i.e.* non sensitiser or sensitiser)—and possibly weak, moderate or strong. Generally there should be good concordance between tests when considered categorically; more caution should be shown in the quantitative comparison of, for example, EC3 values. This latter concern has an impact on the development of QSARs for skin sensitisation.

4.5.6 Mammalian Toxicity

Mammalian toxicity measures are usually related to whole body phenomena, including the processes of absorption, distribution, metabolism and excretion. On the one hand, a compound may affect the organism's integrity at different sites leading to different toxic effects; on the other hand, multiple mechanisms caused by different compounds can lead to the same toxic symptoms. The complexity and multiplicity of the processes and mechanisms involved in mammalian toxicity cause inherent difficulties in their *in silico* modelling. In addition, problems arise from the lack of high quality data suitable for modelling purposes.

Theoretically, data sets for any toxicological endpoint can be subjected to a correlative or statistical analysis to create a relationship between structure and toxicity. If the toxic effect can be related to a known, and ideally single mechanism, such an approach is likely to be successful. However, most toxicological effects are a complex expression of several mechanisms happening in sequential or parallel order leading to highly variable outcomes. Therefore, a statistical correlation between structure and effect without consideration of causal relationships will in most cases lead to flawed models.

For effects that can be attributed to specific physico-chemical properties (*e.g.* water or fat solubility, vapour pressure) or reactivity of a molecule (*e.g.* DNA alkylation, protein binding), a reasonable predictivity for the respective toxicological endpoint (mutagenicity, sensitisation, irritation on eye or skin) can be achieved.[65,66] In contrast, complex endpoints such as acute toxicity or lethality, chronic toxicity, carcinogenicity or reproductive toxicity are amenable to predictions only to a limited extent. Liver toxicity, for example, concerns diverse histopathological observations (cirrhosis, steatosis, fibrosis, *etc.*) which cannot be covered by the same model. To make things even more complicated, significant differences exist in the assessment of toxic effects, the nomenclature of histopathological or clinical–chemical findings, and the interpretation of the biological significance of effects between different examiners.

The quality of toxicological *in vitro* data may vary substantially if the information has been obtained under varying conditions. Some factors responsible for quantitative and even qualitative differences in *in vitro* results have been identified, *e.g.* the use of cell lines with different metabolic capacities, different measured endpoints, and limited solubility or stability of the test substances in the test medium.[67] For genetic toxicity studies *in vitro*, the use of cell lines with insufficient genetic stability can greatly influence the study results. Further, the selection of test substance concentrations is crucial, considering that a number of *in vitro* genotoxicity assays are known to produce false positive results if operated in the cytotoxic concentration range.

The scope of *in vivo* toxicological guideline studies is largely defined in terms of species, number of animals, exposure duration, *etc.* However, these guidelines still allow for many variations within a specific study type. For systemic toxicity studies, rodent species are recommended by the guidelines; therefore most studies are conducted with rats but mice may also be used. More variability is possible regarding the animal strain. Several different rat and mouse strains are available with different susceptibilities, especially with respect to the development of strain-specific tumour types in carcinogenicity studies. Age or weight of animals may also have a relevant influence on study results (especially for short-term studies) because younger animals may be more susceptible to adverse effects. In addition, the route of administration of the test substance may affect the results of a study, *e.g.* for oral dosing, either a bolus exposure (by gavage) or a continuous exposure (administration of the compound *via* feed or drinking water) can result in different toxicities. Furthermore, the selection of dose levels is crucial for the derivation of LO(A)EL (lowest observed adverse effect level) and NO(A)EL (no observed adverse effect level) because they are not statistically derived (while LD_{50} is), but give the definitive doses applied in the study with LO(A)EL or without NO(A)EL effects.

Minimum numbers of animals per group and the number of dose groups are defined by the guidelines. Higher animal numbers and more dose groups would lead to higher statistical reliability such that spontaneous lesions could be better differentiated from effects related to the test substance. However, for animal welfare reasons, only the minimum adequate number of animals should be used.

Due to the high biological variability, the reproducibility of experimental results from non-standardised toxicity studies can be rather low. The high experimental variability observed in rodent cancer studies becomes evident when results from the highly standardised National Toxicology Program (NTP) carcinogenicity studies are compared with less standardised protocols. For a data set of 100 chemicals, a reproducibility of less than 60% was found.[68] This demonstrates the importance of experimental data quality, documented with respect to reliability and relevance, to allow the selection of consistent studies for ITS model building. Poor study planning or limited reporting of study details may prevent their use in *in silico* toxicology.

4.5.6.1   Repeated Dose Toxicity

The RepDose database (www.fraunhofer-repdose.de) has been developed within the framework of the CEFIC Long-range Research Initiative (LRI) to enable (Q)SAR modelling on repeated dose toxicity.[69] Data for about 600 compounds from *circa* 1,700 studies have been extracted from peer-reviewed publications based on the following criteria:

- only organic chemicals—preferentially High Production Volume (HPV) chemicals—with a limited number of defined functional groups were selected; polymers, mixtures (including racemates), salts, metals, *etc.* were excluded;
- only studies using mouse or rat species were used;
- only studies with oral (drinking water, feed, gavage) or inhalation route were used; and
- subacute, subchronic and chronic studies with exposure duration from 14 days to life-span (*circa* 730 days) were used.

RepDose contains chemical structure data in sdf-file format and physico-chemical data (molecular weight, melting and boiling point, solubility in water, log *P*, pKa, vapour pressure) in addition to general study data and study results. The section on study data describes the design of the study and enables a direct comparison of the quality of studies and associated uncertainties. Details about animal strains, number of animals per group and sex are given as well as purity of the test chemical, route and frequency of application, study duration, post-exposure periods and dose groups.

The section on study results features information about effects and target organs. The lowest observed effect levels (LOEL) is given for each observed effect. Furthermore a study LOEL (defined as the concentration where the most sensitive target organ is affected) and, if possible, a study NOEL is given for each study. The quality of the study is assessed and reliability indicators are defined differentiating guideline studies (A), studies with some deficiencies (B), poorly documented studies (C), and special studies with only limited data designed for evaluation of a certain endpoint (D).

Evaluation of *in vivo* tests considers:

- description of the test substance (identity, purity and stability);
- description of the dose finding rationale: description of a range finding study or literature search to verify dose selection (optimal is a minimal dose with no effect, a middle dose with some effect and a maximum dose with adverse effect);
- description of the test procedure including exposure period, post-exposure period and analytical control of exposure concentration;
- data on the test species including the number of individuals tested and information on biology/physiology of test animals (sex, age, feeding, disease control, strains, *etc.*);
- description of the technical procedures including narcosis and fixation methods;
- demonstration that the negative control groups do not show any substance related effect;
- description of the examination extent: targets and organs as well as details of analysis (necropsy, histopathology, urine analysis, haematology, *etc.*);
- description of effects and targets/organs including LOEL, severity and statistical significance (if available consideration of historical data of control groups); and
- description of the methods used for data evaluation, *e.g.* statistical analysis.

The analysis of chemical distribution with respect to their minimal study LOEL in RepDose reveals a normal distribution for oral and inhalation exposure (Figure 4.3). About 80% of the chemicals in RepDose are of moderate toxicity with a study LOEL value between 0.1 and 10 mmol kg^{-1} bw d^{-1}. This corresponds to 14–1400 mgkg^{-1} bw d^{-1} taking the mean molecular weight of all substances in RepDose of 138.2 Da.

Many HPV chemicals have been tested in animal studies before, for example, OECD guidelines were published. Thus, a number of 'old' non-guideline studies are present for this group of substances. The extent of examination has increased in previous decades, so the checklist approach considering examination extent and/or quality of study description would rate all these studies in IUCLID as 'valid with restrictions' or even 'invalid' (see Table 4.3). However, an 'old study' does not automatically mean that the results obtained in the study are of poor quality. Therefore a 'coverage' approach is under investigation to assess to which extent 'old' non-guideline studies can be used for risk assessments under REACH. RepDose offers the opportunity to analyse targets, organs and L(N)OELs of guideline and non-guideline studies and to differentiate those targets/organs that trigger the study L(N)OEL and in consequence the derived no effect level (DNEL) for targets/organs which do not occur at study LOEL. In parallel, Bayesian statistics are under investigation to assess this aspect.

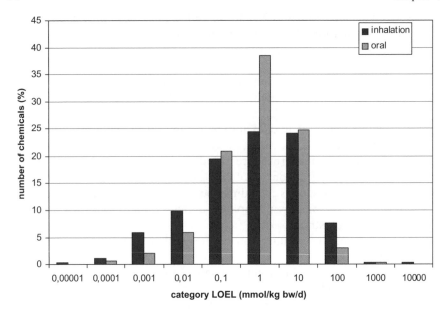

Figure 4.3   Distribution of chemicals according to their study LOELs in RepDose.

4.6   Integration of Data with Distinct Degrees of Reliability

Decision frameworks like ITS require tools to integrate data from different sources and to attribute a quantitative measure of adequacy to both individual test results and the combined WoE result. Two possible methodologies, namely Dempster–Shafer theory and Bayesian statistics, are presented as partly complementary methodologies which enable handling of data quality issues in a quantitative manner.

4.6.1   Dempster–Shafer Theory of Evidence for Information Fusion

The problems concerning how to combine, or fuse, data from multiple sources in order to support decision making have been recently gathered into the research field of *information fusion*. The field originated from the fusing of online sensor data; more recent work has also considered other sources such as databases, simulations or ontologies. Fusion techniques support both human decision-makers and automated decision-making without human intervention. The following definition for this research field has been proposed by Boström *et al.*[70] '*Information fusion is the study of efficient methods for automatically or semi-automatically transforming information from different sources and different points in time into a representation that provides effective support for human or automated decision making.*'

4.6.1.1 Belief Models

Let s_i, $i = 1, \ldots, I$ be different data obtained from I sources. Let $\Theta = \{h_1, \ldots, h_N\}$ be a set of hypotheses under consideration, which can be generally called *sample space*. Such hypotheses may be the presence or absence of a certain characteristic, or a particular value for a parameter measured.

In information fusion systems, the information extracted from each source is represented as a *degree of belief* in an event H from Θ: $x_i(H)$, $i = 1, \ldots, I$. These degrees of belief take values in a real interval and are modelled within a framework of a particular uncertainty theory (probabilities in Bayesian theory, basic probability assignments in Dempster–Shafer's theory of evidence, or membership degrees in fuzzy set theory) taking into account the uncertain nature of the information.[71] The combination of such degrees of belief is performed through a numerical fusion operator $F(x_1, \ldots, x_I)$, and a large variety of such operators has been proposed in the literature. Bloch[72] proposes a classification of these operators providing a guide for choosing an operator in a given problem.

Most processes of combination of beliefs are concerned with adequate modelling of uncertainty since an inadequate belief model can lead to the combination of conflicting and unreliable beliefs. A natural way to deal with this problem is to establish a *reliability* of the beliefs computed within the framework of the model selected and take it into account while building fusion operators.[71] This considers reliability as a second order uncertainty (uncertainty of evaluation of uncertainty) representing a measure of the adequacy of the model used and the state of the environment observed.

4.6.1.2 Representation of Uncertainty

Using Dempster–Shafer's theory of evidence to fuse several sources of information provides an alternative to traditional Bayesian theory for the representation of uncertainty (see Chapter 22). The main characteristic of this framework is that it allows for the allocation of beliefs to subsets of hypotheses, *i.e.* Dempster–Shafer theory does not require an assumption regarding the probability of the individual hypotheses. Another important aspect of this theory is the combination of information obtained from multiple sources of evidence and the modelling of conflicts between them.

Let Θ be the set of all possible hypotheses under consideration and 2^Θ be the set of all subsets of Θ. In the framework of evidence theory,[73–77] information obtained from any source is represented by a *basic probability assignment*, *i.e.* a function m: $2^\Theta \rightarrow [0,1]$ satisfying:

$$m(\Theta) = 0, \tag{4.4}$$

$$\sum_{H \subseteq \Theta} m(H) = 1 \tag{4.5}$$

The concept of basic probability assignment does not refer to probability in the classical sense, but a basic probability assignment $m_i(H)$ expresses the degree of support of the evidential claim from source i that the true hypothesis is in the set H but not in any special subset X of H. Any additional evidence from the same source i supporting the claim that the true hypothesis is in a subset X of H must be expressed by another non-zero value $m_i(X)$.

The original combination rule of multiple basic probability assignments, respectively modelling multiple sources of evidence, is Dempster's rule.[77] This rule of combination is a generalisation of Bayes' rule and it is used for the fusion of independent and equally reliable sources of information. When the basic probability assignments $m_i(H)$ are not equally reliable, several alternative approaches can be followed to include reliabilities into the framework of evidence theory.[78] One of them consists of defining new combination rules (*e.g.* the trade-off rule) as alternatives to Dempster's rule. If a measure of reliability or trust for each belief can be established: R_i in $[0, 1]$, $i = 1, \ldots, I$. R_i is close to 0 if source i is unreliable and is close to 1 if it is reliable. Without loss of generality, reliabilities sum up to 1, that is, $\Sigma_i R_i = 1$. In the trade off combination rule, reliability factors R_i serve as weights assigned to each source:

$$m(H) = \sum_{i=1}^{I} R_i m_i(H) \tag{4.6}$$

This weighted averaging combination rule generalises the averaging operation usually used for probability distributions.

4.6.1.3   Case Study of BCF Classification

This subsection illustrates how knowledge is represented and inference performed using Dempster–Shafer theory of evidence in a representative case study dealing with the classification of chemicals according to their bio-concentration factor (BCF). In the context of REACH, the aim is to provide a measure of predicted BCF and to decide whether that value falls above or below specific BCF regulatory limits. Here, a substance is identified as non-bioaccumulative (*nB*) when its log BCF is lower than 3.3, bioaccumulative (B) when its log BCF is between 3.3 and 3.7, and very-bioaccumulative (vB) when it is higher than 3.7.[18]

To analyse the representative case of 1,2,3,4-tetrachlorobenzene (CAS 634-66-2), experimental BCF data for this compound were collected from two different information sources:

- the EURAS database,[17] which contains two or more BCF values for many chemicals and a reliability score ranging from 1 to 4 assigned on an expert basis; and
- Arnot's database,[40] with more BCF values for each chemical and an overall reliability score ranging from 1 to 3 based on six evaluation criteria.

Table 4.12 Reliability scores and log BCF values for 1,2,3,4-tetra-chlorobenzene obtained from the EURAS[17] and Arnot[40] databases.

Data source	Reliability score	log BCF								
EURAS	2	3.2	3.2							
Arnot	1	3.0	3.0	3.4	3.4	3.4	3.4	3.4	3.6	3.8

Table 4.13 Normalised reliability weights and basic probability assignments for 1,2,3,4-tetrachlorobenzene calculated from the EURAS[17] and Arnot[40] databases.

Data source	Reliability weight	Basic probability assignments			
		$m(\{nB\})$	$m(\{B\})$	$m(\{vB\})$	$m(\{nB, B, vB\})$
EURAS	0.47	0.50	0	0	0.50
Arnot	0.53	0.18	0.55	0.09	0.18

In both cases, the score 1 indicates the highest reliability. Table 4.12 shows the reliability scores and the log BCF values available for 1,2,3,4-tetrachlorobenzene.

First, the reliability scores are converted into reliability weights. The Klimisch score of 2 (*reliable with restrictions*) of the BCF values for 1,2,3,4-tetrachlorobenzene from the EURAS database prompts a reliability weight of 0.9 (almost but not completely reliable). The data from Arnot's database take the maximum reliability score of 1 for 1,2,3,4-tetrachlorobenzene and, therefore, a reliability weight of 1 can be assigned. Finally, the two reliability scores are normalised with the purpose that they sum up to 1, obtaining the normalised reliability weights as shown in Table 4.13.

The frame of discernment Θ in the current case study of chemical bioaccumulation assessment is the set $\{nB, B, vB\}$. Therefore, a basic probability assignment specifies the amount of information supporting that the true hypothesis lays in the subsets $\{nB\}$, $\{B\}$, $\{vB\}$ or $\{nB, B, vB\}$. The latter case, $\{nB, B, vB\}$, denotes lack of evidence (*i.e.* the bioaccumulation category is unknown).

The way to distribute belief values into a basic probability assignment depends on the available data from the source of information. For instance, when a gold standard test method exists for the endpoint in question, then a test validation of the new source can be performed and this information can be used to define the basic probability assignment. Another possibility is to use some information obtained from the distribution of the source data. A third alternative is described as follows. The information in this example consists of a number k of BCF values. As stated in Fernández *et al.*,[73] every single item of evidence falling inside a certain category in k trials can be assigned a belief value equal to $1/(k+2)$. Since EURAS provides two BCF measures for the chemical 1,2,3,4-tetrachlorobenzene, a confidence of $1/4 = 0.25$ can be assigned

Table 4.14 Basic probability assignments obtained for 1,2,3,4-tetra-chlorobenzene after the combination of the EURAS[17] and Arnot[40] databases.

Data source	Basic probability assignments			
	$m(\{nB\})$	$m(\{B\})$	$m(\{vB\})$	$m(\{nB, B, vB\})$
Combined	0.33	0.29	0.05	0.33

to each of these two measures. The remaining amount of information, $2/4 = 0.50$, can be attributed to the subset containing the three individual hypotheses, *i.e.* $\{nB, B, vB\}$. Following the same reasoning, Arnot's database provides nine BCF measures for the same chemical, hence a confidence of $1/11 = 0.09$ can be assigned to each of these nine measures. The remaining amount of information, $2/11 = 0.18$, can be assigned to the subset $\{nB, B, vB\}$.

Analysing the basic probability assignments m_1 and m_2 obtained from the EURAS and Arnot databases, respectively, shown in Table 4.13, reveals that EURAS data gather the supporting evidence on the category nB, although the other two bioaccumulation categories cannot be discarded according to the high uncertainty value equal to 0.50. On the other hand, Arnot's evidence is stronger on the category B. Using Equation (4.6) to combine the two sources of evidence taking into account their corresponding reliability coefficients, the basic probability assignments are obtained, as shown in Table 4.14.

Comparing the final basic probability assignments with the two previous ones, the categories nB and B seem to be equally plausible, although the third alternative, vB, cannot be completely discarded in this case according to the high uncertainty value of 0.33 assigned to the set $\{nB, B, vB\}$.

It is obvious that the results obtained from any combination of partially reliable sources of information will depend largely on their relative reliability weights. In order to analyse this issue in the case of 1,2,3,4-tetrachlorobenzene, Figure 4.4 depicts the belief intervals obtained for the categories nB and B. In the case of category nB, these belief intervals are lower bounded by the values $m(\{nB\})$, and upper bounded by the sum $m(\{nB\}) + m(\{nB, B, vB\})$. In the case of category B, the corresponding intervals are lower and upper bounded by the values $m(\{B\})$ and $m(\{B\}) + m(\{nB, B, vB\}\{nB, B, vB\})$, respectively. Three zones are perfectly distinguishable, and the case discussed here with the reliability weights given in Table 4.13 corresponds to the boundary between zones I and II in Figure 4.4.

4.6.2 Bayesian Networks for Integrating Evidence and Data Quality

Expert systems for integrating chemical and toxicological knowledge and information have to deal with various aspects of uncertainty. Current ITS are largely rule-based and attempts to include uncertainty factors[79,80] have not

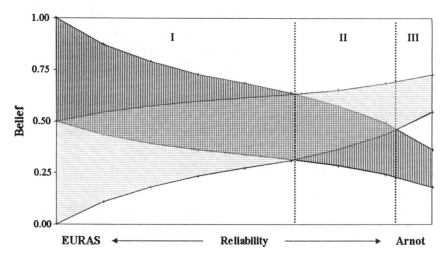

Figure 4.4 Belief intervals for categories nB (vertical lines) and B (horizontal lines) for 1,2,3,4-tetrachlorobenzene, computed along the complete variation of EURAS and Arnot reliability weights. Three zones are perfectly distinguishable: I - reliability weights where both intervals partially overlap, but with a dominance of the category nB; II - reliability weights where both intervals are still partially overlapped, but with a dominance of the category B; and III – reliability weights where the belief intervals corresponding to category B dominate completely over those corresponding to category nB.

been satisfactory.[81] Bayesian networks are an approach to combine information from different sources and handle uncertainty.[81–84]. These graphical networks connect probabilistic quantities through causal relationships. The simplest Bayesian network for the prediction of a chemical endpoint property (C) from a test result (T) can be drawn as follows:

The arrow indicates that the presence of the endpoint property (C) increases the probability of a positive test result (T). This is the direction of the (assumed) causality. Application of Bayes' theorem allows the chemical property to be inferred from the given test result. For *binary* (yes/no) variables, C and T, the calculations are the same as those for the two-by-two table in medical diagnosis.[85,86]

It should be noted that the concept of Bayesian inference, with uncertainty defined in terms of sensitivity (proportion of true positives) and specificity (proportion of true negatives), will distinguish between the predictive power for positive and negative results. A test which has a high sensitivity but a low specificity might give sufficient posterior probability in the case of a positive result, but it may be insufficient to support a negative result.

Furthermore, Bayesian networks allow for the reconciliation of probabilities and costs. These utilities may be monetary but also relate to, for example, animal welfare, social effects or industrial values.[87] Thus, a framework is provided for the evaluation of all kinds of utilities in relation to the uncertainties of tests, datasets, chemical and toxicological properties, *etc.*[81]

4.6.2.1   Uncertainty and Weight-of-Evidence

The term WoE is used in risk assessment in a variety of ways. It can be used to qualitatively assess and combine different lines of evidence in human and ecological risk assessment.[88,89] In a more quantitative way, WoE relates to the logarithm of the likelihood ratio of one hypothesis against another.[81,90] In addition to verbal arguments of WoE, which are always necessary, the assessment requires quantitative probabilities expressed numerically and/or graphically. Evidently, the probabilities for effects of a substance at a specific toxicological endpoint may differ between information from *in vivo* experiments, *in vitro* tests and/or *in silico* predictions, and may positively or negatively impact on the evidence.

Bayesian inference can combine prior information (including expert opinion) and information from testing, and allows for successive updating of the prediction probability when new, additional information is found or generated. In a quantitative WoE procedure, the result of sequentially adding existing information or generating new information will show whether the confidence in the conclusion has increased or decreased.[91,92] It should be noted that the order in which information from different sources is combined does not influence the calculated posterior probability.

Calculation of the uncertainties and WoE for a conclusion in a Bayesian ITS framework requires input on the statistical performance (predictability) of all the sources of information to be taken into account. In its simplest form, the sensitivity and specificity of a test towards a gold standard can be estimated from the performance of the test on given training set(s) of chemicals. The nature or selection of the training set is a formidable problem.

If multiple tests are separately evaluated, a so-called *naïve Bayesian classifier* is used,[83] assuming that the test performances are independent of one another, for each state of the endpoint, C (left network). But the test performances may be correlated, which is indicated by the dashed causality arrow (right network):

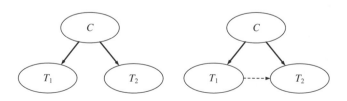

To model test correlation, joint test results must be available, but more often than not there is an unacceptable fraction of missing data. Test correlation may slip in when one is a model developed on largely the same training data, or the tests may essentially measure the same mechanism.

Bayesian networks may account for a prior probability (*e.g.* when a specific class of substances applies). Specific properties can be designated through an indicator property (*e.g.* the presence of a chemical fragment or structure) to model prior dependencies.

In the absence of test performance data or when available data are too scarce to give reasonable estimates, other methods might replace the empirical, or statistical, estimates. As a temporary solution, expert opinion-based estimates of test sensitivity/specificity, and prior probabilities, can also be used until fact-based estimates become available. A big advantage of this quantitative approach is its transparency and consistency, especially when expert judgement is used to temporarily estimate the (inherent) uncertainty of a method.

4.6.2.2 Data Quality Factors in Bayesian Networks

If a study of some chemical does not meet an agreed quality standard (*e.g.* it is not performed under GLP), it may have lower influence in the overall WoE procedure compared with a similar study that was performed under GLP. Likewise, if a guideline study was performed for a substance outside of its applicability domain, this ought to be reflected by a lower posterior probability of the chemical endpoint. Since the statistical methodology should be similar in all cases, quality factors (Q) are introduced to account for the (un)reliability of a specific test result. This is approached by extending the Bayesian network with quality variables ('nodes'). Here is an example of a one-test 'ITS', with one (left) or two (right) quality node(s) attached to the test.

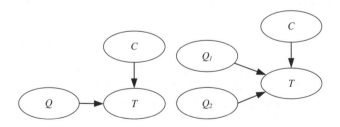

The fulfilment of the quality condition(s) sets the performance indicators of the test (sensitivity, specificity) to their best value. Insufficient quality conditions are supposed to degrade the test performance. The quality weight factors, Q, range from 0 to 1, where a value of 0 makes the information content of the test result worthless (no increase in posterior probability of WoE), and a quality factor of 1 would mean that the test result is not penalised for any omission or defects in the specific test result.

The way in which these quality factors are taken into account in the calculation of the posterior probability is through modification of the specificity (s_p) and sensitivity (s_e) values. The sensitivity of the test, the probability of true positives, taking into account the (combined) quality factor, Q, becomes:

$$s_e(Q) = 0.5 + Q \cdot (s_e - 0.5) \qquad (4.7)$$

This means that, when $Q = 1$ (*i.e.* in case of optimal conditions), the sensitivity is s_e. At the other extreme, when $Q = 0$, the sensitivity becomes 0.5 (*i.e.* the test is as bad as tossing a coin). Intermediate Q values lead to values between the optimal s_e and 0.5 (with $s_e > 0.5$). Similarly, the specificity of the test, the true negative probability, taking into account the quality factors (Q), and with optimal $s_p > 0.5$, becomes:

$$s_p(Q) = 0.5 + Q \cdot (s_p - 0.5) \qquad (4.8)$$

The equations for $s_e(Q)$ and $s_p(Q)$ can be used as a default option and may be modified by expert opinion, *e.g.* about how the quality factor relates to Klimisch[7] codes or how quality factors may differ for sensitivity and specificity. Each quality issue (*e.g.* GLP, domain of applicability, *etc.*) may have its own quality factor, and the relevant quality factors are combined through multiplication:

$$Q = Q_{GLP} \cdot Q_{ApplicabilityDomain} \cdot Q_{...} \qquad (4.9)$$

So, if the Q factor for not performing a study under GLP is 0.9 and the Q factor for applying a methodology outside its applicability domain is 0.3, then the combined Q factor becomes 0.27.

4.6.2.3   *Implementation Issues of Quality Factors in ITS*

The quality factor for a specific test reflects the importance of expert judgement with respect to the quality of an *individual* test result for a given chemical on the continuous scale, as proposed in the previous section.

The main issues that should be reflected in the quality factor of an individual test result are the reliability topics dealt with in the description of the QSAR Prediction Reporting Format,[93,94] (see Chapter 11). These issues are not specific for QSARs, but can be applied to any test method. They are:

- domain of applicability;
- experimental execution of the method; and
- documentation of the result.

It should be noted that general test method information (*e.g.* the predictive power of a method) should not be reflected in this quality factor, as it will already have been taken into account by means of the sensitivity and specificity of the method. Therefore, a method with a very bad performance in terms of predictivity, but which has been applied correctly within the domain of applicability and which was documented adequately would be assigned a quality factor of 1.

In order to systematically and transparently assign a (combined) quality factor to a specific test outcome, it is advisable to create checklists of (method-specific) issues, guiding the user in assessing the quality of an individual test result. A number of issues on these checklists will apply in general (*e.g.* is the test substance within the domain of applicability of the method), but a number of test-specific and endpoint-specific issues should also be part of these checklists. For each method defined as an available source of data within the ITS, it would be necessary to generate method-specific lists of quality issues to consider.

The data quality factor does *not* reflect the overall *adequacy* of the result, as the *adequacy* of the result consists of both the (optimal) statistical performance of the method (*relevance*, expressed as sensitivity and specificity of a method), as well as the quality of the application of the method to the specific substance of interest (*reliability*, expressed as data quality factors). These two aspects should not be confused (see Section 4.2).

The probability of an individual test result calculated using Bayesian inference and taking into account the data quality factors can be seen as a measure of *adequacy*; in other words, a weight factor in a WoE procedure. Bayesian inference furthermore allows the combination of individual test result probabilities, giving the probability of the conclusion of their combined WoE. This probability can be used to determine if the WoE procedure complies with the information requirements set by a specific regulatory framework.

4.6.2.4 One-Test Bayesian Network with Quality Node(s)

This subsection presents the equations needed to calculate the posterior probability and WoE for the one-test Bayesian network to predict a binary chemical property (endpoint), *C*, from binary test information, *T*, given a (total) data quality factor, *Q*.

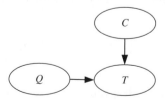

No specialised software is needed to be able to use the network for any input data as the calculation is very simple. The results have been checked using the

Table 4.15   Standard medical diagnosis two-by-two table layout for the test performance: probability of test result, T, given the chemical property, C: $\Pr(T|C)$. s_e and s_p are sensitivity and specificity of the test, respectively.

	$C = C^+$	$C = C^-$
$T = 1$	s_e	$1 - s_p$
$T = 0$	$1 - s_e$	s_p
Total probability	1.0	1.0

review version Hugin Lite® 7.0 (www.hugin.com). First, consider the conditional probability table (CPT) of the test. Without the quality factor, it is the standard table of medical diagnosis; see Table 4.15 *cf.* Campbell and Machin.[86]

The chemical property has two states: C^+ (present) and C^- (absent). A 'positive' test result is denoted as 1, a 'negative' test result as 0. The conditional probability of a test result, given the chemical property, is indicated as $\Pr(T|C)$, when the variables are unspecified. The sensitivity is:

$$s_e = \Pr(T = 1 | C = C^+) \qquad (4.10)$$

and the specificity is:

$$s_p = \Pr(T = 0 | C = C^-) \qquad (4.11)$$

Bayes' theorem for the one-test network with an overall data quality factor can be formulated as:

$$\Pr(C|Q, T) = k \cdot \Pr(T|C, Q) \cdot \Pr(C, Q) \qquad (4.12)$$

which can be interpreted as the posterior probability of the chemical property (given test quality and test result) is proportional to the likelihood of the test result (given the chemical property and the test quality) times the prior probability of chemical property and test quality.

The joint prior probability of chemical property and test quality is assumed to be independent:

$$\Pr(C, Q) = \Pr(C) \cdot \Pr(Q) \qquad (4.13)$$

The posterior 'odds' towards a positive chemical property is:

$$\frac{\Pr(C^+|Q, T)}{\Pr(C^-|Q, T)} = \frac{\Pr(T|C^+, Q)}{\Pr(T|C^-, Q)} \cdot \frac{\Pr(C^+)}{\Pr(C^-)} \cdot \frac{\Pr(Q)}{\Pr(Q)} = \frac{\Pr(T|C^+, Q)}{\Pr(T|C^-, Q)} \qquad (4.14)$$

Table 4.16 Extension of the test performance table (Table 4.15) to include the effect of the quality of the test, Q: probability of the test result given the chemical property and the quality of the test: $\Pr(T|C,Q)$.

	$C = C^+$		$C = C^-$	
	$Q = 1$	$0 \leq Q < 1$	$Q = 1$	$0 \leq Q < 1$
$T = 1$	s_e	$0.5 + Q \cdot (s_e - 0.5)$	$1 - s_p$	$0.5 - Q \cdot (s_p - 0.5)$
$T = 0$	$1 - s_e$	$0.5 - Q \cdot (s_e - 0.5)$	s_p	$0.5 + Q \cdot (s_p - 0.5)$
Total probability	1.0	1.0	1.0	1.0

when assuming non-informative (*fifty–fifty*) prior odds for the chemical property, and the prior probability of the test quality cancels. In this way, the posterior odds are equal to the *diagnostic likelihood ratio*.[85]

From the posterior odds, and using Table 4.16, the posterior probability distribution of the chemical property, when the test is positive is:

$$
\begin{cases}
\Pr(C^+|T = 1) = \dfrac{0.5 + Q \cdot (s_e - 0.5)}{1 + Q \cdot (s_e - s_p)} \\[3mm]
\Pr(C^-|T = 1) = \dfrac{0.5 - Q \cdot (s_p - 0.5)}{1 + Q \cdot (s_e - s_p)}
\end{cases}
\tag{4.15}
$$

and when the test is negative is:

$$
\begin{cases}
\Pr(C^+|T = 0) = \dfrac{0.5 - Q \cdot (s_e - 0.5)}{1 - Q \cdot (s_e - s_p)} \\[3mm]
\Pr(C^-|T = 0) = \dfrac{0.5 + Q \cdot (s_p - 0.5)}{1 - Q \cdot (s_e - s_p)}
\end{cases}
\tag{4.16}
$$

Both posterior distributions are a function of the quality factor, Q. Note that when the test quality is nil (*i.e.* $Q = 0$), all posterior expressions become 0.5. As the *fifty–fifty* prior distribution was also {0.5, 0.5}, we have learned nothing.

As an (artificial) example, consider: $s_e = 0.8$, $s_p = 0.7$, and vary Q from 1 to 0, in two steps. The posterior probabilities for a chemical property being positive, given a positive or negative test result are given in Table 4.17. The posterior probability of the chemical property being negative is 1 minus the positive posterior probability for the same test result.

Quantitative WoE can be defined as the logarithm of the test odds:[81,90]

$$
\mathrm{WoE}_T^+ = 10 \cdot \log_{10}\left(\frac{\Pr(T|C^+, Q)}{\Pr(T|C^-, Q)}\right)
\tag{4.17}
$$

where the common logarithm and multiplication by ten is used to obtain a

Table 4.17 Posterior probability for a chemical property being positive, when the test is either positive or negative, as a function of the data quality factor, Q.

	$Q=1$	$Q=05$	$Q=0$
$\Pr(C^+\vert T=1)$	0.73	0.62	0.50
$\Pr(C^+\vert T=0)$	0.22	0.37	0.50

Table 4.18 Posterior probability for a chemical property being positive, when the test is either positive or negative, as a function of the data quality factor, Q.

	$Q=1$	$Q=0.5$	$Q=0$
$\mathrm{WoE}^+_{T=1}$	4.26	2.11	0
$\mathrm{WoE}^+_{T=0}$	−6.53	−2.34	0

convenient unit of WoE (see Jaworska *et al.*[81] and references therein). In the case of the non-informative prior, the prior odds are 1, in which case this is also the posterior WoE.

Thus, the WoE towards a chemical property depends on both the test result and the test quality. Using the posterior expressions, the WoE toward the chemical being positive are obtained as a function of data quality:

$$\mathrm{WoE}^+_{T=1} = 10 \cdot \log_{10}\left(\frac{0.5 + Q \cdot (s_e - 0.5)}{0.5 - Q \cdot (s_p - 0.5)}\right) \tag{4.18}$$

$$\mathrm{WoE}^+_{T=0} = 10 \cdot \log_{10}\left(\frac{0.5 - Q \cdot (s_e - 0.5)}{0.5 + Q \cdot (s_p - 0.5)}\right) \tag{4.19}$$

For the example of Table 4.17, $s_e = 0.8$, $s_p = 0.7$, the WoE for the chemical property being positive at a given test result is calculated as a function of degrading test quality (Table 4.18). Note that the WoE eventually turns zero, when the test quality becomes worthless.

4.7 Cost-Effectiveness Analysis of Tests and Testing Systems

Testing reveals information on the intrinsic properties of chemicals. This allows for hazard assessment and, if additional information on human or environmental exposure is included, for risk classification of a substance. The new European chemicals regulation, REACH, has defined a set of endpoint-specific information requirements for the registration of chemicals. In addition, the REACH regulation explicitly underlines the need to '*replace, reduce or refine testing on vertebrate animals . . . wherever possible*' and to use animal tests only '*as a last resort*'. This indicates political awareness of animal welfare as a genuine ethical concern. Clearly, generating the required information is costly, where 'costs' denote any type of resources that needs to be invested in order to achieve a required information outcome.

Assuming that improving animal welfare in toxicity testing is a policy objective, a decision is required as to which tests or testing systems are selected if, for a given toxicological endpoint, different testing options are available or are expected to become available in the future. This requires comparing and evaluating available tests and testing systems according to:

- (expected) test information outcome;
- (expected) testing costs; and
- (expected) animal welfare loss.

From the regulator's perspective it is important to know how information targets should be defined in order to maximise social welfare. The decision required from chemical producers or manufacturers is different. Being private decision-makers on a competitive market, they need to know which test or testing system to select for a particular endpoint in order to maximise profits (*i.e.* the difference between revenue and costs).[95] Solving such multi-objective decision problem requires an economic assessment of tests and testing systems that allows:

- weighing of the assessment indicators against each other; and
- taking (limited) testing budgets and minimum information requirements into account.

To provide guidance on how to optimally allocate scarce resources on a set of competing programmes or treatments, cost-effectiveness analysis (CEA) has become a well-established analytic tool, particularly in cases where assessment indicators of interventions are valued in different measurement units.[96–99] Applying CEA to the problem of test selection reveals a cost function for information, denoting additional costs for an additional unit of information outcome. Based on the cost function, a decision-maker can identify the choice set of cost-effective testing strategies for achieving given (minimum) information targets and for animal test replacements. This facilitates the development and implementation of optimal solutions to the test selection problem.

Clearly, CEA application relies on data availability of relevant input parameters. Hence, conclusions drawn from CEA crucially depend on the quality of input data used. To apply CEA to the problem of test selection as described above, data are needed on:

- test information outcome, which is taken to be the 'effect' of testing;
- monetary testing costs; and
- animal welfare.

Table 4.19   Data quality matrix for CEA application.

	Data quality levels		
CEA assessment indicators	*Contextual DQ*	*Intrinsic DQ*	*Accessibility and timeliness*
Information outcome			
Monetary testing costs			
Animal welfare loss			

For each of these assessment indicators different measures may exist. For example, the 'information outcome' of a test or a testing system is commonly expressed in terms of its predictive accuracy, for which different measures have been proposed in the literature (sensitivity, specificity, concordance, *etc.*). Which of these measures to select to perform CEA can be regarded as an issue of data quality assessment: While CEA is usually applied on the basis of best currently available data, data quality analysis is a useful step within the overall analytical process because it stimulates critical reflection about datasets used and supports identification of further research needs. In the literature, different frameworks for data quality assessment have been suggested. Following Strong *et al.*[100] it is proposed to distinguish different 'levels' of data quality assessment in CEA for constructing a 'data quality matrix' as shown in Table 4.19:

- contextual data quality, addressing the adequacy of analytic concepts used for determining required assessment indicators of CEA;
- intrinsic data quality of data selected for indicators used in CEA, addressing the accuracy or 'goodness-of-fit' of datasets (e.g. by means of the indicators discussed in Section 4.3); and
- accessibility and timeliness of data used, addressing the extent to which datasets can easily be retrieved and are sufficiently up-to-date for the task at hand.

Use of the data quality matrix has two main objectives. First, the matrix provides a systematic overview of data quality for every test and testing system included in CEA and the associated assessment indicators for effects and costs. Secondly, it facilitates in the identification of asymmetries in data quality both across quality levels and across tests and testing systems.

A detailed discussion of data quality in CEA can only be performed for a concrete CEA model. In addition, some more general aspects regarding data quality in CEA applications to chemical testing can be identified:

- First, attention should be paid to contextual data quality needs of information outcome measures. This is because standard measures of predictive accuracy do not cover all error probabilities of a test, which is insufficient for regulatory decision-making on the risks of chemicals. Moreover, for *in silico* methods such as SARs and QSARs predictive accuracy depends on the applicability domain of the model, which means that standard predictivity measures cannot readily be compared across different (Q)SARs.[66] Suggestions to arrive at improved concepts for measuring the information outcome of tests and testing systems, including ITS, have been brought forward in the literature.[101–103]

- Second, while there is general agreement that the harm done to experimental animals should be minimised,[1,104] there is no common standard for assessing and measuring that harm. This requires further research for developing appropriate concepts of laboratory animal welfare assessment. Including such concepts in CEA allows for a more sophisticated economic evaluation of different tests and testing systems that addresses animal welfare impacts of toxicity testing.
- Third, accessibility of data on monetary testing costs is particularly low for non-testing methods. One reason is that most non-testing methods still lack formal validation and regulatory acceptance, which prevents industry from using these methods as standard tools. As a consequence, there have been only few efforts towards a systematic assessment of cost data and towards an inventory of cost data quality. This indicates that the design of regulatory policies does not only guide the type of data generated for hazard and risk assessment of chemicals; it also impacts data quality at different levels.

The strength of CEA is to uncover trade-offs between information outcome, animal welfare loss and direct testing costs, and to facilitate the selection of tests and testing systems that perform best under given constraints. How these constraints are set remains to be subject to (policy) preferences. In this sense, CEA serves as a consistent decision-support tool that facilitates risk management of chemicals. Data quality evaluation and improvement is an integral part of the CEA process.

4.8 Acknowledgements

This work was supported by the EU 6th Framework Integrated Project OSIRIS (contract no. GOCE-ET-2007-037017).

References

1. Regulation (EC) No. 1907/2006 of the European Parliament and of the Council of 18 December 2006 concerning the Registration, Evaluation, Authorisation and Restriction of Chemicals (REACH), establishing a European Chemicals Agency, amending Directive 1999/45/EC and repealing Council Regulation (EEC) No. 793/93 and Commission Regulation (EC) No. 1488/94 as well as Council Directive 76/769/EEC and Commission Directives 91/155/EEC, 93/67/EEC, 93/105/EC and 2000/21/EC, *Off. J. Eur. Union*, 2006, **L396**, 1–849.

2. W. Lilienblum, W. Dekant, H. Foth, T. Gebel, J. G. Hengstler, R. Kahl, P. J. Kramer, H. Schweinfurth and K. M. Wollin, *Arch. Toxicol.*, 2008, **82**, 211.

3. C. J. van Leeuwen, G. Y. Patlewicz and A. P. Worth, in *Risk Assessment of Chemicals: An Introduction, ed. C. J. van Leeuwen and T. G. Vermeire,* Springer, Dordrecht, The Netherlands, 2007, pp. 467–509.

4. Organisation for Economic Co-operation and Development, *Guidance Document on the Validation of (Quantitative) Structure–Activity Relationships [(Q)SAR] Models*, OECD, Paris, 2007, OECD Environment Health and Safety Publications Series on Testing and Assessment No. 69, ENV/JM/MONO(2007)2, www.olis.oecd.org/olis/2007doc.nsf/LinkTo/ NT00000D92/$FILE/JT03224782.PDF [accessed March 2010].

5. J. P. Myers, F. S. vom Saal, B. T. Akingbemi, K. Arizono, S. Belcher, T. Colborn, I. Chahoud, D. A. Crain, F. Farabollini, L. J. Guillette, T. Hassold, S. Ho, P. A. Hunt, T. Iguchi, S. Jobling, J. Kanno, H. Laufer, M. Marcus, J. A. McLachlan, A. Nadal, J. Oehlmann, N. Olea, P. Palanza, S. Parmigiani, B. S. Rubin, G. Schoenfelder, C. Sonnenschein, A. M. Soto, C. E. Talsness, J. A. Taylor, L. N. Vandenberg, J. G. Vandenbergh, S. Vogel, C. S. Watson, W.V. Welshons and R. T. Zoeller, *Environ. Health Perspect.*, 2009, **117**, 309.

6. Organisation for Economic Co-operation and Development, *Descriptions of Selected Key Generic Terms Used in Chemical Hazard/Risk Assessment: Joint Project with IPCS on the Harmonisation of Hazard/Risk Assessment Terminology*, OECD, Paris, 2003, OECD Environment Health and Safety Publications Series on Testing and Assessment No. 44, www.who.int/ipcs/ publications/methods/harmonization/definitions_terms/en/ [accessed March 2010].

7. H. J. Klimisch, E. Andreae and U. Tillmann, *Regul. Toxicol. Pharmacol.*, 1997, **25**, 1.

8. P. G. P. C. Zweers and T. G. Vermeire, in *Risk Assessment of Chemicals: An Introduction, ed. C. J. van Leeuwen and T. G. Vermeire,* Springer, Dordrecht, The Netherlands, 2007, pp. 357–374.

9. R. Bowden, *Building Confidence in Geological Models,* Geological Society, London, 2004, Special Publications Vol. 239, pp. 157–173.

10. S. J. T. Pollard, G. J. Davies, F. Coley and M. Lemon, *Sci. Total Environ.*, 2008, **400**, 20.

11. M. F. W. Festing, *TIPS*, 2003, **24**, 341.

12. S. Festing, *Altern. Lab. Anim.*, 2008, **36**, 1.

13. HERA, *HERA Guidance Document Methodology February 2005*, HERA Secretariat, Brussels, 2005, www.heraproject.com/Library.cfm [accessed March 2010].

14. OECD Secretariat, *Manual for Investigation of HPV Chemicals. Chapter 3.1: Data Evaluation, Guidance for Determining the Quality of Data for the SIDS Dossier (Reliability, Relevance and Adequacy)*, OECD, Paris, 2004. www.oecd.org/dataoecd/60/46/1947501.pdf [accessed March 2010].

15. P. Lepper, *Towards the Derivation of Quality Standards for Priority Substances in the Context of the Water Framework Directive,* Fraunhofer Institut for Molecular Biology and Applied Ecology, Schmallenberg, Germany, 2002, Final Report of the Study, Contract No. B4-30401/20001111/30637/MAR/E1.

16. M. Nendza, *Structure-activity Relationships in Environmental Sciences,* Chapman & Hall, London, 1998.

17. CEFIC Long-range Research Initiative (LRI), *EURAS Bioconcentration Factor (BCF) Gold Standard Database*, Cefic, Brussels, 2007, http://ambit.sourceforge.net/euras/[accessed March 2010].

18. C. Zhao, E. Boriani, A. Chana, A. Roncaglioni and E. Benfenati, *Chemosphere*, 2008, **73**, 1701.

19. D. Young, T. Martin, R. Venkatapathy and P. Harten, *QSAR Comb. Sci.,* 2008, **27**, 1337.

20. US EPA, *DSSTox* [online], US Environmental Protection Agency, Washington DC, www.epa.gov/ncct/dsstox/sdf_nctrer.html [accessed March 2010].

21. E. H. W. Heugens, A. J. Hendriks, T. Dekker, N. M. van Straalen and W. Admiraal, *Crit. Rev. Toxicol.*, 2002, **3**, 247.

22. R. P. Lanno, B. E. Hickie and D. G. Dixon, *Hydrobiology*, 1989, **188/189**, 525.

23. T. Braunbeck and H. Segner, *Ecotoxicol. Environ. Saf.*, 1992, **24**, 72.

24. J. Spitsbergen and M. L. Kent, *Toxicol. Pathol.*, 2003, **31**(suppl), 62.

25. M. L. Kent, S. W. Feist, C. Harper, S. Hoogstraten-Miller, J. MacLaw, J. M. Sanchez-Morgado, M. L. Tanguay, G. E. Sanders, J. M. Spitsbergen and C. M. Whipps, *Comp. Biochem. Physiol.*, 2009, **149C**, 240.

26. K. W. H. Kwok, K. M. Y. Leung, G. S. G. Lui, V. K. H. Chu, P. K. S. Lam, D. Morritt, L. Maltby, T. C. M. Brock, P. J. van den Brink, M. St. J. Warne and M. Crane, *Integr. Environ. Assess. Manag.*, 2007, **2**, 49.

27. M. Hrovat, H. Segner and S. Jeram, *Regul. Toxicol. Pharmacol.*, 2009, **54**, 294.

28. T. H. Hutchinson, N. Scholz and W. Guhl, *Chemosphere*, 1998, **36**, 143.

29. ECVAM, *Development of a Quality Assessment Tool for Toxicological Data*, European Centre for the Validation of Alternative Methods, Institute for Health and Consumer Protection, Joint Research Centre, Ispra, Italy, 2008.

30. D. A. Hobbs, M. S. Warne and S. J. Markich, *Integr. Environ. Assess. Manag.*, 2009, **1**, 174.

31. Organisation for Economic Co-operation and Development, *The Report from the Expert Group on (Quantitative) Structure–Activity Relationships [(Q)SARs] on the Principles for the Validation of (Q)SARs*, OECD, Paris, 2004, OECD Environment Health and Safety Publications Series on Testing and Assessment No. 49, ENV/JM/MONO(2004)24,-www.olis.oecd.org/olis/2004doc.nsf/LinkTo/NT00009192/$FILE/JT00176183.pdf [accessed March 2010].

32. J. C. Dearden, M. T. D. Cronin and K. L. E. Kaiser, *SAR QSAR Environ. Res.*, 2009, **20**, 241.
33. Organisation for Economic Co-operation and Development, *Guidelines for Testing of Chemicals No. 107. Partition Coefficient (n-Octanol/Water)*. OECD, Paris, 1981.
34. D. N. Brooke, A. J. Dobbs and N. Williams, *Ecotoxicol. Environ. Saf.*, 1986, **11**, 251.
35. J. de Bruijn, F. Busser, W. Seinen and J. L. M. Hermens, *Environ. Toxicol. Chem.*, 1989, **8**, 499.
36. P. J. Taylor, *Hydrophobic Properties of Drugs,* Pergammon Press, Oxford, 1990.
37. S. H. Unger, J. R. Cook and J. S. Hollenberg, *J. Pharm. Sci.*, 1978, **67**, 1364.
38. M. T. D. Cronin and D. J. Livingstone, in *Predicting Chemical Toxicity and Fate, ed. M. T. D. Cronin and D. J. Livingstone,* CRC Press, Boca Raton, FL, 2004, pp. 31–40.
39. R. Mannhold, G. I. Poda, C. Ostermann and I. V. Tetko, *J. Pharm. Sci.*, 2009, **98**, 861.
40. J. A. Arnot and F. A. P. C. Gobas, *Environ. Rev.*, 2006, **14**, 257.
41. US Environmental Protection Agency, *EPI Suite Version 4.0*, US EPA, Washington DC, 2005.
42. R. F. Rekker, *The Hydrophobic Fragmental Constant,* Elsevier Scientific, New York, 1977.
43. European Commission, *Technical Guidance Document on Risk Assessment in Support of Commission Directive 93/67/EEC on Risk Assessment for New Notified Substances, Commission Regulation (EC) No 1488/94 on Risk Assessment for Existing Substances, and Directive 98/8/EC of the European Parliament and of the Council Concerning the Placing of Biocidal Products on the Market*, European Commission Joint Research Centre, Ispra, Italy, 2003.
44. Organisation for Economic Co-operation and Development, *Guidelines for Testing of Chemicals No. 305. Bioconcentration: Flow-through Fish Test*. OECD, Paris, 1996.
45. J. A. Arnot and F. A. P. C. Gobas, *QSAR Comb. Sci.*, 2003, **22**, 337.
46. D. T. H. M. Sijm, M. G. J. Rikken, E. Rorije, T. P. Traas, M. S. McLachlan and W. J. G. M. Peijnenburg, in *Risk Assessment of Chemicals: An Introduction , ed. C. J. van Leeuwen and T. G. Vermeire,* Springer, Dordrecht, The Netherlands, 2007, pp. 73–158.
47. ILSI HESI/JRC/SETAC-EU, *Workshop on Bioaccumulation Assessments*, Dutch Congress Centre, The Hague, The Netherlands, 2006.
48. T. Parkerton, J. A. Arnot, A. Weisbrod, C. L. Russom, R. A. Hoke, K. Woodburn, T. P. Traas, M. Bonnell, L. P. Burkhard and M. A. Lampi, *Integr. Environ. Assess. Manag.*, 2008, **4**, 139.
49. S. Dimitrov, N. Dimitrova, T. Parkerton, N. Comber, M. Bonnell and O. G. Mekenyan, *SAR QSAR Environ. Res.*, 2005, **16**, 531.

50. G. E. Howe, L. L. Marking, T. D. Bills, J. J. Rach and F. L. Mayer, *Environ. Toxicol. Chem.*, 1994, **13**, 51.
51. G. A. LeBlanc, *Environ. Toxicol. Chem.*, 1984, **3**, 47.
52. F. L. Mayer, M. R. Ellersieck, *Manual of Acute Toxicity*, US Department of the Interior, Washington DC, 1986, Fish and Wildlife Service Resource Publication 160.
53. M. Hrovat, H. Segner and S. Jeram, *Regul. Toxicol. Pharmacol.*, 2009, **54**, 294.
54. T. W. Schultz, *Toxicol. Mech. Methods*, 1997, **7**, 289.
55. J. R. Seward, G. D. Sinks and T. W. Schultz, *Aquat. Toxicol.*, 2001, **53**, 33.
56. C. M. Ellison, M. T. D. Cronin, J. C. Madden and T. W. Schultz, *SAR QSAR Environ. Res.*, 2008, **19**, 751.
57. J. R. Seward, E. R. Hamblen and T. W. Schultz, *Chemosphere*, 2002, **47**, 93.
58. R. Benigni, C. Bossa, A. M. Richard and C. Yang, *Ann. Ist. Super. Sanità*, 2008, **44**, 48.
59. D. Basketter, R. Darlenski and J. W. Fluhr, *Skin Pharmacol. Physiol.*, 2008, **21**, 191.
60. M. T. D. Cronin and D. Basketter, *SAR QSAR Environ. Res.*, 1994, **2**, 159.
61. G. F. Gerberick, C. A. Ryan, P. S. Kern, H. Schlatter, R. J. Dearman, I. Kimber, G. Y. Patlewicz and D. Basketter, *Dermatitis*, 2005, **16**, 157.
62. D. W. Roberts, G. Y. Patlewicz, S. Dimitrov, L. K. Low, A. O. Aptula, P. S. Kern, G. Dimitrova, M. H. I. Comber, R. D. Phillips, J. Niemela, C. Madsen, E. B. Wedeby, P. T. Bailey and O. G. Mekenyan, *Chem. Res. Toxicol.*, 2007, **20**, 1321.
63. D. W. Roberts, A. O. Aptula and G. Y. Patlewicz, *Chem. Res. Toxicol.*, 2007, **20**, 44.
64. I. R. Jowsey, C. J. Clapp, B. Safford, B. T. Gibbons and D. A. Basketter, *Cutan. Ocular Toxicol.*, 2008, **27**, 67.
65. I. Gerner, S. Zinke, G. Graetschel and E. Schlede, *Altern. Lab. Anim.*, 2000, **28**, 665.
66. B. Simon-Hettich, A. Rothfuss and T. Steger-Hartmann, *Toxicology*, 2006, **224**, 156.
67. L. Pohjala, P. Tammela, S. K. Samanta, J. Yli-Kauhaluoma and P. Vuorela, *Anal. Biochem.*, 2007, **362**, 221.
68. E. Gottmann, S. Kramer, B. Pfahringer and C. Helma, *Environ. Health Perspect.*, 2001, **109**, 509.
69. A. Bitsch, S. Jacobi, C. Melber, U. Wahnschaffe, N. Simetska and I. Mangelsdorf, *Regul. Toxicol. Pharmacol.*, 2006, **46**, 202.
70. H. Boström, S. F. Andler, M. Brohede, R. Johansson, A. Karlsson, J. van Laere, L. Niklasson, M. Nilsson, A. Persson and T. Ziemke, *On the Definition of Information Fusion as a Field of Research*, School of Humanities and Informatics, University of Skövde, Sweden, 2007, Technical Report HS-IKI-TR-07-006.

71. G. Rogova and E Bossé, *Information Quality Effects on Information Fusion*, Defence Research and Development Canada, Ottawa, 2008, Technical Report DRDC Valcartier TR 2005-270, 2008, http://pubs.drdc.gc.ca/PDFS/unc72/p529448.pdf [accessed March 2010].

72. I. Bloch, *IEEE Trans. Syst. Man. Cybern.*, 1996, **26**, 52.

73. A. Fernández, R. Rallo and F. Giralt, *Environ. Sci. Technol.*, 2009, **43**, 5001.

74. P. Smets, *Data Fusion in Transferable Belief Model*, in Proceedings of 3rd International Conference on Information Fusion (Fusion 2000), Paris, July 2000, pp. PS21–PS33.

75. E. Lefevre, P. Vannoorenberghe and O. Colot, *IEEE Intern. Fuzzy Syst. Conf. Proc.*, 1999, **1**, 173.

76. I. Bloch, *Pattern Recognit. Lett.*, 1996, **17**, 905.

77. G. Shafer, *A Mathematical Theory of Evidence,* Princeton University Press, Princeton, NJ, 1976.

78. M. C. Florea, A. L. Jousselme, E. Bossé and D. Grenier, *Inform. Fusion*, 2009, **10**, 183.

79. E. Castillo, J. M. Gutiérrez and A. D. Hadi, *Expert Systems and Probabilistic Network Models,* Springer, New York, 1997.

80. S. Parsons and P. McBurney, in *Predictive Toxicology, ed. C. Helma,* CRC Press, Boca Raton, FL, 2005, pp. 135–175.

81. J. S. Jaworska, S. Gabbert and T. Aldenberg, Towards optimization of chemical testing under REACH: A Bayesian network approach to Integrated Testing Strategies, *Regul. Toxicol. Pharmacol.,* 2010, doi:10.1016/j.yrtph.2010.02.003.

82. U. B. Kjaerulff and A. L. Madsen, *Bayesian Networks and Influence Diagrams: A Guide to Construction and Analysis,* Springer, New York, 2008.

83. F. V. Jensen and T. D. Nielsen, *Bayesian Networks and Decision Graphs,* Springer, New York, 2007.

84. J. Pearl, *Probabilistic Reasoning in Intelligent Systems: Networks of Plausible Inference,* Morgan Kaufmann Publishers, San Mateo, CA, 1998.

85. M. S. Pepe, *The Statistical Evaluation of Medical Tests for Classification and Prediction,* Oxford University Press, Oxford, 2003.

86. M. J. Campbell and D. Machin, *Medical Statistics: A Commonsense Approach,* John Wiley & Sons, Chichester, 1993.

87. R. T. Clemen, *Making Hard Decisions,* Duxbury Press, London, 1996.

88. P. M. Chapman, B. G. McDonald and G. S. Lawrence, *Hum. Ecol. Risk Assess.,* 2002, **8**, 1489.

89. G. A. Burton, G. E. Batley, P. M. Chapman, V. E. Forbes, E. P. Smith, T. Reynoldson, C. E. Schlekat, P. J. Den Besten, A. J. Bailer, A. S. Green and R. L. Dwyer, *Hum. Ecol. Risk Assess.,* 2002, **8**, 1675.

90. E. P. Smith, I. Lipkovich and K. Ye, *Hum. Ecol. Risk Assess.,* 2002, **8**, 1585.

91. L. Eriksson, J. S. Jaworska, A. P. Worth, M. T. D. Cronin, R. M. McDowell and P. Gramatica, *Environ. Health Perspect.,* 2003, **111**, 1361.

92. T. G. Vermeire, T. Aldenberg, Z. Dang, G. Janer, J. A. de Knecht, H. van Loveren, W. J. G. M. Peijnenburg, A. H. Piersma, T. P. Traas, A. J. Verschoor, M. van Zijverden and B. Hakkert, *Selected Integrated Testing Strategies (ITS) for the Risk Assessment of Chemicals*, RIVM, Bilthoven, The Netherlands, 2007, RIVM report 601050001/2007, www.rivm.nl/bibliotheek/rapporten/601050001.html [accessed March 2010].

93. E. Rorije, E. Hulzebos and B. Hakkert, *The EU (Q)SAR Experience Project: Reporting Formats. Templates for Documenting (Q)SAR Results under REACH*, RIVM, Bilthoven, The Netherlands, 2007, RIVM Report 601779001/2007, www.rivm.nl/bibliotheek/rapporten/601779001.html [accesses March 2010].

94. QSAR Working Group, *QSAR Reporting Formats and JRC QSAR Model Database* [online], Version 1.2, Toxicology and Chemical Substances Unit, Institute for Health and Consumer Protection, Joint Research Centre, Ispra, Italy, 2007, http://ecb.jrc.ec.europa.eu/qsar/qsar-tools/index.php?c=QRF [accessed March 2010].

95. S. Gabbert and E. C. van Ierland, *Regul. Toxicol. Pharmacol.*, 2010, in press.

96. B. O. Hansen, J. L. Hougaard, H. Keiding and L. P. Osterdal, *J. Health Econ.*, 2004, **23**, 887.

97. P. Dolan and R. Edlin, *J. Health Econ.*, 2002, **21**, 827.

98. A. M. Garber and C. E. Phelps, *J. Health Econ.*, 1997, **16**, 1.

99. M. Johannesson and M. C. Weinstein, *J. Health Econ.*, 1993, **12**, 459.

100. D. Strong, Y. L. Lee and R. Y. Wang, *Commun. ACM*, 1997, **40**, 103.

101. R. Benigni and C. Bossa, *J. Chem. Inf. Model.*, 2008, **48**, 971.

102. F. Yokota, G. Gray, J. K. Hammitt and K. M. Thompson, *Risk Anal.*, 2004, **24**, 1625.

103. Y. P. Zhang, N. Sussman, G. Klopman and H. S. Rosenkranz, *Quant. Struct.-Act. Relat.*, 1997, **16**, 290.

104. W. M. S. Russell and R. L. Burch, *The Principles of Humane Experimental Technique,* Methuen & Co Ltd, London, 1959.

CHAPTER 5

Calculation of Physico-Chemical and Environmental Fate Properties

T. H. WEBB AND L. A. MORLACCI

SRC, Inc., 2451 Crystal Drive, Suite 804, Arlington, Virginia 22202, USA

5.1 Introduction

The physico-chemical properties of a chemical substance influence its biological activity, environmental fate and transport behaviour. Successful modelling of uptake, bioavailability, distribution, exposure potential and toxicity depends in part on the availability of reliable physico-chemical descriptors for the chemicals of interest. Estimated properties can fill data gaps for chemicals that lack a complete set of reliable measured properties. Modern estimation methods can provide the desired data rapidly and cost-effectively. Property estimation is invaluable for the screening level characterisation of chemicals when the measurement of properties is not possible or is impractical, such as when evaluating potential candidates in the design phase before they are synthesised. Estimation programs that allow batch operation can be used in conjunction with high throughput techniques such as combinatorial chemistry to evaluate large numbers of structures in a short timeframe.

A comprehensive review of the programs and computational methods available for physico-chemical property estimation is beyond the scope of this chapter. More complete discussions of estimation methods may be found in the review literature.[1–4] Some common methodologies used to estimate the major

Issues in Toxicology No.7
In Silico Toxicology: Principles and Applications
Edited by Mark T. D. Cronin and Judith C. Madden
© The Royal Society of Chemistry 2010
Published by the Royal Society of Chemistry, www.rsc.org

physico-chemical and environmental fate properties, relevant to toxicity assessment, are presented here in the context of four readily available estimation suites of programs. Each of these is capable of calculating physico-chemical properties for diverse chemical substances from the input of the chemical structure alone. The aim of this chapter is to illustrate how these properties can be successfully calculated and to provide practical advice for evaluating estimated data.

5.2 Getting the Most Out of Estimation Methods

The development of readily available, broadly applicable and easy to use computerised estimation software has greatly increased access to reliable estimated property data for chemical assessment. However, the increased reliability of the available software and the ease with which a user can obtain an answer without understanding the basis for the estimate can foster the temptation to place too much trust in the results. A few basic practices can help increase the accuracy of estimated data. First, it is important to be aware of the capabilities and limitations of the estimation method to ensure that the structural features of the substance to be estimated fall within the design and domain of the program. Secondly, the substance to be estimated must be correctly represented in order for the program to provide accurate results. Thirdly, because some estimation methods are based on correlations between physico-chemical properties, it is important to provide any measured property data that are available to increase the accuracy of subsequent estimations. Finally, all estimated results should be critically evaluated before use.

All estimation programs are limited in the types of chemicals for which they can provide accurate estimations. Most programs are designed to estimate the properties of small, neutral organic molecules and are not capable of providing accurate estimates for substances such as inorganic materials, organometallics, most salts, strongly ionising substances, polymers, nanomaterials or substances with molecular weights of greater than about $1,000\,g\,mole^{-1}$. In addition, some programs may be limited in the types of functional groups or molecular backbones that can be effectively estimated. The limitations of the particular program to be used should be reviewed to establish whether the program is capable of providing reasonable estimates for the structure of interest. Although it is often possible for a program to perform calculations on an inappropriate structure, the results may not be reliable.

Most physico-chemical property estimation programs require the input of a chemical structure, which is often obtained from a chemical name or a Chemical Abstracts Service (CAS) registry number. Obtaining accurate property estimations depends on establishing the correct chemical identity for the substance and choosing the appropriate structure to be estimated (see Chapter 4). This is often easy and obvious, especially when dealing with discrete organic substances. However, it can be quite complicated for complex mixtures such as certain industrial reaction products, natural product extracts or distillation fractions.

Most property estimation methods perform their estimations on discrete structures, so an appropriate structure or structures must be chosen to represent such mixtures. If the substance is a mixture with a dominant component, then this component can often be used as a typical structure. If the substance is a mixture of closely related chemicals, the lowest and highest molecular weight components can sometimes be chosen and estimated to represent the extremes of the mixture.

Occasionally, a substance may contain similar amounts of a number of structurally unrelated components, and the typical or extreme structures may not be readily identified. In this case, it may be necessary to perform complete assessments on several different structures in order to adequately represent the mixture. It can also be challenging to choose an appropriate structure for discrete substances that can be represented by multiple structures. For example, some substances can exist in several different tautomeric forms. An estimation program may calculate different results for each tautomeric structure, even though they represent the same substance. In this case, the structure of the most stable tautomer should be estimated. In contrast, some structural variations may not be important in the context of physical property estimations. Most estimation programs do not recognise the difference between diastereomers or cis/trans isomers of a substance, assuming that the differences in their physico-chemical properties will be negligible. Therefore, it is not usually necessary to distinguish between them. The selection of an appropriate structure for estimation often requires professional judgement.

Several online resources are available to help establish and verify the exact chemical identity of a substance. The most authoritative and extensive source of chemical identity and structural information is the CAS Registry File,[5] which is available online *via* a subscription service. In addition to the Chemical Index Names, synonyms, chemical structures and CAS Registry Numbers (CASRN), the CAS Registry file also contains substance definitions which describe the chemical compositions of many complex mixtures. Two useful alternatives are ChemIDplus Advanced[6] and ChemBioFinder,[7] which are both free and contain chemical names, CASRNs, and structural information for tens of thousands of substances.

Whenever possible, it is important to identify any known physico-chemical properties of a substance before estimating its properties. One widely used estimation approach makes use of correlations between properties that are readily obtainable—either by measurement or estimation—and the property of interest. These empirically derived mathematical relationships allow the calculation of the property of interest as a function of another property or properties. A limitation to the use of such correlations is the potential for the propagation of errors. If the value for the property used is inaccurate, then that error is carried through the calculation and may be magnified in the value of the property of interest; however, the use of quality experimental data will generally increase the accuracy of subsequent estimates. Programs that contain methods based on such correlations, and which can utilise experimental property data, will provide data entry fields or display prompts where the measured data can be entered.

Unfortunately, a significant amount of the measured physical property data found in the literature is inaccurate or inconsistent. Therefore, it is important to evaluate the reliability of any measured properties before using them (see Chapter 4). A number of free databases are available that contain compilations of physico-chemical properties that have already been evaluated for accuracy and reliability such as:

- Physical Properties (PHYSPROP©) database,[8] which is included in the Estimation Programs Interface Suite (EPI Suite™);[9]
- SRC Inc.'s Environmental Fate Data Base[10] (EFDB); and
- Hazardous Substances Data Bank[11] (HSDB).

A more complete list of such sources of physico-chemical properties has been collated by Boethling *et al.*[1]

Once an estimated value for a property has been obtained, it is important to evaluate the reliability of the estimate. Professional judgement, experience with the model and chemical intuition are invaluable in spotting questionable results.

There are several strategies that are useful for evaluating the reliability of estimated data.

- Review any information the program provides on how the value was calculated for the structure evaluated. This is especially useful for fragment constant methods, which can be checked for missing or inappropriately assigned fragments, or to determine if any particular fragment appears to over-influence the results.
- Be aware of any characteristics of the substance of interest that could interfere with the accurate estimation (or measurement) of its properties. Some examples include substances that undergo rapid hydrolysis, substances with surfactant properties, and substances exhibiting unusual intermolecular interactions such as highly fluorinated compounds.
- Estimated data should be evaluated in the context of related properties. For example, substances with high boiling points tend to have low vapour pressures and those with high octanol–water partition coefficient (P) values tend to have low water solubility values. Results that appear to be contradictory may indicate that an estimation method is not performing well and should prompt a closer inspection of the data.
- Be sceptical of extremely high or extremely low values, particularly if the value would not be measurable. Although such results may not be quantitatively accurate, they may still be useful for screening purposes. For example, an estimated vapour pressure of 10^{-20} torr is not likely to be an accurate value; however, it may correctly indicate that the volatility of the chemical is negligible.
- When in doubt, check a suspect value against known values for closely related analogues. Even though the exact quantitative accuracy cannot be verified, the analogue data may indicate whether or not the value is qualitatively reasonable.

Knowing when to trust estimated data is as valuable as knowing how to run the estimation program. These guidelines can help to validate or disqualify suspect data. However, there is not always an easy way to verify the accuracy of estimated data for novel structures. Use of estimated data in these cases may still be worthwhile, as long as the uncertainty about the accuracy of the data is acknowledged and considered in any subsequent analysis.

5.3 Selected Software for the Estimation of Physico-Chemical and Environmental Fate Properties

Over the years, a wide assortment of software has been developed to estimate the various physico-chemical and environmental fate properties of chemicals relevant to assessing chemical exposure and toxicity. However, a comprehensive review of these programs is well beyond the scope of this chapter. Four widely used software packages have been selected for discussion:

- EPI Suite;[9]
- Advanced Chemistry Development Inc. (ACD)/PhysChem Suite;[12]
- SPARC (SPARC Performs Automated Reasoning in Chemistry);[13] and
- ClogP.[14]

All are readily available products that are capable of providing estimated properties for a diverse and varied set of chemicals. Estimation methods within each of these products have been recommended by authorities in the field.[1,2] The selected programs all generate estimates based solely on the chemical structure, although many of the estimations can be improved by providing supporting experimental data. All of them can produce data useful for the hazard and risk characterisation of new and existing chemicals. The Cheminformatics and QSAR Society website (www.qsar.org) provides a more comprehensive list of the property estimation software available. A number of these products are included in recent reviews.[1–3,15–16]

In any comparison of estimation software, the question naturally arises as to which software gives the most accurate results. Although a number of comparative studies have been published,[3,16–19] many report contradictory results, making it difficult to reach a satisfactory conclusion.

Performing comparative studies on estimation software is complicated by the difficulty of compiling a truly independent set of test chemicals. If the set of test chemicals has compounds in common with the training set used to develop the software, the results of the study will be skewed in favour of that software. Since the identities of the chemicals in a program's training set are not always made public, this problem is difficult to avoid. The results of a comparison can also be skewed if the set of test chemicals does not contain a sufficiently diverse range of chemical structures and properties, or if some of the measured properties used in the test set are erroneous.

The information presented in this section and the remainder of the chapter is intended to increase understanding of how these estimation programs work and how they can be used to the best of their potential. Method performance will therefore be treated in general terms throughout this chapter, with an emphasis on increasing accuracy when possible. Performance statistics for each method are generally available in the original literature describing the development of that method.

5.3.1 EPI Suite

The EPI Suite software was developed by the US Environmental Protection Agency (EPA) Office of Pollution Prevention and Toxics and SRC, Inc. It is available on the Internet as a free downloadable program at www.epa.gov/oppt/exposure/pubs/episuite.htm and is incorporated into the Organisation for Economic Cooperation and Development (OECD) (Q)SAR Application Toolbox.

EPI Suite contains 14 programs for estimating physico-chemical properties, environmental fate properties and transport parameters. The main interface allows the user to run all the programs simultaneously, or each module can be run as a standalone program. It also includes the PHYSPROP database of measured physico-chemical and environmental fate properties for about 25,000 chemicals.

EPI Suite utilises fragment constant/group contribution methods and regression equations to estimate the melting point, boiling point, vapour pressure, water solubility, log P, Henry's Law constant, log K_{oc}, bioconcentration factor (BCF) and biodegradability, as well as other properties and models not covered in this chapter. EPI Suite also includes the EPA's ECO-SAR™ program[20] for predicting aquatic toxicity endpoints.

5.3.2 ACD/PhysChem Suite

ACD/PhysChem Suite is a commercial software package available from Advanced Chemistry Development Inc. (www.acdlabs.com) which includes a desktop estimation program with a database of measured physico-chemical properties and an optional subscription to web-based applications (ACD/I-Lab). A free version of ACD/logP is included with the ACD/ChemSketch Freeware (www.acdlabs.com/resources/freeware/). Estimated data generated by the ACD/PhysChem Suite for discrete chemicals are included in the CAS Registry File under the Prop field, available *via* paid subscription. A limited set of ACD estimations are also available free of charge on the ChemSpider website (www.chemspider.com).

ACD/PhysChem Suite's proprietary methods use fragment constant/group contribution methods, regression equations and linear free energy relationships (LFER) to estimate the boiling point, vapour pressure, water solubility, log P, pK_a, log D, K_{oc} and BCF, as well as other properties not covered in this chapter.

5.3.3 SPARC

SPARC was developed by the US EPA's Office of Research and Development and the University of Georgia. SPARC is a freely accessible web-based application located at http://ibmlc2.chem.uga.edu/sparc. SPARC contains a database of measured physico-chemical properties and can estimate the boiling point, vapour pressure, water solubility, log *P*, log *D*, pK_a and Henry's Law constant, as well as other properties not covered in this chapter.

SPARC utilises a 'toolbox' of mechanistic models that can be implemented where needed for a specific query. SPARC has developed resonance models, electrostatic interaction models, solute–solute interaction solvation models and solute–solvent interaction solvation models which can be used to estimate both physico-chemical properties and chemical reactivity. Each of these models are implemented using a combination of methods such as structure–activity relationships (SAR) for structure–activity analysis, LFER to describe thermodynamic and thermal properties, and perturbed molecular orbital theory (PMO) to calculate quantum effects such as delocalisation energy, charge distribution and polarisabilities of π electrons.

5.3.4 ClogP

ClogP was developed by Albert Leo and the Pomona Medicinal Chemistry Project with initial funding from the US EPA, and has undergone extensive upgrades since its initial development.[21] It is available for purchase online from Biobyte Inc. (www.biobyte.com/bb/prod/clogp40.html). Unlike the other software discussed later in this chapter which provide estimations for multiple properties, ClogP is focused exclusively on estimating log *P*. It was one of the earliest log *P* estimation programs to use a fragment constant method. Its updates have kept it current and it remains one of the most widely used programs for estimating log *P*.

5.4 Physico-Chemical Property Estimations

5.4.1 Octanol–Water Partition Coefficient

The octanol–water partition coefficient (K_{ow} or *P*) describes the partitioning behaviour of a substance between two immiscible liquids, octanol and water. It is expressed as the ratio of the concentration of a substance in the octanol phase to the concentration of the substance in the aqueous phase at equilibrium. Since *P* can range from less than 10^{-4} to greater than 10^8, it is typically expressed as a logarithmic value (log K_{ow} or log *P*). Substances with a high log *P* are considered hydrophobic, while substances with a low log *P* are considered hydrophilic.

As a representation of hydrophobicity, *P* has a strong correlation to many environmentally and biologically relevant processes such as environmental distribution, bioavailability, absorption through the skin, transport through biological membranes, partitioning between the tissues of an organism, biological binding effects, biodegradation and biological activity. This wide range of influence makes log P one of the most commonly used descriptors in QSARs. It is used as a key term in regression-derived estimation methods for predicting water solubility, sorption to soil, aquatic bioconcentration, aquatic toxicity, dermal absorption and human toxicity.

Most estimation methods assume that the molecule of interest is in a neutral, unionised state. However, compounds with acidic or basic functional groups may be ionised depending on their pK_a and the pH of the medium. In this case, it may be necessary to adjust the log P to account for the degree of ionisation. This adjusted P is known as the distribution coefficient (D), which is the ratio of the concentration of all species (ionised and unionised) of a substance in the octanol phase to the concentration of all species of the substance in the aqueous phase at equilibrium. For a monoelectrolyte, the distribution coefficient can be calculated for any pH using eqn (5.1) or (5.2).[22]

$$\text{Log } D_{\text{acids}} = \log P + \log[(1 + 10^{pH-pKa})^{-1}] \tag{5.1}$$

$$\text{Log } D_{bases} = \log P + \log[(1 + 10^{pKa-pH})^{-1}] \tag{5.2}$$

Due to the importance of P, a large number of methods have been developed for its estimation and several recent reviews have been published on the subject.[1,17,23] EPI Suite, ACD/PhysChem Suite, ClogP and SPARC all include programs for estimating P. Boethling *et al.* recommend EPI Suite and ClogP.[1]

EPI Suite, ACD/PhysChem Suite and ClogP estimate log P *via* atom/fragment constant contribution methods. This is a broadly applicable approach that has been employed in the estimation of a wide variety of properties. Using this approach, the molecule of interest is deconstructed into its constituent fragments for which numerical contribution factors have been derived. Fragments may be defined by functional groups, atoms or groups of atoms. Correction factors may also be used to account for any other relevant molecular attributes that are not adequately described using molecular fragments and which influence the property being estimated such as steric interactions, hydrogen bonding and polar effects. The property of interest is calculated by first multiplying both the atom/fragment values (f) and the correction factor values (c) by the number of times (n) that either the atom/fragment or the substructure appears in the molecule. The property is calculated as the sum of these quantities as shown in eqn (5.3):

$$Property = \sum n_i f_i + \sum n_j c_j \tag{5.3}$$

Two strategies have been reported for deriving fragment contribution values for log P estimation. Using a 'reductionist' approach, contribution values for each fragment are derived by multiple regression analysis of measured values for a selected set of chemicals known as the training set. Values for the correction factors, representing interactions between more complex substructures, are derived by a separate multiple regression analysis. In contrast, using a 'constructionist' approach, the values for the fragment contributions and the correction factors are derived from experimental measurements of the simplest examples in which they occur in isolation. Advocates of the 'constructionist' approach for estimating log P believe that it better reflects the relevant physical chemical factors affecting log P than do the arbitrary statistics of regression analysis,[21] but both approaches generally appear to provide reasonably and comparably accurate results.[4,17–19]

Fragment constant methods are versatile and flexible, but can fail when the target molecule contains fragments that are not in the program's fragment library. In addition, if a molecule contains a large number of functional groups, they may interact in a complex manner that is not properly accounted for by the correction factors.

EPI Suite's KOWWIN™ program[24] estimates the log P using a 'reductionist' atom/fragment constant method where the values for fragment contributions and the values for the correction factors are derived from separate multiple regression analyses of a database of measured log P values.[25] The atom/fragment values (f) and the correction factor values (c) are each multiplied by the number of times the atom/fragment or substructure appears in the molecule (n), and log P is calculated as the sum of these quantities using a modification of eqn (5.3) shown in eqn (5.4).

$$\log P = \sum n_i f_i + \sum n_j c_j + 0.229 \qquad (5.4)$$

When the KOWWIN program is run in standalone mode and not through the EPI interface, it allows the log P of a substance to be estimated based on the measured log P of a closely related analogue, using what is called an 'experimental value adjusted' method. In this method, the closely related analogue is modified by adding and/or subtracting atom/fragments to construct the compound of interest. The estimated log P is then the sum of the measured log P for the analogue and the values of the fragment modifications. The advantage to this approach is that the calculations are based on the relatively minor structural differences between the target and the analogue, which introduces less uncertainty than constructing the target molecule from basic fragments. This approach is potentially more accurate and is recommended for estimating the P in KOWWIN when a reliable measured value is available for a closely related analogue.

ACD/PhysChem Suite also uses a 'reductionist' atom/fragment constant method, where both the atom/fragment contribution values and the correction factor values are derived from separate multiple regression analyses.[12] If any atom/fragment contribution values or correction factor values are missing from the list of those derived from the regression analyses, estimates of those values can be calculated. The log P is then calculated from these values using eqn (5.3). From log P, ACD/PhysChem Suite can also calculate log D at any pH using its estimation of pK_a.

ClogP estimates the log P using a 'constructionist' atom/fragment constant method, which derives its fragment contributions and correction factors from experimental measurements of the simplest examples in which they occur in isolation. For example, the aliphatic carbon and hydrogen atom fragment values were determined from careful measurements of log P values of methane, ethane and molecular hydrogen.[26] Correction factors are similarly derived from simple representative compounds to account for special attributes such as the effects of certain structural features, fragment interaction effects, electronic effects and intra-molecular hydrogen bonding. Missing fragment values or correction factor values can be derived from existing or calculated values. Log P is calculated from the atom/fragment contribution values and the correction factor values using eqn (5.3).

SPARC calculates log P from the activity coefficients at infinite dilution (γ^{∞}) of the substance in water and wet octanol as shown in eqn (5.5), where -0.82 is the ratio of the molecularities of each phase.[27]

$$\log P = \log \gamma^{\infty}_{water} - \log \gamma^{\infty}_{octanol} + \log(-0.82) \tag{5.5}$$

The activity coefficients are calculated from the free energies of inter-molecular interactions and the Flory–Huggins[28,29] term for excess entropy of mixing. The inter-molecular interactions represented are dispersion, induction, dipole–dipole and hydrogen bonding effects using SAR, LFER and PMO theory. SPARC can also estimate pK_a, which it can use to calculate log D at any pH.

Extreme values of log P are not measurable due to the low concentration of the substance in one phase or the other. Most neutral organic molecules have log P values within the range of -4 to 8. Estimated values that fall outside of this range are not quantitatively reliable and should be taken as an indication that the substance is either strongly hydrophilic or strongly hydrophobic.

Substances that may be expected to readily hydrolyse should be assessed cautiously. If the effect being evaluated involves contact with water (*e.g.* aquatic bioconcentration), it may be necessary to estimate log P of the hydrolysis product instead of the parent compound. If the effect being evaluated does not involve contact with water (*e.g.* dermal absorption), it may be more appropriate to estimate log P of the parent compound.

Log P should not be estimated for surfactants. Instead of partitioning to either the octanol phase or the aqueous phase, surfactants can stabilise octanol–water dispersions. Because of the difficulty in measuring log P for surfactants, few measured data are available for building and evaluating estimation methods for such substances.

Mixtures may be adequately represented by estimating a range of log P values for several components, or by estimating log P of the predominant component. Care must be taken in the selection of representative structures when estimating log P because the most or least desirable values may depend on their implication for the effect that is of interest. For example, the ability to pass readily through biological membranes tends to be greater for substances with log P values of less than 4,[30] while bioconcentration factors (*BCF*s) tend to

be higher for substances with log *P* values in the range of 6–7.[1] Simply estimating either the extreme or the typical values for a mixture may not be sufficient to characterise a mixture.

5.4.2   Water Solubility

Water solubility is the maximum amount of a substance that can be dissolved in
water at a given temperature, pressure and *pH*. It is one of the more important
physico-chemical properties due to its influence on other factors important in
assessing a substance's environmental fate and toxicity. A substance's solubility
in water has a direct effect on its partitioning between water and other media,
thereby influencing its *P*, its air–water partition coefficient (Henry's Law constant), its soil adsorption coefficients and its partitioning between environmental compartments. Water solubility will also affect the potential routes of
exposure to the substance, its bioavailability and its potential for biodegradation. Some useful reviews of water solubility estimation methods can be found
in Mackay[31] and in Boethling *et al.*[1] EPI Suite, ACD/PhysChem Suite and
SPARC all include modules for estimating water solubility.

   EPI Suite includes two programs which each calculate the water solubility by
a different method. The WSKOWWIN™ program[32] uses two regression
equations which relate water solubility (*WS*), log *P*, melting point (*MP*) and
molecular weight (*MW*).[33]

$$\log WS(mol/L) = 0.796 - 0.85 \log P - 0.00728\, MW + Corrections \qquad (5.6)$$

$$\log WS(mol/L) = 0.693 - 0.96 \log P - 0.0092(MP - 25) \\ - 0.00314\, MW + Corrections. \qquad (5.7)$$

The accuracy of the estimations can be improved by entering measured values
for log *P* and melting point into the program. If a measured melting point is not
entered, eqn (5.6) is used. Equation (5.7) is used only when a measured melting
point is entered into the program and usually provides a more accurate estimation than eqn (5.6). When the measured melting point is $\leq 25\,°C$, indicating
that the substance is a liquid, the program automatically removes the melting
point term from eqn (5.7). If the substance is known to be a liquid and does not
have a measured melting point, the accuracy of the estimation can therefore be
improved by manually entering an estimated melting point of 25 °C into the
program. This informs the program that the substance is a liquid and is the only
case where an estimated property should be entered into EPI Suite. According
to the program literature, WSKOWWIN is usually the more accurate method
within EPI Suite and is therefore preferred over WATERNT™ in most cases.

   The WATERNT program[34] in EPI Suite uses a 'reductionist' atom/fragment
constant method, where the fragment contribution values and the correction
factor values are derived by separate multiple regression analyses of a thousand
measured water solubility values. These regression analyses use the same

methodology as in the EPI Suite KOWWIN™ program.[24] The log of the water solubility (log WS) in moles L^{-1} is calculated by multiplying the atom/fragment values (f) and the correction factor values (c) by the number of times the atom/fragment or substructure appears in the molecule (n) and adding them together using eqn (5.8).

$$\log WS = \sum n_i f_i + \sum n_j c_j + 0.24922 \qquad (5.8)$$

The EPI Suite WATERNT program also allows the water solubility of a substance to be estimated based on the measured water solubility of a closely related analogue, using the 'experimental value adjusted' method. In this method, the estimated water solubility is derived from the measured value for the analogue and the relatively minor structural differences between the substance and the analogue, which introduces less uncertainty than constructing the target molecule from basic fragments. This method is not available in the EPI Suite main interface and can only be accessed when running the WATERNT module in standalone mode. This method for estimating water solubility has the potential to be highly accurate, but has been recently developed and, therefore, has not yet been extensively validated.

ACD/PhysChem Suite estimates water solubility at *pH*s ranging from 0 to 14, using separate proprietary equations for various classes of chemicals.[12] The parameters used in its correlation equations are log P, log D, boiling point, molecular volume, refractive index and the number of hydrogen-bond acceptors and donors. The ACD/PhysChem Suite also allows the user to indicate whether the substance is a solid or liquid and enter the measured melting point, which can improve the accuracy of the estimation.

SPARC can estimate the solubility of a substance over a range of temperatures in any solvent, including water, from the activity coefficients at infinite dilution of the substance in that solvent.[27] The activity coefficients are calculated from the free energies of inter-molecular interactions, the crystal energy and the Flory–Huggins[28,29] term for excess entropy of mixing. The inter-molecular interactions represented are dispersion, induction, dipole–dipole and hydrogen bonding effects, and are calculated using SAR, LFER and PMO theory. SPARC also provides the option of entering a measured melting point to improve the accuracy of the estimation.

Mixtures may be represented by estimating the solubilities of several components to define a range. Alternatively, the water solubility of the predominant component may be estimated if one can be identified. If there is no predominant component, it may be more appropriate to estimate the most soluble component to determine the upper solubility limit of the mixture.

Under real world conditions, the apparent solubility of a substance may be affected by its volatility or its tendency to adsorb to suspended organic matter. The water solubility should be evaluated together with Henry's Law constant and K_{oc} to determine if volatilisation and sorption are likely to reduce the amount of the substance that is freely dissolved in the aqueous phase.

Certain substances may not be appropriate for estimation. For example, water solubility cannot be determined for substances that hydrolyse rapidly. In this case, the species in solution are likely to be the hydrolysis products instead of the parent compound. If the effects of the hydrolysis products are of interest, it may be appropriate to estimate their water solubilities in place of those of the parent compound. It is generally inappropriate to estimate the water solubility of surfactants or other self-dispersible substances. These substances form micelles and other colloidal aggregates in water, which makes it impossible to experimentally determine their maximum solubility in water. The absence of reliable experimental solubilities for such substances makes it difficult to evaluate the performance of an estimation method.

5.4.3 Dissociation Constants

Chemicals that act as acids or bases may become ionised to form charged species depending on the *pH*. The dissociation of an acid is shown in eqn (5.9). The equilibrium constant is known as the acid dissociation constant (K_a), which describes the strength of the acid.

$$HA + H_2O \rightleftharpoons H_3O^+ + A^- \quad K_a = \frac{[H_3O^+][A^-]}{[HA]} \qquad (5.9)$$

This constant is typically reported as the pK_a, which is the negative logarithm of the dissociation constant. The ratio of the concentration of ionised to unionised species ($[A^-]/[HA]$) in solution at a particular *pH* can be determined from the pK_a using the Henderson–Hasselbalch eqn (5.10).

$$pH = pK_a + \log\frac{[A^-]}{[HA]} \qquad (5.10)$$

The ratio of the concentration of ionised to unionised species in solution varies with *pH*. When the *pH* equals the pK_a, a substance is exactly 50% ionised. The $[A^-]/[HA]$ ratio will change by one order of magnitude with each *pH* unit difference between the *pH* and pK_a. This ratio is important because charged species often exhibit significantly different behaviour than neutral species. Since estimation methods typically provide estimations for the neutral species, the estimated properties may be misleading for substances that exist predominantly as ionised species and may not represent the actual behaviour of the substance. Charged species tend to have higher water solubility and lower volatility than neutral species, and properties that depend on water solubility and volatility such as log *P*, Henry's Law constant, sorption to soil, *BCF*, bioavailability, dermal absorption and toxicity may be affected as well. The programs in SPARC and ACD/PhysChem Suite are recommended by Boethling *et al.*[1] for the prediction of pK_a.

SPARC calculates the pK_a of a substance by first dividing the structure into units called the reaction centre and the perturber. The reaction centre is the smallest substructure that has the potential to ionise, and the perturber is the substructure attached to the reaction centre that influences, or perturbs, the ionisation behaviour. The pK_a is then calculated as shown in eqn (5.11), where $(pK_a)_c$ describes the ionisation potential of the reaction centre and $\delta_p(pK_a)_c$ is the change in the ionisation behaviour caused by the perturber. This change in ionisation behaviour is described in eqn (5.12), where $\delta_{res}pK_a$, $\delta_{ele}pK_a$, $\delta_{sol}pK_a$, and $\delta_{H\text{-}bond}pK_a$ represent the difference in resonance effects, electrostatic effects, solvation effects and intramolecular hydrogen bonding effects, respectively, between the unionised and ionised states of the reaction centre. These effects are calculated using SAR, LFER and PMO theory.[35]

$$pK_a = (pK_a)_c + \delta_p(pK_a)_c \tag{5.11}$$

$$\delta_p(pK_a)_c = \delta_{res}pK_a + \delta_{ele}pK_a + \delta_{sol}pK_a + \delta_{H\text{-}bond}pK_a \tag{5.12}$$

The ACD/PhysChem Suite calculates the pK_a of a substance using eqn (5.13), where $(pK_a)_c$ represents the dissociation potential of the unsubstituted reaction centre, $\rho_i\sigma_j$ represent electronic interactions from the substituents, and F_m represents correction factors for non-electronic effects.

$$pK_a = (pK_a)_c + \sum \rho_i\sigma_j + \sum F_m \tag{5.13}$$

Sigma and rho values are either taken from experimental databases or calculated. The correction factors are only used when experimental rho values are not available and represent effects caused by structural elements such as steric interactions, resonance, tautomerism, covalent hydration and the presence of multiple ionisation centres.[12]

5.4.4 Vapour Pressure

The vapour pressure of a substance is the pressure exerted by its vapour in equilibrium with its solid or liquid phase. Some of the factors that influence the vapour pressure are molecular weight and intermolecular interactions such as van der Waals forces, dipole–dipole interactions and hydrogen bonding. The vapour pressure is the main descriptor of volatility, and together with Henry's Law constant, is important for determining the partitioning behaviour of volatile chemicals from wet and dry media, including soils, bodies of water, airborne particulates and aerosols. It is a critical parameter for determining a substance's maximum attainable vapour concentration and can also be used to estimate Henry's Law constant.

Estimation methods and software useful for determining the vapour pressures of pure compounds have been recently reviewed.[1,3,36] Boethling *et al.*[1] and Sage and Sage[36] recommend the Antione equation[37] for liquids that boil at <200 °C and the Modified Grain method[38] for high boiling liquids and solids. EPI Suite, ACD/PhysChem Suite, and SPARC all contain programs to predict vapour pressure.

The MPBPWIN™ program[39] in EPI Suite calculates vapour pressure from the normal boiling point using both the Antione equation and the Modified Grain method. These are presented briefly here, and are described in more detail by Sage and Sage[36] and the documentation included with MPBPWIN. The Antoine equation is shown in its condensed form in eqn (5.14), where VP is the vapour pressure of a chemical at temperature T, VP_b is the vapour pressure at the boiling point and T_b is the normal boiling point.

$$\ln \frac{VP}{VP_b} = B\left[\frac{1}{T-C} - \frac{1}{T_b-C}\right] \tag{5.14}$$

The terms B and C are calculated from the boiling point according to eqn (5.15) and eqn (5.16).

$$B = \frac{\Delta H_b(T_b - C)^2}{RT_b^2} \tag{5.15}$$

$$C = -18 + 0.19T_b \tag{5.16}$$

The heat of vapourisation at the boiling point (ΔH_b) is derived from the chemical's calculated dipole moment, the gas constant R and the normal boiling point. The Antoine relationship is most accurate for liquids that boil at temperatures $<200\,°C$ and have vapour pressures $>10^{-2}\,kPa$ at $25\,°C$.

For less volatile liquids, MPBPWIN calculates vapour pressure according to the Grain and Watson adaptation of the Clausius–Clapeyron equation (5.17), where $T_\rho = T/T_b$ and $m = 0.4133 - 0.2575\,T_\rho$. An additional term incorporating the melting point is included for solids.

$$\ln \frac{VP}{VP_b} = \frac{\Delta H_b}{RT_b}\left[1 - \frac{(3 - 2T_\rho)^m}{T_\rho} - 2m(3 - 2T_\rho)^{m-1}\ln T_\rho\right] \tag{5.17}$$

MPBPWIN calculates the vapour pressure according to each of these methods and selects a 'suggested' vapour pressure. Estimations from the Modified Grain method are suggested for solids. For liquids and gases, the suggested vapour pressure is the average of the results from the Antione equation and the Modified Grain method.

Because these estimation methods correlate the vapour pressure to the normal boiling point (and melting point for solids), the accuracy of the estimations are improved by entering measured boiling points and melting points where known. If a measured boiling point is not available, MPBPWIN will use an estimated boiling point. When MPBPWIN is run as a separate program, the vapour pressure can be calculated at a user-selected temperature. If run as part of EPI Suite, the program uses a default temperature of $25\,°C$.

The ACD/PhysChem Suite calculates vapour pressure as part of its boiling point program. The boiling point is calculated according to a proprietary algorithm that expresses boiling point as a linear function of the molar volume. Molar volume is calculated using an additive algorithm including atom and group contributions, and descriptors for atomic and group interactions.[12] The ACD/PhysChem Suite includes a feature that can calculate the vapour pressure as a function of temperature, providing the results as either a graph or a table.

SPARC calculates the vapour pressure (VP) as a function of the sum of the free energies of the intermolecular interactions ($\Delta G_{interactions}$), the gas constant (R) and the temperature (T) according to eqn (5.18).[40]

$$\text{Log } VP = \frac{\Delta G_{interactions}}{2.303RT} + \log T + C \qquad (5.18)$$

The term ($\text{Log} T + C$) represents the change in entropy contribution associated with the transition from the liquid phase to the gas phase, which is more disordered. For solids at standard temperature ($25\,^{\circ}C$), a separate term that includes the melting point is added to account for crystal lattice energy contributions. SPARC also calculates the vapour pressure as a function of temperature and presents the data as a graph or as a data table.

The accuracy of estimated vapour pressures may decrease for low vapour pressures ($<1\,\text{Pa}$). However, there is little practical distinction among vapour pressures lower than $10^{-4}\,\text{Pa}$ when screening chemicals for potential adverse effects. At these very low vapour pressures, volatilisation rates and inhalation exposures are generally expected to be negligible.

The vapour pressure of a mixture is the sum of the partial pressures of each component. For screening purposes, mixtures may be adequately represented by the most volatile component to define an upper limit to the vapour pressure. Alternately, the predominant component may be estimated for mixtures having a high concentration of one component relative to the others.

5.4.5 Boiling Point

The boiling point is the temperature at which the vapour pressure of a liquid is equal to the pressure of the surrounding atmosphere on the surface of the liquid. When the pressure is equal to 1 atmosphere ($101,325\,\text{Pa}$), this temperature is referred to as the normal boiling point. The boiling point directly indicates whether a substance will exist as a liquid or a gas at a given temperature. It has a strong correlation to vapour pressure and influences other properties related to volatility such as Henry's Law constant. Boiling point estimation methods have recently been reviewed.[1,3,41] EPI Suite, ACD/PhysChem Suite and SPARC all predict the normal boiling point.

The group contribution method of Stein and Brown[42] is recommended by Boethling *et al.*[1] and is one of the methods recommended by Lyman.[41] EPI Suite employs an adapted version of this method to estimate boiling point in the MPBPWIN program. A preliminary boiling point (T_b) in Kelvin is first calculated according to eqn (5.19), where g is the group contribution value and

n is the number of times that the group is present in the molecule. The preliminary boiling point is then corrected using eqn (5.20) when the T_b is $\leq 700\,\mathrm{K}$, or eqn (5.21) when the T_b is $> 700\,\mathrm{K}$.[39]

$$T_b(\mathrm{K}) = 198.2 + \sum n_i g_i \tag{5.19}$$

$$T_b(corr) = T_b - 94.84 + 0.5577 T_b - 0.0007705 T_b^2 \tag{5.20}$$

$$T_b(corr) = T_b + 282.7 - 0.5209 T_b \tag{5.21}$$

MPBPWIN uses the group contribution values from Stein and Brown,[42] but has expanded upon the original method to include additional group contribution values and additional correction factors for certain specific types of compounds.

The ACD/PhysChem Suite calculates boiling point according to a proprietary algorithm that expresses boiling point as a linear function of molar volume. Molar volume is calculated using an additive algorithm including atom and group contributions, and descriptors for atomic and group interactions.[12] The ACD/PhysChem Suite includes a feature that can calculate boiling point as a function of pressure, providing the results as either a graph or a table.

SPARC calculates normal boiling point from its vapour pressure model.[40] The boiling point is calculated according to eqn (5.18) by setting the vapour pressure to 1 atmosphere and solving the equation for T. SPARC will also calculate boiling point as a function of pressure, providing the results as either a graph or a table.

Very high estimated boiling points should be interpreted with care. Estimated boiling points for substances with high molecular weights can exceed practically attainable values. Many organic substances will decompose at elevated temperatures without boiling, making it difficult to verify a high estimated value. Although these values might not be measurable, estimations indicating a high boiling point may still be useful for estimating other properties such as the vapour pressure.

Depending on the intended use for the data, the boiling point of a mixture may be approximated either by calculating the boiling point of the lowest boiling component or by calculating the boiling points of several components to define a general boiling range. These strategies for determining the boiling points of mixtures can provide useful data, but it should be remembered that they are simplifications and cannot account for changes in the boiling point due to interactions among the components of a mixture such as azeotropic distillation.

5.4.6 Melting Point

The melting point is the temperature at which a substance changes phase from a solid to a liquid at 1 atmosphere of pressure. In addition to determining the physical state of the material, a melting point also has relevance in predicting toxicity. Pure substances that are liquids or low melting point solids tend to be better absorbed than high melting non-ionic solids by all routes of exposure, because it takes less energy to break the crystal lattice to allow disaggregation of free molecules.[30] The melting point has also been observed to correlate with other physical properties such as boiling point, vapour pressure and water solubility.[43] The melting point has therefore been incorporated as a descriptor in some of the estimation methods for predicting these properties and the use of reliable melting point data can improve the accuracy of results from such methods.

Although the melting point is one of the easiest of the physical properties to measure in the laboratory, it has proven to be one of the most difficult to model accurately for structurally diverse compounds. The melting point of a substance depends in part upon the strength of intermolecular forces such as hydrogen bonding, van der Waals forces, dipole interactions, crystal packing pattern and crystal lattice energy as well as structural parameters such as molecular flexibility and molecular symmetry.[43] Inter-molecular interactions can vary depending on the orientation of the molecules within a solid, which cannot be readily predicted from the structure alone. Several estimation methods for melting point have been recently reviewed.[1,3,43] To date, the methods that have proven to be the most accurate are those designed to estimate melting points for local models, or within a narrow, homologous series of chemicals such as the normal alkanes. Few accepted methods are available that can estimate melting points for diverse sets of chemicals with a reasonable degree of accuracy.

The MPBPWIN program in EPI Suite estimates the melting point using both a modified Joback method[44] and the Gold and Ogle method.[45] It then provides a recommended melting point, which is a weighted average of the two predicted values. Joback's method is an additive group contribution method that correlates the melting point (T_m) to molecular groups or fragments according to eqn (5.22), where g is the group contribution value and n is the number of times that the group is present in the molecule.

$$T_m(\text{K}) = 122.5 + \sum n_i g_i \qquad (5.22)$$

The melting point algorithm in MPBPWIN extends the original Joback method to include additional groups as defined by the Stein and Brown boiling point method[42] and correction factors for specific structures. The Gold and Ogle method correlates the melting point (T_m) to the boiling point (T_b) according to eqn (5.23).

$$T_m = 0.5839 T_b \qquad (5.23)$$

Although averaging the results obtained from these two methods can increase the overall accuracy, significant errors can still occur. In particular, the melting point can be significantly overestimated for certain structures. To compensate for this, MPBPWIN employs a cut-off value of 350 °C, where any estimated melting points above this value are reduced to 350 °C.

The melting point should not be estimated for mixtures because disruptions in the crystal lattice in impure and heterogeneous substances depress the melting point to an unpredictable extent relative to that of a pure substance. These interactions between unlike substances are not incorporated in the available methods. The observed melting point of a mixture will vary with its composition, and the melting point values of the individual components— whether measured or estimated—cannot adequately represent the entire substance.

The results of melting point estimations should be interpreted with caution because significant errors can occur. It has been speculated that the accuracy of melting point estimations may ultimately be increased by combining a variety of approaches, including group contributions, entropic considerations, physical property correlations and topological indices.[3]

5.5 Environmental Fate Properties

5.5.1 Henry's Law Constant

Henry's Law describes the water solubility of a gas as a function of its partial pressure above the solution at equilibrium. For non-reactive and non-dissociating substances with relatively low vapour pressures and water solubilities, a plot of the partial pressure *versus* the mole fraction of the substance in water is a linear function. The slope of the line is the Henry's Law constant (HLC or H) and is generally expressed as P_n/C_W, the partial pressure of the vapour (P_n) over its concentration in water (C_W). HLC indicates the volatility of a substance in solution, and is useful for evaluating environmental partitioning and transport and assessing potential exposures to substances both in solution and in the vapour phase. Substances having high values of HLC tend to evaporate from water, whereas those having low values tend to remain in solution. In addition, vapour phase substances having low values of HLC can partition to airborne water droplets and wash out of the atmosphere when it rains.

Estimation methods for determining HLC have been recently reviewed.[1,16,46] EPI Suite and SPARC can calculate HLC values.

The HENRYWIN™ program[47] in EPI Suite calculates HLC using three different methods. The first method calculates HLC as the ratio of the vapour pressure to the water solubility. Calculation of HLC by this method is preferred only if reliable measured data are available for both the water solubility and the vapour pressure.[1] This calculation should not be used for substances that are highly soluble in water or if measured data are not available for both input properties.

The two remaining estimation methods used in HENRYWIN are contribution methods. The first of these is a bond contribution method that calculates the logarithm of the water–air partition coefficient, log C_W/C_A (where C_A is the concentration in air), according to the method described by

Meylan and Howard.[48] In contrast to group contribution methods, the target molecule is evaluated according to the bond types present (*e.g.* C–H or C–C bonds) rather than the atoms and molecular fragments present. The contributions of each bond are summed according to eqn (5.3), where f_i is the bond contribution value. This water–air partition coefficient is the inverse of the air–water partition coefficient, C_A/C_W, which is the unitless *HLC*. The air–water partition coefficient is multiplied by the gas constant R and the temperature to calculate *HLC* expressed as P_n/C_W.

HENRYWIN also calculates the log C_W/C_A according to the group contribution method of Hine and Mookerjee.[49] This calculation is also performed according to eqn (5.3). The group contribution method is preferred for its increased accuracy over the bond method.[1,16] However, the group contribution method sometimes fails to calculate *HLC* due to missing fragments. The bond method is more broadly applicable, having the ability to evaluate most structures from 59 chemical bond types and 15 correction factors.

Running HENRYWIN as a standalone program (rather than as a part of EPI Suite) allows access to the 'experimental value adjustment' feature. This feature allows the user to enter the structure and a measured Henry's Law constant for an analogue of the target chemical. HENRYWIN then calculates *HLC* based on the experimental value of the analogue and the structural differences between the two chemicals using the bond method. HENRYWIN will also calculate the value of the *HLC* at any temperature between 0 and 50 °C in 5 °C increments. When run as part of EPI Suite, the program uses a default temperature of 25 °C.

SPARC's Henry's Law constant predictor calculates Henry's Law constant as the product of the vapour pressure of the pure solute (VP_i) and the activity coefficient (γ^{∞}_{water}) of the solute in water at infinite dilution (eqn 5.24).[50]

$$HLC = VP_i \gamma^{\infty}_{water} \tag{5.24}$$

The vapour pressure is calculated according to SPARC's model as described previously. The activity coefficient is calculated from the free energy of the solute–solvent and solvent–solvent intermolecular interactions and the Flory–Huggins[28,29] term for excess entropy of mixing. The default solvent is water, but SPARC will calculate the appropriate activity coefficient and *HLC* for any user-entered solvent based on its chemical structure.

Substances that are highly water soluble or dissociate in water do not adhere to Henry's Law. Estimations for these types of substances should be used with caution. For substances that hydrolyse rapidly in water, it should be considered whether it is more appropriate to calculate *HLC* for the parent or for the hydrolysis products, based on the expected hydrolysis rate of the parent and the conditions relevant to the toxicological effects of interest.

Insoluble substances with low vapour pressures may have high estimated values of *HLC*. In this case, a high *HLC* does not necessarily indicate that the compound will volatilise. It is more likely that such a substance will form a separate organic phase.

5.5.2   Soil Sorption Coefficients

The soil sorption coefficient (K_d) is the ratio of the amount of a substance adsorbed per unit weight of the soil or sediment phase to the concentration of the substance dissolved in the aqueous phase at equilibrium. This is typically the value that is measured experimentally. Because the compositions of different soils and sediments can vary widely from one to another, the K_d for a substance will also vary from soil to soil or from sediment to sediment. However, for most neutral organic compounds, the sorption is directly proportional to the organic carbon content of the soil or sediment. When K_d is normalised with respect to the organic carbon content of the soil or sediment, K_{oc} is obtained which exhibits much less variation between different soil or sediment types.

$$K_d = \frac{\mu g\ adsorbed/g\ soil\ or\ sediment}{\mu g/mL\ in\ water} \tag{5.25}$$

$$K_{oc} = \frac{K_d}{\%\ Organic\ Carbon}\,100 \tag{5.26}$$

K_{oc} is the ratio of the amount of the substance adsorbed per unit weight of organic carbon in the soil or sediment phase to the concentration of the substance dissolved in the aqueous phase at equilibrium. K_{oc} will influence a substance's distribution between environmental compartments and can indicate a substance's mobility in soil and its potential to leach through soil to contaminate groundwater. It can also indicate how readily the substance may be removed from wastewater treatment plants by adsorbing to the sludge, as well as the potential of a substance to be removed from environmental waters by adsorbing to suspended sediments. Sorption to suspended sediments in environmental waters may also decrease its bioavailability to aquatic species, thereby mitigating the potential toxic effects of the substance. Strong sorption to soil or sediment may also reduce a substance's bioavailability to microorganisms, thereby reducing its rate of biodegradation in the environment.

Several useful reviews of estimation methods for K_{oc} are available.[1,51,52] EPI Suite and ACD/PhysChem Suite both include modules for estimating K_{oc}.

EPI Suite's KOCWIN™ program[53] calculates K_{oc} by two different methods. The first method is based on the first order simple molecular connectivity index (*MCI* or χ). A molecular connectivity index is a numerical descriptor derived from a method of counting bonds that provides a quantitative characterisation of skeletal variation in a molecule's structure. *MCI*s can encode information describing such attributes as the molecular size, the presence of heteroatoms and the degree of branching, unsaturation and cyclisation of the structure, all of which can be related to its hydrophobicity. KOCWIN uses the first-order simple molecular connectivity index as a descriptor in regression equation (5.27), along with a series of group contribution correction factors for polar compounds.[54]

$$\text{Log}\,K_{oc} = 0.5213\,MCI + 0.60 + \sum P_f N \tag{5.27}$$

$\Sigma P_f N$ is the summation of the products of all applicable correction factor coefficients (P_f) multiplied by the number of times (N) that factor is counted for the structure, with the exception of certain structures for which the factors are only counted once. The relationships of log K_{oc} to *MCI*, and the values of the correction factors for polar compounds, were each derived from separate regression analyses. According to the program literature, the *MCI* approach is the more accurate and preferred method in EPI Suite.

The second method by which EPI Suite's KOCWIN calculates K_{oc} is by two regression equations based on log P.

$$\text{Log } K_{oc} = 0.8679 \log P - 0.0004 \tag{5.28}$$

$$\text{Log } K_{oc} = 0.55313 \log P + 0.9251 + \sum P_f N \tag{5.29}$$

Equation (5.28) is used for non-polar compounds and eqn (5.29) is used for polar compounds that require correction factors (P_f).

ACD/PhysChem calculates K_{oc} by a proprietary method derived from the following regression equation (5.30) based on the log P.[12] ACD/PhysChem can provide K_{oc} estimates at *pH* values ranging from 0 to 14.

$$\text{Log } K_{oc} = 0.544 \log P + 1.377 \tag{5.30}$$

Some substances bind to soils or sediments by mechanisms other than sorption to the organic carbon content. Certain substances, particularly aromatic amines, may form covalent chemical bonds to humic matter in soil or sediment, binding them regardless of the substance's K_{oc}. Other substances may adsorb more strongly to clays, irrespective of the amount of organic carbon present. When assessing a substances ability to adsorb or bind to soil or sediment, these factors should be considered separately from K_{oc}.

5.5.3 Persistent, Bioaccumulative and Toxic (PBT) Chemicals

The past decade has seen increased interest in regulating PBT chemicals.[55–57] PBT chemicals resist environmental degradation and accumulate in the food web at toxic (or potentially toxic) concentrations. Releases of PBT chemicals are of concern, even in small quantities, because their concentrations can increase over time and adversely impact human health and the environment. This section describes some tools that have proven useful for screening-level assessments of potential persistence (P) and bioaccumulation (B).

5.5.3.1 *Persistence*

The persistence of a substance is measured as the time that it is present in the environment before it is destroyed by natural processes. Chemicals may be degraded in the environment by a variety of biotic and abiotic processes including hydrolysis, photolysis, oxidation and biodegradation.

The specific criteria defining a persistent substance depend upon the applicable regulation. For example, the US EPA's PBT rule[55] defines a persistent chemical as having a half-life of 60 days or longer in water, sediment or soil, or a half-life of 2 days or longer in air. Under the European Union's Registration, Evaluation, Authorisation and Restriction of Chemicals (REACH) legislation, different criteria have been established to define persistent and very persistent substances in different environmental compartments.[57] Persistent substances are identified as having a half-life exceeding 60 days in marine water, 40 days in fresh or estuarine water, 180 days in marine sediment, 120 days in freshwater or estuarine sediment, or 120 days in soil. Very persistent chemicals are identified as having a half-life exceeding 60 days in marine, fresh or estuarine water, 180 days in marine, freshwater or estuarine sediments, or 180 days in soils.

Biodegradation is one of the most important degradation processes for chemically stable organic compounds in the water, soil and sediment compartments,[58] and is discussed in this section. A discussion of environmental degradation by abiotic pathways such as hydrolysis, photolysis and atmospheric oxidation can be found in Boethling and Mackay.[2]

Biodegradation is a process by which micro-organisms (typically bacteria and fungi) metabolise organic compounds as a means of energy production and a source of raw material for biomolecular synthesis. In evaluating the biodegradation potential of a substance, a distinction is made between primary biodegradation, which is the conversion of a substance to its first metabolite, and ultimate biodegradation, which is the conversion of a substance to water, carbon dioxide, inorganic salts and biomass. A substance that undergoes rapid primary biodegradation may be converted to degradation-resistant intermediates, thereby reducing the rate of ultimate biodegradation. Therefore, the rate of primary biodegradation does not provide a complete assessment of the biodegradation potential of a substance.

5.5.3.2   Estimating Biodegradation Potential

The biodegradation potential of a chemical is dependent in part on its bioavailability, its solubility and the presence of structural features affecting its ability to undergo typical biomediated transformations. General guidelines and estimation methods for predicting the biodegradation potential for organic chemicals are described by Howard[58] and Boethling et al.[59]

EPI Suite's BIOWIN™ program[60] estimates biodegradation potential using regression-derived fragment constant equations encompassing 36 chemical fragments that influence biodegradability. All of the BIOWIN models use either a linear (5.31) or non-linear (5.32) regression equation to determine the biodegradation potential of a chemical from its structure, either as a qualitative indication of rapid or ready biodegradation or as a semi-quantitative biodegradation timeframe. The biodegradation potential (Y_j) is calculated from the sum of the fragment coefficients (a_n) multiplied by the number of instances of each fragment (f_n), the product of the molecular weight (MW) and the molecular weight regression coefficient (a_m), an equation constant (a_0) and a correction term (e_1).[61] Each model uses its own set of coefficients (a_n and a_m) and constants for these equations.

$$Y_j = a_0 + \sum a_n f_n + a_m M_w + e_1 \tag{5.31}$$

$$Y_j = \frac{\exp(a_0 + \sum a_n f_n + a_m MW)}{\exp(1 + a_0 + \sum a_n f_n + a_m MW)} \tag{5.32}$$

The BIOWIN program in EPI Suite contains seven modules.

- BIOWIN 1 and BIOWIN 2 provide qualitative indications of whether a chemical is likely to undergo rapid biodegradation using eqn (5.31) and eqn (5.32), respectively.
- The expert survey modules BIOWIN 3 and BIOWIN 4 both use eqn (5.31) to predict semi-quantitative biodegradation timeframes for the substance to completely undergo either ultimate or primary biodegradation, respectively.
- The modules BIOWIN 5 and BIOWIN 6 provide qualitative indications of whether a chemical will be likely to pass the Japanese Ministry of International and Trade Industry (MITI) ready biodegradability test (OECD 301 C) using eqn (5.31) and eqn (5.32), respectively.
- BIOWIN 7 predicts biodegradation potential under anaerobic conditions using eqn (5.31).[62]

The BIOWIN models are useful for screening level assessments of biodegradation potential and rate. The complete conversion timeframes from the ultimate biodegradation survey model (BIOWIN 3) have been translated to approximate biodegradation half-lives.[63] Those substances receiving a rating of 'months' or 'recalcitrant' should be examined further for potential persistence.

5.5.3.3 Bioconcentration, Biomagnification and Bioaccumulation

Bioconcentration, biomagnification and bioaccumulation describe the accumulation of a substance in an organism by different routes of exposure. The similarity of these three processes has sometimes led to confusion of the terms. A recent review by Gobas and Morrison[64] describes these in more detail and discusses the various biological processes that affect bioaccumulation.

Bioconcentration is the accumulation of a chemical within an organism from exposure to that chemical in its environment. Bioconcentration is often measured in fish because the exposure levels are easily manipulated by adjusting the concentration of the substance in the water. Fish are exposed to the freely dissolved substance in water through exposed tissues such as the gills and skin. The bioconcentration factor (*BCF*) in fish is expressed as the ratio of the concentration of the substance in the organism to its concentration in the surrounding water.

Biomagnification is the accumulation of a compound in an organism from dietary exposure. The biomagnification factor (*BMF*) is expressed as the ratio of the concentration of the substance in the organism to its concentration in the organism's diet.

Bioaccumulation is the accumulation of a compound in an organism from all routes of exposure, including exposure from the environment and from its diet. Bioaccumulation includes both bioconcentration and biomagnification. For fish, the bioaccumulation factor (*BAF*) is expressed as the ratio of the concentration of the substance in the organism to its concentration in both the surrounding water and in the fish's diet.

Although it is an oversimplification, *BCF* has become accepted as an approximation of the potential for a chemical to bioaccumulate.[55] *BCF* in fish is readily measured and a substantial dataset of measured values is available to support the development of estimation methods.

Fewer measured data are available for *BAF*. *BAF* is more dependent on the identity of the organism and the conditions of its environment than *BCF* in fish. The concentration of the test substance in the diet and in the surrounding environment must be carefully controlled and monitored to get a valid measurement. Because *BAF* measurements tend to be species-specific, the available data cannot readily be adapted for developing estimation methods. Estimations for *BAF* exist, but must rely on assumptions about an organism's trophic level, lipid content, food consumption and growth rates, metabolism and other factors.[64,65]

As with persistence, the specific criteria for identifying a 'bioaccumulative' substance depend upon the applicable regulation. The US EPA's PBT Rule identifies a bioaccumulative substance as having a *BCF* value of $\geq 1,000$.[55] REACH identifies a bioaccumulative substance as having a *BCF* value exceeding 2,000 and a 'very bioaccumulative' substance as having a *BCF* value exceeding 5,000.[57]

5.5.3.4   *Estimating BCF*

Nearly all of the available *BCF* estimation methods calculate the *BCF* in fish as a function of log *P*. For log *P* values ranging from 1 to approximately 7, the log *BCF* increases linearly with increasing log *P*. At higher log *P* values, the measured *BCF* begins to decrease with increasing log *P*. Proposed explanations for this phenomenon include decreased bioavailability for highly hydrophobic compounds due to low water solubility or to increased sorption to dissolved organic matter, colloidal material or particulate matter in the water column.[1]

EPI Suite's BCFBAF™ program[66] uses two regression-derived equations to calculate *BCF* for non-ionic compounds as a function of log *P*.[67] Equation (5.33) is used for log *P* values ranging from 1 to 7, and eqn (5.34) is used for log *P* values greater than 7. For compounds with log *P* values of less than 1, log *BCF* is assigned a value of 0.5.

$$\text{Log } BCF = 0.6598 \log P - 0.333 + \sum Correction\, factors \qquad (5.33)$$

$$\text{Log } BCF = -0.49 \log P + 7.554 + \sum Correction\, factors \qquad (5.34)$$

The ACD/PhysChem Suite calculates *BCF* as a function of log *P* according to eqn (5.35).[12] For ionisable compounds, the program will calculate *BCF* at *pH* values ranging from 0 to 14.

$$\text{Log } BCF = 0.79 \log P - 0.40 \qquad (5.35)$$

5.5.3.5 PBT Profiler

The PBT Profiler[68] was developed by the US EPA and SRC, Inc. as part of the Pollution Prevention (P2) Assessment Framework. It is a free web-based screening tool (www.pbtprofiler.net) for determining potential PBT hazard.

Using a simplified combination of models in EPI Suite, including estimations for vapour pressure, water solubility, log *P* , Henry's Law constant, K_{oc}, *BCF* and biodegradability, the PBT profiler predicts the distribution of a chemical to different environmental compartments (air, water, soil, sediment), the chemical's persistence in each compartment, its *BCF* and its chronic toxicity to fish. The PBT profiler then evaluates the results against US EPA's PBT criteria and determines high, moderate or low potential hazard for P, B and T.

Substances having biodegradation half-lives of less than 60 days are considered to have low persistence. Substances with half-lives between 60 and 180 days are given a rating of moderate, and those with half-lives exceeding 180 days are given a rating of high persistence. Substances with a *BCF* less than 1,000 are rated low for bioaccumulation. Those with a *BCF* between 1,000 and 5,000 are given a moderate rating and those with a *BCF* exceeding 5,000 are given a high rating.

5.6 Conclusions

Physico-chemical property and environmental fate data are essential for performing chemical hazard and risk assessments. These data provide the basis for understanding the fate and transport of a substance in the environment, as well as its bioavailability and potential toxicity. Although reliable measured data are preferred, estimation methods can provide useful information when a full set of measured properties is not available. It is hoped that the information presented in this chapter will help researchers to:

- effectively and accurately generate estimations of physico-chemical and environmental fate properties; and
- understand and successfully interpret the results.

References

1. R. S. Boethling, P. H. Howard and W. M. Meylan, *Environ. Toxicol. Chem.*, 2004, **23**, 2290.
2. R. S. Boethling and D. Mackay, *Handbook of Property Estimation Methods for Chemicals, Environmental and Health Sciences,* CRC Press, Boca Raton, FL, 2000.
3. J. C. Dearden, *Environ. Toxicol. Chem.*, 2003, **22**, 1696.
4. P. H. Howard, W. M. Meylan, in *Quantitative Structure-Activity Relationships in Environmental Science VII,* ed. F. Chen and G. Schüürrman, SETAC, Pensacola, FL, 1997, pp. 185–205.
5. Chemical Abstracts Service (CAS) Registry file, American Chemical Society, http://stnweb.cas.org [accessed March 2010].
6. ChemID Plus Advanced, US National Library of Medicine, http://chem.sis.nlm.nih.gov/chemidplus/ [accessed March 2010].
7. ChemBioFinder, CambridgeSoft Corporation, www.cambridgesoft.com [accessed March 2010].
8. Physical Properties Database (PHYSPROP©), SRC, Inc., Syracuse, NY, www.syrres.com/what-we-do/product.aspx?id = 133 [accessed March 2010].
9. EPI Suite™, v.4.00, US Environmental Protection Agency, Washington DC, 2009, www.epa.gov/oppt/exposure/pubs/episuite.htm [accessed March 2010].
10. Environmental Fate Data Base (EFDB), SRC, Inc., Syracuse, NY, www.srcinc.com/what-we-do/efdb.aspx [accessed March 2010].
11. Hazardous Substances Data Bank (HSDB), US National Library of Medicine, http://toxnet.nlm.nih.gov/ [accessed March 2010].
12. ACD/PhysChem Suite, Advanced Chemistry Development Inc., www.acdlabs.com [accessed March 2010].
13. SPARC (SPARC Performs Automated Reasoning in Chemistry), v.4.2, US Environmental Protection Agency, Washington DC, 2007 http://ibmlc2.chem.uga.edu/sparc [accessed March 2010].
14. ClogP v.4.0, Biobyte Corporation, Claremont, CA, 1993–2003, www.biobyte.com/bb/prod/clogp40.html [accessed March 2010].
15. M. T. D. Cronin and D. J. Livingstone, *Predicting Chemical Toxicity and Fate,* CRC Press, Boca Raton, FL, 2004.
16. J. C. Dearden and G. Schüürrman, *Environ. Toxicol. Chem.*, 2003, **22**, 1755.
17. R. Mannhold, G. I. Poda, C. Ostermann and I. V. Tetko, *J. Pharm. Sci.*, 2009, **98**, 861.
18. Y. Sakuratani, K. Kasai, Y. Noguchi and J. Yamada, *QSAR Comb. Sci.*, 2007, **26**, 109.
19. A. A. Patrauskas and E. A. Kolovanov, *Perspectives in Drug Discovery and Design*, 2000, **19**, 99.
20. ECOSAR™; ECOWIN v.1.00, , US Environmental Protection Agency, Washington DC, 2000, available as part of EPI Suite™, www.epa.gov/oppt/exposure/pubs/episuite.htm [accessed March 2010].

21. A. Leo and D. Hoekman, *Perspectives in Drug Discovery and Design*, 2000, **18**, 19.
22. R. A. Scherrer and S. M. Howard, *J. Med. Chem.*, 1977, **20**, 53.
23. A. Leo, in *Handbook of Property Estimation Methods for Chemicals** Environmental and Health Sciences*, ed. R. S. Boethling and D. Mackay, CRC Press, Boca Raton, FL, 2000, pp. 89–114.
24. KOWWIN™, US Environmental Protection Agency, Washington DC, 2000, available as part of EPI Suite™, www.epa.gov/oppt/exposure/pubs/episuite.htm [accessed March 2010].
25. W. M. Meylan and P. H. Howard, *J. Pharm. Sci.*, 1995, **84**, 83.
26. A. Leo, *Chem. Rev.*, 1993, **93**, 1281.
27. S. H. Hilal, S. W. Karickhoff and L. A. Carreira, *QSAR Comb. Sci.*, 2004, **23**, 709.
28. P. Flory, *J. Chem. Phys.*, 1942, **10**, 51.
29. M. L. Huggins, *J. Am. Chem. Soc.*, 1942, **64**, 1712.
30. S. C. DeVito, in *Handbook of Property Estimation Methods for Chemicals, Environmental and Health Sciences*, ed. R. S. Boethling and D. Mackay, CRC Press, Boca Raton, FL, 2000, pp. 261–278.
31. D. Mackay, in *Handbook of Property Estimation Methods for Chemicals** Environmental and Health Sciences*, ed. R. S. Boethling and D. Mackay, CRC Press, Boca Raton, FL, 2000, pp. 125–139.
32. WSKOWWIN™, US Environmental Protection Agency, Washington DC, 2000, available as part of EPI Suite™, www.epa.gov/oppt/exposure/pubs/episuite.htm [accessed March 2010].
33. W. M. Meylan, P. H. Howard and R. S. Boethling, *Environ. Toxicol. Chem.*, 1996, **15**, 100.
34. WATERNT™, US Environmental Protection Agency, Washington DC, 2000, available as part of EPI Suite™, www.epa.gov/oppt/exposure/pubs/episuite.htm [accessed March 2010].
35. S. H. Hilal and S. W. Karickhoff, *Quant. Struct.-Act. Relat.*, 1995, **14**, 348.
36. M. L. Sage and G. W. Sage, in *Handbook of Property Estimation Methods for Chemicals, Environmental and Health Sciences*, ed. R. S. Boethling and D. Mackay, CRC Press, Boca Raton, FL, 2000, pp. 53–65.
37. C. Antoine, *C. R. Seances Acad. Sci.*, 1888, **107**, 681.
38. C. F. Grain, in *Handbook of Chemical Property Estimation Methods*, ed. W. J. Lyman, W. F. Reehl and D. F. Rosenblatt, McGraw-Hill, New York, 1982, Chapter 14, pp. 1–20.
39. MPBPWIN™, US Environmental Protection Agency, Washington DC, 2000, available as part of EPI Suite™, www.epa.gov/oppt/exposure/pubs/episuite.htm [accessed March 2010].
40. S. H. Hilal, S. W. Karickhoff and L. A. Carriera, *QSAR Comb. Sci.*, 2003, **22**, 565.
41. W. J. Lyman, in *Handbook of Property Estimation Methods for Chemicals, Environmental and Health Sciences*, ed. R. S. Boethling and D. Mackay, CRC Press, Boca Raton FL, 2000, pp. 29–51.
42. S. E. Stein and R. L. Brown, *J. Chem. Inf. Comp. Sci.*, 1994, **34**, 581.

43. M. Tesconi and S. Yalkowski, in *Handbook of Property Estimation Methods for Chemicals, Environmental and Health Sciences*, ed. R. S. Boethling and D. Mackay, CRC Press, Boca Raton, FL, 2000, pp. 3–27.
44. K. G. Joback and R. C. Reid, *Chem. Eng. Commun.*, 1987, **57**, 233.
45. P. I. Gold and G. J. Ogle, *Chem. Eng.*, 1969, **76**, 119.
46. D. Mackay, W. Y. Shiu and K. C. Ma, in *Handbook of Property Estimation Methods for Chemicals, Environmental and Health Sciences*, ed. R. S. Boethling and D. Mackay, CRC Press, Boca Raton, FL, 2000, pp. 69–87.
47. HENRYWIN™, US Environmental Protection Agency, Washington DC, 2000, available as part of EPI Suite™, www.epa.gov/oppt/exposure/pubs/episuite.htm [accessed March 2010].
48. W. M. Meylan and P. H. Howard, *Environ. Toxicol. Chem.*, 1991, **10**, 1283.
49. J. Hine and P. K Mookerjee, *J. Org. Chem.*, 1975, **40**, 292.
50. S. H. Hilal, S. N. Ayyampalayam and L. A. Carriera, *Environ. Sci. Technol.*, 2008, **42**, 9231.
51. W. J. Doucette, in *Handbook of Property Estimation Methods for Chemicals, Environmental and Health Sciences*, ed. R. S. Boethling and D. Mackay, CRC Press, Boca Raton, FL, 2000, pp. 141–190.
52. W. J. Lyman, in *Handbook of Chemical Property Estimation Methods, Environmental Behavior of Organic Compounds*, ed. W. J. Lyman, W. F. Reehl and D. H. Rosenblatt, McGraw-Hill, New York, 1982, Chapter 4, pp. 1–33.
53. KOCWIN™, US Environmental Protection Agency, Washington DC, 2000, available as part of EPI Suite™, www.epa.gov/oppt/exposure/pubs/episuite.htm [accessed March 2010].
54. W. Meylan, P. H. Howard and R. S. Boethling, *Environ. Sci. Technol.*, 1992, **26**, 1560.
55. US Environmental Protection Agency, 40 CFR Part 372, *Federal Register*, 1999, **64**(209), 58666.
56. B. D. Rodan, D. W. Pennington, N. Eckley and R. S. Boethling, *Environ. Sci. Technol.*, 1999, **33**, 3482.
57. Regulation (EC) No. 1907/2006 of the European Parliament and of the Council of 18 December 2006 concerning the Registration, Evaluation, Authorisation and Restriction of Chemicals (REACH), establishing a European Chemicals Agency, amending Directive 1999/45/EC and repealing Council Regulation (EEC) No. 793/93 and Commission Regulation (EC) No. 1488/94 as well as Council Directive 76/769/EEC and Commission Directives 91/155/EEC, 93/67/EEC, 93/105/EC and 2000/21/EC, *Off. J. Eur. Union*, 2006, **L396**, 383.
58. P. H. Howard, in *Handbook of Property Estimation Methods for Chemicals, Environmental and Health Sciences*, ed. R. S. Boethling and D. Mackay, CRC Press, Boca Raton, FL, 2000, pp. 281–310.
59. R. S. Boethling, E. Somme and D. DiFiore, *Chem. Rev.*, 2007, **107**, 2207.
60. BIOWIN™, US Environmental Protection Agency, Washington DC, 2000, available as part of EPI Suite™, www.epa.gov/oppt/exposure/pubs/episuite.htm [accessed March 2010].

61. R. S. Boethling, P. H. Howard, W. Meylan, W. Stiteler, J. Beauman and N. Tirado, *Environ. Sci. Technol.*, 1994, **28**, 459.
62. W. Meylan, R. Boethling, D. Aronson, P. Howard and J. Tunkel, *Environ. Toxicol. Chem.*, 2007, **26**, 1785.
63. D. Aronson, R. Boethling, P. Howard and W. Stiteler, *Chemosphere*, 2006, **63**, 1953.
64. F. A. P. C. Gobas and H. A. Morrison, in *Handbook of Property Estimation Methods for Chemicals, Environmental and Health Sciences*, ed. R. S. Boethling and D. Mackay, CRC Press, Boca Raton, FL, 2000, pp. 189–231.
65. J. A. Arnot and F. A. P. C. Gobas, *QSAR Comb. Sci.*, 2003, **22**, 337.
66. BCFBAF™, US Environmental Protection Agency, Washington DC, 2000, available as part of EPI Suite™, www.epa.gov/oppt/exposure/pubs/episuite.htm [accessed March 2010].
67. W. M. Meylan, P. H. Howard, R. S. Boethling, D. Aronson, H. Printup and S. Gouchie, *Environ. Toxicol. Chem.*, 1999, **18**, 664.
68. PBT Profiler v.1.203, US Environmental Protection Agency and Syracuse Research Corporation (SRC), 2006, www.pbtprofiler.net [accessed March 2010].

CHAPTER 6

Molecular Descriptors from Two-Dimensional Chemical Structure

U. MARAN, S. SILD, I. TULP, K. TAKKIS AND M. MOOSUS

Institute of Chemistry, University of Tartu, Ravila 14A, Tartu 50411, Estonia

6.1 Introduction

The description of chemical processes can be divided, in a broad sense, into two chemical model categories—macroscopic and microscopic. Macroscopic models describe the general properties of systems and processes (*e.g.* transport kinetics of chemicals, relative rates of uptake of chemicals, depletion rates of entities influenced by the chemicals), and do not consider the structural properties of individual molecules. Microscopic models, on the contrary, consider all atoms in the model, *i.e.* structural properties.

Methods for the development of microscopic models can be also divided into two—the so-called fitting procedures and methods applying theory from first principles (*ab initio*). The purpose of a fitting procedure is to find a relationship between a physico-chemical property or chemico-biological activity and the corresponding molecular structure(s). Usually, this property or activity (P) is represented as a mathematical function from the molecular structure which is described with numerical values—descriptors (d): $P = f(d_1, d_2, \ldots, d_n)$.

Expounding and modulation of properties depending on the change in molecular structure is critically linked to the ability of researchers to develop a

Issues in Toxicology No.7
In Silico Toxicology: Principles and Applications
Edited by Mark T. D. Cronin and Judith C. Madden
© The Royal Society of Chemistry 2010
Published by the Royal Society of Chemistry, www.rsc.org

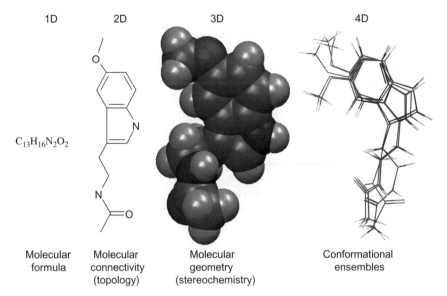

1D   2D   3D   4D

$C_{13}H_{16}N_2O_2$

| Molecular
formula | Molecular
connectivity
(topology) | Molecular
geometry
(stereochemistry) | Conformational
ensembles |

Figure 6.1   The increasing dimensionality, information richness and complexity captured by the molecular descriptors. The example chemical structure is
melatonin (*N*-acetyl-5-methoxytryptamine).

relationship between structure and property/activity. Often those relationships
are qualitative, *i.e.* structure–activity relationships (SAR), derived either from
expert knowledge or from case-based reasoning. Representation of property or
activity as a mathematical function of structure helps to transfer qualitative
ideas into quantitative numerical relationships, often in the form of quantitative structure activity relationships (QSARs). The prerequisite for this is the
representation of information coded in the chemical structure into a numerical
form. This numerical representation—descriptors—depends on the amount of
information used for their calculation. This information can be very simple or
very complex and allows grouping of molecular descriptors according to their
dimensionality (Figure 6.1):

- **One-dimensional (1-D) descriptors** capture information about the chemical
  composition of compounds and, for these, the chemical formula is sufficient to calculate the descriptors.
- **Two-dimensional (2-D) descriptors** extend the complexity and require
  information about the connectivity in molecules that is usually expressed
  in the form of mathematical objects—graphs. Therefore, 2-D molecular
  descriptors are defined to be numerical properties that can be calculated
  from the connection table representation of a molecule, but not atomic co-
  ordinates; they can also include formal charges. It should be emphasised
  here that, in mathematical terms, the graphs are one-dimensional
  objects,[1,2] but as they are realised as chemical objects in higher

dimensional space, the use of 2-D notation for the discussion of molecular descriptors is justified. Two-dimensional descriptors in turn can be formally divided into two—whole-molecule numerical characteristics and structural fragments (subunits) of the molecule.

- **Three-dimensional (3-D) descriptors** extend the complexity further, considering co-ordinates and orientation of atomic nuclei in space, atomic masses and/or atomic radii. Combination of 3-D co-ordinates and information about the electronic structure of molecules allows explicit consideration of charge distribution in the molecules. This can be calculated at various levels of theory starting from empirical methods and ending with high level quantum chemical methods.
- **Four-dimensional (4-D) descriptors** go beyond the single molecule descriptors and consider the effect of molecular ensembles. They incorporate conformational and alignment freedom into the development of descriptors.

The beginning of QSAR dates back to the 19th century. In 1884, Mills developed a QSAR for the prediction of melting and boiling points of homologous series.[3] Similar pioneering work in QSAR followed shortly after the studies of relationships between the potency of local anaesthetics and oil/water partition coefficient,[4] and between narcosis and chain length.[5] A subsequent attempt to link a property to critical structural features was reported in 1925 when Langmuir proposed linking intermolecular interactions in the liquid state to the surface energy.[6] Important contributions to the area were made in the 1930s by Hammett[7,8] and in the 1950s by Taft[9–12] *via* the development of linear free energy relationships (LFER).

Probably the most important milestone in the development of theoretical whole-molecule descriptors dates back to 1947, when the first 2-D descriptors, Wiener index[13] and Platt number[14] were proposed in the framework of the modelling of boiling points of hydrocarbons. It took another 15 years before QSAR as a methodology got a boost after developments in the 1960s, when Hansch and Fujita connected biological activities to the hydrophobic, electronic and steric properties of compounds[15]—today considered cornerstone of modern QSAR. From then it took a further ten years until Randić introduced a descriptor for the characterisation of branching, accelerating further the development of 2-D descriptors.[16]

This chapter concentrates on 2-D whole-molecule descriptors and aspects of their calculation and use. The aim is not to give an all-inclusive overview of theoretical developments but rather to emphasise their practical use in the context of *in silico* toxicology. Section 6.2 provides a mathematical foundation of 2-D descriptors. Section 6.3 helps to navigate among 2-D descriptors and provides definitions of the most commonly used descriptors. Section 6.4 provides an overview of QSAR models involving 2-D descriptors for the prediction of toxicological endpoints. Section 6.5 provides special insight into the interpretability of 2-D descriptors. Section 6.6 collects and presents, in a systematic way, the current state-of-the-art in the arsenal of 2-D descriptor calculators. Finally, Section 6.7 concludes this chapter and Section 6.8 lists the key literature in the field for complementary reading.

6.2 Mathematical Foundation of 2-D Descriptors

The common foundation for the representation of molecular structures and the definition of 2-D descriptors is graph theory, which is a branch of mathematics that studies graphs. A graph is a mathematical structure (*i.e.* ordered pair) consisting of a set of objects (called vertices or nodes) and a set of pair-wise relations (called edges) between these objects. Graphs are usually denoted as G = (V, E), where V is a set of vertices and E is a set of edges representing binary relationships between pairs of vertices. Each edge connects two vertices. Two vertices are called adjacent if there is a common edge connecting them. Graphs can be visualised (see Figure 6.2) as a diagram where vertices are drawn as small circles or dots and the edges as lines connecting the vertices. Graphs may contain cycles. Graphs with cycles are called cyclic and graphs without cycles are called acyclic. The graphs depicted on Figure 6.2 have a very obvious resemblance with the molecular structure representation chemists have used in everyday communication about chemicals and their reactions ever since the work of August Kekulé,[17] who started to use graphs for the representation of molecules in the framework of theory of chemical structure.[i]

Over the past decades, the applications of graph theory to chemistry related problems have created an interesting research area in chemistry called chemical graph theory.[1] In chemical graph theory, different chemical systems can be represented as graphs—molecules, proteins, reactions, polymers, *etc.* All these systems can be viewed as chemical graphs where the relevant objects are atoms,

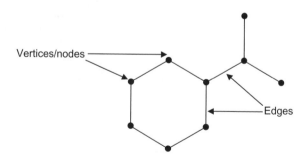

Vertices/nodes

Edges

Figure 6.2 Representation of graphs.

---

[i] August Kekulé is considered the principal founder of the theory of chemical structure. At the same time, several other prominent scientists independently or jointly contributed to the theory: Archibald Scott Couper (independently arrived at the idea of self-linking of carbon atoms); Aleksandr Mikhailovich Butlerov (first to incorporate double bonds into structural formulas); Alexander Crum Brown; Emil Erlenmeyer and Adolphe Wurtz.

amino acids, nucleotides, reactants, intermediates, monomers, *etc.* The connections between these objects may represent covalent bonds, hydrogen bonds, non-bonded interactions, reaction steps, *etc.* Chemical graphs representing molecules are also called molecular graphs. In these graphs, vertices correspond to individual atoms in the molecule and edges correspond to chemical bonds between them. In molecular graphs, hydrogen atoms are often left out (*i.e.* a hydrogen-suppressed graph or hydrogen-depleted graph) to simplify the handling of molecular graphs. Molecular graphs are usually fully connected, meaning that each atom in the molecule has one or more chemical bonds.

Although the connection between molecular structures and graphs is very natural for chemistry, both disciplines have evolved independently and have their own histories. The interface between these disciplines has provided many practical applications to molecular modelling, but also introduced a new terminology from graph theory. Table 6.1 provides a brief glossary of common terms in graph theory and their corresponding terms in chemistry.

There are multiple data structures for the representation of graphs. The most common among them are adjacency and distance matrices (see Table 6.2). The adjacency matrix is a square symmetric matrix containing information about the internal connectivity of vertices in the graph. The rows (i) and columns (j) correspond to the vertices of the graph. The value of a matrix element (i, j) is one if vertices i and j are connected with an edge and zero otherwise.

Table 6.1 Translation between graph theoretical and chemical terms.

Graph theory	*Chemistry*
graph	molecular structure
vertex	atom
degree of vertex	valency of an atom
edge	chemical bond
chain	chain (*e.g.* alkane or alkene)
cycle	cyclic structure (*e.g.* cycloalkane)
graph invariant	topological index or descriptor
path	fragment
sub-graph	substructure
tree	acyclic structure

Table 6.2 Adjacency and distance matrices for a graph corresponding to isopentane.

Adjacency matrix						Distance matrix						
v_{ij}	1	2	3	4	5	v_{ij}	1	2	3	4	5	
1	0	1	0	0	0	**1**	0	1	2	3	2	
2	1	0	1	0	1	**2**	1	0	1	2	1	
3	0	1	0	1	0	**3**	2	1	0	1	2	
4	0	0	1	0	0	**4**	3	2	1	0	3	
5	0	1	0	0	0	**5**	2	1	2	3	0	

The diagonal matrix elements $(i = j)$ have zero values. The distance matrix is a square symmetric matrix where each element (i, j) corresponds to the length of the shortest path between vertices i and j. The length or distance is defined as the number of edges between vertices i and j. These matrices are valuable starting points for calculating numerical characteristics about graphs.

Molecular structures and the corresponding graphs are discrete entities and cannot be used directly in numerical comparisons or mathematical models. In order to use structural information for quantitative predictions it is necessary to translate this information to numbers. A topological index is a single number that can be used to characterise the graph corresponding to a molecular structure. A topological index is called a graph invariant in graph theory. For the given graph, the topological index has the same value regardless of how the graph is depicted or labelled. For example, the number of vertices in the graph (*i.e.* number of atoms) and the number of edges (*i.e.* the number of bonds) can be both considered as the simplest graph invariants. In both cases, the value of a graph invariant does not change when the vertices or edges are counted in a different order. The same property holds for more complex graph invariants. The calculation of topological indices is a straightforward process, but the reverse process for the reconstruction of the molecular structure is generally not possible.

One of the drawbacks of topological indices is degeneracy. A large part of graph invariants are degenerate, because the calculation process loses some of the information available in the graph. This means, in practical terms, that several different graphs may have identical values for certain topological indices. For example, Figure 6.3 shows five graphs that all have the same value for the connectivity index $^1\chi$—see eqn (6.4). This drawback can cause serious problems in predictive QSAR modelling because the descriptor might not be able to discriminate between two isomers having different physico-chemical and biological properties.

6.3 Navigation Among 2-D Descriptors

Over the years the number of topological descriptors has grown vastly. This section provides a short overview of the most common groups of descriptors; this is by no means a rigorous classification scheme but rather considers traditional families of descriptors bearing a certain similarity. At present there is no common and widely accepted classification scheme established to classify all known descriptors. Attempts to distinguish descriptors based on certain mathematical or chemical rules have so far only been of local value, often confined within one workgroup or publication.

6.3.1 Topological Indices

Topological indices (TIs) are often divided into four generations for reasons that are part historical, part mathematical.[18,19] The first generation TIs are based on integer graph properties and the resulting index value is always an

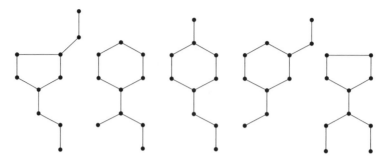

Figure 6.3 Example graphs with the same value for the connectivity index $^1\chi$.

integer such as the Wiener index. Second generation TIs are also calculated from integer graph properties but the resulting index is a real number. Third generation TIs are based on real-number graph properties, also resulting in a real number. This kind of grouping orders TIs in terms of degeneracy and discrimination power. First generation TIs are highly redundant while the third generation TIs have higher discrimination power. The fourth generation involves 3-D structure in TIs and considers the information related to the stereochemistry of molecules. Another way to divide the topological indices is to follow the information they represent;[20] they can be topostructural where TIs are purely calculated from a graph without any additional chemical information, or topochemical where different bond or atom types are considered.

Historically, one of the first and still useful topological indices is the Wiener index (w), which is a sum of bond contributions where the bond contribution is defined as the product of the carbon atoms on each side of the bond.[21,22] This descriptor can be conveniently calculated as the half-sum of all entries in the distance matrix or of all distance sums for i-th row or column corresponding to vertex i.

$$W = \frac{1}{2}\sum_i \sum_j d_{ij} \tag{6.1}$$

Wiener used this simplest topological index for the description of boiling points of paraffins.[13] Several Wiener index modifications and extensions are proposed for various purposes.[22]

Later Zagreb indices were introduced. They are based directly on the graph adjacency matrix,[21,23] where M_1 is defined as the sum of squared vertex degrees and M_2 is the sum of vertex degrees products:

$$M_1 = \sum_i v_i^2 \tag{6.2}$$

$$M_2 = \sum_{ij} v_i v_j \tag{6.3}$$

where v is the vertex degree (number of edges, number of bonds). These descriptor types belong to the first generation, because they result always in integer numbers and thus they rank chemical substances.

In 1975 Randić introduced an index for the characterisation of molecular branching which is currently known as the connectivity index or Randić index.[16,21]

$$^1\chi = \sum_{allbonds} (\delta_i \delta_j)^{-\frac{1}{2}} \tag{6.4}$$

This descriptor looks similar to Zagreb groups M_2, where the inverse square root is applied to vertex degree products in the summation. Later, Kier and Hall extended the Randic connectivity index to several orders (m) and introduced χ as a notation of connectivity indices.[21,24–26] (They also started to use δ instead of v as a notation of vertex degree.)

Thus the Randić connectivity index (eqn 6.4) is the same as the Kier–Hall first order connectivity index; the general equation of connectivity indices is given in eqn (6.5) where n denotes the total number of distinct connected subgraphs, each having m edges.

$$^m\chi = \sum_{j=1}^{s} \prod_{i=1}^{n} (\delta_i)^{-\frac{1}{2}} \tag{6.5}$$

Different orders represent the path length in the molecular graph which is used to calculate the particular topological index. Order 1 indices are summed over paths with length 1 (*i.e.* chemical bonds), order 2 indices are summed over paths with length 2 (*i.e.* fragments consisting of two adjacent bonds), and so on. It is also possible to define a 0th order index. In this case, only individual atoms (vertices) are considered in the calculation of the index. In QSAR modelling orders from 1 to 3 are mostly used, but sometimes they are calculated even up to the 6th order. Example 6.1 illustrates the calculation of connectivity indices of isobutyl methyl ether (shown in Figure 6.4).

Example 6.1 Calculation of connectivity indices of isobutyl methyl ether.

$$\chi = (\delta_1)^{-\frac{1}{2}} + (\delta_2)^{-\frac{1}{2}} + (\delta_3)^{-\frac{1}{2}} + (\delta_4)^{-\frac{1}{2}} + (\delta_5)^{-\frac{1}{2}} + (\delta_6)^{-\frac{1}{2}}$$
$$= (1)^{-\frac{1}{2}} + (3)^{-\frac{1}{2}} + (2)^{-\frac{1}{2}} + (2)^{-\frac{1}{2}} + (1)^{-\frac{1}{2}} + (1)^{-\frac{1}{2}} = 4.99$$
$$^1\chi = (\delta_1 \cdot \delta_2)^{-\frac{1}{2}} + (\delta_2 \cdot \delta_3)^{-\frac{1}{2}} + (\delta_3 \cdot \delta_4)^{-\frac{1}{2}} + (\delta_4 \cdot \delta_5)^{-\frac{1}{2}} + (\delta_2 \cdot \delta_6)^{-\frac{1}{2}}$$
$$= (1 \cdot 3)^{-\frac{1}{2}} + (3 \cdot 2)^{-\frac{1}{2}} + (2 \cdot 2)^{-\frac{1}{2}} + (2 \cdot 1)^{-\frac{1}{2}} + (3 \cdot 1)^{-\frac{1}{2}} = 2.77$$
$$^2\chi = (\delta_1 \cdot \delta_2 \cdot \delta_3)^{-\frac{1}{2}} + (\delta_2 \cdot \delta_3 \cdot \delta_4)^{-\frac{1}{2}} + (\delta_3 \cdot \delta_4 \cdot \delta_5)^{-\frac{1}{2}} + (\delta_1 \cdot \delta_2 \cdot \delta_6)^{-\frac{1}{2}} + (\delta_3 \cdot \delta_2 \cdot \delta_6)^{-\frac{1}{2}}$$
$$= 1 \cdot 3 \cdot 2)^{-\frac{1}{2}} + (3 \cdot 2 \cdot 2)^{-\frac{1}{2}} + (2 \cdot 2 \cdot 1)^{-\frac{1}{2}} + (1 \cdot 3 \cdot 1)^{-\frac{1}{2}} + (2 \cdot 3 \cdot 1)^{-\frac{1}{2}} = 2.18$$

etc.

The Kier–Hall valence connectivity indices are further extensions of connectivity indices. Assigning fixed weights to vertices allows the implicit characterisation of the presence of heteroatoms in the molecule.[21,24–27] In this case

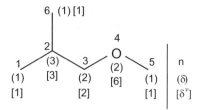

Figure 6.4 Hydrogen-depleted graph of isobutyl methyl ether where connectivity delta (δ) values are in round brackets and valence connectivity delta (δ^v) values are in square brackets.

the valence connectivity delta is expressed as:

$$\delta_i^v = \frac{Z_i^v - h_i}{Z_i - Z_i^v - 1} \tag{6.6}$$

where Z_i^v is the number of valence electrons, Z_i is the count of all electrons and h is the count of hydrogen atoms bonded to atom i.

The valence δ^v index encodes the atomic electronic state because it takes into account the number of valence electrons, the number of core electrons and the number of bonded hydrogens. Kier and Hall used the valence δ^v index to define a new family of molecular valence connectivity indices:

$$^m\chi_t^v = \sum_{j=1}^{s} \prod_{i=1}^{n} (\delta_i^v)^{-\frac{1}{2}} \tag{6.7}$$

where s is the number of connected sub-graphs of type t with m edges and n is the number of vertices in the sub-graph.

Four types of sub-graphs can be used for the calculation of connectivity indices (Figure 6.5):

(i) Path, which is the simplest one and follows the chain of the structure
(ii) Cluster, which accounts for the branching fragments
(iii) Path/Cluster, where first the two types are merged
(iv) Chain, which accounts for closed rings (cycles).

All possible matches of these sub-graphs are searched over the graph (structure) while calculating the particular type of connectivity indices.

These connectivity indices represent a weighted sum over all molecular fragments with the same topology in a molecule. Different orders of valence connectivity indices follow the same mathematical pattern as discussed previously for simple connectivity indices (Example 6.1). Connectivity delta (δ) and valence connectivity delta (δ^v) values are the same for carbon atoms (with sp³ hybridisation state) but different for other atoms such as in the Figure 6.4 where, for oxygen, δ is 2 and δ^v is 6. If the carbon atom is in sp² or sp hybridisation state then δ and δ^v differ too due to the different electron configuration in double and triple bonds.

Order	Type			
m	Path (P)	Cluster (C)	Path/Cluster (PC)	Chain (CH)

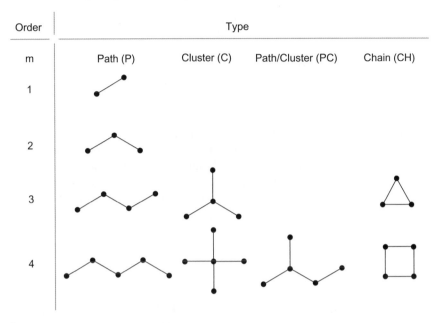

Figure 6.5 Different sub-graphs types (fragments) of different orders.

Unlike all other preceding topological indices, the average distance connectivity index, J (Balaban index), was designed not to increase rapidly with the graph size.[21,28] A similar summation over all edges i–k as in the Randić connectivity index $^1\chi$ is used, but instead of connectivity, distance sums S_i are used, which are less degenerate and unlimited in magnitude. In addition, the index J (eqn 6.8) is normalised for graph size and cyclicity by the means of number of edges E and the cyclomatic number $R = E - N + 1$, where N is the number of vertices.

$$ J = \frac{E}{R+1}\sum_{i-k}(S_i S_k)^{-\frac{1}{2}} \qquad (6.8) $$

As a consequence of this normalisation, for an infinite linear chain (poly-ethylene), $J = \pi$, and for infinite polypropylene, $J = 3\pi/2$.[29] For alkanes, J starts to degenerate when there are 12 carbon atoms in the molecule, and for cyclic graphs, starting from eight carbon atoms in the molecule.[30] An interesting observation is that J orders alkanes similarly to Wiener index.[31,32]

Another set of very useful topological indices are kappa indices of molecular shape and flexibility.[21,26,33,34] The descriptors $^m\kappa$ are defined in terms of the number of graph vertices, which is the number of non-hydrogen atoms, denoted as A, and the number of paths ^mP with length m ($m = 1, 2, 3$) in the hydrogen-depleted molecular graph. Thus, three orders of kappa indices exist:

$$^1\kappa = A(A-1)^2/(^1P_i)^2$$
$$^2\kappa = (A-1)(A-2)^2/(^2P_i)^2$$
$$^3\kappa = (A-1)(A-3)^2/(^3P_i)^2 \text{ when } A \text{ is odd}$$
$$^3\kappa = (A-3)(A-2)^2/(^3P_i)^2 \text{ when } A \text{ is even}$$

(6.9)

Later, Kier extended the kappa topological indices in order to account for the variation in size contribution to shape from atoms other than carbon sp^3. This was achieved by modifying each A and ^mP_i with α, particularly $(A+\alpha)$ and $(^mP_i+\alpha)$; the Kappa shape indices are obtained where α is a parameter derived from the ratio of the covalent radius R_i of the i-th atom relative to the sp^3 carbon atom (R_{Csp3}).[35] Alpha values for most common atomic states are given in Table 6.3 along with atomic radii.

$$\alpha = \sum_{i=1}^{A}\left(\frac{R_i}{R_{Csp3}} - 1\right)$$

(6.10)

Further, Kier introduced a flexibility index which is a combination of kappa indices of path lengths of 1 and 2 where $^1\kappa$ encodes the count of atoms and the relative cyclicity of molecules, $^2\kappa$ encodes branching or the relative spatial density of molecules and the presence of other atoms than sp^3 carbon are encoded into the alpha value modifying each kappa index.[26]

$$\Phi = \frac{(^1\kappa\,^2\kappa)}{A}$$

(6.11)

Table 6.3 Alpha values from covalent radii.[40,42]

Atom valence state	R_i [Å]	α
C(sp^3)	0.77	0.00
C(sp^2)	0.67	-0.13
C(sp)	0.60	-0.22
N(sp^3)	0.74	-0.04
N(sp^2)	0.62	-0.20
N(sp)	0.55	-0.29
O(sp^3)	0.74	-0.04
O(sp^2)	0.62	-0.20
F	0.72	-0.07
P(sp^3)	1.10	0.43
P(sp^2)	1.00	0.30
S(sp^3)	1.04	0.35
S(sp^2)	0.94	0.22
Cl	0.99	0.29
Br	1.14	0.48
I	1.33	0.73

Table 6.4   Shape and flexibility index values for selected compounds.

Structure	$^1\kappa$	$^2\kappa$	$^3\kappa$	Φ
	6.0	3.2	5.33	3.2
	5.96	3.16	5.30	3.14
	4.17	2.22	1.33	1.54
	4.13	2.19	1.31	1.51
	3.41	1.61	0.84	0.91

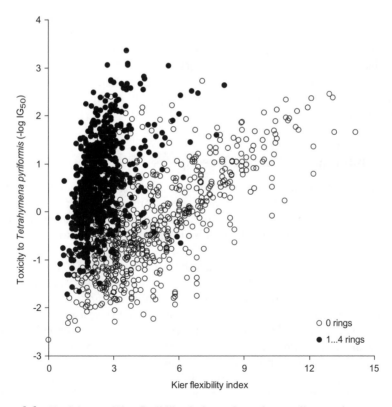

Figure 6.6   Toxicity *vs.* Kier flexibility index coloured according to the presence of rings in the compounds.

Example 6.2 illustrates the calculation of shape and flexibility indices of isobutyl methyl ether (Figure 6.4).

Example 6.2 Calculation of shape and flexibility indices of isobutyl methyl ether.

$$A = 6$$
$$^1P_i = 5; \, ^2P_i = 5; \, ^3P_i = 3$$
$$\alpha = 0 + 0 + 0 - 0.04 + 0 + 0 = -0.04$$
$$^1\kappa = (6 - 0.04)(6 - 0.04 - 1)^2/(5 - 0.04)^2 = 5.96$$
$$^2\kappa = (6 - 0.04 - 1)(6 - 0.04 - 2)^2/(5 - 0.04)^2 = 3.16$$
$$^3\kappa = (6 - 0.04 - 3)(6 - 0.04 - 2)^2/(3 - 0.04)^2 = 5.30$$
$$\Phi = (5.96 \times 3.16)/6 = 3.14$$

The shape and flexibility index values of some six atom structures (without hydrogens) are given in Table 6.4, which indicates how the nature of the structure is reflected in the index values. For example, linear structures are bigger and have a higher conformational freedom, and therefore the indices are also larger. This is also more prominently illustrated with the relationship between toxicity to *Tetrahymena pyriformis* and the Kier flexibility index (Figure 6.6); a diverse dataset of 1,371 compounds was used for the illustration.[36] In most of the cases, the overlap area between two groups contains either cycles with long aliphatic chains or small aliphatic compounds.

6.3.2   Information Content Descriptors

Information theoretic indices (sometimes noted as molecular complexity indices) are derived from the idea of partitioning structural features into disjoint subsets using an equivalence relation. The computations of the descriptors are based on Shannon's theory of information content or complexity.[37]

Information theoretic indices can be calculated based on hydrogen-suppressed or hydrogen-filled graphs. Most common descriptors were introduced by Basak and co-workers and take into account all atoms in the constitutional formula (hydrogens also being included), considering the information content provided by various classes of atoms based on their topological neighbourhood. There are three main types of information indices:[21,38,39] *IC* – mean information content or complexity of a hydrogen-filled graph, with vertices grouped into equivalence classes, where the equivalence is based on the nature of atoms and bonds, in successive neighbourhood groups; *CIC* – complementary information content; and *SIC* – structural information content.

For atomic complexity information, these indices can be calculated as follows:

$$IC_k = -\sum_{i=1}^{k} \frac{n_i}{n} \log_2 \frac{n_i}{n}$$

$$CIC_k = \log_2 n - IC_k \tag{6.12}$$

$$SIC_k = IC_k / \log_2 n$$

where n_i is the number of atoms in the i-th class, n is the total number of atoms in the molecule and k is the number of atomic layers in the co-ordination sphere around a given atom.

The division of atoms into different classes depends upon the co-ordination sphere taken into account and the indices are expressed in different order (k). Generally orders from 0 to 2 are used. Example 6.3 illustrates the calculation of information content indices of orders 0 to 2 of isobutyl methyl ether (Figure 6.7). The division of atomic layers into classes with the corresponding count of elements is given in Figure 6.5. Notably, 0th order atomic layers are just different atom types, whilst first order considers additionally bonded atom types and so on.

Example 6.3 Calculation of information content indices of isobutyl methyl ether

$$IC_0 = -\left(\frac{12}{18} \log_2 \frac{12}{18} + \frac{5}{18} \log_2 \frac{5}{18} + \frac{1}{18} \log_2 \frac{1}{18} \right) = 1.13$$

$$IC_1 = -\left(\begin{array}{l} \frac{12}{18} \log_2 \frac{12}{18} + \frac{2}{18} \log_2 \frac{2}{18} + \frac{1}{18} \log_2 \frac{1}{18} + \frac{1}{18} \log_2 \frac{1}{18} + \\ \frac{1}{18} \log_2 \frac{1}{18} + \frac{1}{18} \log_2 \frac{1}{18} \end{array} \right) = 1.67$$

$$IC_2 = -\left(\begin{array}{l} \frac{6}{18} \log_2 \frac{6}{18} + \frac{1}{18} \log_2 \frac{1}{18} + \frac{2}{18} \log_2 \frac{2}{18} + \frac{3}{18} \log_2 \frac{3}{18} + \\ \frac{2}{18} \log_2 \frac{2}{18} + \frac{1}{18} \log_2 \frac{1}{18} + \frac{1}{18} \log_2 \frac{1}{18} + \frac{1}{18} \log_2 \frac{1}{18} \end{array} \right) = 2.59$$

$$CIC_0 = \log_2 18 - 1.13 = 3.04$$

$$CIC_1 = \log_2 18 - 1.67 = 2.50$$

$$CIC_2 = \log_2 18 - 2.59 = 1.58$$

$$SIC_0 = 1.13 / \log_2 18 = 0.271$$

$$SIC_2 = 2.59 / \log_2 18 = 0.621$$

Other extensions of information content indices exist such as total information content (TIC_k), relative non-structural information content $(RNSIC_k)$, information energy content (I_E) and information divergence (D_{KL}). More

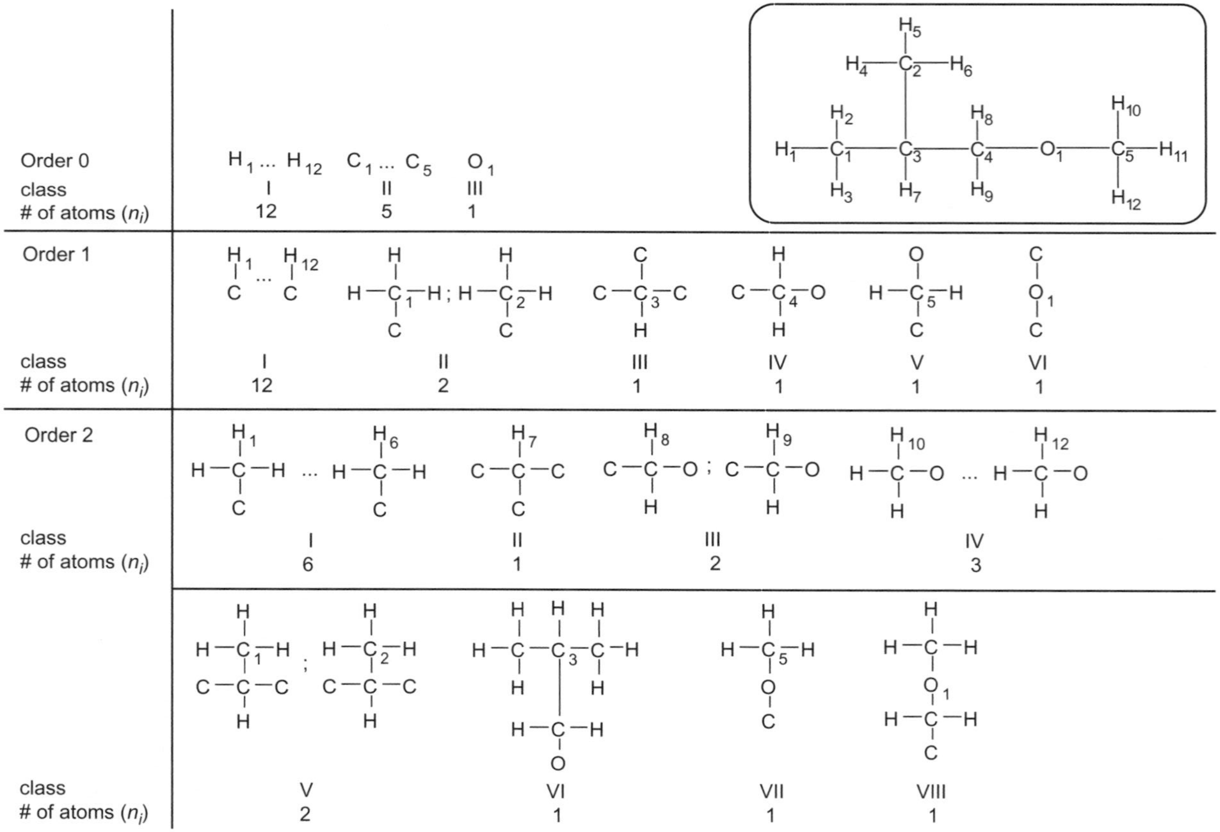

Figure 6.7 Hydrogen-filled graph and atomic layers of orders 0 to 2 of isobutyl methyl ether.

detailed explanation of information content indices can be found in textbooks and handbooks.[21,38]

6.3.3 Electrotopological Descriptors

The topological indices discussed previously are calculated for the whole molecule as a sum over sub-graphs of the molecular graph. In contrast, elec-trotopological state (E-state) indices are computed for each atom in the molecule. The E-state index is a graph invariant that combines the electronic state of the bonded atom within the molecule with its topological nature in the context of the whole molecular skeleton.[21,40–42] It encodes the intrinsic electronic state of the atom as perturbed by the electronic influence of all other atoms in the molecule. The intrinsic state is based on Kier–Hall electronegativity and is calculated according to equation (6.13):

$$I = \frac{(2/N)^2 \delta^v + 1}{\delta} \tag{6.13}$$

where N is the principal quantum number for the valence shell of particular atom, and simple delta and valence delta values are calculated according to molecular connectivity formalism (eqn 6.14):

$$\delta = \sigma - h$$
$$\delta^v = Z^v - h = \sigma + \pi + n - h \tag{6.14}$$

where Z^v is the number of valence electrons, σ is the number of electrons in sigma orbitals, π is the number of electrons in π-orbitals, n is the number of electrons in lone-pair orbitals and h is the number of bonded hydrogen atoms.

Note that here δ^v is slightly different than in Kier–Hall valence connectivity indices (eqn 6.7). While in the case of connectivity indices the differences of atoms among periods are taken into account with δ^v (eqn 6.6), here it is dis-criminated by N (eqn 6.13). Thus, for example, the δ^v values for halogens are the same. For second period atoms, the δ^v values calculated by eqn (6.6) are the same as those calculated by eqn (6.14).

Furthermore, a field influence is taken into account by using the electro-negativities of other atoms to modify the state of each atom within the field of the overall molecular structure. For that, the perturbation on each other are calculated as the difference between intrinsic states of atom i and atom j ($I_i - I_j$) in respect of the topological distance between the atoms. Note that r_{ij} is equal to the usual graph distance d_{ij} plus 1. The perturbation term is expressed in eqn (6.15).

$$\Delta I_{ij} = \frac{I_i - I_j}{r_{ij}^2}, \ j \neq i \tag{6.15}$$

The total perturbation of the intrinsic state of atom i is a sum of individual perturbations, and adding this to the atom intrinsic state, the electrotopological state S_i is expressed in eqn (6.16).

$$S_i = I_i + \sum_j \Delta I_{ij} \tag{6.16}$$

Example 6.4 illustrates the calculation of E-state indices of isobutyl methyl ether. Indices are calculated for each atom together with intrinsic state calculations. The calculation of the perturbations of intrinsic states for each atom of isobutyl methyl ether is given in Example 6.5. The calculation for first carbon atom (C1) is given in more explicit detail.

Example 6.4 Calculations of E-state indices of isobutyl methyl ether.

Atom	N	δ^v	δ	I_i	S_i
1(C)	2	1	1	$I_1 = \dfrac{(2/2)^2 1 + 1}{1} = 2$	$S_1 = 2 + 0.192 = 2.192$
2(C)	2	3	3	$I_2 = \dfrac{(2/2)^2 3 + 1}{3} = 1.33$	$S_2 = 1.33 - 0.661 = 0.669$
3(C)	2	2	2	$I_3 = \dfrac{(2/2)^2 2 + 1}{2} = 1.5$	$S_3 = 1.5 - 0.624 = 0.876$
4(O)	2	6	2	$I_4 = \dfrac{(2/2)^2 6 + 1}{2} = 3.5$	$S_4 = 3.5 + 1.30 = 4.8$
5(C)	2	1	1	$I_5 = \dfrac{(2/2)^2 1 + 1}{1} = 2$	$S_5 = 2 - 0.278 = 1.722$
6(C)	2	1	1	$I_6 = \dfrac{(2/2)^2 1 + 1}{1} = 2$	$S_6 = 2 + 0.130 = 2.130$

Example 6.5 Perturbation of intrinsic states of isobutyl methyl ether.

r_{1-j}	ΔI_{1-j}		r_{2-j}	ΔI_{2-j}		r_{3-j}	ΔI_{3-j}
2	$\Delta I_{1-2} = \dfrac{2 - 1.33}{2^2} = 0.175$		2	$\Delta I_{2-1} = -0.168$		2	$\Delta I_{3-2} = 0.0425$
3	$\Delta I_{1-3} = \dfrac{2 - 1.5}{3^2} = 0.0556$		2	$\Delta I_{2-6} = -0.168$		2	$\Delta I_{3-4} = -0.5$
3	$\Delta I_{1-6} = \dfrac{2 - 2}{3^2} = 0$		2	$\Delta I_{2-3} = -0.0425$		3	$\Delta I_{3-1} = -0.0556$
4	$\Delta I_{1-4} = \dfrac{2 - 3.5}{4^2} = -0.0938$		3	$\Delta I_{2-4} = -0.241$		3	$\Delta I_{3-5} = -0.0556$
5	$\Delta I_{1-5} = \dfrac{2 - 2}{5^2} = 0$		4	$\Delta I_{2-5} = -0.0419$		3	$\Delta I_{3-6} = -0.0556$
	$\displaystyle\sum_j \Delta I_{1-j} = 0.192$			$\displaystyle\sum_j \Delta I_{2-j} = -0.661$			$\displaystyle\sum_j \Delta I_{3-j} = -0.624$

r_{4-j}	ΔI_{4-j}	r_{5-j}	ΔI_{5-j}	r_{6-j}	ΔI_{6-j}
2	$\Delta I_{4-3} = 0.5$	2	$\Delta I_{5-4} = -0.375$	2	$\Delta I_{6-2} = 0.168$
2	$\Delta I_{4-5} = 0.375$	3	$\Delta I_{5-3} = 0.0556$	3	$\Delta I_{6-1} = 0$
3	$\Delta I_{4-2} = 0.241$	4	$\Delta I_{5-2} = 0.0419$	3	$\Delta I_{6-3} = 0.0556$
4	$\Delta I_{4-1} = 0.0938$	5	$\Delta I_{5-1} = 0$	4	$\Delta I_{6-4} = -0.0938$
4	$\Delta I_{4-6} = 0.0938$	5	$\Delta I_{5-6} = 0$	5	$\Delta I_{6-5} = 0$
	$\sum_{j} \Delta I_{4-j} = 1.30$		$\sum_{j} \Delta I_{5-j} = -0.278$		$\sum_{j} \Delta I_{6-j} = 0.130$

In order to use E-state indices in QSAR models, a skeletal superimposition must be performed on the dataset so that the E-state values for corresponding atoms can be used as variables in regression analysis. In order to improve the applicability to QSAR studies, the E-state formalism has been adapted to atom types and hydride groups. This extension removes the need for superimposition and makes it also applicable for more diverse datasets.

In this case, the E-state indices are calculated for a particular atom type (*e.g.* >C<, >C=, >N–, –N=, –O–, =O) or hydride group (*e.g.* >CH–, –CH$_2$–, >NH, –NH$_2$, –OH). For each atom type in a molecule, the E-state index values are summed. As a result, the individual atom positions in a molecule become irrelevant and such E-state descriptors are more usable in QSAR modelling and in other applications in chemistry. The E-state indices contain both electronic and topological structural information, and are thus considered to be information-rich topological indices. A detailed derivation of the E-state index and a comprehensive overview of their use can be found in Kier and Hall's textbook.[41]

6.3.4 Autocorrelation Descriptors

Autocorrelation descriptors or autocorrelation vectors are topological descriptors that encode numerical values assigned to individual atoms in addition to the topological structure of the molecule.[21,43] Originally developed by Moreau and Broto,[44] the first atomic properties used were π-functionality, electronegativity and van der Waals radius.[45] Later, numerous other properties have been adapted such as van der Waals volume, atomic contributions to the water-octanol partition coefficient,[46] charge and atom polarisability,[47] *etc.*[43,48] The individual components of an autocorrelation vector ($C_0, \ldots, C_d$) are calculated according to eqn (6.17) where x is a numeric value assigned to an atom, D_d is a set of atom pairs (i, j) that are separated by the same topological distance d.

$$C_d = \sum_{(i,j) \in D_d} x_i x_j \qquad (6.17)$$

For instance, if one chooses the connectivity delta (δ) as a property (Figure 6.4), then the autocorrelation vector for 5 components of isobutyl methyl ether are calculated as in Example 6.6.

Example 6.6 Calculation of autocorrelation vector of isobutyl methyl ether

$$C_0 = \delta_1 \cdot \delta_1 + \delta_2 \cdot \delta_2 + \delta_3 \cdot \delta_3 + \delta_4 \cdot \delta_4 + \delta_5 \cdot \delta_5 + \delta_6 \cdot \delta_6$$
$$= 1 \cdot 1 + 3 \cdot 3 + 2 \cdot 2 + 2 \cdot 2 + 1 \cdot 1 + 1 \cdot 1 = 20$$
$$C_1 = \delta_1 \cdot \delta_2 + \delta_2 \cdot \delta_3 + \delta_3 \cdot \delta_4 + \delta_4 \cdot \delta_5 + \delta_2 \cdot \delta_6$$
$$= 1 \cdot 3 + 3 \cdot 2 + 2 \cdot 2 + 2 \cdot 1 + 3 \cdot 1 = 18$$
$$C_2 = \delta_1 \cdot \delta_3 + \delta_2 \cdot \delta_4 + \delta_3 \cdot \delta_5 + \delta_1 \cdot \delta_6 + \delta_3 \cdot \delta_6$$
$$= 1 \cdot 2 + 3 \cdot 2 + 2 \cdot 1 + 1 \cdot 1 + 2 \cdot 1 = 13$$
$$C_3 = \delta_1 \cdot \delta_4 + \delta_2 \cdot \delta_5 + \delta_4 \cdot \delta_6$$
$$= 1 \cdot 2 + 3 \cdot 1 + 2 \cdot 1 = 7$$
$$C_4 = \delta_1 \cdot \delta_5 + \delta_5 \cdot \delta_6$$
$$= 1 \cdot 1 + 1 \cdot 1 = 2$$
$$C_5 = 0$$
$$\mathbf{C} = (20, 18, 13, 7, 2, 0)$$

Components of the autocorrelation vector are ordered by the length of d, for the first component, $d = 0$ (the atom's property is multiplied by itself), for the second $d = 1$ (calculated over all bonds), for the third $d = 2$ (calculated over all paths with length 2), and so on to the longest distance in the given molecule.

Naturally, the maximal distances are not the same for all structures, hence usually up to the first eight components are used. In QSAR models, individual components are used as molecular descriptors; however, in the case of artificial neural networks, the whole vector can be used as input for the characterisation of molecular structure. Autocorrelation descriptors can also be extended for 3-D structures by replacing topological distance with Euclidean distance[46] or considering surface properties of grid points on van der Waals surface.[49]

As a result of including numerical properties of atoms, autocorrelation vectors have very low redundancy.[46] Despite a different mathematical formulation, the autocorrelation descriptors can be quite similar to the other topological descriptors. For example, when using connectivity as the atomic property, the second component of autocorrelation vector is the same as Zagreb group's M_2 index (eqn 6.3) or if inverse square roots of connectivity delta values ($\delta^{-0.5}$) are used, then the first order of Kier–Hall's connectivity index (eqn 6.4) is the same as the second component of the autocorrelation vector. This illustrates that topological descriptors can overlap sometimes. In some cases, they may differ only by a constant.

6.4   Examples of Use

Over the years, many different applications of topological descriptors have emerged. The advantage of topological descriptors is that they provide direct and often simple descriptions of molecular structure. It is also argued that chemical topologies do not have a direct mechanistic meaning to toxicity.[50] As such, they are easy to calculate but are often avoided because of difficulties in their interpretation. Despite this, the literature provides numerous uses of topological descriptors in QSAR analysis. From the vast amount of examples, we have selected those where the authors attempt to discuss how topological descriptors modulate the property and provide some reasoning for the physical, chemical or mechanistic interpretation of a descriptor. The selected examples are related to various toxicological and environmental endpoints.

6.4.1 *Pimephales promelas*

Pimephales promelas (fathead minnow) is the most widely used small fish model for regulatory ecotoxicology. Several lethality assay tests exist with different time frames starting from 48 hours up to 30 days. In short-term tests, juvenile fish are used and, in partial life-cycle tests, fathead minnow embryos are used that are younger than 24 hours. The most common is a 96-hour test; more than 500 data points of aquatic toxicity (LC_{50}) are currently available for this duration. A review by Ankley and Villeneuve provides a survey of the past, present and future of this test.[51] Numerous data sets are extensively modelled with QSAR. Some developed models are global, spanning a wide range of chemicals, while other models specifically relate to a certain mode of action (MOA). The most common descriptor in these models is hydrophobicity—log P (logarithm of octanol–water partition coefficient). Several models are based on topological descriptors alone or in combination with other types of descriptors.

For instance, Basak *et al.* studied the narcotic toxic action of 15 industrially important esters using QSAR methods.[52] They obtained a good, one parameter correlation with first-order valence connectivity index ($^1\chi^v$) (eqn 6.18). Comparison with log P resulted in a correlation with slightly lower statistics (eqn 6.19).

$$\log LC_{50} = -1.39 - 0.721\,^1\chi^v$$
$$n = 15; \quad R = 0.938 \tag{6.18}$$

$$\log LC_{50} = -2.68 - 0.519 \log P$$
$$n = 15; \quad R = 0.883 \tag{6.19}$$

This shows that log P is an important characteristic for aquatic toxicity; however, topological descriptors often perform better in the content of small sets of congeneric compounds. In particular the valence connectivity index accounts for theoretical electronic parameters in addition to size and branching. Furthermore, Hall *et al.* showed that even a relatively bigger set of 65 benzene derivates can be successfully modelled with a set of topological descriptors (eqn 6.20).[53]

$$-\log LC_{50} = 5.49 + 1.43\,^1\kappa_\alpha - 2.71\,^1\chi - 1.04\,^6\chi_p^v + 0.68I$$
$$n = 65; \quad R = 0.940; \quad s = 0.26 \tag{6.20}$$

The first two descriptors in this equation are the first-order kappa shape index ($^1\kappa_\alpha$), which reveals that increased molecular branching is the cause of greater toxicity. The first-order molecular connectivity index ($^1\chi$), ranks molecules according to their size and branching. The sixth-order valence connectivity index ($^6\chi^v_p$) encodes information about the arrangement of atoms on the ring. Particularly in the case of benzene derivates, the presence of sixth and higher order paths reduces the toxicity relative to lower order paths. The last descriptor (I) is an indicator variable reflecting the presence of *ortho-* or *para*-dinitro substituents. A similar set of 69 benzene derivates was modelled by Basak *et al.*[54] They resulted in the five parameter linear model given in eqn 6.21.

$$- \log LC_{50} = 5.20 + 0.849 P_9 + 1.80 {}^4\chi^v_{pc} - 0.444 E_{lumo}$$
$$- 0.138\mu - 0.296 {}^{3D}WH \tag{6.21}$$
$$n = 69; \; R = 0.927; \; R_{cv} = 0.914; \; s = 0.287$$

The first two descriptors in this equation are topological, path of length nine (P_9) and *valence* path–cluster connectivity index of order four ($^4\chi^v_{pc}$). The other descriptors are the geometrical 3-D Wiener number (^{3D}WH), the quantum-chemically derived energy of lowest unoccupied molecular orbital (E_{lumo}) and the dipole of the molecule (μ). Due to the diversity of descriptor calculators (see Section 6.8), mixing of different types of descriptors in QSARs is very common nowadays. This leads to the situation where the topological descriptors complement 3-D descriptors and *vice versa*.

Another good example of the combined use of different types of descriptors in QSAR modelling is the study by Papa *et al.*[55] They addressed a structurally heterogeneous set of industrial organic chemicals where 249 compounds were selected for the training set and 200 compounds remained in the test set. The proposed QSAR model (eqn 6.22) combines calculated log P with other theoretical descriptors.

$$- \log LC_{50} = 2.9 + 0.56 AlogP + 0.34 DP03 + 20.8 H8m$$
$$- 0.79 GATS1v - 1.59 R1v \tag{6.22}$$
$$n = 249; \; R^2 = 0.81; \; R^2_{cv} = 0.80; \; R^2_{ext} = 0.72$$

The descriptors in this model are: calculated octanol-water partition coefficient ($AlogP$); Randic molecular profile n03 ($DP03$); H autocorrelation of lag8/weighted on atomic masses ($H8m$); Geary autocorrelation of lag 1 weighted by atomic van der Waals volumes ($GATS1v$); and R autocorrelation of lag 1 weighted by atomic van der Waals volumes ($R1v$). Descriptors other than $AlogP$ were calculated using the DRAGON software package. These four theoretical and topological descriptors ($DP03$, $H8m$, $GATS1v$, and $R1v$) are mainly related to the dimensional features of the chemicals.

Another QSAR model proposed by Eldred *et al.* also combines topological descriptors with surface area descriptors (eqn 6.23).[56] They extracted a relatively big and diverse data set of 375 compounds from the COMPUTOX toxicity database and divided it randomly into training and test set of 287 and 88 compounds respectively. This resulted in an eight parameter linear model.

$$-\log LC_{50} = 1.43 + 0.0409MOMH5 - 2.72FPSA1 + 1.05FNSA2$$
$$+ 0.0572RPCS1 + 1.69CHAA1 + 0.559MOLC9$$
$$+ 0.355NDB13 + 0.144WTPT1$$
$$n = 287; \; R = 0.82; \; R_{ext} = 0.78$$

$$(6.23)$$

The model consisted of one topological descriptor—an average distance sum connectivity (*MOLC9*), which is topological Balaban index *J*. The other descriptors are constitutional (*NDB13*, *WTPT1*), geometrical (*MOMH5*), electrostatic (*CHAA1*) and partial charged surface area related (*FPSA1*, *FNSA2* and *RPCS1*).

6.4.2   *Tetrahymena pyriformis*

Tetrahymena pyriformis is a ciliate protozoa, a freshwater organism common in aquatic environments. Its short life-cycle and easy cultivation in laboratory conditions has made it one of the favourite test organisms among (eco)toxicologists.

Toxic substances affect *Tetrahymena pyriformis* in a number of observable ways such as altering biochemical markers, changing its behaviour and causing ultrastructural modifications. The most popular evaluation method in toxicity studies is to measure the cell growth rate. In optimal conditions, the growth of *Tetrahymena pyriformis* is well characterised and contact with xenobiotics is known to affect this. The inhibition of growth rate is measured at several concentrations of the studied substance and, for each concentration, the cell density is measured. To enable comparison between different chemicals, the concentration of substance needed to reduce the growth rate to a certain percentage, usually 50%, is used. A comprehensive review of *Tetrahymena pyriformis* and its role in toxicological studies has been published by Sauvant *et al.*[57]

The number of studies correlating the structure of compounds to their toxicity towards *Tetrahymena pyriformis via* various experimental or theoretical parameters is quite high. Among all the descriptors used, octanol–water partition coefficient and molecular orbital energies are by far the two most popular one, but topological indices are also fairly common.

In one of the earliest studies, Schultz *et al.* modelled mono- and dinitrogenous heterocyclic compounds.[58] Only one topological descriptor, the

Kier–Hall first order valence connectivity index, was used in the model given in eqn 6.24.

$$-\log IG_{50} = 0.911\,{}^{1}\chi^{v} - 2.969$$
$$n = 23;\ R = 0.974;\ s = 0.23 \tag{6.24}$$

The relationship found between the descriptor and the activity is strong, and the structural features of the data set are thoroughly analysed. Change in several properties such as size, methylation and number of nitrogen atoms is tied to the increase or decrease of IG_{50}. This kind of thoroughness is rare though, as usually topological descriptors in the models are interpreted only briefly as a measure of the molecular size and complexity,[59] or as a measure of hydrophobicity such as in the work of Katritzky *et al.*[60] They studied a set of 97 nitrobenzenes and the resulting five- and three-parameter models contain one topological descriptor each, namely Kier–Hall index ${}^{1}\chi^{v}$ and ${}^{2}\chi$.

$$\log IG_{50} = 13.154 + 0.06 E_{en}^{\min}(C{-}C) + 0.605\,{}^{2}\chi$$
$$- 0.165 E^{SOMO} - 2.731\overline{V_{0}} + 4.654 FNSA_{PNSA}^{(2)}$$
$$n = 97;\ R^{2} = 0.815;\ R_{cv}^{2} = 0.789;\ s = 0.348 \tag{6.25}$$
$$\log IG_{50} = 605.2 - 1.15 E_{en}^{\max}(N{-}O) + 0.562\,{}^{1}\chi^{v} + 0.313 E_{C}^{\min}(C{-}C)$$
$$n = 97;\ R^{2} = 0.754;\ R_{cv}^{2} = 0.736;\ s = 0.348$$

The higher order connectivity index can be found in the models developed by Cronin and Schultz, where the fifth order path molecular connectivity index ${}^{5}\chi^{v}$ is used in the models for the whole set of 166 phenols, and in the model for the slightly reduced set of 151 compounds, from which carboxyl- and amino derivatives had been removed.[61] The authors proposed that ${}^{5}\chi^{v}$ acts only as an indicator variable for nitrophenols since the nitrophenols would encode more topological information and therefore have larger descriptor values compared with other compounds. Subsequently, nitrogen-containing compounds were removed and ${}^{5}\chi^{v}$ stopped to appear in the redeveloped models.

Connectivity indices tend to be the most popular topological descriptors in toxicology studies due to their correlation to hydrophobicity, but others have also been used occasionally. This is the case in the work of Enoch *et al.*, where a set of 250 phenols was studied with multiple linear regression (MLR) and artificial neural network models.[62] The MLR model they developed consisted of one topological descriptor, the E-state index for a carbon atom with two single and a double bond (eqn 6.26).

$$\log IG_{50} = 4.92 + 0.49\log P - 1.25 AHard + 0.22 NH_{don} + 1.15 dssC$$

$$n = 200; \quad R^2 = 0.66; \quad R^2_{test} = 0.72; \quad s = 0.48 \tag{6.26}$$

The artificial neural network model they developed consisted of six descriptors: log P; log D (Distribution coefficient at pH 7.35); Molecular Volume; and three Kier shape indices with statistical parameters of n = 200, $R^2 = 0.71$, $R^2_{test} = 0.73$ and s = 0.45.

6.4.3 *Daphnia magna* and *Vibrio fischeri*

Due to its high sensitivity, easy handling and a high reproduction rate, the water flea, *Daphnia magna*, is used as a representative for the freshwater animals in standard toxicity tests. It is also important in the food chain as a consumer of primitive plant life and a major food source for vertebrate and invertebrate predators. Because of this, one can find a number of publications with different QSAR models for water flea toxicity data. These publications mainly utilise the octanol–water partition coefficient (log P), the energy of the lowest unoccupied molecular orbital (E_{lumo}) and the energy of the highest occupied molecular orbital (E_{homo}) as molecular descriptors.[63–66] At the same time, 2-D descriptors have shown valuable utility in combination with other descriptors and also independently for smaller congeneric datasets.

For instance, to model aquatic toxicity against *Daphnia magna*, Katritzky *et al.* constructed from the literature a diverse data set of 130 compounds containing benzoic acids, benzaldehydes, phenylsulfonyl acetates, cycloalkanecarboxylates, benzanilides and other esters.[67] The statistical evaluation resulted in a model (eqn 6.27) with five descriptors containing one topological descriptor—the Balaban index (J) among others.

$$-\log EC_{50} = 8.968 + 1.227\#Hacc - 0.2478\#ArB - 1.929J$$

$$+ 0.332\log P - 280.3 HDCA - 2/TMSA \tag{6.27}$$

$$n = 130; \quad R^2 = 0.712; \quad R^2_{cv} = 0.676; \quad s^2 = 0.6$$

From the remaining four descriptors in the model, the first two are hydrogen bond related: count of hydrogen acceptor sites ($\#Hacc$); area-weighted surface charge of hydrogen acceptor dependent hydrogen bonding donor atoms ($HDCA$-$2/TMSA$); number of aromatic bonds ($\#ArB$); and hydrophobicity (log P). In the authors' interpretation, the Balaban index increases with the size of the molecule and with the degree of branching and unsaturation. They concluded that branched molecules are less toxic, probably due to their lower membrane penetration abilities.

Papa *et al.* focused on a congeneric set of 29 esters in the modelling of toxicity against *Daphnia magna*.[68] Esters present one of the most important classes among High Production Volume (HPV) chemicals and were collected for this study from the literature. The best four descriptor model (eqn 6.28) was derived with the ordinary least squares method for a training set of 24 compounds and validated externally with five compounds.

$$- \log EC_{50} = -0.193 + 0.0539\,TIC0 - 0.82\,nCp + 0.94n{=}CH2$$
$$n(train) = 24; \ R^2 = 0.879; \ R^2_{cv} = 0.831; \ R^2_{cv-LMO(50\%)} = 0.794; \qquad (6.28)$$
$$s = 0.47; n(test) = 5; \ R^2_{ext} = 0.79$$

The total information content index (*TIC0*) in the equation is a topological index counting the information about neighbourhood symmetry of 0-order. The other descriptors consider number of primary sp^3 carbons (*nCp*) and number of double bonds (*n=CH2*), accounting for size and shape and reactivity, respectively.

A congeneric *Daphnia magna* data set of 12 organic nitro compounds was modelled by Pasha *et al.*[69] Good results were obtained with Kier flexibility index (Φ) and lopping centric index (*Lop*) (eqn 6.29). The authors concluded that, since both descriptors have positive coefficients, it shows that branched and flexible molecules might have higher toxicities. However, a deeper analysis of their work shows that this conclusion can be misleading. The chemicals within the given dataset of nitro compounds are not flexible and toxicities depend on substitution patterns varying from mono- to tri- substitution. In this dataset, the Kier flexibility index is influenced by the increase of substitutions in benzene ring and subsequent changes in electron delocalisation. The lopping centric index is an information content index considering the branching of substituents and is correlated to the electronic effects, because di-nitro benzenes show higher toxicity and have also higher branching, while di-halo nitro benzenes have lower toxicity and lower branching.

$$- \log I_{50} = -11.27 + 1.085\Phi + 11.29\,Lop$$
$$n = 12; \ R^2 = 0.88; \ R^2_{cv} = 0.83; \ s = 0.23 \qquad (6.29)$$

Yu *et al.* compared the performance of various modelling techniques and descriptors (log *P*, valence connectivity index ($^1\chi^v$), linear solvation energy relationship (LSER), Free-Wilson group contribution) for the modelling of aquatic toxicities of *Daphnia magna* and *Vibrio fischeri* of 43 benzene derivatives.[70] For the given data set, the first order valence connectivity index ($^1\chi^v$) provided QSAR models (eqn 6.30 for *Daphnia magna* and eqn 6.31 for *Vibrio fischeri*), with the same statistical characteristics in comparison with multilinear regression approaches and linear regression with log *P* alone. This showed again the utility of topological descriptors in QSARs for congeneric datasets.

$$-\log IC_{50} = 1.25 + 1.03\,{}^{1}\chi^{v}$$
$$n = 43; \; R^2 = 0.72; \; s = 0.39$$
(6.30)

$$-\log IC_{50} = 0.99 + 1.05\,{}^{1}\chi^{v}$$
$$n = 43; \; R^2 = 0.72; \; s = 0.39$$
(6.31)

In the quest for rapid, reproducible and cost-effective assays for toxicity screening and assessment, bacteria have shown their utility over animals, plants and algal assays. Probably the most common is the *Vibrio fischeri* (formerly referred to as *Photobacterium phosphoreum*) bioluminescence inhibition assay.[71] Over the years this has gathered a considerable amount of experimental data[72] which has led to a variety of QSAR modelling efforts, also applying 2-D descriptors as already shown in previous example.

Gombar *et al.* provide a good example of use for molecular shape kappa indexes (${}^{1}\kappa$) for modelling toxicity to *Vibrio fischeri* with a dataset of 46 chlorophenols, chloro-anilines and chlorobenzenes.[73] A single parameter equation (eqn 6.32) shows a considerably good correlation. The authors concluded that the positive coefficient of the kappa shape index of first order confirms that toxicity increases with the increase in the number of chlorine atoms. The authors also reported good correlation between log *P* and kappa shape index for the given data set, which shows that topological descriptors are very often a good replacement for log *P*. The model was further improved (eqn 6.33) by considering directly electronic effects through effective polarisability (α_z) of heteroatoms (O for phenol, N for anilines and Cl for benzenes) and charge on the carbon (C_{z-o}) placed in *ortho*-position from the heteroatom.

$$-\log EC_{50} = -1.573 + 0.871\,{}^{1}\kappa$$
$$n = 46; \; R = 0.829; \; s = 0.367$$
(6.32)

$$-\log EC_{50} = -0.935 + 1.204\,{}^{1}\kappa - 0.421\alpha_z - 3.244 C_{z-o}$$
$$n = 46; \; R = 0.893; \; s = 0.302$$
(6.33)

Dongbin *et al.* measured toxicity to *Vibrio fischeri* for a data set of 16 substituted naphthalenes.[74] They performed stepwise regression analysis and a comparative study of three different pools of descriptors obtained from quantum chemical calculations, topology of chemical structure and LSER parameters. The best model (eqn 6.34) proposed for the molecular connectivity indices consisted of two descriptors.

$$-\log EC_{50} = 5.8454 - 0.4250\,{}^{2}\chi_p + 14211\,{}^{3}\chi_p$$
$$n = 46; \; R^2 = 0.8049; \; s = 0.2477$$
(6.34)

The authors related the Kier–Hall second order path connectivity index (${}^{2}\chi_p$) to molecular volume and the Kier–Hall third order path connectivity index

($^3\chi_p$) to the number of substituting groups. The obvious conclusion was that the compounds with larger volumes have lower toxicity.

Agrawal and Khadikar coupled first order valence connectivity index ($^1\chi^v$) and electronegativity (χ_{eq}) together with some indicator variables to model the toxicity of *Vibrio fischeri* in the set of 39 chemicals acting according to the narcotic mechanism of action (eqn 6.35).[75]

$$- \log EC_{50} = -4.6563 + 0.8867\,^1\chi^v + 2.2796\chi_{eq}$$
$$+ 0.8881\,Ip_3 + 0.888\,Ip_4 + 1.9269\,Ip_5 \quad\quad (6.35)$$
$$n = 39;\ \ R = 0.8616;\ \ s = 0.7075$$

The authors proposed that $^1\chi^v$ distinguishes the degree of unsaturation and the presence of heteroatoms and gives rise to the toxic effects. In addition, the electronegativity indicator variables consider the substitution pattern of monosubstituted compounds (Ip_3), di-substituted compounds (Ip_4) and poly-chlorination (Ip_5). Melagraki *et al.* used the same dataset of 39 heterogeneous compounds and divided it into a training set (29 compounds) and a test set (ten compounds) for modelling.[76] They replaced indicator variables with log P and compared the performance of radial basis function (RBF) neural networks and multilinear regression analyses (MLR). The statistical performance of neural networks was better in comparison with MLR. The three-parameter MLR equation still showed fairly good performance on external validation (eqn 6.36).

$$- \log EC_{50} = -4.0110 + 0.4878\log P + 2.2905\chi_{eq} + 0.4712\,^1\chi^v$$
$$n = 39;\ \ R^2 = 0.7851;\ \ Q^2 = 0.6579; \quad\quad (6.36)$$
$$R^2_{\text{ext}} = 0.8373;\ \ s = 0.5250$$

6.4.4 *Chlorella vulgaris*

Green alga is a phylum providing interesting species for toxicity studies. Frequently used and commonly chosen species for regulatory use are *Pseudo-kirchneriella subcapitata* (previously named *Selenastrum capricornutum*), *Scenedesmus subspicatus* and *Chlorella vulgaris*. Despite this, the amount of publicly available data is small and this results in a scarce selection of *in silico* QSAR models for the prediction of toxicities and application of 2-D descriptors in the modelling process.

Perhaps the most systematic set of studies is provided by Cronin and co-workers in the modelling of toxicity measured from a short-term toxicity assay for *Chlorella vulgaris*.[77] In the first study, Netzeva *et al.* modelled toxicity using a chemically heterogeneous dataset of 65 aromatic compounds containing phenols, anilines, nitrobenzenes, benzaldehydes and other benzenes.[78] They derived a four parameter equation (eqn 6.37) *via* stepwise elimination of descriptors.

$$- \log EC_{50} = -5.40 + 0.40 \log P - 0.23 E_{lumo} + 9.84 A_{max} + 0.20^{0} \chi^{v}$$
$$n = 65; \quad R^{2} = 0.858; \quad Q = 0.843; \quad s = 0.403 \tag{6.37}$$

The authors discussed the role of $\log P$ and $^{0}\chi^{v}$ together, and found that these descriptors reflected the readiness of molecules to penetrate through biological membranes. The Kier–Hall valence connectivity index of 0th order ($^{0}\chi^{v}$) was believed to account for molecular size with a correction for the presence of π- and lone pair electrons. The latter are obviously more important to non-covalent intermolecular interactions such as those occurring with molecular transport through membranes.

In the second publication Cronin *et al.* extended the data set to 91 chemicals,[79] which they split it into a training and a test set with 73 and 18 compounds respectively. In the reported final QSAR (eqn 6.38), one topological descriptor successfully complemented classical $\log P$ and E_{lumo}:

$$- \log(EC_{50}) = -3.03 + 0.9 \log P - 0.156 E_{lumo} - 0.369 \Delta^{1}\chi^{v}$$
$$n(train) = 73; \quad R^{2} = 0.892; \quad R^{2}_{cv} = 0.878; \quad s = 0.496, \tag{6.38}$$
$$n(test) = 18; \quad R^{2}(test) = 0.901$$

where $\Delta^{1}\chi^{v}$ is the difference between $^{1}\chi^{v}$ for the compound in question, and the straight chain alkane with the same molecular formula.

Analysis reveals that the given descriptor correlates modestly with the electrotopological indices for double-bonded oxygen atom and hydrogen bond acceptors, reflecting its valence nature. The analysis of the residuals illustrated that the descriptor contributed to the minimisation of the error associated with compounds that contain double bonded oxygen. The authors concluded that $\Delta^{1}\chi^{v}$ does not encode any particular property or process with the used data set and that rather it is a statistical supplement to the other two descriptors which improves the statistical performance. However, the reason for the development of $\Delta^{1}\chi^{v}$ was to consider non-dispersive interaction forces or non-hydrophobic interaction or polar and ionisable characteristics of molecules using topological descriptors.[80] Considering the analysis described above, it can be hypothesised that the descriptor accounts for the polar characteristics of the molecules.

6.4.5 Toxicities to Rodents and Human

The application of topological descriptors to the modelling of toxicity is not limited to aquatic organisms. Mammalian toxicities have also been modelled. Those organisms are more complex and one of the most widely used test organisms in these studies are rodents, but parallel examples on human toxicity data can be found. This section provides some examples where topological descriptors are involved in the modelling of toxicity on mouse, rat and humans.

Lessigiarska *et al.* investigated acute toxicity in various biological systems, including humans and rodents *in vivo* and *in vitro*.[81] The data was adapted from the MEIC (Multicentre Evaluation of *In vitro* Cytotoxicity) programme and supplemented with toxicity data for rat hepatocytes from the literature. The main attempt was to develop quantitative structure–activity–activity relationships (QSAAR), but at the same time QSAR models for 26 heterogeneous chemicals were derived that included topological descriptors. Among the QSAR models derived for the human endpoints, Kier–Hall second and third order valence connectivity path indexes ($^2\chi_p^v$, $^3\chi_p^v$) were involved. An example for approximate acute blood/serum peak LC_{50} values $^2\chi_p^v$ complements the lowest unoccupied molecular orbital (E_{lumo}) energy (eqn 6.39). The authors proposed that the topological descriptors in the presented models encode the size and shape of the molecules while E_{lumo} addressees chemical reactivity.

$$-\log LC_{50} = 0.258\,^2\chi_p^v - 0.283\,E_{lumo} - 0.878$$
$$n = 26; \ R^2 = 0.821; \ Q^2 = 0.794; \ s = 0.618$$

(6.39)

QSARs were also derived for the toxicity to rat hepatocytes. Single parameter models showed that toxicity to rat hepatocytes is well correlated ($R^2 = 0.763$) with the solubility (log S). At the same time, the average molecular connectivity index ($^0\chi_A$) considerably improves the correlation (eqn 6.40).

$$-\log IC_{50} = -0.405\,\log S - 6.90\,^0\chi_A + 4.19$$
$$n = 25; \ R^2 = 0.907; \ Q^2 = 0.870; \ s = 0.441$$

(6.40)

The rat LD_{50} toxicity was modelled with three descriptors (eqn 6.41). This time, the electrotopological state descriptor (*TIE*) was used and was linked, in the article, to the ability of a molecule to enter into non-covalent intermolecular interactions. Two other descriptors were hydrophilicity factor (*Hy*) and the number of six-membered rings (N_{ring6}).

$$-\log LD_{50} = 0.00382\,TIE - 0.645\,H_y + 0.607\,N_{ring6} - 1.41$$
$$n = 26; \ R^2 = 0.815; \ Q^2 = 0.735; \ s = 0.527$$

(6.41)

Cronin *et al.* modelled the toxic and metabolic effects of 23 aliphatic alcohols on the perfused rat liver with the third order path-cluster molecular connectivity index ($^3\chi_{pc}$).[82] The endpoints they used were extracellular release of the enzymes, where they used glutamate–pyruvate–transaminase (*GPT*), lactate dehydrogenase (*LDH*), glutamate dehydrogenase (*GLDH*) and adenosine triphosphate (*ATP*) determination with a reduction in the amount of intracellular *ATP*. The data set used contains compounds with different saturation and molecular branching. The results (eqn 6.42–6.45) show that descriptors for hydrophobicity, electrophilicity and branching model the hepatotoxic effect of the alcohols very well.

$$\log GPT = 0.576 \log P - 0.193 E_{lumo} - 0.494\,^3\chi_{pc} + 1.19$$
$$n = 23;\ R^2 = 0.836;\ R^2_{cv} = 0.801;\ s = 0.183 \tag{6.42}$$

$$\log LDH = 0.561 \log P - 0.297 E_{lumo} - 0.487\,^3\chi_{pc} + 1.57$$
$$n = 22;\ R^2 = 0.848;\ R^2_{cv} = 0.813;\ s = 0.184 \tag{6.43}$$

$$\log GLDH = 0.399 \log P - 0.037 E_{lumo} - 0.384\,^3\chi_{pc} + 0.579$$
$$n = 19;\ R^2 = 0.846;\ R^2_{cv} = 0.797;\ s = 0.132 \tag{6.44}$$

$$\log 1/ATP = 0.393 \log P - 0.362 E_{lumo} - 0.263\,^3\chi_{pc} + 1.48$$
$$n = 20;\ R^2 = 0.857;\ R^2_{cv} = 0.789;\ s = 0.159 \tag{6.45}$$

The models show that hepatotoxic effects increase with increasing hydrophobicity and electrophilicity, and decrease by the effect of molecular branching. Based on the comparison with previous studies, the authors proposed that molecular branching can be important in terms of the steric hindrance of reactive centres. The authors suggested that increased leakage of *GPT* and *LDH* is associated with cell membrane damage, and that the loss of *GLDH* is associated with mitochondrial membrane damage. QSAR models for *GPT* (eqn 6.42) and *LDH* (eqn 6.43) tend to have similar slope and intercept values, probably due to a similar mechanism of action; the QSAR model for *GLDH* (eqn 6.44) has significantly different coefficients and therefore might have different mechanism of action. The QSAR model (eqn 6.45) for the concentration of *ATP* also demonstrates separate effects for this endpoint. The addition of a double or triple bond to an aliphatic alcohol clearly increases the toxicity relative to the saturated alcohols. This is said to be consistent with the action of unsaturated alcohols as pro-electrophiles and oxidation of such alcohols *via* dehydrogenase makes them act as Michael-type acceptors.

Papa *et al.* used an ordinary least squares method to develop QSAR models for three toxicological endpoints:[83] acute oral toxicity for mouse; inhibition of NADH oxidase; and inhibition of mitochondrial membrane potential. Their data set included compounds used in fragrance materials. The model (eqn 6.46) for the inhibition of mitochondrial membrane potential included two autocorrelation descriptors (*ATS4v* of Broto-moreau and *MATS4m* of Moran) which give information on the distribution of the atomic properties along the topological structure. Particularly for those descriptors, weight properties were van der Waals volumes and atomic mass, which shows that descriptors encode the information related to the dimensions of molecules.

$$\log EC_{50} = 5.79 - 1.48 ATS4v - 0.65 MATS2m$$
$$n = 15; R^2 = 0.917; R_{cv}^2 = 0.886; s = 0.132$$

(6.46)

Bhhatarai and Gramatica investigated the toxicity of per- and poly-fluorinated compounds.[84] The data set contained inhalation LC_{50} data for mice (56 compounds) and rats (52 compounds). The mouse LC_{50} inhalation model (eqn 6.47) involved four descriptors: Moriguchi log P (*MlogP*); third order Kier–Hall valence connectivity index ($^3\chi^v$); frequency of C–C bond at a topological distance of 01 (*F01[C–C]*), and atom-centred fragment descriptor representing hydrogen attached to $C^2(sp^3)/C^1(sp^2)/C^0(sp)$ carbon (*H048*).

The authors proposed that $^3\chi^v$ accounts for the presence of hetero atoms, and double and triple bonds in the compound, whereas the increase in one or the other features increases the value of $^3\chi^v$. *F01[C–C]* represents the total number of C–C bonds and its value increases as the length of alkyl chain increases. Since the correlation constants for both descriptors are positive, increasing chain length and increase in bond order, as well as the presence of the heteroatom contributes to increase in mouse inhalation toxicity.

$$-\log LC_{50} = 4.21 - 1.27\, MlogP + 1.43\,^3\chi^v + 0.38 F01[C\text{–}C] - 1.14 H048$$
$$n = 56; R^2 = 0.7983; R_{cv}^2 = 0.7631; s = 0.717$$

(6.47)

The rat LC_{50} inhalation model (eqn 6.48) also included four descriptors: Balaban-type index from van der Waals weighted distance matrix (*Jhetv*); ratio of multiple path count over path count (*PCR*); Moriguchi log *P*; and a 2-D binary fingerprint descriptor characterising the presence/absence of Cl–Cl at a topological distance of 2 (*B02[Cl–Cl]*).

The first two topological descriptors are the most influential. Their increase also increases the toxicity. *Jhetv* exhibits the bond multiplicity, the heteroatoms and the number of atoms present in the compound. When the number of heteroatoms in a molecule increases, the molecule hydrates better in water and becomes more soluble. The *PCR* accounts for the multiple bonds in the molecule. In fully saturated molecules, its value is one, but in the presence of multiple bonds its value will increase.

$$-\log LC_{50} = 12.76 + 1.87 Jhetv + 11.43 PCR - 0.60 MlogP - 1.41 B02[Cl\text{–}Cl]$$
$$n = 52; R^2 = 0.7814; R_{cv}^2 = 0.7385; s = 0.82$$

(6.48)

Comparison of LC_{50} models (eqn 6.47 and 6.48) for mice and rats shows that some toxicity trends are similar in both organisms, as descriptors related to hydrophobicity (*MlogP*) and electronegativity (*Jhetv* and $^3\chi^v$) were common in both endpoint modelling studies.

Devillers and Devillers reviewed more than 150 models (QSAR, expert systems and interspecies correlations) for estimating the acute toxicity of chemicals to mice and rats.[85] The review included QSARs for congeneric data sets (alcohols, pyrines, substituted benzenes, herbicides, pesticides, fluoroacetamides, barbiturates, *etc.*) and non-congeneric data sets composed of multiple chemical classes. Several reviewed models include topological indices such as connectivity indices, valence connectivity indices and E-state indices.

6.4.6 Soil Sorption

The evaluation of soil mobility of chemicals is one of the primary tasks in estimating ecological consequences that may follow from their distribution in environment. This is often determined with the soil–sediment adsorption coefficient, K_{oc}. This parameter describes the extent to which a chemical is distributed between the solid and solution phases in soil, or between water and sediment in aquatic ecosystems, and indicates whether a chemical is likely to be transported through the soil or would be immobile. Sabljic *et al.* comprehensively evaluated the quality and reliability of the relationships between soil sorption coefficients and octanol–water partition coefficients, showing that for many chemical classes with specific interactions (*e.g.* alkyl ureas, amines, alcohols, organic acids, amides and dinitroanilines) description is inadequate.[86]

A review of more than 200 existing QSARs for the estimation of log K_{oc} has been presented by Gawlik *et al.*[87] and a list of recent QSAR is summarized by Kahn *et al.*[88] Within a variety of theoretical molecular descriptors applied for modelling of log K_{oc}, the molecular connectivity indices have gained high attention. A frequently used descriptor in the estimation of K_{oc} values is the first order molecular connectivity index ($^1\chi$) as comprehensively discussed by Sabljic.[89]

A vivid example relates to the predominantly hydrophobic chemicals—compounds that contain only carbon (C), hydrogen (H) and halogens (F, Cl, Br, I). The single parameter correlation with first order molecular connectivity index ($^1\chi$) is very strong (eqn 6.49).[86] However, a correlation with log P for the same data set has lower performance: $R^2 = 0.887$, $Q^2 = 0.881$ and $s = 0.451$.[86]

$$\log K_{oc} = 0.70 + 0.52\,^1\chi$$
$$n = 81; R^2 = 0.961, Q^2 = 0.959; s = 0.264 \tag{6.49}$$

Sabljic extensively discussed the interpretation of molecular connectivity indexes[89] and concluded that $^1\chi$ has high correlation with the molecular surface area[90] and that it reflects dependence on molecular size.

The situation is different in the case of a non-hydrophobic chemical that possesses specific interactions. To model those chemicals with purely topological descriptors, extension to connectivity indexes is provided by Bahnick and Doucette.[80] The success of the first order molecular connectivity index ($^1\chi$) in the previous example is due to the fact that, for uncharged organic molecules, hydrophobic interactions or London dispersion forces (which increase with the size of molecule) are dominant. For compounds containing polar and/or ionisable functional groups, non-hydrophobic or non-dispersive forces dominate. To account for non-dispersive forces, Bahnick and Doucette designed a new descriptor based on valence connectivity indices. This was achieved *via* removal of the molecular size contribution by subtracting the value of a particular index, for the molecule of interest, from the same index for a molecule formed by replacing highly electronegative atoms by atoms of another type (usually carbon).

$$\log K_{oc} = 0.34 + 0.44\,^1\chi$$
$$n = 56; R = 0.71; s = 0.97 \tag{6.50}$$

$$\log K_{oc} = 0.64 + 0.53\,^1\chi - 2.09\,\Delta^1\chi^v$$
$$n = 56; R = 0.969; s = 0.34 \tag{6.51}$$

The example shows that a large improvement in the regression model (eqn 6.50) is resulting from addition of a non-dispersive forces factor term $\Delta^1\chi^v$ (eqn 6.51), which extends the applicability of the model to include organic chemicals with substantial hydrophilicity.

It has been shown that in many instances, for class specific models, topological descriptors can be successful alternatives to conventional descriptors.[88] A congeneric set of 13 alcohols exhibited a well-defined relationship with the Kier–Hall valence connectivity index of third order, $^3\chi^v$ (eqn 6.52). In the homologous series of alcohols (from methanol to decanol), soil sorption increases with the increase in the size of the molecule. The positive correlation shows an increase in sorption with the increase in the value of the descriptor. Similar correlation with the conventional log P has lower statistical characteristics: $R^2 = 0.8123$, $R^2_{cv} = 0.7405$ and $s = 0.3565$.

$$\log K_{oc} = 0.143 + 1.001\,^3\chi^v$$
$$n = 12; R^2 = 0.9371, R^2_{cv} = 0.9069; s = 0.2064 \tag{6.52}$$

6.5 Interpretation of 2-D Descriptors

The interpretation of molecular descriptors in QSAR models is a significant part of the model development process to understand the meaning of model parameters. Several descriptors are easily interpreted in terms of physico-chemical interactions. However, in general this is complicated because most molecular descriptors are not directly measured properties and can not be physically related to the modelled physical property or biological activity. The interpretation of topological descriptors is complicated because their mathematical foundation comes from graph theory and they are derived using abstract mathematical operations. Although chemists can get quite a clear idea about underlying mechanisms behind modelled activities by analysing structural formulae that resemble graph structures, the topological indices are just plain numbers and are much harder to rationalise and comprehend. This has raised some criticism about topological descriptors concerning the lack of physico-chemical interpretability.

Over several decades, considerable efforts have been made to correlate topological indices with different physico-chemical properties. A large number of models have been developed for the estimation of octanol–water partition coefficient, water solubility, boiling point, melting point, density, soil sorption coefficients and many other properties. There have been several attempts to interpret the meaning of topological indices in such models. So far, for interpretation most efforts have concentrated on the meaning of the molecular connectivity indices and valence connectivity indices.

The seminal work on the interpretation of connectivity indices was made by Randić with the introduction of branching index or $^1\chi$ index.[16] This descriptor was originally derived for the characterisation of molecular branching in hydrocarbons by additively encoding weighted bond contributions between bonded atoms (excluding hydrogens). Correlations with experimental data (*e.g.* boiling point data, enthalpies of formation and vapour pressure) showed that this descriptor is suitable for the modelling of properties which critically depend on molecular size and shape. In addition, a correlation with theoretically calculated total surface area of saturated acyclic hydrocarbons was shown.

Burkhard *et al.* used principal component analysis of topological descriptors, calculated for a set of *n*-alkanes and polychlorinated biphenyls, and found that three principal components accounted for 98% of the variation.[91] These principal components were associated with degree of branching, molecular size and structural flexibility, respectively.

Saxena carried out a systematic study to relate topological descriptors ($^1\chi$, $^2\chi^v$, IC_1) to physico-chemical properties such as molecular refractivity, hydrophobicity (log P), electronic σ parameters and polarity.[92] The analysis of different data sets with substituted benzene and methane derivates concluded that IC_1 index does not have a significant correlation with any of the physico-chemical parameters considered, and that indices $^1\chi$ and $^2\chi^v$ have stronger correlations with molar refractivity than with hydrophobicity. The latter conclusion was also confirmed with a follow-up study with an extended data set.[93]

Sabljic reviewed QSAR models involving topological descriptors for estimating various properties of persistent organic pollutants.[89] The review concluded that topological indices are most efficient in describing and quantifying properties related to non-specific molecular interactions such as hydrophobic and dispersion forces. Low-order connectivity indices are mainly related to molecular size and bulk properties, while the higher order connectivity indices are associated with local structural properties such as degree of branching, substitution patterns on rings, *etc.*

On the other hand, Randic *et al.* argued that one should not expect that a topological index will necessarily have an interpretation in terms of standard physico-chemical concepts such as molecular surface, molecular volume, compactness, flexibility, foldedness, *etc.*[94,95] Instead, the descriptors should have a valid interpretation within the model in which they are used. The interpretation of topological indices within structural chemistry is justified because of the compatible concepts such as atoms that can be viewed as paths of length zero and chemical bonds that can be viewed as paths of length one. Several studies have used concepts from structural chemistry for alternative interpretations.

Kier and Hall have developed an intermolecular accessibility concept for understanding the meaning of molecular connectivity.[96] This concept interpreted the $^1\chi$ index in terms of molecular structure and showed that it characterises the intermolecular accessibility of molecules. The interpretation proposed that each individual bond contribution, $(\delta_i\delta_j)^{-0.5}$ in eqn (6.4) encodes its relative accessibility (*e.g.* topological and electronic availability) to encounter with other molecules in its immediate environment. When the bond contribution value is high, there is an expectation that the bond is relatively accessible to other bonds in the milieu. Conversely, a low value of the bond contribution infers a lower accessibility. They also showed that the same accessibility concept also applies to valence connectivity indices. In this case, the accessibility values can also reflect the polarity of bonds due to the unequal electronegativity .

Randić and Zupan re-evaluated the connectivity index $^1\chi$, Wiener index, Hosoya topological index, and Harary index in terms of bond contributions by differentiating between terminal and interior bond types.[94] The topological descriptors were partitioned into bond additive terms.

Figure 6.8 shows an example with bond contributions for $^1\chi$ and the Wiener index for different pentane derivatives. In the case of $^1\chi$, each bond makes a contribution of $(\delta_i\delta_j)^{-0.5}$ to the magnitude of the connectivity index (see eqn 6.4). The analysis of bond contributions suggested that terminal bonds make larger contributions to $^1\chi$ than the interior bonds. In the case of the Wiener index, the bond contribution was calculated by multiplying the number of carbons at the each side of the bond (see eqn 6.1). However, in this case the situation is reversed and interior bonds make greater contributions than terminal bonds.

Based on the dominance of terminal or interior bonds, topological descriptors can be classified into two groups (*e.g.* 'the connectivity type' or 'the Wiener type'). Descriptors belonging to the first group have stronger correlations with bond additive properties such as boiling points; for other properties, the opposite may be the case. This knowledge was used to develop new modifications for the second group descriptors where terminal groups had higher contributions. In all cases, the modified descriptors had stronger correlations with boiling points than with the original descriptors.

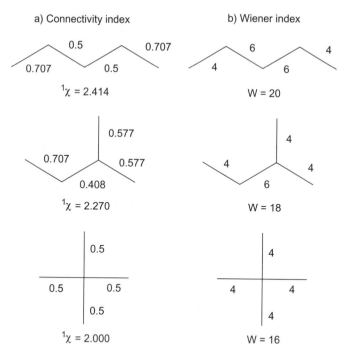

a) Connectivity index   b) Wiener index

$^{1}\chi = 2.414$ $W = 20$

$^{1}\chi = 2.270$ $W = 18$

$^{1}\chi = 2.000$ $W = 16$

Figure 6.8 Bond contribution terms for the connectivity index $^{1}\chi$ and the Wiener index.

Estrada has modified the intermolecular accessibility concept developed by Kier and Hall to use atomic intermolecular accessibility.[97] Equation 6.4 can be rewritten in an equivalent form as $^{1}\chi = \sum\limits_{allbonds} \left(\delta_i^{-0.5} \delta_j^{-0.5} \right)$ and the $\delta^{-0.5}$ terms can be interpreted as the accessible perimeter of an atom from the environment. The accessible perimeter is relative to the carbon atom with degree one and can be expressed in length units. Consequently, the bond contribution terms represented relative bond accessibility areas and the whole $^{1}\chi$ index can be viewed as the relative molecular accessibility area. The interpretation of extended connectivity indices concluded that the second order connectivity indices represent molecular accessibility volumes and the higher order connectivity indices represent hypervolumes.

It is interesting to note that conclusions about the meaning of connectivity indices with different physico-chemical and structural chemistry concepts arrive at similar results. It has been shown that connectivity indices characterise the branching and solvent accessible area of molecular structures. Since the connectivity indices are bond additive they also characterise the molecular size and other non-specific molecular interactions such as dispersion forces and polarisability. In terms of solvation free energy, it has been discussed that topological descriptors do have a relationship with the cavity formation term and consequently modulate related interaction in QSAR modelling of solubility and related properties.[98]

In addition, quantum chemical interpretation of connectivity indices is possible. Stankevich *et al.* found that $^1\chi$ describes π-electron properties of conjugated molecules and can be interpreted as the energy functional depending on the π-electron density.[99] Gálvez has confirmed that the relationship between quantum mechanical parameters from the Hückel molecular orbital approach and molecular connectivity can be established.[100] In conjugated hydrocarbons, the $^1\chi$ represents the measure of the global π-electron energy and also reproduces resonance energies, vibrational energies, bond lengths and the vibrational frequencies of C=C double bonds. Rouvray has discussed the use of the topological (*i.e.* adjacency) matrix in the field of quantum chemistry and showed that it can be used for the calculation of quantum chemical parameters, such as energy eigenvalues, charge densities, and bond orders.[101] The basic reason for the topological dependence of these parameters is the dominance of short range forces.

6.6   Sources for the Calculation of 2-D Descriptors

While the calculation of topological descriptors for small and simple molecules can be performed with the help of paper and pencil, the handling of more complicated molecular structures requires computers in order to get any practical work done. This section presents a brief overview of the software currently available for the calculation of topological descriptors.

The software has been grouped into two classes—open source and proprietary. Open source software can be downloaded from the Internet for free and comes together with source code that implements the algorithm. The source code is very useful for scientists, because this allows study of the descriptor calculation algorithms for better understanding. Proprietary software is often available for a fee from software vendors or developed for in-house use by various research groups or companies. Sometimes free trial versions are provided for demonstration or free licences for academic and non-profit use are provided. Proprietary software usually does not provide explicit information about the implementation of the descriptor calculation algorithms.

6.6.1   Open Source Software

The **Chemistry Development Kit** (CDK) is an open source Java library for chemo- and bio-informatics.[102,103] It has a fairly large and active developer community, and over the years it has implemented a wide array of different molecular descriptors that are useful in the development of QSAR/QSPR models. Among the topological descriptors are: atomic and valence connectivity indices; Kier–Hall kappa molecular shape indices; Petitjean number and shape indices; Wiener indices; Zagreb index; Randić weighted path descriptors; molecular distance edge descriptor; and autocorrelation descriptors.

JOELib2 is an open source chemoinformatics library written in Java.[104] It contains various algorithms for the analysis and mining of chemical structure information. It also includes molecular descriptor calculation algorithms, which are mainly calculated from the topological information of molecular structures including topological charge index, Kier shape indices, graph shape coefficient, Zagreb index, and topological radius and diameter.

6.6.2 Proprietary Software

A wide array of proprietary software exists for the calculation of 2-D descriptors. In most cases there is a certain overlap in the selection of descriptors calculated. Therefore choosing the descriptor calculation software can be quite a difficult task. Often software vendors bundle additional tools (*e.g.* statistical analysis tools, database tools, visualisation tools, *etc.*) with their software which make their product more usable for certain applications.

Accelrys Discovery Studio[105] and **TSAR**[106] packages provide tools for molecular descriptor calculation, including topological descriptors. These include topological descriptors such as chi, kappa and E-state keys.

ADMEWORKS ModelBuilder[107] is a tool for building QSAR/QSPR models for various chemical and biological properties of compounds. It can build both quantitative and qualitative models based on 400 predefined 2-D and 3-D descriptors, and an unlimited number of substructure-related descriptors. Its descriptors originate mostly from the Automated Data Analysis and Pattern Recognition Toolkit (ADAPT), which is a software system for the development of QSAR/QSPR models.[108,109] The topological descriptors included are: molecular distance edge descriptors; E-state indices; molecular connectivity and valence connectivity descriptors; Kier's kappa shape indices; path count and length descriptors; and eccentric connectivity index.

CODESSA Pro provides tools for developing QSAR/QSPR models.[110] It can calculate a large number of molecular descriptors. Among the topological descriptors included are: various information content descriptors; Balaban index; Kier–Hall valence connectivity indices; Kier flexibility and shape indices; Randić connectivity indices; Wiener index; Balaban index; and topological electronic indices.

DRAGON is an application for the calculation of molecular descriptors for QSAR/QSPR modelling, similarity analysis and high-throughput screening of molecule databases.[111] DRAGON calculates a wide array of topological descriptors including: various connectivity and shape indices; information indices; autocorrelations; Balaban indices; Narumi indices; Wiener indices; and Zagreb indices.

Marvin software from ChemAxon provides various tools for chemoinformatics.[112] This software calculates a number of molecular descriptors based on the topological information such as: Balaban index; Wiener indices; Platt index; Szeged index; Randić index; Harary index; and distance degree.

Molconn-Z is widely used software for the calculation of molecular connectivity, shape and information indices for QSAR analyses.[113] Among them are Kier–Hall molecular connectivity indices, Kier's kappa shape indices, E-State indices and many other topological indices.

Molecular Operating Environment (MOE) includes tools for QSAR modelling.[114] MOE calculates various topological descriptors[115] including: Kier–Hall molecular connectivity indices; Kier's kappa shape and flexibility indices; E-State indices; Zagreb index; Balaban index; Wiener path and polarity numbers; and topological diameter and radius.

MOLD² is software for the rapid calculation of a large and diverse set of descriptors encoding 2-D chemical structure information.[116] The topological descriptors included are: autocorrelation descriptors; Balaban indices; Randić connectivity indices; Wiener indices; Zagreb indices; detour indices; and topological charge indices.

MOLGEN-QSPR is an integrated software tool for structure generation, descriptor calculation and regression analysis in combinatorial chemistry.[117,118] Among topological descriptors are: Wiener index; Zagreb indices; Randić connectivity indices; Kier and Hall indices; solvation connectivity indices; Kier shape and molecular flexibility indices; Balaban index; Basak information content indices; and Hosoya Z-index.

TAM is a program for the calculation of topological indices in QSPR/QSAR studies including: Wiener's index; a modified Wiener's index; Balaban's index; a modified Balaban's index; the information–theoretical index; a modified information–theoretical index; Schultz's index; a modified Schultz's index; and vertex–connectivity indices.[119]

TOPIX software calculates approximately 130 topological and structural descriptors, including simple constitutional descriptors, fragment descriptors, and various connectivity and information theory based topological indices.[120]

6.7   Conclusions

This chapter presents a short review of topological descriptors covering computational background, interpretation and examples of their use in toxicology modelling. In short, topological descriptors are an attractive choice in modelling. They can be applied when working with large and complex systems where 3-D and quantum mechanical methods would be impractically expensive.

The ease of calculation comes from the fact that they are derived from graph theory. This simple and mathematically well-studied basis enables both easy development of new indices and fast computation of existing ones. The unfortunate side of such simplicity is that the number of existing topological descriptors has grown almost incomprehensibly large. Although choice is generally regarded as a positive aspect, the amount of indices to study can give newcomers into the field lot of extra work when selecting the most suitable ones.

On the other hand, the graph theoretical ancestry of topological descriptors has also received some criticism of being unfathomable and distant from chemistry. The claims that graph invariants are impossible to sensibly interpret in a chemical context have gained ground due to the abundance of works where the discussion on topological descriptors used is non-existent, inadequate or sometimes even misleading. However, this criticism does not appear to be constructive. Admittedly, thorough and conclusive interpretation of graph-theoretical parameters is not simple. It usually requires total dissection of the calculation formula as well as some knowledge of the underlying theory and property modelled, but as can be seen from the interpretation section of this chapter, it is not impossible. Older and more established descriptors have already been exhaustively studied and their connection to certain physical parameters appears to be consistent. Probably the best example here would be connectivity indices and their correlation with hydrophobicity and lipophilicity to such an extent that they are, in some instances, replacing the more traditional log *P*.

As in everything else, topological descriptors are in the process of evolution. Despite some sporadic criticism, they are being used, developed and interpreted. Less useful ones are becoming obsolete and eventually this will lead to the selection of a well-characterised, reliable set. The descriptors require further work, but the results are rewarding and in our opinion this gives reason to be optimistic.

6.8 Literature for In-depth Reading

This chapter does not give an all-inclusive overview of 2-D descriptors. For more information, readers are advised to consult the following:

- T. Balaban, *Chemical Applications of Graph Theory*, Academic Press, New York, 1976.
- J. Devillers and A. T. Balaban, *Topological Indices and Related Descriptors in QSAR and QSPR*, Gordon & Breach, New York, 1999.
- M. V. Diudea, *QSPR/QSAR Studies by Molecular Descriptors*, Nova Science, Huntington, NY, 2001.
- M. Karelson, *Molecular Descriptors in QSAR/QSPR*, John Wiley & Sons, New York, 2000.
- L. B. Kier and L. H. Hall, *Molecular Connectivity in Chemistry and Drug Research*, Academic Press, New York, 1976.
- L. B. Kier and L. H. Hall, *Molecular Connectivity in Structure Activity Analysis*, Research Studies Press, Letchworth, UK, 1986.
- L. B. Kier and L. H. Hall, 'The molecular connectivity chi index and kappa shape indexes in structure–property modeling, in *Reviews of Computational Chemistry*, ed. D. B. Boyd and K. Lipkowitz, Wiley-VCH, USA, 1991, vol. 2, pp. 367–422.
- L. B. Kier and L. H. Hall, *Molecular Structure Description: The Electro-topological State*, Academic Press, San Diego, CA, 1999.

- T. I. Netzeva, 'Whole molecule and atom-based topological descriptors, in *Predicting Chemical Toxicity and Fate*, ed. M. T. D. Cronin and D. J. Livingstone, CRC Press Boca Raton, FL, 2004, pp. 61–83.
- R. Todeschini and V. Consonni, *Molecular Descriptors for Chemoinformatics*, Wiley-VCH, Weinheim, Germany, 2nd edn, 2009.
- N. Trinajstić, *Chemical Graph Theory*, CRC Press, Boca Raton, FL, 2nd edn, 1992.

6.9   Acknowledgements

Financial support from the Estonia Ministry of Education and Research (grant SF0140031Bs09) and the Estonian Science Foundation (grants 7099 and 7153) is gratefully acknowledged.

References

1. N. Trinajstić, *Chemical Graph Theory,* CRC Press, Boca Raton, FL, 1992.
2. M. Gordon and W. B. Temple, in *Chemical Applications of Graph Theory,* ed. A. T. Balaban, Academic Press, New York, 1976, 299–332.
3. E. J. Mills, *Philosophical Magazine*, 1884, **17**, 173.
4. H. Meyer, *Arch. Exper. Pathol. Pharmakol.*, 1899, **42**, 109.
5. E. Overton, *Studien über die Narkose zugleich ein Beitrag zur allgemeinen Pharmacologie*, Verlag Gustav Fischer, Jena, Germany, 1901.
6. I. Langmuir, *Colloid Symp. Monogr.*, 1925, **3**, 48.
7. L. P. Hammett, *Chem. Rev.*, 1935, **17**, 125.
8. L. P. Hammet, *Physical Organic Chemistry,* McGraw-Hill, New York, 1940.
9. R. W. Taft, *J. Am. Chem. Soc.*, 1952, **74**, 2729.
10. R. W. Taft, *J. Am. Chem. Soc.*, 1952, **74**, 3120.
11. R. W. Taft, *J. Am. Chem. Soc.*, 1953, **75**, 4231.
12. R. W. Taft, *J. Am. Chem. Soc.*, 1953, **75**, 4538.
13. H. Wiener, *J. Am. Chem. Soc.*, 1947, **69**, 17.
14. J. R. Platt, *J. Chem. Phys.*, 1947, **15**, 419.
15. C. Hansch and T. Fujita, *J. Am. Chem. Soc.*, 1964, **86**, 1616.
16. M. Randić, *J. Am. Chem. Soc.*, 1975, **97**, 6609.
17. A. T. Balaban and F. Harary, in *Chemical Applications of Graph Theory,* ed. A. T. Balaban, Academic Press, New York, 1976, 1–4.
18. A. T. Balaban and O. Ivanciuc, in *Topological Indices and Related Descriptors in QSAR and QSPR,* ed. J. Devillers and A. T. Balaban, Gordon & Breach, New York, 1999, pp. 21–57.
19. A. T. Balaban, in *QSPR/QSAR Studies by Molecular Descriptors,* ed. M. V. Diudea, Nova Science, Huntington, NY, 2001, pp. 1–30.
20. S. C. Basak, G. D. Grunwald, B. D. Gute, K. Balasubramanian and D. Opitz, *J. Chem. Inf. Comput. Sci.*, 2000, **40**, 885.
21. R. Todeschini and V. Consonni, *Molecular Descriptors for Chemoinformatics,* Wiley-VCH, Weinheim, Germany, 2009.

22. I. Lukovits, in *QSPR/QSAR Studies by Molecular Descriptors,* ed. M. V. Diudea, Nova Science, Huntington, NY, 2001, pp. 31–38.
23. I. Gutman, B. Rušćić, N. Trinajstić and C. Wilcox, *J. Chem. Phys.*, 1975, **62**, 3399.
24. L. H. Hall and L. B. Kier, in *Topological Indices and Related Descriptors in QSAR and QSPR,* ed. J. Devillers and A. T. Balaban, Gordon & Breach, New York, 1999, pp. 307–360.
25. L. B. Kier and L. H. Hall, *Molecular Connectivity in Structure Activity Analysis,* Research Studies Press, Letchworth, UK, 1986.
26. L. H. Hall and L. B. Kier, in *Reviews of Computational Chemistry*, ed. D. B. Boyd and K. Lipkowitz, Wiley-VCH, USA, 1991, Vol. 2, pp. 367–422.
27. L. B. Kier and L. H Hall, *Molecular Connectivity in Chemistry and Drug Research,* Academic Press, New York, 1976.
28. A. T. Balaban, *Chem. Phys. Let.*, 1982, **89**, 399.
29. A. T. Balaban, N. Ionescu-Pallas and T. S. Balaban, *MATCH Commun. Math. Comput. Chem.*, 1985, **17**, 121.
30. A. T. Balaban and L. V. Quintas, *MATCH Commun. Math. Comput. Chem.*, 1983, **14**, 213.
31. A. T. Balaban, in *Topology in Chemistry: Discrete Mathematics of Molecules,* ed. D. H. Rouvray and R. B. King, Horwood Publishing, Chichester, 2002, pp. 89–122.
32. A. T. Balaban, D. Mills and S. C. Basak, *MATCH Commun. Math. Comput. Chem.*, 2002, **45**, 5.
33. L. B. Kier and L. H. Hall, in *Topological Indices and Related Descriptors in QSAR and QSPR,* ed. J. Devillers and A. T. Balaban, Gordon & Breach, New York, 1999, pp. 455–489.
34. L. B. Kier, *Quant. Struct.-Act. Relat.*, 1985, **4**, 109.
35. L. B. Kier, *Quant. Struct.-Act. Relat.*, 1986, **1**, 1.
36. I. Kahn, S. Sild and U. Maran, *J. Chem. Inf. Model.*, 2007, **47**, 2271.
37. C. A Shannon, *Bell Syst. Technol. J.*, 1948, **27**, 379.
38. S. C. Basak, in *Topological Indices and Related Descriptors in QSAR and QSPR,* ed. J. Devillers and A. T. Balaban, Gordon & Breach, New York, 1999, pp. 563–593.
39. S. C. Basak, A. T. Balaban, G. D. Grunwald and B. D. Gute, *J. Chem. Inf. Comput. Sci.*, 2000, **40**, 891.
40. L. B. Kier and L. H. Hall, in *Topological Indices and Related Descriptors in QSAR and QSPR,* ed. J. Devillers and A. T. Balaban, Gordon & Breach, New York, 1999, pp. 491–562.
41. L. B. Kier and L. H. Hall, *Molecular Structure Description: The Electrotopological State,* Academic Press, San Diego, CA, 1999.
42. L. H. Hall and L. B. Kier, *J. Chem. Inf. Comput. Sci.*, 1995, **35**, 1039.
43. J. Devillers, in *Topological Indices and Related Descriptors in QSAR and QSPR,* ed. J. Devillers and A. T. Balaban, Gordon & Breach, New York, 1999, pp. 595–612.
44. G. Moreau and P. Broto, *Nouv. J. Chim.*, 1980, **4**, 359.
45. G. Moreau and P. Broto, *Nouv. J. Chim.*, 1980, **4**, 757.

46. P. Broto, G. Moreau and C. Vandycke, *Eur. J. Med. Chem.*, 1984, **19**, 66.
47. H. Bauknecht, A. Zell, H. Bayer, P. Levi, M. Wagener, J. Sadowski and J. Gasteiger, *J. Chem. Inf. Comput. Sci.*, 1996, **36**, 1205.
48. P. Broto and J. Devillers, in *Practical Applications of Quantitative Structure-Activity Relationships (QSAR) in Environmental Chemistry and Toxicology,* ed. W. Karcher and J. Devillers, Kluwer Academic Publishers, Dordrecht, The Netherlands, 1990, pp. 105–127.
49. M. Wagener, J. Sadowski and J. Gasteiger, *J. Am. Chem. Soc.*, 1995, **117**, 7769.
50. M. T. D. Cronin and T. W. Schultz, *J. Mol. Struct. (Theochem.)*, 2003, **622**, 39.
51. G. T. Ankley and D. L. Villeneuve, *Aquat. Toxicol.*, 2006, **78**, 91.
52. S. C. Basak, D. P. Gieschen and V. R. Magnuson, *Environ. Toxicol. Chem.*, 1984, **3**, 191.
53. L. H. Hall, E. L. Maynard and L. B. Kier, *Environ. Toxicol. Chem.*, 1989, **8**, 783.
54. S. C. Basak, B. D. Gute, B. Luić, S. Nikolić and N. Trinajstić, *Comput. Chem.*, 2000, **24**, 181.
55. E. Papa, F. Villa and P. Gramatica, *J. Chem. Inf. Model.*, 2005, **45**, 1256.
56. D. V. Eldred, C. L. Weikel, P. C. Jurs and K. L. E. Kaiser, *Chem. Res. Toxicol.*, 1999, **12**, 670.
57. M. P. Sauvant, D. Pepin and E. Piccinni, *Chemosphere*, 1999, **38**, 1631.
58. W. T. Schultz, L. B. Kier and L. H. Hall, *Bull. Environ. Contam. Toxicol.*, 1982, **28**, 373.
59. J. C. Dearden, M. T. D. Cronin, W. T. Schultz and D. T. Lin, *Quant. Struct.-Act. Relat.*, 1995, **14**, 427.
60. A. R. Katritzky, P. Oliferenko, A. Oliferenko, A. Lomaka and M. Karelson, *J. Phys. Org. Chem.*, 2003, **16**, 811.
61. M. T. D. Cronin and T. W. Schultz, *Chemosphere*, 1996, **32**, 1453.
62. S. J. Enoch, M. T. D. Cronin, T. W. Schultz and J. C. Madden, *Chemosphere*, 2008, **71**, 1225.
63. E. Zvinavashe, T. Du, T. Griff, H. H. J. van den Berg, A. E. M. F. Soffers, J. Vervoort, A. J. Murk and I. M.C.M. Rietjens, *Chemosphere*, 2009, **75**, 1531.
64. Y. H. Zhao, G. D. Ji, M. T. D. Cronin and J. C. Dearden, *Sci. Total Environ.*, 1998, **216**, 205.
65. J. Davies, R. S. Ward, G. Hodges and D. W. Roberts, *Environ. Tox. Chem.*, 2004, **23**, 2111.
66. G. Hodges, D. W. Roberts, S. J. Marshall and J. C. Dearden, *Chemosphere*, 2006, **63**, 1443.
67. A. R. Katritzky, S. H. Svalov, I. S. Stoyanova-Svalova, I. Kahn and M. Karelson, *J. Tox. Environ. Health.*, 2009, **72**, 1181.
68. E. Papa, F. Battaini and P. Gramatica, *Chemosphere*, 2005, **58**, 559.
69. F. A. Pasha, M. M. Neaz, S. J. Cho, M. Ansari, S. K. Mishra and S. Tiwari, *Chem. Biol. Drug. Des.*, 2009, **73**, 537.
70. R. L. Yu, G. R. Hu and Y. H. Zhao, *J. Environ. Sci.*, 2002, **14**, 552.

71. S. Parvez, C. Venkataraman and S. Mukherji, *Environ. Int.*, 2006, **32**, 265.
72. K. L. E. Kaiser, *Environ. Health Perspect.*, 1998, **106**, 583.
73. V. K. Gombar and K. Enslein, *In Vitro Toxicol.*, 1988/1989, **2**, 117.
74. D. B. Wei, A. Q. Zhang, Z. B. Wei, S. K. Han and L. S. Wang, *Ecotox. Environ. Safety.*, 2002, **52**, 143.
75. V. K. Agrawal and P. V. Khadikar, *Bioorg. Med. Chem.*, 2002, **10**, 3517.
76. G. Melagraki, A. Afantitis, H. Sarimveis, O. Iglessi-Markopoulou and A. A Alexandridis, *Mol. Div.*, 2006, **10**, 213.
77. A. D. P. Worgan, J. C. Dearden, R. Edwards, T. I. Netzeva and M. T. D. Cronin, *QSAR Comb. Sci.*, 2003, **22**, 204.
78. T. I. Netzeva, J. C. Dearden, R. Edwards, A. D. P. Worgan and M. T. D. Cronin, *J. Chem. Inf. Comput. Sci.*, 2004, **44**, 258.
79. M. T. D. Cronin, T. I. Netzeva, J. C. Dearden, R. Edwards and A. D. P. Worgan, *Chem. Res. Toxicol.*, 2004, **17**, 545.
80. D. A. Bahnick and W. J. Doucette, *Chemosphere*, 1988, **17**, 1703.
81. I. Lessigiarska, A. P. Worth, T. I. Netzeva, J. C. Dearden and M. T. D Cronin, *Chemosphere*, 2006, **65**, 1878.
82. M. T. D. Cronin, J. C. Dearden, J. C. Duffy, R. Edwards, N. Manga, A. P. Worth and A. D. P. Worgan, *SAR QSAR Environ. Res.*, 2002, **13**, 167.
83. E. Papa, M. Luini and P. Gramatica, *SAR QSAR Environ. Res.*, 2009, **20**, 767.
84. B. Bhhatarai and P. Gramatica, *Chem. Res. Toxicol.*, 2010, **23**, 528.
85. J. Devillers and H. Devillers, *SAR QSAR Environ. Res.*, 2009, **20**, 467.
86. A. Sabljic, H. Güsten, H. Verhaar and J. Hermens, *Chemosphere*, 1995, **31**, 4489.
87. B. M. Gawlik, N. Sotiriou, E. A. Feicht, S. Schulte-Hostede and A. Kettrup, *Chemosphere*, 1997, **34**, 2525.
88. I. Kahn, D. Fara, M. Karelson, P. Andersson and U. Maran, *J. Chem. Inf. Model.*, 2005, **45**, 95.
89. A. Sabljic, *Chemosphere*, 2001, **43**, 363.
90. A. Sabljic, *Model. Environ. Sci. Technol.*, 1987, **21**, 358.
91. L. P. Burkhard, A. W. Andrew and D. E. Armstrong, *Chemosphere*, 1983, **12**, 935.
92. A. K. Saxena, *Quant. Struct.-Act. Relat.*, 1995, **14**, 31.
93. A. K. Saxena, *Quant. Struct.-Act. Relat.*, 1995, **14**, 142.
94. M. Randiæ and J. Zupan, *J. Chem. Inf. Comput. Sci.*, 2001, **41**, 550.
95. M. Randiæ, A. T. Balaban and S. C. Basak, *J. Chem. Inf. Comput. Sci.*, 2001, **41**, 593.
96. L. B. Kier and L. H. Hall, *J. Chem. Inf. Comput. Sci.*, 2000, **40**, 792.
97. E. Estrada, *J. Phys. Chem. A.*, 2002, **106**, 9085.
98. A. R. Katritzky, A. A. Oliferenko, P. V. Oliferenko, P. Petrukhin, D. B. Tatham, U. Maran, U. A. Lomaka and W. E. Acree Jr, *J. Chem. Inf. Comput. Sci.*, 2003, **43**, 1794.

99. I. V. Stankevich, M. I Skvortsova and N. S. Zefirov, *J. Mol. Struct. (Theochem.)*, 1995, **342**, 173.

100. J. Gálvez, *J. Mol. Struct. (Theochem.)*, 1998, **429**, 255.

101. D. H. Rouvray, in *Chemical Applications of Graph Theory*, ed. A. T. Balaban, Academic Press, New York, 1976, 175–221.

102. C. Steinbec, Y. Han, S. Kuhn, O. Horlacher, E. Luttmann and E. Willighagen, *J. Chem. Inf. Comp. Sci.*, 2003, **43**, 493.

103. http://sourceforge.net/projects/cdk [accessed February 2010].

104. http://sourceforge.net/projects/joelib [accessed February 2010].

105. http://accelrys.com/products/discovery-studio/ [accessed February 2010].

106. http://accelrys.com/products/accord/desktop/tsar.html [accessed February 2010].

107. www.fqs.pl/Chemistry_Materials_Life_Science/products/admeworks_modelbuilder [accessed February 2010].

108. A. J. Stuper, W. E. Brugger and P. C. Jurs, *Computer-Assisted Studies of Chemical Structure and Biologial Function,* Wiley, New York, 1979.

109. http://research.chem.psu.edu/pcjgroup/adapt.html [accessed March].

110. www.codessa-pro.com [accessed March].

111. www.talete.mi.it/dragon.htm [accessed March].

112. www.chemaxon.com/product/marvin_land.html [accessed March].

113. www.edusoft-lc.com/molconn/ [accessed March].

114. www.chemcomp.com/software.htm [accessed March].

115. www.chemcomp.com/journal/descr.htm [accessed March].

116. H. Hong, Q. Xie, W. Ge, F. Qian, H. Fang, L. Shi, Z. Su, R. Perkins and W. Tong, *J. Chem. Inf. Model.*, 2008, **48**, 1337.

117. A. Kerber, R. Laue, M. Meringer and C. Rücker, *MATCH Commun. Math. Comput. Chem.*, 2004, **51**, 187.

118. http://molgen.de/?src = documents/molgenqspr.html [accessed March].

119. M. Vedrina, S. Markovic, M. Medic-Saric and N. Trinajstić, *Comp. Chem.*, 1997, **21**(6), 355.

120. www.lohninger.com/topix.html [accessed March].

CHAPTER 7

The Use of Frontier Molecular Orbital Calculations in Predictive Reactive Toxicology

S. J. ENOCH

School of Pharmacy and Chemistry, Liverpool John Moores University, Liverpool L3 3AF, UK

7.1 Introduction

Molecular descriptors utilised in QSAR modelling can be roughly divided into three categories—hydrophobic, steric and electronic. It is the electronic class of descriptors that falls into the domain of quantum chemical calculations, with a number of chemical properties involved in the toxicity of chemicals being modelled by such descriptors. One of the primary tenets of a good quality QSAR model is the ability to relate the model to the mechanism of toxic action[1] and thus it is important to outline the mechanistic factors that quantum chemical descriptors aim to encapsulate. Therefore, the aim of this chapter is to outline the use of such descriptors in the modelling of chemical reactivity and thus covalent bond formation. The ability to model chemical reactivity is important as covalent bond formation between a biological macromolecule, and an exogenous chemical plays a key role in a number of toxicological endpoints. The chapter highlights how, within well defined mechanistic domains, simple and interpretable descriptors can be utilised for such endpoints. In addition, an outline of the importance of frontier molecular orbital theory and its relationship to chemical reactivity is presented. Finally, the

Issues in Toxicology No.7
In Silico Toxicology: Principles and Applications
Edited by Mark T. D. Cronin and Judith C. Madden
© The Royal Society of Chemistry 2010
Published by the Royal Society of Chemistry, www.rsc.org

chapter aims to illustrate the considerations that must be taken into account if the results of such calculations are going to be of practical use in regulatory environments.

7.2 Mechanistic Chemistry

Industrial and pharmaceutical chemicals are able to elicit toxicity across a range of ecotoxicity and human endpoints *via* a variety of mechanisms. These mechanisms can be grouped into a number of classes but for the purpose of this chapter can be thought of simply as[2–5] mechanisms acting *via* the formation of a covalent bond between a biological nucleophile and the chemical; and mechanisms involving non-covalent interactions (*e.g.* general disruption of membranes or cellular processes, receptor binding, disruption of hormonal balance, *etc.*).

The ability of a chemical to partition into cellular membranes and thus be bioavailable to the organism in sufficient concentration to be able to cause a toxic response is important for chemicals acting *via* both non-covalent and (the majority of) covalent mechanisms. Within the differing mechanisms, however, additional chemical factors become important in the determination of how toxic a given chemical will be. For example, with regard to acute aquatic toxicity, chemicals within the polar narcosis mechanistic domain are relatively more toxic than those within the non-polar narcosis domain due to the presence of a chemical dipole.[5,6] Thus knowledge of the chemistry underlying a toxicity mechanism will assist in elucidating effects. In addition, also for fish acute toxicity, chemicals acting *via* specific covalent mechanisms of action are often relatively more toxic than those acting *via* non-covalent (general) mechanisms such as narcosis. This is due to the ability of specific toxicants to form irreversible chemical bonds with membranes such as the gill and destroy their function. Thus, to identify such toxicants, one needs to direct attention to their chemistry. In particular, the ability to form a covalent chemical bond depends upon the presence, and electrophilicity, of a reactive sub-structure within a chemical.[7,8] Quantum chemical descriptors have been utilised in QSAR studies for the mechanisms by which the electronic properties of a given chemical control its toxicity.

The (organic chemistry) mechanisms by which a biological nucleophile and an exogenous electrophilic chemical form a covalent bond are governed by frontier molecular orbital theory. In a reaction controlled by the frontier orbitals, the most likely product will be formed by a reaction between the highest occupied molecular orbital (HOMO) of the nucleophile and the lowest unoccupied molecular orbital (LUMO) of the electrophile (Figure 7.1). Importantly, reactions that are orbital controlled proceed at faster rates the closer in energy the HOMO of the nucleophile and the LUMO of the electrophile are to one another (all other factors such as steric effects being equal).

Frontier molecular orbital theory also states that the most likely site of attack within a given acceptor molecular orbital (such as the LUMO) will occur

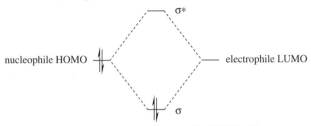

anti-bonding orbital formed between the HOMO of the
nucleophile and LUMO of the electrophile

σ*

nucleophile HOMO

electrophile LUMO

σ

bonding orbital formed between the HOMO of the
nucleophile and LUMO of the electrophile

Figure 7.1 Simplified molecular orbital diagram showing how the HOMO of the
nucleophile and the LUMO of the electrophile interact to produce new
bonding (occupied) and anti-bonding (unoccupied) molecular orbitals.

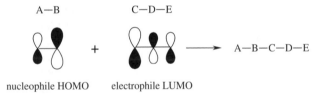

A—B   C—D—E

+   ⟶   A—B—C—D—E

nucleophile HOMO   electrophile LUMO

Figure 7.2 Example of how the maximum electron density in the HOMO of the
nucleophile and the LUMO of the electrophile controls the site of covalent
bond formation in frontier molecular orbital controlled chemical reactions.

at the position where the electron density is at its maximum (this can be
visualised as the atom with the largest atomic orbital contribution to the
molecular orbital). Thus, the new chemical bond will form between the atom in
the nucleophile with the maximum electron density in the HOMO and the
atom of the electrophile with the maximum electron density in the LUMO
(Figure 7.2). It is important to realise that steric hindrance at the site of attack
(usually in the electrophile) can alter the atom at which the covalent bond
forms. In such cases it becomes important to inspect the shape and energy of
the next lowest unoccupied molecular orbital (LUMO + 1) as, if this orbital is
sufficiently close in energy to the LUMO, then the chemical reaction is likely to
occur *via* an interaction with this orbital.

7.3 Commonly Utilised Quantum Mechanical Descriptors and Levels of Computational Theory

There is a rich history of the use of quantum chemistry calculations to determine molecular orbital properties. Inevitably in the past four decades, these

methods have been computerised and are now commonplace. A review of the available computational theories and their relative merits is outside the scope of this chapter (for a detailed overview of the commonly used computational theories refer to Koch and Holthausen,[9] Foresman and Frisch,[10] and Leach[11]).

Over the decades a large number of calculation methods have been developed from simplistic (not requiring a computer) to highly parameterised and computationally expensive. In terms of this chapter, two widely applied computational levels of theory will be discussed semi-empirical Austin Model 1 (AM1); and density functional theory (DFT). It is worth noting that these two methods represent opposing ends of the computational spectrum in that calculations using AM1 are extremely fast whilst those using DFT are significantly slower (Table 7.1). The trade-off being that DFT calculations offer a significantly better representation of the underlying electronic structure than AM1 and thus, in certain circumstances, may provide a more accurate representation of the properties of the molecule.[9–11]

Given the benefits of speed, the majority of descriptors utilised in the predictive toxicology literature have been calculated using the AM1 methodology. Descriptors calculated at this level of theory have been applied in the modelling of large databases containing non-generic chemicals acting primarily *via* non-covalent mechanisms such as narcosis.[12,13] In addition, such descriptors have also been shown to be useful in the modelling the toxic effects associated with smaller databases of congeneric chemicals acting *via* covalent mechanisms.[14–17] In contrast there has been significantly fewer studies investigating the use of DFT descriptors for non-covalent mechanisms,[18] with one study showing there to be no statistical benefit in the use of the more computationally demanding methods.[19] It is perhaps unsurprising, therefore, that the use of DFT-derived descriptors has focussed primarily towards the modelling of both congeneric and mechanistically related categories of covalently reactive chemicals.[20,21] Inspection of the literature reveals several key descriptors have been shown to be of use in predictive toxicology (Table 7.2).[22]

7.4 Descriptors Derived from Frontier Molecular Orbitals

Values associated with frontier molecular orbitals are the most commonly used—and mechanistically interpretable—descriptors available from quantum mechanical calculations. These descriptors have been used to offer an insight into the electronic effects that play an important role in a given chemical's toxicity, especially those acting *via* covalent bond formation. If one considers frontier molecular orbital theory (FMO)[23] and the fact that the HOMO of the biological nucleophile interacts with the LUMO of the electrophile (exogenous chemical), then it is clear how and why such descriptors are important in predictive toxicology. The general premise of FMO theory being that the closer the energy of the HOMO (E_{HOMO}) of the nucleophile and the energy of the LUMO (E_{LUMO}) of the electrophile are to one another the more facile the

Table 7.1 Commonly used computational methods in predictive toxicology for molecular orbital calculations.

Method	Example level of theory	Advantages	Disadvantages	Software (website)
Semi-empirical	AM1/RM1/PM6	• Computationally efficient • Calculations can be performed rapidly for large databases of chemicals. • No additional parameters (e.g. basis sets) need to be considered	• Underlying mathematics limited to the valance electrons. • Significant simplifications made in order to ensure computational efficiency	MOPAC (http://openmopac.net) AMPAC (www.semichem.com)
Ab initio	Hartree–Fock theory (HF)/Density functional theory (DFT)	• Significantly more complete mathematical description of the underlying electronic structure.	• Computationally demanding. • Limited to smaller numbers of chemicals. • Consideration of an appropriate basis set must be made (usually 631Gd is utilised).	Gaussian (www.gaussian.com) ADF (www.scm.com)

Table 7.2 Examples of descriptors derived from the frontier molecular orbitals.

Name	Definition	Mechanistic interpretation
E_{LUMO}	Energy of the lowest unoccupied molecular orbital	Ability to accept electrons, *i.e.* electrophilicity and the ease of reduction
E_{HOMO}	Energy of the highest occupied molecular orbital	Ability to donate electrons, *i.e.* nucleophilicity and the ease oxidation
μ	Chemical potential (negative of electronegativity) $\mu = (E_{LUMO} + E_{HOMO})/2$	Opposite of electronegativity
η	Chemical hardness $\eta = (E_{LUMO} - E_{HOMO})/2$	Measure of chemical hardness: useful in modelling hard–soft chemical interactions
σ	Chemical softness $\sigma = 1\text{-}\eta$	Measure of chemical softness: useful in modelling hard–soft chemical interactions
ω	Electrophilicity $\omega = \mu^2/2\eta$	Measure of chemical electrophilicity: useful in modelling frontier orbital controlled chemical reactivity
A_{max}	Maximum atomic acceptor super-delocalisability within a molecule	Measure of the atom with the greatest ability to accept electron density within an aromatic system: useful in modelling reactivity within aromatic systems
D_{max}	Maximum atomic donor super-delocalisability within a molecule	Measure of the atom with the greatest ability to donate electron density within an aromatic system
A_N	Atomic acceptor super-delocalisability for atom N	Measure of atom N's ability to accept electron density within an aromatic system
D_N	Atomic donor superdelocalisability for atom N	Measure of atom N's ability to donate electron density within an aromatic system
$A_{orbital}$	One-orbital delocalisability associated with a given orbital (*e.g.* LUMO or LUMO + 1)	Measure of the atomic contribution

reaction will be (and thus the compound has an increased potential for covalent interactions that may lead to toxicity). This, of course, assumes all other factors such as steric effects within a molecule (*i.e.* the shielding of a reactive group by other substituents) are equal. In practice one usually calculates E_{LUMO} for a series of electrophiles, with the assumption that the lower the calculated value the more electrophilic the chemical is likely to be. Increased electrophilicity leads to increased reactivity and hence potentially greater toxicity. It is generally considered that the biological nucleophile remains constant for a given series of chemicals, although this clearly is not the case if one starts to consider differing mechanisms of action and biological endpoints. Thus it is important to consider chemicals acting *via* a single covalent mechanism of action when performing such analyses.[7]

7.4.1 Use of HOMO and LUMO in Modelling Chemical Reactivity

In order to illustrate the relationship between reactivity and E_{LUMO} in covalent toxicity mechanisms, it is useful to first consider previously published reactivity data for a set of simple chemicals undergoing Michael addition (a classic organic chemistry mechanism) with the model nucleophile 4-(dimethylamino)pyridine.[24] These data highlight frontier molecular orbital theory as it can be seen that as E_{LUMO} increases (thus becoming less stable) then the rate of constant, K_{Nu}, decreases (meaning a slower reaction) (Table 7.3).

An important consideration when utilising quantum mechanical descriptors to model covalently reactive chemicals is in understanding the applicability domain, *i.e.* the area of chemical space occupied by a group of chemicals with the same mechanism of action (see Chapter 12). For example, Table 7.3 shows that the reactivity of methyl vinyl sulfone is poorly modelled by its E_{LUMO} value (which suggests it should be significantly more reactive than acrolein). A possible explanation for this is that the AM1 method deals poorly with sulfur and thus perhaps sulfur compounds should be excluded from the model (*i.e.* they fall outside of the applicability domain of the model).

In addition to considering the presence of chemicals that are not well modelled by *in silico* techniques, one also needs to be aware of chemicals that are not modelled well due to additional factors that influence reactivity (and ultimately toxicity). For example, inspection of the reactivity data for crotonaldehdye highlights the influence of substitution at the β-carbon atom, with the additional methyl group resulting in a significant reduction in reactivity. In contrast, the corresponding E_{LUMO} value does not significantly change (compare acrolein to crotonaldehdye in Table 7.3). This is as one would expect given that the methyl group at the β-carbon has no π-orbital with which to influence the LUMO. However, there is clearly a steric effect on experimental reactivity. This serves as an important consideration when using quantum mechanical descriptors in modelling, *i.e.* one should not expect electronic descriptors to model non-electronic effects such as steric effects or entropy. If one removes

Table 7.3 Reactivity (Log K_{Nu}) versus E_{LUMO} values for a series of Michael acceptors.

Name	SMILES	Log K_{Nu} (taken from ref. 24)	E_{LUMO} (eV) (AM1)
acrolein	C=CC=O	0.38	−0.0051
methyl vinyl ketone	C=CC(=O)C	−0.26	−0.0025
methyl vinyl sulfone	C=CS(=O)(=O)C	−1.08	−0.018
methacrolein	C=C(C)C=O	−1.74	−0.0033
methyl acrylate	C=CC(=O)OC	−1.84	−0.00051
crotonaldehyde	C/C=C/C=O	−2.10	−0.0049
acrylonitrile	C=CC#N	−2.39	0.0018
acrylamide	C=CC(=O)N	−3.21	0.0061
N,*N*-dimethylacrylamide	C=CC(=O)N(C)C	−3.74	0.016

these two chemicals for these reasons, then the correlation coefficient between reactivity and E_{LUMO} improves considerably ($r^2 = 0.48$ to 0.74).

The above discussion has utilised reactivity data in order to illustrate some important points about the use of E_{LUMO} values to model electrophilicity (and thus chemical reactivity). Clearly it is important to demonstrate that the same trends exist in the toxicity data if E_{LUMO} values will be of use in the modelling of such data.

It is important to remember that the modelling of toxicity data is frequently complicated by the need to consider chemical partitioning and thus hydrophobicity. However, examination of 40-hour toxicity data to the ciliated protozoan, *Tetrahymena pyriformis* (used as a measure of acute aquatic toxicity), for an equivalent series of polarised alkenes (acrolein, methyl vinyl ketone and methyl acrylate) illustrates that the same trends exist with toxicity decreasing when the polarising group is varied from an aldehyde to a ketone to an ester.[25] This confirms that relative acute toxicity decreases as the reactivity of the functional group decreases. These data also illustrate the significant differences associated with substitution at the β-carbon atom of the alkenes. A more in-depth review of the numerous previous QSAR studies for the prediction of acute aquatic toxicity—many of which involve the use of E_{LUMO} to model electrophilicity—can be found in ref. 26.

In addition to being used to model electrophilicity, E_{LUMO} values have also been suggested to be useful in the modelling of the ability of a chemical to undergo metabolic reduction to a covalently reactive species.[27] It has been hypothesised that the inclusion of E_{LUMO} values in such modelling accounts for the ability of a chemical to accept electrons and hence the ease with which it can undergo reduction. The mechanistic rationale for the use of E_{LUMO} in this fashion is clearly well-defined and similar to the mechanistic rationale used to justify its use in the modelling of covalent bond formation. As previously, the use of E_{LUMO} to model such effects requires a careful definition of the applicability domain. Such models are usually restricted to a single chemical class such as aromatic nitro compounds. As might be expected given the utility of E_{LUMO} in the modelling of reductive bioactivation, E_{HOMO} values have been utilised to model the ability of chemicals to be oxidatively bioactivated to covalently reactive species.[28,29]

7.4.2 Use of Chemical Potential, Hardness and Electrophilicity in Modelling Chemical Reactivity

Several quantum chemical properties related to the frontier molecular orbitals have also been suggested to be useful in understanding nucleophilic–electrophilic reactions; these being chemical potential (μ), chemical hardness (η) and chemical electrophilicity (ω).[30–32] These descriptors are all derived from a density functional theory understanding of the mechanistic factors governing chemical reactivity.[33] Given this derivation, it is perhaps unsurprising that these descriptors are typically calculated at significantly more complex levels of computational theory than the descriptors discussed thus far (usually at least

Hartree–Fock theory coupled with a reasonable sized basis set; the discussion of Hartree–Fock theory and basis sets is outside the scope of this chapter and interested readers are referred to the excellent text by Leach[11] for further reading). The additional complexity has resulted in these descriptors being successfully used in the modelling of chemical reactivity and toxicity.[18,20,21,34–36] As with the use of E_{LUMO} and E_{HOMO} values calculated at the AM1 level of theory, the successful use of μ, η and ω still requires a well-defined applicability domain based on the considerations already discussed.

This need for a careful consideration of the applicability domain based not only on the reactivity but also the mechanism of toxicity was illustrated by comparing the electrophilic indices for two skin sensitising chemicals (both assigned to the Michael mechanistic domain).[21] For these two chemicals, the study demonstrated that the electrophilicity index could be used to predict the skin sensitising potentials of alkene derivatives within the Michael domain. Thus, inspection of the electrophilicities of 5,5-dimethyl-3-methylene-dihydro-2(3H)-furanone and ethyl acrylate would be expected to suggest that they have similar skin sensitising potentials (Table 7.4). However, this is not the case with the furanone being significantly more toxic than the acrylate. The authors of the study suggested that the furanone is out of the domain of the model due to the additional reactivity (and thus toxicity) gained upon release of the ring strain (compared with the acrylate) during the covalent bond formation with the biological nucleophile (Figure 7.3).[21]

Table 7.4 Electrophilicities and skin sensitisation potentials of 5,5-dimethyl-3-methylene-dihydro-2(3H)-furanone and ethyl acrylate.

Chemical	Electrophilicity index (eV)	Skin sensitising potential (EC3)
5,5-dimethyl-3-methylene-dihydro-2(3H)-furanone	1.49	1.8
ethyl acrylate	1.49	28.0

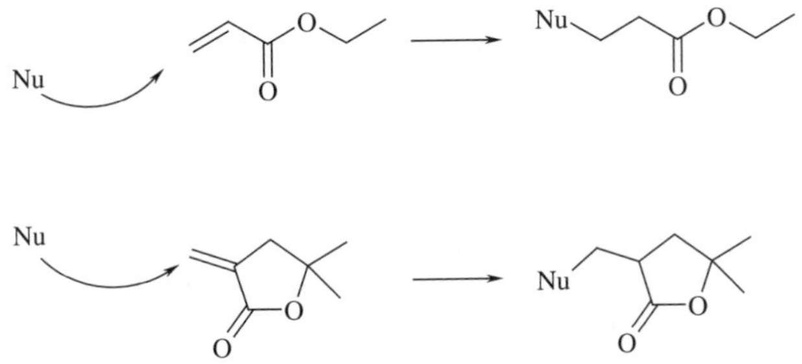

Figure 7.3 Ring strain release in 5,5-dimethyl-3-methylene-dihydro-2(3H)-furanone upon covalent bond formation with a biological nucleophile.

In addition to the need to consider the applicability domain of chemicals within a single mechanistic domain carefully, a recent study also highlighted that fact that differing mechanistic domains have differing relationships between the electrophilicity index and toxicity.[36] This finding is in keeping with similar work which also highlighted that differing relationships between E_{LUMO} and toxicity existed depending upon whether the chemical domain being modelled applied to aromatic or aliphatic chemicals.[26,37] This is an important concept to be aware of, as it indicates that relationships should be investigated and applied only within defined chemical domains.

7.4.3 Use of Superdelocalisability in Modelling Chemical Reactivity

The discussion of the importance of the frontier molecular orbitals and descriptors derived from them has thus far focussed on using the entire molecular orbital. Going beyond this theory, a number of authors have made use of the superdelocalisability theory introduced by Fukui *et al.*[38] This theory utilises the occupied and unoccupied molecular orbitals to generate donor (nucleophilic) and acceptor (electrophilic) delocalisability descriptors for aromatic systems. The superdelocalisability descriptors were devised to model the contributions of each atom within an aromatic system to the stabilisation energy upon formation of a charge interaction with an incoming nucleophile (or electrophile) in the transition state. These descriptors have been suggested to be useful as both global reactivity descriptors and site-specific descriptors capable of suggesting the most reactive centre in a given aromatic molecule.[14,39–42]

The mechanistic rationale for their usage is that the atom within an aromatic molecule with the maximum value of acceptor (or donor) delocalisability corresponds to the site most likely to undergo attack by a nucleophile (or electrophile). As with all the electronic descriptors discussed, these values assume that all other factors such as steric hindrance at the site of attack remain (relatively) constant. Inspection of the acceptor delocalisabilities of the aromatic atoms within dinitrochlorobenzene (DNCB) and tetrachloroisophthalonitrile (TCPN) (Table 7.5) reveals them to predict the most reactive atoms that are required for these chemicals to react *via* the S_NAr mechanism correctly (which has been suggested to be the mechanism of action for both chemicals).[7] In the case of DNCB, the maximum acceptor delocalisability (A_{max}) corresponds to the chloro-substituted carbon atom (C_1), which has the activating nitro groups *ortho* and *para* to it (note that the intermediate formed during the S_NAr reaction is maximally stabilised by groups in these positions). In the case of TCPN, three equally reactive sites are suggested (C_1, C_3 and C_5); again these are the chloro-substituted carbon atoms which have the activating cyano groups in the *ortho* and *para* positions. Crucially, the chloro-substituted carbon atom with both cyano groups meta to it (C_6) is the least reactive site. In addition, maximum aromatic delocalisability (A_{max}) values have also been utilised as global measures of reactivity.[39–42] A comparison of the skin

Table 7.5 Acceptor delocalisabilities and one-orbital acceptor delocalisabilities for dinitrochlorobenzene, tetrachloroiso-phthalonitrile, acrolein and chloroethane.

Name	Structure	Atom number	Acceptor delocalisability (A_N)	One-orbital acceptor delocalisability (A_{LUMO})
dinitrochlorobenzene		1	−0.56	32.58
		2	−0.54 (A_{max})	17.63
		3	−0.52	0.63
		4	−0.53	23.90
		5	−0.50	21.33
		6	−0.48	1.03
tetrachloroisophthalonitrile		1	−0.58 (A_{max})	25.26
		2	−0.52	17.89
		3	−0.583 (A_{max})	0.00
		4	−0.52	17.94
		5	−0.58 (A_{max})	25.23
		6	−0.56	0.00
acrolein		1	−0.47	38.62 (15.32)
		2	−0.44	19.62 (28.39)
		3	−0.49 (A_{max})	23.93 (40.11)
		4	−0.21	17.80 (16.07)
chloroethane		1	−0.42	0.95
		2	−0.49 (A_{max})	53.52
		3	−0.14	42.03

sensitising potentials of DNCB and TCPN with the values of A_{max} suggest that TCPN is a stronger skin sensitiser than DNCB based on it being more electrophilic (and thus reactive).[2]

A similar analysis of the acceptor delocalisabilities for the aliphatic chemical acrolein reveals that the highest acceptor value does not correspond to the most likely site of attack (Table 7.5). In this system, the acceptor delocalisability value corresponds to attack at the carbonyl carbon (atom 3), whereas experimentally it has been shown that acrolein undergoes Michael addition at the terminal β-carbon.[43] This difference between aliphatic and aromatic delocalisability values serves to illustrate the difficulties and risks associated with using a descriptor designed for a specific type of system (aromatic systems for delocalisabilities) and applying it to alternate systems without a thorough analysis of its predictivity and applicability (which based on acrolein would suggest that superdelocalisabilities are not applicable to the Michael mechanistic domain).

The shape and relative atomic electron densities of the frontier molecular orbitals can also provide useful information about reactive mechanisms of toxic action.[44,45] Considering that such mechanisms are orbital controlled, it is clear that in order for a chemical to act *via* a given reactive mechanism it must possess an unoccupied frontier orbital of the correct shape for a biological nucleophile to donate into. Importantly, the relative atomic electron densities pinpoint the most likely position of attack, as frontier molecular orbital theory dictates that the reaction between nucleophile and electrophile will occur at the position where orbital overlap is maximised. This information has been encapsulated in the so-called one-orbital superdelocalisabilities (the acceptor delocalisability is donated here as A_{LUMO} and A_{LUMO+1}) which use only information relating to the frontier molecular orbitals in the calculation of the index (in contrast to the previously discussed delocalisabilities which use information derived from all occupied and unoccupied orbitals).

Inspection of the comparable A_{LUMO} values for the chemicals in Table 7.5 shows the same trends as A_{max} for DNCB and two of the three sites previously highlighted for TCPN. In contrast to the A_{max} values for TCPN, substitution at carbon 3 is suggested not to be favourable. A significant improvement in reactive site prediction was calculated for acrolein, with the A_{LUMO} value correctly identifying the terminal β-carbon as the most reactive site. Interestingly, inspection of the A_{LUMO+1} values (shown in italics and parentheses in the A_{LUMO} column in Table 7.5) suggests that the most favourable site of attack into this orbital is at the carbonyl carbon. This is in keeping with the mechanistic hypothesis that steric hindrance at the terminal β-carbon results in a Schiff base mechanism becoming favoured.[2,45] A_{LUMO} can also be used to predict the site of attack in mechanisms involving interactions with σ orbitals, such as the S_N2 mechanism for the reaction of chloromethane with a nucleophile. Such an analysis shows that, in keeping with experimental data, the halogen substituted carbon atom is the most reactive site (Table 7.5). In contrast to the full superdelocalisability descriptors, one-orbital descriptors are not (on their own) good indicators of reactivity between chemicals. It has been suggested that such descriptors need to be weighted by the energy of the orbital from which they have been derived in order to be comparable between chemicals.[44]

7.4.4 Orbital Descriptor Summary

The ability to analyse the electron density of frontier molecular orbitals and suggest potential mechanisms (and the likely associated site of attack) is of use in the formation of chemical categories. Given the difficulty in assigning chemicals to mechanistic domains with any degree of confidence, it is clear that any additional information that the electronic structure can offer to aid such assignments is likely to be of use. The above analysis demonstrates that the shapes of the frontier molecular orbitals of the electrophile are a powerful tool for such assignments. One must remember, however, that an appreciation of steric hindrance at the most likely site of attack needs to guide such analyses for the reasons discussed.

7.5 Isomers and Conformers

It is important to consider the potential geometrical isomers when calculating descriptors based on quantum mechanics as the energies (and shapes) of the orbitals (including the HOMO and LUMO) can be altered. Clearly, if the energies of the orbitals change, then descriptors calculated from them will also vary.

A good illustration the effect of isomerisation on the E_{HOMO} and E_{LUMO} can be seen by examining dimethyl fumarate and dimethyl maleate (Table 7.6). The E_{HOMO} values are almost identical in energy, whilst in contrast the E_{LUMO} values are significantly different. This difference is easy to rationalise when one considers the differing geometrical strain on the two isomers. In the case of dimethyl fumarate, the entire extended π-system is planar; whilst in dimethyl maleate only one of the ester groups can be planar with the alkene π-system as the other is twisted out of the plane due to steric effects. As might be expected, the fully planar π-system leads to a more stable LUMO orbital in dimethyl fumarate compared to dimethyl maleate. In contrast, the E/Z isomerisation between the two forms of heptadienal makes very little difference to the E_{HOMO} or E_{LUMO}. As before, this can be understood if one considers the orbitals

Table 7.6 Calculated E_{HOMO} and E_{LUMO} and associated descriptor values for dimethyl fumarate, dimethyl maleate, 2E,4E-heptadienal and 2Z, 4Z-heptadienal.

Name	SMILES	E_{HOMO} (eV)	E_{LUMO} (eV)
dimethyl fumarate	COC(=O)/C=C/C(=O)OC	−11.44	−0.99
dimethyl maleate	COC(=O)\C=C/C(=O)OC	−11.42	−0.53
2E,4E-heptadienal	C\C=C\C=C\C=O	−9.48	−0.59
2Z,4Z-heptadienal	C\C=C/C=C\C=O	−9.52	−0.61

involved; in both isomers, achieving fully planar π-systems results in the near identical orbital energies.

This simple illustration of the potential variability in isomer orbital energies (and thus any descriptors calculated from them) highlights the need to ensure that the chemical structures used in calculations are accurate (as far as possible). In addition, given that quantum mechanical descriptors are nearly always calculated using the ground state geometry, it also highlights the potential issue of conformational flexibility and how the presence of multiple conformers may (or may not) affect descriptor values. In terms of modelling reactive toxicology, conformer flexibility is likely to be less important than ensuring that calculations are performed on the correct geometrical isomer. This is because chemicals are likely to react in (or very close) to their ground state geometries when in free aqueous solution. In contrast, a conformer other than the ground state for a chemical undergoing an enzyme mediated covalent reaction may be important. In such cases, quantum mechanical descriptor values should be calculated on this alternate conformer.

7.6   Chemical Category Formation and Read-Across

The examples discussed in this chapter have highlighted the fact that trends in covalent bond formation and thus reactivity can be understood using calculations probing the electronic structure of a series of chemicals. Importantly, this chapter also discusses the fact that descriptors derived from quantum chemical calculations (Table 7.1) can only be expected to model reactivity in which the electronic effects of the system are the dominant factor. One should not expect steric effects at the reactive site to be dealt with by any of the descriptors discussed. This concept of descriptor applicability is extremely important in the use of chemical categories and read-across that have been proposed to be of use in regulatory toxicology (see Chapters 14–16 for further details).[46] A number of recent publications have demonstrated the utility of computationally derived descriptors to model reactivity (and thus toxicity) within well-defined mechanistic domains.[18,20,21,36,47] Importantly, these studies showed that calculations can only help elucidate those trends in toxicity data that are predominantly influenced by the electronic structure of the chemical of interest.

7.7   Quantum Chemical Calculations

All calculations in this chapter were carried out using the Austin Model 1 semi-empirical method as implemented in MOPAC2009 (from http://open-mopac.net/MOPAC2009.html), which is freely available to academics and non-profit organisations. Calculations utilised the following keywords in order to generate the required orbital information:

- VECTORS EIGEN. These two keywords when present together ensure that the orbital eigenvectors and eigenvalues for the seven highest occupied and seven lowest unoccupied molecular orbitals are printed to the output file. This information is useful in the visualisation of the atomic orbital contributions to each of the molecular orbitals. It is used in the calculation of both types of super delocalisabilities.

- SUPER. This keyword calculates acceptor and donor superdelocalisabilities as well as the one-orbital delocalisabilities for the HOMO-1, HOMO, LUMO and LUMO + 1.

7.8 Conclusions

This chapter has outlined how frontier molecular orbitals can be utilised to offer mechanistic insights into the nucleophilic–electrophilic chemistry involved in covalently mediated toxicology. The ability to understand such interactions is of high importance given the number and nature of the adverse effects that can be attributed to the formation of a covalent bond between an exogenous chemical and a biological macromolecule.

It is clear from the discussions that quantum mechanically derived descriptors are able to support previously assigned mechanisms of action. In addition, such calculations can also help confirm mechanistic hypotheses for new chemicals. Such analyses are clearly important if chemicals are going to be assigned to mechanism-based chemical categories with confidence. In addition to supporting mechanistic hypotheses, the discussions have also highlighted how frontier molecular orbitals and the electronic information within them can be used to predict trends in chemical reactivity. This ability to predict chemical electrophilicity and thus reactivity has been demonstrated to be extremely useful in the prediction of reactive toxicity.

7.9 Acknowledgements

The funding of the European Union 6th Framework CAESAR Specific Targeted Project (SSPI-022674-CAESAR) and the European Chemicals Agency (EChA) Service Contract No. ECHA/2008/20/ECA.203 is gratefully acknowledged.

References

1. J. A. Vonk, R. Benegni, M. Hewitt, M. Nendza, H. Segner, D. van de Meent and M. T. D. Cronin, *ATLA*, 2009, **37**, 557.
2. D. W. Roberts, G. Patlewicz, P. S. Kern, F. Gerberick, I. Kimber, R. J. Dearman, C. Ryan, D. A. Basketter and A. O. Aptula, *Chem. Res. Toxicol.*, 2007, **20**, 1019.
3. A. O. Aptula, G. Patlewicz and D. W. Roberts, *Chem. Res. Toxicol.*, 2005, **18**, 1420.

4. T. W. Schultz, G. D. Sinks and M. T. D. Cronin, in *Quantitative Structure-Activity Relationships in Environmental Sciences VII,* ed. F. Chen and G. Schuurmann, SETAC, Florida, 1997, p. 329–337.

5. H. J. M. Verhaar, C. J. van Leeuwen and J. L. M. Hermens, *Chemosphere,* 1992, **25**, 471.

6. C. L. Russom, S. P. Bradbury and S. J. Broderius, *Environ. Toxicol. Chem.,* 1997, **16**, 948.

7. A. O. Aptula and D. W. Roberts, *Chem. Res. Toxicol.,* 2006, **19**, 1097.

8. T. W. Schultz, R. E. Carlson, M. T. D. Cronin, J. L. M. Hermens, R. Johnson, P. J. O'Brien, D. W. Roberts, A. Siraki, K. B. Wallace and G. D. Veith, *SAR QSAR Environ. Res.,* 2006, **17**, 413.

9. W. Koch and M. C. Holthausen, *A Chemist's Guide to Density Functional Theory,* Wiley-VCH, Weinheim and Chichester, 2000.

10. J. B. Foresman and A. Frisch, *Exploring Chemistry with Electronic Structure Methods,* Gaussian, Pittsburgh, PA, 1996.

11. A. R. Leach, *Molecular Modelling: Principles and Applications,* Pearson Education, Harlow, 2001.

12. M. T. D. Cronin, A. O. Aptula, J. C. Duffy, T. I. Netzeva, P. H. Rowe, I. V. Valkova and T. W. Schultz, *Chemosphere,* 2002, **49**, 1201.

13. S. J. Enoch, M. T. D. Cronin, T. W. Schultz and J. C. Madden, *Chemosphere,* 2008, **71**, 1225.

14. R. Purdy, *Sci. Total Environ.,* 1991, **109/110**, 553.

15. M. T. D. Cronin, B. W. Gregory and T. W. Schultz, *Chem. Res. Toxicol.,* 1998, **11**, 902.

16. R. Benigni, L. Conti, R. Crebelli, A. Rodomonte and M. R. Vari, *Environ. Mol. Mutagen.,* 2005, **46**, 268.

17. R. Benigni, A. Giuliani, R. Franke and A. Gruska, *Chem. Rev.,* 2000, **100**, 3697.

18. D. R. Roy, R. Parthasarathi, V. Subramanian and P. K. Chattaraj, *QSAR Combi. Sci.,* 2006, **25**, 114.

19. T. I. Netzeva, A. O. Aptula, E. Benfenati, M. T. D. Cronin, G. Gini, L. Lessigiarska, U. Maron, M. Vracko and G. Schuurmann, *J.Chem. Inf. Model.,* 2005, **45**, 106.

20. D. R. Roy, R. Parthasarathi, B. Maiti, V. Subramanian and P. K. Chattaraj, *Bioorg. Med. Chem.,* 2005, **13**, 3405.

21. S. J. Enoch, M. T. D. Cronin, T. W. Schultz and J. C. Madden, *Chem. Res. Toxicol.,* 2008, **21**, 513.

22. G. Schuurmann, in *Predicting Chemical Toxicity and Fate,* ed. M. T. D. Cronin and D. J. Livingstone, Taylor and Francis, London, 2004, p. 85–149.

23. M. G. Moloney, *Structure and Reactivity in Organic Chemistry,* Blackwell, 2008.

24. C. K. M. Heo and J. W. Bunting, *J. Org. Chem.,* 1992, **57**, 3570.

25. T. W. Schultz and J. W. Yarbrough, *SAR QSAR Environ. Res.,* 2004, **15**, 139.

26. M. T. D. Cronin, in *Quantitative Structure-Activity Relationship (QSAR) Models of Mutagens and Carcinogens,* ed. R. Benigni, CRC Press, Boca Raton, FL, 2003, p. 235–282.

27. A. K. Debnath and C. Hansch, *Environ. Mol. Mutagen.*, 1992, **20**, 140.

28. A. K. Debnath, R. L. L. Decompadre, A. J. Shusterman and C. Hansch, *Environ. Mol. Mutagen.*, 1992, **19**, 53.

29. A. K. Debnath, R. L. L. Decompadre, A. J. Shusterman and C. Hansch, *Environ. Mol. Mutagen.*, 1992, **19**, 37.

30. R. G. Parr, R. A. Donnelly, M. Levy and W. E. Palke, *J. Chem. Phys.*, 1978, **68**, 3801.

31. R. G. Parr and R. G. Pearson, *J. Am. Chem. Soc.*, 1983, **105**, 7512.

32. R. G. Parr, L. V. Szentpaly and S. Liu, *J. Am. Chem. Soc.*, 1999, **121**, 1922.

33. R. G. Parr and W. Yang, *J. Am. Chem. Soc.*, 1984, **106**, 4049.

34. L. R. Domingo, M. J. Aurell, P. Perez and R. Contreras, *J. Phys. Chem. A*, 2002, **106**, 6871.

35. L. R. Domingo, P. Perez and R. Contreras, *Tetrahedron*, 2004, **60**, 6585.

36. S. J. Enoch, D. W. Roberts and M. T. D. Cronin, *Chem. Res. Toxicol.*, 2009, **22**, 1447.

37. M. T. D. Cronin and T. W. Schultz, *Chem. Res. Toxicol.*, 2001, **14**, 1284.

38. K. Fukui, T. Yonezawa and C. Nagata, *Bull. Chem. Soc. Jpn.*, 1954, **27**, 423.

39. M. T. D. Cronin, N. Manga, J. R. Seward, G. D. Sinks and T. W. Schultz, *Chem. Res. Toxicol.*, 2001, **14**, 1498.

40. T. W. Schultz, *Chem. Res. Toxicol.*, 1999, **12**, 1262.

41. S. Karabunarliev, O. G. Mekenyan, W. Karcher, C. L. Russom and S. P. Bradbury, *Quant. Struct.-Act. Relat.*, 1996, **15**, 302.

42. S. Karabunarliev, O. G. Mekenyan, W. Karcher, C. L. Russom and S. P. Bradbury, *Quant. Struct.-Act. Relat.*, 1996, **15**, 311.

43. T. W. Schultz, J. W. Yarbrough, R. S. Hunter and A. O. Aptula, *Chem. Res. Toxicol.*, 2007, **20**, 1359.

44. M. Karelson, V. S. Lobanov and A. R. Katritzky, *Chem. Rev.*, 1996, **96**, 1027.

45. S. J. Enoch, J. C. Madden and M. T. D. Cronin, *SAR QSAR Environ. Res.*, 2008, **19**, 555.

46. K. van Leeuwen, T. W. Schultz, T. Henry, B. Diderich and G. D. Veith, *SAR QSAR Environ. Res.*, 2009, **20**, 207.

47. A. O. Aptula, S. J. Enoch and D. W. Roberts, *Chem. Res. Toxicol.*, 2009, **22**, 1541.

CHAPTER 8

Three-Dimensional Molecular Modelling of Receptor-Based Mechanisms in Toxicology

J. C. MADDEN AND M. T. D. CRONIN

School of Pharmacy and Chemistry, Liverpool John Moores University, Byrom Street, Liverpool L3 3AF, UK

8.1 Introduction

The interaction between a toxicant and an organism (human or environmental species) may be relatively easy to describe using simple descriptors such as those for global hydrophobicity or molecular size. These interactions include, for example, generalised perturbation of cell membranes inducing narcosis, or direct chemical reaction leading to irritation. The generation and use of descriptors relating to physico-chemical properties and molecular topology are discussed in Chapters 5 and 6 respectively. This aim of this chapter is to discuss the importance of considering three-dimensional (3-D) factors in the prediction of certain toxicological endpoints (*e.g.* receptor-mediated effects) that are not amenable to description with standard two-dimensional (2-D) descriptors alone.

This chapter introduces the nature of 3-D descriptors, *i.e.* those associated with the spatial orientation of the molecular features (steric requirements, lipophilicity and electrostatic interactions) of a chemical and their use in predicting toxicity. Four-dimensional (4-D) and higher-dimensional descriptors used in quantitative structure–activity relationships (QSAR) accounting for not only the 3-D structure of the molecule but also different possible conformations,

Issues in Toxicology No.7
In Silico Toxicology: Principles and Applications
Edited by Mark T. D. Cronin and Judith C. Madden
© The Royal Society of Chemistry 2010
Published by the Royal Society of Chemistry, www.rsc.org

protonation states, tautomeric forms, stereoisomers and induced-fit models are not explicitly dealt with here, though it is recognised that ligands and targets are flexible, complex structures with adaptable conformations.[1]

The approaches discussed in this chapter are applicable to modes and mechanisms of action where toxicity is induced by a complex interaction between a chemical and a biological regulatory molecule (*e.g.* an enzyme, receptor or DNA molecule). Toxicity resulting from such interactions often relates to low dose, chronic effects with endocrine disrupting effects being a good example. Other effects, however, can be acute such as rapid and permanent inhibition of acetylcholinesterase (see Chapter 14).

Chronic effects may be difficult to measure and may require long-term experimentation which is costly in terms of expense and animal use. Prediction of such toxicities is therefore important, but the nature of the complex biochemical mechanisms involved also makes these much more difficult endpoints to model. High quality data, particularly binding affinity data, are vitally important here from which to develop models. Note that a substance may bind productively, non-productively, inhibit the activity of the target or prevent binding of an endogenous ligand. Endpoint data themselves may not always be useful for several reasons:

- a specific endpoint may be the result of more than one pathway;
- in certain cases (*e.g.* foetal exposure), the timing of the toxic insult may lead to very different toxic outcomes or no observed toxicity; and
- factors affecting external exposure, bioavailability or metabolism of the parent may give misleading information on potential toxicity, *i.e.* the toxic effect may be elicited by a metabolite with the parent compound being unable to bind to the receptor.

The descriptors employed in predicting receptor-mediated effects may include simple representations of molecular features such as structural alerts or fragment-based approaches. More complex and computationally expensive methods that can be used include 3-D QSAR, pharmacophore generation and docking approaches.

The use of 3-D approaches to predict toxicity is a highly complex, rapidly evolving area of interest. It is fortunate that the importance of specific 3-D interactions has long been recognised in the field of drug design. Hence, there is a range of both established and newer technologies that can be usefully employed in predicting toxicological endpoints. This chapter introduces some of these methods and provides examples of endpoints, such as endocrine disruption, hERG channel binding, affinity for cytochrome P450 enzymes, *etc.* where the application of 3-D approaches has been shown to be successful.

8.2 Background to 3-D Approaches

Simple QSAR approaches may be limited to the consideration of global features of molecules, or sub-structures thereof; for example representing the partition coefficient of the whole molecule (log P) or the contribution to hydrophobicity of an individual substituent (π). Two-dimensional approaches are similarly restricted in that they provide only limited information on the overall size and shape of molecules. In reality, molecules such as toxicants are 3-D entities and, to predict toxicity relating to binding within specific regions of biological macromolecules (*e.g.* receptors), it is essential to consider these interactions at the 3-D level. Conceptually it is easier to understand the effects and relative potencies of xenobiotics if it can be visualised how changes in structure can change the interaction at the receptor site. Hence 3-D modelling is a promising approach to solving receptor-based problems in toxicology and understanding the mechanisms involved. The approaches described in this chapter have generally been derived from drug design, where they are accepted methods to predict receptor-binding phenomena.

Traditional approaches to considering the 3-D properties of a molecule were based on the application of calculated grids or fields of interaction around it. A molecule could be placed (computationally) within a 3-D grid, and points of interaction between the molecule and the grid could be translated into individual descriptors which could then be incorporated into regression analysis. Such techniques are widely employed and integrate statistical methodologies, such as partial least squares (PLS), to make the high number of potential descriptors more manageable (see Section 8.4.1).

A range of probe molecules with different hydrophobic or hydrophilic moieties can be 'rolled' across the surface of the molecule in order to provide information not just on the spatial orientation of the molecule, but also on the organisation of different molecular features (*e.g.* the water accessible surface area or regions of hydrophobic interaction). This allows more detail to be gained of region-specific properties of molecules, giving a more accurate estimation of binding interactions.

The development of the Comparative Molecular Field Analysis (CoMFA) program in 1988 was a breakthrough in the use of 3-D QSAR.[2] CoMFA is an example of a ligand-based approach, *i.e.* knowledge concerning the receptor is surmised from information obtained from known ligands. The information relates to the 3-D shape of the ligand and its potential for electrostatic and hydrophobic interaction with complementary sites on the receptor. The methods require information (*e.g.* binding affinities) for an appropriate number of substrates for which the biologically relevant conformation can be generated. Interactions are highly conformation dependent; determining the correct conformation for each substrate can be computationally expensive and can be a source of error in generating models. Ideally the range of activities should cover at least three orders of magnitude.

The converse of ligand-based approaches is receptor-based approaches–also referred to as structure-based or target-based approaches. These require structural information on the receptor; potential ligand interactions can then be determined by 'fitting' ligands to the known receptor. X-ray crystallography, nuclear magnetic resonance (NMR) and electron microscopy can be used to determine the 3-D structure of proteins. Unfortunately, many proteins are not compatible with these techniques (due to problems with size, stability, solubility, *etc.*). Therefore relatively few 3-D structures are available for receptors of interest; this limits the application of these methods. However, proteins with common amino acid sequences tend to form similar 3-D structures. This can be exploited in homology (comparative) modelling, in which the known structure of one protein is used to predict the structure of another with a similar amino acid sequence. The number of resolved 3-D structures for proteins is steadily increasing. This provides increasing opportunity for the development of models based on the true receptor or on a homology model.

Three-dimensional approaches have been used successfully in drug development for many years and the number of 3-D QSAR studies reported in the literature has risen significantly, particularly in the past decade. This undoubtedly reflects the expansion of computational power and the development of myriad approaches to incorporating 3-D information into predictions of biological interaction. However, it is only recently that the bespoke application of these methods for toxicology has been considered (see Section 8.6.3). It is not possible here to provide a complete overview of these approaches in detail (readers are referred to recent literature for further information[1,3,4]). Some examples of computational approaches and their application to solving problems in receptor-mediated toxicology are given in the following sections.

8.3 Modelling Approaches for Receptor-Mediated Toxicity

8.3.1 Rejection Filters, Structural Alerts and Fragment-Based Approaches

The main focus of this chapter is on 3-D approaches for predicting toxicity. However, these are often computationally expensive and therefore not suitable for preliminary screening of large databases of potential toxicants. In such cases, less refined screens may be useful in rejecting those unlikely to elicit an effect or in selecting those of potential concern for further evaluation.

At the most basic level, cut-off values for physico-chemical properties have been proposed for classification purposes (*e.g.* for use as rejection filters). Examples include:

- classifying toxicants with log P values >4.5 as having the potential to bioaccumulate;[5]
- using Lipinski's 'rule of fives' to identify compounds likely to show poor absorption characteristics and therefore possibly low bioavailability *in vivo*;[6] and
- using molecular weight cut-off values of less than 94 or greater than 1,000 Da as limits beyond which compounds are unlikely to show oestrogen receptor binding[7] (see Section 8.6.2).

At a more refined level, structural alerts may be used to identify compounds likely to elicit a particular toxicity. Structural alerts are key 2-D fragments associated with an activity or toxicity, and identification of such a feature in a compound may indicate the potential for that compound to elicit the associated effect. However, the overall likelihood of the toxic event occurring can be modulated by other features of the molecule. For example, bulky substituents adjacent to an important functional group may sterically hinder interaction; elsewhere the parent molecule may be insoluble in the relevant biophase or metabolically unstable, hence obviating the effect *in vivo*. Tong *et al.*[7] successfully used structural alerts as part of their 'four-phase' system to predict oestrogen receptor binding.

There are cases where identification of structural alerts associated with toxicity have been formalised into a range of expert systems software. For example the Derek software from Lhasa (updated to Derek Nexus, https://www.lhasalimited.org) is an expert knowledge base system that applies structure–activity relationships and other expert knowledge rules to derive a reasoned conclusion about the potential toxicity of a query chemical. Expert systems have been demonstrably successful in predicting certain toxicity endpoints. They are discussed in greater detail in Chapter 19 and therefore are not considered further here.

Small 2-D molecular fragments are seldom sufficient to model the complex processes of receptor binding. They may, however, account for the covalent (permanent) binding of a chemical within a receptor that may result in inhibition of activity (see Chapters 7, 13 and 24).

There are other examples of using 2-D QSAR descriptors to predict receptor-binding phenomena, particularly within distinct categories of compounds. For example, in a study of 120 aromatic chemicals, Schultz *et al.*[8] identified three structural criteria related to xenoestrogen activity and potency:

(i)   the hydrogen bonding ability of the phenolic ring mimicking the A ring of a steroid structure;
(ii) a hydrophobic centre of similar shape and size to the B and C rings; and
(iii) a hydrogen bond donor mimicking the hydroxyl of the D ring.

Structure–activity relationships for individual chemical classes demonstrated that certain features are associated with activity for one group of compounds, whereas different features were associated with activity for other groups. For example, within the phenols class, the relative activity was dependent on the bulk of the non-phenolic moiety. Previous studies have also shown the phenolic hydroxyl moiety facilitates binding via hydrogen bond donor ability. This is affected by the nature of *ortho*-substituent(s); hence the use of relatively simple parameters can be rationalised in terms of what is known about oestrogen receptor binding and can aid prediction of potency.

Fragments are also applied in the freely available OECD (Q)SAR Application Toolbox (discussed further in Chapter 16).[9] In version 1.1 of the OECD (Q)SAR Toolbox DNA binding is profiled using 19 categories, each of which is defined by 2-D structural alerts that are necessary for covalent interaction with DNA (readers should note that the DNA binding profiler within the Toolbox is being updated and expanded in a new version). Oestrogen receptor binding is profiled using a series of alerts derived from the 'four-phase' scheme of Tong *et al.*[7] Justification for the inclusion of each of the structural fragments in the profilers is based on evidence from the literature, references for which are given in the guidance notes accompanying the Toolbox.[10]

These relatively trivial approaches for capturing molecular phenomena associated with receptor binding have some utility in identifying potential binders, eliminating non-binders and assisting in the formation of categories of molecules for further of investigation and possible read-across. They have the advantage of being able to screen large databases rapidly, but lack the sophistication and detail of 3-D models.

8.3.2 Introduction to 3-D Methods

The advantage of 2-D approaches to predicting toxicity is that a representation of the molecule's structure is sufficient to generate the descriptors. Three-dimensional molecular descriptors are calculated from the 3-D structure of the compound and, as such, are much more computationally demanding.

Molecular models need to be built and an energy minimised conformation generated. Descriptors generated from these structures are critically dependent on the conformation of the molecule used. Determining a reasonable conformation on which to base the estimations can be problematic; procedures to obtain minimum energy conformations may result in a local minimum and many interactions are based on induced-fit mechanisms; hence the conformation generated *in silico* may not be representative of the biologically relevant conformation *in vivo*. The computational expense of generating minimum energy conformations can be prohibitive if large data sets are of interest. Molecular mechanics approaches can be used to generate an approximate conformation relatively rapidly; more computationally time-consuming calculations based on molecular orbital theory can then be used to optimise the conformations (Chapter 7 covers the generation and use of parameters obtained from molecular orbital theory).

Three-dimensional QSAR relies on the alignment of the ligands and the assumption that they interact with the same regions of the receptor. The alignment of ligands may be driven by knowledge of the ligands themselves or of the target biomolecule. The problems of aligning diverse, flexible structures and consideration of which conformation is the true biologically active conformer are discussed in Section 8.4.5.

The results of the geometry optimisation process are typically stored in molecular spreadsheets. These allow for the import, export, manipulation and storage of results for calculations of additional descriptors, alongside the structural information for the compound. The following sections provide an introduction to the application of 3-D modelling approaches to predict toxicity for complex endpoints.

8.3.2.1   Molecular Dimensions

Descriptors for molecular dimensions can be used to describe several key features of molecules in 3-D space. These are important as it is often the distance between specific atoms or functional groups within a molecule that correlates with their ability to elicit a response. Overall dimensions (*e.g.* the maximum or minimum diameter for a compound) can also determine specific fit within, and binding to, a receptor. Global size and shape characteristics have been proposed as cut-offs for certain toxicological endpoints (*i.e.* exceeding a particular size) may render the chemical unable to pass through the relevant physiological membrane and is therefore associated with lack of toxicity.

Such factors were considered above for simple descriptors, but the same is applicable to size and shape characteristics determined in three dimensions; this information can be used alongside other cut-off values. For example, indicators for limited bioaccumulation potential have been developed as part of the REACH guidance, wherein (with appropriate caveats) a chemical may be considered as potentially not bioaccumulative if the following factors are considered the average maximum diameter (Dmax aver) is greater than 17 Å; the molecular weight is greater than 1,100 (Da); the maximum molecular length is greater than 43 Å (4.3 nm); and the logarithm of the octanol–water partition coefficient is greater than 10.[11]

A recent report for the Environment Agency into the use of molecular dimensions by Brooke and Cronin compared[11] a range of different packages capable of computing molecular dimensions.[11] The packages considered were:

- OASIS: http://oasis-lmc.org
- MOE: www.chemcomp.com
- TSAR: http://accelrys.com/products/informatics/
- Mol2Mol: www.gunda.hu/mol2mol/index.html
- SPARTAN: www.wavefun.com/products/spartan.html

The report includes comments on the general performance and ease of use of the packages, together with details of the various descriptors that can be generated and their interpretation in terms of modelling. The descriptors include maximum and minimum diameters; ellipsoidal volumes; moments of inertia; distances between atoms; and molecular surface area and others. These descriptors can be employed not only to provide cut-off values in terms of the overall likelihood of toxicological effect, but can also provide information on the potential of specific atoms or groups (*e.g.* hydrogen bond donor/acceptor groups) to bind to the corresponding site of a target molecule. These calculations bridge the gap from 2-D to 3-D as they include a consideration of the different conformers of molecules which is, in itself, an important point.

The report demonstrates that very different values for descriptors can be obtained where different, energetically feasible, conformers are generated. This suggests that a number of issues must be borne in mind in using 3-D descriptors. First, when investigating the 3-D structure of (large) flexible molecules, conformational analysis must be considered. Secondly, that as dimensions of the different conformers varied widely, it would be safe to assume that other properties would also vary in a similar manner. This has implications, for example, when using a single lowest energy conformation—such as in quantum chemical calculations (discussed in Chapter 7).

If descriptors vary according to the conformer studied, a strategy must be developed for using the various values obtained, *i.e.* should the lowest, highest or median value be used? This decision may be influenced by the purpose for which the model is developed, *i.e.* does the precautionary principle need to be applied as in risk assessment scenarios or is there scope for more tolerance where the models are designed for general screening purposes? Despite these caveats, molecular dimensions can provide useful information concerning the 3-D structure of compounds that may correlate to toxic potential.

8.3.2.2 COMPACT

The Computer-Optimised Parametric Analysis of Chemical Toxicity (COMPACT) approach has been used to evaluate toxicity potential of chemicals in terms of identifying those which act as substrates for cytochrome P450 (and are thereby metabolically activated to a toxicant, *i.e.* potential carcinogenic activity) or which interact with the aryl hydrocarbon receptor.[12] A compound only interacts with cytochrome P450 enzymes if it has the correct shape and electronic distribution. The approach identifies molecular conformation and electronic structure characteristics necessary for interaction with the enzyme/ receptor. This information is then used to predict the likelihood of interaction and therefore the toxicity of the compound. This approach is discussed further in Chapter 19.

8.4 Ligand-Based Approaches

8.4.1 GRID

The GRID concept was originally developed by Goodford[13] to generate contour surfaces at given energy levels to indicate sites of potential interaction between a ligand and its target. The GRID program, now marketed by Molecular Discovery (www.moldiscovery.com) is used to determine energetically favourable binding sites on molecules of known structure (*i.e.* where the 3-D crystal structure of a target is known) by generating molecular interaction fields (MIFs). The output from GRID contains a large number of variables; thus combination with a program such as GOLPE (Generating Optimal Linear

PLS Estimations, www.miasrl.com/golpe.htm) can be used to reduce the data matrix.

In terms of drug design, GRID and related approaches have been used to design ligands with appropriate features and so generate new leads.[14] Within toxicology, the approach could be used to determine likely binding interaction between a potential toxicant and its receptor. Surprisingly, there have been relatively few illustrations of its use in toxicology, with it being applied mainly in drug design. However, there is potential for this method to assist in the modelling of receptor-based effects.

8.4.2 Comparative Molecular Field Analysis (CoMFA) and Comparative Molecular Shape Indices Analysis (CoMSIA)

The CoMFA software marketed by Tripos (www.tripos.com) is probably the most commonly implemented ligand-based 3-D QSAR method. As mentioned above, ligand-based approaches rely on information obtained from a set of aligned ligands (*e.g.* molecules from a database aligned to a template molecule) based on a common substructure. Steric and electrostatic interaction energies are calculated using the interaction of a probe atom with each molecule on the points of a grid which surrounds the molecule. The grid is designed to be large enough to enclose all superimposed ligands in all directions.

CoMFA uses a Lennard-Jones function for steric interaction potential and a Coulomb function for electrostatic potential, whereas CoMSIA uses a Gaussian function to generate steric, electrostatic, hydrophobic and hydrogen bond donor or acceptor fields. Results can be viewed as contour maps identifying regions of favourable or unfavourable interactions. The differences in steric and electrostatic interaction fields generated for each ligand should correlate with the relative biological activity of that ligand, enabling visualisation of the factors affecting potency. Integration with other applications (*e.g.* the QSAR with CoMFA software from Tripos) enables information to be stored in molecular spreadsheets from which predictive models may be built.

There has been relatively wide application of the CoMFA technology in receptor-based toxicological prediction, especially in the area of endocrine disruption. Cronin and Worth report a number of CoMFA models for oestrogenic receptor binding and several models for androgenic receptor binding, noting that there were a larger number not considered in the report.[15]

The CoMFA approach is a promising tool for investigating receptor-based interactions in toxicology but, due to its reliance on alignment of optimised structures, is computationally expensive.

8.4.3 VolSurf and Almond

The VolSurf software marketed by Tripos (www.tripos.com) reads or generates 3-D molecular interaction fields which it then converts into simple descriptors to quantify size, shape, volume, surface/volume ratio, polarity and hydrophobicity of molecules, *etc.* Various probe atoms can be used to generate GRID molecular fields. The models do not require molecular alignment; therefore the descriptors generated are not sensitive to alignment rules. These descriptors are chemically intuitive and have found application in the prediction of absorption, distribution, metabolism and excretion (ADME) related properties of both drugs and toxicants alike. The importance of incorporating ADME properties into overall predictions of toxicity is discussed further in Chapter 21.

The software contains models for predicting solubility, absorption, blood–brain barrier permeation and other ADME properties. As with many existing models, these were largely developed for the purposes of drug design and consideration needs to be given to the chemical space of the models with respect to their application for industrial chemicals.

The Almond software, also marketed by Tripos, similarly generates 3-D descriptors without the need for molecular alignment, making the modelling process more rapid. The descriptors represent the internal geometrical relationships of pharmacophoric regions of the ligands and are therefore useful for predicting specific ligand-receptor interactions important in determining activity.

A GRID force field is used to produce molecular interaction fields, the information from which is transformed to generate descriptors that are independent of the molecule's location within the grid—GRid-INdependent Descriptors (GRIND). Descriptors containing redundant information are removed. The program also generates shape descriptors, generated from a shape probe that can be used to investigate ligand-receptor shape complementarity when developing QSAR models.

The software contains many other features for the development of 3-D QSAR models and can be readily integrated with other software.

8.4.4 Hologram QSAR (HQSAR)

HQSAR is also marketed by Tripos. The software uses a combination of molecular holograms and PLS to generate structure–activity relationships. Hologram QSAR is a technique whereby fragments of molecules are arranged to form a molecular hologram, such that 3-D information is implicitly encoded from input 2-D structures. Molecular holograms are extensions of 2-D fingerprints (including information on branched and cyclic fragments and stereochemistry). HQSAR encodes all possible fragments and constituent subfragments, and therefore encodes region-specific information on fragments (*i.e.* this can detect the influence of moving a substituent molecule around a ring system).

As the analysis does not require alignment of molecules or conformational analysis, it is much faster and therefore applicable to larger datasets than other

3-D approaches. This approach appears to be little used in toxicology, but could offer the possibility of screening large databases and inventories rapidly.

8.4.5 Pharmacophore Modelling

A pharmacophore is a 3-D representation of the arrangement of key features of a molecule, indicating the relative spatial orientation between them. Key features may include hydrogen bond donor or acceptor groups, hydrophobic regions, presence of positively or negatively charged or ionisable groups, or excluded volumes (*i.e.* regions where the presence of any group hinders binding/ activity possibly due to steric constraints at the target site). Pharmacophores are not real molecular entities, but indicate the association of functional groups in 3-D.

The common (pharmacophoric) features shared by a set of ligands should interact with complementary sites on the biological target. Molecular alignment is performed for a set of energy-minimised ligands that interact with the same biological target. As the structures are potentially diverse this alignment can be problematic, *i.e.* it may be possible to align some of the features but not all, and it may not be known *a priori* which are the most important to align. Relative biological activities or binding energies of ligands can be included into the analysis to give a weighting, taking account of optimal interactions.

As with most modelling techniques, a reasonable range of activity values will provide a better model. Whilst 'inactives' can potentially be included, this should be done with caution as inactivity may be due to many reasons not related to pharmacophore fit (*e.g.* metabolism of parent or inability to reach the target site). Identifying features likely to be important in binding to the target (and their spatial orientation) allows prediction of toxicity or ranking of possible order of toxicity for unknown compounds.

Development of a pharmacophore is computationally expensive as it requires energy minimisation and conformational analysis of flexible ligands as well as alignment of key molecular features; these may be aligned in many ways and must then be matched to the various conformers using least squares fitting.[3] As such this is a relatively 'low throughput' approach.

Use of pharmacophores is particularly useful if the target's structure is unknown. A potential structure (pseudo-receptor) can be surmised from corresponding interaction with the elements of the pharmacophore. Conversely if the target is known, then a pharmacophore can be derived from this.

Pharmacophore generation programs include:

- CATALYST (DS Catalyst): http://accelrys.com
- GALAHAD: www.tripos.com
- GASP: www.tripos.com
- PHASE: www.inteligand.com
- MOE: www.chemcomp.com

Field-based pharmacophores use alignment based on the molecular fields exhibited by the molecule rather than assigning features to functional groups; this represents what the target receptors 'sees' in terms of charge distribution and shape rather than the underlying structural skeleton.[3]

Pharmacophores have been utilised for oestrogen receptor binding but few other toxicological endpoints. This is certainly a technique that could be used more widely, as it is clear and transparent and can be related to specific toxicological mechanisms of action.

8.5 Receptor-Based approaches

These techniques rely on using information from known receptor structures to determine interaction with potential ligands. As mentioned previously, this is problematic as the techniques used to determine 3-D structure (X-ray crystallography, NMR, electron microscopy) are not amenable to the study of many proteins, limiting the number of known structures available.

The Protein Data Bank (www.pdb.org) contains a repository of experimentally determined protein structures that is continually being updated. Where the structures are known, several programs are available that enable docking of ligands into the receptor site to determine the ligand–receptor interactions. Such programs typically use scoring functions to quantify protein–ligand interactions relating to consideration of steric, polar, entropic and solvation contributions. Examples of software that can be used for docking ligands in protein targets include:

- Surflex-Dock: www.tripos.com
- FlexX: www.biosolveit.de/software/
- Glide: www.schrodinger.com
- GOLD: www.ccdc.cam.ac.uk/products/life_sciences/gold/
- AutoDock: http://autodock.scripps.edu/
- FRED: www.eyesopen.com/products/applications/fred.html

Whilst application of this method is currently limited, the number of receptors for which the structure has been ascertained is increasing. Consequently, these methods will become increasingly important in predicting specific receptor-mediated toxicity in future.

8.5.1 Homology (Comparative) Modelling

Lack of 3-D structural information is a limiting factor in receptor-based methods. However, this has been in part overcome by the use of homology modelling. The concept, introduced above, is based on the premise that proteins with similar amino acid sequences will adopt similar 3-D structures; hence structural identity of one protein can be used to predict the structure of another related protein. This is another example of a technology with a background in drug design that has found successful application in predictive toxicology.

Three-dimensional protein structures that have been determined or computationally predicted can be stored and shared with other modellers, leading to a rapid expansion in available models. Homology models are developed iteratively, *i.e.* as more protein structures are experimentally determined, these can be used to improve the homology models.

The SWISS-MODEL (http://swissmodel.expasy.org) is an automated protein structure homology-modelling server. The SWISS-MODEL repository developed by the Swiss Institute of Bioinformatics provides a very useful resource in the form of a database of protein structure models generated by the automated homology-modelling pipeline SWISS-MODEL and is updated regularly. The Protein Model Portal (http://proteinmodelportal.org) and ModBase (http://modbase.compbio.ucsf.edu) also contain protein models.[16]

As with receptor-based models, homology modelling is being increasingly employed in prediction of toxic effects as more 3-D protein structures are resolved. Selected examples of the application of these approaches are discussed below.

8.6 Examples of the Application of 3-D Approaches in Predicting Receptor-Mediated Toxicity

8.6.1 Homology Models

The cytochrome P450 system for metabolism has been studied intensely in terms of drug development, partly because of concerns of drug–drug interactions. In such cases co-administration of two or more drugs with affinity for the same metabolising enzymes can result in a prolongation of effect or bioaccumulation (and therefore potential toxicity) because each drug competes with the other(s) for the enzyme. Enzymes can also be responsible for the metabolism of innocuous parent compounds into toxic metabolites. For example, the cytochrome P450 isoforms CYP1A1, CYP1A2, CYP2A6, CYP2B1 and CYP2E1 catalyse the activation of pro-carcinogenic environmental pollutants into carcinogenic species.[17] Prediction of metabolism is therefore of great interest in predicting toxicity overall. De Groot reported on developments in predicting metabolism of specific compounds by cytochrome P450 enzymes.[18] References are given for pharmacophore models for human P450s and the available crystal structures for bacterial, fungal and human P450s. These resolved structures can be used to develop homology models, allowing for the prediction of the affinity of ligands with P450 enzymes.

The human ether-a-go-go-related gene (hERG) potassium channel has received much attention of late. This is due to the potential of certain compounds to block the hERG channel causing prolongation of the cardiac QT interval that may result in sudden death. This has led to the high profile withdrawal of certain drugs. Prediction of hERG channel blockade has therefore become a priority in toxicity prediction. Farid *et al.* created a homology model of the homo-tetrameric pore domain of hERG using the crystal structure of the bacterial potassium channel, KvAP, as a template.[19] This enabled docking of known channel blockers into the model providing a means to investigate the interactions involved. Molecular docking and molecular simulations have also been investigated by Du *et al.*, who reported on a model for the prediction of hERG channel blockade of potential use on evaluating cardiac liability of new compounds.[20] A review of *in silico* receptor-based and ligand-based studies for the investigation of ligand-hERG interactions has also been published.[21]

2,3,7,8-Tetrachlorodibenzo-*p*-dioxin (TCCD) and similar compounds of environmental concern have been associated with tumour promotion, dermal toxicity, immunotoxicity, developmental and reproductive toxicity, and enzyme inhibition. TCCD is suspected of interacting with the aryl hydrocarbon receptor (AhR). A homology model of AhR has also been developed to investigate the interaction of TCCD and similar compounds with this receptor.[1]

Nuclear receptors mediate the effects of endogenous ligands (*e.g.* hormones) to regulate gene expression; as such they are involved in a wide range of physiological regulatory processes, disruption of which can have manifold effects. The oestrogen receptor (ER) is the most studied of these receptors as it is recognised that many exogenous compounds (*e.g.* natural products and synthetic chemicals) may act as substrates for these receptors and therefore disrupt physiological function. The effects can influence homeostasis of many underlying processes in humans and environmental species including reproductive effects, developmental toxicity, immune response, neurological effects and susceptibility to cancer. Chemicals with the capacity to interact with the ER and other nuclear receptors are referred to as endocrine-disrupting chemicals (EDCs). The potential toxicity of these compounds is a highly contentious issue, partly because many suspected EDCs are high production volume, economically important chemicals which have the potential to elicit toxic effects at low doses.[22] The computational approaches applied in studies of nuclear receptor binding have been reviewed by Ai *et al.*[23]

8.6.2 NCTR Four-Phase Approach

In response to concern over the effects of endocrine-disrupting chemicals (EDCs), the United States Environmental Protection Agency (US EPA) set up the Endocrine Disrupting Screening and Testing Advisory Committee (EDSTAC) to evaluate the endocrine disrupting potential of 58,000 chemicals, most of which have no biological data (originally the figure was 87,000 but many of these were polymers and therefore excluded). The National Center for Toxicological Research (NCTR) devised a 'four-phase' approach to identify compounds with the greatest potential to elicit endocrine disrupting effects with the aim to prioritise testing.[7] In order to develop models, a robust training set of 232 chemicals was obtained using a validated rat oestrogen receptor (ER) binding assay.

The four-phase system involved consideration of the following factors:

(i) Rejection filters: Compounds were required to have a molecular weight between 94 and 1,000 Da and to possess at least one ring structure.

(ii) Compounds were assigned as being likely to be active/inactive based on a series of criteria. Three structural alerts (phenolic ring, steroid structure and diethylstilbestrol) were used to detect likely actives, along with seven pharmacophore models (using the bound ligand–oestrogen receptor crystal structure of known binders as a template) and a decision tree classification. Only compounds classified as inactive by all methods were rejected at this stage.

(iii) The CoMFA software was used to predict binding affinity quantitatively; compounds with higher binding affinity were considered of higher concern and therefore given higher priority for further evaluation.

(iv) A rule-based decision-making system was devised that used the associated criteria to determine those compounds with higher priority for testing.

This system allowed potential ER binders to be prioritised and the methodology was subsequently extended to consider chemicals with potential to bind to the androgen receptor.

8.6.3 VirtualToxLab

As indicated above, EDCs are of great concern to public and regulatory authorities. Hence, there have been many studies into the prediction of endocrine-disrupting effects. Some of the newer technologies are encompassed within VirtualToxLab from Biograph (www.biograf.ch), which predicts endocrine-disrupting effects of chemicals by simulating their interaction with proteins known or suspected to cause adverse effects. It is based purely on thermodynamic considerations. Currently, VirtualToxLab comprises 12 validated models for the aryl hydrocarbon, oestrogen α/β, androgen, thyroid α/β, glucocorticoid, liver X, mineralocorticoid and peroxisome proliferator-activated receptor γ as well as for the enzymes CYP450 3A4 and 2A13.[1] The program simulates induced fit of ligands to the receptor quantifying the interactions. It takes account of thermodynamically relevant processes such as ligand desolvation energy, entropic cost and binding affinity. Results may indicate a high affinity for an individual target (*e.g.* a strong singular response for AhR or CYP binding) or an overall high toxic potential if the chemical is predicted to bind to several nuclear receptors and therefore display a wide range of endocrine-disrupting effects. Toxic potential can be classified on a scale ranging from benign to low, moderate, high or extreme potential. The technique is very computationally expensive as it requires fitting of flexible ligands to different receptors.

Toxicity prediction is based on interaction with these macromolecules; however, no consideration is given to potential mitigating ADME considerations (*i.e.* the ability of the chemical to reach the target site). It is recommended that the information is linked to predictions for ADME effects for a more accurate estimation of true potential to elicit toxic effects.

Whilst many 3-D approaches may have originally been developed for drug design applications, the examples above demonstrate how this methodology can be usefully applied in predicting receptor-mediated toxicity elicited by other types of compounds. It is anticipated that, despite the current caveats and limitations, there will be an expansion in the application of such techniques in toxicology.

8.7 Advantages and Disadvantages of 3-D Methods

The discussion above has identified several techniques of differing levels of complexity which can be applied to the prediction of receptor-based toxicity. As with all modelling approaches, there are advantages and disadvantages to each method and these should be considered carefully.

Clearly, for all approaches there is a need for high quality data, preferably generated within the same laboratory and ideally covering at least three orders of magnitude. *In vitro* data (for binding affinity) are essential as *in vivo* outcomes can be influenced by ADME properties and do not usually reflect equilibrium conditions; hence this does not provide reliable information solely related to the interaction of the ligand with the receptor.[4]

Simpler 2-D models (*e.g.* structural alerts) can be applied to large data sets, but they do not provide details of 3-D interactions. Conversely 3-D models provide much more specific information but can only be applied to small data sets. Three-dimensional models are also highly reliant on the conformation selected. Conformational analysis is computationally expensive and does not guarantee that the biologically relevant conformation will be generated. A pragmatic approach will be required to use the most appropriate technique for a given problem taking into account computational expense and the degree of accuracy required for a given study.

8.8 Conclusions and Future Outlook

There will be an increasing interest in receptor-based modelling as *in silico* toxicology tackles the more complex, chronic endpoints associated with long-term, low dose exposure. The techniques available for modelling these processes range from relatively trivial filters and alerts to very complex—and computationally expensive—approaches derived from developments in drug design.

The technology will no doubt be implemented in different ways, *e.g.* in terms of predicting toxicity in risk assessment; the 'four-phase' system is a good illustration of this. Simple filters can be used to rapidly screen large databases and inventories. For compounds not rejected at this initial stage, structural alerts and fragments may identify potential hazards, but these techniques can also be used also be used to group molecules together into relevant categories. Once a suspect molecule or a group/category of molecules has been identified, then the more computationally expensive approaches can be implemented (*e.g.* the use of pharmacophore generation packages or CoMFA routines).

The implementation of these approaches will require expert knowledge and modelling skills. This is clearly an area where expertise from those working in drug development can be employed and closer collaborations between the different research areas would be beneficial.

A final word on implementation is likely to be how these methods can be validated. Currently such computational 3-D approaches are not considered by the OECD Principles for the Validation of (Q)SARs nor mentioned in detail in regulatory guidance. However, they offer real possibilities for predicting complex toxicological events and hence their application should be encouraged in future.

8.9 Acknowledgements

The funding of the European Union 6th Framework Programme ReProTect Integrated Project (LSHB-CT-2004-503257) is gratefully acknowledged.

References

1. A. Vedani and M. Smiesko, *Altern. Lab. Anim.*, 2009, **37**, 477.
2. R. D. Cramer III, D. E. Patterson and J. D. Bunce, *J. Am. Chem. Soc.*, 1988, **110**, 5959.
3. A. R. Leach, V. J. Gillet, R. A. Lewis and R. Taylor, *J. Med. Chem.*, 2010, **53**, 539.
4. W. Sippl, in *Recent Advances in QSAR Studies*, ed. T. Puzyn, J. Leszczynski and M. T. D. Cronin, Springer, London, 2010, pp. 103–125.
5. T. Feijtel, P. Kloepper-Sams, K. den Haan, R. van Egmond, M. Comber, R. Heusel, P. Wierich, W. Ten Berge, A. Gard, W. de Wolf and H. Niessen, *Chemosphere*, 1997, **34**, 2337.
6. C. A. Lipinski, F. Lombardo, B. W. Dominy and P. J. Feeney, *Adv. Drug. Deliv. Rev.*, 1997, **23**, 3.
7. W. Tong, H. Fang, H. Hong, Q. Xire, R. Perkins and D. M. Sheehan, in *Predicting Chemical Toxicity and Fate*, ed. M. T. D. Cronin and D. J. Livingstone, CRC Press, Boca Raton, FL, 2004, pp. 285–314.
8. T. W. Schultz, G. D. Sinks and M. T. D. Cronin, *Environ. Toxicol.*, **17**, 14.
9. www.oecd.org/document/23/0,3343,en_2649_34379_33957015_1_1_1_1,00.html [accessed March 2010].
10. www.oecd.org/dataoecd/23/40/41881208.pdf [accessed March 2010].
11. D. N. Brooke and M. T. D. Cronin, Calculation of Molecular Dimensions Related to Indicators for Low Bioaccumulation Potential, Environment Agency, Bristol, 2009, Science Report.
12. D. F. V. Lewis, C. Ioannides and D. V. Parke, *Mutagenesis*, 1990, **5**, 433.
13. P. J. Goodford, *J. Med. Chem.*, 1985, **28**, 849.
14. J. Nilsson and H. Wikstrom, *J. Med. Chem.*, 1997, **40**, 833.

15. M. T. D. Cronin and A. P. Worth, *QSAR Comb. Sci.*, 2008, **1**, 91.
16. C. N. Cavasotto and S. S. Phatak, *Drug Discov. Today*, 2009, **14**, 676.
17. G. R. Wilkinson, in *Goodman and Gilman's: The Pharmacological Basis of Therapeutics*, ed. J. G. Hardman and L. E. Limbird, McGraw Hill, New York, 10th edn 2001, pp. 3–29.
18. M. J. de Groot, *Drug Discov. Today*, 2006, **11**, 601.
19. R. Farid, T. Day, R. A. Friesner and R. A. Pearlstein, *Bioorg. Med. Chem.*, 2006, **14**, 3160.
20. L. Du, M. Li, Q. You and L. Xia, *Biochem. Biophys. Res. Comm.*, 2007, **355**, 889.
21. L. Du, M. Li and Q. You, *Curr. Top. Med. Chem.*, 2009, **9**, 330.
22. H. Hong, H. Fang, Q. Xie, R. Perkins, D. M. Sheehan and W. Tong, *SAR QSAR Environ. Res.*, 2003, **14**, 373.
23. N. Ai, M. D. Krasowski, W. J. Welsh and S. Ekins, *Drug Discov. Today*, 2009, **14**, 486.

CHAPTER 9

Statistical Methods for Continuous Measured Endpoints in In Silico Toxicology

P. H. ROWE

School of Pharmacy and Chemistry, Liverpool John Moores University, Byrom Street, Liverpool L3 3AF, UK

9.1 Continuous Measured Endpoints (Interval Scale Data)

This chapter deals with various regression methods used for the prediction of endpoints described as 'interval scale'. These are endpoints recorded as values that form a scale of measurement with high and low values representing the extremes of behaviour, *e.g.* toxicity or another quantitative measure of potency. The values are also continuous, *i.e.* any value between the highest and lowest is logically possible. Finally, a step upwards of one unit along the scale of measurement is defined in size, *e.g.* the interval between LD_{50} values of 5 and 6 mg is defined precisely.

9.2 Regression Analysis Models in QSAR

For a general introduction to regression analysis see Rowe (2007).[1] Regression models assume there is a linear relationship between a potency outcome and one (or more) chemical descriptors. The methods are illustrated using the toxicity data for phenols that act by polar narcosis (compounds with IDs 101-273 in Appendices 1.1 and 1.2 for *Tetrahymena pyriformis*).

Issues in Toxicology No.7
In Silico Toxicology: Principles and Applications
Edited by Mark T. D. Cronin and Judith C. Madden
© The Royal Society of Chemistry 2010
Published by the Royal Society of Chemistry, www.rsc.org

9.2.1   Simple and Multiple Regression

9.2.1.1   Simple Linear Regression

In the *T. pyriformis* dataset, toxicity is recorded as log IGC_{50}^{-1} (the logarithm of the reciprocal of the millimolar concentration causing 50% inhibition of growth). In Figure 9.1, toxicity is related to hydrophobicity recorded as the logarithm of the octanol–water partition coefficient (log P) (data in Appendix 1.2). Log P was calculated using the KOWWIN software (see Chapter 5). The latter is referred to as the 'predictor'. Simple linear regression uses just one predictor.

A line is fitted through the scatter of points and this can be described by an equation in the standard form for a straight line:

$$y = a + bx \tag{9.1}$$

where:

y is the predicted toxicity value for a particular substance
x is the value of log P for the relevant molecule
a is the intercept on the vertical axis
b is the gradient of the line.

For the data in Figure 9.1 the regression equation is:

$$Toxicity = -1.03 + 0.637 \ \log P \tag{9.2}$$

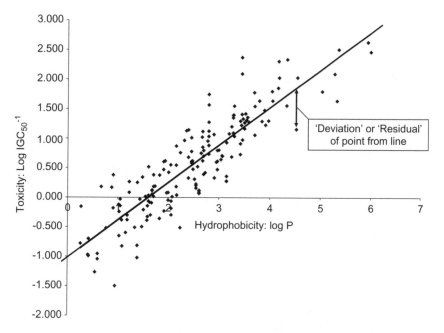

Figure 9.1   Simple regression prediction of toxicity from lipophilicity (log P).

P for log *P* < 0.001;

$$R^2 = 77.6\%$$

The constant (-1.03) in eqn (9.2) indicates toxicity if the predictor had a value of zero. This may be a hypothetical extrapolation if zero is outside the real range of values for the chemical descriptor (which is marginally the case with these data).

The term '$+0.637$ log *P*' in eqn (9.2) means that there is a positive relationship and that an increase in log *P* of one unit is associated with an increase in toxicity of 0.637 units.

The data that are used to establish the regression model are referred to as the 'training' or 'learning' set. Other molecules for which toxicity has not been experimentally assessed can be assigned a predicted toxicity using the model. Thus the toxicity of an untested molecule with a log *P* of 2.0 would be predicted as below from eqn (9.2):

$$Toxicity = -1.03 + 0.637 \times 2.0 = 0.24$$

9.2.1.2   *Multiple Linear Regression*

Multiple linear regression (MLR) extends the above idea and allows the use of several predictors by adding extra terms—one for each descriptor. In eqn (9.3), the energy of the lowest unoccupied molecular orbital (E_{LUMO}; data in Appendix 1.2) has been added.

$$Toxicity = -1.08 + 0.648 \ \log P - 0.469 \ E_{LUMO} \qquad (9.3)$$

P for both predictors < 0.001;

$$R^2_{(adj)} = 82.9\%$$

Account is now taken of the influence of both log *P* and E_{LUMO}. There is still seen to be a positive effect of log *P* and additionally a negative relationship with E_{LUMO} can be seen. A unit increase in E_{LUMO} is associated with a reduction in toxicity of 0.469 units. Note that the coefficient describing the responsiveness of toxicity to log *P* changes very little (0.637 to 0.648) with the addition of the extra predictor. This is a desirable feature but cannot automatically be relied upon (see Section 9.2.2.4).

With a single predictor, the relationship to toxicity can be represented as a straight line graph as in Figure 9.1. With two predictors, the relationship has to be represented as a sloping plane (Figure 9.2). Any given compound can be plotted onto a flat plane according to its values of log *P* and E_{LUMO}. Two compounds are indicated; compound A has a low value of log *P* and a high value for E_{LUMO}, whereas B has a high log *P* and a low E_{LUMO} value. The positions of the two compounds are then projected upwards onto a sloping

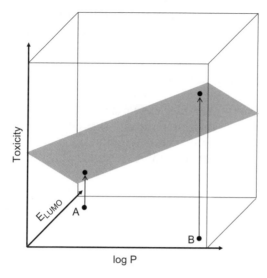

Figure 9.2   A plane fitted to toxicity data in multiple linear regression (MLR).

plane defined by the regression equation (note you are looking at the under-side of the plane). Both properties of A are associated with low toxicity and those for B with greatest toxicity, and so the points A and B project onto the lowest and highest corners of the regression plane.

With three or more predictors, it is no longer possible to represent the relationship as a simple diagram such as Figures 9.1 or 9.2, and the relationship is described as following a hyper-plane.

9.2.2 Data Properties

MLR is a robust procedure and will generally produce sensible results so long as it is not used with grossly inappropriate data. However, the following points do need to be borne in mind.

9.2.2.1 Linearity

Regression is based on an assumption of straight line relationships between toxicity and each explanatory variable. If data from a markedly non-linear relationship are submitted to regression analysis, an equation will be produced but may be highly misleading. Sections 9.2.4.5 and 9.2.6.1 respectively describe how such problems may be detected and resolved.

9.2.2.2 Normal Distribution

Technically, any parameter that is to be included in a regression analysis should follow a normal distribution, but in reality practical problems only arise with severe departure from normality.

- Data that fall into two (or more) distinct groups of higher and lower values with few examples of intermediate values (described as 'polymodal') are hazardous. Such data suggest that the compounds may fall into chemically distinct groups and any variation in toxicity may be due to these chemical differences rather than whatever property is represented by the descriptor.
- Isolated extreme outliers may exert excessive leverage (see Section 9.2.4.6).
- Data that are very strongly positively skewed (asymmetrical distribution with a long tail of high values) are quite common and may include high values that effectively constitute outliers which may exert excessive leverage.

Severe positive skewing can often be corrected by taking the logarithm of the original data (log transformation).

9.2.2.3   Data Variance

The various descriptors used as predictors are likely to have differing standard deviations (SD). For most purposes this is irrelevant. The only caution is that MLR equations need to be interpreted with care. Each predictor has its own coefficient and it is tempting to think that those with the largest coefficients have the greatest influence. However, this is not necessarily true because a predictor with a low coefficient may have a very large SD and would then have a large effect on the predicted toxicity. If it is desired to make the coefficients easily interpretable, the descriptors can be re-scaled so they all have a mean of zero and an SD of ± 1.0. This is achieved by eqn (9.4):

$$Scaled\ value = (Original\ value - Mean)/\text{SD} \qquad (9.4)$$

Mean and SD refer to the mean and SD for that particular descriptor. Data scaled in this way are sometimes referred to as 'Z scores'.

9.2.2.4   Collinearity

Some of the greatest problems with MLR arise when there are strong correlations among the descriptors being considered as possible predictors. The problem is referred to as 'collinearity'—or 'multicollinearity' if several are involved.

One problem with collinear predictors is that the coefficients attached to each predictor become unstable. In Section 9.2.1.2 it was noted that the coefficient attached to log P was almost unchanged when E_{LUMO} was added as a further predictor. This is highly desirable and will be the case in the absence of excessive collinearity. However, log P and solubility (data in Appendix 1.2) are strongly correlated ($r = -0.940$) and, if regression is carried out using these two predictors, the coefficient attached to log P changes from 0.637 to 0.177 creating the impression that toxicity is much less responsive to changes in log P than is actually the case. This can make interpretation of the model difficult or misleading. The instability leads to very high values for the standard error of these coefficients reflecting how poorly they are estimated in the presence of collinearity.[2] In extreme cases, even the sign of a coefficient may change giving a completely misleading impression of the effect of that particular factor.

Another problem is that collinear predictors may be dismissed as lacking statistical significance even when they actually have good predictive power. The reason for this effect is described in Section 9.2.4.4.

Collinearity problems also arise where the value of one descriptor is a linear combination of two (or more) of the others, *e.g.* one of the descriptors might be equal to the sum of two others.

Where collinearity is present, one of the following will be necessary:

- Prior to attempting to fit an MLR, exclude sufficient descriptors so that those retained lack excessive correlation.
- Exclude strongly correlated descriptors as part of the fitting process.
- Replace MLR with one of the techniques described in Sections 9.3 or 9.4.
- In the extreme case of perfect collinearity, the regression calculation simply cannot be executed and an error will be generated.

9.2.3   Least Squares Fit

Figure 9.1 illustrates the 'deviation' from the regression line for one substance—the vertical distance between the point and the fitted line. The goodness of fit of a regression line to the data is assessed by calculating the deviations for all substances and squaring these. The overall lack of fit is then measured as the sum of the squared deviations. The best fitting line is that which minimises this value, hence the term 'least squares fit'. Fitting for multiple regression follows the same logic; the coefficients are adjusted in the equation so as to minimise the sum of squares of the differences between the toxicities predicted by the equation and the experimentally measured values.

9.2.4   Regression Statistics, Diagnostics and Significance Testing

When a statistical package is used to produce an MLR model, all (or most) of the following results should be generated:

9.2.4.1   Regression Equation

Whatever statistical package is used a regression equation will be produced.

9.2.4.2   Standard Deviation of the Errors

Section 9.2.3 emphasised the errors in fitting a line to the points as measured by the vertical deviations between the points and the line. For the best fitting line these errors always average to zero and the standard deviation among the errors reflects how good the fit is; a high standard deviation reflects a poor fit. Statistical packages report an estimate of this under varying labels—'S' in Minitab software (www.minitab.com) or 'Standard Error of the Estimate' in SPSS (www.spss.com/software). In this chapter, the term 'S' is used.

S is of limited value as a measure of the quality of the fit as it takes no account of the range of toxicity values. If the SD among the toxicity values is 10 and $S = 1$, this would indicate a reasonable fit, but if the toxicity values themselves had an SD only slightly greater than unity, the same value of S would indicate virtually no fit at all. R-squared (next section) overcomes this limitation.

S can be very informative if the practical imprecision in the determination of toxicity values is known. Any case where S is less than this practical imprecision is almost certainly a warning that the model has been over-fitted (see next section).

9.2.4.3   R-squared and R-squared(adjusted)

R-squared (R^2) is a more interpretable measure of the closeness of the predicted toxicity values to the true values. It takes account of both the SD among the prediction errors (S) and the range of values in the toxicity data. A value of 100% indicates that all the values are fitted exactly by the equation and 0% that there is no fit at all. Some packages quote R-squared as a percentage and others as a decimal; thus an R-squared of 70% is the same as one of 0.7.

With real toxicity data there will always be deviations between the measured and predicted values which do not arise from the influence of any of the remaining unused descriptors. Nonetheless, most of these descriptors will have some chance correlation with the residual deviations and so their incorporation with appropriate coefficients into the regression equation will increase the fit to the data and increase R-squared. However, if this equation is then used to predict the toxicity of a series of new substances the extra term in the equation will simply add random noise and reduce the precision of the predictions. This characteristic pattern where extra terms improve the fit to the training set but reduce the ability to predict values for new substances is referred to as 'over-fitting'. Topliss and Edwards[3] have explored, in some detail, the relationship between over-fitting and the size of the initial pool of descriptors, from which a sub-set may be trawled. Notice that descriptors that reflect the real causes of toxicity do not suffer from this problem.

Because of the problem of potential over-fitting, the generally preferred measure of fit is 'R-squared(adjusted)' [$R^2_{(adj)}$]. This is like R-squared but includes a penalty factor for each descriptor used. The penalty factor reflects the average improvement in fit that would be expected to arise from the addition of a descriptor consisting of random values and so R-squared(adjusted) has no systematic tendency to rise with the addition of irrelevant descriptors. In practice, R-squared(adjusted) is reasonably stable and the incorporation of meaningless 'predictors' causes only moderate random fluctuations in this statistic (except with very small data sets).

With log P as the sole predictor of toxicity, R-squared(adjusted) was 77.6% and this increased to 82.9% when E_{LUMO} was added. The five percentage point improvement would probably be accounted practically as well as statistically significant (see next section for a discussion of practical as opposed to merely statistical significance).

9.2.4.4   Significance Testing for Equation as a Whole and for Individual Predictors

A P value is provided which tests the complete regression equation. Statistical significance can, as usual, be claimed if $P < 0.05$, but this provides a very weak test. Any regression equation that does not comfortably meet this criterion is unlikely to be of any real utility as a predictive tool.

P values are also provided for each of the individual descriptors incorporated into the model. The necessity for the additional P-values arises because of the danger that a regression equation may have real predictive power solely because a limited sub-set of the predictors are truly correlated with toxicity while others are simply adding random noise to the predictions. The latter are referred to as 'redundant'. The statistical testing of individual predictors can be thought of as being achieved by calculating the increase in S if each predictor were in turn removed from the regression equation. A large increase in S indicates that the predictor had been making a major contribution to the fit and so it should be retained. However, below a certain point, the increase may be so slight that there is no convincing evidence that it had been contributing any new information beyond that already available from the other predictors.

In a case where two predictors correlate with toxicity but are also strongly collinear, both may be declared non-significant. The reason for this is that, being so correlated, they must largely carry the same information. If one of these is excluded from the regression equation, little information will cease to be available as the other predictor can provide it and S will increase very little, leading to a non-significant result for the first predictor. Unfortunately, the same logic applies to the second predictor and it is also declared non-significant. So, although the two predictors do carry useful information they are both declared non-significant! However, if either predictor is discarded and the reduced equation is tested, the remaining descriptor will be found to be significant as its removal would now reduce the effectiveness of the model. This means that any routine for selecting a set of predictors that involves starting with a large set and removing non-significant candidates must carry out the removals one at a time. Wholesale ejections could result in the loss of a collinear set of descriptors, one of which should have been retained.

It would be very limiting to rely solely on tests of statistical significance without also considering practical significance. Statistical tests take account of both the strength of the descriptor/toxicity relationship and the number of substances in the training set. Consequently, sufficiently large data sets make it possible to discern very weak predictors as being statistically significant. However the inclusion of such descriptors may improve the predictive power of an MLR model to only a trivial extent. A balanced view is required that takes account of the improvement in predictive power provided by the descriptor along with a consideration of its interpretability and the biological and chemical feasibility that it would influence toxicity. Uninterpretable and/or mechanistically unconvincing descriptors that increase R-squared(adjusted) by a fraction of a percent are not practically significant, even if formally statistically significant.

An extreme note of caution needs to be issued concerning *P* values in the context of toxicity prediction. If a very large number of descriptors have been screened for correlation with toxicity and a small but select band incorporated into a regression equation, the *P* value for the overall equation and those for the individual predictors are bound to be grossly optimistic. If the *P* value for one of the above predictors is quoted as 0.05 that tells you that any single set of random numbers would have only a 5% chance of correlating this strongly with toxicity. That information is perfectly correct, but in the case considered, a single set of such numbers was not taken; what was taken was (possibly) hundreds of sets from which were chosen the ones most strongly correlated. The chances that a selected set of random numbers would show this degree of correlation may be nearer 100% than 5%. The problem of statistical testing of allegedly predictive models is followed up in Section 9.2.9.

9.2.4.5 Residuals

A 'residual' is the difference between the observed toxicity of a substance and its predicted toxicity (in fact the same as the 'deviation' shown on Figure 9.1 when just one predictor is being used).

Figure 9.3a shows an extreme case of a non-linear relationship. If such a chemical descriptor were to be included in a linear regression, the residuals would show a distinctive pattern (see Figure 9.3b). Moving from low to high values of the predictor, the residuals indicate first over-estimation, then under- and finally back to over-estimation. This non-random pattern is diagnostic of the non-linear relationship present in this case. A relationship with the opposite curvature would result in residuals that indicated under-, over- and then under-estimation. More complex curvilinear relationships might arise, but would be difficult to detect amid the random scatter of most quantitative structure activity relationship analysis outcomes.

A series of 'residual plots' can be obtained by plotting the residuals against each of the predictors in turn. Where there is a linear relationship between the outcome and the predictor, the residuals should show a random scatter of points.

One of the residual plots for the relationship between toxicity and E_{LUMO} (Section 9.2.1.2) does give cause for concern (Figure 9.4a). Unfortunately, it is

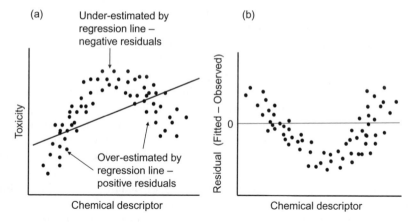

Figure 9.3   A non-linear relationship revealed by the pattern of residuals. (a) Endpoint versus descriptor (b) Residual versus descriptor.

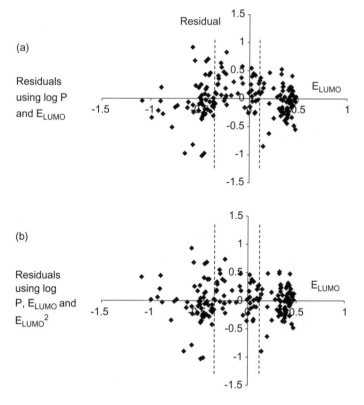

Figure 9.4   Detection and correction of a non-linear relationship between toxicity and
E_{LUMO} by inclusion of $E_{LUMO}{}^2$. (a) Linear fit to E_{LUMO}. (b) Quadratic fit.

possible to draw vertical lines that isolate a central section with a strong pre-ponderance of positive residuals which is balanced by an excess of negative residuals to the left and right, suggesting a non-linear relationship with E_{LUMO}. Section 9.2.6.1 suggests a possible solution to this problem.

Residuals can also be used to detect 'heteroscedacity'. Residuals should be similar in magnitude at all levels of the endpoint being fitted. This is referred to as 'homoscedacity'. A common problem is that the residuals may be propor-tional to the value of the endpoint; thus they are low for low values of the endpoint but then steadily increase (an example of heteroscedacity). Figure 9.5 shows an example—much greater scatter at one end of the graph that at the other.

Heteroscedacity can complicate the process of defining a best fitting line. Using the normal least squares criterion, a molecule with a toxicity that devi-ates by a certain amount from the fitted line makes the same contribution to the sum of squares regardless of whether it has a low or high toxicity. However, more weight ought to be given to deviations for low toxicity compounds as they are subject to less random fitting error. Figure 9.5 includes the best fit (solid line) as defined by the normal least squares criterion; it is clearly questionable as it runs systematically above the majority of the points where toxicity is low. This poor fit still minimises the sum of squares because the calculation is

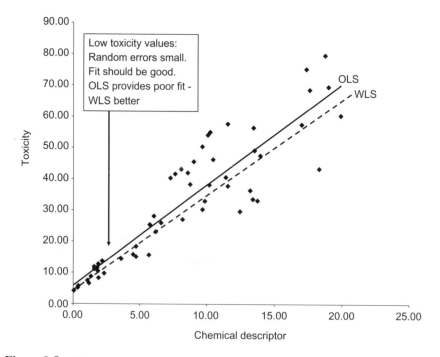

Figure 9.5   Heteroscedastic data fitted by ordinary least squares (OLS) or weighted least squares (WLS) regression.

dominated by the large errors at the top of the graph. Fox[4] discusses how severe heteroscedacity must be for it to become a real problem.

There are two possible solutions to this problem. The first is transformation of the toxicity values such as the log transform (Section 9.2.2.2), which commonly removes the form of heteroscedacity mentioned above. An alternative solution is to replace the standard least squares criterion ('ordinary least squares') with 'weighted least squares'. With this approach each substance's squared deviation is weighted so as to increase those that should be allowed greater influence. In the example quoted above, dividing each squared deviation by either the toxicity value or the toxicity-squared may suitably increase the influence of the deviations for low toxicity molecules. The weighted squared deviations are then summed to produce a 'weighted sum of squares' which is then minimised. The broken line in Figure 9.5 indicates the fit achieved by weighting according to the square of the toxicity and it provides a better fit for the low toxicity compounds. The full algorithm is available from the Liverpool John Moores University website (www.staff.ljmu.ac.uk/phaprowe/rsc/wls.xls).

9.2.4.6 Outlying Points (Outliers)

Statistical analysis may indicate outlying points (or outliers). These may be points (*i.e.* compounds in a QSAR) that have a large residual and should be checked for possible data entry errors. Alternatively they may have an extreme value for one or more of the predictors and therefore have a large effect on the regression equation. The latter are said to have 'high leverage'. Here the concern would be that the apparent relationship between toxicity and the predictors might be due to a small group of high leverage points. As a test, the removal of these points should not unduly alter the regression equation.

Outliers can result from QSAR for a variety of other reasons as well. Dearden *et al.* described a number of reasons for compounds appearing as outliers including poor data, incorrect data transfer, alternative mechanisms of action or simply a poor model.[5]

The question becomes whether an outlier should be removed. It is tempting to remove such compounds, and historically this has been done. Such removal is seldom acceptable and, if performed, must be carefully recorded. Possibly acceptable is the removal of outliers should they act by a different mechanism of action—and that there is good evidence to support this assumption. Thus, with the *T. pyriformis* data set utilised above, there would be good reason to omit compounds with other mechanisms of action. Obviously, following removal of outliers, the equation should be re-calculated and the statistical analysis repeated.

By analogy, analysis of outliers can also be used to identify compounds with different mechanisms of action. Thus, whilst these compounds may be seen as problematic, they are also useful and have been shown to provide new mechanistic knowledge and insight.[6] For an illustration of this process, the reader is referred to Enoch *et al.* who analysed the *T. pyriformis* data in more detail.[7]

9.2.5 Categorical Variables

A categorical descriptor is one that categorises molecules rather than acting as a continuous measure. An example would be a descriptor for whether a molecule does/does not contain a particular atom or functional group. Where such a descriptor is dichotomous, it can be coded into the regression by representing the two possibilities by two convenient numbers (traditionally 0 and 1). There may be more than two categories, *e.g.* the ring type for a series of molecules might be benzene, pyridine or pyran. These can also be rendered as 'indicator' or 'dummy' variables, but they cannot simply be coded as 1, 2 and 3. The ring types are classifications with no natural order and if we coded the rings as above, there is absolutely no logical reason why toxicity would increase (or decrease) in a stepwise manner as we proceeded along the series. There is also no reason to believe that toxicity would change in steps of equal size as we proceeded along the sequence.

The solution is to use n-1 indicator variables to describe n categories. For the example above, we would use two descriptors, one to encode 'Has a pyridine ring?' and the second for 'Has a pyran ring?'. These would both be encoded as 0 = false and 1 = true. Any benzene-based compounds would be coded with both set to zero (*i.e.* 'none of the above'). This arrangement can model any pattern of toxicities among the three ring types. For example Fillipatos *et al.*[8] used precisely this kind of approach to distinguish between groups of substances.

9.2.6 Non-linear Models and Interaction

9.2.6.1 Non-linearity

Section 9.2.4.5 described a problem case of non-linear relatedness. Toxicity may fit better to a quadratic curve rather than a straight line. To test this, both E_{LUMO} and its squared value are entered into the regression as predictors. The two values can then jointly model (or at least approximate) a range of curvilinear relationships. When the squared term (E_{LUMO}^2) is added, the regression equation becomes:

$$Toxicity = -1.02 + 0.664 \log P - 0.628 \quad E_{LUMO} - 0.660 E_{LUMO}^2 \tag{9.5}$$

P for all predictors <0.001;

$$R^2_{(adj)} = 84.2\%$$

The extra term is statistically significant, but whether an increase in R-squared(adjusted) from 82.9 to 84.2% could be adjudged practically significant is questionable. Figure 9.4b shows a more random distribution of residuals for the new equation, suggesting that a better fit to the shape of the relationship between toxicity and E_{LUMO} has been achieved. Draper and Smith[9] give more detail of non-linear regression.

The quadratic function mentioned above is a useful but very non-specific approach. A more rational approach would be to use knowledge of underlying biological and chemical mechanisms to select more specific mathematical relationships between descriptors and endpoints that could be incorporated into non-linear regression models.[10]

9.2.6.2   Interaction

Interaction arises when the responsiveness of the endpoint to one predictor depends on the value of another. For example, the regression eqn (9.3) suggests that toxicity changes by $+0.648$ units for every one unit increase in log P. However, the responsiveness of toxicity to log P may not be constant. It could be that for compounds with a low value of E_{LUMO}, a one unit increase in log P would cause a greater change in toxicity, whereas for compounds with high E_{LUMO} values, toxicity was less sensitive to log P. In such a case the figure of $+0.648$ would represent average responsiveness over the full range of E_{LUMO} values. Interaction between two predictors can be tested for by creating a new descriptor that is the product of the two terms (log $P \times E_{LUMO}$). This is then added an additional descriptor. The equation becomes:

$$Toxicity = -1.09 + 0.649 \log P - 0.351 E_{LUMO} \\ -0.0442 \log P \times E_{LUMO} \tag{9.6}$$

P for both individual predictors <0.001
P for interaction $=0.430$

$$R^2_{(adj)} = 82.9\%$$

The interaction falls far short of significance and it can be concluded that the two predictors act essentially independently. The interaction term does not need to be added to the equation.

9.2.7   Descriptor Selection

Starting from a large set of descriptors all of which are potential predictors, the removal of those that are simply irrelevant is generally not particularly challenging, but groups of collinear predictors that have not been thinned out at an earlier stage present a greater difficulty.

9.2.7.1   Best Sets

Some statistical packages have utilities that will assess reasonably large sets of descriptors (up to 31 in Minitab) looking at all possible combinations of (say) two, three or four predictors to find those providing the best fit. This procedure should select just one descriptor from each collinear set, as any more from the same set would add little new information.

9.2.7.2 Stepwise Selection

For larger descriptor sets, stepwise procedures will either start with the best single predictor and gradually build up a larger set, always adding the candidate that most improves the model, or start with a large set and reduce it by removing the least informative member. Routines may also mix both approaches. Theoretically such an approach could miss the ideal combination. This could arise if the early stages of model building incorporate a descriptor that is not part of the ideal set. However, there is little practical evidence of significantly sub-optimal models being selected.

9.2.7.3 Genetic Algorithms

Genetic algorithms offer another way to tackle large descriptor sets. These mimic the processes of biological evolution. They start with an initial pool consisting of several sets of descriptors that are selected purely randomly. Most of these sets will produce very poor regression models, but a proportion will, by chance, contain some descriptors that are correlated with toxicity. All candidate descriptor sets in the pool are assessed and those least able to model the toxicity data are rejected. New candidates are then created by either mutation or breeding from the retained cases. A mutation is a copy of an existing member of the pool where one descriptor has been replaced by a randomly selected alternative. Breeding consists of taking half the descriptors in one member of the pool and adding half those from another. The models are then re-assessed and, if any of the newly created models are superior to the originals, they displace the inferior ones. This process is then repeated many times, each cycle tending to improve the quality of the pool.

There may be some data sets where stepwise procedures can never find the optimum solution. A particular descriptor may be the best one to add when the model contains only one or two terms; it is therefore incorporated but it might not form part of the optimum model with three or four terms. A genetic algorithm might overcome this problem as it starts out with full-sized models and never passes through the smaller stages. While there can be no guarantee that a genetic algorithm will find the true optimum model, there is at least a theoretical case that it could outperform stepwise procedures in some cases.

9.2.7.4 General Comments

All automated procedures are subject to the criticism that they select descriptors using purely statistical criteria without reference to other aspects such as the ability to interpret them.

9.2.8   Interpretation of MLR Models

Any MLR equation is capable of interpretation, at least on a superficial level. Equation (9.3) tells us that for every unit increase in log P, toxicity will rise by 0.648 and for every unit increase in E_{LUMO}, it will decrease by 0.469. However, the possibility of a deeper interpretation will depend on the particular descriptors chosen. In this case both are well understood; log P reflects hydrophobicity, so more hydrophobic molecules tend to be more toxic. This makes perfectly good biological sense in terms of access to cellular components *via* penetration of lipid membranes. Increasing toxicity with lower E_{LUMO} values (*i.e.* increased molecular electrophilicity) also makes biological sense, as electrophilicity is associated with non-specific protein binding (see Chapter 7). If QSAR is going to be anything more than a conjuring brick, the selection of interpretable descriptors is highly desirable.

9.2.9   Testing Predictive Power— Cross Validation and Training, Test and Evaluation Data Sets

There is a great deal of sloppy thinking concerning the predictive power of MLR and most other modelling techniques. The development of an MLR involves three stages (as in Figure 9.6):

- Choose MLR as opposed to some other modelling technique.
- Select a set of descriptors that correlate with toxicity.
- Calculate the coefficients for each descriptor that will enter into the regression equation.

A particularly weak test is 'Leave-One-Out' at a time (LOO) cross-validation. This involves removing one compound and using the reduced data set to

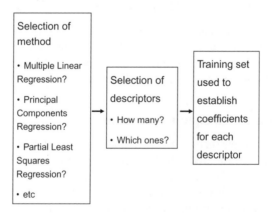

Figure 9.6   The full process of model development, all of which should be tested if a claim is to be made that a model has predictive power.

calculate the coefficients and then an equation based on these is used to predict the toxicity of the one removed compound. This is then repeated for all compounds and the prediction errors are used to calculate a 'cross-validated R-squared' value which will be lower than simple R-squared. However, the full data set would have contaminated the whole process by informing the selection of both the technique and the descriptor set. All that would be tested by LOO cross-validation would be the final calculation of the coefficients. For a more detailed critique of this important topic, the reader is referred to Tropsha *et al.*[11]

It is possible with a large collection of entirely random descriptors and a full data set to choose MLR and select a small sub-set of descriptors showing chance correlation with toxicity. LOO cross-validation would then test only the calculation of the regression coefficients. The reduced data sets would differ only slightly from the full set which is already known to produce a regression equation capable of fitting the toxicity values. The result would inevitably be apparent evidence of predictive power.

A meaningful test must be based upon the ability to predict the toxicity of compounds that have had no chance to influence any of the three stages in model development. This can be achieved either by obtaining toxicity data for a new 'test' set of substances after the model has been finalised, or by dividing an initial data set into a training set and a ring-fenced 'evaluation' set with the latter carefully insulated from model development and only used to assess a finalised model. Once all three stages of model building have been completed using the training set, the MLR can be used to predict the toxicities of the test/evaluation set in an uncontaminated manner. Any descriptor that showed only chance correlation with toxicity within the training set would have no special likelihood of correlating within the test set and should be revealed as a false prophet.

9.3   Principal Components Regression (PCR)

Principal components regression (PCR) and partial least squares (PLS) regression are two techniques that can allow the retention of sets of collinear descriptors within a regression model without the (sometimes) rather arbitrary selection of just one descriptor from each collinear group.

9.3.1   What are Principle Components?

For the purposes of this section, the two descriptors log *P* and *solubility* (data in Appendix 1.2) are considered but they are used in the scaled form of *Z*-scores (see Section 9.2.2.3). These are referred to as log *P(Sc)* and *Solub(Sc)*.

Log *P(Sc)* and *Solub(Sc)* are strongly negatively correlated with $r = -0.94$ (see Figure 9.7). For any group of correlated descriptors, a series of 'principal components' (PCs) can be calculated. The first PC is shown as line AB in Figure 9.7. It looks superficially like a regression line, but is selected by different

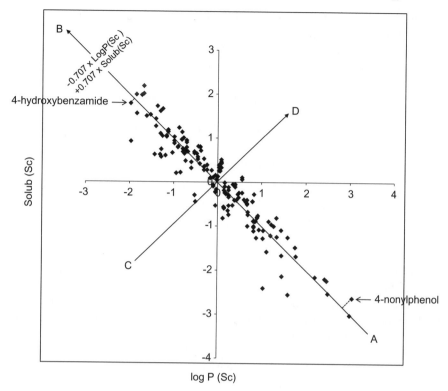

Figure 9.7 The principal components for log *P* and solubility.

criteria and is positioned slightly differently. The placement of the first PC is determined as follows:

- The PC can be considered as a new scale of measurement with lowest values at end A of the line and highest at end B.
- The point corresponding to each molecule is projected perpendicularly onto the new line. This is shown for two molecules as dotted lines. The points at which these lines hit the PC establish the PC scores for the molecules. Thus 4-nonylphenol has a low score and that for 4-hydro-xybenzamide is much higher.
- The PC is arranged so that it follows the longest axis of the scatter of points. Mathematically, the criterion for this is that the PC scores among the molecules achieve the greatest possible SD.

The second PC is line CD. The criterion for the placement of this PC is that the scores arising by projection of the points onto this new line have zero correlation with those from the first PC. In practice, this means that it follows a much shorter axis of the cluster. It can be seen that projection of the molecules onto this line would yield scores of much lower variability.

It is possible to calculate as many PCs as there are molecules or descriptors (whichever is the lesser).

The output from a PC analysis will normally include the equation used to calculate the scores for each PC. In the current case, the first PC is scored as:

$$PC1 = -0.707 \log P(Sc) + 0.707 \, Solub(Sc) \qquad (9.7)$$

Thus, for the two very contrasting molecules highlighted in Figure 9.7:

> PC1 score for 4−nonylphenol
> $= -0.707 \times 3.057 + 0.707 \times -2.633 = -4.023$
> PC1 score for 4−hydroxybenzamide
> $= -0.707 \times -1.983 + 0.707 \times 1.783 = 2.663$

These PC scores are readily interpretable. The low value for the first compound indicates high log P and low solubility, and the second has the reverse pattern. Presumably these have low and high polarities respectively and the PC score is essentially a measure of polarity. The PC scores for the molecules can, in effect, be used as a new predictor variable and will replace all the descriptors that were used to create the PC.

Because of the high degree of correlation between log P and solubility, the scores for the first PC effectively summarise most of the information carried by both descriptors. Thus the low PC score for 4-nonylphenol carries the information that this molecule must have both a low solubility and a high log P, whereas 4-hydroxybenzamide must have a high solubility and a low log P. If two descriptors were perfectly correlated, the first PC score would convey all the information from both descriptors, and then as correlation becomes weaker, the proportion of the information summarised declines. Part of the output from PC analysis is a measure of the proportion of the total information carried by each PC.

PCs can also be produced for groups of more than two descriptors. With three, the points could be plotted as a cloud in three dimensions. The first PC would follow the most extended axis of the cloud, the second would then be at right angles to that oriented along the next most extended axis and the third is then forced to occupy the direction of least variation. The same principal applies with more than three descriptors, but is no longer easily represented pictorially.

The first PC always carries the highest proportion of the information present in the initial descriptors, followed by the second, *etc.* Large sets of descriptors can generate equally large numbers of PCs. However, the first few will generally account for most of the real information, with the latter ones reflecting little other than noise. In the case of log P and solubility, the first PC explains 97% of the variability in the two descriptors, with the second accounting only for the remaining 3%. It would almost certainly be reasonable to allow the scores for PC1 to replace both descriptors and discard the second PC.

9.3.2 Applying Principal Components Regression

To illustrate the technique, it will be assumed that only log P, solubility, molecular mass, surface area and volume (data in Appendix 1.2) are available as predictors. Log P and solubility form one strongly collinear group and mass, surface area and volume another.

9.3.2.1 Establishing a Regression Model

Section 9.3.1 showed that 97% of the variability in log P and solubility could be summarised by their first PC and the second can probably be discarded. The scores from the first PC for log P and solubility are called 'LgPS1'.

For the other three descriptors, the first PC encapsulates 83.1% of the information and the inclusion of the second raises this to 99.3%. In this case, the first two PCs would probably be considered, but not the third. These two PCs for mass, surface area and volume are called MSV1 and MSV2. The weights of the descriptors as they contribute to these first two PCs are:

$$MSV1 = +0.513 \, Mass + 0.593 \, Surface \, area + 0.621 \, Volume$$

$$MSV2 = +0.593 \, Mass - 0.488 \, Surface \, area - 0.229 \, Volume$$

The coefficients for the three contributors to MSV1 are all positive and of similar magnitude. Thus this PC can be interpreted as a simple measure of molecular size. Greater mass, volume and surface area all lead to greater values of MSV1. With MSV2, mass is positively weighted and volume and surface area are negative. Highest scores will therefore be associated with molecules that are heavy (high mass) but small (low surface area and volume). It seems to be a measure of density.

Regression analysis relating toxicity to LgPS1, MSV1 and MSV2 shows MSV1 to be non-significant ($P = 0.797$). Removal of this term leads to the reduced equation:

$$Toxicity = 0.572 - 0.526 \, LgPS1 + 0.120 \, MSV2 \tag{9.8}$$

P value for LgPS1 < 0.001; for MSV1 $= 0.002$

$$R^2_{(adj)} = 83.3\%$$

9.3.2.2 Interpretation

LgPSV1 is a marker of polarity and has a negative coefficient, so the interpretation is that the least polar compounds tend to be the most toxic.

MSV1 was a marker of molecular size and explained the highest proportion of variability in the three size descriptors, but was evidently not usefully correlated with toxicity. The second PC (MSV2) is clearly significant with a positive coefficient. It seemed to be a marker of density—denser molecules being associated with greater toxicity. It is a general characteristic of PC regression that the first PC will be the most informative (and often the most easily interpretable) component, but it will not necessarily be the most effective predictor.

Either log *P* or solubility could have been used almost equally well as a predictor, and any decision to select one and exclude the other would have been largely arbitrary. Here, both have been allowed to contribute and the subsequent regression equation remains interpretable. The three size parameters have ended up as a single predictor with no arbitrary elimination of descriptors and the regression equation is interpretable as indicating that higher density molecules are likely to be more toxic.

9.3.2.3   Assessment

The R-squared(adjusted) value (83.3%) is about the same as that from previous models.

The potential advantages of using PC scores rather than the original descriptors are that:

- a small number of PC scores may carry most of the information present in a large number of descriptors, avoiding the need to exclude descriptors; and
- because the PC scores are uncorrelated, they do not interact among themselves thus avoiding the problems we have seen with collinear descriptors (see Section 9.2.2.4).

The greatest hazard with PC regression is that PC scores may not always be as easily interpretable as those seen above. The first PC is often reasonably interpretable, especially when it is derived from a single group of collinear descriptors. Subsequent PCs may be a lot less interpretable. If large numbers of descriptors including several separate collinear groups are included in a single PC analysis, the PCs may be almost impossible to interpret.

9.4   Partial Least Squares (PLS) Regression

A theoretical objection to PC regression is that the calculation of the components is based entirely on correlations among the potential predictors with no view as to any possible correlation of the resultant PC scores with toxicity. In the PLS approach, a series of components is again calculated, but the calculations take account of both correlation among groups of predictors and their joint correlation with the toxicity endpoint. With PC analysis, the first component simply emphasises a group of descriptors that are collinear whereas the first component in PLS will emphasise a group of descriptors that are both collinear and correlated with toxicity. Unlike PC regression, it would be expected that the first component would be the most effective predictor of toxicity, followed by the second, *etc.*

PLS can potentially generate as many components as PC analysis. The more components that are used, the more closely the data can be fitted, but this may degenerate into over-fitting. It is therefore necessary to monitor the process by cross-validation. In the final stage of PLS regression, the scores from the appropriate number of components will be incorporated into a predictive equation, with suitable weighting for each component. All the components will contain the same set of variables with different patterns of weighting. Consequently the final model generated by PLS can be expressed as a multiple regression-like equation containing all the original descriptors.

9.4.1 Applying PLS

The following illustration of PLS employs the seven previously used descriptors (log *P*, solubility, LUMO, LUMO-squared, mass, surface area and volume). Using just the first component, cross-validated R-squared is 71.5% and this rises to 85.4% with two. After that, additional components change R-squared by only fractions of a percent, so two components are adequate. The first two components are then combined to give the equation:

$$
\begin{aligned}
Toxicity = {}& -0.758 + 0.343 \ \log P - 0.310 \ Solubility \\
& - 0.259 \ E_{LUMO} - 0.280 \ E_{LUMO}^2 \\
& + 0.00211 \ Mass - 0.00252 \ Surface \ area \\
& - 0.000333 \ Volume
\end{aligned}
\tag{9.9}
$$

9.4.2 Interpretation

The model is interpretable as previously. Log *P* and solubility are again weighted oppositely with toxicity negatively linked to polarity. The two E_{LUMO} terms show negative correlation with toxicity and the size parameters again suggest a positive relationship between toxicity and density.

Used in this restrictive manner, PLS produces a useful and interpretable model. Problems arise when PLS is used as a receptacle for vast numbers of descriptors. The reduction of the data to a limited number of components should preserve the real predictive power of such models without over-fitting, but there is no guarantee that one would easily be able to interpret the model.

9.4.3 Assessment

In summary, PLS possesses the same strengths and weaknesses as PC regression. Both offer a way to use sets of collinear descriptors without the need for arbitrary pruning. Both can lead to interpretable models if the list of descriptors incorporated is sensibly restricted. Simplistically applied to interminable sets of descriptors, both are likely to generate uninterpretable models. PLS has a theoretical advantage over PC regression as it constructs its components with a view to their predictive power whereas PC analysis simply aims to summarise the descriptors as efficiently as possible.

9.5   Conclusions

There are a variety of statistical techniques that are applicable to continuous data and which have often been used in QSAR analysis. The techniques described in this chapter (*e.g.* regression analysis, PCR, PLS) can be applied to develop interpretable models for QSAR analysis. The importance of understanding the diagnostic statistics and limitations of the models is illustrated and, whilst a QSAR modeller or user may not be a statistician, they should have an appreciation of these issues. There are many more places for the reader to go to find more detailed accounts of statistical analysis for QSAR; in particular Eriksson *et al.* provide a good background to the diagnostics[12] and Livingstone[13] a well-written overview of all methods (including those applicable to Chapter 10).[3] Reference can also be made to the excellent free guidance available from the Organisation for Economic Cooperation and Development (OECD).[14]

References

1. P. H. Rowe, *Essential Statistics for the Pharmaceutical Sciences,* Wiley, Chichester, 2007.
2. W. D. Berry, *Understanding Regression Assumptions,* Sage, Newbury Park CA, 1993.
3. J. G. Topliss and R. P. Edwards, *J. Med. Chem.,* 1979, **22**, 1066.
4. J. Fox, *Regression Diagnostics,* Sage, Newbury Park, CA, 1991.
5. J. C. Dearden, M. T. D. Cronin and K. L. E. Kaiser, *SAR QSAR Environ. Res.,* 2009, **20**, 241.
6. R. L. Lipnick, *Sci. Total Environ.,* 1991, **109/110**, 131.
7. S. J. Enoch, M. T. D. Cronin, T. W. Schultz and J. C. Madden, *Chemosphere,* 2008, **71**, 1225.
8. E. Fillipatos, A. Tsantili-Kakoulidou and A. Papadaki-Valirake, in *Trends in QSAR and Modelling 92,* ed. C.G. Wermuth, ESCOM, Amsterdam, 1993.
9. N. R. Draper and H. Smith, *Applied Regression Analysis,* Wiley, New York, 1998.
10. H. Kubinyi, *QSAR: Hansch Analysis and Related Approaches,* VCH, Weinheim, 1993.
11. A. Tropsha, P. Gramatica and V. K. Gombar, *QSAR Comb. Sci.,* **22**, 69.
12. L. Eriksson, J. Jaworska, A. P. Worth, M. T. D. Cronin, R. M. McDowell and P. Gramatica, *Environ. Health Perspect.,* 2003, **111**, 1361.
13. D. Livingstone, *A Practical Guide to Scientific Data Analysis,* John Wiley and Sons, Chichester, 2009.

14. Organisation for Economic Co-operation and Development, *Guidance Document on the Validation of (Quantitative) Structure–Activity Relationships [(Q)SAR] Models*, OECD, Paris, 2007, OECD Environment Health and Safety Publications Series on Testing and Assessment No. 69, ENV/JM/MONO(2007)2, www.olis.oecd.org/olis/2007doc.nsf/LinkTo/NT00000D92/$FILE/JT03224782.PDF [accessed March 2010].

CHAPTER 10

Statistical Methods for Categorised Endpoints in In Silico Toxicology

P. H. ROWE

School of Pharmacy and Chemistry, Liverpool John Moores University, Byrom Street, Liverpool, L3 3AF, UK

10.1 Ordinal and Nominal Scale Endpoints and Illustrative Data

The statistical analyses that can be applied to *in silico* models for continuous toxicity data are described in Chapter 9. Elsewhere in toxicology, ordinal data provide a scale of measurement covering a range of outcomes. However, these scales are discontinuous (commonly coded as simple integer values) and in this case the intervals between scores progressing along the scale are undefined in size. Toxicities recorded as '1 = little/none', '2 = moderate' or '3 = high' would be ordinal data as they form an ordered sequence but the steps between values are undefined. Where only two outcomes are recorded (*e.g.* 'non-toxic' and 'toxic'), the data are reduced to a simple categorisation and are now described as 'nominal'. This chapter covers techniques (discriminant analysis, logistic regression and k-nearest neighbours) used to predict toxicities recorded as ordinal or nominal data.

Issues in Toxicology No.7
In Silico Toxicology: Principles and Applications
Edited by Mark T. D. Cronin and Judith C. Madden
© The Royal Society of Chemistry 2010
Published by the Royal Society of Chemistry, www.rsc.org

The skin sensitisation data set is used to provide illustration (refer to Appendices 2.1 and 2.2). The main endpoint used is 'LLNA binary class' (data in Appendix 2.2). This is a simple dichotomisation and substances will be referred to as 'non-sensitising' or 'sensitising'. For just one technique (ordinal logistic regression), the endpoint 'LLNA class' (data in Appendix 2.2) is used which includes the classes 'non', 'weak', 'moderate', 'strong' and 'extreme' sensitiser. The full data set contains such a heterogeneous set of compounds that it can be fitted only very poorly. To allow some possibility of a useful fit, a reduced and more homogeneous data set is used to illustrate the methods. This reduced data set consists of those compounds that contain no elements other than carbon, hydrogen or halogens. There are 31 such compounds (Table 10.1).

10.2 Discriminant Analysis

Discriminant analysis is based on arranging compounds in two, three or more dimensional space based upon their chemical properties in the hope that substances with similar biological properties will cluster together. A line or plane is then constructed so that it cuts through this space. Ideally all compounds which fall into one classification (*i.e.* toxic or non-toxic) will lie on one side of the line or plane and those of the other class on the other. In reality, a few compounds will probably lie on the wrong side and therefore be misclassified.

10.2.1 Discriminant Function

A discriminant function is a new scale of measurement that looks similar to a principal component (see Section 9.3.1). Figure 10.1 shows a simple example using two descriptors. The function is indicated as the arrow with lowest values at the lower end increasing towards the top. Each compound can be projected perpendicularly onto this line yielding a discriminant score. Note that although the toxic compounds are generally associated with higher values of both descriptors, there is no value for either descriptor alone that would cleanly cut off the toxic from the non-toxic substances. While the discriminant function may look like a principal component, the criterion for its placement is quite different. This new function is positioned so as to achieve the best discrimination between the two groups. There are two criteria for 'best' discrimination:

(i) The mean scores for the two groups should be as different as possible.
(ii) Those compounds that belong in the same class should have the lowest possible variation among their discriminant scores.

It is therefore the ratio of the variance between the two groups to that within the groups that is maximised.

Table 10.1 Reduced table of skin sensitisation data. Includes only substances containing no elements other than carbon, hydrogen and halogens.

Name	LLNA class	LLNA binary class	E_{LUMO}	Molecular mass	Molecular volume	$E_{LUMO}\_Std^a$	$Volume\_Std^a$
1-bromobutane	Non	0	0.8284	137.0	80.5	0.1257	−1.6545
1-bromodocosane	Moderate	1	0.8251	389.6	314.0	0.1218	1.9811
1-bromododecane	Weak	0	0.8259	249.3	184.3	0.1227	−0.0383
1-bromoeicosane	Moderate	1	0.8248	361.5	287.7	0.1214	1.5713
1-bromoheptadecane	Moderate	1	0.8251	319.4	248.9	0.1218	0.9680
1-bromohexadecane	Moderate	1	0.8257	305.4	236.1	0.1225	0.7679
1-bromohexane	Weak	0	0.8277	165.1	106.3	0.1248	−1.2530
1-bromononane	Non	0	0.8266	207.2	145.2	0.1235	−0.6466
1-bromooctadecane	Weak	0	0.8251	333.5	262.1	0.1218	1.1723
1-bromopentadecane	Moderate	1	0.8248	291.4	223.1	0.1214	0.5664
1-bromotetradecane	Moderate	1	0.8252	277.3	209.9	0.1219	0.3599
1-bromotridecane	Weak	0	0.8253	263.3	196.9	0.1220	0.1584
1-bromoundecane	Weak	0	0.8261	235.2	171.3	0.1229	−0.2399
1-chlorohexadecane	Moderate	1	1.5059	260.9	231.3	0.9129	0.6938
1-chloromethylpyrene	Extreme	1	−1.0794	250.7	164.4	−2.0913	−0.3481
1-chlorononane	Non	0	1.5073	162.7	140.9	0.9146	−0.7142
1-chlorooctadecane	Weak	0	1.5060	289.0	257.4	0.9130	1.1002
1-chlorotetradecane	Weak	0	1.5063	232.9	205.5	0.9134	0.2915

1-iodododecane	Weak	0	0.4611	296.3	190.2	-0.3012	0.0533
1-iodohexadecane	Weak	0	0.4606	352.4	242.1	-0.3018	0.8613
1-iodohexane	Non	0	0.4631	212.1	112.5	-0.2988	-1.1562
1-iodononane	Weak	0	0.4617	254.2	151.3	-0.3005	-0.5521
1-iodooctadecane	Non	0	0.4605	380.5	267.9	-0.3019	1.2640
1-iodotetradecane	Weak	0	0.4608	324.3	216.2	-0.3016	0.4588
7,12-dimethylbenz[alpha] anthracene	Extreme	1	-0.8200	256.4	187.6	-1.7898	0.0136
7-bromotetradecane	Weak	0	0.7743	277.3	213.2	0.0628	0.4118
benzo[a]pyrene	Extreme	1	-1.1107	252.3	164.6	-2.1277	-0.3453
benzyl bromide	Strong	1	-0.0500	171.0	94.8	-0.8951	-1.4317
chlorobenzene	Non	0	0.1547	112.6	74.9	-0.6572	-1.7410
hexane	Non	0	3.7357	86.2	85.7	3.5040	-1.5726
R(+) limonene	Weak	0	1.1952	136.3	122.5	0.5519	-1.0003

[a] E_{LUMO}_Std and Volume_Std are the Z score standardised values (see Section 9.2.2.3) of E_{LUMO} and Volume.

10.2.2   Cut-off Values

Some cut-off value of the discriminant function is then selected in order to split the compounds into two classes. This is shown as a short heavy line crossing the discriminant function in Figure 10.1. Any compound lying at a point on the dashed line would project onto the cut-off point and so this line acts as the discriminatory surface. Compounds above/to the right of this boundary will be predicted to be toxic and those below/to the left will be predicted as non-toxic. Note that the discriminant score allows perfect classification, which was not achievable using either descriptor alone.

While the calculation of the discriminant function is fairly straightforward, there are several methods to decide upon the placement of the cut-off point. To illustrate the various approaches, consider a larger data set as in Figure 10.2. This shows 100 non-toxic and 20 toxic compounds. The discriminant function is shown as an arrow with the toxic compounds generally producing higher discriminant scores, but with some overlap. Two triangles point to the mean discriminant score for both groups of compounds.

10.2.2.1   Using Chemical Similarity Alone

The simplest way to set the cut-off is to calculate the mid-point between the two mean scores. Line (a) is drawn through that point. Any chemical producing a score above this line is then chemically more similar to the toxic molecules and any below are more similar to the non-toxic ones. Classification is commonly achieved by using position relative to this simple midpoint line.

10.2.2.2   Adding in Prior Likelihood

Among the 120 chemicals tested only a minority (20) were toxic. This observation needs to be interpreted carefully. If the compounds tested were a random selection from all those that might have been studied, then we can use the observed 5:1 ratio as an estimate for the general ratio of non-toxic to toxic substances. However, in the real world, the selection of molecules for testing could have been biased in either direction. If it is reasonable to assume that among the generality of compounds of interest the 5:1 ratio does apply, then this additional information should be used. This 5:1 ratio is referred to as a 'prior likelihood'.

Previously line (a) was used to divide substances into two classes as only chemical similarity was considered. But because of the imbalance in the two groups of molecules, the general vicinity of this line contains many more non-toxic than toxic compounds and any new compound that falls into this region is more likely to be non-toxic than toxic.

Line (b) is drawn at the point where the density of toxic and non-toxic substances will be equal if we can rely on the prior likelihood that any new substance will have odds of 5:1 against being toxic. Compounds are then compared to this new boundary. The territory between lines (a) and

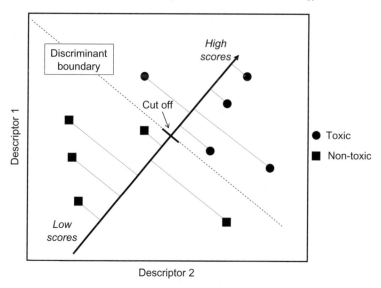

Figure 10.1 The underlying principal of discriminant analysis.

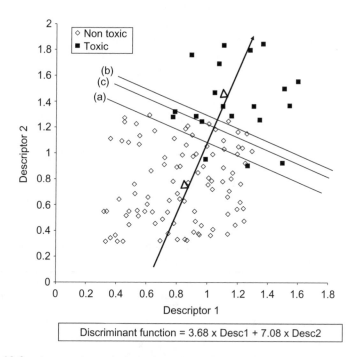

Figure 10.2 Approaches to the calculation of a cut-off point for discriminating between hypothetical toxic and non-toxic substances.

(b) is dominated by non-toxic substances and moving the line will mean that these would now all be classified as non-toxic and so the total number of cases of misclassification will be reduced—in fact to the minimum possible value.

If it is felt that the 5:1 ratio of non-toxic to toxic substances observed in the sample selected is likely to be biased and that there is no real idea of the prior likelihood of toxicity, then it may be necessary to revert to simple chemical similarity (line a). This is mathematically equivalent to using a prior likelihood set at 50:50.

10.2.2.3   Adding in Judgements of the Relative Costs of Misclassification

It might be judged that misclassifying a toxic substance as non-toxic would be more damaging than declaring a non-toxic one to be toxic (or *vice versa*). If it is felt that the former (false negative) was twice as damaging as the latter (false positive), then this relative cost of misclassification can also be added into the calculation of the placement of the boundary. Note that 'cost' is not necessarily financial cost; it can be any measure of harm done. If this is to be taken into account, then allowance must also be made for prior likelihood.

If the relative costs of misclassification are 2:1 (as above), then we should move the boundary so as to play safe and arrange for marginal substances to be classified as toxic. In Figure 10.2 this means moving the boundary to position (c). This is closer to (a) and will increase the total number of cases of misclassification, but on balance, total costs will be reduced by avoiding the more expensive misclassifications even if a greater number of less expensive ones are now generated.

- Using the prior likelihood alone minimises the total number of cases of misclassification.
- Adding in the cost factor minimises the total cost of misclassification.

Table 10.2 summarises the outcomes for the three classification criteria. Costs are calculated allowing one unit of cost per false positive and two per false negative.

Table 10.2   Summary of the outcomes when using different classification criteria for discriminant analysis.

Method	False positives	False negatives	Total mis-classifications	Cost of mis-classification
Chemical similarity alone	15	2	17	19
Prior likelihood	1	7	8	15
Prior likelihood and relative cost	5	4	9	13

10.2.2.4 Generating the Discriminant Function and Setting the Cut-Off Value in Statistical Packages

Simple statistical packages such as Minitab and SPSS are limited and default to using simple chemical similarity as the basis for the classification cut-off. Minitab can be coaxed to take account of both prior likelihood and relative costs, while SPSS has a limited ability to use the former and none for the latter. Instructions for using Minitab are available on the Liverpool John Moores University website (www.staff.ljmu.ac.uk/phaprowe/rsc/DiscriminantDetails.pdf).

10.2.3 Applying the Technique to the Reduced Set of Skin Sensitisation Data

Using the descriptors for the energy of the lowest unoccupied molecular orbital (E_{LUMO}) and calculated molecular volume (Vol) (data in Appendix 2.2) in the skin sensitisation spreadsheet as the descriptors allows some degree of separation between sensitisers and non-sensitisers, although discrimination is weak. Figure 10.3 presents the data as a scatter plot of the two descriptors.

10.2.3.1 Producing a Discriminant Function

The discriminant function (DF) is shown as Equation 10.1.

$$DF = -1.0597\,E_{LUMO} + 0.0129\,Vol \tag{10.1}$$

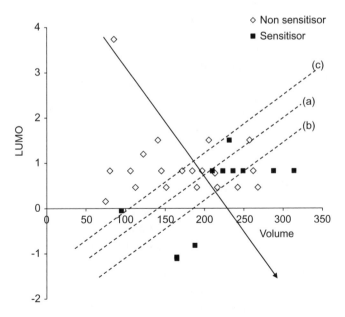

Figure 10.3 Using E_{LUMO} and volume to discriminate sensitising from non-sensitising substances

The ratio of the coefficients for the two descriptors in the above function is 82.15:1. In fact, any function that maintained this ratio would work equally well. Various different statistical packages may therefore generate functions that look different from the above, but so long as they maintain this ratio, they will simply re-scale the scores. The function is indicated as an arrow on Figure 10.3 with the direction of the arrow indicating that discriminant values are lowest at the top left hand end (high E_{LUMO}, low *Vol*) and increase to the bottom right (low E_{LUMO}, high *Vol*). The fact that the above function provides more than just chance separation of the two categories is attested to by a P value of 0.025. It is also possible to generate P values for the individual predictors within the model to allow the removal of any that are redundant. This is similar to the procedure for multiple linear regression (MLR) discussed in Section 9.2.4.4. Unfortunately, the simpler statistical packages are very parsimonious in their provision of these statistics.

10.2.3.2 Classification Using Cut-off Values

Consider first classification by simple chemical similarity. Using the function above, the non-sensitising and sensitising substances produce mean discriminant scores of 1.21 and 2.44 respectively and the mid-point between these is 1.825. Any compound yielding a score less than this value will be more similar to the group of non-sensitisers while those with higher scores resemble the sensitisers. The dotted line (a) in Figure 10.3 indicates the discriminatory surface based on this simple criterion. If this line is used to model the compounds, 14 of the 20 non-sensitisers are correctly classified with six incorrect, and nine of the 11 sensitisers are correct (two incorrect).

With the reduced data set, 11 out of 31 substances (35.5%) are sensitisers. If it can be assumed that this figure is a reasonable estimate of the prior likelihood of any molecule within the relevant group being a sensitiser, then this additional information should be incorporated. Using this empirical prior likelihood, the cut-off value shifts from 1.825 to 2.419 and the boundary moves to (b) in Figure 10.3.

Finally a misclassification cost can also be added. It will be assumed that misclassification of a sensitiser as a non-sensitiser is three times as damaging as the reverse error. This shifts the cut-off value to 1.321 and the new boundary is shown as (c) in Figure 10.3.

10.2.3.3 Error Rate and Predictive Ability

The overall fit achieved to the data can be expressed in what is sometimes referred to as a 'confusion table'. Where simple chemical similarity was used, the results were as in Table 10.3. The figures of nine and 14 represent those substances that were correctly predicted, and six and two are the numbers misclassified. Thus 23/31 (74.2%) of cases were correctly fitted.

In general, the proportion correctly fitted will increase if prior likelihood based on the observed data is incorporated. In the current case, small numbers and chance distribution of the data actually reduces the proportion correct to 22/31. One would always expect the proportion correctly predicted to decline if

Table 10.3 Confusion table for prediction of sensitisation by discriminant analysis using E_{LUMO} and volume as predictors.

		Observed property	
		Sensitiser	Non-sensitiser
Predicted property	Sensitiser	9	6
	Non-sensitiser	2	14

relative costs are incorporated, as minimisation of the number of incorrect assignments is no longer the goal.

The number of molecules successfully fitted can create an overly optimistic impression of goodness-of-fit unless this is contrasted with the number that would be expected by blind guessing. Tabachnick and Fidell show how this chance proportion should be estimated.[1] Using prior likelihood to carry out classification (as above), one would expect 35.5% of cases (11/31) to be classified as sensitisers and 35.5% of these would (by chance) fit that category (3.9 cases). Similarly 64.5% (20) would be classed as non-sensitisers and 64.5% of these would fit that category (12.9 cases). Thus in total, 16.8 would be guessed correctly. The achieved figure of 22 is something of an improvement on that but hardly stunning.

Leave-one-out (LOO) cross-validation can be used to avoid undue optimism. This operates in a similar way to that described in Section 9.2.9. Each compound is omitted one at a time and the discriminant function and cut-off value are re-calculated without that substance and then its group membership is predicted. Re-analysing the data using simple chemical similarity and LOO cross validation reduces the number of correct assignments only moderately (from 23 to 22/31). However it should be noted that all 31 substances were used to select E_{LUMO} and *Vol* as descriptors, so the real predictive power of any such approach is very uncertain.

10.2.3.4 *Interpretability of Discriminant Analysis*

The model generated is readily interpretable. Sensitisers tend to have high discriminant scores which, in turn, implies an association with low E_{LUMO} values and high Vol.

10.2.3.5 *Using Greater Numbers of Predictors and Predictor Selection*

The method was illustrated using only two descriptors, as this allows a simple pictorial presentation of the method; however the method can be extended to three or more descriptors. With three, the data are arranged in three-dimensional (3-D) space and the discriminant surface will then consist of a plane cutting the space into two. With more than three descriptors, the data have to be imagined in multi-dimensional hyper-space.

If *molecular mass* (data in Appendix 2.2) is used as an additional descriptor, the number of correctly modelled substances rises to 25/31, but there is then evidence of over-fitting as this model is more sensitive to LOO cross-validation (success rate falls to 22/31 correct). The 3-D model confirms the previous observation that sensitisation is associated with low E_{LUMO} and high *Vol* values, but suggests a further association with low molecular mass.

When many descriptors are available, collinearity is likely to raise its ugly head. As before (Section 9.2.2.4), there is no point in trying to use collinear groups of descriptors as they simply repeat the same information and destabilise models. One solution is to select a limited number of descriptors that avoid excessive collinearity using techniques such as 'best sets', 'step-wise selection' or genetic algorithms (see Section 9.2.7). Alternatively principle components analysis may be used to create uncorrelated PC scores (see Section 9.3). Discrimination may then be achieved by plotting (say) the first *versus* the second PC scores. Similar comments apply to the selection of descriptors for all the other modelling methods (logistic regression, *etc.*) mentioned in the rest of this chapter.

10.2.4   More Complex Patterns of Outcomes

Not all data will fall into just two categories that are best discriminated by a boundary in the form of a straight line. Figure 10.4 shows various patterns that might be encountered.

In the simplest case (Figure 10.4a), discrimination among three classes can still be achieved using a single discriminant function but with two cut-offs among the scores. In Figure 10.4b we would need two discriminant functions, each with a cut-off value creating its own boundary (boundaries shown as dotted lines). Even where there are just two categories, they may not be efficiently discriminated by a simple linear boundary (Figure 10.4c). In that case a non-linear boundary can be created by using a quadratic rather than simple linear function (similar to Section 9.2.6.1). The situation shown in (Figure 10.4d) is virtually intractable by discriminant analysis and is referred to as containing an 'embedded' class. Various methods have been proposed to develop QSAR models for embedded classes including cluster significance analysis[2] and embedded cluster modelling[3] but not discussed further in this chapter.

10.3   Logistic Regression

Logistic regression is similar to MLR (see Section 9.2). It uses a regression equation and the predictors are continuous variables, but the outcome is categorised rather than measured.

10.3.1   The Logistic Model

The probability that a given molecule will be a sensitiser could be treated as a dependent variable to be predicted from the value of a chemical descriptor

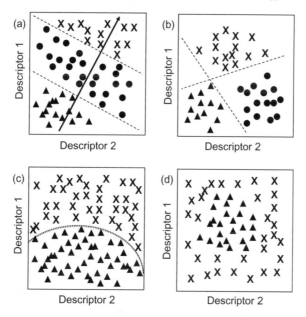

Figure 10.4 Discriminant analysis with more complex data structures. (a) Three classes separated by a single discriminant function. (b) Three classes requiring two discriminant functions. (c) Two classes requiring a quadratic function. (d) Two classes difficult to separate by discriminate analysis.

using a regression equation as in Section 10.2. In the example shown (Figure 10.5a), the greater the value of the descriptor, the more likely the molecule is to be a sensitiser. However, the regression line would extend indefinitely at both ends and include values of less than zero and more than one which are logically impossible as probabilities can only range from zero (no chance) to one (certain to occur). Figure 10.5b shows a more realistic relationship where the probability of being a sensitiser increases along with the value of the descriptor, but above a certain value of the descriptor, sensitisation is virtually certain and the graph flattens off. Similarly, there is a lower limit below which sensitisation is virtually excluded and the graph is again flat.

Thus, regression techniques should not be applied directly to predict the probability of an event because of these non-linear relationships. However, if we apply a suitable mathematical transformation to the probability, the relationship may be linearised. A commonly used transformation is the 'Logit' (log odds) of the probability. The 'odds' of an event occurring is defined as the ratio between the probability that the event will occur and the probability that it will not. Hence, if P is the probability that the event will occur:

$$Odds = P/(1-P) \tag{10.2}$$

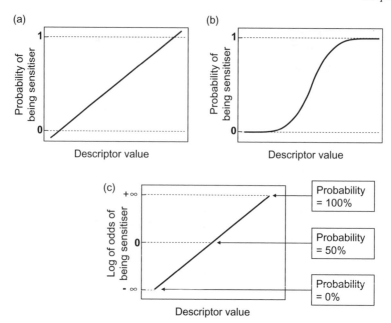

Figure 10.5 The Logit transform for linearising the relationship between the probability of a substance being toxic and a chemical descriptor. (a)Linear regression. (b) Non-linear regression. (c) Logistic regression.

The Logit of the probability of such an event is then defined as the natural log of these odds and so:

$$\text{Logit}(P) = \text{Ln}[P/(1-P)] \tag{10.3}$$

Table 10.4 shows that, in the most extreme case of a molecule with no possibility of sensitisation, the probability and its odds are zero and the log of the odds is therefore minus infinity. At the other extreme, a molecule that is certain to sensitise has a probability of one and then the odds and the Logit are plus infinity. So, while probabilities are restrained between values of zero and one, the Logit of the probability can meaningfully take any value from minus to plus infinity. In the neutral case of a molecule that has a 50:50 chance of being a sensitiser, the odds take a value of one and the Logit is zero. This is shown in Figure 10.5c.

10.3.2 Fitting a Logistic Model

Fitting the available data using a model such as that shown in Figure 10.5c is complicated by the fact that each substance is categorised as either a sensitiser or non-sensitiser. These can be recorded as values of one and zero, but we end up with data consisting of just these two values and no intermediate values and a special method of fitting is required.

Table 10.4 Relationship between the probability of an event and the Logit of its probability.

Situation	Probability = P	Odds = P/(1-P)	Logit = Log odds
Certain not to be sensitiser	0	0/1 = 0	Minus infinity
Neutrality – cannot predict	0.5	0.5/0.5 = 1	0
Certain to be sensitiser	1.0	1/0 = Infinity	Plus infinity

A regression equation is set up in the usual form:

$$\text{Logit}(P) = Constant + coefficient \times descriptor \qquad (10.4)$$

Because of the use of the Logit transform, such equations are given the special name of 'logistic regressions'.

To assess the fit of the regression equation to the data, each chemical is taken in turn and the Logit of the probability that it will be a sensitiser is determined using its value for the descriptor. Since a value of zero equates to the point of exact balance (Table 10.4), any positive value for the Logit indicates a greater than 50% likelihood of being a sensitiser and a negative value indicates that it is more likely to be a non-sensitiser. In essence, optimisation then consists of adjusting the values in the regression equation so that, (as far as possible) those chemicals that are known to be sensitisers produce positive Logits, and those that are not produce negative values. Technically, the fitting does not use the usual 'least squares' approach; instead it uses a criterion called 'maximum likelihood'. This means selecting a model such that the likelihood of observing the set of outcomes obtained here is as high as possible.[4]

Unlike least squares fitting, maximum likelihood fitting cannot be achieved in a single step. Instead an iterative approach is used. In the initial step, a first attempt at a fit is established. The constant and coefficients in the equation are then adjusted to achieve greater likelihood. These adjustments are repeated over and over, increasing the likelihood each time. Eventually a point is reached where further iterations produce minimal increases in likelihood. The computer algorithm will include a threshold below which these improvements are considered trivial and the process will terminate.

Equation (10.4) can be modified to make it a form of multiple regression, taking account of two or more descriptors:

$$\text{Logit}(P) = Constant + coefficent_1 \times descriptor_1$$
$$+ coefficent_2 \times descriptor_2 \ldots etc \qquad (10.5)$$

10.3.3 Fitting the Reduced Set of Skin Sensitisation Data and Assessing the Fit

E_{LUMO}, molecular mass and molecular volume are used as predictors for the likelihood that a substance will be a sensitiser using the reduced data set.

10.3.3.1. Producing a Logistic Regression Equation

The equation is:

$$Logit = 1.610 - 6.116 \, E_{LUMO} - 0.0900 \, Mass + 0.131 \, Vol \qquad (10.6)$$

P for $E_{LUMO} = 0.045$
P for $Mass = 0.051$
P for $Vol = 0.035$

Mass produces a P value that is technically just outside statistical significance, but it will be retained for illustrative purposes. In agreement with the conclusions from discriminant analysis, low values of E_{LUMO} are associated with greater likelihood of sensitisation, and a combination of low mass and high volume (low density) also seem to make sensitisation more likely.

10.3.3.2   Calculating the Likelihood that a Given Substance will be a Sensitiser

Using eqn (10.6), it is possible to take each compound and calculate a Logit value and hence the probability that it will be a sensitiser. This is illustrated below for one compound (chlorobenzene):

$$
\begin{aligned}
Logit(P) &= 1.610 - 6.116 \times 0.1547 - 0.0900 \times 112.56 + 0.131 \times 74.908 \\
&= 1.610 - 0.946 - 10.130 + 9.813 \\
&= 0.347
\end{aligned}
$$

The Logit of the probability is then converted to the probability (P) using eqn (10.7):

$$P = \text{Exp}(Logit)/[\text{Exp}(Logit) + 1] \qquad (10.7)$$

Thus in our example:

$$
\begin{aligned}
P &= \text{Exp}(0.347)/[\text{Exp}(0.347) + 1] = 1.415/2.415 = 0.586 \\
&= 58.6\% \text{ probability that it will be a sensitiser.}
\end{aligned}
$$

For any given compound, it will be predicted to be a sensitiser if its probability is greater than 50%; otherwise non-sensitisation is predicted. For chlorobenzene, the probability is (just) greater than 50% and so our prediction is that it will be a sensitiser. Unfortunately this particular prediction conflicts with the observed data. It is clearly foolhardy to rely on any prediction where the calculated likelihood of category membership is so close to 50%.

10.3.3.3. *Assessing the Effectiveness of the Model*

The overall effectiveness of the predictions can be illustrated using a graph such as Figure 10.6. Logit values are shown along the horizontal axis and the corresponding probability values are shown up the vertical. The sigmoidal graph shows that relationship. The key value of zero for the Logit, which corresponds to 50% probability, is shown as a dotted line. Each compound is superimposed on the graph by plotting its calculated Logit value on the horizontal axis with its vertical position fixed at either one or zero according to whether it is or is not observed to be a sensitiser. Twenty-two compounds have negative Logit values and are therefore predicted to be non-sensitisers. In practice, 17 of these are observed to fit this prediction, but five were found to cause sensitisation. Among the nine compounds predicted to be sensitisers, six fitted this category but three did not. Table 10.5 summarises the outcomes. The overall success was 23/31 cases (74.2%).

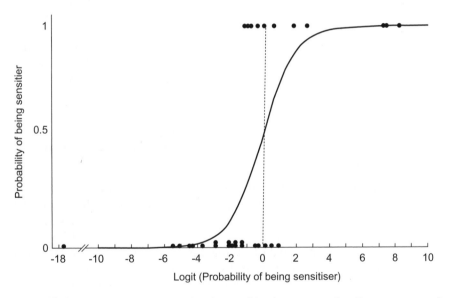

Figure 10.6 Binary logistic regression for sensitisation status using E_{LUMO}, *mass* and *volume*.

Table 10.5 Confusion table for prediction of sensitisation by binary logistic regression using E_{LUMO}, mass and volume as predictors.

		Observed property	
		Sensitiser	*Non-sensitiser*
Predicted property	Sensitiser	6	3
	Non-sensitiser	5	17

It is a general assumption that probabilities close to 50% will lead to unreliable predictions whereas those closer to zero or a 100% should be more certain. It is noticeable that all eight compounds that were misclassified had Logit values close to zero and therefore their probabilities for sensitisation were in the region of 50%. All the compounds for which a clearer prediction was possible were correctly classified.

10.3.4 Taking Account of Prior Likelihood and Relative Costs of Misclassification in Logistic Regression

The relative frequencies of the two (or more) categories present in the training data are automatically taken into account in logistic regression. Thus there is no need to add prior likelihood as a separate consideration in the way that was described for discriminant analysis (see Section 10.2.2.2).

In the description above it is assumed that, for binary categorisations, predicted class membership will simply be that for which likelihood is greater than 50%. Differing costs of misclassification could easily be accommodated by changing this criterion. For example, if it was felt that predicting a toxic molecule to be non-toxic would be three times as damaging as the reverse pattern, the criterion for a declaration of non-toxicity would be that the probability for this must be at least 75% (the point at which there will be a 3 : 1 ratio among the molecules).

10.3.5 Ordinal Logistic Regression—More Than Two Grades of Outcome

Logistic regression for dichotomised outcomes (an outcome did/did not arise) is referred to as 'binary logistic regression'. The technique can be extended to cover ordinal outcomes. These are situations where outcomes are classified into three or more groups and these groups have a natural order. The 'LLNA class' column in the sensitisation data base (data in Appendix 2.2) provides a more finely graded outcome measure recorded as 'none', 'weak', 'moderate', 'strong' and 'extreme'. These can be given suitable numerical codes such as 1–5.

10.3.5.1 Producing a Regression Model

Logistic regression then produces a series of equations that allow the calculation of the likelihood that a compound would fall into any given sensitisation category. The first equation calculates likelihood of membership of the lowest category, the next calculates likelihood of membership within either of the two lowest categories, and then the next does the same for membership within the three lowest and so on. Because of the cumulative nature of these predictions, the likelihood always increases working up through the series. If it is desired to calculate the likelihood of membership of a given category (say the second), then this can be calculated as the difference between the likelihood for categories 1 or 2 minus that for 1 alone.

With the reduced set of compounds, ordinal logistic regression generates the following coefficients:

Constant $(1) = -5.619$
Constant $(2) = -2.674$
Constant $(3) = 0.1373$
Constant $(4) = 2.155$
E_{LUMO} 5.782 $P < 0.001$
Mass 0.07384 $P = 0.003$
Vol -0.1041 $P = 0.001$

10.3.5.2 Calculating Likelihoods of Category Membership

As an example, 1-chlorononane has values for E_{LUMO}, mass and volume of 1.5073, 162.73 and 140.87 respectively. Hence the Logit(probability) of membership of the first category (non-sensitiser) is calculated using the first constant:

$$\text{Logit(Prob)} = -5.619 + 5.782 \times 1.5073 + 0.07384 \times 162.73$$
$$- 0.1041 \times 140.87 = 0.448$$

$$\text{Probability of membership of first category} = \text{Exp}(0.448)/[\text{Exp}(0.448) + 1]$$
$$= 0.610$$

Probability of membership within categories 1 or 2 (non or weak) is calculated as above, but replacing the first constant (-5.619) with the second (-2.674) yielding a probability of 0.967. The probability of membership within classes 1, 2 or 3 is 0.998 and for any of the first four classes is 0.9998. No further constant was provided (or necessary) as membership of one of the five categories must be 1.0.

The probability of membership of the second category can be calculated as the difference between that for categories 1 or 2 and that for 1 alone:

$$\text{Probability of membership of category } 2 = 0.967 - 0.610 = 0.357.$$

The probabilities of membership of the single categories 3, 4 or 5 are similarly calculated as 0.031, 0.0018 and 0.0002 respectively. These results, along with those for two strongly contrasting molecules, are shown in Figure 10.7.

If compounds are predicted as most likely to belong within any of the classes where that likelihood exceeds 25% (an arbitrary figure), 1-chlorononane would be predicted as being a member of categories 1 or 2, 1-bromopentadecane would probably belong in categories 2 or 3, and 1-chloromethylpyrene very likely belongs to category 5 (extreme). In all three cases, the observed categories agree with these predictions (see Figure 10.7).

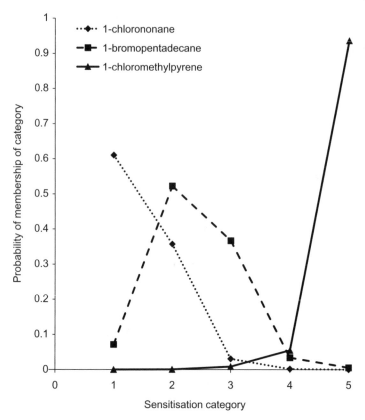

Figure 10.7 Logistic regression for sensitisation recorded on an ordinal scale scored 1 (none) to 5 (extreme). Observed sensitisation categories for 1-chlorononane, 1-bromopentadecane and 1-chloromethylpyrene are 1, 3 and 5 respectively.

10.3.5.3 Assessment of the Model

A spreadsheet (www.staff.ljmu.ac.uk/phaprowe/rsc/ordinal_logist.xls) shows similar calculations for all 31 compounds; it also identifies for each substance any category for which there is a 25% (or greater) probability of membership. For all substances, there are only one or two categories to which it is likely to belong. In all but four cases, these predicted categories include the true value for that compound ($27/31 = 87.1\%$ success).

It is not fair to compare this success rate against others quoted in this chapter as a different endpoint is being used, but the use of an endpoint recorded using five categories does emerge as considerably more powerful than the simple dichotomisation. The P values for E_{LUMO}, mass and volume were 0.045, 0.051 and 0.035 respectively when using the binary procedure but are now <0.001, 0.003 and 0.001 respectively. Markedly higher levels of statistical significance have been achieved. This illustrates the damaging loss of information when data are unduly simplified. The ordinal form of the data can offer considerably greater statistical power to identify which descriptors are related to the biological endpoint.

10.4   k-Nearest-Neighbour(s)

k-Nearest-neighbours (k-NN) models are conceptually very simple and have the advantage of making minimal statistical assumptions.

10.4.1   Distance Matrix

The method is based on the calculation of the distance between pairs of compounds. 'Distance' is generally defined in the simple Euclidean sense using space with as many dimensions as the number of descriptors considered. For two descriptors, the substances could be plotted onto a graph of one descriptor *versus* the other (as in Figure 10.8). Any two points on the graph would be vertically and horizontally displaced by distances D_1 and D_2. The distance between the points would then be that of the hypotenuse of a right-angled triangle which would be:

$$Distance = \sqrt{(D_1^2 + D_2^2)} \tag{10.8}$$

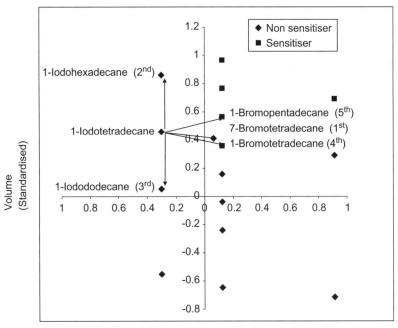

Figure 10.8   *k*-Nearest neighbour (k-NN) fitting of sensitiser status for 1-iodote-tradecane using Z score standardised volume and E_{LUMO}.

If further descriptors are used, the geometry becomes three (or more) dimensional and additional terms are simply added within the brackets of eqn (10.8).

It is essential that distance is determined using the descriptors after transformation to Z scores. Without this, descriptors may be recorded in different units and the one with the greatest standard deviation would meaninglessly predominate over the others. The distances for all pairs of compounds are then set out in a matrix that looks like a mileage chart used by motorists. This can then be used to identify the closest neighbours to each substance.

10.4.2   Prediction

The assumption is made that any compound is likely to belong to the same class as that of its closest neighbours. The number of neighbours inspected is represented as 'k'. For binary classifications, k is usually set to an odd number so that the set of nearest neighbours is guaranteed to contain a simple majority of one class or the other. With $k = 1$, prediction will be open to chance effects and greater numbers are likely to represent the surrounding territory more reliably. However, if k is too large, some of the compounds included will be excessively distant from the target molecule. Normal practice is therefore to try a range of values of k looking for an optimum. This process should be carried out on a cross-validated basis, although with this technique, LOO approaches are logically impossible.

10.4.3   Applying k-NN Prediction to the Sensitisation Data

Figure 10.8 zooms in to show just part of the skin sensitisation data as a graph of volume *versus* E_{LUMO} with each descriptor standardised as Z scores. As an example of class prediction, consider 1-iodotetradecane which is a non-sensitiser. Simple Euclidean distances are indicated from this molecule to its five nearest neighbours (1st to 5th). Prediction based on $k = 1$ would take account only of the nearest molecule (7-bromotetradecane), which is a non-sensitiser, leading to a correct prediction. Using $k = 3$, introduces two more molecules which are also non-sensitisers, giving a unanimous 3/3 correct prediction again. Raising k to 5 adds two sensitisers, and the prediction of non-sensitiser still stands but now only by a 3/5 majority. Using $k = 1$, 3 or 5 results in respectively 21, 22 and 19 correct predictions out of 31 cases. The number of compounds is too small to say anything definitive, but using $k = 3$ seems a reasonable compromise. (For the full details of the analysis see www.staff.ljmu.ac.uk/pha-prowe/rsc/kNN.xls).

The number of descriptors can be increased beyond two. For example, mass might be added which would leave the compounds positioned in 3-D space.

k-NN models may provide a useful predictive tool, but they have very limited interpretability. The above model suggests some form of relationship of E_{LUMO} and volume to sensitisation, but its nature is unknown.

10.5 Choice of Method for Categorised Data

Discriminant analysis makes extensive assumptions about the data. Among these it is assumed that the individual predictors follow normal distributions. For example, the E_{LUMO} values among substances that are sensitisers should be normally distributed. However, additionally there is a requirement for 'multivariate normality'. This means that any dimension through the predictors will yield scores that are normally distributed. Thus, any principle component of the predictors for (say) the sensitisers should produce normally distributed scores, and furthermore any dimension however oriented should fit this pattern. Failure to meet this assumption can affect estimates of statistical significance and prediction success rates. Logistic regression has the considerable advantage that it makes no assumptions about the data distribution.

There is evidence that, in some circumstances, dichotomised data may be more successfully modelled by logistic regression than by discriminant analysis.[5,6]

When just two predictors are being used, discriminant analysis does have the slight advantage that its method of working can be easily grasped from an illustration such as Figure 10.1. Logistic regression is (sadly) less visual in nature.

k-NN analysis makes no assumptions about the patterns in which substances are distributed within the data space. For example, it would be able to make predictions about embedded classes (Figure 10.4d) for which discriminant analysis and logistic regression have little to offer. However, it lacks the clear interpretability of discriminant analysis and logistic regression.

On balance, logistic regression should probably be the first thing to try and it is worth remembering that endpoints expressed on a multi-point ordinal scale are likely to carry considerably more information than dichotomisations.

10.6 Conclusions

A number of statistical methods have been widely applied by the QSAR community to predict the membership of a compound to a particular toxicological class. These are particularly appropriate for (semi-)qualitative endpoints such as the presence or absence of an effect or gradings (*e.g.* non, low, moderate or high, *etc.*) This chapter has focussed on three such methods—discriminant analysis, logistic regression and k-nearest neighbours. More details and further methods can be found throughout the published literature and in commercial software. For good introductory reading material, readers are referred elsewhere in the first instance.[7,8,9]

References

1. B. G. Tabachnick and L. S. Fidell, *Using Multivariate Statistics,* Pearson, Boston, 2007.
2. J. W. McFarland and D. J. Gans, *Quant. Struct.-Act. Relat.*, 1994, **13**, 11.

3. A. P Worth and M. T. D. Cronin, *Quant. Struct.-Act. Relat.*, 1999, **18**, 229.
4. J. J. Hox, *Multilevel Analysis,* Lawrence Erlbaum, Mahwah, NJ, 2002.
5. S. E. Fienberg, *The Analysis of Cross-Classified Categorical Data,* Springer, New York, 2007.
6. S. J. Press and S. Wilson, *J. Am. Stat. Assoc.*, 1978, **73**, 699.
7. A. P. Worth and M. T. D. Cronin, *J. Mol. Struct. (Theochem.)*, 2003, **622**, 97.
8. D. Livingstone, *A Practical Guide to Scientific Data Analysis,* John Wiley and Sons, Chichester, 2009.
9. Organisation for Economic Co-operation and Development, *Guidance Document on the Validation of (Quantitative) Structure–Activity Relationships [(Q)SAR] Models*, OECD, Paris, 2007, OECD Environment Health and Safety Publications Series on Testing and Assessment No. 69, ENV/JM/MONO(2007)2.

CHAPTER 11

Characterisation, Evaluation and Possible Validation of In Silico Models for Toxicity: Determining if a Prediction is Valid

M. T. D. CRONIN

School of Pharmacy and Chemistry, Liverpool John Moores University, Byrom Street, Liverpool L3 3AF, UK

11.1 Introduction

In silico models for the prediction of toxicity—as well as physico-chemical and fate properties—may be used in a variety of situations. These range from reducing the hazard, through the design of safer new compounds, assessing existing compounds for hazard and identifying priorities for assessment.[1] However, for predictions from any *in silico* model to be useful, there must be some assessment of the validity of the prediction. This means whether the prediction can be considered to be reliable and the degree of confidence (if any) the user may place in the predicted value. This process is complex for any type of prediction and will involve consideration of the model, its performance and whether the model is appropriate to make a prediction. Thus it will require the collation of 'evidence' regarding the model and its ability to make predictions,

Issues in Toxicology No.7
In Silico Toxicology: Principles and Applications
Edited by Mark T. D. Cronin and Judith C. Madden
© The Royal Society of Chemistry 2010
Published by the Royal Society of Chemistry, www.rsc.org

as well as 'evidence' as to whether it is appropriate to be applied in the particular context being considered.

For regulatory acceptance of a prediction (normally of toxicity but also of physico-chemical or fate properties), it is likely that the model itself, as well as the prediction, should be assessed. This is a formal process of documentation, description and evaluation. Whilst the concepts and techniques are likely to be useful for the identification and development of a new compound (*i.e.* in R&D), the level of documentation and formal reporting required is likely to be less than for regulatory use of a prediction. This chapter aims to describe the process by which a (Q)SAR may be described and evaluated with a view to validating it for regulatory use; it directs readers to relevant sources of information and illustrates how the information can be applied.

11.2 A Very Brief History

The possibility of using *in silico* approaches to predict toxicity in response to regulatory pressure has been promoted for at least two decades.[2,3] This stimulus has been supplemented by ethical and financial pressures to reduce animal use in toxicity testing, societal pressure for safer chemicals and a healthier environment. The result has been a strong development in predictive toxicology methods and applications.

There is no single piece of legislation that is accountable for this direction being taken, but historically the Toxic Substances Control Act (TSCA) in the United States, as well as consideration of the effects of High Production Volume (HPV) chemicals have encouraged the development and use of (Q)SARs and categories.[2,3] More recently assessment of the Domestic Substance List (DSL) in Canada, the Registration, Evaluation, Authorisation and restriction of Chemicals (REACH) legislation and the ban on testing of cosmetic ingredients in the European Union have led to the need for the application of existing methods and development of new methods.[4]

The context for using QSARs specifically for REACH is well described by Worth.[5] It is deemed possible to use data from (Q)SAR models instead of experimental data if each of four main conditions is fulfilled:

 (i) The model used is shown to be scientifically valid.
 (ii) The model used is applicable to the chemical of interest.
 (iii) The prediction (result) is relevant for the regulatory purpose.
 (iv) Appropriate documentation on the method and result is given.

The requirement for approaches to satisfy, in a regulatory context, these four criteria was recognised early on.[6] A large number of projects and initiatives were started as a result of the prospect of the REACH legislation. Some of these were at the national level, other across the EU and others at an international level crossing continental boundaries (*e.g.* the OECD approaches and Toolbox described in Chapter 16). One particular project was to provide approaches and guidance on how to assess the validity of a (Q)SAR; this resulted in 'OECD Principles for the Validation, for Regulatory Purposes, of (Quantitative) Structure-Activity Relationship Models' (referred to hereafter as the OECD Principles), which were agreed through the Organisation for Economic Co-operation and Development (OECD) in November 2004.[7,8]

11.3 OECD and European Chemicals Agency Documents and Further Background Papers and Sources of Information

This chapter provides an overview and some pointers for the interested reader. It cannot, and is not intended to, be wholly comprehensive regarding the use of the appropriate methods and approaches to assess the validity of a (Q)SAR prediction; it is only a starting point. It is strongly recommended that readers acquaint themselves with the information provided in OECD and European Chemicals Agency (EChA) guidance as well as other key references.[5,7–10] The purpose of this chapter is merely to introduce the concepts; much has been taken from the guidance (particularly the definitions, OECD Principles, OECD Checklist (Table 11.1), *etc.*), and the application of this process should use the original source information and guidance.

11.4 Definition of Terms

A number of terms are defined and explained below. Where noted these definitions are taken from the relevant guidance.[7–9] If no reference is made to these documents then readers must assume that they are the author's interpretation of the definition and, as such, have no legal status—they are provided for illustration only.

11.4.1 Characterisation

The term 'characterisation of a (Q)SAR' is not an official regulatory phrase. It is used in this context to mean the description of a (Q)SAR, *i.e.* elucidation (where possible) of the algorithm, compounds and endpoint. This has required a terminology to be developed and also that the person performing the characterisation is familiar with (Q)SAR principles and methods (many of which can be found in this book). The characterisation of the model is one of the first steps in fulfilling the information requirements of the OECD Principles. A document has been developed describing the characterisation process;[11] much of this information can also be found in the OECD Guidance.[8]

11.4.2 Evaluation

Again, the term 'evaluation of a (Q)SAR' is not an official regulatory phrase. In this context it is taken to mean the consideration of the (Q)SAR in terms of its mechanistic significance (if any), statistical robustness and predictivity, applicability domain and any other relevant aspect of its description. The evaluation of a (Q)SAR is in no way independent of the characterisation and also forms the basis of the fulfilment of the requirements for the OECD Principles. The guidance associated with the OECD Principles, summarised below, provides an ideal framework for both characterisation and evaluation of *in silico* models.

11.4.3 Validation, Reliability and Relevance

There can be some ambiguity in the use of the word validation according to the context within which it is used. Here the OECD definitions of the terms 'validation', 'reliability' and 'relevance' are used:

According to the OECD guidance, the term 'validation' is defined as follows:[8]

'Validation: The process by which the reliability and relevance of a particular approach, method, process or assessment is established for a defined purpose.'

The terms 'reliability' and 'relevance' are defined as follows:

'Reliability: Measures of the extent that a test method can be performed reproducibly within and between laboratories over time, when performed using the same protocol. It is assessed by calculating intra- and inter-laboratory reproducibility and intra-laboratory repeatability.'

'Relevance: Description of relationship of the test to the effect of interest and whether it is meaningful and useful for a particular purpose. It is the extent to which the test correctly measures or predicts the biological effect of interest. Relevance incorporates consideration of the accuracy (concordance) of a test method.'

These are general definitions, and with regard to (Q)SARs, the assessment of reliability places a greater emphasis on the accuracy of the (Q)SAR predictions for different chemicals rather than on the reproducibility of the (Q)SAR within and between laboratories (which should be irrelevant for a computational method). In addition, the assessment of the relevance of the (Q)SAR model must be made with respect to the expected molecular interactions and pathways by which each causes the biological effect.

11.4.4 Valid Prediction

According to the above definitions, a (Q)SAR can be characterised, evaluated and undergo a process of validation. The model can then be used to make a prediction and the prediction assessed as to whether it is valid. For a (Q)SAR prediction to be adequate for a given regulatory purpose, the estimate should be generated by a valid model, and the model should be applicable to the chemical of interest with the necessary level of reliability.[5] How does this ensure a prediction is correct? A better line of thinking is to consider whether a prediction is 'valid'. Currently there is no accepted (for regulatory purposes) definition of what may constitute a valid prediction, but clearly it would be for a compound that falls within the applicability domain (in terms of structure, physico-chemical properties, metabolism and mechanism/mode of action—see Chapter 12) of a (Q)SAR that has undergone a validation procedure.

This process of selecting a model to achieve a valid prediction is shown as a workflow in Figure 11.1.

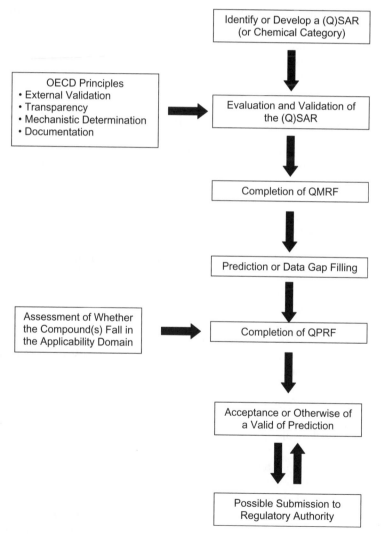

Figure 11.1 General workflow for producing a valid prediction.

11.4.5   A Comment on Validation of (Q)SARs

One of the key words in the title of the OECD Principles is 'validation'. This has caused some confusion in the past due to the different uses of the word 'validation'. For instance, in the context of *in vitro* tests for toxicology, validation represents a formal process whereby an *in vitro* test may be deemed acceptable for the replacement of an *in vivo* assay.[12] For a chemoinformatician, validation is the process whereby the statistical robustness of a QSAR is assessed. In the terms of the OECD Principles (and this chapter), the term validation is taken to mean somewhere between these two meanings. It is not the intention that the process of validation of a (Q)SAR (as described here) is to replace an animal test formally. Neither is it merely a process of the assessment statistical robustness, it is much more than that. Therefore, the term validation in the context of this chapter is taken to include the formal processes to characterise, describe and evaluate a (Q)SAR or category in terms of its biological data, chemistry and statistical robustness. The formal definitions are given in the key texts and further defined below.[7–9]

11.5   OECD Principles for the Validation of (Q)SARs

The OECD Principles are described briefly below. However, before attempting validation of a (Q)SAR, it is recommended that readers become acquainted with the OECD Guidance Document which provides considerably more detail.[8]

At the heart of the validation of a (Q)SAR are the five Principles agreed in 2004 with the purpose being:

'To facilitate the consideration of a (Q)SAR model for regulatory purposes, it should be associated with the following information:

1. A defined endpoint
2. An unambiguous algorithm
3. A defined domain of applicability
4. Appropriate measures of goodness-of-fit, robustness and predictivity
5. A mechanistic interpretation, if possible.'

In addition to their formal definition in the OECD Guidance Document,[8] much has already been written about these five Principles.[13–21] Some brief comments are provided below along with the types of questions that their consideration should stimulate.

The easiest way to interpret the information below, in the first instance, is with reference to the checklist provided in Table 11.1 and also in the illustrative case studies provided in Section 11.8. Table 11.1 is checklist for the OECD Principles which has been adapted from the OECD Guidance Document. The purpose is to allow the user to assess the Principles against these measurable criteria.

Table 11.1   Check list for providing guidance on the interpretation of the
OECD Principles for the Validation of (Q)SARs (adapted from
OECD guidance[8]).

Principle	Considerations: Is the following information available for the model?
1) Defined endpoint	
1.1	A clear definition of the scientific purpose of the model (*i.e.* does it make predictions of a clearly defined physico-chemical, biological or environmental endpoint?)
1.2	The potential of the model to address (or partially address) a clearly defined regulatory need (*i.e.* does it make predictions of a specific endpoint associated with a specific test method or test guideline?)
1.3	Important experimental conditions that affect the measurement and therefore the prediction (*e.g.* sex, species, temperature, exposure period, protocol)
1.4	The units of measurement of the endpoint
2) Defined algorithm	
2.1	In the case of a SAR, an explicit description of the substructure, including an explicit identification of its substituents
2.2	In the case of a QSAR, an explicit definition of the equation, including definitions of all descriptors
3) Defined domain of applicability	
3.1	In the case of a SAR, a description of any limits on its applicability (*e.g.* inclusion and/or exclusion rules regarding the chemical classes to which the substructure is applicable)
3.2	In the case of a SAR, rules describing the modulatory effects of the substructure's molecular environment
3.3	In the case of a QSAR, inclusion and/or exclusion rules that define the following variable ranges for which the QSAR is applicable (*i.e.* makes reliable estimates): a) descriptor variables b) response variables
3.4	A (graphical) expression of how the descriptor values of the chemicals in the training set are distributed in relation to the endpoint values predicted by the model
4A) Internal performance	
4.1	Full details of the training set given, including details of: a) number of training structures b) chemical names c) structural formulae d) CAS numbers e) data for all descriptor variables f) data for all response variables g) an indication of the quality of the training data
4.2	a) an indication whether the data used to the develop the model were based upon the processing of raw data (*e.g.* the averaging of replicate values) b) if yes to a), are the raw data provided? c) if yes to a), is the data processing method described?

Table 11.1   (*Continued*)

Principle	*Considerations: Is the following information available for the model?*
4.3	An explanation of the approach used to select the descriptors, including: a) the approach used to select the initial set of descriptors b) the initial number of descriptors considered c) the approach used to select a smaller, final set of descriptors from a larger, initial set d) the final number of descriptors included in the model
4.4	a) a specification of the statistical method(s) used to develop the model (including details of any software packages used) b) if yes to a), an indication whether the model has been independently confirmed (*i.e.* that the independent application of the described statistical method to the training set results in the same model)
4.5	Basic statistics for the goodness-of-fit of the model to its training set (*e.g.* r^2 values and standard error of the estimate in the case of regression models)
4.6	a) an indication whether cross-validation or resampling was performed b) if yes to a), are cross-validated statistics provided, and by which method? c) if yes to a), is the resampling method described?
4.7	An assessment of the internal performance of the model in relation to the quality of the training set and/or the known variability in the response
4B) Predictivity 4.8	An indication whether the model has been validated by using a test set that is independent of the training set
4.9	If an external validation has been performed (yes to 4.8), full details of the test set, including details of: a) number of test structures b) chemical names c) structural formulae d) CAS numbers e) data for all descriptor variables f) data for all response variables g) an indication of the quality of the test data
4.10	If an external validation has been performed (yes to 4.8): a) an explanation of the approach used to select the test structures, including a specification of how the applicability domain of the model is represented by the test set b) was the external set *sufficiently large and representative* of the training data set? c) a specification of the statistical method(s) used to assess the predictive performance of the model (including details of any software packages used) d) a statistical analysis of the predictive performance of the model (*e.g.* including sensitivity, specificity, and positive and negative predictivities for classification models)

Table 11.1   (*Continued*)

Principle	Considerations: Is the following information available for the model?
	e) an evaluation of the predictive performance of the model that takes into account the quality of the training and test sets, and/or the known variability in the response
	f) a comparison of the predictive performance of the model against previously defined quantitative performance criteria
5) **Mechanistic interpretation**	
5.1	In the case of a SAR, a description of the molecular events that underlie the properties of molecules containing the substructure (*e.g.* a description of how substructural features could act as nucleophiles or electrophiles, or form part or all of a receptor-binding region)
5.2	In the case of a QSAR, a physicochemical interpretation of the descriptors that is consistent with a known mechanism of (biological) action
5.3	Literature references that support the (purported) mechanistic basis
5.4	An indication whether the mechanistic basis of the model was determined *a priori* (*i.e.* before modelling, by ensuring that the initial set of training structures and/or descriptors were selected to fit a pre-defined mechanism of action) or *a posteriori* (*i.e.* after the modelling, by interpretation of the final set of training structures and/or descriptors)

11.5.1   A Defined Endpoint

The purpose of this Principle is to assess whether the (Q)SAR has been derived from 'suitable' data. Thus, the types of questions that that need to be answered include:

- What is the endpoint that is modelled and the method by which the data were determined?
- Is the endpoint of relevance to a regulatory decision or, for instance, product development?
- Are the data taken from a consistently performed assay? This is very much a question that requires consideration of data and data set quality (see Chapters 3 and 4 respectively).
- Is the protocol(s) used to derive the assay reliable according to current knowledge, *e.g.* a current OECD or equivalent accepted assay?

The OECD Guidance Document provides copious illustrations for many types of endpoints as to the issues that can support the conclusion of a 'defined endpoint'.

11.5.2 An Unambiguous Algorithm

The purpose of this Principle to assess whether the (Q)SAR can be defined (for instance written down) in a clear manner so that it is transparent and portable. Some types of QSAR algorithm (*e.g.* regression analysis) lend themselves very well to this definition whereas, for others (*e.g.* neural networks), it is more difficult to define the unambiguous algorithm. This goes beyond a simple question of where the equation for a QSAR is defined, but also considers the availability of the chemicals on which it is based and its description. Thus, the types of questions that need to be answered include:

- For a SAR, is a direct relationship between chemical structure(s) and/or sub-structure(s) and the effect defined?
- For a category, is it defined in terms of chemical structures and (possibly) physico-chemical domain?
- Is the (Q)SAR known and/or can it be derived?
- For a QSAR, is some type of mathematical model present in a clear and interpretable manner?

If the relevant questions above are satisfied, then the availability of the following information should be assessed as appropriate:

- the chemical structures (or their identification), effect values which are modelled and the values for the descriptors in the (Q)SAR;
- a clear description of how the descriptors were obtained (*i.e.* calculation or experimental measurement) such that this step could be repeated;
- a clear description of the test and training sets and, if outliers were removed, a clear description and justification for this;
- a description of the statistical model(s) used to explore the relationships between the effects and descriptors, as well as any variable selection; and
- the statistical parameters describing the model fit and predictivity (this is expanded upon in the fourth Principle).

The OECD Guidance Document provides useful information relating to common statistical methods to assist in the answering of these questions. Readers will also find the descriptions of common statistical methods in Chapters 9 and 10 of assistance in clarifying these issues.

11.5.3 A Defined Domain of Applicability

The purpose of this Principle is to encourage the definition and description of the applicability domain of a (Q)SAR. This is vital for users to have confidence that any chemical for which they make a prediction falls within the applicability domain of the model. The definition of term 'applicability domain' of a model and how it can be defined is provided in Chapter 12 (readers are also referred to other papers[22,23]).

Applicability domains can be defined in a number of ways, each of which is usually required, *e.g.*

- structural domain (chemical structures and fragments);
- physico-chemical property domain—this may be the domain of the descriptors in the (Q)SAR itself or could be solubility or other physico-chemical property cut-offs (which may not themselves form part of the model); and
- mechanistic domain.

To determine if these aspects of the domain have been described adequately, the following types of questions that require answers include:

- Has a recognised method (*e.g.* AMBIT—described fully in Chapters 12 and 17) been used to define the domain of the (Q)SAR?
- Has the training set of the (Q)SAR been described in terms of the descriptor domain, chemical structures, physico-chemical properties and/ or mechanisms?
- Have any limits of the domains been implemented?
- Are any properties, fragments or mechanisms specifically included in, or excluded from, the domain?

The OECD Guidance Document provides detailed information on how to describe the applicability domain of a model. It is recognised that such definition has not been regularly performed and, for historical models (see the case studies below), it will be absent. Readers are also referred to Chapter 12 for practical assistance in domain definition.

The assessment of a QSAR can be performed initially with the yes/no checklist in Table 11.1. In the author's opinion this should not be viewed as a pass or fail checklist, and it is unlikely there will be many 'perfect' QSARs that tick all the boxes. However, it assists in the appropriate characterisation of a QSAR. Therefore it is important that there is a suitable format to capture the information generated. This can be achieved by the QSAR Model Reporting Format (QMRF) described below.

11.5.4 Appropriate Measures of Goodness-of-fit, Robustness and Predictivity

This Principle allows for the statistical fit and predictivity of a (Q)SAR to be assessed. There are many methods to assess statistical fit of a model and using the appropriate method will depend on the type of model assessed. Models based on categorical (qualitative) data require different diagnostic statistics from quantitative data (see Chapters 9 and 10). In addition, there are a variety of methods to achieve 'external' validation, *i.e.* prediction of the activities of compounds not included in the model—thus providing an estimate of predictivity.[24] There should be some realism and pragmatism with regard to finding true test sets, *i.e.* compounds not included in a model; this may not be possible for every model due to data availability. Thus sub-sampling or cross-validation may be required.

In order to fulfil the criteria for this principle, the following questions should be asked:

- Are full details of the training and (if available) test set given?
- Is there access to the raw data?
- Is the model development procedure, especially with regard to variable selection for QSARs, described and transparent?
- If compounds (such as outliers) have been removed from a QSAR, is this recorded and accompanied with a reasonable justification?
- Is the statistical method along with statistics for goodness of fit and predictivity described?
- Was the external validation (if performed) appropriate in terms of data quality, fit into the applicability domain and number of compounds?

The OECD Guidance Document provides full and unambiguous definitions of many statistical criteria for describing a (Q)SAR.[8] As well as this source of information, readers should also look at the discussion and debate regarding the correct use of statistical terms from Aptula *et al.*,[25] Gramatica *et al.*[26] and Schüürmann *et al.*;[27] in addition, Chapters 9 and 10 are intended to provide valuable background information.

11.5.5   A Mechanistic Interpretation, if Possible

The purpose of this Principle is to provide a mechanistic rationale and basis for any model. This is important as it will increase confidence in the model that it is a causal, not casual, relationship. This aspect of the Validation Principles is by far the most controversial with some researchers resisting the call to build mechanistic models (and indeed there may be occasions when they are not desirable, such as when mechanisms are not known). Despite this, every attempt should be made to develop mechanistically interpretable models. In order to fulfil the principle, the following questions should be asked:

- Is there a known mechanistic basis to the (Q)SAR or category that is supported in the scientific literature?
- Were descriptors used in a model determined before the modelling process (*i.e.* an empirical selection on a mechanistic basis) or as a result of statistical analysis?

The OECD Guidance Document provides further illustration and examples of how descriptors can be interpreted mechanistically.[8] This is probably the most subtle of all the OECD Principles and requires some knowledge of the underlying interaction between the chemical and the material (biological or otherwise) on which it is acting. Mechanistic interpretability goes beyond simply stating that a descriptor in a model relates to a property (*e.g.* the common type of statement 'the inclusion of log P in the model indicates that hydrophobicity is important for biological activity'); instead it should say why the property is important in terms of the effect being modelled (*e.g.* the previous statement could be complemented by saying (for non-polar narcosis—Section 11.8.1 Case Study 1) 'hydrophobicity is a known determinant of narcotic activity as it controls the capability of the chemical to accumulate in biological membranes').

11.6 Examples of Validation of (Q)SARs from the Literature

As well as a number of examples of the OECD Principles being applied which are available in OECD documents (OECD guidance and reports), an increasing number of literature reports have attempted to describe (Q)SARs for toxicity in terms of the OECD Principles. A number of these reports are summarised in Table 11.2. Reader are strongly encouraged to consider at least a selection of these papers (*e.g.* for relevant endpoints) as they continue to learn about the validation procedures.

11.7 Describing a QSAR: Use of the QSAR Model Reporting Format

In order to characterise and describe (Q)SARs, a QSAR Model Reporting Format (QMRF) has been proposed. The OECD Guidance Document states that the (Q)SAR Model Reporting Format (QMRF) is a framework for structuring and summarising key information about a model in order to provide the end-user with details on:

a) The source of the model (including the developer, where known), *e.g.* reference for a literature model or expert system, *etc.*
b) Model type, *e.g.* structural alert (SAR) or regression based, neural network, *etc.,* QSAR, expert system, *etc.*
c) Model definition, *e.g.* the algorithm
d) The development of model, *e.g.* the training set of chemicals, descriptors considered and selected
e) The validation of the model, *e.g.* the predictivity of the test set
f) Possible applications of the model.

Whilst the QMRF has not been formally agreed upon and is likely to undergo a series of evolutions, Chapter R.6 of the EChA Guidance Document[9] provides a detailed description of the type of information that will be included. This will

Table 11.2 Examples of the application of the OECD Principles to toxicological endpoints

Endpoint	Reference
Acute fish toxicity[a]	OECD Expert Group report[7]
QSARs for atmospheric degradation[a]	OECD Expert Group report[7]
QSARs for mutagenicity and carcinogenicity[a]	OECD Expert Group report[7]
Multi-CASE model for *in vitro* chromosomal aberrations[a]	OECD Expert Group report[7]
Multi-CASE and MDL models for human NOEL[a]	OECD Expert Group report[7]
ECOSAR[a]	OECD Expert Group report[7] Hulzebos *et al.*[10]
BIOWIN[a]	OECD Expert Group report[7] Hulzebos *et al.*[10]
DEREK[a]	OECD Expert Group report[7] Hulzebos *et al.*[10]
RIVM[a]	OECD Expert Group report[7]
DEREK skin sensitisation rulebase[a]	OECD Expert Group report[7]
Japanese METI biodegradation model[a]	OECD Expert Group report[7]
Rat oral chronic toxicity models in TOPKAT[a]	OECD Expert Group report[7]
Derek for Windows model for skin sensitisation[b]	OECD guidance[8]
Multi-CASE Model for *in vitro* chromosomal aberrations in mammalian cells[b]	OECD guidance[8]
Fish acute neutral organics 96-hour (Q)SAR, a constituent of ECOSAR[b]	OECD guidance[8]
CATABOL for biodegradation[b]	OECD guidance[8]
BIOWIN for biodegradation[b]	OECD guidance[8]
Skin permeability (permeability coefficient, maximum flux)	Bouwman *et al.*[13]
Oestrogen receptor binding	Liu *et al.*[14]
Acute fish toxicity	Pavan *et al.*[15]
Oestrogen gene activation	Saliner *et al.*[16]
Acute fish toxicity	Vracko *et al.*[17]
Various including acute toxicity, mutagenicity, carcinogenicity, and other health effects	Saiakhov and Klopman[18]
Night-time degradation rates with the nitrate radical	Papa and Gramatica[19]
Mutagenicity	Gramatica *et al.*[20]
Skin sensitisation	Roberts *et al.*[21]

[a]A total of six principles for validation were considered in these reports, these were originally termed the 'Setúbal Principles' and were translated into the five OECD Principles.
[b]The description of the (Q)SARs is given in the context of the QSAR Model Reporting Format.

include, in the electronic version, drop-down menus for many of the options. The current structure of the QMRF is extremely helpful in guiding the model developer or user through the OECD Principles. The QMRF should, therefore, be considered at the same time as the OECD Principles. Within the QMRF there are also opportunities for free text entry, so that more details can be given on the description of the model and, for instance, mechanistic interpretation.

The QMRF comes in a number of formats; readers are referred to the website (http://ecb.jrc.ec.europa.eu/qsar/) of the former European Chemicals Bureau— now the Computational Toxicology Group of the Institute for Health and

Consumer Protection—at the European Commission's Joint Research Centre for assistance.

The QMRF operates *via* a Microsoft® Excel file with various worksheets within an overall Excel workbook. One Excel workbook is needed for each (Q)SAR model. An easy way to record the information regarding a (Q)SAR into a QMRF is through QMRF Editor—available through the JRC's website and also from the developer IdeaConsult (http://ambit.sourceforge.net) in Sofia, Bulgaria. The QMRF Editor ultimately produces a PDF file which can be saved or printed. A number of example QMRFs are available from the JRC's website (http://ecb.jrc.ec.europa.eu/qsar/).

Another application that will produce a QMRF (as an Excel file) is through the QSAR generator function of the OECD Toolbox (see Chapter 16). This is extremely helpful, although it is the author's experience that the QMRF from the OECD Toolbox may need some of the information entering manually such as the nature of the test, types of descriptors, *etc.* It is also restricted to those models (or categories) created in the Toolbox.

A further bonus of the QMRF is that it forms an electronic method of storing QSARs. The JRC has on its website (noted above) a QSAR model database which contains a growing number of QMRFs and also the related QSARs. When fully built, this will allow searching of QSARs by the various items they contain (*e.g.* endpoint, compounds, descriptors, *etc.*).

11.8   Case Studies: Application of the OECD Principles

A small selection of QSAR models are evaluated below to illustrate how the OECD Principles can be applied and the types of questions that should be addressed. These case studies have been chosen because they are published models and represent a range of modelling approaches and philosophies as well as a number of challenges for their assessment by the OECD Principles. The three papers span three decades of science and very different approaches to modelling and technologies. To understand some of the comments made in these brief assessments, readers are referred to the original papers.

11.8.1   Case Study 1: QSAR for Non-Polar Narcosis

In 1981, Professor Hans Könemann published what became a seminal paper[28] relating the 14-day toxicity to the guppy (LC_{50} in molar units) of 50 neutral organic compounds to the logarithm of the octanol–water partition through the following equation:

$$\text{Log } 1/LC_{50} = 0.871 \text{ log } P - 4.87$$
$$n = 50, \quad r = 0.988, \quad s = 0.237$$

(11.1)

where:

n is the number of compounds
r is the correlation coefficient
s is the standard error.

The original paper by Könemann is rightly considered to be one of the most important in stimulating the use of QSAR to predict acute environmental toxicity.[28] Whilst it is very simple, it provides the basis for predicting the acute toxicity of a large number of chemicals to fish. Its importance, significance and continued relevance are undeniable and it is interesting to consider it through the OECD Principles, particularly in the context of when it was developed. There was no computational method available in 1981 when the paper was published to deal with chemical structures or calculate log P (although fragment methods existed), software to calculate a regression equation was in its infancy.

Those readers interested in the application of the OECD Principles to the prediction of acute fish toxicity are referred to Table 11.2, and the examples reported in the OECD Expert Group's report[7] and by Pavan *et al.*,[15] as well as the example QMRF published on the JRC's website. The information in these articles will supplement that given briefly below.

11.8.1.1   A Defined Endpoint

Könemann gives the full method for assessment of acute fish toxicity; the endpoint is the lethal concentration (LC_{50}). The tests were undertaken using the guppy (*Poecilia reticulata*). This is a method that was relevant to regulatory decisions at that time and is also relevant at the present time. Whilst the method is described, it clearly predates any OECD guideline (so is best thought of as an in-house method) and has not been performed to Good Laboratory Practice (GLP) standards.

The methods imply, although do not explicitly state, that the tests were all performed in the same lab, *i.e.* that of the author. Therefore the protocol can be considered to be consistent. An interesting aspect to the data measurement is that toxicity was measured either at 14 or 7 days. In terms of data quality (see Chapter 4) this may be considered to be inconsistent.

11.8.1.2   An Unambiguous Algorithm

A clear and unambiguous algorithm is provided (presented as eqn 11.1 above). This relates the inverse of the molar LC_{50} to log P using linear regression analysis.

With regard to the training set of compounds, the full data set is provided along with the values for toxicity and log P. Chemical names only are provided, with some ambiguities (*e.g.* trichloroethane and tetraethane); this study predates the development of the SMILES notation and CAS numbers were not regularly used at that time.

The log P values were calculated using the Rekker method.[29] This was the only usable method at the time of the study, but would now be considered difficult to repeat. The user of such a model may wish to use a more modern method to calculate log P (*e.g.* KOWWIN – see Chapter 5).This analysis has been repeated using predictions from KOWWIN and no significant difference was found with the original equation.

The software used to calculate the regression analysis is not reported, but it is unlikely it could be obtained or utilised today anyway. This analysis has been repeated and the same results obtained. There is not an issue over variable selection as only one descriptor was used. The statistical analysis is discussed in more detail with regard to the fourth OECD Principle.

One last note is with regard to outliers. Careful reading of the original paper indicates that a number of compounds were removed from the analysis as their toxicity values did not fit the QSAR. There may be reasons for this (*e.g.* water solubility), but their removal may, at least in part, be the reason for the excellent statistical fit of the QSAR.

11.8.1.3 A Defined Domain of Applicability

Unsurprisingly there is no mention in this paper of the term 'domain of applicability'. However, these chemicals are described as 'unreactive' industrial organic compounds and anaesthetics, although chemical groups are specifically included or excluded. Whilst no software (*e.g.* AMBIT) has been used to assess the applicability domain, there is no reason why this could not be easily applied by users of the model. This is also clearly a case where knowledge of the mechanistic domain (*e.g.* see Ellison *et al.*[30] for more information on the non-polar narcosis domain) could be incorporated into the model to increase its coverage.

11.8.1.4 Appropriate Measures of Goodness-of-fit, Robustness and Predictivity

By modern standards, appropriate measures of goodness-of-fit and predictivity are not provided. However, since the raw data are provided, this is easily and rapidly achieved; this procedure was performed and it was able to repeat successfully the claims for goodness-of-fit and provide all other relevant measures.

There is no estimate of predictivity in Könemann's analysis.[28] This concept was not really considered at that time and this should not be considered a criticism of the model. It would be a relatively easy matter to find further data not considered in the model and make predictions for them. Alternatively the data set could be analysed using a cross-validation and/or sub-sampling approach.

11.8.1.5 A Mechanistic Interpretation, if Possible

The original article by Könemann does not specifically mention the QSAR to be top non-polar narcosis,[28] or the context in which we now know it can be placed (for a recent analysis of this context see the paper by MacKay *et al.*[31]). Thus, since its publication, it is possible to place a strong mechanistic interpretation on this model in terms of these chemicals acting by a narcotic mode of action (reversible disruption of membranes) and more specifically by the nonpolar narcosis mechanism.[31,32]

The strong relationship of toxicity with a descriptor for hydrophobicity is entirely consistent with the mechanistic hypothesis that toxicity is brought about by accumulation of the toxicant in membranes. Since its publication this paper has been supported by many other studies.[33]

11.8.1.6 Summary

The QSAR model developed by Könemann[28] provides a simple and transparent method to predict the toxicity to fish of chemicals falling within the nonpolar narcosis domain. Whilst the techniques used are relatively out-dated, the method is easy to update with more recent descriptors and techniques without changing the model and its meaning.

11.8.2 Case Study 2: Identification of Structural Features Associated with Carcinogenicity

The paper by Cunningham and co-workers[34] published in 1992 is one of many describing the basis of models in the MultiCASE software. In comparison with the first case study, it provides an interesting illustration of how far methods to develop *in silico* approaches to predict toxicity had progressed.

This paper describes an approach to identify, automatically, structural fragments within a molecule that are related statistically to carcinogenicity. The fragments are defined and grouped together into 26 biophores (*i.e.* fragments or structures associated with carcinogenicity). When these structural fragments are combined together, they can be used in the form of descriptors in a QSAR approach. This has been (part of) the basis of the MultiCASE approach and applied to many endpoints. For further information on the MultiCASE type of algorithm see some of the publications of Professor Klopman[35] and Chapter 19. For an overview of the MultiCASE type of products (*e.g.* MC4PC, CASETOX, META and several others), please see the MultiCASE Inc. website (www.multicase.com). An additional source of information can be gained from the developers' own consideration of their *in silico* approaches according to the OECD Principles.[18]

11.8.2.1 A Defined Endpoint

The endpoint modelled by Cunningham *et al.*[34] is carcinogenicity; specifically, the dose which, if administered chronically for the standard lifespan of the species, will halve the probability of remaining tumourless throughout that period (TD_{50}) was considered. These values were considered for the rat and taken from the Carcinogenicity Potency DataBase (CPDB) for 745 chemicals (383 carcinogens, 14 marginally active carcinogens, 348 non-carcinogens). This endpoint is clearly of considerable relevance for regulatory decisions.

It is interesting to consider the database further. The CPDB is a well-respected database of high quality values. These values have been compiled from, in theory, standardised bioassays performed by the US National Toxicology Program. Whilst the exact protocol is not provided (nor is it given in the context of OECD guidelines or GLP), it would be easy to obtain the original source data from the original references or, more likely now, through databases such as ISSCAN through DSSTox (see Chapter 3). Therefore it can be assumed that the data quality is reasonably high.

One peculiarity about this type of modelling approach is the use of the so-called 'CASE units' which are a conversion of the original units on a scale of 0 to 100. It should, however, be possible to convert back from the CASE units to the original units.

11.8.2.2. An Unambiguous Algorithm

The algorithm on which the model is based could be considered as being difficult to determine as it is contained within a long table of structural fragments. At the end of the day, users of this algorithm would find it very difficult to reconstruct the algorithm without the bespoke software of the authors and model developers. There must be a sense of realism and pragmatism about this issue: commercial vendors of software cannot give all their algorithms away and still maintain the resources to continue to build models. Strictly speaking, it is possible to determine how a prediction could be made from the structural fragments without necessarily requiring the full unambiguous model.

The data on which the model was built are not given though the reference is (*i.e.* to the CPDB as stated for the first principle). It is not obvious whether the data set represents all the data available in the reference, or if some have been selected. Likewise, the removal of outliers during the modelling process is not well described, leading readers to assume no outliers were removed.

With regard to the descriptors, there is a reliance on structural fragments (which could in theory be coded through the use of SMARTS strings) and molecular orbital properties, the calculation for which is relatively poorly defined. Thus, as noted for the first principle, it would be very difficult to repeat the modelling process without the developer's software.

The statistical properties of the model are only moderately addressed—see the fourth Principle.

11.8.2.3 A Defined Domain of Applicability

The data set (TD_{50} values or descriptors) and the domain of applicability are not formally defined by Cunningham *et al.*[34] The latter was an issue not considered explicitly even in the late 1990s. Whilst the applicability domain is not defined in terms of physico-chemical properties or structures, the developer's software should be applauded as, for many years, it has automatically searched query molecules to determine if they contain any fragments unknown within the training set. This goes at least some way to providing a method to ascertain whether that query compound does fall within the (structural) applicability domain of the training set. Again, the developer's software is required to obtain this information.

11.8.2.4 Appropriate Measures of Goodness-of-fit, Robustness and Predictivity

As noted above, the training set for the model is not provided explicitly, though the source references and hence the raw data are. The authors present a good statistical fit for their model. Results from a cross-validation study confirm the predictivity of the model, although it may now be possible to perform a full external validation study as more data should have been made available.

With regard to the statistical modelling, details are provided of how the model was created. It does require a statistical evaluation of all fragments to identify the most significant. This variable selection process is performed on a statistical basis.

11.8.2.5 A Mechanistic Interpretation, if Possible

A full mechanistic evaluation has not been provided by Cunningham et al.;[34] however due to the complexity of the model and the number of descriptors, this would be a very difficult and lengthy process (probably beyond the reasonable length of a journal article). The authors do, however, attempt to place the fragments in the context of genotoxic and non-genotoxic carcinogens. Some fragments are clearly based on electrophilic groups; others associated with carcinogenicity are not electrophilic and hence may be related to non-genotoxic mechanisms. This approach of selecting fragments through statistical rather than mechanistic means has advantages and disadvantages. The ability to build models rapidly and not be restricted by current mechanistic knowledge (or lack of it) is a good attribute; however, not being able to contextualise all fragments mechanistically may reduce the user's confidence in them.

11.8.2.6 Summary

This model requires specialist software for efficient operation. Given that caveat, it complies with the OECD Principles reasonably well although more clarity over the mechanistic interpretation and precise data set would be preferable.

11.8.3 Case Study 3: Identification of Structural Features Associated with Skin Sensitisation

The third case study considered is the most recent and is published from the author's laboratory.[36] It represents a different aspect of the (Q)SAR spectrum from the other case studies in that it simply presents chemistry encoded by SMARTS strings. A SMARTS string is a two-dimensional (2-D) description of a fragment within a chemical structure. Enoch *et al.*[36] presented SMARTS strings to describe mechanistic organic chemistry, with over 50 strings being used to describe six main mechanisms. In this way it goes beyond SAR or structural alerts; however it is this type of information that will be used to form categories of chemicals (and hence allow read-across). It is included here not only due to the author's personal interests, but also because it represents a different modelling approach and should illustrate some of the problems associated with this approach when applying the OECD Principles.

The important issue to note with the approach of Enoch *et al.*[36] is that it would be incorrect to consider it to be a model; it really forms a simple version of what is termed in the OECD Toolbox a 'profiler' (see Chapter 16). The fragments represent areas of chemistry that are known to promote skin sensitisation. These fragments have been derived from a knowledge of chemistry rather than the known test results of sensitisers.

11.8.3.1   A Defined Endpoint

The structural fragments defined by the SMARTS strings are derived from mechanistic organic chemistry. The mechanisms have, in part, been derived from the analysis of results from the Local Lymph Node Assay (LLNA), which has considerable regulatory significance. It is important to note that this is not a method to predict skin sensitisation directly, but an approach to assist in the identification of hazard and also group chemicals into mechanistic categories.

With regard to the source LLNA data, they come from a well-recognised database for this endpoint.[37,38] These data can be considered to be of good quality. Whilst the endpoint itself has an OECD Guideline, and many of the data were performed to GLP, some data pre-date the existence of the OECD Guideline. The data are a compilation from different testing laboratories, so there may be some inconsistency across the database.

11.8.3.2   An Unambiguous Algorithm

Strictly speaking there is no algorithm for this predictive approach. However, this paper reports the structural fragments in the form of SMARTS strings associated with various mechanisms. These fragments are clearly defined and can be obtained electronically from the authors. In addition, the authors can provide a very simple piece of free software code to run the structural fragments.

For this study, the training and test set of compounds are supplied as both chemical names and CAS numbers. No outliers have been removed from the analysis. A small number of local descriptors have been calculated to illustrate the role of electrophilicity within a mechanism—the method is clearly defined using AM1 optimisation and *ab initio* calculation of molecular orbital properties. All calculations were performed using the commercially available Gaussian 03 software.

11.8.3.3 A Defined Domain of Applicability

The domain is mostly implicit in the fragments as they are descriptions of structural chemistry. However, it is not possible to determine if a query molecule contains a fragment that is not known by this mechanistic approach.

The data set (*i.e.* the LLNA data) has been defined, thus this could be used to define an applicability domain with, for example, the AMBIT software. Also, the experimental applicability domain could be defined (*e.g.* in terms of solubility cut-offs, *etc.*).

11.8.3.4 Appropriate Measures of Goodness-of-fit, Robustness and Predictivity

A limited assessment of the performance of this approach is provided. This has been performed in terms of the training set (*i.e.* data from Gerberick *et al.*[37]) and also a test/validation set (Roberts *et al.*[38]). Both sets of chemicals are available (including the raw data). It may be possible to provide a more detailed statistical analysis, but this rather misses the point; this is not intended to be a predictive algorithm but more a method to define hazard, identify risk, and develop groups or categories. The method and philosophy of this approach is clearly stated, *i.e.* to describe, for computational purposes, the mechanistic organic chemistry associated with skin sensitisation.

11.8.3.5 A Mechanistic Interpretation, if Possible

The real strength of this approach is its foundation in mechanistic organic chemistry. Thus this can be considered to be a wholly mechanistic model, where the reaction chemistry has led the development of the fragments—there is no statistical bias in their creation. The mechanistic chemistry knowledge being applied has been developed over the past two decades and is well-supported in the literature and through *in vitro*, *in vivo* and *in chemico* testing (see, for instance, the excellent review by Patlewicz *et al.*[39] and the paper by Cronin[40]).

11.8.3.6 Summary

The SMARTS strings defined by Enoch *et al.*[36] describe fragments associated with reactive chemistry which, in turn, can be related to endpoints such as skin sensitisation. There is no algorithm as such, though these fragments do provide a valid method to identify potentially hazardous compounds and provide a basis for groupings of molecules.

11.9   Is the Prediction Valid?

As noted in the introduction, the concern of a (Q)SAR user should be whether the prediction is valid, rather than the formal validation of an *in silico* model. It must be stressed that simply because a prediction may be classed as being 'valid', it does not mean that it is correct. It simply could imply that greater confidence can be given to the predicted value. The assessment of whether a valid or reliable prediction has been achieved is described in detail by Worth.[5]

In order to determine whether a prediction from an *in silico* model is valid, the following criteria should be assessed.

- Is the (Q)SAR reliable, statistically robust and well-documented, *etc*? This will result from the assessment of the model according to the OECD Validation Principles (see Section 11.8) and reporting the model using a QMRF (see Section 11.7).
- Does the query compound fall within the applicability domain? Whether or not the query compound falls within the applicability domain of the endpoint and model must be assessed. Methods to assess the applicability domain are described in Chapter 12.

11.9.1   QSAR Prediction Reporting Format (QPRF)

In order to provide adequate documentation, the QSAR Prediction Reporting Format (QPRF) has been proposed. The QPRF is a harmonised format for sharing key information on substance-specific QSAR predictions, including their reliability. Guidelines for the QPRF are given on the JRC website (http://ecb.jrc.ec.europa.eu/qsar/). The QPRF allows the user of the model to include information on

1. The substance, *i.e.* query compound. This includes chemical identifiers and structure.
2. General information regarding the prediction process. This includes the date of the QPRF, author and contact details.
3. The prediction of the effect for the query compound. This will facilitate consideration of the scientific validity of the model (as defined by the OECD Principles) and hence the reliability of the prediction. Detailed information on the model is stored in the corresponding QMRF. It should be remembered that the QMRF and the QPRF are complementary, and a QPRF should always be associated with a defined QMRF. An important issue relates to the fourth OECD Principle; this is to comment on the uncertainty of the prediction for this chemical, taking into account relevant information (*e.g.* variability of the experimental results).
4. A final, optional, section is on the adequacy of the prediction. This is to facilitate considerations about the adequacy of the (Q)SAR prediction (result) estimate. A (Q)SAR prediction may or may not be considered

adequate ('fit-for-purpose'), depending on whether the prediction is sufficiently reliable and relevant in relation to the particular regulatory purpose.

This information is summarised from the published QPRF to which readers are referred for more complete detail.

11.10   Conclusions and Recommendations

A framework exists to assess the reliability of *in silico* models for toxicity and the validity of the predictions. It is important to remember that this is not a process to validate a (Q)SAR formally in the same manner as an *in vitro* assay, but it is intended to characterise and evaluate the models. The 'acceptability' of QSARs for any purpose (product development/prioritisation, *etc.*) should be guided by consideration of the OECD Principles. The rigidity by which these are applied depends on why the QSAR is being used, *i.e.* it will be important to consider other factors than just regulatory use.

Thus, the following recommendations can be made:

- (Q)SARs should be assessed (for regulatory use) according to the OECD Principles.
- (Q)SARs should be reported, following assessment by the OECD Principles, using the QMRF.
- Assessment according to the OECD Principles will always be, to a certain extent, subjective and requires expertise in all areas of *in silico* toxicology.
- Predictions need to be assessed for validity.
- Predictions should be reported (for regulatory use) using the QPRF.

11.11   Acknowledgements

This project was sponsored by the Department for Environment, Food and Rural Affairs (Defra) through the Sustainable Arable Link Programme. This work was also supported by the EU 6th Framework Integrated Project OSIRIS (www.osiris-reach.eu; contract no. GOCE-ET-2007-037017).

References

1. M. T. D. Cronin, in *Recent Advances in QSAR Studies. Methods and Applications*, ed. T. Puzyn, J. Leszczynski and M. T. D. Cronin, Springer, Dordrecht, The Netherlands, 2010, pp. 3–11.
2. M. T. D. Cronin, J. S. Jaworska, J. D. Walker, M. H. I. Comber, C. D. Watts and A. P. Worth, *Environ. Health Perspect.*, 2003, **111**, 1391.

3. M. T. D. Cronin, J. D. Walker, J. S. Jaworska, M. H. I. Comber, C. D. Watts and A. P. Worth, *Environ. Health Perspect.*, 2003, **111**, 1376.
4. G. Schaafsma, E. D. Kroese, E. L. J. P. Tielemans, J. J. M. van de Sandt and C. J. van Leeuwen, *Reg. Toxicol. Pharmacol.*, 2009, **53**, 70.
5. A. P. Worth, in *Recent Advances in QSAR Studies. Methods and Applications,* ed. T. Puzyn, J. Leszczynski and M. T. D. Cronin, Springer, Dordrecht, The Netherlands, 2010, pp. 367–382.
6. J. S. Jaworska, M. Comber, C. Auer and C. J. van Leeuwen, *Environ. Health Perspect.*, 2003, **111**, 1358.
7. Organisation for Economic Co-operation and Development, The Report from the Expert Group on (Quantitative) Structure-Activity Relationship ([Q]SARs) on the Principles for the Validation of (Q)SARs, OECD, Paris, 2004, OECD Environment Health and Safety Publications Series on Testing and Assessment No. 49, ENV/JM/MONO(2004)24, www.olis.oecd.org/olis/2004doc.nsf/LinkTo/NT00009192/$FILE/JT00176183.pdf [accessed March 2010].
8. Organisation for Economic Co-operation and Development, *Guidance Document on the Validation of (Quantitative) Structure–Activity Relationships [(Q)SAR] Models*, OECD, Paris, 2007, OECD Environment Health and Safety Publications Series on Testing and Assessment No. 69, ENV/JM/MONO(2007)2, www.olis.oecd.org/olis/2007doc.nsf/LinkTo/NT00000D92/$FILE/JT03224782.PDF [accessed March 2010].
9. European Chemicals Agency, *Guidance on Information Requirements and Chemical Safety Assessment,* EChA, Helsinki, 2010, http://guidance.echa.europa.eu/guidance_en.htm#GD_METH [accessed March 2010].
10. E. Hulzebos, D. Sijm, T. Traas, R. Posthumus and L. Maslankiewicz, *SAR QSAR Environ. Res.*, 2005, **16**, 385.
11. A. P. Worth, T. Hartung and C. J. van Leeuwen, *SAR QSAR Environ. Res.*, 2004, **15**, 345.
12. A. P. Worth and M. Balls, *Altern. Lab. Anim.*, 2002, **30**(Suppl 2), 15.
13. T. Bouwman, M. T. D. Cronin, J. G. M. Bessems and J. J. M. van de Sandt, *Hum. Exp. Toxicol.*, 2008, **27**, 269.
14. H. X. Liu, E. Papa and P. Gramatica, *Chem. Res. Toxicol.*, 2006, **19**, 1540.
15. M. Pavan, T. I. Netzeva and A. P. Worth, *SAR QSAR Environ. Res.*, 2006, **17**, 147.
16. A. G. Saliner, T. I. Netzeva and A. P. Worth, *SAR QSAR Environ. Res.*, 2006, **17**, 195.
17. M. Vracko, V. Bandelj, P. Barbieri, E. Benfenati, Q. Chaudhry, M. Cronin, J. Devillers, A. Gallegos, G. Gini, P. Gramatica, C. Helma, P. Mazzatorta, D. Neagu, T. Netzeva, M. Pavan, G. Patlewicz, M. Randic, I. Tsakovska and A. Worth, *SAR QSAR Environ. Res.*, 2006, **17**, 265.
18. R. D. Saiakhov and G. Klopman, *Toxicol. Mech. Meth.*, 2008, **18**, 159.
19. E. Papa and P. Gramatica, *SAR QSAR Environ. Res.*, 2008, **19**, 655.
20. P. Gramatica, P. Pilutti and E. Papa, *SAR QSAR Environ. Res.*, 2007, **18**, 169.
21. D. W. Roberts, A. O. Aptula, M. T. D. Cronin, E. Hulzebos and G. Patlewicz, *SAR QSAR Environ. Res.*, 2007, **18**, 343–365.

22. T. I. Netzeva, A. P. Worth, T. Aldenberg, R. Benigni, M. T. D. Cronin, P. Gramatica, J. S. Jaworska, S. Kahn, G. Klopman, C. A. Marchant, G. Myatt, N. Nikolova-Jeliazkova, G. Y. Patlewicz, R. Perkins, D. W. Roberts, T. W. Schultz and D. T. Stanton, *Altern. Lab. Anim.*, 2005, **33**, 155.

23. J. Jaworska, N. Nikolova-Jeliazkova and T. Aldenberg, *Altern. Lab. Anim.*, 2005, **33**, 445.

24. L. Eriksson, J. Jaworska, A. P. Worth, M. T. D. Cronin, R. M. McDowell and P. Gramatica, *Environ. Health Perspect.*, 2003, **111**, 1361.

25. A. O. Aptula, N. G. Jeliazkova, T. W. Schultz and M. T. D. Cronin, *QSAR Comb. Sci.*, 2005, **24**, 385.

26. P. Gramatica, *QSAR Comb. Sci.*, 2007, **26**, 694.

27. G. Schüürmann, R. U. Ebert, J. W. Chen, B. Wang and R. Kühne, *J. Chem. Inf. Model.*, 2008, **48**, 2140.

28. H. Könemann, *Toxicology*, 1981, **19**, 209.

29. R. Rekker, *The Hydrophobic Fragmental Constant*, Elsevier, Amsterdam, 1977.

30. C. M. Ellison, M. T. D. Cronin, J. C. Madden and T. W. Schultz, *SAR QSAR Environ. Res.*, 2008, **19**, 751.

31. D. Mackay, J. A. Arnot, E. P. Petkova, K. B. Wallace, D. J. Call, L. T. Brooke and G. D. Veith, *SAR QSAR Environ. Res.*, 2009, **20**, 393.

32. G. D. Veith, D. J. Call and L. T. Brooke, *Can. J. Fish. Aquat. Sci.*, 1983, **40**, 743.

33. B. I. Escher and J. L. M Hermens, *Environ. Sci. Technol.*, 2002, **36**, 4201.

34. A. R. Cunningham, G. Klopman and H. S. Rosenkranz, *Mut. Res.*, 1998, **405**, 9.

35. G. Klopman, *Quant. Struct.-Act. Relat.*, 1992, **11**, 176.

36. S. J. Enoch, J. C. Madden and M. T. D. Cronin, *SAR QSAR Environ. Res.*, 2008, **19**, 555.

37. G. F. Gerberick, C. A. Ryan, P. S. Kern, H. Schlatter, R. Dearman, I. Kimber, G. Patlewicz and D. Basketter, *Dermatitis*, 2005, **16**, 157.

38. D. W. Roberts, G. Patlewicz, S. D. Dimitrov, L. K. Low, A. O. Aptula, P. S. Kern, G. D. Dimitrova, M. I. H. Comber, R. D. Phillips, J. Niemelä, C. Madsen, E. B. Wedebye, P. T. Bailey and O. G. Mekenyan, *Chem. Res. Toxicol.*, 2007, **20**, 1321.

39. G. Patlewicz, A. O. Aptula, D. W. Roberts and E. Uriarte, *QSAR Comb. Sci.*, 2008, **27**, 60.

40. M. T. D. Cronin, in *Recent Advances in QSAR Studies. Methods and Applications*, ed. T. Puzyn, J. Leszczynski and M. T. D. Cronin, Springer, Dordrecht, The Netherlands, 2010, pp. 305–325.

CHAPTER 12

Developing the Applicability Domain of In Silico Models: Relevance, Importance and Methods

M. HEWITT AND C. M. ELLISON

School of Pharmacy and Chemistry, Liverpool John Moores University, Byrom Street, Liverpool L3 3AF, UK

12.1 Introduction

Given the problems and limitations of animal testing, the development and utilisation of *in silico* approaches to predict biological activity, toxicity and absorption, distribution, metabolism and excretion (ADME) endpoints continues to grow year on year. As introduced previously (Chapter 2), (quantitative) structure-activity relationship [(Q)SAR] models are at the forefront of these alternative approaches and have been developed for a plethora of physico-chemical and toxicological endpoints.[1] As a consequence, the reliability of predictions from such models is of increasing concern, particularly when these predictions are being used in a regulatory setting.[2] The need to ensure that a prediction generated by a given model is valid has brought about a multitude of requirements and methods aimed at characterising/evaluating a model. These are discussed in detail in Chapter 11.

Issues in Toxicology No.7
In Silico Toxicology: Principles and Applications
Edited by Mark T. D. Cronin and Judith C. Madden
© The Royal Society of Chemistry 2010
Published by the Royal Society of Chemistry, www.rsc.org

12.1.1   Definition of the Term 'Applicability Domain'

In order to ensure the appropriate use of (Q)SAR models and their predictions, investigations into a model's applicability domain are required. The applicability domain defines the constraints of the training set compounds of a (Q)SAR model, allowing a user to choose the most suitable model or use a given model within its own predictive capacity. Ensuring the appropriate use of a (Q)SAR model is key, as it is easy for end users to make a prediction blindly without knowledge of the model's limitations, possibly resulting in erroneous predictions. To prevent such errors and to ensure the most relevant model is selected, the applicability domain is used to detail the most appropriate prediction space of a model (*i.e.* the chemicals to which the model can be applied). It does this by defining the interpolation region of a model (the region where predictions can be reliably made),[3] which can be defined using a number of different approaches (discussed later in this chapter).

As there are many ways in which to define the applicability domain of a (Q)SAR model, obtaining a standard and universal definition is difficult. However, an ECVAM (European Centre for the Validation of Alternative Methods) workshop held in 2005, which was given the task of reviewing the state-of-the-art of methods for applicability domain definition, agreed upon a general definition. The published report from that meeting states '*the applicability domain of a (Q)SAR model is the response and chemical structure space in which the model makes predictions with a given reliability*'[4] whereby the 'response' is the effect being modelled (*e.g.* toxicity). Despite the publication of a common definition, numerous others are present within the published literature resulting in the applicability domain being an often misunderstood and misinterpreted subject.

In an effort to maximise the appropriate use of (Q)SAR models and to ensure reliable predictions, an international workshop was held in 2002 in Setúbal, Portugal. From this workshop, six principles were proposed for the assessment of the validity of a (Q)SAR model (termed the Setúbal Principles). Following subsequent rewording, these principles formed what are now known as the Organisation for Economic Co-operation and Development (OECD) QSAR Validation Principles (see Chapter 11). As such, the applicability domain of a model is now considered a core requirement in validating both a (Q)SAR model and the reliability of the predictions which it yields.

12.2   Regulatory Context

As a result of various pieces of legislation regarding chemicals detailed in previous chapters, heavy demands are being placed upon the (Q)SAR community to deliver models suitable for regulatory use.[5] As stated previously, one of the biggest hurdles to overcome is to provide users/regulators with sufficient evidence to prove that the (Q)SAR model in question is relevant and fit-for-purpose. Currently, in addition to the standard statistical evaluation of the predictions,[6–8] examining a model's applicability domain is a crucial process in providing this information.

12.2.1 Assigning Confidence and Predicting Outside the Domain

The role of the definition of the applicability domain is to indicate to the user that a model is relevant to make a prediction for any given query compound. If a query compound is within the applicability domain of the model, then increased confidence may be placed on the prediction made by that model. Likewise, if a query compound is found to be outside a model's applicability domain, then reduced levels of confidence need to be placed on that prediction. Therefore, the applicability domain of a model expresses the scope (applicability) of a model together with its limitations.[4]

It is important to stress that the applicability domain is not a measure of performance. Making a prediction outside a model's applicability domain does not automatically mean that prediction is not accurate. It simply indicates that the model used to make that prediction is not designed to do so and is being operated outside its intended area of relevance (*i.e.* prediction *via* extrapolation). However, in many cases, predicting outside a model's applicability domain does result in increased prediction error.[9–11]

12.3 Applicability Domain Definition

12.3.1 Strategies to Define the Applicability Domain

Although the core concept of what an applicability domain is has been present within the (Q)SAR community for many years, it is only within the past decade that the term 'applicability domain' was coined and particular emphasis was placed upon its definition and utilisation.[12–16] A key factor in this was the adoption of the OECD (Q)SAR Validation Principles at policy level by OECD member countries in November 2004.[17]

Since the Setúbal workshop, interest in applicability domain definition has grown considerably and, as a result, applicability domain definition is now a very active area of (Q)SAR research.[5,9,18–22] However, despite numerous methodologies being proposed, it is generally accepted that there is no single universal technique to define the applicability domain.[3–4,16,23,24] In the same way that using multiple (Q)SAR models built using different methodologies and descriptors can be more revealing, considering multiple approaches to define the applicability domain can also be beneficial. As each of the proposed applicability domain approaches gives more information, a stepwise approach combining multiple methods has been suggested to be beneficial.[25,26]

The main consideration of a modeller, therefore, should not be which methods should be used in domain definition, but how many methods can be implemented with the information available. The greater the number of methods used, the more information will be obtained and therefore the more thorough the definition of the applicability domain. However, over-definition of the domain may result in an excessively strict domain where only the training compounds of a model are found within it, rendering the model useless. A balance is therefore needed between using sufficient information to define the domain adequately and an understanding of when a domain has been over-defined.

Building on the strategies suggested previously for applicability domain definition,[25,26] an adapted scheme is represented in Figure 12.1.

Although potentially limited by the availability and quality of data (*i.e.* information as to a chemical's mechanism of action, metabolic profile, *etc.*), it is likely that a stepwise approach will be beneficial, providing more information to the modeller and a more robust and comprehensive definition of the applicability domain.

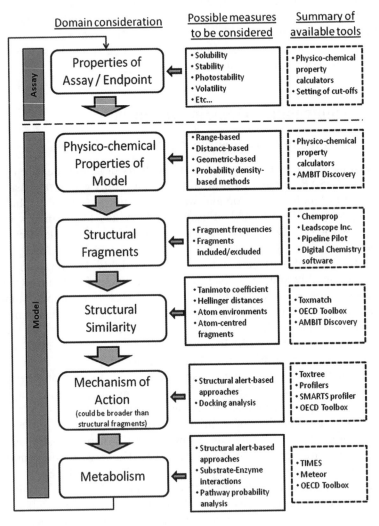

Figure 12.1   Schematic of a stepwise applicability domain definition workflow.

12.3.2 Acknowledging the Assay/Endpoint Domain

Often overlooked in applicability domain studies is the domain (or constraints) of the assay being used to generate activity data. In the same way that a query chemical can be assessed to see if fits within the constraints of a training set of chemicals (see sections 12.3.4 and 12.3.5), the suitability of a particular assay should also be considered. For example, if a query compound is very poorly soluble it is likely to show reduced toxicity. In the *Tetrahymena pyriformis* growth impairment assay,[27] compounds that are poorly soluble show little or no toxicity as they are not able to enter the aqueous media.[18] Similarly, other factors affecting bioavailability (*e.g.* chemical volatility) can also affect the reliability of an assay and the confidence that can be placed upon it.

It is therefore important for applicability domain studies to begin with an assessment of assay suitability. More specifically, consider the ability of a particular assay to perform reliably when given a specific query compound. If this query compound is associated with properties that make it unsuitable for the assay used to generate the toxicity data, then that compound should be considered outside the domain of the assay. Using a (Q)SAR model to perform a prediction of activity is only likely to result in erroneous results, as the predictions are based on results from an assay that is unsuitable for assessing the activity potential of the given query compound (see Chapter 4).

12.3.3 Dependence of Domain Definition on the Modelling Method

When attempting to define the applicability domain of any model, an important factor that needs to be taken into consideration is the method by which the model was developed. For example, a model based on structural alerts that can make qualitative (active/inactive) predictions is very different from a model based on linear regression with the ability to make quantitative predictions. These differences are important as different information is required and used from the training set to build these models (*i.e.* chemical structure in the structural alert model *versus* physico-chemical properties in the linear regression model). The format of these variables defines the nature of the resultant model and, likewise, is crucial in the definition of the model's applicability domain. Therefore the approach used to define the domain will depend on what information is available from the training data. For example, structural similarity could be considered more appropriate than descriptor-based methods for structural alert models as no descriptors are used in the building of the model. It is important to understand that, although several methodologies exist for defining the applicability domain, not all of them may be appropriate for any given model.

12.3.4 Approaches for Defining the Applicability Domain

As highlighted in Figure 12.1, there are many approaches available to investigate and define the applicability domain of a model. A stepwise approach, taking into account as many of these techniques as possible, is the most likely to yield a robust assessment. Each of these approaches is discussed in detail below.

12.3.4.1   Descriptor-Based Approaches

In this approach, an applicability domain is defined based on one or more physico-chemical descriptors on which the model has been built; this may also include the response variable (property being modelled). In this way, the applicability domain takes the form of an interpolation region that is defined by the training data. Any query compound falling within this interpolation region would be considered within the applicability domain of the model, as shown in Figure 12.2.

There are four major approaches to defining an interpolation region in one- or *n*- dimensional space, where *n* is the number of descriptors in the model. Each of these approaches is discussed below. Further details on all of these methods can be found in the literature.[3,4]

12.3.4.1.1 Range-Based Methods. This is the simplest method for approximating the interpolation region and is based upon the ranges observed in descriptor values. For example, the applicability domain of a simple linear regression model based on the logarithm of the octanol–water partition coefficient (log P) and the energy value of the lowest unoccupied molecular orbital (E_{LUMO}) could be described in terms of the numeric range of the two descriptors.

12.3.4.1.2 Distance-Based Methods. A second approach is to consider the distance between data points using one of several distance measures as

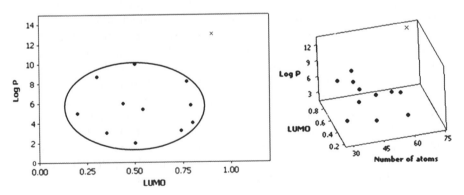

Figure 12.2 Plots showing the basic principle behind descriptor-based applicability domain methods. A number of descriptors (two and three in this example) can be used to define a region in descriptor space (the applicability domain). Query compounds falling outside of this region (X) can be easily identified.

detailed below. These approaches calculate the distance from a query compound to that of the training set compounds.[28] This is more complex than the simple range-based methods based on descriptors as the position of the query compound is established. Numerous distances can be calculated, including distance to the mean and the average, maximum and minimum distances between the query compound and training compounds. Three of the most common distance methods are Euclidean, Mahalanobis and City Block distances. The Hotellings test, leverage[29] and distance methods have also been recommended for assessing the applicability domain of (Q)SAR models.[3]

12.3.4.1.3 Geometric-Based Methods. Another method of defining the applicability domain is to define the smallest convex area which contains the training set data (termed a convex hull). The query compounds can then be plotted onto this *n*-dimensional space to see if they fall within the domain, as exemplified in Figure 12.3. Here, a convex hull is created that contains all training data points plus two test compounds. It is clear that one of the compounds is within the applicability domain and the other is not, as it falls outside the convex hull. Efficient convex hull algorithms are available for two and three dimensions, but the complexity of the calculations rapidly increases in higher dimensions.

12.3.4.1.4 Probability Density-Based Methods. The final method for estimating the interpolation region is based upon an estimation of the

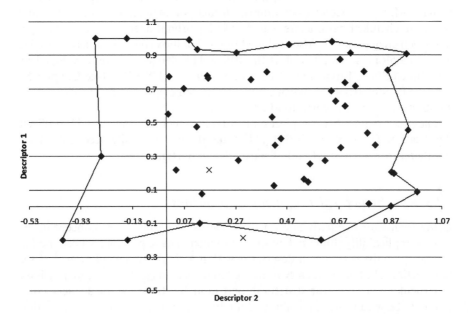

Figure 12.3 An example two-dimensional convex hull. Training compounds are represented as (◆) and the two test compounds are given as (X).

probability density function for the training data using either parametric or non-parametric methods.[4] The non-parametric approach is often favoured as it does not make any assumptions about the data distribution. Here probability density is estimated solely from the data using kernel density estimation or mixture density methods. Probability density methods are also attractive as they are the only methods able to identify internal regions of empty space within the convex hull, leading to more accurate mapping of the applicability domain.

12.3.4.1.5 Available Tools for Descriptor-Based Applicability Domains. Obviously for a descriptor-based approach, a prerequisite is the calculation of numerous physico-chemical descriptors. Therefore, any software package which is able to calculate such descriptors is important. At present there are a multitude of commercially available software tools which are able to generate numerous descriptors. These commercially available resources are discussed elsewhere (see Chapters 5–7).

A freely downloadable piece of software known as the OECD (Q)SAR Application Toolbox was released by the OECD in March 2008 (see Chapter 16).[30] The Toolbox is a software application intended to be used by governments, the chemical industry and other stakeholders in filling gaps in (eco)-toxicity data needed for assessing the hazards of chemicals. A key feature of the software is the grouping of chemicals into chemical categories. One method of doing this is based upon physico-chemical descriptors, which can be calculated by the software. Selected descriptors can be calculated including three-dimensional (3-D) parameters (including molecular diameters, volumes, *etc.*) and quantum chemical parameters (*e.g.* molecular orbital energies, atomic charges and electronegativities). Also included in the Toolbox, but also freely available as a standalone software tool, is the EPISUITE software suite, developed by the United States Environment Protection Agency (US EPA) (see Chapter 5). The predictions from this software include parameters such as log P, aqueous solubility and bioconcentration factor (BCF).

The free availability of pieces of software such as the OECD (Q)SAR Application Toolbox and EPISUITE make these tools easily accessible. They are therefore widely used and of particular importance.

12.3.4.2 Structural Fragment Approaches

Using fragments explicitly and on their own is a relatively new concept in the field of applicability domain definition. Fragments have previously been used in similarity analyses (*i.e.* as a descriptor within a similarity calculation,[31–37] see next section), but have recently been used in their own right[9,38–40] and may have a particular relevance for structural alert models.[9] In order for a query compound to be within the applicability domain, all the structural fragments of that query compound must be present within the training set compounds. Therefore, if the query compound contains an exotic and unique functionality not present within the training set chemicals, it would be highlighted as being outside the applicability domain of the model as illustrated in Figure 12.4.

12.3.4.2.1 Tools for Fragment Generation. At present there are a number of commercially available tools for the creation of structural fragments such as:

- Chemprop (www.ufz.de);
- Leadscope Inc. (www.leadscope.com);
- Pipeline Pilot (http://accelrys.com);
- Digital Chemistry (www.digitalchemistry.co.uk); and
- MOE (www.chemcomp.com).

However, few of these programs have the in-built capability to sufficiently compare and attempt to match the fragments produced from training and test sets—highlighting how much this approach is still in its infancy. The one exception is Pipeline Pilot, where it is possible to add further nodes to a pipeline to screen for novel molecules (or in this case fragments) within a file as compared to a library of other molecules (or fragments) defined by the user. The user can therefore search the library of fragments produced by the training data for the fragments produced from a test compound. If there are no novel fragments within the test compound, then it is within the applicability domain of the model.

12.3.4.3 Structural Similarity Approaches

One approach that is increasing in popularity is that of structural similarity. This can be differentiated from the structural fragment approach as structural similarity considers the molecule as a whole, providing a global measure of similarity, whereas the structural fragment approach considers only component fragments. Therefore, all fragments within a query compound may be present in the training set but their 3-D arrangement within the query compound may be unique, making it dissimilar in a global sense to any training compound. When considering toxicological endpoints, the effect (toxicity) can be a response triggered by a specific structural feature within the molecule such as a particularly reactive functional group. Similarly, when considering receptor-binding effects, a complex arrangement of structural features is often required within a molecule to elicit an effect. Therefore, structural similarity approaches are able to consider the structural diversity of a training set and can be used to identify whether a query compound is sufficiently similar to the training set and hence whether the query compound is within the applicability domain (see Figure 12.5).

A quantitative structural similarity measure can be calculated using one of numerous methods. One of the most utilised methods codes molecular structure as a pattern of bit sets, within a bit string (or fingerprint).[41] Each bit set within the string corresponds to a unique structural fragment found within the

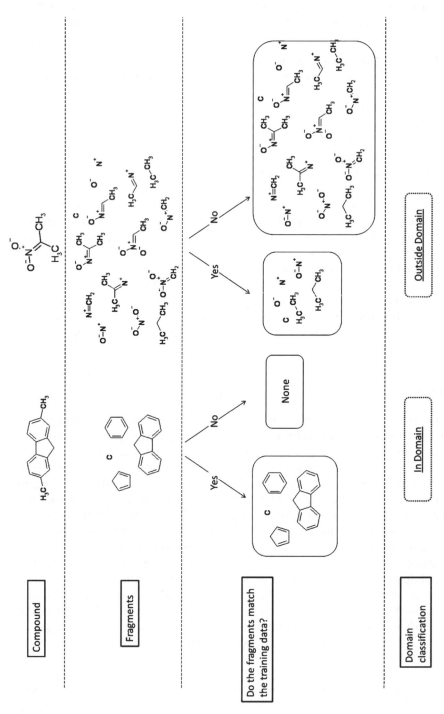

Figure 12.4 Example of how fragments can be explicitly used to classify compounds as within or outside of the domain of applicability.

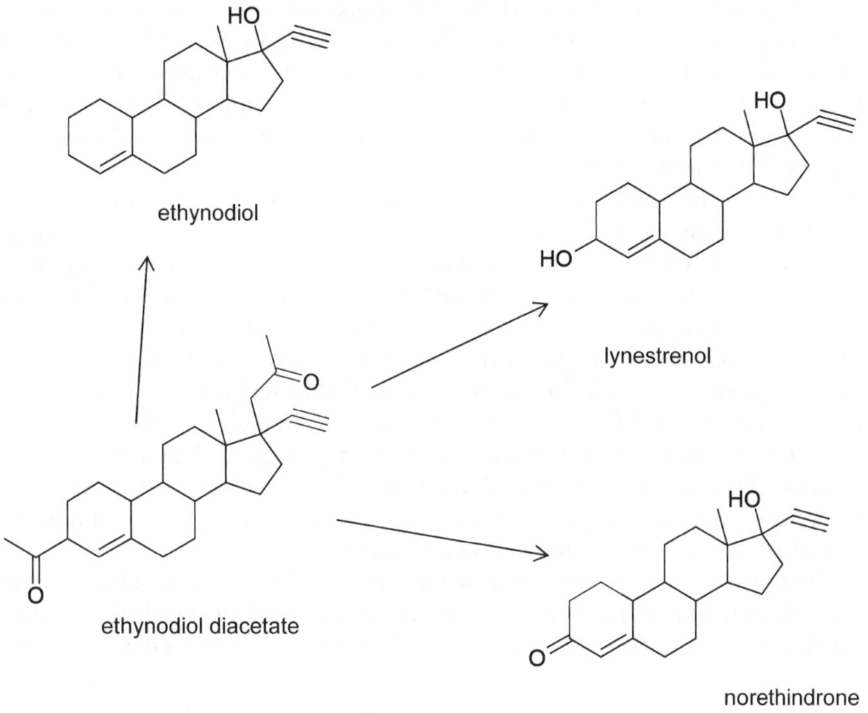

Figure 12.5   Use of structural similarity to find structurally similar chemicals from the training set for a given query compound (ethynodiol diacetate). The identification of similar compounds in the training set suggests the query compound is within the applicability domain of the model.

training set compounds.[33,42] A similarity measure can then be employed to quantify the similarity between two bit strings, such as the Tanimoto coefficient.[3] Therefore, the similarity of a query compound to all training set compounds can be investigated and, if not sufficient, would be classified as being outside the applicability domain. Of the many similarity approaches available, it is the comparison of bit strings that is the most utilised and, as a consequence, has been implemented in numerous software applications.[33]

Numerous other methods are also available which yield a quantitative similarity measure,[43] including both atom environments and atom centred fragments. Both of these approaches operate in a similar manner, considering the fragments surrounding a particular atom and then connected atoms, to a pre-defined level.[3,44] This differs from methods such as the Tanimoto method, as it takes into account the global arrangement of specific structural fragments within a molecule and, as such, is a more exacting measure of structural similarity.

12.3.4.3.1 Tools for Structural Similarity-based Applicability Domains. As discussed previously, the similarity between a query compound and training set compounds can be used to define the applicability domain. In addition to commercial tools, two freely available software tools are able to perform structural similarity analysis using various similarity measures. First, one of the most utilised is the OECD (Q)SAR Application Toolbox. The Toolbox is able to perform similarity calculation based upon one of four similarity coefficients (Tanimoto, Dice, Ochiai and Kulczynski-2). Similarity in molecular structure can be analysed using one of four similarity methods (atom pairs, topological torsions, atom centred fragments and natural functional groups).

A second software tool is Toxmatch.[45] Toxmatch was developed by Ideaconsult Ltd (Sofia, Bulgaria) under the terms of a European Commission Joint Research Centre contract. The software is freely available as a service to scientific researchers and anyone with an interest in the application of computer-based estimation methods in the assessment of chemical toxicity. Comparable to the OECD (Q)SAR Application Toolbox, Toxmatch is able to generate structural similarity measures between a query compound and training set compounds using one of several methods. These include distanced based methods (such as Tanimoto and Euclidean distances), fingerprint methods and also atom environment approaches (see Chapter 17).

The similarity indices generated by these pieces of software (a value between 0 and 1, or 0 and 100) can then be used to screen query chemicals to ascertain whether they are within the applicability domain of a given model.

12.3.4.4   Mechanism of Action Informed Approaches

A different approach to defining the applicability domain is to consider mechanism of action. When taking into consideration toxicological effects, the mechanism of action is defined as the molecular sequence of events leading from the absorption of an effective dose of a chemical to the production of a specific biological response in the target organ.[46] If the mechanism of action of a chemical is known, it is possible to identify chemicals acting *via* a single mechanism (*e.g.* polar narcosis) and build a (Q)SAR for that mechanism.[47–49] Given the differing characteristics of compounds acting *via* different mechanisms of action (*i.e.* potency, stability, *etc.*), developing a model for a single mechanism is often beneficial – yielding more predictive, although restricted, models.[50] If mechanistic information is lacking, mechanism of action may be assigned using numerous approaches—a commonly used approach being the use of structural alerts based on expert knowledge.[51] For a query compound to be within the applicability domain of such a model, it must inherently act *via* the same mechanism of action (Figure 12.6). Any mechanistic outlier would be classed as outside the applicability domain. It must be noted that mechanism of action is often unknown, which thus prevents the definition of a mechanistic domain.

12.3.4.4.1   Tools for Mechanism of Action-Based Applicability Domains. Although this is still very much an emerging approach in terms of its relation

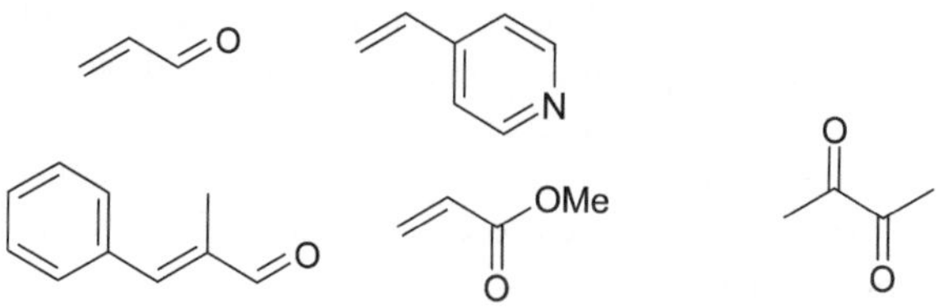

Compounds within the Michael
acceptor domain

Schiff-Base Former
OUTSIDE of Michael
acceptor domain

Figure 12.6 Example of a mechanistic domain and outlying chemical. Although this chemical is similar to some of the Michael addition chemicals it belongs to a different mechanism of action (Schiff base formation).

to applicability domain studies, there are a small number of software tools that have been developed to predict the mechanism of action for a set of chemicals. Currently, the vast majority of these approaches, if not all, are based upon the presence (or absence) of specific structural components that are related to a particular mechanism of action. These classifiers, which are based on expert knowledge, consider each chemical and assign a mechanism of action based solely upon molecular structure. As such, these methods are very useful when the mechanism of action of a compound is unknown (*i.e.* a new query compound). Again, there are a small number of freely available tools with this capability. First, there is the OECD (Q)SAR Application Toolbox. One of the shaping features of this software is its ability to form chemical categories based on mechanism of action. As such, multiple mechanism classifiers are present within the Toolbox, including classifiers for protein binding and DNA binding mechanisms of action, in addition to the well established Verhaar[52] and Cramer[53] classification schemes.

Similarly, a second freely available software tool, Toxtree,[54] has been developed by Ideaconsult Ltd (Sofia, Bulgaria) and which is capable of making mechanism of action predictions. This software again contains the Verhaar classification scheme[52] in addition to other classification schemes including the Cramer rules[53] and, from a simple SMILES input, will predict the mechanism of action (see Chapter 17). Therefore, software packages such as these are able to provide important information, crucial in the definition of the mechanistic aspect of the applicability domain.

12.3.4.5 Metabolism-Based Approaches

A further consideration relating to the applicability domain of a model is the role of metabolism and the effect of this on the reliability of a prediction. Although metabolism is not accounted for in the majority of (Q)SAR models it is, of course, important. If a query compound is metabolised, then a prediction based on the parent compound would be incorrect as the toxicity of the metabolite(s) must also be considered. The extent to which a compound undergoes metabolism can therefore be used as a measure of prediction confidence. Predictions for compounds which are not metabolised can be associated with high confidence, whereas predictions for readily metabolised compounds would be associated with lower levels of confidence. This rather simple approach attempts to correlate prediction confidence to the extent of metabolism, as shown in Figure 12.7.

As detailed metabolic studies are often lacking, metabolites often need to be predicted. Modelling systems containing metabolic simulators are under development to take metabolism into consideration. Computational software including the OECD (Q)SAR Application Toolbox,[30] METEOR,[55] CATABOL[56] and TIMES[56,57] are now available which can model metabolic pathways including abiotic transformation, enzyme-catalysed phase I and phase II reactions, and reactions with protein nucleophiles.

Although metabolism cannot be used to define an applicability domain in isolation, it is an important consideration. Compounds that are readily metabolised should be flagged when assessing the applicability domain as the (reactive) metabolites may not be within the scope of the model.

Another angle of the metabolic domain is the fact that the compounds within the training set may themselves undergo metabolic activation. To fully define the metabolic domain, therefore, the metabolic pathways and the products of the training set also need to be taken into consideration. As current methods for predicting metabolism have been shown to have problems,[58,59] obtaining all this information is likely to be difficult if not impossible. In order to explore and

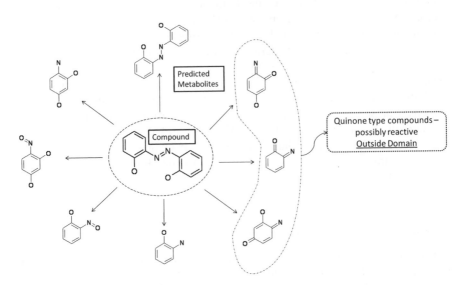

Figure 12.7   Example showing that compounds with the potential to produce metabolites could be considered outside the domain of applicability. Metabolites shown here have been created using the skin metabolism simulator within the OECD (Q)SAR Application Toolbox.

define fully the metabolic domain information relating to a metabolite's stability, the relative proportion and abundance are also required. However, the current state-of-the-art tools are lacking in these respects. Therefore, although in theory the metabolic domain is an important aspect to define, the process is often too difficult to put into practice.

12.3.4.5.1 Tools for Metabolism-Based Applicability Domains. It is important to acknowledge that applicability domains cannot be based solely on metabolism. Nevertheless, information relating to the metabolism of a query compound can be vital in assessing the reliability of a (Q)SAR prediction. Although predicting metabolism is difficult and is dependent upon the organism in question, a small number of software tools have been developed to predict metabolites.

Again one of the most significant freely available tools for metabolism prediction is the OECD (Q)SAR Application Toolbox. As stated previously, a key feature of the Toolbox is to form chemical categories; as part of this functionality, the Toolbox contains metabolism profilers able to simulate metabolism and predict metabolites in a number of different media (gastrointestinal tract, liver, skin) as well as metabolites produced in a microbial environment.

12.3.5 Available Tools for Applicability Domain Analysis

In addition to the software tools discussed above, which all yield information that can be used in the assessment of the applicability domain, there are a small number of software tools that have been specifically designed to assess whether a query compound (or set of compounds) falls within the applicability domain based on a given training set. A freely available example is the AMBIT Discovery software tool.[60] This was developed within the Bulgarian Academy of Sciences and is available online or for download (see Chapter 17).

Through the import of a text (.csv) file containing the compound information (name, SMILES *etc.*) and descriptor data, the software is able to estimate the applicability domain of the training set. A second file containing the query compound (or set of compounds) is then imported and these are screened to see if they are within the applicability domain of the training set. This assessment can be descriptor-based, in which case one of four methods can be used (ranges, Euclidean distance, city-block distance or probability density) or a structure-based assessment of the applicability domain can be performed using a fingerprint method as discussed in Section 12.3.4.3. Following analysis, each training and query compound is stated as being within or outside the applicability domain. The results can be displayed in a number of ways (*i.e.* graphical or tabular) or exported for further analysis.

AMBIT Discovery can be used to define the applicability domain of a model using one of the methods available in the software. However, if the applicability domain is multifaceted and in effect multidimensional, then all aspects of the domain must be considered (*i.e.* structural, physico-chemical descriptor, mechanistic and metabolitic). Although there is a debate as to the relative importance of each aspect of the applicability domain (structural, mechanistic, *etc.*), most people agree that the confidence one can place on a prediction will increase as each of the applicability domain approaches are considered. Therefore, it seems prudent to consider, where possible, each of the five applicability domain aspects in turn (descriptor, structural fragment, structural similarity, mechanism of action and metabolism).

12.4 Case Study

Although the process of applicability domain definition must be tailored for each individual dataset and the information that is available (mechanism of action, likely metabolites, *etc.*), this chapter considers an example dataset and, using some of the software tools discussed previously, defines a workflow to define the applicability domain.

12.4.1 Example Data: *Tetrahymena pyriformis*

12.4.1.1 Dataset

A dataset containing the toxicity of 350 compounds to *Tetrahymena pyriformis* (IGC_{50}) was used in this case study (see Appendix 1). As in most (Q)SAR studies, the available data are split into training data [used to develop the (Q)SAR model] and test data (compounds for which predictions are made, which may also be used to validate model performance). In this case the split between training and test set data was 80 : 20. In order to gain a representative sample, the dataset was ranked according to endpoint value (toxicity) and every fifth compound entered the test set. This resulted in a training set of 280 compounds and a test set of 70 compounds.

12.4.1.2 (Q)SAR Model

To simulate a real-world situation whereby a modeller would have a dataset (training and possibly also test data) and a (Q)SAR model, a widely accepted two-descriptor model for the toxicity of phenols was used based on log P and E_{LUMO}.[61] This model was rebuilt for the current training set yielding a regression equation (12.1). Log P was calculated using the KOWWIN module of the EPISUITE software, developed by the US EPA, whilst E_{LUMO} was calculated using TSAR for Windows, developed by Accelrys Software Inc.[62]

$$\text{Log } IGC_{50}^{-1} = 0.474 \log P - 0.389 E_{LUMO} - 0.637 \quad (12.1)$$

$$R^2_{(adj)} = 0.58; \ s = 0.51; \ F = 192$$

It is important to stress that the training set compounds used to derive this model are diverse in both structure and mechanism of action. This is the causal factor resulting in the reduced statistical fit seen in the above model. Although not sufficient for the practical modelling of toxicity, this model and the datasets used in this chapter are well suited to a discussion on the applicability domain.

12.4.2 Definition of the Applicability Domain for the Case Study

As already discussed in this chapter, there are many ways to define the applicability domain. A stepwise method taking into consideration a number (if not all) of these approaches is likely to be beneficial, following the workflow shown earlier in Figure 12.1.

Using the *T. pyriformis* case study dataset and the above workflow as a guide, the applicability domain was considered using each of the five approaches. It is not possible to fully explore all the methods in this chapter, but this case study serves as an example highlighting the different approaches.

12.4.2.1 Physico-Chemical Descriptor Domain

The descriptor domain is possibly the simplest approach to applicability domain definition. It involves the analysis of the descriptors that are already available to the user, *i.e.* used within the (Q)SAR model. Of course, as discussed previously, there are numerous ways in which the domain of these values can be quantified (*i.e.* range, distance, geometric or probability density-based). It is worth noting that the descriptor domain is not limited to those descriptors used within a (Q)SAR model and can include any descriptor of interest with relevance to the endpoint in question. However, this case study uses the (Q)SAR model discussed previously (see Section 12.4.1, eqn 12.1), which was based on two physico-chemical descriptors (log P and E_{LUMO}).

12.4.2.1.1 Range-Based Applicability Domain. The simplest approach is to merely state the minimum and maximum values for each descriptor for the training set compounds. If a query compound has descriptor values within this range, then it will be considered within the applicability domain. Table 12.1 shows the descriptor ranges for log P and E_{LUMO} for the training and test set.

Table 12.1 clearly shows the test set to be within the log P domain. However, not all the test compounds are within the E_{LUMO} domain as their E_{LUMO} values are not within those of the training set. Closer inspection of the data reveals that only one compound (1-phenyl-2-propanol) falls outside this E_{LUMO} domain. Therefore, using this approach, this compound must be highlighted as being outside the applicability domain of the model. Indeed, a comparison between the observed toxicity (log $IGC_{50}^{-1} = -0.62$) and that predicted by the (Q)SAR model (0.08) shows the prediction to be poor. It is therefore for useful to define the domain in this manner; compounds outside the domain often have poor predictions and therefore it is advantageous to have reduced confidence in the prediction.

Table 12.1 Ranges of descriptors used in the case study (Q)SAR model and
whether the test set is within the applicability domain of the
model.

	Log P	*LUMO*
Training set	− 0.87 to 5.99	− 2.53422 to 0.543965
Test set	0.07 to 5.36	− 1.452615 to 0.568566
Are all test compounds within the domain?	Yes	No

12.4.2.1.2 Geometric-Based Applicability Domain. A similar approach to
that of ranges is the definition of an applicability domain using geometric-
based methods. Although this approach can increase exponentially in complexity when considering more than three dimensions, this approach is quite
simple when only two variables are used. As described previously, a convex
hull can then be plotted graphically defining the applicability domain. Figure
12.8 shows a convex hull for the case study datasets, based again on log P
and E_{LUMO}.

Using the geometric approach, Figure 12.8 reveals two test compounds (α,α-
dimethylbenzenepropanol and 2-methyl-1-phenyl-2-propanol) to be outside the
applicability domain as these compounds fall outside the convex hull generated
using the descriptor values for log P and E_{LUMO}. Looking at the experimental
and predicted toxicity values of these compounds suggests that this is indeed
likely due to the large residuals between these values; α,α-dimethylbenzene-
propanol has an experimental toxicity value of − 0.07 and a prediction of 0.56.
Likewise, 2-methyl-1-phenyl-2-propanol has an experimental value of -0.41 and
a prediction of 0.31—a difference of almost one log unit. This shows that this is
also a useful approach to defining the domain.

12.4.2.2 Structural Fragment Domain

The structural fragment domain was assessed using Pipeline Pilot. The training
data were fragmented and then stored as a library of fragments to be used later.
For the test data, a pipeline was constructed which was able to fragment the
molecules (input as SMILES files) and also find any novel fragments within the
test data that were not present within the training data. Each individual
compound within the test set was run through the pipeline individually so that
the output of novel and non-novel fragments for each molecule could be stored
separately. It was then possible to see how many fragments for each test
molecule did or did not match the training data (as shown in Table 12.2). Test
compounds with any number of unmatched fragments were considered outside
the domain.

It can be seen from Table 12.2 that 14 of the compounds were classified as
being outside the domain. This includes α,α-dimethylbenzenepropanol, which
had also been classified as out of the domain by the descriptor-based methods.
As discussed previously, it is advantageous to have reduced confidence in the

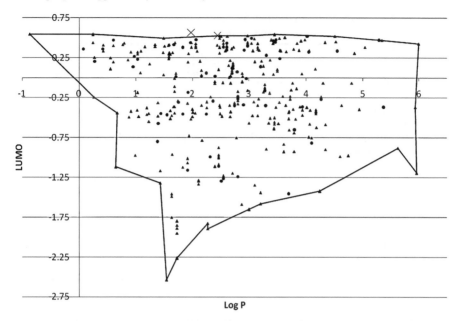

Figure 12.8 Plot of the convex hull defining the applicability domain found based on log P and E_{LUMO} for the training set ($\blacktriangle$). Test set is shown by ($\bullet$), with the compounds outside the domain denoted as (X).

prediction for this compound as it has a high residual. However, of the remaining 13 compounds, only three also have high residuals (*i.e.* >0.5) (4-hydroxyphenylaceticacid, 1.259; 4-hexylresorcinol, 0.647; 4-heptylox-yphenol, 0.65). The remaining ten compounds are well predicted by the model.

This highlights how strict the structural domain can be. Any difference in structure between the test and training compounds will force the test compound outside the domain. However, these changes in structure do not necessitate an alteration in the overall chemistry of a molecule and therefore the activity of the compound may still be well predicted by the model.

12.4.2.3 Structural Similarity Domain

The next stage of domain definition is to consider the structural similarity between the training and the test sets. This was performed using two different tools—Toxmatch and AMBIT Discovery.

12.4.2.3.1 Structural Similarity Using Toxmatch. Using the Toxmatch software, a quantitative measure of structural similarity was calculated between each query compound and the training set using the Hellinger distance (atom environments) method. In order to elucidate the applicability domain, the number of training compounds with a similarity greater than 0.6

Table 12.2 Compounds classified as being outside the applicability domain according to the structural fragment method.

Name	Number of fragments which:	
	Match training set	Do not match training set
4-hydroxy-3-methoxybenzylamine	3	1
4-hydroxyphenylaceticacid	2	1
4-nitrosophenol	2	1
1-phenyl-2-butanone	1	1
3-phenyl-1-propanol	1	1
α,α-dimethylbenzenepropanol	1	1
5-phenyl-1-pentanol	1	1
4-benzyloxyphenol	3	1
4-sec-butylphenol	2	1
n-butylbenzene	1	1
4-hexylresorcinol	2	1
4-heptyloxyphenol	2	1
4-(tert)octylphenol	2	1
2-ethylhexyl-4'-hydroxybenzoate	2	1

Table 12.3 Results of structural similarity analysis between query and training compounds.

Number of structural matches (similarity measure >0.6)	Frequency
0	0
1–5	46
6–10	18
≥ 11	6

was found for each query compound. For a query compound to be within the applicability domain, it must have at least one match with a similar training set compound. Table 12.3 shows the number of compounds and the number of structural matches.

Analysis of the results shows that all of the 70 query compounds had at least one structurally similar match within the training set, making them within the domain. Although this is a simple approach, it does show that none of the query compounds are drastically structurally dissimilar to those used to derive the model. It could be argued that a single similarity match may not be enough to classify a query compound as within the applicability domain as it may be that this single chemical is itself unique within the training data. Therefore a closer examination of the results is called for. Table 12.4 shows compounds with a single structural match in more detail.

As can be seen from Table 12.4, the majority of these compounds with a single structural match actually have reasonable predictions with small associated residuals. However, one query compound (5-amino-2-methoxyphenol) has a significantly larger residual of − 1.19, indicating that the toxicity of this compound is significantly under-predicted. Analysis of the data shows that this

Table 12.4 Detailed results of structural matches between query and training compounds.

Compound	Experimental log IGC_{50}^{-1}	Predicted log IGC_{50}^{-1}	Residual (Pred-Exp)
1-bromo-2,3-dichloro-benzene	1.300	1.509	0.209
2,3,5,6-tetrafluorophenol	1.167	0.845	− 0.322
2-bromo-4-methylphenol	0.599	0.766	0.167
2-chloro-5-methylphenol	0.393	0.620	0.227
3-chloro-4-fluorophenol	1.131	0.595	− 0.536
5-amino-2-methoxyphenol	0.450	− 0.743	− 1.193
Pentafluorophenol	1.638	1.057	− 0.581

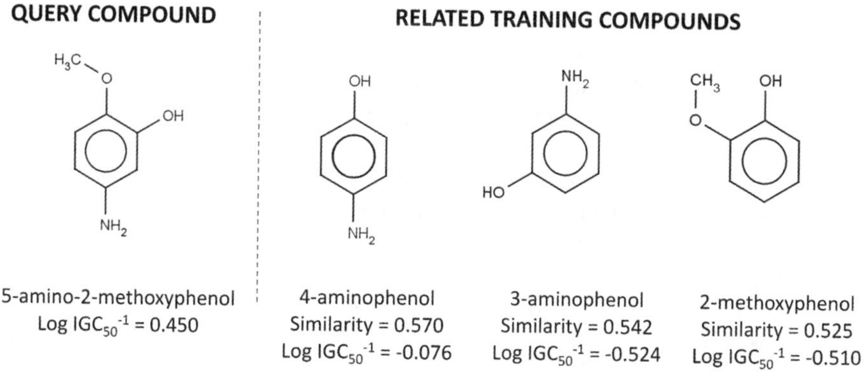

QUERY COMPOUND   **RELATED TRAINING COMPOUNDS**

5-amino-2-methoxyphenol
Log IGC_{50}^{-1} = 0.450

4-aminophenol
Similarity = 0.570
Log IGC_{50}^{-1} = -0.076

3-aminophenol
Similarity = 0.542
Log IGC_{50}^{-1} = -0.524

2-methoxyphenol
Similarity = 0.525
Log IGC_{50}^{-1} = -0.510

Figure 12.9 Structures, similarity indices and toxicities of three training compounds compared with the query compound 5-amino-2-methoxyphenol.

compound is matched with only one training compound (3-hydroxy-4-methoxybenzylalcohol). However, further analysis reveals that there are multiple structurally related compounds within the training set just falling short of the 0.6 similarity threshold. These compounds are shown in Figure 12.9.

It is clear from Figure 12.9 that each of the three related training compounds is structurally similar to the query compound. However, inspection of the toxicities of these compounds reveals that the three training compounds are considerably less toxic. In contrast to the query compound, which contains both methoxy and amino functional groups, each training compound contains only one of these functionalities, which accounts for the reduction in toxicity. The only conclusion which can be drawn is that this chemical should be classed as outside the domain of the model.

The lack of any commonly accepted 'rules' highlights the subjective nature of applicability domain definition. As, in most cases, one would choose to err on the side of caution, all query compounds with only a single matched training compound may also be classed as out of the domain. In effect, each of these compounds is sitting on the boundary of the domain, being similar to a single training compound, and may therefore suggest that the boundary of the domain should not be definite—possibly a fuzzy boundary is more appropriate. In this situation, compounds could still be definitively in or out of the domain, but ones nearing the boarder of the domain would need to be examined and a professional judgement made upon their status (as has occurred in this case).

12.4.2.3.2 Structural Similarity Using AMBIT Discovery. A number of the methods considered above (descriptor and structurally-based) are available within the specialised applicability domain tool known as AMBIT Discovery (version 0.04).[60] This freely available software can be used to assess the fit of a set of query compounds within a domain defined using a set of training compounds, by applying one of several available methods.

Based on structural similarity (calculated using fragments and fingerprints), AMBIT Discovery was used to assess how many of the 70 test compounds were within the applicability domain of the training set (280 compounds). Chemical structures were imported as SD files (other formats are available). If a distance method based on physico-chemical descriptors is utilised, the descriptor values must also be imported.

Using the fragment method within AMBIT Discovery, a test compound with fragments unmatched to those of the training set, will be classed as out of the applicability domain. In other words, the query compound contains a unique arrangement of atoms and therefore may possess unique chemistry or act *via* an alternative mechanism of action to the training set compounds. Analysis of the case study compounds revealed 11 test compounds as being outside the applicability domain, each with unique unmatched fragments.

12.4.2.4   *Mechanistic Domain*

Fortunately, the mechanism of action for each compound in the *Tetrahymena pyriformis* dataset used in this study has been previously derived and is freely available in the literature.[61] However, if this mechanistic information were not available there are, as discussed previously, software packages available which can be used to predict mechanistic class. These pieces of software have seen rapid development in recent years and the information they generate can be of great use. Software including the OECD (Q)SAR Application Toolbox or Toxtree could be used to predict this information. However, given that only selected mechanistic classifiers (*e.g.* protein binding) are currently available within these pieces of software, experimentally supported or expert derived mechanism classification would be preferential.

As with all these approaches for defining the applicability domain, anything outside the training set/model domain is classified as being outside the applicability domain. Therefore, when considering mechanisms of action, a differing mechanism of action (between the training and test data) is clearly a valid premise for exclusion. When considering toxicity, as the mechanism of action determines the reactivity (hence potency) of a chemical, the mechanism of action is particularly important. It has been seen that many outliers in (Q)SAR models are actually compounds acting *via* differing mechanisms of action, usually those with more reactive electrophilic mechanisms.[63] Therefore the identification of these mechanistic outliers is crucial.

Figure 12.10 shows the experimental *versus* predicted log IGC_{50}^{-1} values for the case study training set, using the (Q)SAR from section 12.4.1 (eqn 12.1). In addition, the mechanism of action of each chemical is also displayed to ascertain any relationship between prediction performance and mechanism of action.

It is clear from Figure 12.10 that the accuracy of the predictions differs according to the mechanism of action. Evidently, the compounds acting *via* more reactive mechanisms are poorly predicted, falling further from the regression line. This is clearly demonstrated by the pro-electrophiles, which are very poorly predicted. For this reason, training sets containing multiple mechanisms of action, although maximising the scope of the model, often lead to a reduction in model performance (as observed in this example). As the model considered here is primarily built using data from narcotic chemicals (both polar and non-polar narcosis), small numbers of compounds acting *via* other mechanisms of action

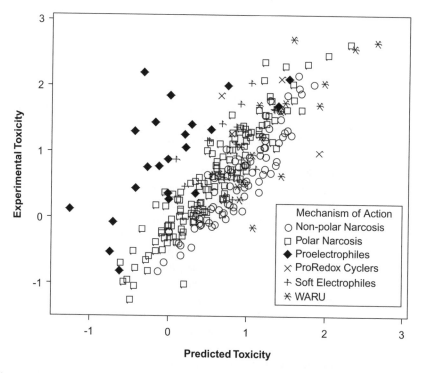

Figure 12.10 Plot of the experimental *versus* predicted log IGC_{50}^{-1} toxicity and mechanism of action for each training set compound.

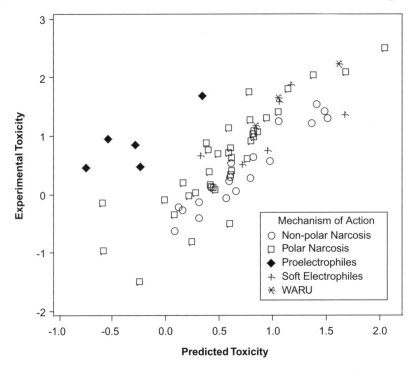

Figure 12.11   Plot showing experimental *versus* predicted log IGC_{50}^{-1} toxicity and mechanism of action for each test set compound.

effectively add noise to the predictions, reducing performance. It is therefore beneficial to restrict the training set to a single mechanism of action. Despite the resultant model being restricted in terms of applicability domain, its predictions are likely to be more accurate. Any test compound acting *via* an alternative mechanism will be outside the model domain.

When examining the test data (Figure 12.11), it can be seen that the more reactive mechanisms are the most poorly predicted. This is probably due to the fact that the majority of the compounds on which the model is built are thought to act *via* narcosis. The under-representation of the more reactive mechanisms has led to the mechanistic domain being dominated by narcosis.

It is evident from Figure 12.11 that the predictive performance between mechanisms varies greatly. The more electrophilic compounds (particularly the pro-electrophiles) show considerably increased prediction error compared with the narcotic compounds. In this case study at least, the reactive mechanisms should be considered individually, leaving the narcotic-acting compounds which, in turn, could be further subdivided into polar and non-polar mechanisms.

It is evident from this example that combining mechanisms of action into a single model can be dangerous and the level of confidence one can place in a prediction differs according to mechanism in question. It may therefore be considered important to contemplate only using one mechanism of action when building a (Q)SAR model to ensure that the mechanistic domain can be easily defined.

12.4.2.5 Metabolic Domain

Although it cannot be denied that metabolism plays an important part in assessing the applicability domain of the model, current metabolic simulators are not developed sufficiently to allow for a satisfactory evaluation. As discussed previously, to fully define the metabolic domain is difficult and the results unreliable. For this reason the metabolic domain has not been fully evaluated for these data. However, for more complex species such as mammals, metabolism is likely to play a much more important role and should be considered in greater depth.

As an example of how the metabolic approach may be used, one of the pro-electrophiles highlighted from the mechanistic approach was submitted to the microbial metabolism simulator within the OECD (Q)SAR Application Toolbox. The Toolbox predicted the possibility of nine metabolites resulting from 5-amino-2-methoxyphenol (Figure 12.12)

Once the metabolites have been predicted, the difficult decision remains as to how this information should be used. One possibility is to start the domain

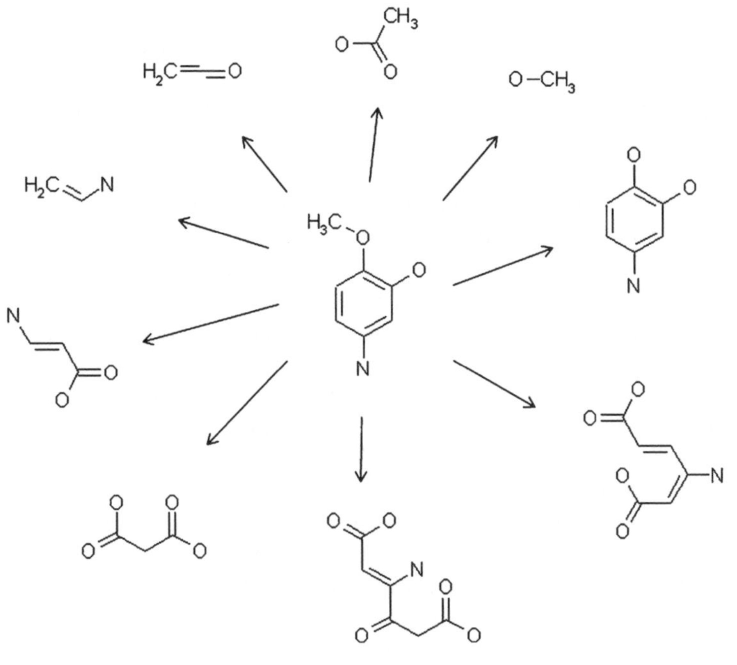

Figure 12.12 Metabolites predicted for 5-amino-2-methoxyphenol by the OECD (Q)SAR Application Toolbox.

definition process again for each of the metabolites. However, considering that these predicted metabolites may not even be produced in an actual assay, this could be a laborious task that results in the acquisition of very little knowledge. Another approach could be to look at the metabolites produced from the training data and see how they compare to those produced from a query compound. To compare the two sets of metabolites, similarity measures could be used or novel metabolites produced by the query compound (as compared with the training data) could be the defining criteria. Both of these methods are again, at present, fairly time-consuming tasks and, considering that the metabolic predictions themselves could be questionable, may be of little value. Therefore, further development of the tools available for metabolism prediction is necessary before this aspect of the domain can be fully defined.

Another point for contemplation is the fact that, if a query compound is within the structural, similarity and mechanistic domain of a model, it shares many structural features in common with the training data. As metabolism predictions are based on structure, structurally similar compounds are therefore likely to be predicted to follow similar metabolic pathways. If this is the case, then the metabolic domain may already be implicitly defined in previous stages of domain definition.

12.4.3   Applicability Domain Conclusions

As introduced in this chapter and utilised in this case study, there are many ways in which an applicability domain can be characterised. These approaches greatly differ in complexity from simple descriptor ranges to complex structural similarity measures. The results of each approach when applied to the case study are summarised in Table 12.5, which lists those compounds defined as outside the applicability domain.

Table 12.5 highlights the complexity of defining the applicability domain, with each approach giving different results. The interpretation of these results is rather context dependent. If the model is being developed for use within a regulatory decision framework (*e.g.* to evaluate the toxicity of a new chemical), it would be prudent to class all these chemicals as out of the applicability domain, reducing the risk of making inaccurate predictions.

In contrast, one could take all the information as a whole to make a more informed decision. Inspection of Table 12.5 reveals five compounds (shown in bold) which have been deemed as being outside the applicability domain in two or more domain approaches.

For example, 5-amino-2-methoxyphenol is highlighted as being outside the domain by both the Toxmatch structural similarity approach and by the AMBIT Discovery similarity approach. Interestingly, this compound is also classified as a pro-electrophile and is, therefore, also outside the mechanistic applicability domain. This is, of course, not surprising since there is an overlap between the structural and mechanistic domains, as mechanism is dependent on chemical structure. It therefore comes of little surprise that this compound was

Table 12.5 Summary of compounds falling outside the applicability domain of the training set.

Approach used	Compounds outside the applic-ability domain
Descriptor-based (ranges)	1-phenyl-2-propanol
Descriptor-based (geometric)	**α,α-dimethylbenzenepropanol**
	2-methyl-1-phenyl-2-propanol
Structural fragment approach (fragments created using Pipeline Pilot)	**4-hydroxy-3-methoxybenzylamine**
	4-hydroxyphenylaceticacid
	4-nitrosophenol
	1-phenyl-2-butanone
	α,α-dimethylbenzenepropanol
	3-phenyl-1-propanol
	5-phenyl-1-pentanol
	4-benzyloxyphenol
	4-sec-butylphenol
	n-butylbenzene
	4-hexylresorcinol
	4-heptyloxyphenol
	4-(*tert*)octylphenol
	2-ethylhexyl-4'-hydroxybenzoate
Structural similarity-based (Toxmatch – Hellinger distances)	**5-amino-2-methoxyphenol**
AMBIT Discovery (using missing fingerprints)	2,6-dibromo-4-nitrophenol
	4-bromo-2-nitrophenol
	3-chlorobenzophenone
	3-ethoxy-4-hydrobenzaldehyde
	3-chloro-4-fluorophenol
	4-chloro-3-ethylphenol
	3-acetamidophenol
	4-benzyloxyphenol
	4-hexylresorcinol
	5-amino-2-methoxyphenol
	4-hydroxy-3-methoxybenzylamine
Mechanism-based	**Electrophilic mechanisms**

found to be particularly poorly predicted with an observed toxicity (log IGC_{50}^{-1}) of 0.45 compared with a value of -0.74 predicted by the (Q)SAR model. Neither descriptor-based approach showed this compound to be outside the applicability domain. On closer inspection, this compound was also found to have a very low log *P* value of just 0.07. Only one training compound (2,4-diaminophenol) had a log *P* value which was lower at -0.87. In addition, locating 5-amino-2-methoxyphenol on the convex hull (Figure 12.9) reveals this compound to be close to the boundary of the domain, therefore suggesting that this compound should be considered as outside the applicability domain.

Similarly, α,α-dimethylbenzenepropanol was identified as being out of the applicability domain by both the geometric descriptor-based approach as well as the structural fragment approach. Unsurprisingly, the prediction for this compound was poor with an observed toxicity (log IGC_{50}^{-1}) of -0.62 compared with a predicted value of 0.08. This compound, in addition to possessing unique structural fragments, was also found to be outside the descriptor domain. This was a result of its E_{LUMO} value of 0.5686, which is outside the E_{LUMO} range defined by the training set compounds (-2.534 to 0.544).

The presence of complementary information in this manner is ideally suited to an informed approach whereby a conclusion can be reached with maximal confidence. The synergy of a multifaceted approach such as this ensures a robust yet sensitive assessment of the applicability domain which can be tailored to fit the user's demands.

A more challenging situation is where a compound is highlighted as being outside the applicability domain by one method only. For example, in the case study above, 4-nitrosophenol is shown to be outside the structural fragment domain but within the domain as defined by the alternative approaches. In this situation, reaching a conclusion is much more difficult. A logical starting point in tackling this dilemma is to consider the nature of the model for which the applicability domain is being defined. By doing so, the relative importance of the result can become clear. For example, if a prediction was made using a traditional (Q)SAR model based upon physico-chemical descriptors, the structural fragment domain is less crucial. However, if the prediction were made using a structural alert model, the structural fragment domain becomes of primary importance as it has direct implications for the validity of the prediction.

In addition to considering the nature of the model, it is also important to consider the requirements of the user. This allows the level of scrutiny to be set in keeping with the user's requirements. For example, if the predictions are to be used in a regulatory setting then the level of scrutiny required will be very high. In this case, any compound outside the applicability domain (irrespective of method) is likely to be classed as such. However, if the model is being used in a purely academic exercise, more evidence is likely to be required before a compound is eliminated as being outside the applicability domain.

It must be stressed that, because a compound is assigned as out of the domain by a given method, it may still yield an accurate prediction. For example, 3-chlorobenzophenone is classified as being outside the applicability domain by AMBIT Discovery (missing fingerprints method), but shows a good prediction of 1.41 compared with the observed toxicity of 1.55—a difference of only 0.14 log units. The accuracy of this prediction suggests that this compound should not be removed from the analysis and, as stated previously, the structural domain has little bearing upon the (Q)SAR model prediction in this case. This type of retrospective analysis is only possible when observed toxicity/activity data are available for a given chemical but, in this case, retrospective analysis has allowed possible weaknesses in certain methodologies to be identified.

The methods of domain definition are numerous and the analysis of their results, in many cases, is clearly far from simple. However, when used together in a stepwise approach, the applicability domain can be defined more accurately than when one approach is used in isolation. With knowledge of both the user requirements and model architecture, the results from several domain assessments can be used to define a robust and fit-for-purpose applicability domain which can be used to ensure maximal prediction confidence.

12.5 Recommendations

Given the reluctance to accept (Q)SAR predictions for regulatory purposes and the complexities of model validation, the applicability domain is an important means of assessing the confidence that can be placed on a given prediction. Despite the confusion which has ensued as a result of the many ways in which the applicability domain can be defined, its place within the (Q)SAR paradigm cannot be argued. In addition, with its inclusion within the OECD (Q)SAR Validation Principles, the consideration of applicability domain is desirable by all (Q)SAR modellers irrespective of the model's prospective use.

Despite the complexities involved, the applicability domain of a model is crucial in assuring that predictions are made sensibly with a given level of confidence. Whether it is the selection of the most appropriate model to make a prediction or an attempt to ascertain the confidence that can be placed in a given prediction, the applicability domain must be considered. The information gathered from applicability domain investigations can be considerable and may also be fed back into the development stages to further refine the model (*i.e.* extend the structural breadth of the training set, take into account metabolic activation, *etc.*).

In light of the multifaceted nature of defining the applicability domain and the widespread uncertainty concerning its use, the following recommendations should be considered:

- **Consider the role of the model.** Is the (Q)SAR model being used to predict physico-chemical properties as an academic exercise or toxicological predictions as part of a regulatory risk assessment? Circumstance can dictate the complexity of applicability domain studies. A simple range-based approach may be sufficient in certain cases.
- **Where possible, consider multiple approaches.** As seen in the case study, different approaches to applicability domain definition yield differing answers. Therefore, considering the applicability domain from a number of angles is beneficial as it will provide a more robust overview.
- **Role in (Q)SAR development.** Although applicability domain studies are most often conducted once a (Q)SAR model has been created, they can also have a place in their development. Applicability domain studies can be used to ensure maximal scope for a model *via* the broadening of training data in terms of chemical structure, mechanism of action, *etc.* Similar methods can also be used in intelligent test set design, ensuring a suitable and robust set of chemicals are selected, which are both within the applicability domain of the model and also representative of the training data.
- **Make use of the available software tools.** There are many freely available software tools (as highlighted in this chapter) which can aid the

characterisation and screening of applicability domains. Such tools provide a wealth of options to the modeller and also allow potential users of (Q)SAR models to assess their query compounds in the same manner, ensuring consistency and maximal confidence in (Q)SAR predictions.

- **Restricting the applicability domain of a model is not a setback.** Although detailed applicability domain studies may restrict the applicability of a (Q)SAR model, they also greatly increase the confidence in its predictions. It is better to make confident predictions for a small number of compounds than predictions with unknown confidence for many compounds. Compounds that fall outside the applicability domain of model A can be used themselves to develop model B, C, *etc.,* each with their own predictive domains.

- **The future.** With the increasing pressure to explore and utilise alternative non-testing methods, the use of *in silico* modelling approaches for toxicity prediction is expanding. Having a defined domain of applicability is one of the five OECD Principles for the validation of (Q)SARs and it is therefore essential for models to be associated with an applicability domain if they are to be used to reduce the requirements of animal testing. Recent years have seen an explosion of applicability domain studies, utilising an array of methods. However, certain types of *in silico* tools still have no accepted method of domain definition that results in a useful domain (*i.e.* a domain that enables the user to put increased confidence in correct predictions and low confidence in incorrect predictions). One major class of *in silico* tools to fall into this category are expert systems that employ the use of structural alerts. It is likely that the future will see further methods being developed and applied to the definition of applicability domains for the less traditional forms of (Q)SAR models (*e.g.* structural alerts), aiding the appropriate use of *in silico* methods.

12.6 Acknowledgements

The funding of Lhasa Ltd, Leeds, England and the EU 6th Framework Integrated Project OSIRIS (www.osiris-reach.eu; contract no. GOCE-ET-2007-037017) is gratefully acknowledged.

References

1. R. E. Hester, R. M. Harrison and M. Balls, *Alternatives to Animal Testing,* Royal Society of Chemistry, Cambridge, UK, 2006.
2. J. Tunkel, K. Mayo, C. Austin, A. Hickerson and P. Howard, *Environ. Sci. Technol.,* 2005, **39**, 2188.
3. J. Jaworska, N. Nikolova-Jeliazkova and T. Aldenberg, *Altern. Lab. Anim.,* 2005, **33**, 445.

4. T. I. Netzeva, A. P. Worth, T. Aldenberg, R. Benigni, M. T. D. Cronin, P. Gramatica, J. S. Jaworska, S. Kahn, G. Klopman, C. A. Marchant, G. Myatt, N. Nikolova-Jeliazkova, G. Y. Patlewicz, R. Perkins, D. W. Roberts, T. W. Schultz, D. T. Stanton, J. J. M. van de Sandt, W. D. Tong, G. Veith and C. H. Yang, *Altern. Lab. Anim.*, 2005, **33**, 155.

5. E. Zvinavashe, A. J. Murk and I. M. C. M. Rietjens, *Chem. Res. Toxicol.*, 2008, **21**, 2229.

6. P. Gramatica, *QSAR Comb. Sci.*, 2007, **26**, 694.

7. P. Gramatica, P. Pilutti and E. Papa, *SAR QSAR Environ. Res.*, 2007, **18**, 169.

8. P. Gramatica, E. Giani and E. Papa, *J. Mol. Graphics Modell.*, 2007, **25**, 755.

9. C. M. Ellison, S. J. Enoch, M. T. D. Cronin, J. C. Madden and P. Judson, *Altern. Lab. Anim.*, 2009, **49**, 533.

10. M. Hewitt, M. T. D. Cronin, J. C. Madden, P. R. Rowe, C. Johnson, A. Obi and S. J. Enoch, *J. Chem. Inf. Comput. Sci.*, 2007, **47**, 1460.

11. H. Dragos, M. Gilles and V. Alexandre, *J. Chem. Inf. Model.*, 2009, **49**, 1762.

12. M. T. D. Cronin, *Altern. Lab. Anim.*, 2002, **30**(Suppl. 2), 81.

13. L. Eriksson, J. Jaworska, A. P. Worth, M. T. D. Cronin, R. M. McDowell and P. Gramatica, *Environ. Health Perspect.*, 2003, **111**, 1361.

14. O. Mekenyan, S. Dimitrov, P. Schmieder and G. Veith, *SAR and QSAR Environ. Res.*, 2003, **14**, 361.

15. W. Tong, Q. Xie, H. Hong, L. Shi, H. Fang and R. Perkins, *Environ. Health Perspect.*, 2004, **112**, 1249.

16. A. Tropsha, P. Gramatica and V. Gombar, *QSAR Comb. Sci.*, 2003, **22**, 69.

17. A. P. Worth, A. Bassan, J. De Bruijin, A. G. Saliner, T. I. Netzeva, G. Patlewicz, M. Pavan, I. Tsakovska and S. Eisenreich, *SAR QSAR Environ. Res.*, 2007, **18**, 111.

18. C. M. Ellison, M. T. D. Cronin, J. C. Madden and T. W. Schultz, *SAR QSAR Environ. Res.*, 2008, **19**, 751.

19. T. W. Schultz, M. Hewitt, T. I. Netzeva and M. T. D. Cronin, *QSAR Comb. Sci.*, 2007, **26**, 238.

20. J. C. Dearden, M. T. D. Cronin and K. L. E. Kaiser, *SAR QSAR Environ. Res.*, 2009, **20**, 241.

21. A. Pery, A. Henegar and E. Mombelli, *QSAR Comb. Sci.*, 2009, **28**, 338.

22. E. Papa, S. Kovarich and P. Gramatica, *QSAR Comb. Sci.*, 2009, **28**, 790.

23. N. Nikolova-Jeliazkova and J. Jaworska, *Altern. Lab. Anim.*, 2005, **33**, 461.

24. I. Tetko, I. Sushko, A. K. Pandet, H. Zhu, A. Tropsha, E. Papa, T. Oberg, R. Todeschini, D. Fourches and A. Varnek, *J.Chem. Inf. Model.*, 2008, **48**, 1733.

25. S. Dimitrov, G. Dimitrova, T. Pavlov, N. Dimitrova, G. Patlewicz, J. Niemela and O. Mekenyan, *J.Chem. Inf. Model.*, 2005, **45**, 839.

26. European Chemicals Agency, *Guidance on Information Requirements and Chemical Safety Assessment. Chapter R.7a. Endpoint specific guidance,*

EChA, Helsinki, 2008, http://guidance.echa.europa.eu/docs/guidance_
document/information_requirements_en.htm [accessed March 2010].

27. T. W. Schultz, *Toxicol. Meth.*, 1997, **7**, 289.
28. R. W. Stanforth, E. Kolossov and B. Mirkin, *QSAR Comb. Sci.*, 2007, **26**, 837.
29. M. Pavan, T. I. Netzeva and A. P. Worth, *SAR QSAR Environ. Res.*, 2006, **17**, 147.
30. Organisation for Economic Co-operation and Development, *OECD Quantitative Structure–Activity Relationships [(Q)SARs] Project*, OECD, Paris, www.oecd.org/document/23/0,3343,en_2649_34379_33957015_1_1_1_1,00.html [accesssed March 2010].
31. J. Batista and J. Bajorath, *J. Chem. Inf. Model.*, 2007, **47**, 59.
32. J. Batista, J. W. Godden and J. Bajorath, *J. Chem. Inf. Model.*, 2006, **46**, 1937.
33. D. R. Flower, *J. Chem. Inf. Comput. Sci.*, 1998, **38**, 379.
34. N. Nikolova and J. Jaworska, *QSAR Comb. Sci.*, 2003, **22**, 1006.
35. R. P. Sheridan, B. P. Feuston, V. N. Maiorov and S. K. Kearsley, *J. Chem. Inf. Comput. Sci.*, 2004, **44**, 1912.
36. R. P. Sheridan, M. D. Miller, D. J. Underwood and S. K. Kearsley, *J. Chem. Inf. Comput. Sci.*, 1996, **36**, 128.
37. P. Willett, *Drug Discov. Today*, 2006, **11**, 1046.
38. M. Casalegno, G. Sello and E. Benfenati, *J. Chem. Inf. Model.*, 2008, **48**, 1592.
39. S. A. Kulkarni and J. Zhu, *SAR QSAR Environ. Res.*, 2008, **19**, 39.
40. R. Kühne, R. Ebert and G. Schüürmann, *J. Chem. Inf. Model.*, 2009, **49**, 2660.
41. S. Weaver and M. P. Gleeson, *J. Mol. Graphics Modell.*, 2008, **26**, 1315.
42. J. M. Barnard and G. M. Downs, *J. Chem. Inf. Comput. Sci.*, 1997, **37**, 141.
43. A. Maunz and C. Helma, *SAR QSAR Environ. Res.*, 2008, **19**, 413.
44. R. Kühne, R. Ebert and G. Schüürmann, *J. Chem. Inf. Model.*, 2006, **46**, 636.
45. The Toxtmatch software can be downloaded from ex-European Chemicals Bureau section of the website of the Institute for Health and Consumer Protection (at the European Commission Joint Research Centre), http://ecb.jrc.ec.europa.eu/qsar/qsar-tools/index.php?c = TOXMATCH [accessed March 2010].
46. C. J. Borget, T. F. Quill, L. S. McCarty and A. M. Mason, *Toxicol. Appl. Pharmacol.*, 2004, **201**, 85.
47. D. W. Roberts, A. O. Aptula and G. Patlewicz, *Chem. Res. Toxicol.*, 2006, **19**, 1228.
48. D. W. Roberts, A. O. Aptula and G. Patlewicz, *Chem. Res. Toxicol.*, 2007, **20**, 44.
49. D. W. Roberts, G. Patlewicz, P. S. Kern, G. F. Gerberick, I. Kimber, R. J. Dearman, C. A. Ryan, D. A. Basketter and A. O. Aptula, *Chem. Res. Toxicol.*, 2007, **20**, 1019.
50. R. Benigni and C. Bossa, *J. Chem Inf. Model.*, 2008, **48**, 971.

51. S. Enoch, J. C. Madden and M. T. D. Cronin, *SAR QSAR Environ. Res.*, 2008, **19**, 555.
52. H. J. M. Verhaar, C. J. Vanleeuwen and J. L. M. Hermens, *Chemosphere*, 1992, **25**, 471.
53. G. M. Cramer, R. A. Ford and R. L. Hall, *Food Cosmet. Toxicol.*, 1978, **16**, 255.
54. The Toxtree software can be downloaded from the ex-European Chemicals Bureau section of the website of the Institute for Health and Consumer Protection (at the European Commission Joint Research Centre), http://ecb.jrc.ec.europa.eu/qsar/qsar-tools/index.php?c = TOXTREE [accessed March 2010].
55. C. A. Marchant, K. A. Briggs and A. Long, *Toxicol. Mech. Methods*, 2008, **18**, 177.
56. O. Mekenyan, S. Dimitrov, N. Dimitrova, G. Dimitrova, T. Pavlov, G. Chankov, S. Kotov, K. Vasilev and R. Vasilev, *SAR QSAR Environ. Res.*, 2006, **17**, 107.
57. S. Dimitrov, L. K. Low, G. Patlewicz, G. Dimitrova, M. Comber, R. D. Phillips, J. Niemela, P. T. Bailey and O. Mekenyan, *Int. J. Toxicol.*, 2005, **24**, 189.
58. L. J. Jolivette and S. Ekins, *Adv. Clin. Chem.*, 2007, **43**, 131.
59. H. Y. Li, J. Sun, 12. W. Fan, 12. F. Sui, L. Zhang, Y. J. Wang and Z. G. He, *J. Comput.-Aided Mol. Des.*, 2008, **22**, 843.
60. Ambit Discovery can be downloaded from http://ambit.sourceforge.net/ [accessed March 2010].
61. M. T. D. Cronin, A. O. Aptula, J. C. Duffy, T. I. Netzeva, P. H. Rowe, I. V. Valkova and T. W. Schultz, *Chemosphere*, 2002, **49**, 1201.
62. TSAR for Windows, Accelrys Software Inc., http://accelrys.com.
63. S. J. Enoch, M. T. D. Cronin, T. W. Schultz and J. C. Madden, *Chemosphere*, 2008, **71**, 1225.

CHAPTER 13

Mechanisms of Toxic Action in In Silico Toxicology

D. W. ROBERTS

School of Pharmacy and Chemistry, Liverpool John Moores University, Byrom Street, Liverpool L3 3AF, UK

13.1 Introduction

The activity of a compound in any biological endpoint is a function of its chemical identity. The relationship between chemical identity and activity may be based on:

- **Molecular shape** (interpreted in a greater than three-dimensional context to include, besides spatial dimensions, distribution of dipoles and hydrophobic/hydrophilic sub-structural fragments). Such relationships, very common in drug activity, usually correspond to receptor-mediated mechanisms and are encountered in toxicology in, for example, endocrine disruption (see Chapter 8).
- **Physical properties** (in particular partitioning properties). Such relationships are encountered in narcosis modes of action in aquatic toxicity (*vide infra*), where the relevant partitioning is between aqueous and hydrophobic phases.
- **Reaction chemistry properties**. Such relationships are the basis of skin sensitisation potency, and are important in mutagenicity and aquatic toxicity.

Issues in Toxicology No.7
In Silico Toxicology: Principles and Applications
Edited by Mark T. D. Cronin and Judith C. Madden
© The Royal Society of Chemistry 2010
Published by the Royal Society of Chemistry, www.rsc.org

It is possible to develop quantitative structure–activity relationships (QSARs) without knowing the nature of the relationship between chemical identity and activity, but QSARs based on mechanistic understanding are preferable. Among their advantages are greater transparency and a greater ability to define the applicability domain in terms of mechanistic criteria.[1] In cases where the mechanism of the toxic endpoint is sufficiently well understood, it can be modelled mathematically and a QSAR derived from such modelling can be referred to as a QMM (quantitative mechanistic model).[2,3]

In this chapter, toxic mechanisms and mechanism-based approaches to their modelling are discussed, the emphasis being on aquatic toxicity and skin sensitisation.

13.2 Modes and Mechanisms

When considering mechanisms of toxic action, it is useful to distinguish between modes of action, biological mechanisms and chemical mechanism.

The mode of action relates to the way in which the toxic endpoint is expressed (*e.g.* morbidity, narcosis, hypersensitivity). The biological mechanism relates to the biological processes by which the toxic effect is produced. For example in skin sensitisation, there is a cascade of processes including cytokine and chemokine release, Langerhans cell migration to the lymph node, antigen presentation and recognition, and clonal expansion of allergen-responsive T-lymphocytes.[4] The cascade of biological processes is triggered by the toxic chemical, to an extent depending on the identity of the chemical. The chemical mechanism refers to the chemical processes by which the toxic chemical triggers the biological mechanism leading to the toxic effect. The mechanism of action forms an integral part of what is termed the 'adverse outcome pathway' described in Chapter 14.

Some toxic endpoints have several modes of action and biological mechanisms, others are less diverse. A selection of examples is shown in Table 13.1 where a comparison of acute aquatic toxicity and skin sensitisation illustrates the point.

13.2.1 Acute Aquatic Toxicity

In acute aquatic toxicity to fish, there are several modes of action such as:

- **General narcosis** – the fish become lethargic, to the point of narcosis. For a short time after this, the toxicity is reversible if the fish are transferred to fresh water, but over a longer period morbidity occurs. This mode of action applies for a diverse variety of compounds, such as aliphatic and aromatic hydrocarbons and halogenated hydrocarbons, alcohols, ketones,[5] and is often referred to as baseline toxicity.[6]
- **Polar narcosis** – similar to general narcosis, but the fish become hyper-active rather than lethargic, before narcosis sets in.[7] This mode of action applies to many phenols,[8] aromatic nitro- and amino-compounds,[9] and anionic surfactants.[10]

Table 13.1 Mechanism and mode of action: definitions.

	Definition	Examples			
		Skin sensitisation	Aquatic toxicity		Electrophilic toxicity
			Polar narcosis	General narcosis	
Chemical mechanism of action	What the toxic chemical and the biological chemicals do together *in vivo*	Electrophiles react with nucleophilic groups on proteins, *e.g.* formation of Michael adducts	2-D solvation of chemical in membrane	3-D solvation of chemical in membrane	Reaction with nucleophilic proteins in membrane or in cytosol
Biological mechanism of action	Events at cellular or sub-cellular level	Antigenic modified proteins expressed by Langerhans cells, then T-cell proliferation in lymph nodes	Reversible (short term) perturbation of membrane function	Reversible (short term) perturbation of membrane function	Irreversible disruption of cell function
Mode of action	What happens to the organism	Becomes sensitised	Hyperactivity, then narcosis	Lethargy, then narcosis	Death

- **Electrophilic toxicity** – morbidity follows hyperactivity, without intermediate reversible narcosis.[1] This mode of action applies, as the name implies, to electrophilic compounds. However, not all electrophiles exhibit this mode of action. If they are of relatively low reactivity and/or of relatively high hydrophobicity, they may be more toxic as narcotics and will act as such.

Each of these modes of action, and others such as redox recycling, corresponds to its own biological mechanism, and each biological mechanism corresponds to different chemical mechanisms. Thus the chemical mechanism underlying general narcosis is thought to be partitioning of the toxicant into the membrane in a three-dimensional (3-D) way such that the toxicant moves freely as a solute in all directions in the membrane.[6] For polar narcosis, the chemical mechanism is thought to be partitioning into the membrane in a two-dimensional (2-D) way such that the toxicant has a head group associated with the membrane head groups and can only move freely in directions parallel to the plane of the membrane.[6] These partitioning mechanisms are shown in Figure 13.1. The toxicity depends on the membrane–water partition coefficient, which can be modelled by the octanol–water partition coefficient (normally as its logarithm, log P), but the relationships between log P (membrane–water) and log P (octanol–water) differ depending whether the partitioning is two-dimensional or three-dimensional. Using log P values for artificial membranes and water, the QSARs for polar narcosis and general narcosis become more similar,[11,12] but are still statistically different and the two toxic modes can be distinguished.[6]

For electrophilic toxicity, the toxicant reacts with nucleophilic groups on cellular proteins or peptides, consequently disrupting cell function. Toxicity can be correlated with measured reactivity parameters such as rate constants[13] or their equivalents,[14] or with computed reactivity parameters based on substituent constants or molecular orbital calculations,[15] with little or no contribution from hydrophobicity. There are various electrophile–nucleophile mechanisms, *vide infra*.

In aquatic toxicity, the various modes of action are not mutually exclusive. Thus an electrophilic compound with a polar group will have an electrophilic toxicity, a polar narcosis toxicity and a general narcosis toxicity. If the compound acts alone, it will express its highest toxicity, *i.e.* as the concentration in a toxicity assay is increased, 50% of the subjects will exhibit the toxic effect when the concentration reaches the lowest of the compound's EC50 values. However if it is a component of a mixture, a compound may contribute a lower toxicity. Figure 13.2 illustrates the point for compounds able to act both as polar and general narcotics.

13.2.2 Skin Sensitisation

In skin sensitisation there is only one mode of action. In response to the sensitising compound, a clonally expanded population of circulating T-lymphocytes able to recognise antigens resulting from that compound is produced.

Generalised structure of a membrane (hydrocarbon chains shown only in centre of diagram)

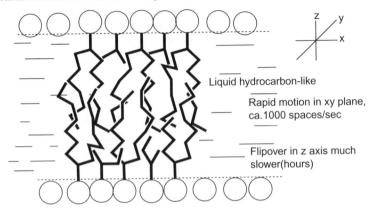

Liquid hydrocarbon-like

Rapid motion in xy plane,
ca.1000 spaces/sec

Flipover in z axis much
slower(hours)

Behaviour of solutes in a membrane

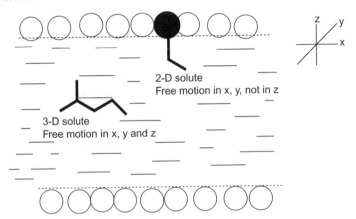

2-D solute
Free motion in x, y, not in z

3-D solute
Free motion in x, y and z

Figure 13.1   Solute partitioning into a membrane.

The biological mechanism may be very briefly summarised as follows: the sensitiser reacts, usually as an electrophile with reactive (usually nucleophilic) groups on proteins, forming antigenic modified proteins which are processed by Langerhans cells, resulting in migration of the Langerhans cells to the lymph node where they present the sensitiser-derived antigens to naïve T-cells, leading to T-cell proliferation and development of the sensitised state.[4] There are some variations in the biological mechanism, depending on the nature of the sensitiser, in the way the migration of Langerhans cells is induced and in the nature of the T-cell populations that are stimulated to proliferate.[16]

The reaction of the sensitiser with reactive groups on proteins constitutes the chemical mechanism of sensitisation, and consequently skin sensitisation has several chemical mechanisms, corresponding to the reaction mechanisms at the protein reaction step. Similarly, electrophilic aquatic toxicity also has several mechanisms.

Narcotic (N) partitioning from water (W) into membrane M

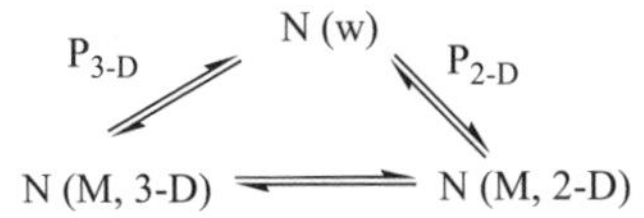

General narcosis $\log [N]_{M,3\text{-}D} = \log [N]_W + \log P_{3\text{-}D}$

Setting $[N]_{M,3\text{-}D} = EC50$
and rearranging: $pEC50_{3\text{-}D} = \log P_{3\text{-}D} - \underbrace{\log [N50]_{M,3\text{-}D}}_{\text{Constant}}$

Similarly…

Polar narcosis $pEC50_{2\text{-}D} = \log P_{2\text{-}D} - \underbrace{\log [N50]_{M,2\text{-}D}}_{\text{Constant}}$

Figure 13.2 Theoretical basis of narcosis QSARs. The symbol p represents the operator-log If $pEC50_{3\text{-}D} > pEC50_{2\text{-}D}$, the compound behaves as a general narcotic. If $pEC50_{3\text{-}D} < pEC50_{2\text{-}D}$, the compound behaves as a polar narcotic.

13.3 Reaction Mechanistic Domains

There are various organic reaction mechanisms whereby chemicals can react covalently with proteins. The major electrophilic domains encountered in skin sensitisation are (see Chapter 7):[1]

- Michael acceptor;
- SNAr;
- SN2;
- Schiff base;
- Acyl transfer; and
- Non-reactive, non-proreactive

These reaction mechanistic domains are also the major ones encountered in reactive toxicity to aquatic organisms. Identification criteria for these domains are shown in Figure 13.3. Other mechanistic domains that are probably relevant—at least in skin sensitisation—are S_N1[3,17] and free radical binding to proteins.[18] It is important to note that many compounds exhibiting reactive toxicity do so not by reacting directly with proteins but by being converted—either metabolically or abiotically—to electrophiles belonging to one of the above classes.[1,19] Such compounds are referred to as pro-electrophiles. Compounds in the non-reactive, non-proreactive domain are not skin sensitisers, and in aquatic toxicity they act as narcotics.

13.4 Mechanism-Based Reactivity Parameters for Electrophilic Toxicity

The identity of the biological nucleophiles involved in skin sensitisation and aquatic toxicity is not known. Possibilities range from any nucleophilic group

Mechanistic domain　　　　**Protein binding reaction**　　　　**Modified protein**

Michael acceptors　　　X $\diagdown\diagdown$ Nu—Protein ⟶ X $\diagdown\diagdown^{Nu}$ Protein

Identification characteristics. Double or triple bond with electron-withdrawing substituent X, such as -CHO, -COR, -CO$_2$R , -CN, -SO$_2$R, -NO$_2$...Includes *para* quinones and *ortho* quinones, often formed by oxidation of para and ortho di-hydroxy aromatics acting as pro-Michael acceptors. X can also be a heterocyclic group such as 2-pyridino or 4-pyridino.

S$_N$Ar electrophiles

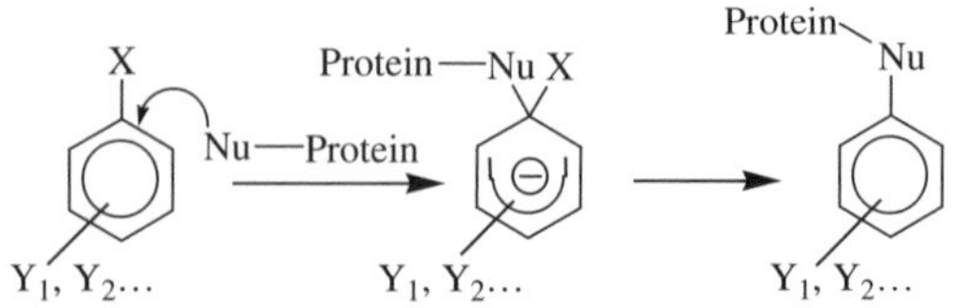

Protein—Nu X ⟶ Protein—Nu

$Y_1, Y_2...$

Identification characteristics. X = halogen or pseudohalogen, Y's are electron withdrawing groups (at least two) such as -NO$_2$, -CN, -CHO, -CF$_3$, -SOMe, -SO$_2$Me, ring fused nitrogen...One halogen is too weak to act as a Y, but several halogens together can activate.

S$_N$2 electrophiles　　　X— $\frown$ Nu—Protein ⟶ —Nu $\diagdown$ Protein

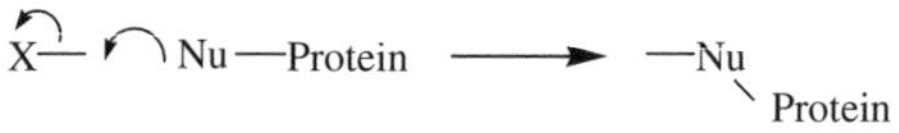

Identification characteristics. X = halogen or other leaving group, e.g. OSO$_2$(R or Ar), OSO$_2$O(R or Ar) bonded to primary alkyl, benzylic, or allylic carbon. OR and NHR or NR$_2$ do not usually act as leaving groups, but can do so if part of a strained 3-membered ring (e.g. epoxides, ethylenimine and substituted derivatives).

Schiff base formers　　　O= $\frown$ NH$_2$—Protein ⟶ =N—Protein

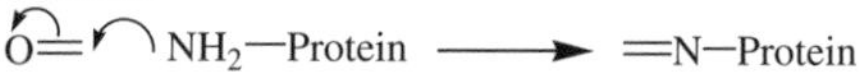

Identification characteristics. Reactive carbonyl compounds such as aliphatic aldehydes, some α,β- and α,γ-diketones, α-ketoesters. Not simple monoketones and aromatic aldehydes. Other hetero-unsaturated systems can behave analogously, e.g. C-nitroso compounds, thiocarbonyl compounds (C=S), cyanates and isocyanates, thiocyanates and isothiocyanates.

Acylating agents

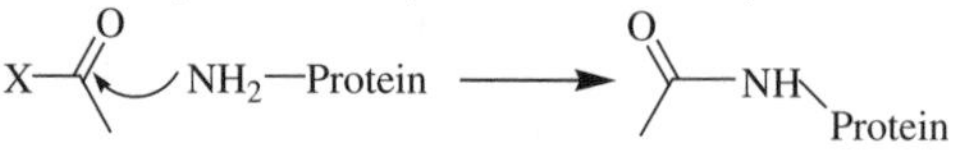

X— $\overset{O}{\diagup}$ —NH$_2$—Protein ⟶ $\overset{O}{\diagup}$ —NH $\diagdown$ Protein

Identification characteristics. X = halogen, or other group (e.g. -OC$_6$H$_5$) such that XH is acidic enough for X$^-$ to act as a good leaving group. Includes anhydrides, cyclic or non-cyclic. X = -Oalkyl does not qualify, except when part of a strained lactone ring, e.g. β-propiolactone (but not γ-butyrolactone). Analogous reactions can occur with attack at sulphonyl S, phosphoryl P and thioacyl C.

Figure 13.3　Identification criteria for reaction mechanistic domains.

on any protein being able to participate in the toxic mechanism to a small number of highly reactive proteins 'designed' by evolution to act as *in vivo* electrophile scavengers. In this situation the successful use in QSARs of rate constants from *in chemico* experiments with model nucleophiles, or of calcu-lated reactivity parameters (which may be regarded as equivalent to rate constants with virtual nucleophiles) depends on the relevant *in vivo* rate constant being correlated with the *in chemico* or *in silico* rate constant. It is therefore necessary to consider whether, to what extent and in what circumstances such correlations can be assumed to be apply. This issue is best analysed in the context of the Swain–Scott relationship,[20] and the hard and soft acids and bases (HSAB) concept.[21]

13.4.1 The Swain–Scott Relationship and the HSAB Concept

In the context of electrophile–nucleophile reactions, it is convenient to consider the HSAB principle against the background of the Swain–Scott relationship, which predates the HSAB concept. Swain and Scott argued[20] that nucleophiles could be assigned a nucleophilicity parameter *n* and that a parameter *s* could be assigned to electrophiles to quantify their susceptibility to changes in *n*. On that basis the Swain–Scott relationship was written originally as:

$$\text{Log}\left(\frac{k_{E/N}}{k_{E/\text{water}}}\right) = ns \qquad (13.1)$$

where:

$k_{E/N}$ is the rate constant for reaction of the electrophile E with the nucleophile N
$k_{E/\text{water}}$ is the rate constant for reaction of the electrophile E with water.

Water may be regarded as the reference nucleophile, $k_{E/\text{water}}$ quantifying the intrinsic reactivity of the electrophile. Taking methyl bromide as the reference electrophile and assigning an *s* value of 1, Swain and Scott measured rate constants at 25 °C and from these derived *n* values for a series of nucleophiles and *s* values for a series of electrophiles (mainly reacting by the S_N2 mechanism). Some *n* and *s* values for various nucleophiles and electrophiles are shown in Table 13.2.

It has been pointed out by Loechler that, towards the hard end of the hard–soft spectrum, deficiencies in the Swain–Scott equation become apparent.[23]

Table 13.2 Swain–Scott parameters[a] for some nucleophiles and electrophiles

Nucleophile	n	Electrophile	s
Aniline	4.49	Ethyl tosylate	0.66
n-Butylamine	5.13	Benzyl chloride	0.87
HS⁻	5.08	Methyl Bromide	1.00
Cysteine	5.08	Epichlorohydrin	0.93
Cysteine ethyl ester	5.24		
Serum albumin	5.37		
HO⁻	5.94		
Acetate ion	2.89		

[a]These values are for reactions in water, with water as the reference nucleophile being assigned an *n* value of zero. A parallel set of *n* and s values based on methanol as the reference nucleophile and methyl iodide as the reference electrophile has also been established for reactions in methanol.[22]

Thus the methyl diazonium ion, a hard S_N2 methylating agent, is more selective for oxygen-centred nucleophiles of DNA than for nitrogen-centred nucleophiles, although with less hard electrophiles, the nitrogen-centred nucleophiles are the more reactive. Loechler argued that n values of nucleophiles are not completely constant, but may vary according to the electrophile;[23] in particular, when the nucleophile and electrophile are well-matched on the hard-soft spectrum, the nucleophile may exhibit an enhanced n value.

Despite this and other deficiencies, the Swain–Scott equation (13.1) provides a useful framework for considering the issues involved in trying to model an *in vivo* nucleophile whose precise identity may not be known, with model nucleophiles *in chemico*. The key questions are:

1. If the relative reactivity of two electrophiles E1 and E2, having s values s_1 and s_2 respectively, is determined with a model nucleophile N_m, is this relative reactivity the same for the *in vivo* nucleophile N_{iv}?
2. If not, what is the difference?
3. How can the difference be minimised?

These questions can be addressed by rearranging the Swain–Scott equation to derive:

$$\text{Relative reactivity } in\ vivo = \log(k_{N_{iv}/E1}/k_{N_{iv}/E2})$$
$$= n_{iv}(s_1 - s_2) + \log(k_{water/E1}/k_{water/E2}) \quad (13.2)$$

$$\text{Relative reactivity to model} = \log\left(\frac{k_{N_{iv}/E1}}{k_{N_{iv}/E2}}\right)$$
$$= n_{iv}(s_1 - s_2) + \log\left(\frac{k_{water/E1}}{k_{water/E2}}\right) \quad (13.3)$$

For the two expressions to be identical, it is necessary that either $s_1 = s_2$ or $n_{iv} = n_m$. If this condition is not met, there is a difference of $(n_{iv} - n_m)(s_1 - s_2)$. This difference, being dependent on the identities of the electrophiles, is not constant, so it will tend to weaken a reactive toxicity QSAR in which reactivity is based on a model nucleophile. Depending how much is known or can be inferred about the *in vivo* nucleophile, selecting a model nucleophile with a similar n value (if the n value of the *in vivo* nucleophile is not known this can be approached by aiming to match pKa and position on the hard-soft scale) will minimise $(n_{iv} - n_m)$ and hence reduce $(n_{iv} - n_m)(s_1 - s_2)$. Working within the same reaction mechanistic domain, or within sub-domains (categories), will minimise differences in s values, and hence reduce $(n_{iv} - n_m)(s_1 - s_2)$.

13.5 Role of Hydrophobicity in Modelling Electrophilic Toxicity

Whether, and if so to what extent, hydrophobicity contributes to electrophilic toxicity depends on the *in vivo* location of the nucleophiles involved in the toxic

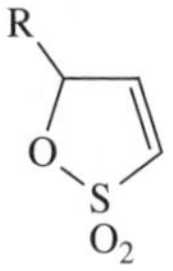

Scheme 13.1   Alk-1-ene-1,3-sultones

mechanism. Although, as pointed out above, little if anything is known definitely about the nature of the *in vivo* nucleophiles, some inferences can be drawn from QSAR findings.

Several QSARs have been developed for electrophilic and pro-electrophilic aquatic toxicity.[13–15,24] These show little or no dependence on hydrophobicity. The inference is that the nucleophiles involved are in an aqueous environment, presumably the cytoplasm inside cells.[24] The slight dependence on hydrophobicity that is sometimes observed may reflect differences between pure water and cytoplasm (a viscous non-ideal solution of organic and inorganic compounds) in terms of partitioning properties.

In skin sensitisation the situation is somewhat more complex. For S_N2 electrophiles, QSAR models for both guinea pig and mouse data require both reactivity and hydrophobicity parameters.[25–28] This suggests that the *in vivo* reaction occurs in a lipid (presumably membrane) environment. Similarly, mouse data for sensitisers in the Schiff base domain (aldehydes and ketones) are correlated with a combination of reactivity and hydrophobicity parameters.[2] For the Michael acceptor domain, a QSAR based on reactivity is not improved by a hydrophobicity parameter for mouse sensitisation data.[29] However, for guinea pig sensitisation data on long chain (total carbon number C12–C16) alk-1-ene-1,3-sultones (Scheme 13.1), shown to act as Michael acceptors,[30] potency is correlated with hydrophobicity.[25] For the S_NAr domain, there is no evidence for any dependence on hydrophobicity in guinea pigs or mice.[31] The overall inference is that there are at least two types of *in vivo* nucleophile involved in skin sensitisation—the first located in membranes and the second in cytoplasm. It can be further inferred that the cytoplasmic nucleophiles are softer than the membrane nucleophiles, since it is the softer electrophilic domains that show no dependence of potency on hydrophobicity. The log P dependence observed with the long chain Michael acceptors in guinea pigs may be because their greater tendency to partition into the membrane environment outweighs their lower reactivity with the membrane nucleophiles.

13.6   Conclusions

A chemistry-based mechanistic approach to prediction of toxicity seems promising for several toxic endpoints including aquatic toxicity and skin sensitisation.

Presented with a new compound, the first step is to classify it into its reaction mechanistic domain. One domain is the 'unreactive' domain, populated by predicted narcotics (aquatic toxicity) and non-sensitisers. For several mechanistic domains there are corresponding pro-electrophilic sub-domains. For example, many skin sensitisers such as hydroquinone and 3-alkyl/alkenyl catechols (active components of poison ivy) are thought to act as pro-Michael acceptors.[32] Domain classification may often be possible by inspection of structure, but inevitably in some cases a confident prediction may not be possible. In such situations, experimental work will be needed to determine the reaction chemistry—in particular to determine if the compound is electrophilic or pro-electrophilic and the nature of the reactions.

Having assigned the compound to its reaction mechanistic applicability domain, the next step is to quantify its reactivity/hydrophobicity relative to known toxicants in the same mechanistic applicability domain. These properties may sometimes be confidently predictable from structure using physical organic chemistry approaches such as linear free energy relationships based on substituent constants or on molecular orbital parameters. In other cases it will be necessary to perform physical organic chemistry measurements such as determination of reaction kinetics and measurement of partition coefficients. If there are sufficient known toxicants in the same mechanistic applicability domain, it should be possible to develop a QSAR by means of which toxicity of new compounds can be predicted. If not, read-across can be applied to make more approximate predictions.

Having assigned the compound to its reaction mechanistic applicability domain and quantified its reactivity/hydrophobicity relative to known toxicants in the same domain, QSAR or mechanistic read-across can be applied to predict the toxicity.

13.7   Acknowledgements

This project was sponsored by the Department for Environment, Food and Rural Affairs (Defra) through the Sustainable Arable Link Programme (Grant LK0984).

References

1. A. O. Aptula and D. W. Roberts, *Chem. Res. Toxicol.*, 2006, **19**, 1097.
2. A. O. Aptula, D. W. Roberts and G. Patlewicz, *Chem. Res. Toxicol.*, 2006, **19**, 1228.
3. D. W. Roberts, G. Patlewicz, P. Kern, F. Gerberick, I. Kimber, R. Dearman, C. Ryan, D. Basketter and A. O. Aptula, *Chem. Res. Toxicol.*, 2007, **20**, 1019.
4. D. A. Basketter and I. Kimber, *J. Appl. Toxicol.*, 2009, **25**, 545.
5. H. Könemann, *Toxicology*, 1981, **19**, 209.
6. D. W. Roberts and J. F. Costello, *QSAR Comb. Sci.*, 2003, **22**, 226.

7. C. L. Russom, S. P. Bradbury, S. J. Broderius, D. E Hammermeister and R. A. Drummond, *Environ. Toxicol. Chem.*, 1997, **16**, 948.
8. J. Saarikoski and M. Viluksela, *Ecotoxicol. Environ. Safety*, 1982, **6**, 501.
9. E. R. Ramos, W. H. J. Vaes, H. J. M. Verhaar and J. M. L. Hermens, *Environ. Sci. Pollut. Res.*, 1997, **4**, 83.
10. G. Hodges, D. W. Roberts, S. J. Marshall and J. C. Dearden, *Chemosphere*, 2006, **64**, 17.
11. W. H. J. Vaes, E. U. Ramos, H. J. M. Verhaar and J. L. M. Hermens, *Environ. Toxicol. Chem.*, 1998, **17**, 1380.
12. B. I. Escher, R. Eggen, E. Vye, U. Schreiber, B. Wisner and R. P. Schwartzenbach, *Environ. Sci. Technol.*, 2002, **36**, 1971.
13. J. Hermens, F. Busser, P. Leeuwangh and A. Musch, *Toxicol. Environ. Chem.*, 1985, **9**, 219.
14. D. W. Roberts, T. W. Schultz, A. M. Wolf and A. O. Aptula, *Chem. Res. Toxicol.*, 2010, **23**, 228.
15. T. W. Schultz, T. I. Netzeva, D. W. Roberts and M. T. D. Cronin, *Chem. Res. Toxicol.*, 2005, **18**, 330.
16. R. J. Dearman, N. Humphreys, R. A. Skinner and I. Kimber, *Clin. Exp. Allergy*, 2005, **35**, 498.
17. H. Alenius, D. W. Roberts, Y. Tokura, A. Lauerma, G. Patlewicz and M. S. Roberts, in *Drug Discovery Today: Disease Mechanisms, ed. M. S. Roberts,* Elsevier, Amsterdam, 2008, pp. 211–220.
18. T. Redeby, U. Nilsson, T. M. Altamore, L. Ilag, A. Ambrosi, K. Broo, A. Börje and A.-T. Karlberg, *Chem. Res. Toxicol.*, 2010, **23**, 203.
19. A.-T. Karlberg, M. A. Bergström, A. Börje, K. Luthman and J. L. G. Nilsson, *Chem. Res. Toxicol.*, 2008, **21**, 53.
20. C. G. Swain and C. B. Scott, *J. Am. Chem. Soc.*, 1953, **75**, 141.
21. R. G. Pearson, *J. Am. Chem. Soc.*, 1963, **85**, 3533.
22. R. G. Pearson, H. Sobel and J. Songstad, *J. Am. Chem. Soc.*, 1968, **90**, 319.
23. E. L. Loechler, *Chem. Res. Toxicol.*, 1994, **7**, 277.
24. D. W. Roberts, in *Aquatic Toxicology and Environmental Fate: Eleventh Volume. ASTM STP 1007*, ed. W. Suter II and M. A. Lewis, ASTM, Philadelphia, 1989, pp. 490–506.
25. D. W. Roberts and D. L. Williams, *J. Theor. Biol.*, 1982, **99**, 807.
26. D. W. Roberts, B. F.J. Goodwin, D. L. Williams, K. Jones, A. W. Johnson and C. J. E. Alderson, *Food Chem. Toxicol.*, 1983, **21**, 811.
27. D. W. Roberts and D. A. Basketter, *Contact Derm.*, 1990, **23**, 331.
28. D. W. Roberts, A. O. Aptula and G. Patlewicz, *Chem. Res. Toxicol.*, 2007, **20**, 44.
29. D. W. Roberts and A. Natsch, *Chem. Res. Toxicol*, 2009, **22**, 592.
30. D. W. Roberts, D. L. Williams and D. Bethell, *Chem. Res. Toxicol*, 2007, **20**, 61.
31. D. W. Roberts, *Chem. Res. Toxicol.*, 1995, **8**, 545.
32. D. W. Roberts and J-P. Lepoittevin, in *Allergic Contact Dermatitis. The Molecular Basis*, ed. J.-P. Lepoittevin, D. A. Basketter, A. Goosens and A-T. Karlberg, Springer, Heidelberg, 1997, pp. 81–111.

CHAPTER 14

Adverse Outcome Pathways: A Way of Linking Chemical Structure to In Vivo Toxicological Hazards

T. W. SCHULTZ

The University of Tennessee, College of Veterinary Medicine, Department of Comparative Medicine, 2407 River Drive, Knoxville, TN 37996-4543, USA

14.1   Categories in Hazard Assessment

There are many national and international efforts to assess the hazards or risks of chemical substances—in particular industrial organic compounds—to humans and the environment. It is generally agreed that establishing the adequacy of information for each endpoint to be evaluated is the first step in the assessment process. If adequate information is not available, which is often the case, additional data are required to complete the assessment. For a number of reasons (including time, resources and animal welfare) testing, especially *in vivo* is not always the first option to obtain additional data. In fact, *in vivo* testing is listed as the option of last resort in the European Union's Registration, Evaluation, Authorisation and restriction of Chemicals (REACH) legislation.[1] One approach to filling data gaps in an assessment, which is gaining favour, is to consider closely related chemicals as a group or 'chemical category' rather than as individual chemicals.[2]

Issues in Toxicology No.7
In Silico Toxicology: Principles and Applications
Edited by Mark T. D. Cronin and Judith C. Madden
© The Royal Society of Chemistry 2010
Published by the Royal Society of Chemistry, www.rsc.org

In its widest context, a chemical category can be thought of as a group of chemicals whose physico-chemical properties, human health and/or environmental toxicological properties, and/or environmental fate properties are likely to be similar or follow a regular pattern. As such, a chemical category can be expressed by a matrix consisting of the chemicals included in the category and corresponding sets of chemical properties and toxicological endpoint data.[2,3]

The most attractive aspect to using categories in the assessment of organic chemicals is that they provide a means of evaluating all members of a category for common toxicological behaviour or consistent trends among data for an endpoint. A further advantage of using categories in hazard assessment is that the identification of a consistent pattern of toxic effects within a category increases the confidence in the reliability of the results for all the chemicals in the category, especially when compared with evaluation of data on a compound-by-compound basis. As a result, forming a category and then using measured data on a few category members to estimate the missing values for one or more untested category members is generally accepted as a common sense application for data gap filling.[3] The net result is that when the category approach is used to fill data gaps, whether by read-across or quantitative structure–activity relationship (QSAR) modelling, there is a greater likelihood that the estimated value will be accurate and accepted by the regulatory community.[2] However, all this is predicated on *a priori* placement of the chemical into the correct category. Thus, category formation is a topic of paramount importance to predictive toxicology.

The challenge in grouping chemicals for hazard assessment is placing them into groups that are meaningful for a particular toxicity endpoint. This challenge is the result of the discontinuity between chemical and toxicological spaces, which are often endpoint related. Substances which are 'similar' in molecule structure are often dissimilar in terms of toxic action, including the ability to elicit a particular hazard endpoint as well as potency within a particular hazard endpoint. The consequence is that toxicological surfaces covering the chemical universe appear as a 'rugged landscape' made up of relatively smooth surfaces punctuated by sharp changes in directions and angles. Modelling such landscapes is easiest and more accurately done by treating each surface as a separate category with a separate structural domain. A means of segregating structural domains or 'forming a toxicologically meaningful category' is therefore required.

Chemicals, historically, have been grouped together using expert judgement, which typically reflected a common feature exhibited by the chemicals included in the group. Well-studied examples are chemicals that react covalently with nucleophilic sites on proteins.[4] Commonalities in structure between members of a group lead to the elucidation of structural alerts. Such structural alerts are then used to identify the category and define the molecular structural limits of the domain of that category. For example, the category of chemicals which act by Michael type nucleophilic addition to thiol is defined by the structural alert for a 'polarised α,β-unsaturated group'.[5] However, Michael addition includes a number of sub-categories where potency is consistent within the sub-category but varies between sub-categories.[5,6] The classic example of the difference between subcategories is the difference in fish acute toxicity between acrylates and methacrylates, with the former being 3,000 times more toxic than the latter (see Chapter 7).[7] Such examples suggest a more systematic approach to refining categories or sub-categorisation may be required to better incorporate potency differences into toxicological predictions to assure a greater likelihood that the estimated value will be accurate.

Additionally, while categorisation schemes allow for the grouping of chemicals into toxicologically meaningful groups (whether a category or sub-category), other exclusion rules based on additional properties may be relevant. For example, when grouping chemicals for acute aquatic toxicity) according to water solubility or hydrophobicity, it could be relevant to exclude chemicals with low water solubility.

14.2   Filling Data Gaps in a Category

The data in a matrix that makes up a category can be used to fill data gaps by one of two techniques: read-across or QSAR modelling. With either technique, interpolation between category members is preferred to extrapolation. Read-across is the simpler technique. It predicts endpoint information for one chemical by using data from the same endpoint from another chemical or group of chemicals, which are considered to be 'similar'.[6,8] When the category is of limited size, data gaps may be filled by read-across.[2] If the group of closely related chemicals is large, typically greater than ten, then read-across is termed the category approach.[2] The key here is that endpoint information for several chemicals is used to predict the same endpoint for an untested chemical. If the number of tested compounds within the category is limited (as few as one), the approach is modified and termed the analogue approach.[2] The difference in the latter case is fewer endpoint values are used to predict the same endpoint for an untested analogue. The rationale for having two approaches to read-across is that the restricted analogue approach is generally open to more uncertainty than the broader category approach. It is intuitive that confidence in predictions increases with increasing numbers of tested chemicals included in the category (see Chapter 15).

When the category is of sufficient size and the Organisation for Economic Co-operation and Development (OECD) Principles[9] are met, data gaps may be filled by a QSAR model. When the QSAR model is a simple linear regression (*i.e.* when for a given category endpoint, the category members are related by an increasing, decreasing or constancy in an effect, which can be related to a single property), the term 'trend analysis' is used to describe the model for the category.[2]

The data matrix in Table 14.1 demonstrates how data gaps in a category can be filled. An examination of the data in Table 14.1 reveals that chemicals A, B, D and E form a category as structure S1 and property P1 are common to all four compounds and related to property P2 and toxicities T1 and T2, while

Table 14.1 Structure/property/toxicity matrix for five compounds (A–E).

	Compound				
	A	*B*	*C*	*D*	*E*
Structure S1	+	+	−	+	+
Property P1	+	+	−	+	+
Property P2	1	2	3	4	5
Toxicity T1	+	+	−	+	?
Toxicity T2	10	15	5	?	30

compound C is not a member of the category. Two data gaps are also observed—T1 for compound E and T2 for compound D. For the category identified, the relationships between structure S1 and/or property P1 and toxicity T1 are qualitative. Using the relation with P1, measured data for compounds A, B and D can be used to fill the data gap in T1 for compound E. This is performed by read-across (*i.e.* read-across from the measured compounds A, B and D to the untested compound E). For the category, the relationship between P2 and T2 is quantitative and modelled as [Toxicity T2 = 5.0 (Property P2) + 5.0]. Using this QSAR model, the potency of T2 for compound D is predicted to be 25.

14.3 Mechanism of Toxic Action

Inherent to forming a toxicologically relevant chemical category is that the compounds in the category are identified based on the hypothesis that the properties of the chemicals in the category will show coherent trends in their toxicological effects. This hypothesis typically presumes a 'common mechanism of action'.[10] Equally, the absence of a common mechanism of action is also a factor in deciding if a chemical would be unlikely to be a member of a category. On the surface, the common mechanism of action seems intuitive and supported by the literature.[11] When the chemicals in a category have a single mechanism of action, the category concept creates a practical and powerful approach to describe the structural requirements of that mechanism (see Chapter 13). Just as the confidence in the category is significantly greater when the number of tested chemicals is greater, it is also increased when it is known that the members of the category share a common mechanism of action. However, current knowledge of toxicological categories and category formation is limited. In part, limitation is due to the depth and breadth of available databases needed to support category formation and the intricacy of the hazard endpoint being evaluated.

Examination of the literature shows that, in mammalian pharmacology and toxicology, the term 'mechanism of action' denotes the sequence of events leading from the absorption of an effective dose of a chemical to the production of a specific biological response in the target organ.[10] According to Borgert *et al.*,[10] understanding a chemical's mechanism requires appreciation of the causality and temporal relationships between the steps leading to a particular toxic endpoint, as well as the steps that lead to an effective dose of the chemical at the relevant biological target(s). They further noted that, to define a mechanism of action of a chemical, the experimental data need to be sufficient to form conclusions regarding the following:

- metabolism and distribution of the chemical in the organism or population, and subsequent modulating influence on the dose delivered to the molecular site(s) of action;
- molecular site(s) of action;
- biochemical pathway(s) affected by the chemical's action on the molecular target and resulting perturbations of those pathways;
- cellular- and organ-level consequences of affecting the particular biochemical pathway(s);
- target organ(s) or tissue(s) in which the molecular site of action and biochemical effect occur;
- physiological response(s) to the biochemical and cellular effects;
- target organ-response(s) to the biochemical, cellular, and physiological effects;
- overall effect on the organism;
- ecological effects, or the overall effect on the population or ecosystem;
- causality and temporal relationships between the mechanistic steps; and
- dose–response parameters associated with each step.

Meeting this definition of a mechanism of action requires an exceptionally large amount of high quality data, which only can be attained for a very limited number of compounds. As such, it is currently out of reach for the vast majority of industrial organic compounds. Clearly, one cannot impose a mechanism of action definition to forming chemical categories for toxicological assessment and expect to make progress in the very near future.

14.4   Toxicologically Meaningful Categories (TMCs)

While a category can be thought of as a group of chemicals whose physicochemical, toxicological and fate properties are likely to be similar or follow a regular pattern, it is unlikely that the members of a category will be the same for any two hazard endpoints. Returning to the example of electrophilic reactivity and Michael-type nucleophilic addition, if for simplification we limit the discussion to acrylates and compare acute toxicity in excess of baseline, it can be seen that those acrylates that are part of a toxicologically meaningful category (TMC) for fish acute toxicity are those with significant aquatic solubilities (*i.e.* those with ten or fewer carbon atoms). In contrast, those acrylates that are part of a TMC for mammalian inhalation acute toxicity are those with significant vapour pressures (*e.g.* those with five or fewer carbon atoms). While the 'acrylate' category is the same for both toxicity endpoints, membership of the

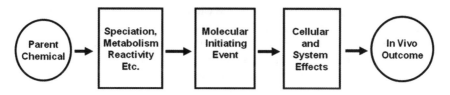

Figure 14.1 Conceptual framework for the molecular initiating event.

category is endpoint-specific; therefore for the category to be 'toxicologically meaningful', it must be defined differently for the two hazard endpoints.

It must be remembered that the concept of a toxicologically-based category linked to chemical structure and the ideas of read-across or QSAR modelling for data filling are coupled to the general explanation of the category concept. The historical description of read-across and QSAR modelling in toxicity are the same. Many of the well-studied toxicity QSARs have applicability domains which can be mechanistically derived from experimental data and quantify critical events along the pathway.[12]

Forming categories for acute endpoints such at LC_{50} and EC_{50} values derived from dose–response relationships is well-established, especially when developed along the OECD validation principle of mechanistic plausibility.[9] Categories based on chemical reactivity (or lack thereof)—especially electro(nucleo)philic interactions where the applicability domains are based on conventional organic chemistry—in particular are well-recognised,[4] especially when augmented by some information on mechanisms or modes of toxic action. Forming such categories has arisen from the conceptual framework described earlier,[13] specifically the notion of the molecular initiating event (MIE) (Figure 14.1).

TMCs are typically formed as a result of a common chemical reactivity mechanism, biological mechanism or mode of toxic action. These are typically based on alterations of basic cellular processes, enzyme function and receptor activity, or based on molecular similarity. The problem lies in that, when one moves from a common chemical reactivity-based category to a biologically-based category, and even more so to molecular similarity-based category, confidence in whether the chemical in question truly belongs to the category to which it is assigned diminishes.

The first reduction in confidence is a reflection of the fact that, while it is relativity easy to form TMCs for those hazard endpoints where covalent chemical reactivity is the MIE, it is more difficult to form such categories where the MIEs are non-covalent molecular interactions. This increase in difficulty is because particular cellular and systems effect data are often required to delineate the category. The second reduction in confidence is a reflection of the fact that there is no accepted best way of defining molecular or structure similarity and that small changes in structure may result in large changes in toxicity.

In most cases, well-established TMCs have been formed for acute hazard endpoints where the chemical interactions at the molecular level are dominated by a single MIE. This is particularly true when the endpoint is measured under steady-state conditions where the numerous variables that impact kinetic factors are limited. In this respect, the short-term LC_{50} assays based on air exposure to rodents and water exposure to fish, which achieve steady-state rapidly, have endpoints where TMCs are well-accepted. In contrast, the single oral dose LD_{50} assay with mammals, while an acute test, is typically considered a transitory endpoint which often does not represent a steady state. Forming categories for this endpoint is fraught with problems.

14.5   The Concept of the Adverse Outcome Pathway

The concept of the adverse outcome pathway is introduced as a substitute for the more stringent mechanism of action approach to form a TMC. It is the means to join the idea of a structural domain and an explanation of why a specific biological effect occurs. In this regard, the pathway is aimed at placing chemicals into categories based on data that are more manageable than that required for a 'mechanism of action' (see above), yet still having a category that is useful in filling data gaps in a transparent and mechanistically plausible manner.

Each adverse outcome pathway is a set of chemical, biochemical, cellular, physiological, behavioural, *etc.* responses which characterise the biological effects cascade resulting from a particular MIE. The term 'adverse outcome pathway' has been selected so not to cause confusion with the term 'Toxicity Pathway', which is used by the US National Research Council in its document, *Toxicity Testing in the Twenty-first Century: A Vision and a Strategy*,[14] where the focus is on 'omics' and high throughput *in vitro* data.

In this chapter, the adverse outcome pathways concept is offered as a means of defining TMCs. Again, a TMC can be thought of as a group of chemicals whose toxicological properties are likely to be similar or follow a regular pattern. The structure of the chemicals within the category characterises the applicability domain of the category. For many chronic endpoints, different chemical interactions with biological systems lead to the same hazard endpoint or outcome (*e.g.* NOEC). Pathways are intended to provide a means to hypothesise, test and demonstrate the mechanistic plausibility of a category for a particular endpoint. Accordingly, the hazard endpoint in question is inherently part of the pathway or may be the outcome of the pathway.

An adverse outcome pathway is a description of plausible causal linkages which illustrate how MIEs result in other critical biological effects quantified at the cellular, tissue, organ and whole animal levels of observation. A chemical can have a variety of interactions with biological systems. The MIE is the seminal, and typically the first, critical event of the sequence of events that are essential to the induction of the outcome, as hypothesised in the postulated pathway; it is also measurable. This concept is illustrated in Figure 14.2.

Whilst the adverse outcome pathway approach to forming categories has arisen from the framework described earlier,[13] specifically the notion of the MIE

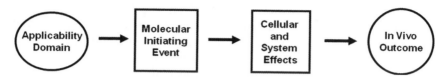

Figure 14.2 The basic outcome pathway.

and toxicity resulting from soft electro(nucleo)philic interactions (see Figure 14.1), its biological foundation can be traced to the modes of action studies of McKim *et al.* and their view of the fish acute toxicity syndrome.[15] Defining these syndromes requires less data than a mechanism of action approach and can be viewed as one of the first attempts to develop pathways for categorising chemicals for toxicity assessment. The fish acute toxicity syndromes were based on biochemical and/or physiological effects of exposure selected as key responses measured *in vivo* from exposure to model chemicals.[12] The mode of action as presented by McKim and co-workers looks only at *in vivo* data.[16] Syndromes do not address the MIE nor use other data (*e.g. in vitro*).

Binding to acetylcholinesterase is an example of a MIE which leads to a toxicologically meaningful group, the acetylcholinesterase inhibitors.[12] These MIEs cause changes in normal chemical and biochemical processes, which lead to a progression of biological effects, some of which are 'specific endpoints'. These can be defined in terms of their scientific significance to the pathway. When these significant biological effects are linked together and grouped to the various levels of biological organisation (cell, tissue, organ, *etc.*), a hazard endpoint can be traced back through a series of critical events to one or more initiating events—hence the term pathway.

Structural alerts are atom-based fragments which, when present in a molecule, are an indication that a compound can be placed into a particular category. They are best known for their ability to indicate the potential for genotoxicity. Structural alerts are the basis of forming categories for soft electro(nucleo)philic interactions where they predict MIEs, which are very useful in predicting selected hazard endpoints.[13] However, as noted above, these alerts in isolation cannot be expected to form TMCs for more elaborate hazard endpoints where damage is a result of multiple events (*e.g.* repeated dose toxicity), accumulation over time (*e.g.* neural toxicity) or particular to a life stage of the organism (*e.g.* developmental toxicity). To predict those longer-term effects that result from multiple factors, it is necessary to integrate MIEs with biological effects, which measure how effects are propagated at the different biological levels. The integration of MIEs with measurements of key biological processes is the pathway which links molecular level effects of chemicals to the outcomes at the organism or population level.

By addressing a sequence of issues such as chemical specificity, the molecular site of action(s), the MIE, the affected biochemical pathway(s), the cellular- and organ-level consequence(s), the target organ(s) or tissue(s), the key physiological response(s), the key target organ-response(s) and the key organism response(s), an outcome pathway can be formulated and anchored at the beginning by molecular structure and at the end by the particular *in vivo* outcome. Chemicals that follow the pathway and elicit the same key responses along the pathway are likely to be in the same TMC.

A TMC is formed by developing an experimentally-based understanding of the pathway including the MIE and, as necessary, verification of membership of the pathway as provided by assessment with *in vitro* assays. Thus, the pathway becomes the basis for establishing the biological and mechanistic relevance for data generating and gathering exercises proposed in any Integrated Testing Strategy (ITS). Using SMiles ARbitrary Target (SMARTS) strings or similar *in silico* methodology, the structural domain for a pathway can be captured computationally. This knowledge can then be applied to search for compounds in a chemical inventory likely to elicit toxicity *via* that particular adverse outcome pathway. In this way, endpoint-specific pathway information is used to identify likely members of a TMC. This membership information can be used to prioritise testing or to direct data gap filling without testing.

14.6   Weak Acid Respiratory Uncoupling

One well-studied adverse outcome pathway to fish mortality is for 'weak acid respiratory uncoupling'.[16,17] Uncoupling of oxidative phosphorylation disrupts and/or destroys the union between oxygen consumption and ATP production. There are several types of uncoupling of which weak acid uncoupling is the best studied, especially with regard to 2,4-dinitrophenol and pentachlorophenol.[18,19] The decoupling of oxidation from phosphorylation is explained by the chemiosmotic hypothesis.[20] Briefly, a hydrogen ion gradient is formed across the inner mitochondrial membrane as a result of electron transport. The hydrogen ions are used to generate ATP *via* the ATP-synthetase enzyme complex, which is membrane bound. The weak acids short-circuit the hydrogen ion gradient by providing an alternative means of transporting the hydrogen ions across the membrane.[21] While respiration continues to pump hydrogen ions across the inner membrane into the mitochondrial matrix, no ion gradient is formed because the weak acid dissipates the ions and no ATP is synthesised. The result is that uncoupling chemicals induce a metabolically futile wasting of energy by stimulating resting respiration, while at the same time decreasing ATP yield.[22]

The fish acute toxicity weak acid uncoupling pathway can be characterised by the following:

1) There is no metabolism of the chemical and distribution of the chemical in the organism is based on partitioning, which is influenced by ionisation.
2) The molecular site of action is the inner mitochondrial membrane.
3) The molecular initiating event is non-covalent perturbation.
4) The biochemical pathways affected by the chemical's action are varied. Resulting disruptions of these pathways lead to non-specific

inhibition of cellular function. Oxidative phosphorylation loss of the hydrogen ion gradient leads to reduction in ATP formation.

5) The cellular- and organ-level consequences of affecting the ATP-production are reversible.
6) The target organ(s) or tissue(s) in which the molecular site of action and biochemical effect occur are diverse, though tissues with high energy demands are particularly susceptible.
7) A key physiological response to the biochemical and cellular effects is increased oxygen consumption (*i.e.* increased rate of metabolism).
8) There is no key target organ–response to the biochemical, cellular and physiological effects.
9) The overall effect on the organism is reversible which, if left untreated, leads to death.

The weak acid respiratory uncoupler pathway targets the basic cellular process of energy production, which in eukaryotes is mitochondrial-based and is largely independent of biological organisation above the cellular level. Therefore, an *in vitro* assay or series of assays monitoring oxygen consumption and ATP production would be sufficient to verify membership of the category. In this example, a TMC is formed from physico-chemical properties. Terada describes the structural alerts for weak acid uncouplers as phenols with ionisation constants (pKa) between 4 and 3 and hydrophobicity (logarithm of the octanol–water partition coefficient, log *P*) between 3 and 8 as defining the membership of this category.[23]

The basic pathway for the process of weak acid respiratory uncoupling is illustrated in Figure 14.2. It can be argued that the available *in vivo* data, mechanistic understanding and applicability domains for weak acid respiratory uncoupling are such that the pathway concept is unnecessary. However, this example illustrates how the pathway concept may be applied to forming TMCs though the use of simple structural features and physicochemical properties. Moreover, by using the pathway one can establish biological or mechanistic relevance for any sets of methods proposed in an Integrated Testing Strategy to support inclusion of a chemical in a particular TMC.

14.7 Respiratory Irritation

A generic acute fish mortality pathway for 'respiratory irritation'[24] can be developed. This pathway was so named because the gill is the target organ; however, protein binding is the MIE. The fish 'respiratory irritation' pathway can be described as follows:

1) The chemical is either a direct-acting electrophile or a chemical that can be abiotically or biotically transformed to an electrophile.
2) The molecular sites of action are specific nucleophiles, typically either thiol or amino-moieties in proteins, but the reactions are non-selective.

3) The molecular initiating event is covalent perturbation of proteins.
4) The biochemical pathways affected by the chemical's action are varied and resulting disruptions of those pathways lead to general inhibition of cellular functions.
5) The cellular- and organ-level consequences of affects are irreversible.
6) The target organ(s) or tissue(s) is the gill, typically the first tissue exposed to the electrophile.
7) The key physiological response to the biochemical and cellular effects is general hypoxia.
8) The key target organ response to the biochemical, cellular and physiological effects is sloughing of the gill epithelium.
9) The key organism responses to the biochemical, cellular and physiological effects are a sharp reduction in blood oxygen level and asphyxiation, which quickly leads to death.

The crucial factor in the pathways described above is that different MIEs give rise to similar biochemical, cellular, physiological, behavioural, *etc.* responses and the same regulatory outcome. While the pathway has, historically, been referred to as 'respiratory irritation', the strategy here is to utilise the knowledge that covalent perturbation of proteins is a series of nearly 50 covalent reactions, or MIEs.[25,26] While these reactions are typically selective and well grounded in organic chemistry, their atomic targets—sulfur and nitrogen—are ubiquitous and the molecular targets non-specific. For example, mitochondria from cells treated with either an iodoacetyl-containing electrophile, where conjugation with the nucleophile is accompanied by the expulsion of the iodo group (nucleophilic substitution of haloaliphatics) or a maleimido-containing electrophile, where conjugation with the nucleophile is addition to the vinylene group (Michael type nucleophilic addition), a total of 1,255 different cysteine adducts representing 809 proteins were identified; moreover, these two thiol-targeting mechanisms exhibit different selectivity.[27] By convention, each electrophilic mechanism is given its own category rather than denoting them as sub-categories of the protein binding.

Since protein binding pathways are chemical reactivity mechanism-specific, the molecular initiating event in the above generic pathway is covalent perturbation of proteins by a particular mechanism. However, regardless of the specific reactive mechanism, the *in vivo* biological events are qualitatively similar. In a chemical mechanism of a reactivity-based pathway (especially for simple hazard endpoints such as acute aquatic toxicity), the resulting biochemical, cellular, physiological and/or behavioural symptoms[15,28] have response-to-response coefficients, which may be regarded as equivalent; hence, quantifying the electro(nucleo)philic interaction by simple *in chemico* reactivity methods[29] predicts the potency of the endpoints and calibrates the QSAR models.[13]

As shown in Figure 14.3 for this hazard endpoint and pathway(s), several structural domains (one for each specific MIE) are related to the same outcome by way of the same series of biological effects. In this case the biological effects

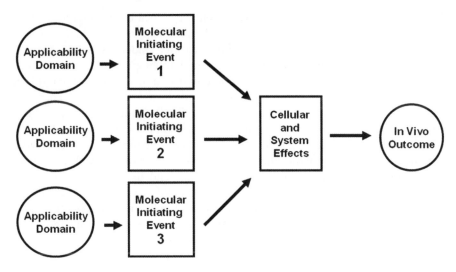

Figure 14.3   Outcome pathway(s) for different MIEs and the same biological effects.

along the pathway contribute little to defining the applicability domains, so without understanding the MIEs, meaningful categories or sub-categories are difficult to form and toxic potency is not easily linked to alterations in chemical structure.

In the example above, TMCs are formed from the MIE. As with the respiratory uncoupler example, the structural domain of each MIE can be defined and thus inclusion of a chemical in a particular domain can be confirmed.

14.8   Skin Sensitisation

A similar pathway strategy (one based on knowledge of protein binding mechanisms) can be utilised to demonstrate and test mechanistic plausibility for skin sensitisation. Skin sensitisation (type IV contact allergy) is an endpoint which is conceptually difficult because of the increased complexity of the pathway and individual mechanisms. There is general agreement that any substance that irreversibly binds to proteins has the potential to be a sensitiser.[30] Yet, in order to determine if a chemical will be an *in vivo* sensitiser, other issues such as bioavailability, epidermal dendritic cell activation and T-cell proliferation may have to be considered.[31] The latter reflects the fact that, while our current knowledge of dermal immunogenicity is in certain contexts broad, our ability to measure all facets of immunotoxicity is wanting. For example, it is well-known that a chemical must undergo a number of processes before it induces skin sensitisation.[32] From a biological perspective the expression of skin sensitisation is a non-antibody mediated reaction, which requires a high degree of co-ordination between several cell types located in different organs and is typically divided into two steps.[32] The first step or the induction phase entails the chemical (allergen) penetrating the outer epidermis of the skin and forming a complex with a dermal protein (a hapten–protein complex). The complex must then be processed by epidermal dendritic cells, which subsequently mature and migrate out of the skin to the local lymph nodes. There the complex is presented and recognised by naïve T-cells leading to cell proliferation and the generation of allergen-specific memory T-cells, some of which subsequently recirculate. The second step, the challenge phase, occurs following subsequent contact with the allergen. The hapten–protein complex is again formed and subsequently taken up by epidermal dendritic cells and other cells. Subsequently, activated memory T-cells secrete cytokines, which induce release of inflammatory cytokines and conscription of T-cells and inflammatory cells from the circulating blood. These inflammatory cells migrate to the epidermis and elicit the characteristic local inflammatory response.

For *in vivo* testing, several methods are reported under OECD Test Guideline No. 406.[33] However perhaps the currently most established method is the Local Lymph Node Assay (LLNA) where cellular proliferation in the draining lymph nodes is measured after repeated topical application of the test chemical on mice ears.[34] This assay is standardised, has undergone validation, and it is published by OECD as a method for regulatory use.[35]

In vitro assays for sensitisation using cell or tissue cultures typically have as their endpoint a single event, which occurs during the induction phase (especially the stimulation of dendritic cells).[36] Measurement of the expression of certain surface markers (*e.g.* MHC class I and class II molecules), adhesion molecules (*e.g.* CD54), co-stimulatory molecules (*e.g.* CD86) or measurement of secretion of specific cytokines (*e.g.* IL-1β) by cells are areas of active research,[31] although no robust assays have yet been developed. Among the more recent *in vitro* assays is one based on the stimulation of antioxidant response element (ARE) dependent gene activity in a recombinant cell line.[37] The antioxidant response element (also known as electrophile response element, EpRE) is a DNA element present in many phase II detoxification genes. The ARE-based assay offers a straightforward means of measuring electrophilic reactivity at the cellular level of organisation. As such it is a logical complement to *in chemico* measurements of reactivity.[30]

Despite the complexity of sensitisation and the LLNA endpoint, and the current limitations of measuring relevant events, sufficient information is available to develop a pathway approach to skin sensitisation. Since the fundamental chemical basis of skin sensitisation is quite well understood, a generic sensitisation pathway for protein binders can be described as follows:

1) A chemical may be a direct-acting electrophile or metabolism may convert a non-reactive compound (a pro-electrophile) to a reactive metabolite.
2) The molecular sites of action are nucleophilic sites in proteins (*e.g.* cysteine and lysine residues).

3) The MIE is the covalent perturbation of dermal proteins, which is irreversible.

4) The biochemical pathways affected by the chemical's action on the molecular targets are incompletely known, but include stimulation of antioxidant response elements.

5) The cellular-level consequence is antigen expression by epidermal dendritic cells. The organ-level consequence formation is proliferation of allergen-specific memory T-cells in the lymph nodes.

6) The target organ is skin with the requirement of intact local lymph nodes and the target tissues are immune cells.

7) The key physiological response to the cellular effects is acquisition of sensitivity.

8) The key organism response to the biochemical, cellular and physiological effects is dermal inflammation upon receiving the chemical challenge.

9) The overall effect on the organism is allergic contact dermatitis.

Although measurement of an event along a skin sensitisation pathway may provide evidence that a chemical can induce sensitisation *via* a particular MIE, it is highly unlikely that the event will achieve the same predictive accuracy as an *in vivo* method. However, measurements of several events, especially events that are postulated to be essential to the induction of the adverse outcome (as hypothesised in the pathway), are likely to achieve greater predictive accuracy.[38] To support an event as being critical to a pathway, there needs to be experimental data which characterise and measure the event, and demonstrate the event to be potency related to the adverse outcome.

In skin sensitisation, a limited number of events are currently measurable and have databases of sufficient size and diversity to demonstrate their importance in the pathway. All of these available measurable events are associated with the induction phase of sensitisation and several are related to potency. Since the work of Landsteiner and Jacobs,[39] there has been a growing agreement that the main potency determining step in skin sensitisation is the formation of a stable covalent association with carrier protein or formation of the hapten–protein complex.[40] The development of ITS based on *in chemico* and *in vitro* assays selected to be relevant to this adverse outcome pathway is an active field of research.

14.9 Acetylcholinesterase Inhibition

Another well-documented adverse outcome pathway is for insecticides such as organophosphate esters where the toxicity is attributed to the inhibition of acetylcholinesterase.[41] Briefly, acetylcholinesterase is the enzyme responsible for the degradation of the neural transmitter acetylcholine, which is involved in synaptic conduction of nerve impulses. Inhibition of acetylcholinesterase leads to the accumulation of acetylcholine at the synaptic receptor and a resulting continuous stimulation of cholinergic nerves.

An acute fish mortality pathway for 'acetylcholinesterase inhibition'[24] can be described as follows:

1) The chemical is either a direct-acting hard electrophile or a chemical that can be metabolised to such an electrophile.
2) The molecular sites of action are specific serine hydroxyl groups on acetylcholinesterase, but the reaction is non-selective.
3) The molecular initiating event is covalent perturbation (*i.e.* phosphorylation) of acetylcholinesterase.
4) The biochemical pathways affected by the chemical's action are destruction of the neural transmitter acetylcholine. Disruptions of the pathways lead to accumulation of acetylcholine until new acetylcholinesterase is synthesised.
5) The cellular- and organ-level consequences of affects are exceedingly slow to reverse so as to be irreversible in acute exposures.
6) The target organ(s) or tissue(s) in which the molecular site of action and biochemical effect occurs are mainly effector tissues (*i.e.* glands and muscles).
7) The key physiological responses to the biochemical and cellular effects are initial stimulation leading to increased glandular secretions and tremors. This is followed by desensitisation leading to general muscle weakness, and a sharp reduction in blood oxygen level.
8) The key target organ–response to the biochemical, cellular and physiological effects is a reduction in cardiac and cardiovascular function, as well as neuromuscular junction damage.
9) The key organism responses to the biochemical, cellular and physiological effects are lethargy, generalised weakness and coma, which is followed by death.

The crucial factor in the pathways described above is that of inhibiting acetylcholinesterase. Because there is an experimentally-based understanding of the pathway, measurement of the MIE is sufficient to ascribe membership to the category for this hazard. As needed, verification of membership of the category can be provided by assessment with a single *in vitro* assay assessing acetylcholinesterase inhibition. The structural domain of this effect can be captured computationally to identify further chemicals likely to elicit acute toxicity *via* acetylcholinesterase inhibition.

The pathway for inhibition of acetylcholinesterase takes the form of the basic pathway illustrated in Figure 14.2. One can argue that the available *in vivo* data, mechanistic understanding and applicability domains for acetylcholinesterase inhibition are such that a pathway is not necessary. Regardless, this example points out how the adverse outcome pathway concept may be applied to forming TMCs based on a defined toxicophore.

14.10 Receptor Binding Pathways for Phenolic Oestrogen Mimics

Another example of a scenario where more than one structural domain (one for each MIE) is related to the same outcome by way of the same series of biological processes (as illustrated in Figure 14.2) is for oestrogen receptor binding. The outcomes from oestrogen receptor binding chemicals are typically reproductive and developmental effects.[42] In these cases, binding to the receptor is the MIE. Endocrine disruption *via* oestrogen receptor binding in the environment has been linked to industrial chemicals including phenols, phytoestrogens and steroid hormones.[42] Well-documented examples of impairment of reproduction in fish include the feminisation of male fish upon exposure to oestrogenic steroids and phenolic compounds,[43] especially when exposed during early life stages.[44] Based in part on the work of Schmieder and co-workers,[45] a pathway resulting in fish reproductive impairment from oestrogen receptor binding has been developed. In this example:

1) A chemical may bind to the oestrogen receptor (*e.g.* cyclic compound with a molecular weight of less than 500).
2) The molecular site of action is the A-site of the oestrogen receptor (*i.e.* a non-hindered phenolic residue is present).
3) The MIE is binding at the A site in the oestrogen receptor.
4) The biochemical pathways affected by the chemical's action on the molecular target are hormonally linked, with resulting perturbations being reversible.
5) The cellular-level consequence is upregulation of oestrogen transcription.
6) The target organs are several, but especially the liver and gonads.
7) The physiological responses to the cellular effects are induction of vitellogenin synthesis in the liver and conversion of testicular tissue to ova tissue.
8) The anatomical response to the biochemical, cellular and physiological effects is the feminisation of male fish.
9) The overall effect on the organism is reproductive impairment.

A number of events are currently measurable in reproductive impairment in fish induced by oestrogen receptor binding. The results of some of these measurements have been recorded in databases of sufficient size and diversity to demonstrate their importance in the pathway. Moreover, oestrogen receptor binding assays can be combined with cellular- and/or tissue-level transcription assays to form part of the ITS for reproductive toxicity.

14.11 Using Pathways to Form TMCs and Reduce Testing

In using adverse outcome pathways, especially when coupled with tiered testing approaches, the number of *in vivo* tests that need to be performed can be reduced markedly. This is especially true if chemicals are selected intelligently to assure proper chemical coverage of the applicability domain and to reduce redundancy of testing. In such a scenario, the selection of chemicals to be tested in a higher order assay is based on the results of a lower order assay. This means that the higher order data cover the relevant aspect of the chemical universe of the lower order assay in a thorough and non-redundant manner, but with fewer tests.

Returning to the example of skin sensitisation, examination of the literature shows that measurable events along the skin sensitisation pathway for Michael acceptor electrophilic chemicals include:

- protein binding assessed qualitatively by structural evidence for protein binding categories;[4,46]
- chemical reactivity measured by some *in chemico* reactivity method (thiol or cysteine reactivity being correlated with Michael-type addition; amino or lysine reactivity being correlated with Schiff-base formation);[47,48]
- stimulation of an antioxidant response element (ARE) measured by an *in vitro* method (*e.g.* molecular engineered ARE promoter with a luciferase reporter in Hepa1C1C7 cells);[37]
- quantification of T-cell proliferation by measurement in the local lymph node assay; and[35]
- observed inflammation measured by a further *in vivo* toxicity assay (*e.g.* Guinea Pig Maximisation Test).[33]

The published data for the well-established LLNA only cover 210 chemicals.[49] These data verify the LLNA as a surrogate for the more elaborate maximisation assays.[32] The compounds reported with LLNA data do not, so far, cover the full spectrum of electrophilic reactivity. They are skewed towards certain mechanisms, as the database was not created from a mechanistic point of view. For example, of the chemicals tested in LLNA, about 25% represent the Michael-type nucleophilic addition mechanism for forming protein adducts;[6] all these chemicals have a polarised α,β-unsaturated substructure fitting the structural alert for Michael-type addition.[4] With noted exceptions, due either to reduced availability as a result of solubility limitations or the need of metabolic activation, all show graded *in chemico* reactivity with the target nucleophile glutathione.[6] In addition, all demonstrated a graded photo response in the Hepa1C1C7-based ARE luciferase reporter system.[37] The chemicals with published test data from the LLNA represent less than half of the structural domain for the Michael-type addition mechanism of protein binding.[5] It can be argued that this lack of a systematic database makes the pathway approach to forming toxicologically significant categories even more appealing, as it does provide for transparency and mechanistic understanding for interpolation within a particular mechanism of protein binding.

In the skin sensitisation example, the TMCs are formed from structural features which have an organic chemical reaction basis. By using intelligent testing strategies, the structural domain of each chemical mechanism can be defined and key steps along the pathway (*e.g.* of the ARE response) can be quantified, thus reducing the need for chemical-by-chemical *in vivo* testing.[50]

A similar, but more intricate, argument can be presented for receptor binding pathways for oestrogen mimicking compounds. In this case measurable events along the pathway include:

- oestrogen binding measured quantitatively in competitive binding assays using radiolabelled [^{3}H]-17-β- estradiol;[45,51]
- altered gene expression as a result of oestrogen binding measured quantitatively in a molecular engineered oestrogen binding promoter linked to a bioreporter system;[52,53]
- vitellogenin induction in fish liver slice assay analysing for vitellogenin mRNA utilising real time reverse transcriptase polymerase chain reaction; and[45]
- conversion of testicular tissue to ova tissue measured morphometrically (from histological section) in whole fish assays.[54]

The energy and steric constraints imposed by a receptor characterise the domain of chemical structures which can bind to it. In the case of the oestrogen receptor, it is known to have a 'dynamic and plastic character' rather than the lock-and-key interaction that characterises more specific receptors.[55] Although only a small portion of the chemical universe has structures which are able to bind, the oestrogen receptor is sufficiently non-specific to permit binding with a structurally diverse group of compounds. The presumption for these various chemicals interacting with the oestrogen receptor is offered on the basis of 'sub-pockets' within the receptor. A review of the recent literature reveals a growing agreement that there are three primary oestrogen receptor binding sub-pockets (typically referred to sites A, B, and C), each with different hydrogen bonding requirements. Understanding of chemical interaction with the oestrogen receptor is historically based on information gained from steroidal compounds, which interact at both the A and B site within the receptor. However, it is also known that chemicals that contain only one hydrogen-bonding group (*e.g.* *tert*-octylphenol or 4-aminobiphenyl) bind and cause subsequent gene activation.[52–53,56] For such chemicals, there will be a possibility to be able to form meaningful categories based on sub-pocket binding rather than listing the binding selectivity as sub-categories of oestrogen receptor binding. As stated for other examples, intelligent chemical selection can be used to demonstrate the key steps along the pathway (*e.g.* vitellogenin induction) in a way that the structural range of a given inventory is completely covered. This will mean that there is no need to test every compound in that inventory and raise the possibility of computational screening to highlight potential binders.

The two examples presented above show how an understanding of the events along the pathway (including the MIE) provides a transparent, mechanistically defensible, biologically relevant, and adverse outcome specific means of forming TMCs and at the same time reduce testing.

14.12   Hazards with Elaborate Datasets

When considering longer-term hazards, such as repeated dose or chronic toxicity and reproductive toxicity (including parental and offspring effects as well as neurological deficits), forming TMCs becomes a more daunting task. Data reported for these longer-term hazards are elaborate, typically based on a 'controlled vocabulary' list of possible symptoms.[57] ToxRefDB, which includes *in vivo* data for 309 structures (typically pesticide active ingredients) lists more than 22,000 vocabulary items; while some are quantitative, most are qualitative with many of these symptoms being reversals (*i.e.* increase *versus* decrease).[57] Symptoms reported include clinical chemistry, haematological and urinalysis, as well as pathological findings (*i.e.* gross, neoplastic and non-neoplastic). It is intuitive that all these symptoms cannot be related to the same *in vivo* hazard. Since exposure for chronic effects is over a longer duration (*e.g.* 28 days or more in the case of repeat dose toxicity), the *in vivo* damage is likely to be cumulative. The reported hazard endpoints such as a no observable adverse effect level (NOAEL) or lowest observable adverse effect level (LOAEL) are the 'critical effects', which vary as to incidence, target organ, severity, *etc.*[58] For example, in examining the ToxRefDB, Martin and co-workers identified 26 high incident tissue-specific pathology endpoints.[58] From a TMC standpoint, these tissue-specific pathology endpoints should be assumed to be separate. However, the pathways leading to them may not be linear and more than one MIE may lead to the same tissue-specific pathology.

Multi-generational reproductive tests typically characterise parental and offspring systemic toxicity as well as reproductive symptoms.[59] On examining ToxRefDB, Martin and co-workers identified 19 high incident parental, offspring and reproductive effects.[59] Again, from a TMC standpoint these endpoints should be assumed to be separate. Similarly, developmental toxicity is a complex endpoint.[60] Historically, developmental toxicity has been defined as an adverse effect induced prior to adult life. Manifestations are typically recorded as type, degree and/or incidence of abnormal development.[57] Final manifestations reported in developmental toxicity studies often include death, malformation, growth retardation and functional deficit. Sensitivity to developmental toxicants is often dependent on the genotype of the zygote. Moreover, expression of embryo- or feto-toxicity is life-stage dependent (*i.e.* the particular stage of development at the time of exposure). In mammals, where maternal metabolism may dramatically impact embryo/foetal exposure, dose–response relationships are very important to developmental effects. All these factors make the elucidation of pathways leading to elaborate *in vivo* outcomes more challenging.

14.13   Pathways for Elaborate Hazard Endpoints

The discussion above focused on endpoints and pathways where the MIEs are well-studied and the downstream biological effects form an orderly progress of events for which experimental data are known. In other words, the pathway is well-documented. However, if the pathways approach is to be useful for categorising all organic chemicals and for as many regulatory endpoints as possible, it must be flexible enough to be adapted to form categories for more elaborate hazards where, as noted, the effects may be cumulative or life stage dependent. These pathways may have multiple MIEs leading to the same *in vivo* outcome, which may be modelled by various *in vitro*. Collecting data for these elaborate hazards is characteristically expensive and time-consuming; therefore, these endpoints are inclined to be data sparse, especially for industrial organic substances.

Even when the targeted *in vivo* outcome is symptom-limited (*e.g.* rat, liver, non-neoplastic lesion), there are likely to be multiple pathways. Multiple types of lesions such as necrosis, steatosis, cholestasis, phospholipidosis and hypertrophy are observed in hepatotoxicity.[61] A structurally diverse series of chemicals has been linked to a single lesion; for example, acetominophen, allyl alcohol, carbon tetrachloride, cyclophosphamide, diethyl nitrosamine, dimethyl formamide and monocrotaline are all related to liver narcosis.[62] Moreover, some chemicals cause more than one type of lesion; for example carbon tetrachloride has been linked to both liver necrosis and steatosis.[63]

The integration of conventional *in vivo* toxicological data with 'omics' and other molecular, biochemical and *in vitro* data (*i.e.* 'systems biology'[64]) is not a new concept.[65] While much attention has been given to toxicogenomics,[62–63,66] the uses of *in vitro* assays in high throughput screening schemes have also been proposed as 'systems' data, which can be linked to *in vivo* outcomes *via* toxic mechanism.[67] Such correlations have been demonstrated for hepatotoxicity.[68] Furthermore, such *in vitro* test results are a major part of the ToxCast™ program.[69]

It has been shown that separate MIEs may be related to the same *in vivo* outcome, but by way of a different series of necessary chemical, cellular and system effects. Each pathway is consequently defined by both the initial chemical interaction and the necessary biological events measurable at various levels of organisation (cell, organ, *etc.*), which lead to the ultimate *in vivo* outcome. The critical step is to identify which *in vitro* assays are key to separating one pathway from another, when both pathways lead to the same *in vivo* outcome. The pathway approach provides a means of identifying both the MIE and the key biological events that allow TMCs to be formed. This approach, while not limited to *in vivo* hazards with elaborate datasets, has the potential to be useful for modelling such hazards as they will permit critical events or symptoms to be linked to alterations in chemical structure. The concept of adverse outcome pathways for elaborate datasets, which include multiple MIEs and multiple key *in vitro* effects, is shown in Figure 14.4. A logical extension, which reflects the potential intricacy of pathways especially for elaborate data sets, would be a combination of the events illustrated in Figures 14.3, 14.4 and 14.5.

Predictive modelling for elaborate hazard endpoints (particularly chronic health-related effects) in the foreseeable future will be limited by the lack of systematic databases. Moreover, data identifying the pathway and measuring the

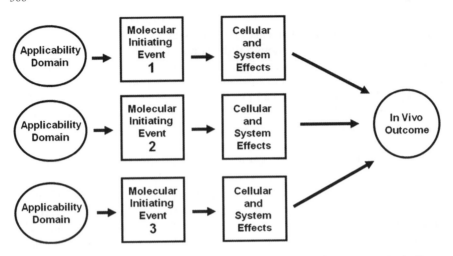

Figure 14.4   Outcome pathway(s) for different MIEs and different biological effects.

key events along the pathway are scattered and fragmentary. All this makes the development and verification of pathways more of a challenge. It is envisaged that the initial formations of TMCs will often require categories based on narrow applicability domains. Whilst no examples of pathways for elaborate data are presented, I am aware of several international initiatives which are collecting symptom data from chronic health studies either from the literature or by strategic testing with a standardised protocol. The result will be that other symptom-based datasets (*e.g.* ToxRefDB[58]) will be available in the near future. This will certainly allow the concept of the outcome pathway to be explored further.

The conventional opinion is that there are so many MIEs and so many system effects that the combinatorial aspects of the problem will preclude the development of useable approaches in the near future. However, a possible useful approach, and the one proposed here, would be to constrain this difficulty with logical boundaries along the pathway. The point being that once the target chemical and hazard outcome are fixed at the beginning and end of a pathway, the number of possible pathways linking the chemical to the endpoint should be manageable—especially as there is knowledge of the key events along a particular pathway.

14.14   Positive Attributes of the Adverse Outcome Pathway Approach

There are many positive features of adapting the pathway approach to forming toxicological categories. Important among these is that the outcome pathway:

- allows a shift from chemical-by-chemical animal testing to hypothesis development, testing and the demonstration of mechanistic plausibility;

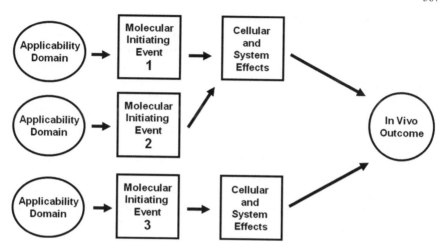

Figure 14.5 Outcome pathway(s) for different MIEs and various biological effects.

- provides a basis for chemical extrapolation by interpolation within a toxicologically relevant chemical category;
- provides a means (a) to compare across the different levels of biological organisation, (b) to consider life form and life stage at the time of exposure, and (c) for species to species extrapolation;
- shifts the emphasis to intrinsic chemical activity and key, or critical, biological events and away from statistical parameters, especially a fixation on the fit and predictivity of the model;
- avoids mixing data from multiple mechanisms which can cause the same *in vivo* outcome.

14.15 Conclusions: Basic Elements in Developing a Pathway

It is important to remember that mechanistic plausibility is the key to acceptance of a pathway and thus any prediction that comes from the toxicological category resulting from this pathway. This requires an understanding of key processes or critical events measured along the pathway, but not a detailed elucidation of the pathway (as required by the mechanism of action approach used in drug design). In the end, the pathway has to be based on chemistry that has biological relevance. The key or critical system measurements must have *in vivo* relevance, especially the MIE. It is important to understand the linkages and scaling factors as the pathway moves up the levels of biological organisation, especially for events which depend on potency in the *in vivo* outcome. It is advantageous to identify triggers (particularly *in vitro*) along the pathway that provide a means of eliminating a chemical from further testing. In the end, the pathway—or at least the key events along the pathway—must be simple enough to be manageable, but complex enough to be useful.

The first issue in establishing an outcome pathway is the identification of the most probable MIE(s) for a chemical or its major metabolites. The question of MIE must be answered first because it is a certainty that predictive methods will lead to nonsense predictions unless we first identify the chemistry in terms of biomolecular interactions. The second issue is to forecast the types of outcomes which might be observed under a variety of exposure scenarios for a variety of test organisms, a variety of life stages and with various testing methods. Briefly, one wants the biology to describe the experimental conditions necessary for an experimentalist to observe specific effects. For example, many anticancer drugs target cell proliferations and cell differentiation, yet targeting these events in an embryo can lead to developmental toxicity.

In summary, an outcome pathway anchored at the start by molecular structure and at the finish by a specific *in vivo* outcome can be devised by addressing a series of topics, which may include but not be limited to:

- the molecular site of action(s);
- the MIE;
- the affected biochemical pathway(s);
- the cellular-, tissue- and organ-level targets and consequence; and
- key physiological and organism response(s).

Chemicals which follow a particular pathway as demonstrated by their ability to elicit the same key responses (*e.g. in vitro*) along the pathway are likely to be members of the same TMC. A recent paper has illustrated these concepts further.[70]

References

1. Corrigendum to Regulation (EC) No 1907/2006 of the European Parliament and of the Council of 18 December 2006 concerning the Registration, Evaluation, Authorisation and Restriction of Chemicals (REACH), establishing a European Chemicals Agency, amending Directive 1999/45/EC and repealing Council Regulation (EEC) No 793/93 and Commission Regulation (EC) No 1488/94 as well as Council Directive 76/769/EEC and Commission Directives 91/155/EEC, 93/67/EEC, 93/105/EC and 2000/21/EC (Official Journal of the European Union L396 of 30 December 2006), *Off. J. Eur. Union*, **L136**, 29.5.2007, 3–280.
2. Organisation for Economic Co-operation and Development, *Guidance on Grouping of Chemicals*, OECD, Paris, 2007, OECD Environmental Health and Safety Publications Series on Testing and Assessment No. 80, ENV/JM/MONO(2007)28, www.olis.oecd.org/olis/2007doc.nsf/LinkTo/NT0000426A/$FILE/JT03232745.PDF [accessed March 2010].
3. K. van Leeuwen, T. W. Schultz, T. Henry, B. Diderich and G. D. Veith, *SAR QSAR Environ. Res.*, 2009, **20**, 207.
4. A. O. Aptula and D. W. Roberts, *Chem. Res. Toxicol.*, 2006, **19**, 1097.

5. T. W. Schultz, J. W. Yarbrough, R. S. Hunter and A. O. Aptula, *Chem. Res. Toxicol.*, 2007, **20**, 1359.
6. T. W. Schultz, K. Rogers and A. O. Aptula, *Contact Derm.*, 2009, **60**, 21.
7. C. L. Russom, R. A. Drummond and A. D. Hoffman, *Bull. Environ. Toxicol.*, 1988, **41**, 589.
8. S. J. Enoch, M. T. D. Cronin, T. W. Schultz and J. C. Madden, *Chem. Res. Toxicol.*, 2008, **21**, 513.
9. Organisation for Economic Co-operation and Development, *Guidance Document on the Validation of (Quantitative) Structure-Activity Relationship [(Q)SAR] Models*, OECD, Paris, 2007, OECD Environmental Health and Safety Publications Series on Testing and Assessment No. 69, ENV/JM/MONO(2007)2, www.olis.oecd.org/olis/2007doc.nsf/LinkTo/ NT00000D92/$FILE/JT03224782.PDF [accessed March 2010].
10. C. J. Borgert, T. F. Quill, L. S. McCarty and A. M. Mason, *Toxicol. Applied Pharm.*, 2004, **201**, 85.
11. J. Wiltse, *Crit. Rev. Toxicol.*, 2005, **35**, 727.
12. S. P. Bradbury, T. R. Henry and R. W. Carlson, in *Practical Application of Quantitative Structure–Activity Relationships (QSAR) in Environmental Chemistry and Toxicology*, ed. W. Karcher, J. Devillers, Kluwer Academic, Dordrecht, The Netherlands, 1900, pp. 295–315.
13. T. W. Schultz, R. E. Carlson, M. T. D. Cronin, J. L. M. Hermens, R. Johnson, P. J. O'Brien, D. W. Roberts, A. Siraki, K. D. Wallace and G. D. Veith, *SAR QSAR Environ. Res.*, 2006, **17**, 413.
14. National Research Council, *Toxicity Testing in the Twenty-first Century: A Vision and a Strategy*, National Academic Press, Washington DC, 2007.
15. J. M. McKim, S. P. Bradbury, S. P. and G. J. Niemi, *Environ. Health Perspect.*, 1987, **71**, 171.
16. J. M. McKim, P. K. Schmieder, R. W. Carlson, E. P. Hunt and G. J. Niemi, *Environ. Toxicol. Chem.*, 1987, **6**, 295.
17. M. Cajina-Quezada and T. W. Schultz, *Aquatic Toxicol.*, 1990, **17**, 239.
18. W. F. Loomis and F. Lippmnn, *J. Biol. Chem.*, 1948, **172**, 807.
19. S. G. McLaughlin and J. P. Dilger, *Physiol. Rev.*, 1980, **60**, 825.
20. P. Mitchell, *Biol. Rev.*, 1966, **41**, 445.
21. F. H. Blaikie, S. E. Brown, L. M. Samuelsson, M. D. Brand, R. A. Smith and M. P. Murphy, *Biosci. Rep.*, 2006, **26**, 231.
22. K. B. Wallace and A. A. Starkov, *Annu. Rev. Pharmacol. Toxicol.*, 2000, **40**, 353.
23. H. Terada, *Environ. Health Perspect.*, 1990, **87**, 213.
24. J. M. McKim, P. K. Schmieder, G. J. Niemi, R. W. Carlson and T. R. Henry, *Environ. Toxicol. Chem.*, 1987, **6**, 313.
25. G. Dupuis and C. Benezra, *Allergic Contact Dermatitis to Simple Chemicals,* Marcel Dekker, New York, 1982.
26. C. K. Smith and S. A. M. Hotchkiss, *Allergic Contact Dermatitis–Chemical and Metabolic Mechanisms,* Taylor & Francis, London, 2001.
27. H. L. Wong and D. C. Leibler, *Chem. Res. Toxicol.*, 2008, **21**, 796.

28. C. L. Russom, S. P. Bradbury, S. J. Broderius, D. E. Hammermeister and R. A. Drummond, *Environ. Toxicol. Chem.*, 1997, **16**, 948.

29. T. W. Schultz, J. W. Yarbrough and E. L. Johnson, *SAR QSAR Environ. Res.*, 2005, **16**, 313.

30. F. Gerberick, M. Aleksic, D. Basketter, S. Casati, A.-T. Karlberg, P. Kern, I. Kimber, J.-P. Lepoittevin, A. Natsch, J. M. Ovigne, R. Costanza, H. Sakaguchi and T. W. Schultz, *Altern. Lab. Anim.*, 2008, **36**, 215.

31. I. R. Jowsey, D. A. Basketter, C. Westmoreland and I. Kimber, *J. Appl. Toxicol.*, 2006, **26**, 341.

32. I. Kimber, D. A. Basketter, G. F. Gerberick and R. J. Dearman, *Int. Immunopharmcol.*, 2002, **2**, 201.

33. E. V. Buehler, *Arch. Dermatol.*, 1965, **91**, 171.

34. I. Kimber, R. J. Dearman, D. A. Basketter, C. A. Ryan and G. F. Gerberick, *Contact Derm.*, 2002, **47**, 315.

35. Organisation for Economic Co-operation and Development, *Guidelines for Testing of Chemicals No. 429. Skin Sensitisation: The Local Lymph Node Assay*, OECD, Paris, 2002.

36. C. A. Ryan, B. C. Hulette and G. F. Gerberick, *Toxicol. In vitro*, 2001, **15**, 43.

37. A. Natsch and R. Emter, *Toxicol. Sci.*, 2008, **102**, 110.

38. A. Natsch, R. Emter and G. Ellis, *Toxicol. Sci.*, 2009, **107**, 106.

39. K. Landsteiner and J. L. Jacobs, *J. Exp. Med.*, 1936, **64**, 625.

40. D. W. Roberts and A. O. Aptula, *J. Appl. Toxicol.*, 2008, **28**, 377.

41. D. J. Ecobichon, in *Casarett & Doull's Toxicology the Basic Science of Poisons*, ed. C. D. Klaassen, McGraw-Hill, New York, 6th edn, 2001, pp. 763–810.

42. J. P. Sumpter, *Acta Hydrochem. Hydrobiol.*, 2005, **33**, 9.

43. M. Seki, H. Yokota, M. Maeda, H. Tadokoro and K. Kobayashi, *Environ. Toxicol. Chem.*, 2003, **22**, 1507.

44. R. van Aerle, N. Ponds, T. H. Hutchinson, S. Maddix and C. R. Tyler, *Ecotoxicology*, 2002, **11**, 423.

45. P. K. Schmieder, M. A. Tapper, J. S. Denny, R. C. Kolanczyk, B. R. Sheedy, T. R. Henry and G. D. Veith, *Environ. Sci. Technol.*, 2004, **38**, 6333.

46. M. P. Payne and P. T. Walsh, *J. Chem. Infom. Comput. Sci.*, 1994, **34**, 154.

47. G. F. Gerberick, J. D. Vassallo, R. E. Bailey, J. G. Chaney, S. W. Morall and J.-P. Lepoittevin, *Toxicol. Sci.*, 2004, **81**, 332.

48. A. Natsch, H. Gfeller, H. Rothaupt and G. Ellis, *Toxicol. In vitro*, 2007, **21**, 1220.

49. G. F. Gerberick, C. A. Ryan, P. S. Kern, H. Schlatter, R. J. Dearman, I. Kimber, G. Y. Patlewicz and D. A. Basketter, *Dermatitis*, 2005, **16**, 157.

50. S. J. Enoch, J. C. Madden and M. T. D. Cronin, *SAR QSAR Environ. Res.*, 2008, **19**, 555.

51. S. Laws, S. Carey, W. Kelce, R. Cooper and L. E. Gray, *Toxicology*, 1996, **112**, 173.

52. T. W. Schultz, G. D. Sinks and M. T. D. Cronin, *Environ. Toxicol.*, 2003, **17**, 14.
53. J. Sanseverino, M. E. Eldridge, A. C. Layton, J. P. Easter, J. W. Yarbrough, T. W. Schultz and G. S. Sayler, *Toxicol., Sci.*, 2009, **107**, 122.
54. J. M. Zha, Z. J. Wang, N. Wang and C. Ingersoll, *Chemosphere*, 2007, **66**, 488.
55. J. Katzenellenbogen, R. Muthyala and B. S. Katzenellenbogen, *Pure Appl. Chem.*, 2003, **75**, 2397.
56. E. L. Hamblen, M. T. D. Cronin and T. W. Schultz, *Chemosphere*, 2003, **52**, 1173.
57. US Environmental Protection Agency, ToxRefDB Program, www.epa. gov/ncct/toxrefdb/ [accessed March 2010].
58. M. T. Martin, R. S. Judson, D. M. Reif, R. J. Kavlock and D. J. Dix, *Environ. Health Perspect.*, 2009, **113**, 392.
59. M. T. Martin, E. Mendez, D. G. Corum, R. S. Judson, R. J. Kavlock, D. M. Rotroff and D. J. Dix, *Toxicol. Sci.*, 2009, **110**, 181.
60. T. B. Knudsen, M. T. Martin, R. J. Kavlock, R. S. Judson, D. J. Dix and A. Singh, *Reprod. Toxicol.*, 2009, **28**, 209.
61. H. J. Zimmerman, *Hepatotoxicity,* Lippincott/Williams & Wilkins, Philadelphia, 1999.
62. J. F. Waring, G. Cavet, R. A. Jolly, J. McDowell, H. Dai, R. Ciurlionis, C. Zhang, R. Stoughton, P. Lum, A. Ferguson, C. J. Roberts and R. G. Ulrich, *Environ. Health Perspect.*, 2003, **111**, 863.
63. N. Zidek, J. Hellmann, P.-J. Kramer and P. G. Hewitt, *Toxicol. Sci.*, 2007, **99**, 289.
64. T. Ideker, T. Galiski and L. Hood, *Annu. Rev. Genomics Hum. Genet.*, 2001, **2**, 343.
65. M. Waters, G. Boorman, P. Bushel, M. Cunningham, R. Irwin, A. Merrick, K. Olden, R. Paules, J. Selkirk, S. Stasiewicz, B. Weis, B. van Houten, N. Walker and R. Tennant, *Environ. Health Perspect.*, 2003, **111**, 811.
66. M. R. Fielden, R. Brennan and J. Gollub, *Toxicol. Sci.*, 2007, **99**, 90.
67. A. A. Rabow, R. H. Shoemaker, E. A. Sausville and D. G. Covell, *J. Med. Chem.*, 2002, **45**, 818.
68. P. J. O'Brien, W. Irwin, D. Diaz, E. Howard-Cofield, C. M. Krejsa, M. R. Slaughter, B. Gao, N. Kaludercic, A. Angeline, P. Bernardi, P. Brain and C. Hougham, *Arch. Toxicol.*, 2006, **80**, 580.
69. D. J. Dix, K. A. Houck, M. T. Martin, A. M. Richards, R. W. Setzer and R. J. Kavlock, *Toxicol. Sci.*, 2007, **95**, 5.
70. G. T. Ankley, R. S. Bennett, R. J. Erickson, D. J. Hoff, N. W. Hornung, R. D. Johnson, D. R. Mount, J. W. Nichols, C. L. Russom, P. K. Schmeider, J. A. Serrano, J. E. Tietge and D. L. Villeneuve, *Env. Toxicol. Chem.*, 2010, **29**, 730.

CHAPTER 15

An Introduction to Read-Across for the Prediction of the Effects of Chemicals

S. DIMITROV AND O. MEKENYAN

Laboratory of Mathematical Chemistry, Bourgas Prof. Assen Zlatarov University, Yakimov St, 1, 8010 Bourgas, Bulgaria

15.1 Read-Across

The process of the assessment of a property of one chemical based on the same property for one or a few different chemicals considered to be similar in some context is known as read-across or the analogue approach.[1,2] There is no practical limitation on the type of properties that can be the subject of read-across including:

- physico-chemical properties;
- environmental fate;
- toxicological; and
- ecotoxicological effects.

From a formal methodological point of view, read-across is not usually recommended for predicting physico-chemical properties that may be used as explanatory variables for predicting more complex environmental or (eco)-toxicological effects. On the other hand, grouping of chemicals to estimate the properties of untested chemicals has been routine practice in chemical engineering and physical chemistry for several decades.

Issues in Toxicology No.7
In Silico Toxicology: Principles and Applications
Edited by Mark T. D. Cronin and Judith C. Madden
© The Royal Society of Chemistry 2010
Published by the Royal Society of Chemistry, www.rsc.org

15.2 Basis of Performing Read-Across

Knowledge discovery traditionally starts with observation and the collection of facts, followed by an attempt to establish order in the collected empiric knowledge. This process is always conducted in a framework of research to express knowledge in formats useful for mutual understanding and further discovery such as essays, reports, formulae, models and theories. The last two formats, models and theories, are the most powerful methods to represent the essential aspects of a studied phenomenon.

Classification is among the most important tools involved in the model and theory building. This is a process of clustering empiric knowledge with the aim of establishing an order through the grouping of objects into categories whose members share some perceivable similarity within a given context. In some cases, a meaningful classification may initiate the generation of new knowledge. The periodic system of chemical elements attributed to Mendeleyev is an example of a very useful tool, leading to the discovery of new elements, creating a framework to comprehend experimental errors and provide new understanding of existing elements. The first two conclusions made by Mendeleyev demonstrate how close he was to today's concept of read-across:[3]

1. The elements, if arranged according to their atomic weights, exhibit an evident periodicity of properties.
2. Elements which are similar with respect to their chemical properties have atomic weights which are either of nearly the same value (*e.g.* platinum, iridium, osmium) or which increase regularly (*e.g.* potassium, rubidium, caesium).

At the time the Periodic Law of the Chemical Elements was discovered, the description and classification of organic chemicals was little developed. The approach used in inorganic chemistry, representing a compound by the enumeration of the chemical elements, was impractical in organic chemistry. In 1861, Butlerov put forward a theory that provided a tool for classifying and understanding the properties of organic chemicals based on the concept of molecular structure.[4] His principal assumption was that the chemical structure of substances defines their physical and chemical properties and, as a consequence, chemicals with similar molecular structure and functionalities possess similar properties. Nowadays, practically any textbook in organic chemistry is divided into chapters according to this principle.

The fundamental discoveries made by Mendeleyev and Butlerov over one and half centuries ago laid the foundations for the performance of read-across far before the approach received its official definition and name:

'Read across is a technique used to predict endpoint information for one chemical by using data from the same endpoint from another chemical which is considered to be similar in some way (on the basis of structural similarity and similar properties and/or activities)'.[2]

In this case, similarity should be understood in a broad sense—not only structural similarity, but also similarity of properties (weight, solubility, volatility, partitioning), mechanism of interaction with different chemical groups including specific biogenic molecules (lipids, proteins, DNA), test species (phyla, classes), *etc.*

Compared with other category approaches such as trend analysis and quantitative structure–activity relationships [(Q)SARs], read-across operates with the most limited set of experimental data—only for one or a few analogues. As a consequence, the success of read-across depends strongly on the quality of these data and the appropriate selection of the characteristics and measures of similarity, as well as the thresholds used to identify analogous chemicals. The selection of similar chemicals is endpoint specific. It has to be based on theoretical knowledge (driving forces, bioavailability, *etc.*) and empirical facts related to the phenomena responsible for the effect studied.

A parallel look at other branches of science reveals that read-across is closely related to the approaches known as the combination of predictions. Combining predictions has a long history and probably started with the claim by Laplace in 1818 that '*in combining of the results of these two methods, one could obtain a result whose probability low of error will be more rapidly decreasing*' (as quoted by Clemen 1989[5]). In 1878 Galton also appreciated the value of combining predictions.[6] Nowadays these approaches are used broadly in psychology, economics, statistics and other activities.[5,7]

15.3 Practical Aspects of Read-Across

The successful application of read-across depends strongly on:

- the approach selected to collate information relating to similar chemicals;
- the type and quality of the data; and
- the formalism of data treatment used to predict the endpoint for the non-tested target chemical.

The type of data, measure of similarity and mathematical formalism are mutually dependent.

15.3.1 Types of Data

Four types of data (nominal, ordinal, interval and ratio; see also Chapter 9) can be distinguished, each one providing more information than the previous one. Thus, interval data could be converted into ordinal data by losing part of the information, but the opposite conversion is not possible.

- **Nominal.** Nominal data normally come in the form of a set of values that have no meaning in either a numerical or ordering (hierarchical) sense. Each value is essentially a label or qualitative descriptor (*nominal* comes from the Latin *nomen* meaning a *name*) and is chosen from a set of nonoverlapping categories. It is possible to count, but not order or measure, nominal data. An example of nominal data is the classification of chemicals into groups such as carbohydrates, lipids, amino acids, *etc.*

- **Ordinal.** Ordinal data normally come in the form of a set of ordered values with a meaning in the order. In this respect, the relative order can be used for comparative purposes but not for arithmetic operations. They can be expressed in both numeric and string forms, though the distinction between neighbouring points is not necessarily equal. An example of ordinal type of data is the classification of chemicals according to their skin sensitisation potential based on the Local Lymph Node Assay (LLNA) classification scheme of extreme, strong, moderate and non sensitisers.[8]

- **Interval.** An interval scale is a scale of measurements where the distance between any two adjacent units of measurement (or 'intervals') is the same, but the zero point is chosen arbitrarily. Interval data allow for the quantification and comparison of the magnitudes of differences between measured items. An example of interval scale is temperature measured in degrees Celsius.

- **Ratio.** A ratio scale affords the most informative level of measurement. The factor which clearly defines a ratio scale is that it has a true zero point. The simplest examples of a ratio scale are the measurement of mass, distance and concentration.

Depending on the type of data, read-across will be qualitative (for nominal or ordinal data) or quantitative (for interval or ratio data). When data for analogous chemicals belong to different scales, the application of read-across requires conversion of data into the least informative scale.

15.3.2 Mathematical Formalism

The simplest version of read-across avoids any processing of data. It simply ascribes the available data for some (reference) chemical to another (target) chemical that lacks such data. In this case, all uncertainty associated with the data of the reference chemical will be transferred to the target chemical. In addition, special attention should be paid also to the measure of similarity used to select the data-rich chemical as an appropriate analogue as any dissimilarity that is not taken into account could significantly affect the quality of the filling of the data gap.

Data processing is necessary when data from more than one reference chemical are used to make a prediction for a target chemical. In this case, from a formal point of view, the read-across approach utilises methods also known as the 'combination of predictions'. The mathematical formalism used to combine predictions and/or data is determined by the type of data:

- **Maximum/minimum function.** This returns the maximum or minimum value of a set of data as the argument (or predicted value). Arguments can be numbers, named ranges or arrays. The function is applicable to ordinal,

interval and ratio types of data only. Depending on the context of the
maximum or minimum value of the predicted endpoint, it can also be
described or thought of as a 'worst or best scenario assessment'.

- **Mode.** The mode is the most frequently occurring value in a set of discrete
data. There can be more than one mode if two or more values are equally
common. The mode is applicable to all types of data.
- **Median.** The median is the value halfway through the ordered data set,
above and below which there are an equal number of data values. For a set
of interval or ratio data, the median is the middle value for an odd number
of data or the arithmetic mean of the two middle values for an even
number of data. It is generally a good descriptive measure of the data
location which works well for skewed data or data with outliers. The
median is applicable to ordinal, interval and ratio types of data.
- **Average.** The arithmetic, harmonised or geometric mean are used and are
defined as follows:

Arithmetic mean (average)

$$\bar{y} = \frac{\sum\limits_{n=1}^{N} y_n}{N} \tag{15.1}$$

Harmonic mean (sub-contrary mean):

$$\bar{y}_H = \frac{N}{\sum\limits_{n=1}^{N} \frac{1}{y_n}} \tag{15.2}$$

Geometric mean:

$$\bar{y}_G = \left(\prod_{n-1}^{N} y_n \right)^{1/N} \tag{15.3}$$

Average values depend on all the data, which may include outliers. They are
especially useful as being representative of the whole sample for use in
subsequent calculations. Average values are applicable for interval and ratio
types of data.

- **Linear or quadratic combinations.**

$$y_L = \mathbf{b}^{\mathrm{T}} \mathbf{y} + c \tag{15.4}$$

$$y_Q = \mathbf{y}^{\mathrm{T}} \mathbf{A} \mathbf{y} + \mathbf{b}^{\mathrm{T}} \mathbf{y} + c \tag{15.5}$$

where:

y_L or y_Q is a prediction based on the linear or quadratic combination
$\mathbf{y}$ is a vector of the combined data and/or predictions

A, **b** and *c* are parameters (weights) of the combination.

The values of combination parameters **A**, **b** and *c* can be assigned on the basis of expert or empirical information, or could be estimated *via* regression analysis. Depending on the specificity of the studied phenomenon, additional restrictions could be added for the combination parameters. For example, in the linear combination if the intercept is assumed to be zero ($c = 0$), the combination is transformed into the weighted average of single forecasts or observations. If the latter are unbiased then the combination parameters **b** could be selected such that they sum to unity ($\mathbf{b}^\mathrm{T} = 1$). The arithmetic mean, eqn (15.1) is a special case where $\mathbf{b}^\mathrm{T}1 = 1$, $c = 0$ and all elements of **b** are equal. It should be noted that, in the case of extrapolation, read-across based on eqn (15.4) and eqn (15.5) provides better results compared with the traditional mean value. Both linear and quadratic combinations are applicable to interval and ratio type of data.

15.3.3 Chemical and Structural Similarity

Similarity plays a crucial role in grouping approaches, one of which is read-across. Similarity (and grouping) of chemicals can be estimated on the basis of single dichotomy symptoms or by making use of a set of symptoms. Although there is an intuitive understanding about the similarity of chemicals, attempts to formalise this understanding with universal practical validity for hazard assessment are not, at present, advanced. The origin of the problem is encapsulated in the interpretation of the similarity measures.

The similarity of chemicals depends on a variety of factors. It may be accounted for by molecular topology (atom connectivity), chemistry (atom and bonds types), the presence of functional or specifically acting groups, physicochemical properties, mechanism of interactions, endpoints, *etc.* (see Chapters 5–7 for a detailed discussion of these descriptors).

Common approaches to encode molecular characteristics defining similarity include their representation in a binary bit-string (fingerprint) format. Such fingerprints contain a number of bits, each of them representing the presence (1) or absence (0) of some molecular feature. Discrete variables that are described by more than two values can be represented by a binary bit-string, using a bit for each possible value. Continuous variables can be represented by defining ranges of values and then assigning a bit to each range. If the multiplicity of the features (*i.e.* the number present) needs to be accounted for, then the information is encoded in 'holographic' fingerprints (holograms). In principle, there is no limitation on the type and number of features that can be coded in the fingerprints or holograms; however, from a practical point of view (especially for read-across), coding together features with different meaning is not recommended. The latter means that a formal equivalency among topology, atom and bond types, physico-chemical properties, mechanisms of action, *etc.* is assumed. In addition, if the number of bits of one of these features dominates over the number of bits of remaining features, the similarity measure will be biased by the dominating feature thus diminishing the impact of the remaining characteristics of the chemicals. In this respect, it is more useful to determine similarity based on features with similar meaning, thus allowing better interpretation of the similarity itself—structural similarity (accounting for topology and type of atoms and bonds), property similarity (based on physico-chemical properties), mechanistic similarity, *etc.*

A quantitative measure of the relatedness between two chemicals can be provided by coefficients of similarity or dissimilarity (as a distance). In general, similarity between chemicals is estimated by the number of matches or overlap with respect to one or more molecular characteristics. Dissimilarity is estimated by the same approach, but the number of mismatches or differences is accounted for. In most cases, the values assessing similarity can be described by a coefficient ranging from 0 to 1 or it can be normalised to this range. The zero-to-one range provides a simple tool to convert between similarity and dissimilarity coefficients, mainly by subtraction from unity. A large number of similarity coefficients providing different values for similarity have been defined.[9] Some coefficients have been reinvented by different authors and a variety of different names are found for the same coefficients. Certain coefficients provide identical ranking of chemicals (monotonic coefficients), although the actual coefficient values are different. Conversely, some pairs of coefficients exhibit very low correlations suggesting that they reflect different aspects of molecular properties.

The short discussion here has emphasised the lack of a single best coefficient of similarity. The answer to the question 'What does 83% similarity mean?' can be obtained easily by analysing the formula used to calculate similarity, features accounted for and the type of coding (fingerprints or holograms). In most cases, it means that 83% of the features compared are common to both chemicals and 17% of the features are present in only one of the chemicals. Unfortunately, the question 'Does the 17% difference indicate that both chemicals are different from a chemical or biochemical point of view?' is meaningless because the measure of similarity is statistical; it only accounts for the number of mismatches and not for the chemistry hidden behind these mismatches. If the only difference in a set of chemicals having the same topology is one atom, then the measure of structural similarity will be the same for all pairs of chemicals no matter the chemical specificity of this atom (*e.g.* Mg, O, S, Zn, or Hg). In this respect, it was confirmed experimentally that the replacement of an oxygen atom with $-NH-$ or $-CH_2-$ in three series of thermolysin inhibitors resulted in analogues 1,000 times more potent[10] this would not be captured by similarity approaches.

The absence of a unique measure of chemical similarity and the difficulties associated with the chemical interpretation of the measured values indicate that similarity should be considered as a context dependent parameter. From a practical point of view, it is more useful if the grouping of similar chemicals is performed in a stepwise manner applying different measures of similarity as this will account for the specificity of different factors affecting similarity, consecutively—topology, functionalities, properties, mechanisms, *etc*. A consecutive approach to grouping (or sub-categorisation) of chemicals allows users to:

- start by grouping chemicals with less conservative conditions thus allowing an initial collection of a large number of potential analogues;
- prune the initial set of analogues consecutively by applying different measures of similarity, starting with more general and ending up with more specific molecular features; and
- select a small number of very similar analogues in the context of the phenomenon studied.

The main benefit of the consecutive approach is that it allows investigation of the effect of different molecular features on the phenomenon studied. If the approach is not applied formally and the user is trying to understand the causality between molecular features and effect, then it could be a premise for the generation of knowledge and arguments supporting the appropriate selection of analogues. No doubt, the final result will be the same if all measures of similarity are logically 'ANDed' and applied simultaneously. Obtaining analogues in a single step, however, risks losing information about the significance of different molecular features on the phenomenon studied as well as the possibility that the resultant grouping could, to some extent, be an artefact.

15.4. Example: Read-Across for the Prediction of the Toxicity of 4-Ethylphenol to *Tetrahymena pyriformis*

In this example, the acute toxicity to *Tetrahymena pyriformis* of 4-ethylphenol (CAS registry number 123-07-9) is predicted using grouping techniques to allow for read-across. This illustrates the capabilities of read-across to reproduce the well-known experimental toxicity for this chemical log $(1/\text{IGC}_{50})$, where IGC_{50} is in mol L^{-1}.

It is known that acute toxicity of non-reactive chemicals such as 4-ethylphenol is strongly correlated with their hydrophobicity. On the other hand, it is also known that alkylphenols demonstrate toxicity in excess of non-polar narcotics. In this respect, the analogues used to assess the toxicity of 4-ethylphenol by read-across should not be non-polar narcotics or reactive chemicals, and should also have a hydrophobicity as close as possible to the target chemical.

From a chemical point of view, the best analogues to the target are expected to be 3-ethylphenol (CAS 620-17-7) and 2-ethylphenol (CAS 90-00-6) followed by other 2-, 3- or 4-alkylphenols. The hydrophobicity and toxicity of the alkylphenols used for the predictions are given in Table 15.1.

Table 15.1 Alkylphenols and their 2-D structure, hydrophobicity (log P) and toxicity [log(1/IGC$_{50}$)].

Name	2-D structure	log P [EPI Suite, v4.0, 2009]	log (1/IGC$_{50}$) in mol L^{-1}
3-Methylphenol		2.06	2.89
4-Methylphenol		2.06	2.92
2-Ethylphenol		2.55	3.16
3-Ethylphenol		2.55	3.26
4-Ethylphenol (target)		**2.55**	**3.21**
4-Propan-2-ylphenol		2.97	3.47
4-Propylphenol		3.04	3.64
4-*tert*-Butylphenol		3.42	3.91
4-Butan-2-ylphenol		3.46	3.98
4-*tert*-Pentylphenol		3.91	4.23
4-Nonylphenol		5.99	5.47

In this example, the toxicity and hydrophobicity data, search engines and categorisation approaches implemented in the Organisation for Economic Co-operation and Development (OECD) (Q)SAR Application Toolbox (Version 1.1) are used (see Chapter 16).[11] Hereafter, only analogues having observed toxicity in the version of the Toolbox utilised are considered.

Based on traditional read-across (which is in concordance with the formal definition of the approach), one should start by searching for analogues based on structural similarity (topology, type of atoms, type of bonds, hybridisation, hydrogen atoms, *etc.*). Table 15.2 illustrates the number and type of analogues

Table 15.2   Similarity threshold, total number and type of analogues identi-
                 fied.

Similarity threshold	Number of analogues	Type of analogues
50%	94	Alkylphenols • 3-Methylphenol • 4-Methylphenol • 2-Ethylphenol • 3-Ethylphenol • 4-Propan-2-ylphenol • 4-Propylphenol • 4-*tert*-Butylphenol • 4-Butan-2-ylphenol • 4-*tert*-Pentylphenol • 4-Nonylphenol Aldehydes Anilines Aryl halides Carboxamides Esters Hydroquinones Nitriles
70%	17	Alkylphenols • 4-Methylphenol • 3-Ethylphenol • 4-Propylphenol Aldehydes Anilines Hydroquinones Nitriles
80%	2	Alkylphenols • 3-Ethylphenol • 4-Propylphenol

obtained (all alkylphenols are listed), depending on the selected similarity threshold. The Dice coefficient calculated on the basis of atom centred fragments (first neighbours accounting for atom type, hybridisation and attached H-atoms) was used as the measure of similarity.[9]

The results shown in Table 15.2 reveal that the attempt to find the best analogues in a single step failed. At the 80% threshold, the approach identifies only two analogues (3-ethylphenol and 4-propylphenol). Even at a 70% threshold, one of the most appropriate analogues from chemical perspective (2-ethylphenol) is not found.

Bearing in mind the insufficiency of experimental data, the restrictive search for the most similar chemicals for read-across would reveal few chemicals and some important analogues are omitted. Such problems can be resolved if the selection of analogues is performed in a stepwise manner. This includes initial grouping of chemicals with relatively low threshold of similarity (50%) followed by sub-categorisations with gradually increasing thresholds of similarity.

Table 15.3   Results of structural similarity categorisation (50% threshold for the Dice coefficient) followed by sub-categorisations based on 'Organic functional groups' and ECOSAR classifications in the OECD (Q)SAR Application Toolbox.

Analogues based on sub-categorisation by Organic functional groups		*Subsequent classification using ECOSAR*
1	4-Hydroxypyridine	Phenols
2	Benzene-1,3-diol	Phenols
3	**Benzene-1,4-diol**	**Quinone/Hydroquinone**
4	Phenol	Phenols
5	**4-Ethylpyridine**	**Neutral**
6	3-Methylphenol	Phenols
7	4-Methylphenol	Phenols
8	2-Ethylphenol	Phenols
9	3-Ethylphenol	Phenols
10	4-Propan-2-ylphenol	Phenols
11	4-Propylphenol	Phenols
12	4-*tert*-Butylphenol	Phenols
13	4-Butan-2-ylphenol	Phenols
14	4-*tert*-Pentylphenol	Phenols
15	4-Nonylphenol	Phenols

The current exercise shows that, at a certain threshold (*i.e.* 70% in this example), the statistical measure of similarity is unable to discriminate 'chemically correct' analogues from those which are 'chemically dissimilar'. Hence, it was decided to make the first categorisation step at relatively low similarity threshold (50%), followed by subsequent sub-categorisations.

Following use of the 50% similarity threshold, an effective elimination of dissimilar chemicals can be achieved by sub-categorisation based on chemical or biochemical knowledge such as organic functional groups, ability to bind to proteins or DNA, or mode of action. Table 15.3 illustrates the results of sub-categorisations based on 'Organic functional groups' as defined in the OECD (Q)SAR Application Toolbox.[11] This selects chemicals containing functional groups that are present in the target chemical and which do not contain any other predefined functional groups.

As can be seen from Table 15.3, sub-categorisation based on 'Organic functional groups' correctly identifies all alkylphenols as candidates to be analogues of 4-ethylphenol and only a few dissimilar chemicals are also included. An attempt to verify the toxic potency of these chemicals can be undertaken by the subsequent application of the ECOSAR classification implemented in OECD (Q)SAR Application Toolbox.[12,13] The ECOSAR classification identified that one chemical (4-ethylpyridine) belongs to the class of 'Neutral' chemicals and one (benzene-1,4-diol) belongs to the class of 'Quinone/Hydroquinone' chemicals. The remaining 12 analogues of 4-ethylphenol are structurally similar, have the same functional groups and similar toxic mechanism of action.

To assess the toxicity of 4-ethylphenol, it is necessary to select one or a few analogues that have the same or very similar hydrophobicity to the target chemical. Two chemicals (2-ethylphenol and 3-ethylphenol) have practically the same hydrophobicity as the target chemical. Their averaged toxicity log $(1/IGC_{50})$ can be used as surrogate of the toxicity of 4-ethylphenol. The result of data-gap filling *via* the read-across examples described here practically coincides with the actual toxicity value of the target chemical and, more importantly, the result was found *via* an approach that is in agreement with current knowledge on the acute toxicity of chemicals.

15.5 Conclusions

Read-across has been shown to be an effective and practical tool for assessing the toxicological hazards of substances for which little or no data are available. The overall conclusion is that the read-across is context dependent in terms of identifying analogues and its relevance needs to be justified on a case-by-case basis. In this respect, the usefulness of the approach is highly dependent on an insight into the mechanisms of studied phenomenon.

The following are some general methodological recommendations on how to apply read-across:

- The identification of potential analogues for read-across should be performed by a stepwise sub-categorisation.
- The process of categorisation should start with less conservative grouping followed by sub-categorisation based on gradually increasing mechanistic understanding for the studied phenomenon.
- A small number of very similar analogues should be selected, in the context of the studied phenomenon, to make the ultimate read-across prediction.

There is no need to justify the usefulness of the read-across and it is well described in more detail in Chapters 14, 16 and 17. However, it is necessary to find ways to make the implementation of this approach easy and efficient for hazard assessment and the legislation of chemicals. The development and release of OECD (Q)SAR Application Toolbox v1.0, initiated by OECD, and the forthcoming v2.0 supported by the European Chemicals Agency (EChA) is a cornerstone in this approach.

References

1. Organisation for Economic Co-operation and Development, *Guidance on Grouping of Chemicals*, OECD, Paris, 2007, OECD Environmental Health and Safety Publications Series on Testing and Assessment No. 80, ENV/JM/MONO(2007)28, www.olis.oecd.org/olis/2007doc.nsf/LinkTo/NT00 00426A/$FILE/JT03232745.PDF [accessed March 2010].

2. European Chemicals Agency, *Guidance on Information Requirements and Chemical Safety Assessment. Chapter R.6. QSARs and grouping of chemicals*, EChA, Helsinki, 2008, http://guidance.echa.europa.eu/docs/guidance_document/information_requirements_en.htm [accessed March 2010].
3. D. Mendelejeff, *Z. für Chem.*, 1869, **12**, 405.
4. A. M. Butlerov, *Z. Chem. Pharm.*, 1861, **4**, 549.
5. R. T. Clemen, *Int. J. Forecast.*, 1989, **5**, 559.
6. F. Galton, *J. Anthropol. Inst. Gr. Brit. Irel.*, 1878, **8**, 132.
7. J. S. Armstrong, *Principles of Forecasting: A Handbook for Researchers and Practitioners,* Kluwer Academic, Norwell, MA, USA, 2001.
8. European Centre for Ecotoxicology and Toxicology of Chemicals, *Contact Sensitization: Classification According to Potency,* ECETOC, Brussels, 2003, Technical Report No. 87.
9. P. Willett, *J. Chem. Comput. Sci.*, 1998, **38**, 983.
10. H. Kubinyi, *J. Braz. Chem. Soc.*, 2002, **13**, 717.
11. Toolbox Downloads, Getting started, January 07, 2010, http://toolbox.oasis-lme.org/?section = downloads.
12. EPI Suite™ v4.0, US Environmental Protection Agency, Washington DC, 2009, www.epa.gov/opptintr/exposure/pubs/episuitedl.htm [accessed March 2010].
13. US Environmental Protection Agency, *User's Guide for the ECOSAR Class Program, MS-Windows Version 1.00*, US EPA Risk Assessment Division, Washington DC, www.docstoc.com/docs/7851157/Ecological-Structure-Activity-Relationships-(ECOSAR)-ECOSAR-Users-Guide-(PDF)/ [accessed March 2010].

CHAPTER 16

Tools for Category Formation and Read-Across: Overview of the OECD (Q)SAR Application Toolbox

R. DIDERICH*

Organisation for Economic Co-operation and Development (OECD), ENV/EHS, 2 rue André Pascal, 75775 PARIS CEDEX 16, France

16.1 The OECD (Q)SAR Project: From Validation Principles to an Application Toolbox

As part of its Environment Health and Safety Programme, the Organisation for Economic Co-operation and Development (OECD) has been active on the subject of the use of (quantitative) structure–activity relationships [(Q)SARs] for regulatory purposes since 1991. Early activities aimed to improve understanding among countries on the performance of (Q)SARs and their possible use in regulatory settings. A number of review documents were published between 1992 and 1994, e.g. on physico-chemical properties, biodegradation, aquatic toxicity and general base-set endpoints for new industrial chemicals.[1–4]

No major activities on (Q)SARs were launched at the OECD between 1994 and 2003 due to divergent needs in member countries. Indeed, while some

---

*The views expressed in this chapter are the sole responsibility of the author and do not necessarily reflect the views of OECD.

Issues in Toxicology No.7
In Silico Toxicology: Principles and Applications
Edited by Mark T. D. Cronin and Judith C. Madden
© The Royal Society of Chemistry 2010
Published by the Royal Society of Chemistry, www.rsc.org

member countries relied heavily on (Q)SAR results for some regulatory processes (*e.g.* assessment of new chemicals in the United States following pre-manufacture notices[5]), others had adopted legislation allowing them to request experimental test results for all relevant regulatory endpoints (*e.g.* notification of new chemicals in the EU following the seventh amendment of the Dangerous Substances Directive[6] or the assessment of existing chemicals in the EU under the Existing Substances Regulation[7]).

Since 1998, many member countries as well as industry have significantly increased their efforts to assess the hazards of existing chemicals either through voluntary programmes such as the US High Production Volume (HPV) Chemicals Challenge Programme[8] or the International Council of Chemical Associations' (ICCA) HPV Chemicals Initiative,[9] or through new legislation such as the Canadian Environmental Protection Act[10] or the EU REACH Regulation.[11]

These efforts have also led to increased international co-operation (*e.g.* within the OECD HPV Chemicals Programme) and to increased need for non-testing methods to fill data gaps to save time, resources and test animals. A new (Q)SAR project was therefore set up at OECD in 2003 with the specific aim of improving regulatory acceptance of (Q)SARs.

The first two activities of the new OECD (Q)SAR Project consisted of:

- The development of 'OECD Principles for the Validation, for Regulatory Purposes, of (Q)SAR Models', and a corresponding guidance document. The development of such a set of agreed principles was considered important, not only to provide regulatory bodies with a scientific basis for making decisions on the acceptability of data generated by (Q)SARs, but also to promote the mutual acceptance of (Q)SAR models by improving the transparency and consistency of (Q)SAR reporting.[12] A detailed description of these principles is provided in Chapter 11.
- The description of experience in member countries in applying (Q)SAR models. This report included a brief description of the activity of OECD member countries with the aim of enhancing the regulatory acceptance of (Q)SAR models and expanding the opportunities for future application of the models.[13]

The third work item of the OECD (Q)SAR project was the development of the (Q)SAR Application Toolbox—computer software which aimed to make practical (Q)SAR approaches readily available and accessible. The philosophy of the Toolbox is based on the 'chemical category' concept. A chemical category is a group of chemicals whose physico-chemical and toxicological or ecotoxicological properties and/or environmental fate properties are likely to be similar or to follow a regular pattern because of their similar chemical structure. Using this so-called category approach, not every chemical needs to be tested for every endpoint because the available test results for the members of the category allow an estimation of the results for the untested endpoints. Guidance on the formation and use of chemical categories was developed at the

OECD and published in 2007.[14] The concept of chemical categories is outlined in detail in Chapters 14, 15, 17 and 18.

The concept of chemical categories was chosen as the background philosophy for the Toolbox because it is a straightforward application of QSAR methodology, which is used successfully in a number of chemical review programmes such as:

- **OECD HPV Chemicals Programme**. Guidance on the formation of chemical categories is part of the OECD Manual for Investigation of HPV Chemicals and numerous categories have been assessed within this programme.[15,16]
- **US HPV Challenge Programme.** Guidance on the formation of chemical categories is part of the official guidance for this programme and the concept has led to considerable savings of resources and test animals.[17,18]
- **EU REACH Regulation**. The concept of chemical categories is specifically mentioned in the REACH Regulation as a method of data gap filling and the OECD guidance document was implemented in the REACH guidance on Information Requirement and Chemical Safety Assessment.[19]

A first version of the OECD (Q)SAR Application Toolbox was made publicly available in March 2008. A project to further develop the Toolbox was launched in October 2008, with the aim of releasing version 2.0 in October 2010 and version 3.0 in October 2012.[20]

16.2 Workflow of the Toolbox

The seminal features of the Toolbox are:

1. Identification of relevant structural characteristics and potential mechanism or mode of action of a target chemical.
2. Identification of other chemicals that have the same structural characteristics and/or mechanism or mode of action.
3. Use of existing experimental data to fill the data gap(s).

The Toolbox has six work modules which are used in a sequential workflow. This workflow is meant to mimic the thought process of assessors evaluating the hazards of chemicals as well as to be compatible with the workflow of forming chemical categories as outlined in the OECD *Guidance on Grouping of Chemicals*.[14] Summary information on the six modules is provided in Table 16.1.

The identification of structural characteristics and potential mechanisms or modes of action is achieved with a set of 'profilers' in the module 'Profiling'. These profilers identify structural alerts involved in specific reactions or binding mechanisms relevant for different regulatory endpoints. For example the profiler 'Protein binding' identifies structural alerts that have the potential to

Table 16.1   Sequential workflow of the Toolbox.

Module	Summary background information
Chemical Input	This module provides the user with several means of entering the chemical(s) of interest or target chemical(s). Since all subsequent functions are based on chemical structure, the goal here is to make sure the molecular structure assigned is the correct one.
Profiling	This module provides the user with the means to 'profile' the target chemical(s) to identify relevant structural characteristics and potential mechanisms or modes of action. Examples of 'profilers' are 'protein binding' or DNA binding, identifying structural alerts for covalent binding with protein or DNA.
Endpoints	This module provides the user with an electronic process of retrieving results for regulatory endpoints (*e.g.* data on environmental fate, ecotoxicity or mammalian toxicity) which are stored in the Toolbox. This data gathering can be executed in a global fashion (*i.e.* collecting all data on all endpoints) or on a more narrowly defined basis (*e.g.* collecting data for a single or limited number of endpoints).
Category Definition	This module provides the user with several means of grouping chemicals into a (eco)toxicologically meaningful category that includes the target molecule(s). This is the critical step in the workflow and several options are available in the Toolbox to assist the user in refining the category definition *via* sub-categorisation.
Filling Data Gaps	This module provides the user with three options for making an endpoint-specific prediction for the untested chemical(s), in this case the target molecule(s). These options, in increasing order of complexity, are by read-across, by trend analysis, and through the use of QSAR models.
Report	The final module provides the user with a downloadable written audit trail of what functions the user performed using the Toolbox to arrive at the prediction.

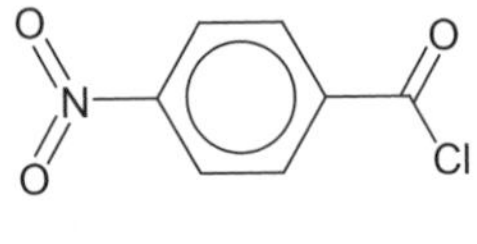

Scheme 16.1   4-Nitrobenzoyl chloride.

covalently bind with proteins. Such a profiler is relevant for grouping chemicals for a number of endpoints, *e.g.* skin sensitisation or acute aquatic toxicity. Table 16.2 and the references contained therein outlines a selection of profilers which are implemented in version 1.1 of the Toolbox released in December 2008.

The workflow of the Toolbox can be illustrated with a simple example. For the chemical 4-nitrobenzoyl chloride (see Scheme 16.1), no experimental data on skin sensitisation are available.

There is growing agreement that all organic chemicals must react covalently with skin proteins in order to behave as skin sensitisers. Therefore, the

Table 16.2 Summary information on selected profilers from version 1.1 of the Toolbox.

Profiler	Summary background information
Protein binding	The protein binding categorisation scheme includes 38 categories such as haloalkanes, isocyanates, isothiocyanates, diketones, aldehydes, acyl halides, alkyl sulfates, sulfonates, *etc.* Each category is presented by defined 2-D structural alerts that are responsible for the protein binding. The associated mechanisms are in accordance with the existing knowledge on electrophilic interaction mechanisms of various structural functionalities.[21] This classification scheme is particularly relevant for skin and respiratory sensitisation and acute aquatic toxicity, but also for chromosomal aberration and acute inhalation toxicity. It is built on conventional organic chemical mechanism and as such is qualitative in character.
Verhaar Classification	Utilising an acute toxicity data collection for guppies, Verhaar and colleagues delineated the classes of inert, less inert, reactive, and specifically acting chemicals, and provided the chemical rules for discrimination of the first three.[22] This classification scheme is specifically relevant for acute aquatic toxicity endpoints.
OASIS Acute Toxicity MOA	This profiler also classifies chemicals for their acute aquatic toxicity mode of action. This is a behavioural mode of action. It was originally developed by the US EPA and is based on a set of chemicals tested with fathead minnow.[23] This classification scheme is specifically relevant for acute aquatic toxicity endpoints.
ECOSAR classification	ECOSAR is a program developed by the US EPA which contains structure–activity relationships (SARs) used to predict the aquatic toxicity of chemicals based on their similarity of structure to chemicals for which the aquatic toxicity has been previously measured.[24] SARs are developed for chemical classes based on measured test results. To date, over 150 SARs have been developed for more than 50 chemical classes. The ECOSAR classification profiler in the Toolbox retrieves the class to which a chemical belongs. This classification scheme is specifically relevant for aquatic toxicity endpoints.
DNA binding	The DNA binding categorisation scheme is based on the model of Ames mutagenicity and was developed by Mekenyan and colleagues.[25] The scheme includes 19 categories. Each category is defined by 2-D structural alerts that are a necessary condition for a chemical to covalently interact with DNA. Definition of these alerts was justified by their interaction mechanisms with DNA, found in the literature.[26] This classification scheme is particularly relevant for Ames mutagenicity.
Benigni/Bossa rulebase	The Benigni/Bossa rulebase for mutagenicity and carcinogencity was developed as a module (plug-in) to the Toxtree software.[27] The structural alerts (SAs) from the rulebase have been included as a profiler in the Toolbox. The list of SAs refers mainly to the knowledge on the action

Table 16.2   (*continued*)

Profiler	Summary background information
	mechanisms of genotoxic carcinogenicity (thus they also apply to the mutagenic activity in bacteria), but includes also a number of SAs flagging potential nongenotoxic carcinogens.[28]
BfR rulebases for skin and eye irritation/ corrosion	The rulebase for skin and eye irritation and corrosion developed by the German Federal Institute for Risk Assessment (BfR) and collaborators has been implemented into the Toolbox as a set of structural alerts and physical-chemical exclusion criteria.[29,30]
BIOWIN biodegrad-ability category	The BIOWIN biodegradability categorization scheme is based on the structural alerts used by the MITI Biodegradation Probability Model developed for the US EPA.[31] This profiler identifies structural alerts relevant for biodegradation.
Cramer classification	The Cramer classification scheme (tree) is probably the best-known approach for structuring chemicals in order to make an estimation of a Threshold of Toxicological Concern (TTC). TTC is a concept that aims to establish a level of exposure for all chemicals below which there would be no appreciable risk to human health. The tree relies primarily on chemical structures and estimates of total human intake to establish priorities for testing. The procedure uses recognised pathways for metabolic deactivation and activation, toxicity data, and the presence of a substance as a component of traditional foods or as an endogenous metabolite.[32]
	Substances are classified into one of three classes:
	Class 1 contains substances of simple chemical structure with known metabolic pathways and innocuous end products which suggest a low order of oral toxicity.
	Class 2 contains substances that are intermediate. They possess structures that are less innocuous than those in Class 1, but they do not contain structural features that are suggestive of toxicity like those in Class 3.
	Class 3 contains substances with a chemical structure that permit no strong initial impression of safety and may even suggest a significant toxicity.
ER binding	Estrogen receptor (ER) binding is a molecular initiating event much like protein binding. It is an endpoint where several comprehensive databases exist, which has lead to the development of several approaches for using (Q)SARs to predict ER binding and possible subsequent endocrine disruption. The ER binding profiler in the Toolbox is based on the 'four phase' assessment that includes Comparative Molecular Field Analysis (CoMFA) and the Common Reactivity Pattern Approach (COREPA).[33–37]
Superfragment profiling	An extended (super) fragment consists of a combination of simple fragments that are in such close proximity that their solvation behaviour (evidenced by octanol–water log *P*) is markedly affected. The simplest definition of an extended or super fragment is the 'largest electronically-connected

Table 16.2 (*continued*)

Profiler	Summary background information
	substructure'. At present the software is restricting the candidates for extended fragments to those separated by one or two 'isolating carbons', designating these Y–C–Y and Y–C–C–Y respectively. Isolating carbons are locations in the compound that act as high threshold barriers to movement of electrons. All remaining connected group of atoms are then designated as 'polar fragments', *i.e.* simple fragments.

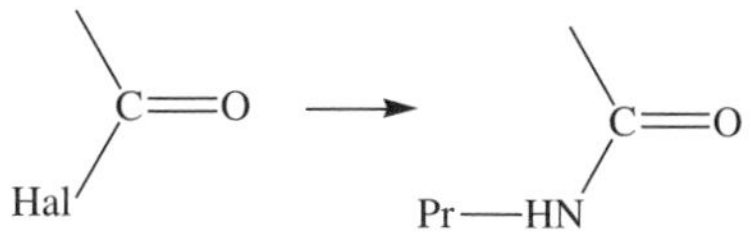

Hal = F, Cl, Br

Scheme 16.2

mechanism by which organic chemicals bind with proteins is relevant to grouping chemicals that may be skin sensitising agents.[21] The profiler 'protein binding' of the Toolbox identifies 4-nitrobenzoyl chloride as having the potential to conjugate with protein *via* nucleophilic substitution of acyl halides, as illustrated in Scheme 16.2.

The Toolbox can then scan its databases to find other chemicals that can potentially bind to protein by the same mechanism and for which experimental results on skin sensitisation are available. Table 16.3 outlines the identified chemicals and their experimental results for skin sensitisation.

All identified chemicals are skin sensitisers. By read-across, it can be predicted that the target chemical, 4-nitrobenzoyl chloride, is also a skin sensitiser. The same approach can be applied for several other endpoints of regulatory relevance. A few representative use scenarios are outlined in the following sections.

16.3 Example Use Scenarios for Regulatory Application

The examples outlined below are largely based on the examples described in the OECD *Guidance Document for using the (Q)SAR Application Toolbox to Develop Chemical Categories according to the OECD Guidance on Grouping of Chemicals*.[38] Furthermore a set of training materials with additional examples and case studies has been developed and published on the website of the OECD (Q)SAR project.[19]

Table 16.3 Experimental skin sensitisation results for chemicals with the same protein-binding mechanism as 4-nitrobenzoyl chloride.

Name	Structure	Experimental results
Benzoyl chloride		LLNA[a] Positive EC3 <10%[b]
3,5,5-Trimethylhexanoyl chloride		LLNA Positive EC3 = 2.7%
Isononanoyl chloride		LLNA Positive EC3 = 2.7%
Nonanoyl chloride		LLNA Positive EC3 = 1.8%
Hexadecanoyl chloride		LLNA Positive EC3 = 8.8%
Octadecanoyl chloride		LLNA Positive EC3 <10%

[a]Local Lymph Node Assay.
[b]Concentrations of test chemical required to provoke a three-fold increase in lymph node cell proliferation.

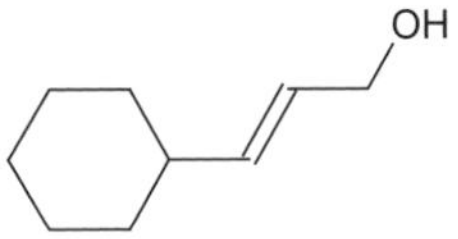

Scheme 16.3   3-Cyclohexyl-2-propen-1-ol.

16.3.1   Filling Data Gaps Using a Simple Analogue Approach

In this example, the Toolbox is used to estimate the toxicity to *Tetrahymena pyriformis* of 3-cyclohexyl-2-propen-1-ol (see Scheme 16.3) *via* the identification of relevant analogues.

Several profilers (Protein binding, OASIS Acute Toxicity Mode of Action, ECOSAR classification) identify structural and mechanistic properties relevant for acute toxicity to aquatic organisms (see Table 16.2).

For 3-cyclohexyl-2-propen-1-ol, the three profilers retrieve the following results:

- Protein binding: no binding
- OASIS Acute Toxicity Mode of Action: α,β-unsaturated alcohols
- ECOSAR classification: vinyl/allyl alcohols.

With those profiling results, the Toolbox can identify chemicals with the same profile for which experimental results are available. Eighteen analogues, as listed in Table 16.4, are identified by the Toolbox.

All the chemicals identified are structurally very similar. Only one chemical having a triple bond (2-penten-4-yn-1-ol) and an aromatic compound (3-phenyl-2-propen-1-ol) could be considered as structural outliers. Using all available results, the Toolbox can estimate the toxicity of 3-cyclohexyl-2-propen-1-ol using a simple linear regression, with Log *Kow* estimated by KOWWIN as the dependent variable.[39] The result is outlined in Figure 16.1. The estimated 48h-IGC$_{50}$ for *Tetrahymena pyriformis* is 39.9 mg L^{-1}.

Among the structural outliers, 2-penten-4-yn-1-ol indeed has a higher toxicity than would be estimated from the regression. Eliminating both chemicals from the set of analogues improves the linear regression (r^2 increases from 0.895 to 0.965), but the estimated result for 3-cyclohexyl-2-propen-1-ol changes only marginally to 41.0 mg L^{-1}.

This example shows that it is very easy to identify analogues of a chemical having the same structural and mechanistic properties and use the available experimental results to fill the data gap for the target compound. It should be noted that this approach is endpoint dependent and that different sets of profilers are relevant for grouping chemicals to fill data gaps for the endpoints of interest (see also Table 16.2).

16.3.2   Building Chemical Categories

As shown above, it is very easy to use the Toolbox to group chemicals with the same profile so that existing experimental results can be used to fill data gaps.

Table 16.4 Experimental results for chemicals with the same profile as 3-cyclohexyl-2-propen-1-ol.

Chemical name	Structure [a]	Experimental result Tetrahymena pyriformis 48h-IGC_{50} (mg L^{-1})[b]
2-Propen-1-ol		4.83E+03
2-Butene-1,4-diol		1.43E+04
2-Methyl-2-propen-1-ol		3.30E+03
3-Methyl-2-buten-1-ol		1.50E+03
3-Buten-2-ol		8.09E+02
2-Buten-1-ol		2.13E+03
1-Penten-3-ol		1.93E+03
3-Penten-2-ol		2.16E+03
2-Penten-1-ol		1.11E+03
2-Penten-4-yn-1-ol		2.91E+02
1-Hexen-3-ol		6.47E+02

Table 16.4 (*continued*)

Chemical name	Structure [a]	Experimental result *Tetrahymena pyriformis* 48h-IGC$_{50}$ (mg L^{-1})[b]
2-Hexen-1-ol		2.96E + 02
2-Hepten-1-ol		1.02E + 02
2-Octen-1-ol		5.86E + 01
2-Nonen-1-ol		2.42E + 01
2-Decen-1-ol		1.16E + 01
2-Undecen-1-ol		4.18E + 00
3-Phenyl-2-propen-1-ol		1.61E + 02

[a]Cis/trans isomers are not distinguished in this table.
[b]All experimental results are from the same publication.[40]

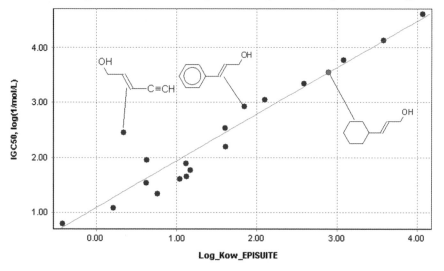

Trend analysis evaluation, making a linear approximation, based on 'Current subcategory', containing 18 data points in 18 analogue structures, Predicted target value: 39.9 mg/L

Figure 16.1   Estimate for 48 h - IGC50 for Tetrahymena pyriformis for 3-cyclohexyl-2-propen-1-ol by linear regression with results available for 18 analogs.

Furthermore, the different profilers are relevant for different endpoints. When building a chemical category for regulatory submission, the aim is to build a category for which data gaps can be filled for as many regulatory endpoints as possible, *i.e.* as mechanistically consistent as possible. Of course the more restrictive the conditions for a chemical to belong to a category, the lower the number of members of the category and the less useful the category will be for regulatory purposes. Alternatively, categories can be built that are meant to allow for data gap filling for only a few endpoints, in which case the mechanistic consistency is required only for those profilers relevant for the chosen regulatory endpoints. Furthermore, it is possible to identify sub-categories within categories, *i.e.* having different profiling results for some profilers. Hence, data gap filling would only be possible within those sub-categories. This is illustrated in the following two examples.

16.3.2.1   *Mechanistically Consistent Category*

Starting with a target chemical such as propanoic acid, 3-mercapto, methyl ester (Scheme 16.4), the most pronounced profiling results are:

- Protein binding: disulfide formation
- ECOSAR classification: esters AND thiols (mercaptans)
- DNA binding: no binding
- Cramer rules: low (Class 1).

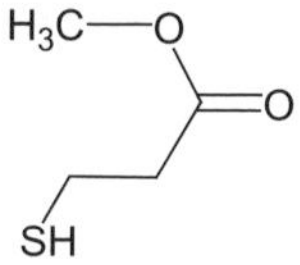

Scheme 16.4 Propanoic acid, 3-mercapto-, methyl ester.

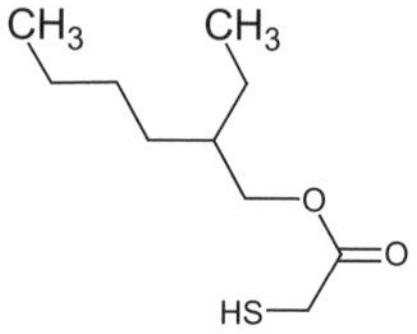

Scheme 16.5 Acetic acid, mercapto-, 2-ethylhexyl ester.

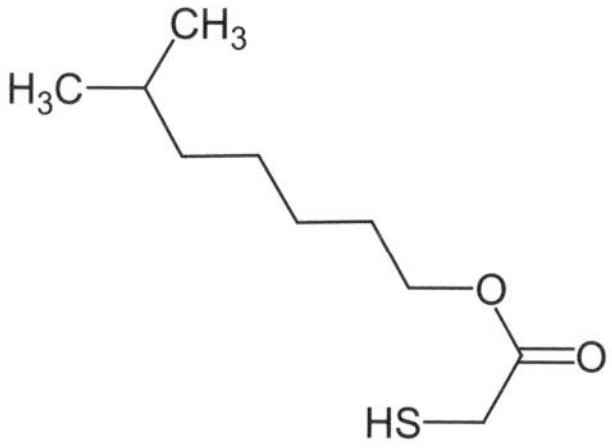

Scheme 16.6 Acetic acid, mercapto-, isooctyl ester.

In the OECD HPV Chemicals List, only two additional chemicals are found that have the same profiling results, *i.e.* acetic acid, mercapto-, 2-ethylhexyl ester and acetic acid, mercapto-, isooctyl ester (see Schemes 16.5 and 16.6 respectively). Together these three chemicals form a mechanistically robust category and read-across could be performed for most regulatory endpoints.

16.3.2.2 Sub-categories

Starting again with the same target chemical, but allowing chemicals to belong to either the ECOSAR classification 'Esters' OR 'Thiols (mercaptans)', a total of 15 chemicals is identified (see Table 16.5). The other profiling results remain identical for all 15 chemicals.

The ECOSAR classifications are specifically relevant for aquatic toxicity. Hence two sub-categories can be formed—one for chemicals with the ECOSAR classification 'Esters AND Thiols (mercaptans)' and one with the ECOSAR classification 'Thiols (mercaptans)'. For aquatic toxicity, read-across or trend analysis could still be performed within one of the sub-categories. For other regulatory endpoints (*e.g.* genotoxicity), the category could still be considered

Table 16.5   Chemicals with differing ECOSAR classifications.

Name	SMILES	ECOSAR classification
Propanoic acid, 3-mercapto-, methyl ester		Esters, Thiols (mercaptans)
Acetic acid, mercapto-, 2-ethylhexyl ester		Esters, Thiols (mercaptans)
Acetic acid, mercapto-, isooctyl ester		Esters, Thiols (mercaptans)
2-Mercaptoethanol		Thiols (mercaptans)
Methanethiol	H_3C—SH	Thiols (mercaptans)
Ethanethiol		Thiols (mercaptans)
2-Propanethiol		Thiols (mercaptans)
2-Propanethiol, 2-methyl-		Thiols (mercaptans)
1-Propanethiol		Thiols (mercaptans)
1-Butanethiol		Thiols (mercaptans)
2-Butanethiol		Thiols (mercaptans)

Table 16.5 (*continued*)

Name	SMILES	ECOSAR classification
1-Octanethiol		Thiols (mercaptans)
tert-Nonanethiol		Thiols (mercaptans)
1-Dodecanethiol		Thiols (mercaptans)
tert-Dodecanethiol		Thiols (mercaptans)

consistent for all 15 category members as the relevant profiling results are consistent for all 15 chemicals.

16.3.3 Evaluating *ad hoc* Categories

Experience from the OECD HPV Chemicals Programme has shown that grouping of chemicals most often depends on which chemicals are manufactured by the consortium of companies elaborating the hazard assessment. This means that, in practice, chemical categories are often constructed in an *ad hoc* manner and not by a systematic review of all possible chemicals that could belong to a category.

The Toolbox allows the user to evaluate the robustness of such *ad hoc* categories. Such an evaluation can be performed in two steps:

1. Verification of the consistency between members of the category of mechanisms of action or relevant structural features.
2. Verification of the consistency of available experimental data among the members of the category.

While the second step can of course be done manually without the help of the Toolbox, the profilers can in particular be used for the first step. This is outlined in the following example.

Table 16.6 lists a group of secondary amines, together with relevant profiling results for all the chemicals.

All chemicals in this group have identical profiling results except for dimethylamine, 2-methylaminoethanol and morpholine. Both 2-methylaminoethanol and morpholine have a superfragment. The presence or absence of a superfragment relates to the solvation behaviour of the chemical and is an indication that the chemical may be an outlier for aquatic toxicity compared with the other chemicals in this group. Hence the read-across from or to these two chemicals to or from the other chemicals in the group should be performed with care. In addition, differing results have been identified by the skin and eye irritation profiler and the same caution should be exerted. It could be noted though that both aliphatic amines alerts and alkylalkanolamines alerts are indicators of potential irritation.[29,30]

For dimethylamine, only the Verhaar classification is different. By analysing the classification tree, it becomes apparent that the classification differs because the octanol–water partition coefficient (log P) of the compound is less than 0. No specific acting mechanism has been identified by this scheme. Compounds with a log P <0 are not classified by the Verhaar scheme due to the unrealistically high effect concentrations that will be predicted using narcosis type QSARs.[22] Nevertheless, it is considered unlikely that they would exhibit acute toxic action towards biota in aqueous environments because of this. This is confirmed by the fact that classifications consistent with other secondary amines have been proposed by the other profilers relevant for aquatic toxicity, *i.e.* ECOSAR classification, OASIS Acute Toxicity MOA and Protein binding.

As indicated above, the verification of the consistency of available experimental data among the members of the category can be performed manually. Furthermore, the resident databases of the Toolbox have been incorporated into the Toolbox as they have been donated. The OECD does not formally provide any quality assurance of data within the Toolbox and therefore users have to decide whether the available data complies with their own quality standards. Nevertheless, the Toolbox allows users to quickly scan the available results and verify their consistency.

For example, for the *ad hoc* category of secondary amines, Ames mutagenicity results are available for nine out of ten chemicals. Using the 'read-across' function to fill the data gap for the tenth chemical (*i.e.* 2-propen-1-amine, N-ethyl-2-methyl-), the results are displayed graphically as shown in Figure 16.2. All the available results are negative (-1 in Figure 16.2), including for 2-methylaminoethanol and morpholine, confirming the consistency of the category regarding Ames mutagenicity.

The same verification can be performed for any other endpoint for the full category or a sub-category.

Table 16.6 Profiling results for secondary amines.

Chemical name and structure	Profiler	Result
Dimethylamine	BfR rulebase for eye irr./ corr.	Aliphatic amines
	BfR rulebase for skin irr./ corr.	No alert identified
	Benigni/Bossa rulebase	No structural alert identified
	DNA binding	No binding
	Cramer rules	High (Class III)
	ECOSAR classification	Aliphatic amines
	OASIS Acute Toxicity MOA	Narcotic amines
	Protein binding	No binding
	Superfragment profiling	No superfragment
	Verhaar scheme	Class 5 (not possible to classify according to these rules)
Diethylamine	BfR rulebase for eye irr./ corr.	Aliphatic amines
	BfR rulebase for skin irr./ corr.	No alert identified
	Benigni/Bossa rulebase	No structural alert identified
	DNA binding	No binding
	Cramer rules	High (Class III)
	ECOSAR classification	Aliphatic amines
	OASIS Acute Toxicity MOA	Narcotic amines
	Protein binding	No binding
	Superfragment profiling	No superfragment
	Verhaar scheme	Class 1 (narcosis or baseline toxicity)
2-Methylamino-ethanol	BfR rulebase for eye irr./ corr.	Group CN (C, H, O, N)
	BfR rulebase for skin irr./ corr.	Alkylalkanolamines
	Benigni/Bossa rulebase	No structural alert identified
	DNA binding	No binding
	Cramer rules	High (Class III)
	ECOSAR classification	Aliphatic amines
	OASIS Acute Toxicity MOA	Narcotic amines
	Protein binding	No binding
	Superfragment profiling	Has superfragment, C(O)CNR[*]
	Verhaar scheme	Class 5 (not possible to classify according to these rules)
	BfR rulebase for eye irr./ corr.	Group CN (C, H, O, N)
	BfR rulebase for skin irr./ corr.	Group CN (C, H, O, N)

Table 16.6   (*continued*)

Chemical name and structure	Profiler	Result
Morpholine	Benigni/Bossa rulebase	No structural alert identified
	DNA binding	No binding
	Cramer rules	High (Class III)
	ECOSAR classification	Aliphatic amines
	OASIS Acute Toxicity MOA	Narcotic amines
	Protein binding	No binding
	Superfragment profiling	Has superfragment, C1CNCCO1
	Verhaar scheme	Class 5 (not possible to classify according to these rules)
Di-n-propylamine	BfR rulebase for eye irr./ corr.	Aliphatic amines
	BfR rulebase for skin irr./ corr.	No alert identified
	Benigni/Bossa rulebase	No structural alert identified
	DNA binding	No binding
	Cramer rules	High (Class III)
	ECOSAR classification	Aliphatic amines
	OASIS Acute Toxicity MOA	Narcotic amines
	Protein binding	No binding
	Superfragment profiling	No superfragment
	Verhaar scheme	Class 1 (narcosis or baseline toxicity)
2-Propanamine, N-(1-methylethyl)-	BfR rulebase for eye irr./ corr.	Aliphatic amines
	BfR rulebase for skin irr./ corr.	No alert identified
	Benigni/Bossa rulebase	No structural alert identified
	DNA binding	No binding
	Cramer rules	High (Class III)
	ECOSAR classification	Aliphatic amines
	OASIS Acute Toxicity MOA	Narcotic amines
	Protein binding	No binding
	Superfragment profiling	No superfragment
	Verhaar scheme	Class 1 (narcosis or baseline toxicity)
1-Propanamine, 2-methyl-N-(2-methylpropyl)-	BfR rulebase for eye irr./ corr.	Aliphatic amines
	BfR rulebase for skin irr./ corr.	No alert identified
	Benigni/Bossa rulebase	No structural alert identified
	DNA binding	No binding
	Cramer rules	High (Class III)
	ECOSAR classification	Aliphatic amines

Table 16.6 (*continued*)

Chemical name and structure	Profiler	Result
	OASIS Acute Toxicity MOA	Narcotic amines
	Protein binding	No binding
	Superfragment profiling	No superfragment
	Verhaar scheme	Class 1 (narcosis or baseline toxicity)
1-Butanamine, N-butyl-	BfR rulebase for eye irr./ corr.	Aliphatic amines
	BfR rulebase for skin irr./ corr.	No alert identified
	Benigni/Bossa rulebase	No structural alert identified
	DNA binding	No binding
	Cramer rules	High (Class III)
	ECOSAR classification	Aliphatic amines
	OASIS Acute Toxicity MOA	Narcotic amines
	Protein binding	No binding
	Superfragment profiling	No superfragment
	Verhaar scheme	Class 1 (narcosis or baseline toxicity)
2-Propen-1-amine, N-ethyl-2-methyl-	BfR rulebase for eye irr./ corr.	Aliphatic amines
	BfR rulebase for skin irr./ corr.	No alert identified
	Benigni/Bossa rulebase	No structural alert identified
	DNA binding	No binding
	Cramer rules	High (Class III)
	ECOSAR classification	Aliphatic amines
	OASIS Acute Toxicity MOA	Narcotic amines
	Protein binding	No binding
	Superfragment profiling	No superfragment
	Verhaar scheme	Class 1 (narcosis or baseline toxicity)
1-Pentanamine, N-pentyl-	BfR rulebase for eye irr./ corr.	Aliphatic amines
	BfR rulebase for skin irr./ corr.	No alert identified
	Benigni/Bossa rulebase	No structural alert identified
	DNA binding	No binding
	Cramer rules	High (Class III)
	ECOSAR classification	Aliphatic amines
	OASIS Acute Toxicity MOA	Narcotic amines
	Protein binding	No binding
	Superfragment profiling	No superfragment
	Verhaar scheme	Class 1 (narcosis or baseline toxicity)

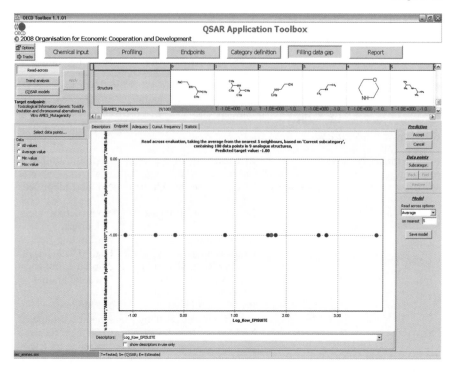

Figure 16.2   Available Ames-mutagenicity results for the ad hoc category of secondary amines.

16.4   Conclusions and Outlook

The OECD (Q)SAR Application Toolbox was developed with the aim of making QSAR methodologies readily accessible and to improve their regulatory acceptance. The focus of the Toolbox is on building toxicologically meaningful categories. Its strength lies in its ability to systematically identify chemicals which have identical structural or mechanistic characteristics so that:

- they can be grouped into chemical categories; and
- the existing experimental results for some of the members of the category can be used to fill data gaps for other members of the category.

While these features work very well for some endpoints in the first versions of the Toolbox, further developments of the Toolbox will focus on expanding this concept to other regulatory endpoints. This will be most challenging for complex endpoints such as repeated dose toxicity, developmental toxicity or reproductive toxicity. The concept of adverse outcome pathways as described in Chapter 14 is expected to play a determining role in developing grouping methods for these endpoints.

References

1. Organisation for Economic Co-operation and Development, *Application of Structure–Activity Relationships to the Estimation of Properties Important in Exposure Assessment*, OECD, Paris, 1993, OECD Environment Monographs No. 67, OECD/GD(93)125, www.oecd.org/document/2/0,3343,en_2649_34379_42926338_1_1_1_1,00.html [accessed March 2010]

2. Organisation for Economic Co-operation and Development, *Structure–Activity Relationships for Biodegradation*, OECD, Paris, 1993, OECD Environment Monographs No. 68, OECD/GD(93)126, www.oecd.org/document/2/0,3343,en_2649_34379_42926338_1_1_1_1,00.html [accessed March 2010].

3. Organisation for Economic Co-operation and Development, *Report of the OECD Workshop on Quantitative Structure Activity Relationships (QSARs in Aquatic Effects Assessment*, OECD, Paris, 1992, OECD Environment Monographs No. 58, OECD/GD(92)168, www.oecd.org/document/2/0,3343,en_2649_34379_42926338_1_1_1_1,00.html [accessed March 2010].

4. Organisation for Economic Co-operation and Development, *US EPA/EC Joint Project on the Evaluation of (Quantitative) Structure Activity Relationships*, OECD Environment Monographs No. 88, OECD, Paris, 1994, OECD/GD(94)28, www.oecd.org/document/2/0,3343,en_2649_34379_42926338_1_1_1_1,00.html [accessed March 2010].

5. Toxic Substances Control Act 1976. Section 5: Pre-manufacture notification for new chemical substances.

6. Council Directive 92/32/EEC of 30 April 1992 amending for the seventh time Directive 67/548/EEC on the approximation of the laws, regulations and administrative provisions relating to the classification, packaging and labelling of dangerous substances, *Off. J. Eur. Commun.*, 1992, **L154**, 1–29.

7. Council Regulation (EEC) No 793/93 of 23 March 1993 on the evaluation and control of the risks of existing substances, *Off. J. Eur. Commun.*, 1993, **L084**, 1–15.

8. US Environmental Protection Agency, *High Production Volume Chemicals Challenge Program*, www.epa.gov/hpv/ [accessed March 2010].

9. International Council of Chemical Associations, *ICCA HPV Tracking System*, www.iccahpv.com/hpvchallenge/HPVChallenge.cfm [accessed March 2010]

10. Canadian Environmental Protection Act 1999. Statutes of Canada 1999. Chapter 33. Act Assented to 14 September 1999.

11. Regulation (EC) No 1907/2006 of the European Parliament and of the Council of 18 December 2006 concerning the Registration, Evaluation, Authorisation and Restriction of Chemicals (REACH), *Off. J. Eur. Union*, 2007, **L136**, 3–128.

12. Organisation for Economic Co-operation and Development, *Guidance Document on the Validation of (Quantitative) Structure–Activity Relationships*

[(Q)SAR] Models, OECD, Paris, 2007, OECD Environment Health and Safety Series on Testing and Assessment No. 69, ENV/JM/MONO(2007)2, www.olis.oecd.org/olis/2007doc.nsf/LinkTo/NT00000D92/$FILE/ JT03224782.PDF [accessed March 2010].

13. Organisation for Economic Co-operation and Development, *Report on the Regulatory Uses and Applications in OECD Member Countries of (Q)SAR Models in the Assessment of New and Existing Chemicals*, OECD, Paris, 2006, OECD Environment Health and Safety Series on Testing and Assessment No. 58, ENV/JM/MONO(2006)25, www.oecd.org/document/2/ 0,3343,en_2649_34379_42926338_1_1_1_1,00.html[accessed March 2010]

14. Organisation for Economic Co-operation and Development, *Guidance on Grouping of Chemicals*, OECD, Paris, 2007, OECD Environmental Health and Safety Publications Series on Testing and Assessment No. 80, ENV/JM/ MONO(2007)28, www.olis.oecd.org/olis/2007doc.nsf/LinkTo/NT0000426A/ $FILE/JT03232745.PDF [accessed March 2010]

15. Organisation for Economic Co-operation and Development, *Manual for Investigation of HPV Chemicals*, OECD, Paris, 2009, www.oecd.org/ document/7/0,3343,en_2649_34379_1947463_1_1_1_1,00.html [accessed March 2010].

16. Organisation for Economic Co-operation and Development, *HPV Database*, http://www.oecd.org/linklist/0,3435,en_2649_34365_2734144_1_1_1_1,00. html#41477034 [accessed June 2010].

17. US Environmental Protection Agency, *Development of Chemical Categories in the HPV Challenge Program*, www.epa.gov/chemrtk/pubs/general/categuid.htm [accessed March 2010].

18. US Environmental Protection Agency, *Status and Future Directions of the High Production Volume Challenge Program*, US EPA, Washington DC, 2004, EPA 743-R-04-001.

19. European Chemicals Agency, *Guidance on Information Requirements and Chemical Safety Assessment*, EChA, Helsinki, 2010, http://guidance.echa.europa.eu/guidance_en.htm#GD_METH [accessed March 2010].

20. Organisation for Economic Co-operation and Development, *OECD Quantitative Structure–Activity Relationships [(Q)SARs] Project*, www. oecd.org/env/existingchemicals/qsar [accessed March 2010].

21. S. D. Dimitrov, L. K. Low, G. Y. Patlewicz, P. S. Kern, G. D. Dimitrova, M. H. Comber, R. D. Phillips, J. Niemela, P. T. Bailey and O. G. Mekenyan, *Int. J. Toxicol.*, 2005, **24**, 189.

22. H. J. M. Verhaar, C. J. van Leeuwen and J. L. M. Hermens, *Chemosphere*, 1992, **25**, 471.

23. C. L. Russom, S. P. Bradbury, S. J. Broderius, R. A. Drummond and D. E. Hammermeister, *Environ. Toxicol. Chem.*, 1997, **16**, 948.

24. US Environmental Protection Agency, *Ecological Structure Activity Relationships, ECOSAR*, v. 1.00a, January 2009, www.epa.gov/oppt/newchems/ tools/21ecosar.htm [accessed March 2010]

25. O. Mekenyan, S. Dimitrov, R. Serafimova, E. Thompson, S. Kotov, N. Dimitrova and J. Walker, *Chem. Res.Toxicol.*, 2004, **17**, 753.
26. R. Serafimova, M. Todorov, S. Kotov, E. Jacob, N. Aptula and O. Mekenyan, *Chem. Res. Toxicol.*, 2007, **20**, 662.
27. European Commission Joint Research Centre, Toxtree version 1.60, http://ecb.jrc.ec.europa.eu/qsar/qsar-tools/index.php?c = TOXTREE [accessed March 2010].
28. R. Benigni, C. Bossa, N. Jeliazkova, T. Netzeva and A. Worth, *The Benigni/Bossa Rulebase for Mutagenicity and Carcinogenicity – A Module of Toxtree*, Office for Official Publications of the European Communities, Luxembourg, 2008, JRC Scientific and Technical Reports, EUR 23241 EN, 2008, http://ecb.jrc.ec.europa.eu/documents/QSAR/EUR_23241_EN.pdf [accessed March 2010]
29. I. Gerner, M. Liebsch and H. Spielmann, *Altern. Lab. Anim.*, 2005, **33**, 215.
30. J. D. Walker, I. Gerner, E. Hulzebos and K. Schlegel, *QSAR Comb. Sci.*, 2005, **24**, 378.
31. J. Tunkel, P. H. Howard, R. S. Boethling, W. Stiteler and H. Loonen, *Environ. Toxicol. Chem.*, 2000, **19**, 2478.
32. G. M. Cramer, R. A. Ford and R. L. Hall, *Food Cosmet. Toxicol.*, 1978, **16**, 255.
33. T. W. Schultz, R. E. Carlson, M. T. D. Cronin, J. L. M. Hermens, R. Johnson, P. J. O'Brien, D. W. Roberts, A. Siraki, K. D. Wallace and G. D. Veith, *SAR QSAR Environ. Res.*, 2006, **17**, 413.
34. M. T. D. Cronin and A. P. Worth, *QSAR Comb. Sci.*, 2008, **27**, 91.
35. W. Tong, W. D. Fang, H. Hong, Q. Xie, R. Perkins and D. M. Sheehan, in *Predicting Chemical Toxicity and Fate*, ed., M.T.D. Cronin and D.J. Livingstone, CRC Press, Boca Raton, FL, 2004, p. 285–314.
36. P. K. Schmieder, G. Ankley, O. Mekenyan, J. D. Walker and S. Bradbury, *Environ. Toxicol. Chem.*, 2003, **22**, 1844.
37. R. Serafimova, M. Todorov, D. Nedelcheva, T. Pavlov, Y. Akahori, M. Nakai and O. Mekenyan, *SAR QSAR Environ. Res.*, 2007, **18**, 1.
38. Organisation for Economic Co-operation and Development, *Guidance Document for Using the (Q)SAR Application Toolbox to Develop Chemical Categories According to the OECD Guidance on Grouping of Chemicals*, OECD, Paris, 2009, OECD Environment Health and Safety Publications Series on Testing and Assessment No. 102, ENV/JM/MONO(2009)5, www.olis.oecd.org/olis/2009doc.nsf/LinkTo/NT00000CAA/$FILE/JT032-60091.PDF [accessed March 2010]
39. KOWWIN™ v. 1.67a, 2008, available as part of EPI Suite™ version 4.0, 2009, www.epa.gov/oppt/exposure/pubs/episuite.htm [accessed March 2010].
40. T. W. Schultz, *Toxicol. Meth.*, 1997, **7**, 289.

CHAPTER 17
Open Source Tools for Read-Across and Category Formation

N. JELIAZKOVA,[a] J. JAWORSKA[b] AND A. P. WORTH[c]

[a] Institute for Parallel Processing, Bulgarian Academy of Science, Sofia, Bulgaria; [b] Procter & Gamble, Modeling & Simulation, Biological Systems, Brussels Innovation Center, Belgium; [c] European Commission Joint Research Centre, Institute for Health and Consumer Protection, Ispra 21027 (VA), Italy

17.1 Introduction

17.1.1 Current Regulatory Guidance for Category Formation and Read-Across

The purpose of grouping chemicals into categories is to be able to identify the group's common properties and to extend the use of existing measured data to untested chemicals that are category members or analogues. Apart from waiving the need for a complete test, knowledge of the properties of compounds in a category may facilitate the decision on the nature and scope of the tests to be performed.

Regulatory guidance documents published by the European Chemicals Agency (EChA)[1] and the Organisation for Economic Co-operation and Development (OECD)[2] distinguish between the category approach and the analogue approach as techniques for grouping chemicals, while the term 'read-across' is used for assigning (predicted) endpoint values to chemical compounds where such information is missing (filling data gaps).

Issues in Toxicology No.7
In Silico Toxicology: Principles and Applications
Edited by Mark T. D. Cronin and Judith C. Madden
© The Royal Society of Chemistry 2010
Published by the Royal Society of Chemistry, www.rsc.org

A chemical category is defined as a group of chemicals whose physico-chemical and human health and/or environmental toxicological properties and/or environmental fate properties are likely to be similar, or follow a regular pattern as a result of structural similarity (or other similarity characteristic). Categories of chemicals are built based on the hypothesis that the properties of a series of chemicals in a category will show coherent trends in their physico-chemical properties, and more importantly, in their toxicological (human health/ecotoxicity) effects or environmental fate properties. It is anticipated that, if one or more consistent trends can be identified, it can be assumed that the category hypothesis is valid and that the effects are due to a common underlying mechanism of action. A prerequisite for category robustness is the inclusion of a sufficient number of chemicals to detect trends across endpoints and to develop hypotheses for specific endpoints. If the number of chemicals being grouped is very limited, the approach is known as the 'analogue approach'. In this case, trends in the properties are not apparent, but read-across may still be applied—mostly on the basis of expert judgement, and with some supporting information to substantiate the read-across.

Once the chemicals have been grouped and analogues identified on the basis of a given hypothesis, one needs to link the category members with existing data. Clearly not all analogue candidates or members of the category will have experimental data and only the candidates with reliable data will be useful. The EChA guidance recommends that missing data should be filled by read-across, trend analysis or quantitative structure–activity relationships (QSARs).

The term 'read-across' is used to refer to the practice of using endpoint information from one chemical to predict the same endpoint for one or more chemicals selected during the analogue or category identification process. Read-across can be used when applying either the category or analogue approach; the distinction between the two being rather artificial depending on the number of source chemicals used. A source chemical (or chemicals) is used to make an estimate, while target chemical(s) are those for which the endpoint is being estimated. Read-across is considered more robust when used within the category approach, mainly because of the larger amount of data being analysed. At this stage a rationale for read-across needs to be tested and verified, and it is heavily reliant on expert knowledge.

A rigorous computer implementation of the expert knowledge involved is still a challenge and human decisions are necessary for crucial steps. One of the possible reasons is that the basic hypothesis of similarity assessment—that similar compounds are likely to have similar biological activities—has known exceptions (the 'similarity paradox'), which can usually only be resolved on a case-by-case basis. Recognising the fact that similarity is not an absolute, but a relative concept, provides a different view of chemical similarity assessment. Objects cannot be similar in absolute terms, but only with respect to some property of the object. In the same way, chemicals can only be similar with respect to some measurable key feature. Given the 'similarity paradox', it is unlikely that a single similarity measure exists which will be universally appropriate; instead similarity is best related to a given endpoint or mechanism of action. Identifying and using the pertinent characteristics of a chemical as a basis for decisions on the similarity of the toxicological effects and fate effects of chemicals offers a systematic approach towards a fully computerised chemical grouping.

The category approach and read-across between chemical analogues have been extensively used in the OECD High Production Volume (HPV) and European Union chemical management programmes.[2] While there is a common understanding that the development of QSAR models should comply with established statistical procedures, read-across and category formation are often used on an *ad hoc* basis and by applying expert judgement.[3] The REACH guidance attempts to address this weakness, for example, by listing available commercial and open source computational tools and by describing a stepwise approach towards developing and documenting chemical grouping.[4] Rosenkranz and Cunningham present a controversial view regarding the categorisation of HPV chemicals[5] by arguing that '*traditional organic chemical categories do not encompass groups of chemicals that are predominantly either toxic or non-toxic across a number of toxicological endpoints or even for specific toxic activities*'. In recognition of this issue, the EChA guidance foresees the development of 'subcategories' in case a general category is not adequate for all endpoints. The guidance includes a number of exceptions and examples of complex cases, *e.g.* when a chemical can be identified as member of more than one category and therefore manifest effects typical of any of those categories. The guidance introduces concepts such as 'breakpoint' chemicals as the chemicals where the trend within a category changes, *etc.* This requires a supporting mechanistic hypothesis.

17.1.2 Needs and Challenges for Computational Approaches in Category Formation and Read-Across

Given the large number of chemicals with data gaps in their toxicological profiles, reliance on expert judgements to identify suitable chemicals for read-across is likely to be slow, inconsistent and non-transparent. Computer tools can play a useful role in overcoming these shortcomings. Building on the guidance mentioned above, a number of computer functionalities facilitating the category approach can be identified. In particular, computer-based approaches can:

- provide access to diverse data on chemicals including chemical structures, physico-chemical properties, human health endpoints, and environmental toxicity and fate properties;
- query datasets to identify chemicals with similar structures or/and other characteristics of similarity;
- help in formulating a hypothesis for chemical grouping;
- provide an interface to examine trends and regular patterns of properties, either globally, or for specific endpoints;
- execute the search strategy;

- provide information that helps to 'validate' the category based on the mechanism of action;
- help in defining category boundaries to make a decision on whether or not a chemical can be considered a category member;
- estimate the presumed toxicological or fate effect(s) expected for a category (presence and absence or certain hazards, as well as a trend across the category); and
- define a procedure for assigning the presumed toxicological or fate effect(s) to an untested chemical(s), given a category or a list of analogues.

The critical issue facing computational tools is the need to formalise the category hypothesis and to define the category boundaries. Such formalisation will allow for simplification of the verification process. The concept of the applicability domain of a category, defined as 'structural requirements and ranges of physico-chemical, environmental fate, toxicological or ecotoxicological properties within which reliable estimations can be made for the (sub)category members' needs to be refined in greater detail (see Chapter 12). For example, there may be a trend of increasing acute aquatic toxicity with increasing chain length from C2 up to a carbon chain length of C12, after which no aquatic toxicity is seen because the water solubility has decreased with increasing chain length. Thus, the applicability domain for aquatic toxicity would be C2 to C12.

The decision whether or not a given read-across assessment is appropriate is based on expert knowledge. This presents a significant challenge for the rigorous and routine implementation of read-across in a computational tool. The required functionalities for a computational tool with read-across support can be divided into two groups—the first focused on verification of the hypothesis and the second concerned with the actual derivation of the predicted toxicological effect:

1. **Verification of the hypothesis**.
   a. Calculate various similarity measures between chemical compounds, both structural and based on other characteristics; provide means to generate descriptors used in similarity calculations.
   b. Identify common structural fragments between chemical compounds and take into account their interaction.
   c. Provide an underlying rationale using a specific similarity measure for assigning the predicted endpoint value to untested chemicals (*e.g.* common functional group, common precursor or breakdown product, reactivity, common mechanism of action).
   d. Provide means to embed and reuse expert knowledge for similarity assessment
2. **Estimation of the toxicological effect**. The exact procedure depends on the number of source chemicals (used to make an estimate) and target chemicals (the chemicals for which endpoint estimates are being made).
   a. For a single source chemical, the only decision that can be taken is to assume the toxicological effect of the target chemical is the same as that of the source chemical.

b. With more than one source chemical, the endpoint value of the target chemical can be obtained by averaging, taking the value of the most representative chemical, or predicted *via* a QSAR model built from source chemicals (internal QSAR model). Trend analysis is essentially the same as building a model from the source chemicals.

c. Prediction by an external QSAR model (a model built with chemicals different to the source chemicals) is also an option, though the applicability of such a model in each particular case should be justified.

Finally, the reporting of the category or analogue approach is a necessary feature of the computational tools that aim to facilitate these tasks. The guidance proposes reporting in a matrix, consisting of the category members and corresponding set of properties and category endpoints.

While the specified functionalities in computational tools are becoming increasingly available, the ultimate utility of these tools depends heavily on the data used for formulating category and read-across proposals. It is expected that more biological data generated by systems biology approaches will be better suited to formulate mechanistic hypotheses than by currently available data and that could be further formalised. Before this can happen there are challenges that can be considered short-term, often related to improved accessibility and management of information. Namely they are the diversity of data sources, providing information on chemical structure, diversity in terms of distributed and disparate resources, data quality and incompatible representations that are often ill-suited for a computerised analysis.

This chapter discusses how these challenges can be addressed starting with the principles of open science, open source software solutions, and why open data standards are of crucial importance.

17.2 Open Source, Open Data and Open Standards in Chemoinformatics

17.2.1 Open Source

The term 'open source' is used to denote software whose source code is published and made publicly available, and which grants the rights to copy, modify and redistribute the source code without fees or royalties. The Open Source Initiative (OSI) provides a list of ten criteria that software must comply with in order to be OSI certified.[6]

Open source software provides many opportunities, of which just not having to pay for it is the weakest argument. Naturally, it facilitates the work of developers who need the option to change and adapt the source code of a module they use in order to integrate it into their projects. Different stakeholders (*e.g.* regulatory authorities and industry) benefit from the transparent and unified access to algorithms, models and data. Open source also lowers the user barrier, facilitates dissemination activities and enables the reproducibility of models and results in scientific research. Open source projects are known to solve problems with faulty software in a short time since bugs are much more easily found and improvements are much more easily made if everyone can examine and work with the source code. Thus peer-reviewed code and community driven support facilitates continued maintenance and a high level of reliability, but has the drawback that the fixes and updates may not always align with time-critical business requirements. These issues are usually addressed by various established and emerging business models, providing commercial support for open source software.

The terms 'community open source' and 'commercial open source' are used to denote the two extremes of open source projects run by a community of volunteers and by profit-making entities, respectively. Nowadays open source software is widely used in both academic and commercial environments, ranging from simple standalone and web applications to full-featured open source solutions relying on open source for all the components from operating system to database management, user interface, data processing and network connectivity. The economic motivation behind open source, and its impact on innovation and the IT market have recently drawn much attention and the effects on system integrators, system vendors, individual employees, customers and society have been extensively discussed.[7–11] Notably, the O'Reilly Open Source Convention (OSCON) motto at its annual conference for the discussion of free and open source software OSCON 2009 was 'Open for Business'.[12]

17.2.1.1 Open Source in Chemoinformatics

Traditionally, chemoinformatics software has been developed and distributed on a closed source and commercial basis; however, internet and modern communication tools are rapidly changing this trend. The timeline for the evolution of chemoinformatics toolkits since 1995 includes the introduction of open source chemoinformatics packages such as:

- The Chemistry Development Kit (2000);[13–15]
- OpenBabel[16] (2001);
- Joelib (2002, although no longer supported); and
- Pybel (Python wrapper for OpenBabel—2005).[17]

Rational Discovery became an open source RDKit in 2006,[18] and most recently Cinfony,[19] offering a common application programming interface (API) to several chemoinformatics toolkits.

The Chemistry Development Kit (CDK) is an open source and open development chemoinformatics library written in Java. CDK is now maintained and further enhanced by more than 50 developers from academic and industrial institutions all over the world. It is used in more than ten different academic and industrial projects worldwide. The library provides methods for many common tasks in molecular informatics, including two-dimensional (2-D) and three-dimensional (3-D) rendering of chemical structures, I/O routines, SMILES parsing and generation, ring searches, isomorphism checking, structure diagram generation, 3-D builder, QSAR module, *etc.*[20] CDK is distributed under the LGPL license (see below for an explanation). A number of well-known high-quality software products are known to rely on and/or be interoperable with CDK (*e.g.* Bioclipse, JChemPaint, Jmol, JOELib, Jumbo, NMRShiftDB, Nomen, SafeBase™, SENECA, QueryConstructor, Evince, KNIME, Toxtree, Toxmatch, AMBIT and others). Most of these programs are entirely open source, although Evince is a good example of commercial software that uses CDK internally.

These chemoinformatics packages offer comparable functionalities and are generally used by end users for data transformation and analysis, or by developers to build end user applications on top of them. For instance:

- JChemPaint is widely used for 2-D chemical diagram editing;
- Jmol provides 3-D chemical animation/viewing;
- Jumbo enables the conversion of full chemical documents (in any format) into Chemical Markup Language (CML);[21] and
- Nomen provides IUPAC name parsing.

Prominent examples are Bioclipse,[22] Taverna workbench and KNIME workflow.[23]

Bioclipse, a free, open source workbench for the life sciences provides advanced functionality mainly in chemoinformatics, with bioinformatics planned for the near future. Some major components include a brand new chemical editor for SWT (JChemPaint), interactive 3-D visualisation of molecules (Jmol), a Molecules Table capable of reading large files, and a powerful backbone in chemoinformatics provided by the Chemistry Development Kit (CDK) library.

The Taverna workbench is a free software tool for designing and executing scientific workflows, created by the myGrid project, and funded through OMII-UK. Taverna allows users to integrate many different software tools including web services from many different domains such as chemistry, music and social sciences. Bioinformatics services include those provided by the National Centre for Biotechnology Information, The European Bioinformatics Institute, the DNA Databank of Japan (DDBJ), SoapLab, BioMOBY and EMBOSS.

Open Babel[17] is a chemical toolbox, offering conversion between over 90 chemical file formats. In addition, it supports SMARTS searching, descriptor calculations, fingerprints and conformer generation. It is an open, collaborative project allowing anyone to search, convert, analyse or store data from molecular modelling, chemistry, solid-state materials, biochemistry or related areas. Open Babel is distributed under the LGPL license. RDKit is a software suite for chemoinformatics,[18] computational chemistry and predictive modelling, offering similar functionalities and distributed under BSD license. It is written in C + + and Python.

Another interesting development is OSCAR3 (Open Source Chemistry Analysis Routines),[24] a software for the semantic annotation of chemistry papers developed by the Unilever Cambridge Centre for Molecular Informatics. The modules OPSIN (a name to structure converter) and ChemTok (a tokeniser for chemical text) are also available as standalone libraries. Crystal structure coordinates are automatically retrieved from the web, aggregated and made available by CrystalEye.[25]

17.2.1.2 Open Source Licences

Open source software tools are distributed under various types of open source licence. How to choose the right one from different types of licence has always been an important issue for free/open-source software developers and authors. An inappropriate licence could lead to an unintended result such as restraining the development of the software and/or the data collection, or even giving rise to legal dispute (*e.g.* copyright or patent cases). Therefore it is recommended that developers, authors and users consult the most recent information on the current open source and open data licences in order to make the most appropriate choice.

The most popular licences among 72 possible licence types are the GNU General Public Licence or GPL (55%) and the GNU Lesser General Public Licence or LGPL (10%). Copyleft is the term created in the GNU General Public Licence to allow the freedom of copying, redistributing and modifying the free software in the copyrights world. The licences using copyleft method are called 'copyleft licences'. Designed to be 'opposite' to copyright, copyleft licences tend to make the program free by '*requiring all the modified and extended versions of the program to be free as well*'.[26] In other words, the software and all derivative works must be made free.[27] This efficiently prevents the unwanted transformation of free software into proprietary software, which is very encouraging for non-commercial software developers. However, in recent years, copyleft licenses (*e.g.* GNU GPLv2) have been surrounded by debates due to allegations of their so-called 'infectious' or perpetuation characteristic. The perpetuation characteristic of copyleft licences is problematic to developers when the code needs to be linked from other (proprietary) code, which makes the other code also copyleft. The choice among copyleft, non-copyleft or semi-copyleft licences depends on the purpose of the software development.

Typically for most developers, the copyleft licences are adopted for standalone programs or software that stay forever free (including the derivative works). Non-copyleft licences are chosen for programs which are likely to be linked with other (proprietary) programs, or developed in the future as proprietary software.

The GNU Lesser General Public Licence (formerly the GNU Library General Public Licence) or LGPL is an open source software licence published by the Free Software Foundation (FSF). It was designed as an alternative to the strong-copyleft GPL. The LGPL places copyleft restrictions on the program itself, but does not apply these restrictions to other software that merely links with the program.

17.2.2 Open Standards

An open standard is a specification (a set of rules or formats) which is agreed and accepted by a community. An open standard is characterised by the following:

- it is accompanied by explicit documentation, enabling independent implementation that results in interchangeable and interoperable software;
- it is openly accessible and freely implemented *via* either open source or proprietary code; and
- its development is based on a consensus between potential vendors and users.

The internet provides a unique example of what society can achieve by adopting common standards. Internet Engineering Task Force (IETF) working groups have the responsibility for developing and reviewing specifications intended as Internet Standards.[28] The standardisation process starts by publishing a Request for Comments (RFC)—a discourse prepared by engineers and computer scientists for peer review or to convey new concepts or information. IETF accepts some RFCs as Internet Standards *via* its three-step standardisation process. If an RFC is labelled as a Proposed Standard, it needs to be implemented by at least three independent implementations, further reviewed, and after correction, becomes a Draft Standard. With a sufficient level of technical maturity, a Draft Standard can become an Internet Standard. Organisations such as the World Wide Web consortium and OASIS support collaborations of open standards for software interoperability.[29,30] While recently some authors have argued that the standardisation process is less than ideal and does not always endorse the best technical solutions,[30] the existence of the internet itself, based on compatible hardware and software components and services, is a demonstration of the opportunities offered by collaborative innovation, flexibility, interoperability, cost-effectiveness and freedom of action.

Historically, the chemoinformatics world has been driven by *de facto* standards, developed and proposed by different vendors (*e.g.* SMILES and MOL/SDF are proprietary standards, developed by respective companies). Recently, the development of IUPAC International Chemical Identifier (InChI) is a positive example of adopting an open standardisation procedure,[31] but without a requirement similar to IETF procedure of at least two independent implementations. Chemical Markup Language[21] is an extension of XML, designed to support a wide range of chemical objects such as molecules, reactions, spectra and analytical data, computational chemistry, chemical crystallography and materials, and is being developed by Peter Murray-Rust and Henry Rzepa at the Unilever Centre for Molecular Informatics, Cambridge, UK. Blue Obelisk[32,33] is an informal organisation founded in 2002 by a group of chemists and software developers in order to encourage open source, open standards and open data approaches in chemoinformatics. Blue Obelisk has initiated an open specification for SMILES,[34] provides a chemoinformatics algorithm dictionary,[35] a descriptors dictionary[36] and a physico-chemical properties repository; it emphasises the agreement of common terminology, data and metadata as crucial prerequisites for interoperability.

17.2.3 Open Content, Open Data and Open Collaboration

'Open data' represents the philosophy and practice requiring that certain data are freely available to everyone without restrictions from copyright, patents or other mechanisms of control. The Open Knowledge Definition provides a list of licences that are conformant with its principles.[37] Peter Murray-Rust in his recent publication in *Nature*, 'Chemistry for everyone',[38] commented that *'Unlike astronomers, geoscientists and biologists, chemists have no global data-collection projects; their data are usually published in many different online journals and then collated by hand into CAS'*. Currently, this trend is changing slightly due to the availability of public repositories such as PubChem, emolecules, *etc.*, but the lack of common protocols for data access complicates the online integration of data resources. Social and commercial resistance of openness undoubtedly has its reasons and the right balance has yet to be found.

OpenTox[39] is an European Commission Seventh Framework Programme (FP7) funded project which is making progress toward a solution of technical problems of interoperability by developing a common API for (Q)SAR modelling and data access, based on a recent web service technology known as REST.[40] An important aspect of the openness is its collaborative nature,[41] from collaborative software development to modern social networking, with tools such as wikis, blogs, virtual collaborative environments and a chemical 'blogosphere' being particularly active. Open, integrated semantic systems are perceived as the future of IT tools for chemistry.[40]

Another example of an open collaboration is the Pistoia Alliance, a joint activity of pharmaceutical companies aiming at sharing and defining pen standards for data and technology interfaces in the life science research industry.[42]

There are several recent reviews on the adoption of open source chemoinformatics applications.[40,43,44] This chapter is not an attempt to repeat this work, but rather to highlight the opportunities that the openness offers and to focus on open source applications, specifically developed to address some of the challenges of REACH.

17.3 Descriptions of the Tools Suitable for Category Formation and Read-Across

17.3.1 Toxtree

Toxtree is a flexible and user-friendly open source application that places chemicals into categories and predicts various kinds of toxic effect by applying decision tree approaches. Toxtree was commissioned by the European Commission's Joint Research Centre (JRC) and developed by Ideaconsult Ltd (Sofia, Bulgaria). It is distributed under the GPL licence. Plug-ins are currently being developed by several different organisations. The latest release of Toxtree version 1.60 (July 2009) includes the following plug-ins:[45–49]

- Structural alerts for the *in vivo* micronucleus assay in rodents;[50]
- Cramer rules with extensions: this plug-in is a copy of the original plug-in, plus minor extensions. Like the Cramer plug-in, this plug-in works by assigning compounds to Class I, II or III according to the original Cramer rules and some extra ones. Several compounds were classified by Munro in 1996 as Class I or Class II compounds according to the Cramer rules,[51] even though Munro reported low NOEL values upon oral administration (indicating relatively high toxicity). To overcome such misclassifications, five rules have been introduced to capture the possible toxicity (and thus Class III assignment) of these compounds;
- Structural alerts for the identification of Michael acceptors: this plug-in contains a series of structural alerts able to identify Michael acceptors.[52]

Toxtree can be applied to datasets from various compatible file types. User-defined molecular structures are also supported—they can be entered as SMILES strings or by using the built-in 2-D structure diagram editor. Toxtree was designed with flexible capabilities for future extensions in mind (*e.g.* other classification schemes that could be developed at a future date). New decision trees with arbitrary rules can be built with the help of a graphical user interface or by developing new plug-ins.

17.3.1.1   Case Study using Toxtree

In addition to making predictions for individual chemicals, Toxtree can be used to profile the toxicological hazard or mechanistic group of a set of chemicals. In this example, structural alerts for the identification of Michael acceptors were applied to a skin sensitisation dataset of 210 structures (accessible from Toxmatch, see below). The 19 structural alerts are documented by Schultz *et al.*[52] The identifying structural characteristic is an acetylenic or olefinic moiety attached to a neighbouring electron-withdrawing group. The 210 structures are most easily processed in batch mode (*e.g.* by using a csv file in which the first column consists of the SMILES and the second column the chemical names). One of the structures, streptomycin sulfate, is not processed because its SMILES is too long, so a total of 209 predictions are generated. The results obtained are summarised in Table 17.1.

On the basis of these numbers, various statistics can be calculated such as the sensitivity ($26/54 \times 100 = 48\%$), specificity ($138/155 \times 100 = 89\%$) and positive predictivity ($26/43 \times 100 = 60\%$). These do not appear to be very good performance statistics. However, when the alerts were tested against an independent test set of 27 compounds by Schultz *et al.*,[52] the alerts made correct predictions

Table 17.1 Results obtained using Toxtree to identify Michael acceptors.

	Known to react via Michael addition	*Known not to react via Michael addition*	*Total*
Predicted to be a Michael acceptor	26 (true positives)	17 (false positives)	43
Predicted not to be a Michael acceptor	28 (false negatives)	138 (true negatives)	166
Total	54	155	209

for all chemicals. Taken together, the findings indicate that the structural alerts are necessary, but perhaps not sufficient, for the identification of Michael acceptors. In addition, they may not be described in sufficient detail to allow unambiguous implementation.

17.3.2 Toxmatch

Toxmatch is a user-friendly software application for exploring structural similarity commissioned by the JRC and developed by Ideaconsult Ltd. Structural similarity assessment is currently a common practice in read-across, category formation, and developing and validating (Q)SARs. Toxmatch provides several endpoint specific similarity measures and descriptors are selected using a training set in combination with data mining methods. A computerised analysis of similarity is based on numerical representation of the compound and a measure (similarity index) between these representations for multiple compounds. The current version of Toxmatch (v 1.06; October 2008) implements the similarity indices listed in Table 17.2.

Toxmatch is able to calculate a limited range of descriptors (examples include logarithm of the octanol–water partition coefficient, molecular weight, energy of the highest occupied molecular orbital, *etc.*) and to extract descriptors imported from a file. Two structural similarity approaches are encoded into Toxmatch. These are based on fingerprints and atom environments.

Based on the view that similarity assessment should be related to the endpoint of interest, Toxmatch suggests different approaches based on the extent of available information:

- The knowledge-based approach is recommended where a chemical understanding of a mechanism exists. For this purpose, Toxmatch integrates the available Toxtree rule bases, estimating several different types of toxicological effects (see Section 17.3.1).
- If expert defined rules are not available, the recommended Toxmatch approach is to use similarity assessment by use of molecular descriptors and structural features that reflect the particular mechanism.
- In other cases, the similarity assessment relies again on descriptors and structural similarity, with a preceding selection of the relevant representation performed by feature selection methods.

Table 17.2 Similarity indices implemented in Toxmatch.

Index	Formula
Distance-like similarity indices	
General definition	$D_{AB}(k,x) = [k(Z_{AA} + Z_{BB})/2 - xZ_{AB}]^{1/2}, D_{AB} = [0, \infty)$
Euclidean Distance Index $(k=x=2)^a$	$D_{AB}(k,x) = [Z_{AA} + Z_{BB} - 2Z_{AB}]^{1/2}$
Correlation-like similarity indices	
General definition	$V_{AB}(k,x) = (k-x)Z_{AB}D_{AB}^{-2}(k,x), V_{AB} = [0,1]$
Hodgkin–Richards index	$H_{AB} = 2Z_{AB}[Z_{AA} + Z_{BB}]^{-1}$
Tanimoto index	$T_{AB} = 2Z_{AB}[Z_{AA} + Z_{BB} - Z_{AB}]^{-1}$
Cosine-like similarity index or Carbó index^b	$C_{AB} = Z_{AB}[Z_{AA}Z_{BB}]^{-1/2}$

Note: Where Z is the similarity matrix, A and B are the two molecules being compared.
$^a D_{AB}$ varies in the interval $[0,\infty]$, values close to zero imply a greater similarity between the compared objects. Two compared objects are identical, when $D_{AB} = 0$.
$^b C_{AB}$ varies in the interval $[0,1]$. The nearer the value to unity, the more similar the objects being compared, whilst a value a value approaching zero indicates the two objects are dissimilar. The exact value of one is obtained when the two objects are identical.

Toxmatch allows the user to exploit the similarity information in two ways that are useful for read across:

- to estimate activity of a chemical of interest by consideration of the activities of similar compounds (nearest neighbours); or
- by clustering/grouping similar chemicals together on the basis of similarity values.

The first approach results in a prediction of activity based on the weighted average of the activity values of the *k* nearest neighbours, *i.e.* the activity of the most similar (closest) chemicals are averaged proportionately and used to estimate the activity of the chemical of interest. In this case the most similar chemicals are the source chemicals and the chemical of interest is the target chemical. The process of relating these source chemicals to a target chemical can be regarded as 'many-to-one read across'. The actual set of *k* most similar compounds will depend on the similarity measure. The weights are proportional to the pairwise similarities (*e.g.* the activity value of the most similar compound has the largest weight and *vice versa*). In order to predict the dependent variable (activity), the measured activity values should be available for the training set. Two values are reported for each compound—averaged similarity to the *k* nearest neighbours and predicted activity value.

The second approach is to perform a classification into groups of activity based on similarity values. The read-across is slightly different in that the source chemicals are binned into one group and the similarity measure provides the means to define the likelihood that the target chemical falls into one or another bin. The procedure also relies on *k* nearest neighbours and classifies the target compound into the group where the majority of the *k* most similar compounds belong (see Chapter 10 for more details on this method). For this purpose, activity groups should be available for the training set (*e.g.* potency classes or some other grouping). The values reported are 'probability to belong to a group' (m/k, where m is the number of compounds in the group) and the 'predicted group'.

Toxmatch makes use of a supervised algorithm based on nearest neighbours and selected similarity measures for prediction and classification. Supervised learning methods take into account the toxicity information. For this reason, several example datasets were also included within Toxmatch 1.05; two for human health endpoints and two for environmental endpoints. The aquatic toxicity dataset is a copy of the DSSTox Fathead Minnow Acute Toxicity dataset (April 2006 update) from the US Environmental Protection Agency (US EPA). It contains 617 chemicals with information pertaining to:[53,54]

- organic chemical class assignments (ChemClass_FHM);
- acute toxicity in fathead minnow (LC50_mg);
- dose–response assessments (LC50_Ratio, ExcessToxicityIndex);
- behavioural assessments (FishBehaviourTest);
- joint toxicity MOA evaluations of mixtures (MOA_MixtureTest), and
- additional MOA evaluation of fish acute toxicity syndrome (Fish-AcuteToxSyndrome) in rainbow trout.

The bioconcentration factor (BCF) dataset comprises log BCF values for 610 non-ionic chemicals.[55] The dataset was used to develop the BCFwin model in the US EPA's EPISuite™. The skin sensitisation dataset contains Local Lymph Node Assay (LLNA) data for 209 chemicals.[56] In addition to the numeric LLNA EC3% values, a qualitative classification of potency category and reaction domain is reported for each compound, where potency category is one of non-sensitiser, weak sensitiser, moderate sensitiser, strong sensitiser, extreme sensitiser.[57] Reaction domain is one of Michael acceptors, SNAr electrophiles, SN2 electrophiles, Schiff base formers, acylating agents, non-reactive and special cases.[58] The skin irritation dataset consists of 72 chemicals, labelled according to EU and Global Harmonisation Standards (GHS) classification for skin irritation potential (NI: not irritating; MI: mild irritation; R38: irritating). The data arise from several sources including the US EPA TSCA (Toxic Substances Control Act) inventory and the European Centre for Ecotoxicology and Toxicology of Chemicals (ECETOC) databank.[59]

Toxmatch allows the user to categorise datasets into groups (*e.g.* potency categorisation for skin sensitisation, a set of different ranges for a BCF endpoint value, a mechanism of action, *etc.*). These can subsequently be used to classify new chemicals into groups of potency, endpoint value ranges or mechanism of action, based on similarity values. Examples of the application of Toxmatch in the formation of chemical categories and the application of read-across are given by Enoch *et al.*,[60] in which predictions are made of teratogenicity potential, and in Patlewicz *et al.*,[61] in which predictions are made of

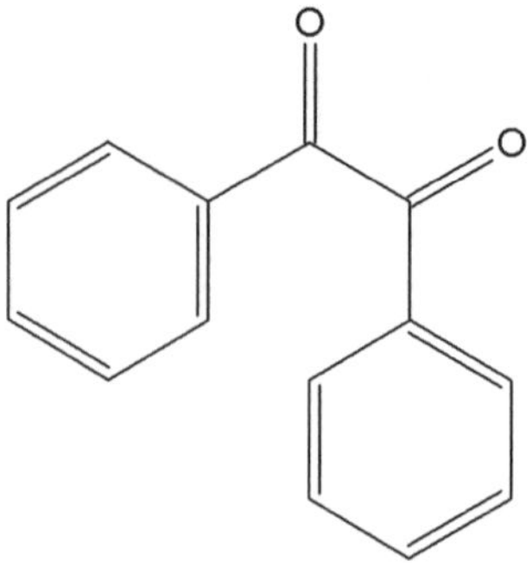

Scheme 17.1

skin sensitisation (mechanistic grouping and potency categorisation) and BCF values.

An example of the application of Toxmatch to a skin sensitisation dataset is given and summarised by Patlewicz *et al.*[61] The first step is to open the pre-loaded LLNA dataset on skin sensitisation with mechanism of action information.[61] This serves as the model training set. Then the chemical of interest, for which a read-across prediction is needed, is inserted (Scheme 17.1). The 2-D structure can be drawn using the available structure-drawing editor, or the SMILES can be entered [c1ccccc1C(=O)C(=O)c1ccccc1]. The structure then appears in the interface as a test set and a similarity assessment can be conducted with respect to the 'training set', *i.e.* the preloaded LLNA dataset.

Using the Tanimoto similarity index applied to nearest neighbour finger-prints, the resulting similarity matrix is a single column depicting all the Tanimoto index values for each of the 209 chemicals in the training set. The chemicals that are most similar to the test (target) chemical can be identified on the basis of their colour and their Tanimoto value. Three analogues were extracted together with their Tanimoto values (Scheme 17.2).

The three analogues are all purported to react *via* a Schiff base mechanism, as based on the original classification. This suggests that the chemical of interest is also likely to behave as a Schiff base former. LLNA data for the three analo-gues varied in terms of their similar sensitising potencies (EC3 values). The most similar compound with a similarity index of 0.866 had an experimental EC3 value of 1.3% (*i.e.* a moderate sensitiser), that with a similarity index of 0.62 had an EC3 of 29% (*i.e.* a weak sensitiser) and that with a similarity index of 0.6 had an EC3 of 3% (*i.e.* a moderate sensitiser). Overall, these results suggest that the target chemical could be a moderate sensitiser. A closer examination of the analogues, taking into account their differing molecular weights, enables a log(1/EC3) or pEC3 (the negative logarithm of the molar EC3) to be computed which shows that all three compounds have potencies in the range 0.8–2.0. Taking an average of all three values leads to an estimated pEC3 of 1.48 and thus a predicted EC3 of 6.9%, *i.e.* a moderate sensitiser.

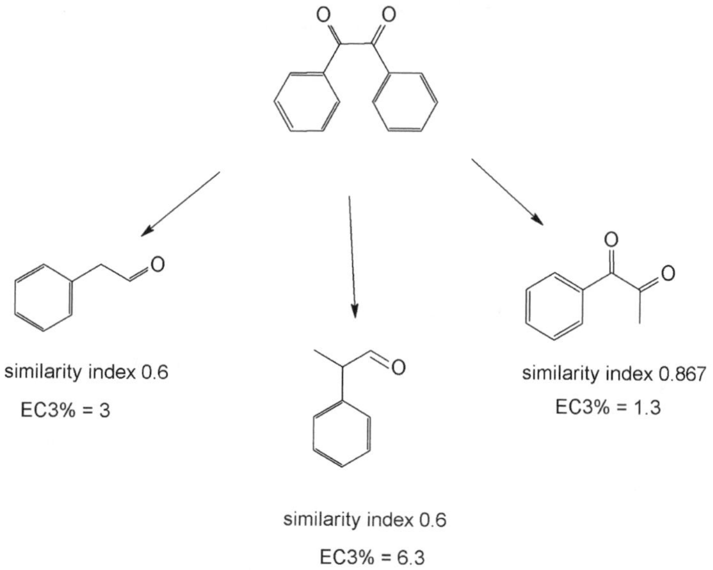

Scheme 17.2

Thus, the target chemical is predicted to be a moderate sensitiser in the LLNA, acting *via* a Schiff base mechanism.[61]

17.3.3 AMBIT

AMBIT is open source software for chemoinformatics data management developed with funding from industry *via* a project funded by the European Chemical Industry Council (CEFIC) Long-range Research Initiative (LRI). It is distributed under Lesser General Public Licence (LGPL).

AMBIT2 software consists of a database and functional modules allowing a variety of queries and data mining of the information stored in the database. AmbitXT is a user-friendly application with a graphical user interface, based on AMBIT2 modules, and is also distributed under LGPL. AmbitXT provides a set of functionalities to facilitate evaluation and registration of the chemicals for REACH. AmbitXT introduces the concept of workflows, allowing users to be guided step-by-step towards a particular goal, and provides workflows for analogue identification and assessment of Persistence, Bioaccumulation and Toxicity (PBT). The software is a standalone application, with an option to install the database on a server.

17.3.3.1 Modules within AMBIT

AMBIT is organised in several modules with a well-defined dependency as shown in Table 17.3.

Table 17.3 Modules within AMBIT.

Module	Description
AmbitXT	GUI application
AmbitXT plugin: database search and analogue identification	AmbitXT plugin, allowing various database queries and analogues identification.
AmbitXT plugin: category building	AmbitXT plugin for analogues identification
AmbitXT plugin: database tools	AmbitXT plugin for database import and management
AmbitXT plugin: database administration	AmbitXT plugin for database administration activities
AmbitXT plugin: REACH PBT assessment	AmbitXT plugin, implementing an workflow for REACH compliant PBT assessment.
ambit2-base	Base classes, without chemoinformatics functionality
ambit2-core	Core classes, with chemoinformatics functionality
ambit2-hashcode	Hashcodes
ambit2-smarts	SMARTS parser
ambit2-db	Database functionality
ambit2-smi23d	Wrapper for Smi23d executables https://czcc-gzid.svn.sourceforge.net/svnzoot/cicc-gzid/cicc-gzid/smi23ol/tzunk
ambit2-mopac	Wrapper for OpenMOPAC
ambit2-ui	User interface
ambit2-dbui	Database user interface
ambit2-workflow	Workflow module
ambit2-namestructure	Chemical name to structure convertor, based on OPSIN package http://sourceforge.net/projects/oscar3-chem/files/
ambit2-model	Similarity calculation, feature selection and QSAR model development
ambit2-taglibs	JSP tags
PubChem utilities	PubChem access utilities
ambit2 REST web services	Allows query of AMBIT database by REST style web services.

17.3.3.2 Database

The AMBIT database is a relational database consisting of a number of repositories for compounds, properties, QSAR models, users, references, as well as a number of tables containing pre-processed information which allow faster substructure and similarity queries. The current implementation based on the MySQL Database functionality is provided by *ambit2-db* module and shown in Table 17.4.

17.3.3.3 Chemical Compounds

The chemical compounds are stored in the table *chemicals* and assigned a unique number. If connectivity is available, a unique SMILES, as well as InChI and molecular formula are generated and stored. The database supports multiple 3-D structures per compound, either coming from different inventories, or generated by external programs and imported into the database. The chemical

Table 17.4 Tables in AMBIT2 database.

Table	Description
Chemical structures	
chemicals	Chemical compounds
structure	Chemical structures, conformers
history	Previous versions of chemical structures
Inventories	
src_dataset	Datasets
struc_dataset	Lookup table for structures, belonging to a dataset
Identifiers, descriptors, properties	
catalog_references	References
properties	Property definition (name, reference, units)
property_values	Numerical property values or links to string values
property_string	String values
property_tuples	Tuples of properties
tuples	Tuples per dataset
template	Templates
template_def	Template definition (which properties belong to a template)
dictionary	Templates hierarchy
Queries	
query	Queries
query_results	Structures per query
sessions	Sessions
Users support	
user_roles	Roles, assigned to users
roles	User roles
users	Users
Quality assessment support	
quality_chemicals	Quality labels of structures and properties
quality_labels	
quality_pair	
quality_structure	
Pre-processed data for substructure, similarity and SMARTS queries	
fp1024	Pre-processed fingerprints for pre-screening and similarity search
fp1024_struc	
sk1024	Pre-processed fragments for accelerating SMARTS searches
atom_distance	Pre-processed data for atom environments similarity
atom_structure	
Schema version	
version	Database version

structures are stored in the table *structure* as a compressed text, where supported formats are SDF, MOL and CML. The choice of text format makes the database transparent and easy to use by external software. Support of multiple formats is motivated by the need to keep the data in the original format. If the original format is not one of the above formats, it is converted to MOL. Support of internal formats will be extended in future releases.

17.3.3.4   Data Provenance

The database provides a means to identify the origin of the data, *i.e.* the specific inventory from which a compound originated. An inventory is identified by its name and reference (table *src_dataset*). Each compound might belong to multiple inventories (table *struc_dataset*), thus allowing users to select the compounds of interest for specific regulatory purposes. Moreover, the data provenance indicator can distinguish between different conformations; for example, in cases where a particular conformation of a compound comes from one inventory and a different conformation comes from another inventory.

Updates of the chemical structures are recorded and subsequent versions are stored in the *history* table. While importing structures from a file, they are stored in its original format in the structure table. If the structure is subsequently updated as a result of a specific calculation (*e.g.* 3-D conversion) or another structure import step (*e.g.* updated version of the original file), the new version will be stored and become currently available, while the previous version will be moved to the history table.

17.3.3.5   Quality Assurance

The discrepancy between structures available in chemical databases presents a challenge for AMBIT as a data integration platform. In order to raise the awareness of possible incorrect structures that might be imported from external sources, AMBIT allows quality labels to be assigned to each chemical record, as follows:

- **Manual verification by expert(s).** Any user can assign quality labels and explain the reasoning for the assignment (table *quality_structure*). The reasons can include discrepancies between registry numbers, names and structure, expert knowledge, manual comparison with external sources, *etc.*
  - *'OK'* – The structure is correct.
  - *'ProbablyOK'* – Most probably the structure is correct, but some issues still need to be verified.
  - *'Unknown'* – Not possible to assign a definite label.
  - *'ProbablyERROR'* – Most probably there is an error
  - *'ERROR'* – The structure is definitely wrong.
- Automatically verified by comparing the structures available under the same chemical compound entry (*e.g.* imported from different sources) – table *quality_chemicals*
  - *'Consensus'* – All structures under the same chemical compound entry are the same.
  - *'Majority'* – The majority of structures under the same chemical compound entry are the same, but there are a small number of structures which differ from the majority.

    ○ *'Ambiguous'* – There is no majority of equal structures under the same chemical compound entry (*e.g.* structures come from three different sources and all the three structures are different).

    ○ *'Unconfirmed'* – The structure comes from a single source and it is impossible to make a comparison.

    ○ *'Unknown'* – No information about the structure (*e.g.* no connectivity).

17.3.3.6 Examples of Quality Assurance in AMBIT

The automatic comparison with three other sources of chemical structures revealed several discrepancies in structures in the dataset for skin sensitisation, two of which are illustrated below. The first one is the chemical with CAS 55-55-0. In addition to the data set, supplied with the book, the structure with the same CAS number is retrieved from three different sources, namely ChemId-Plus,[62] Chemical Structure Lookup Service and the EINECS[63] list. The structure provided in the data set is incomplete and is imported as such into the AMBIT database. An automatic verification identifies if there are discrepancies between structures, which should represent the same chemicals and assigns quality labels for the chemicals—'Consensus', 'Majority', 'Ambiguous', 'Unconfirmed', 'Unknown' (as explained above). The same automatic verification procedure also assigns consensus labels for every structure, labelling structures as follows:

- *'OK'* – If there are multiple structures for a chemical and all of them are the same.
- *'ProbablyOK'* – If there are multiple structures for a chemical and the structure belongs to the set where most of the structures are the same.

This is the case with the example in Figure 17.1. The structures from EINECS, ChemIDPlus and Chemical Structure Lookup Service can be considered the same, apart from the differences in charges, while the first structure is truncated. As a minority structure, it is assigned the 'ProbablyERROR' label to indicate there are discrepancies with the other structures representing this chemical.

In the second example the structure with CAS 39236-46-9 provided in the skin sensitisation data set has an erroneous structure—ethyl substitution occurs on the wrong nitrogen in a ring due to an incorrect SMILES string in the data set supplied with the book (see Figure 17.2).

17.3.3.7 Identifiers, Descriptors and Properties

The database schema is designed to provide unified storage for an arbitrary number of text fields (*e.g.* registry numbers or names) and numerical properties (*e.g.* descriptors, experimental data). The properties are not predefined, but stored in the database on demand. Thus, the AMBIT database is ready to incorporate any number of chemical compounds, identifiers, descriptors and experimental data.

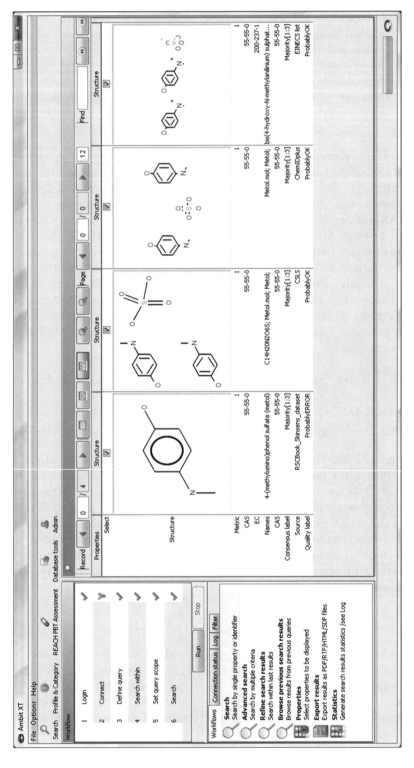

Figure 17.1   AMBIT representations for chemical with CAS 55-55-0.

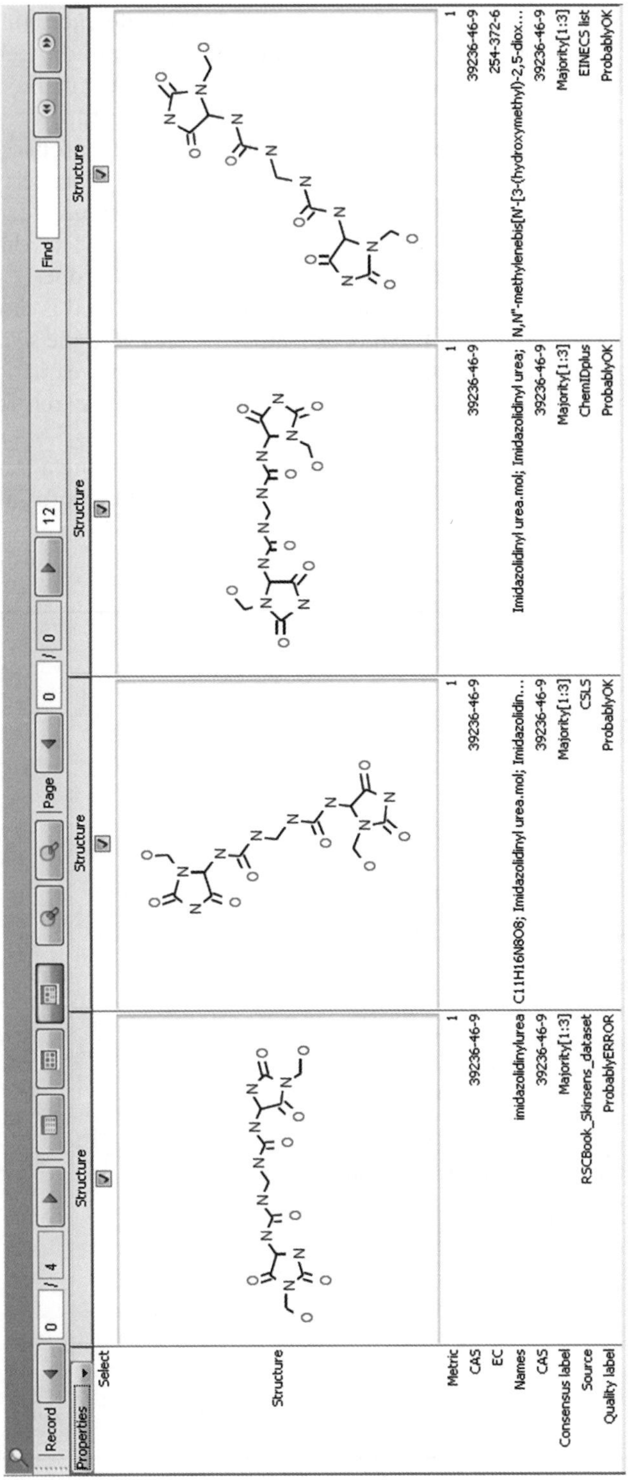

Figure 17.2 C AMBIT representations for chemical with CAS 39236-46-9.

A property (table *properties*) is identified by a name and reference, thus allowing properties with coinciding names but originating from different sources to be distinguished (*e.g.* log *P* calculated internally by different methods and log *P* imported from an external file). Every newly added property or descriptor is added to a *properties* table, with information about the property/ descriptor name, units, alias and reference. The reference for a property, imported from a file is the name of the file itself, while the reference for a descriptor contains the name of the software used for calculation. The alias usually contains a copy of the name, except in cases where the property is recognised as a specific type of registry number or a chemical name. In this case, the alias is assigned a fixed value (*e.g.* CasRN or Names). Fields with the same meaning but different names can be assigned the same alias to facilitate queries (*e.g.* species field same across all endpoints in order to be able to search for species); see Table 17.5.

The flat list of properties provides flexible storage, though presenting a long list of properties and descriptors in the user interface might be confusing.

Table 17.5   Templates.

Template: ambit2.mopac.DescriptorMopacShell
Properties:
'NO. OF FILLED LEVELS'
'TOTAL ENERGY'
'FINAL HEAT OF FORMATION'
'IONIZATION POTENTIAL'
'ELECTRONIC ENERGY'
'CORE-CORE REPULSION'
'MOLECULAR WEIGHT'
'EHOMO'
'ELUMO'
Template: Benigni/Bossa rulebase (for mutagenicity and carcinogenicity)
Properties:
Structural Alert for genotoxic carcinogenicity
Structural Alert for nongenotoxic carcinogenicity
No alerts for carcinogenic activity
Potential *S. typhimurium* TA100 mutagen based on QSAR
Unlikely to be a *S. typhimurium* TA100 mutagen based on QSAR
Potential carcinogen based on QSAR
Unlikely to be a carcinogen based on QSAR
For a better assessment a QSAR calculation could be applied.
Error when applying the decision tree
Template: ambit2.descriptors.FunctionalGroupDescriptor
Properties:
'Alkyl carbon'
'Nitro group'
'Halogen'
'Amide'
'Sulfonate'
'Sulfonamide'
'Sulfonamide'

Table 17.6   An excerpt view of ontology.

Template	Relationship	Parent template
Endpoints		
Identifiers		
Dataset	Top level templates	
Descriptors		
Ecotoxic effects	is_a	Endpoint
Human health effects	is_a	Endpoint
Physicochemical effects	is_a	Endpoint
Short-term toxicity to algae (inhibition of the exponential growth rate)	is_a	Ecotoxic effects
Toxicity to birds	is_a	Ecotoxic effects
CAS number	is_a	Identifier
RSCBook_Skinsens_dataset.sdf	is_a	Dataset
org.openscience.cdk.qsar.descriptors.molecular. HBondAcceptorCountDescriptor	is_a	Descriptor
org.openscience.cdk.qsar.descriptors. molecular.HBondDonorCountDescriptor	is_a	Descriptor
Verhaar scheme	is_a	Descriptor

Templates (tables *template* and *template_def*) allow properties to be organised in groups.

Templates themselves can be organised hierarchically, with the help of table *dictionary*. The database is distributed with a set of default templates including top level templates 'Endpoints', 'Identifiers, Datasets and Descriptors' and a number of endpoints—according to the EChA classification of endpoints.[1] Convenience view *ontology* combines the templates with its hierarchical organisation. An excerpt of this view is shown in Table 17.6.

By default, properties imported from a file with chemical compounds belong to the data set of origin, but can be moved to any user selected group.

Quality labels can also be assigned to any property value stored in the database (table *quality_labels*):

- *'OK'* – The value assigned to property is correct.
- *'ProbablyOK'* – Most probably the value is correct, but some issues still need to be verified.
- *'Unknown'* – Not possible to assign a definite label.
- *'ProbablyERROR'* – Most probably there is an error.
- *'ERROR'* – The value assigned to this property is definitely wrong.

17.3.3.7.1   Queries.   The results of the searches performed by a user are stored in the *query* and *queryresults* tables. As well as providing the ability to record user actions, this enables subsequent browsing of query results and combining queries with arbitrary logic.

17.3.3.8   Search Methods

17.3.3.8.1   Exact Structure, Fixed Sub-structures, Similarity and SMARTS.
The core substructure search functionality (graph isomorphism) is provided
either by the CDK chemoinformatics library or by a faster algorithm implemented in AMBIT. Substructure searching is a non-deterministic polynomial-
time problem, which means that the complexity of the algorithm increases
rapidly with the size of the molecule. To speed up substructure searching in
large datasets, pre-calculated fingerprints are used to identify structures
potentially containing the substructure. The AMBIT database and software
combines this technique with fast relational database queries, which results in
very fast substructure searching in large datasets. In addition, fingerprints are a
standard tool for representing chemical structures to assess structural similarity
by calculating Tanimoto coefficient between two fingerprints.

17.3.3.8.2   Similarity.   Fingerprint generation was based on the fingerprint
implementation by an open source chemoinformatics library, CDK, and follows the ideas of Daylight fingerprint theory which state:[64]

1) For a given molecule all possible paths of a predefined length (default is 7)
   are generated.
2) The path is converted into a set of bits *via* specific mathematical procedure
   (hash function).
3) The set of bits thus produced is added (with a logical OR) to the fingerprint. AMBIT uses 1024 bit fingerprints by default.

The Tanimoto coefficient is calculated as Tanimoto $N_A \cap N_B/(N_A + N_B - N_A \cap N_B)$, where N_A is the number of bits 'on' in fingerprint A, N_B is the
number of bits 'on' in fingerprint B, and $N_A \cap N_B$ is the number of bits 'on' in
both fingerprints. Since Tanimoto distance is a pairwise measure and here the
objective is to assess the similarity to the set of molecules, we generate a consensus fingerprint, which is again 1024 bit fingerprint where each bit is set.

Atom environments (AE) can be regarded as fragments,[65,66] surrounding
each atom in a molecule, up to a predefined level. The calculation procedure is
as follows. First, atom types to be included in the generation of AEs are
selected. Thirty-four atom types are used (as listed in Table 17.7), which are

Table 17.7   Atom types used to generate atom environments.

H	C.default	N.sp2	P3	F	I
Hplus	Cplus.sp2	Nplus	P4	F-	I-
Hminus	Cminus.sp2	Nplus.sp3	S2	Cl	Misc
C.sp3	Caromatic.sp2	O.sp2	S2-	Cl-	
C.sp2	Cminus	Oplus	S4	Br	
C.sp	N	Ominus	S	Br-	

very similar to the Sybyl atom types recommended by Bender *et al.*[66] The choice is based on the available atom type parameterisation in the CDK library. Next, a vector of length $(34 *L + 1)$ is constructed for each atom, where L is the maximum level for generating atom environments and $L = 3$ by default. Thirdly, for each atom, neighbours at level 1, 2, 3 are identified and corresponding counts stored in the vector. An example of a string representation of the result for a single atom (C.sp2) is:

C.sp2, 0, 0, 0, 0, 1, 0, 0, 0, 0, 0, 0, 0, 0, 0, 0, 1, 0, 0, 1, 0, 0, 0, 0, 0, 0, 0, 0, 0, 0, 0, 0, 0, 0, 0, 0, 0, 0, 1, 0, 0, 1, 0, 0, 0, 0, 0, 0, 0, 0, 0, 0, 0, 0, 0, 0, 0, 0, 0, 0, 0, 0, 0, 0, 0, 0, 0, 0, 0, 0, 0, 0, 0, 0, 0, 2, 0, 0, 0, 0, 0, 0, 0, 0, 0, 0, 0, 0, 0, 0, 0, 0, 0, 0, 0, 0, 0, 0, 0, 0, 0, 0, 1

Note that if there are several C.sp2 atoms with the same neighbours up to third level in the molecule, they will have the same string representation. This representation will be referred to as a 'fragment'. AEs could be compared by Tanimoto distance (see above), where NA is the number of fragments in molecule A, NB is the number of fragments in molecule B, and NA ∩ NB is the number of common fragments between the two molecules. Here, we take the average Tanimoto distance for the nearest neighbours instead of defining a consensus fingerprint. For each molecule, the similarity measure is the averaged Tanimoto distance between the molecule and its five nearest molecules.

17.3.3.8.3 Substructure Search. The implementation of substructure searching allows only fixed structures (*e.g.* no wildcards for atoms and bonds). The structure is drawn using a structure diagram editor (JChem-Paint[67]) and submitted as a hydrogen-depleted structure, which may present difficulties in distinguishing certain types of functional groups (*e.g.* aldehydes *vs.* carbonyls, amines *vs.* nitro groups).

AMBIT also allows querying the database by using the Smiles ARbitrary Target Specification (SMARTS) language,[68] accelerated with certain pre-processed information for structural features, stored in the database. The SMARTS specification was originally developed and maintained by Daylight Inc.,[69] but is supported by many more commercial and open source software suites.[70–72] A list of predefined functional groups and their SMARTS definitions are available, but formulating more complex queries requires knowledge of SMARTS. The SMARTS line notation allows extremely precise and transparent substructural specification and atom typing. SMARTS expressions for atoms and bonds can be combined by logical operators to form more complex queries. Recursive SMARTS allow detailed specification of an atom's environment. For example, the more reactive (with respect to electrophilic aromatic substitution) ortho and para carbon atoms of phenol can be defined as: [$(c1c([OH])cccc1),$(c1cc-c([OH])cc1)]. Atoms that are in an environment where (the atom is connected to an aliphatic oxygen) and where (the atom is connected to two sequential aliphatic carbons) are as expressed as [$(*O);$(*CC)].

Various queries and their combinations on properties, inventories and quality labels are available. A query can be restricted to search for compounds within a specified data set within a previous search results query.

Table 17.8   The CDK library based descriptors.

ALogP and Molar Refractivity	*Largest Chain*
Atomic Polarizabilities	Largest Pi System
Amino Acids Count	Largest Aliphatic Chain
Aromatic Atoms Count	Moments of Inertia
Aromatic Bonds Count	Petitjean Number
Atom count	Petitjean Shape Indices
BCUT	Rotatable Bonds Count
Bond Polarizabilities	Lipinski's Rule of Five
Bond Count	Topological Polar Surface Area
Charged Partial Surface Area	Vertex adjacency information magnitude
Gravitational Index	WHIM
Hydrogen Bond Acceptors	Wiener Numbers
Hydrogen Bond Donors	XLogP
Kier and Hall kappa molecular shape indices	Zagreb Index

17.3.3.9   3-D Structure Generation

The AMBIT module *ambit2-smi23d* integrates the open source 3-D coordinate generation smi23d for generation of an initial 3-D structure from a connectivity matrix.[73] The initial structure is further optimised by OpenMOPAC 7.1,[74] which is embedded into the *ambit2-mopac* module.

17.3.3.10   Molecular Descriptors

AMBIT provides facilities to calculate and store descriptors for all chemical structures in the database, as well as specification of search criteria based on descriptor values. The CDK library based descriptors[36] are shown in Table 17.8.

17.3.3.11   Workflow Engine

A workflow engine is a software application which manages and executes modelled business processes. In general, the models can be edited by non-programmers using workflow editors. The workflow models can be as simple as a series of sequential steps, but can also be complex, including many conditions and loops. The Workflow Management Coalition provides standards for defining workflows in a XML based format.[75]

Given the importance of support for workflows in AMBIT, a number of existing open source workflow engines were evaluated for their suitability to be embedded into the AmbitXT application. The final decision of embedding micro-workflow is based on a trade-off between simplicity and available functionalities.[76] AmbitXT is entirely based on a micro-workflow engine, providing an extensible platform for workflow-based wizards and facilitating recording of user actions. It allows the steps of a workflow to be defined with very low granularity, down to a single line of code. The workflow below has a very high granularity. Each step from a workflow of high granularity can be represented as a workflow of steps with a finer granularity, allowing for flexibility and uniform tracking of every action performed.

17.3.3.12 Workflow for Analogue Identification

The workflow will consist of the following steps:

1) Definition of the starting structure or set of structures. The structure(s) can be defined as:
   - Identifiers (*e.g.* CAS, EINECS number, name).
   - Structure—represented as SMILES, MOL, SDF—drawn manually by the structure diagram editor, available in AMBIT or drawn using externally installed ISIS/Draw software, copied to the system clipboard and then pasted into AMBIT user interface.
2) Basic analogue search consists of a similarity search (hashed fingerprints compared by Tanimoto distance by default).
3) The results are displayed in the Structure browser. The user can decide to restrict the forthcoming queries within the set of selected structures.
4) Substructure search by user-defined fragment.
5) The results are displayed in the Structure browser. The user can decide to restrict forthcoming queries within the set of selected structures.
6) Further filtering of the results by conducting additional compound profiling based on experimental and calculated data (log P, $Dmax$, other 2-D and 3-D descriptors chosen by the user).
7) The results are displayed in the Structure browser. The user can decide to restrict the forthcoming queries within the set of selected structures.
8) The selected structures are grouped into typical chemical classes or by clustering, allowing the user to inspect small groups of analogues and derive the final decision of the query compound(s).
9) The system proposes to calculate the final value by average, min–max, Euclidean distance to user selected properties.

17.3.3.13 Workflow for REACH PBT and vPvB Assessment

If the tonnage exceeds 10 tonnes/year, REACH requires a set of information including a PBT (Persistent, Bioaccumulative and Toxic) and vPvB (Very Persistent and Very Accumulative) Assessment for every substance to be registered and not exempted. If the necessary information is available, the REACH PBT & vPvB Assessment allows a straightforward, user-friendly and quick assessment. An important goal is to rapidly identify those REACH substances that are not PBT or vPvB. In addition, those substances identified as potentially PBT or vPvB can immediately be investigated in a higher tier assessment to find out what is necessary as a next step. Such higher tier assessments are very often time-consuming and costly; they are preferably avoided or the strict registration deadlines will not be met due to an ongoing

PBT assessment. Because the assessment is performed transparently and always in the same way, it allows a standardised PBT & vPvB Assessment throughout the company, independent of the personal judgments of an assessor. Printing the result sheet (*e.g.* as a PDF file) allows proper documentation of the PBT & vPvB Assessment.

Only organic substances can be assessed. This workflow should not be applied to inorganic or organometallic substances, polymers and mixtures. PBT assessment is visually organised in five pages:

- definition of the substance;
- persistency check;
- bioaccumulation check;
- toxicity check; and
- presenting the final results.

17.3.3.14   Population of AMBIT Database with Data

The following datasets are imported and distributed with the AMBIT database:

- EINECS list;[63]
- Bioconcentration factor dataset;[77]
- ECETOC aquatic toxicity data;[78]
- Local Lymph Node Assay (LLNA) data;[56] and
- ECETOC skin irritation data.[59]

The data are imported using the standard data import functionality. The EINECS list is publicly available at the JRC site and consists of 100,204 chemicals.[63] There has been extensive verification of EINECS structures in order to improve their reliability, based on comparison of structures with matching registry numbers and available from public sources. Quality labels have been assigned, as explained above.

The bioconcentration factor data set is distributed without structural information and chemical compounds are identified only by CAS numbers and chemical names. Structures have been retrieved from publicly available sources and imported into the database. Datasets 3–5 consist of relatively small numbers of compounds and presumably contain high quality structures, manually checked by experts before making them publicly available.

17.3.3.15   Analogue Identification with SMARTS According to the Wu et al. Framework.[79]

17.3.3.15.1   Example 1.   In this example we illustrate formalisation of an expert written query *via* SMARTS. The search strategy for compound **1** (Scheme 17.3) is shown in compound **2** (Scheme 17.4), where R and R' are small alkyls of 1–2 carbons such as methyl and ethyl groups. The n is 1–2

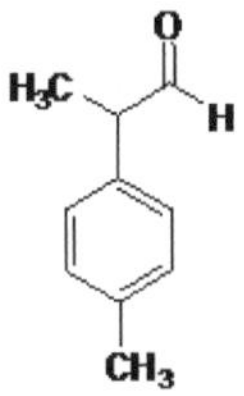

1 (CAS# 99-72-9)

Scheme 17.3

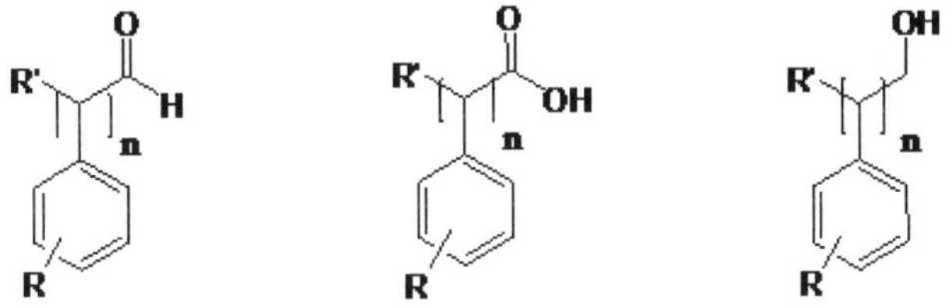

2   3   4

Scheme 17.4

carbons. When n is 2, the R′ group is connected to the carbon next to the aldehyde functional group. R may be at any of five positions on the aromatic ring. Additionally, if searches around **2** do not bring back sufficient data, it may be appropriate to consider carboxylic acid **3** and alcohol **4** (Scheme 17.4), which would be the potential metabolites of **1** *via* the *C*-oxidation and reduction pathways. In the compound **3** and **4**, the R and R′ groups and n should be the same as that of the compound **1**.

The following can be noted:

1) The whole strategy can be written as the following SMARTS [$(c1cc([$([CH3]),$([CH2][CH3])])ccc1[$(C(C)C(=O)),$(CC(C)C(=O))]), $(c1c([$([CH3]),$([CH2][CH3])])cccc1[$(C(C)C(=O)),$(CC(C)C(=O))]), $(c1([$([CH3]),$([CH2][CH3])])ccccc1[$(C(C)C(=O)),$(CC(C)C(=O))])] and was executed as one single query.

2) The search did not yield any hits, and after consultation with experts, the SMARTS searches were relaxed to allow for arbitrary substituents at all positions. The relaxed SMARTS and the new query was c1ccccc1-[$(C(C)C(=O)),$(CC(C)C(=O))].

3) The relaxed search yielded the following hits: 103-95-7, 93-53-8, 80-54-6.
   The chemical assessed for skin sensitisation potential is evaluated for its potential to act as an electrophile towards nucleophilic groups on skin

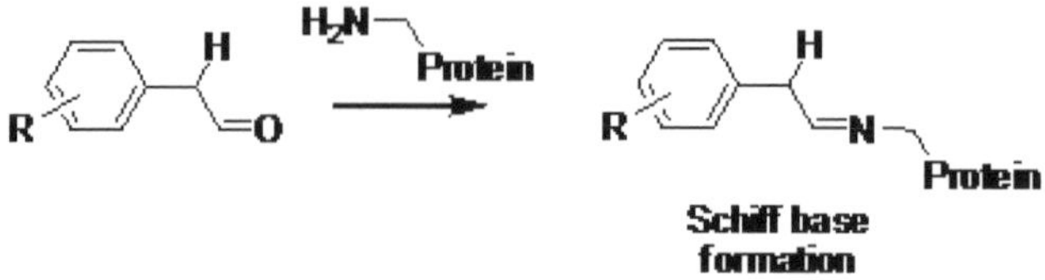

Schiff base
formation

Scheme 17.5

proteins. The aldehydes which cause skin sensitisation are most likely to form Schiff bases *via* the carbonyl group of the aldehyde that reacts with the amine group of protein. So far, this provides the best rationale even though it is not definitively established. The analogue candidates found for 99-72-9 (namely the compounds with CAS RN 103-95-7, 93-53-8, 80-54-6) are aliphatic aldehydes with a *para*-substituted aromatic ring. From a structural features and reactivity point of view, these aldehydes have similar reactivity towards to the nucleophiles (Scheme 17.5). The substituents such as methyl and ethyl groups at the *para*-position would not significantly change the electronic nature and reactivity of these molecules. In contrast, the steric hindering groups such as isopropyl, *tert*-butyl and isobutyl groups may have steric effects and tend to affect the physico-chemical properties and skin absorption or penetration ability. These factors may affect their toxic potential.

4) By considering the above, 99-72-9 (our target chemical) is more likely to be close to 93-53-8 (Moderate) rather than 103-95-7 and 80-54-6 (Weak) because the steric effects of the substituents of the latest two are likely to lower absorption/penetration ability and therefore skin sensitisation potential.

Next, the analogue search *via* SMARTS was compared with a standard fingerprints based similarity search (Figure 17.3). The results are ordered by Tanimoto distance. Low Tanimoto distance for molecules that could be considered as appropriate analogues are discussed above. They would be missed if the traditional 0.75 cut-off were used. Similarity, searching also retrieves chemicals that differ in reactivity from the target (e.g. 579-07-7).

In general, a similarity based search yields a mix of suitable and not suitable chemicals for read-across and needs to be evaluated based on chemistry expertise. In homogenous data sets, searches by SMARTS and similarity will give similar results—though this is not to be expected when searching diverse, large libraries. As such, a SMARTS based search is far more efficient, and read-across more transparent and verifiable.

17.3.3.15.2 Example 2. Chemical class: alpha-beta unsaturated ester – 2 hydroxypropyl methacrylate

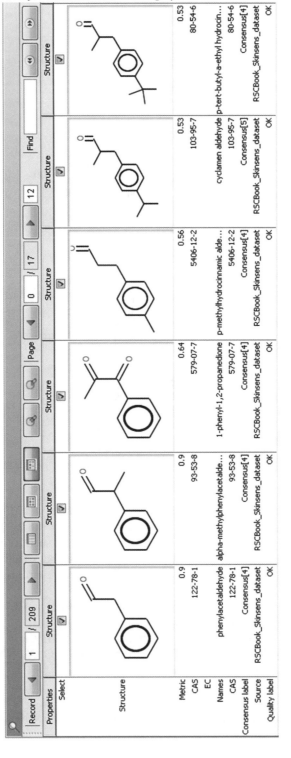

Figure 17.3 Results from fingerprint based structural similarity search for CAS 99-72-9.

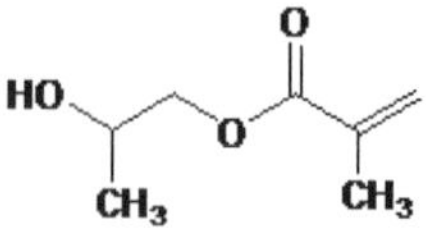

Scheme 17.6

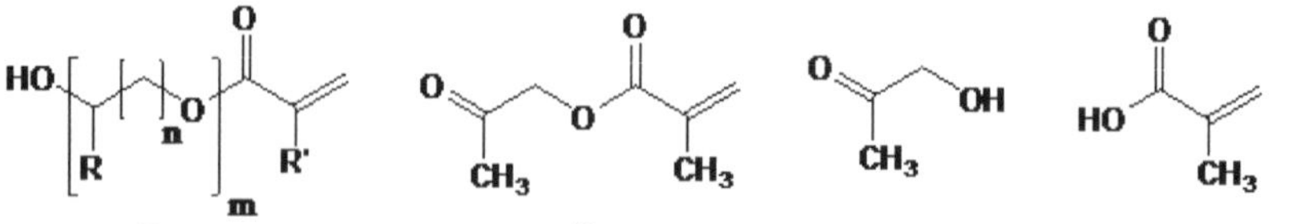

Scheme 17.7

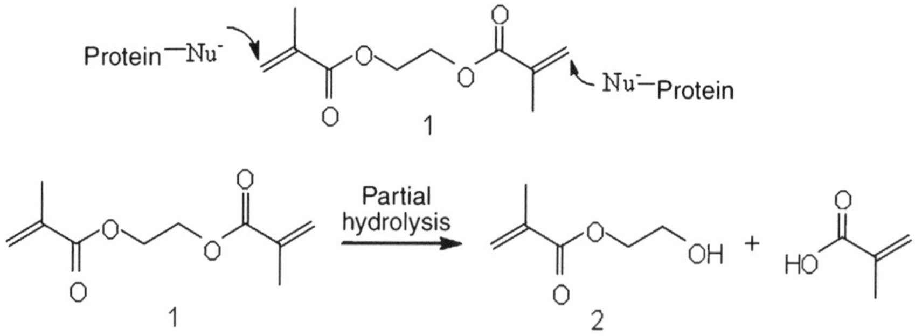

Scheme 17.8

The reasonable search strategy for compound **1** (Scheme 17.6) is shown in Scheme 17.7 via compounds **2**, **3**, **4**. For compound **1** n can be 1–3 carbons. When n is 1, m can be 1–2. R and R' can be small alkyl such as methyl, ethyl groups. Additionally, if searches around **2** do not bring back sufficient data, it may be appropriate to consider metabolites **3** and ester hydrolysis products **4** and **5** (Scheme 17.8).

Step 1: The initial SMARTS strategy was captured in C(=C)C(=O)O [$(CCCCO),$(CCCO),$(CCO)] ($n = 1,3$) , $m = 1$ with no hits.

The searches for **3**, **4** and **5** were formulated as **3** [CH3]C(=O)COC(=O)C(=C)[CH3], **4** [CH3]C(=O)C[OH] and **5** [OH]C(=O)C(=C)[CH3], respectively, and also generated no hits.

Step 2: The revised strategy contained only a methyl acrylate substructure OC(=O)C=C and yielded:

ethylene glycol dimethacrylate O=C(OCCOC(=O)C(=C)C)C(=C)C (97-90-5).

This compound (**1**) can be considered as suitable analogue for following reasons:

1) It has the structural features and reactivity as well as major metabolic pathway partially similar to the target (2-hydroxyethyl acrylate).
2) This analogue contains two Michael acceptor units which could react with nucleophilic sites of proteins.
3) This analogue could convert to a metabolite, 2-hydroxyethyl methyl acrylate (**2**), which has very similar structural feature to the target, *via* a partial hydrolysis pathway.

Ethylene glycol dimethacrylate is a weak sensitiser. In addition data on 2-hydroxypropyl methylacrylate (**2**) indicate it is a non-sensitiser. Given that ethylene glycol dimethacrylate has two Michael acceptor units, it is conceivable that it is more reactive than our target. Thus the overall conclusion is that the 923-26-2 is a non-sensitiser, which agrees with experimental data.

A similarity search for 923-26-2 points to 818-61-1, with a Tanimoto coefficient equal to 0.74 between the two compounds. However, read-across would be inappropriate in this case. Both are the alpha–beta unsaturated esters, which are capable of undergoing the Michael addition reaction to cause the skin sensitisation (*via* Michael acceptors reacting with the nucleophilic site of protein). However, from a structural features and reactivity point of view, the first compound (818-61-1) is a slightly stronger Michael acceptor due to the lack of a methyl group at the alpha-carbon position. Additionally, it contains a primary alcohol which could be transformed to the corresponding aldehyde *via* an oxidation pathway. This aldehyde is also a potential skin sensitiser. The 818-61-1 compound has a methyl substitution on the alpha-carbon of the olefinic moiety. This methyl group may decrease the electron withdrawing ability of unsaturated group and increase steric effects for the Michael addition reaction. Overall, the 923-26-2 compound is most likely to have lower reactivity than 818-61-1. This hypothesis is confirmed by the experimental data since 818-61-1 is a moderate sensitiser.

17.3.4 AmbitDiscovery

AmbitDiscovery is a software tool for assessing structural similarity to a set of chemicals, and for applicability domain assessments. It calculates the similarity of a chemical to a set in the following ways. The first approach is based on distances (Euclidean, City block) suitable when similarity is assessed in descriptors space. Several pre-processing data functionalities are available to deal with linearly dependent descriptors (rotation *via* Principal Component Analysis, PCA), centring to remove effects of different numerical values ranges of the descriptors before PCA. In AmbitDiscovery, there is also a means to calculate the probability of a chemical belonging to a set represented by probability density function. This probability density function is fully non-parametric and was described by Jaworska *et al.*[3] To assess structural similarity, the user has a choice of Daylight fingerprints and two methods to quantify similarity (missing fragments and 1-Tanimoto).

Another approach to assess structural similarity is based on atom environments. Similarity using AEs can be compared by the Hellinger distance when representing the set as the consensus fingerprint or k-NN average (n = 5) of 1-Tanimoto scores.

17.4   Summary and Conclusions

This chapter summarises the requirements and challenges for computational tools to support category formation and read-across. A brief overview of the open source, open data and open standards approaches in chemical informatics is presented. The opportunities afforded by 'openness' are highlighted, with an emphasis on open source applications specifically developed to address the challenges posed by the EU chemicals regulation, REACH. The identification of analogues is a process of assessing structural, metabolic, reactivity and physico-chemical similarity between target and read-across chemical. Traditionally, this assessment has been performed on a case-by-case basis. Computational tools implementing various ranges of similarity metrics can facilitate identification of suitable analogues in a more efficient and consistent manner.

The major challenges for computational tools to support category formation and read-across centre around formalisation of expert knowledge and improved management of toxicological data. Open source, open data and open standards approaches in chemical informatics are a way forward to improved management of chemical data. Therefore, further development of the freely accessible, transparent and inter-operable tools will promote the uptake of non-testing methods in toxicology. Their increasing availability and user-friendliness should eventually contribute to the acceptance of reliable and reproducible non-testing data for use in chemical hazard assessments. To illustrate the potential of open source chemoinformatics applications, the Toxtree, Toxmatch and AMBIT tools are described with illustrative analogue identification case studies.

17.5   Acknowledgements

AMBIT software was developed within the framework of CEFIC LRI project *EEM-9* 'Building blocks for a future (Q)SAR decision support system: databases, applicability domain, similarity assessment and structure conversions' and extended under a subsequent CEFIC LRI contract for developing AmbitXT.

References

1. European Chemicals Agency, *Guidance on Information Requirements and Chemical Safety Assessment, Chapter R.6. QSARs and grouping of chemicals*, EChA, Helsinki, 2008, http://guidance.echa.europa.eu/docs/guidance_document/information_requirements_en.htm [accessed March 2010].

2. Organisation for Economic Co-operation and Development, *Guidance on Grouping of Chemicals*, OECD, Paris, 2007, OECD Environmental Health and Safety Publications Series on Testing and Assessment No. 80, ENV/JM/MONO(2007)28, www.olis.oecd.org/olis/2007doc.nsf/LinkTo/NT0000426A/$FILE/JT03232745.PDF [accessed March 2010].

3. J. Jaworska and N. Nikolova-Jeliazkova, *SAR QSAR Environ. Res.*, 2007, **18**, 195.

4. A. Worth and G. Patlewicz, *A Compendium of Case Studies that Helped to Shape the REACH Guidance on Chemical Categories and Read Across*, European Commission Joint Research Centre, Ispra, Italy, 2007, EUR 22481 EN, http://ecb.jrc.ec.europa.eu/DOCUMENTS/QSAR/EUR_22481_EN.pdf [accessed March 2010].

5. H. S. Rosenkranz and A. R. Cunningham, *Regul. Toxicol. Pharmacol.*, 2008, **33**, 313.

6. www.opensource.org

7. D. Riehle, *IEEE Comp.*, 2007, **40**, 25.

8. http://ec.europa.eu/enterprise/sectors/ict/files/2006-11-20-flossimpact_en.pdf

9. F. Letellier, *Open Source Software: The Role of Nonprofits in Federating Business and Innovation Ecosystems*, presented at AFME 2008, http://flet.netcipia.net/xwiki/bin/download/Main/publications-fr/GEM2008-FLetellier-SubmittedPaper.pdf [accessed March 2010].

10. J. Bitzer and P. J. H. Schröder, *The Impact of Entry and Competition by Open Source Software on Innovation Activity*, EconWPA, 2005, Industrial Organization 0512001, http://ideas.repec.org/p/wpa/wuwpio/0512001.html [accessed March 2010].

11. J. Feller, B. Fitzgerald and S. A. Hissam, *Perspectives on Free and Open Source Software*, MIT Press, Cambridge, MA, 2005.

12. http://en.oreilly.com/oscon2009

13. http://sourceforge.net/apps/mediawiki/cdk/index.php?title = Main_Page

14. C. Steinbeck, Y. Han, S. Kuhn, O. Horlacher, E. Luttmann and E. L. Willighagen, *J. Chem. Inf. Comput. Sci.*, 2003, **43**, 493.

15. C. Steinbeck, C. Hoppe, S. Kuhn, M. Floris, R. Guha and E. L. Willighagen, *Curr. Pharm.*, 2006, **12**, 211.

16. http://openbabel.org

17. N. M. O'Boyle, C. Morley and G. R. Hutchison, *Chem. Central J.*, 2008, **2**, 5.

18. www.rdkit.org

19. N. M. O'Boyle and G. R. Hutchison, *Chem. Central J.*, 2008, **2**, 24.

20. http://sourceforge.net/apps/mediawiki/cdk/index.php?title = Features

21. http://cml.sourceforge.net

22. www.bioclipse.net

23. http://taverna.sourceforge.net/
24. http://sourceforge.net/projects/oscar3-chem/
25. http://wwmm.ch.cam.ac.uk/crystaleye/
26. www.gnu.org/copyleft/
27. www.ietf.org
28. www.w3.org
29. www.oasis-open.org
30. J. Day, *Patterns in Network Architecture: A Return to Fundamentals,* Pearson Education, Boston, 2008.
31. www.iupac.org/inchi/download/index.html
32. R. Guha, M. T. Howard, G. R. Hutchison, P. Murray-Rust, H. Rzepa, C. Steinbeck, J. Wegner and E. L. Willighagen, *J. Chem. Inf. Model.*, 2006, **46**, 991.
33. www.blueobelisk.org
34. www.opensmiles.org
35. http://qsar.sourceforge.net/dicts/blue-obelisk/index.xhtml
36. http://qsar.sourceforge.net/dicts/qsar-descriptors/index.xhtml
37. www.opendefinition.org/licenses
38. P. Murray-Rust, *Nature*, 2008, **451**, 648.
39. http://opentox.org
40. L. Richardson and S. Ruby, *RESTful Web Services*, O'Reilly Media, 2007.
41. T. B. Kepler, M. A. Marti-Renom, S. M. Maurer, A. K. Rai, G. Taylor and M. H. Todd, *Aust. J. Chem.*, 2006, **59**, 291.
42. www.pistoiaalliance.org
43. M. H. Todd, *Chem. Central J.*, 2007, **1**, 3.
44. W. J. Geldenhuys, K. E. Gaasch, M. Watson, D. D. Allen and C. J. van der Schyf, *Drug Discov. Today*, 2006, **3–4**, 127.
45. G. M. Cramer, R. A. Ford and R. L. Hall, *J. Cosmet. Toxicol.*, 1978, **16**, 255.
46. H. J. M. Verhaar, C. J. van Leeuwen and J. L. M. Hermens, *Chemosphere*, 1992, **25**, 471.
47. J. D. Walker, I. Gerner, E. Hulzebos and K. Schlegel, *QSAR Comb. Sci.*, 2005, **24**, 378.
48. I. Gerner, M. Liebsch and H. Spielmann, *Altern. Lab. Anim.*, 2005, **33**, 215.
49. R. Benigni, C. Bossa, N. Jeliazkova, T. Netzeva, and A. Worth, *The Benigni/Bossa Rulebase for Mutagenicity and Carcinogenicity – A Module of Toxtree,* Office for Official Publications of the European Communities, Luxembourg, 2008, JRC Scientific and Technical Reports, EUR 23241 EN, http://ecb.jrc.ec.europa.eu/documents/QSAR/EUR_23241_EN.pdf [accessed March 2010].
50. R. Benigni, C. Bossa, O. Tcheremenskaia and A. Worth, *Development of Structural Alerts for the in vivo Micronucleus Assay in Rodents*, Office for Official Publications of the European Communities, Luxembourg, 2009, EUR 23844 EN.
51. I.C. Munro, R. A. Ford, E. Kennepohl and J. G. Sprenger, *Food Chem. Toxicol.*, 1996, **34**, 829.

52. T. W. Schultz, J. W. Yarbrough, R. S. Hunter and A. O. Aptula, *Chem. Res. Toxicol.*, 2007, **20**, 1359.

53. C. L. Russom, S. P. Bradbury, S. J. Broderius, D. E. Hammermeister and R. A Drummond, *Environ. Toxicol. Chem.*, 1997, **16**, 948.

54. www.epa.gov/ncct/dsstox/sdf_epafhm.html

55. W. M. Meylan, P. H. Howard, R. S. Boethling, D. Aronson, H. Printup and S. Gouchie, *Environ. Toxicol. Chem.*, 1999, **18**, 664.

56. G. F. Gerberick, C. A. Ryan, P. S. Kern, H. Schlatter, R. J. Dearman, I. Kimber, G. Patlewicz and D.A. Basketter, *Dermatitis*, 2005, **16**, 157.

57. I. Kimber, D. A. Basketter, M. Butler, A. Gamer, J. L. Garrigue, G. F. Gerberick, C. Newsome, W. Steiling and H.W. Vohr, *Food Chem. Toxicol.*, 2003, **41**, 1799.

58. D. W. Roberts, G. Patlewicz, P. S. Kern, F. Gerberick, I. Kimber, R. J. Dearman, C. A. Ryan, D. A. Basketter and A. O. Aptula, *Chem. Res. Toxicol.*, 2007, **20**, 1019.

59. European Centre for Ecotoxicology and Toxicology of Chemicals, *Skin Irritation and Corrosion: Reference Chemicals Data Bank*. ECETOC, Brussels, 1995, Technical Report, Vol. 66, pp. 1–247.

60. S. J. Enoch, M. T. D. Cronin, J. C. Madden and M. Hewitt, *QSAR Comb. Sci.*, 2009, **28**, 696.

61. G. Patlewicz, N. Jeliazkova, A. Gallegos-Saliner and A. P. Worth, *SAR QSAR Environ. Res.*, 2008, **19**, 397.

62. http://chem.sis.nlm.nih.gov/chemidplus/

63. http://ecb.jrc.ec.europa.eu/qsar/information-sources/

64. www.daylight.com/dayhtml/doc/theory/theory.finger.html

65. L. Xing and R. C. Glen, *J. Chem. Inf. Comput. Sci.*, 2002, **42**, 796.

66. A. Bender, H. Y. Mussa, R. C. Glen and S. Reiling, *J. Chem. Inf. Comput. Sci.*, 2004, **44**, 170.

67. http://sourceforge.net/apps/mediawiki/cdk/index.php?title = JChemPaint

68. www.daylight.com/dayhtml/doc/theory/theory.smarts.html

69. www.daylight.com

70. http://openbabel.sourceforge.net/wiki/SMARTS

71. http://cdk.sourceforge.net/

72. http://www-ra.informatik.uni-tuebingen.de/software/joelib/

73. http://cicc-gzid.svn.sourceforge.net/viewcc/cicc-gzid/cicc-gzid/smi23d

74. OpenMOPAC 7.1, www.openmopac.net

75. www.wfmc.org/wfmc-standards_framework.html

76. http://sourceforge.net/projects/micro-workflow/

77. www.euras.be/eng/project.asp?ProjectId = 92

78. European Centre for Ecotoxicology and Toxicology of Chemicals, *Aquatic Toxicity (EAT) Database. Supplement to ECETOC. Aquatic Hazard Assessment II*, ECETOC, Brussels, 2003, Technical Report No. 91.

79. S. Wu, K. Blackburn, J. Amburguy, J. Jaworska and T. Federle, *Regul. Toxicol. Pharmacol.*, 2010, **56**, 67.

CHAPTER 18

Biological Read-Across: Mechanistically-Based Species–Species and Endpoint–Endpoint Extrapolations

M. T. D. CRONIN

School of Pharmacy and Chemistry, Liverpool John Moores University, Byrom Street, Liverpool L3 3AF, UK

18.1 Introduction

The use of surrogate, or alternative, species in the risk assessment of chemicals is seen as one of the main methods of reducing the *in vivo* use of (higher) animals in toxicology. This relies on there being a relationship (either direct or indirect) on the results from the different toxicity tests. There may be many uses of developing such relationships such as to provide range-finding information, identify hazardous chemicals or confirm other hypotheses (*e.g.* predictions from *in silico* methods).[1,2]

The purpose of this chapter is to discuss the possibilities for using information from alternative species to make an assessment of the effects to a similar or higher species. In particular, this includes direct correlations between effects in different species for the same endpoint as well as more speculative considerations of extrapolation between endpoints. There is an emphasis on including these species–species extrapolations in a mechanistic framework. Thus, they could be applied within a suitable mechanistic grouping or category.

Issues in Toxicology No.7
In Silico Toxicology: Principles and Applications
Edited by Mark T. D. Cronin and Judith C. Madden
© The Royal Society of Chemistry 2010
Published by the Royal Society of Chemistry, www.rsc.org

With this in mind, they may support and extend category formation. Their use is illustrated through some attempts to predict fish acute toxicity from data from similar—and aquatic non-fish—species. Use of an aquatic species as an illustration does not, of course, mean that mammalian toxicity endpoints cannot be treated in this manner; readers are encouraged to apply these techniques to a broad range of endpoints.

18.2 Extrapolation of Toxicological Information from One Species to Another

The premise of this section is that a similar toxicological effect will be caused in different species. There may, of course, be a change in magnitude of the effect (*e.g.* the concentration causing acute effects) and other species-specific issues will come into play. Species-specific issues include metabolism (activation or deactivation), clearance, distribution, cellular defence mechanisms, *etc.* As such, it is often useful to derive a formal relationship between the activities of (the same) chemicals to different species. These relationships provide a means of extrapolation and, if the statistics are considered, will give an estimate of the strength of a particular relationship and relative sensitivity. The next section illustrates various general methods of making these extrapolations of the effects between species. Various terms can be applied (note these are loose definitions and should not be interpreted in a regulatory context).

18.2.1 Inter-Species Relationships of Toxicity

For an inter-species relationship of toxicity, a direct correlation is usually sought between the toxicity potency of the same chemicals to different species.[3] Such relationships have occasionally in the past been termed 'quantitative activity–activity relationships' (QAARs).[4] Regression analysis is an ideal tool in this case to present the inter-species relationship of toxicity as it is clear and provides an unambiguous model. Typically, for a toxic potency, such models have the form:

$$C_1 = aC_2 + c \qquad (18.1)$$

where:

C_1 is the concentration causing the toxic effect to the species to be replaced (usually the higher species)
C_2 is the concentration causing the toxic effect to the alternative species
a is the regression coefficient
c is the constant.

Usually inter-species relationships of toxicity are thought of in terms of lethal (or equivalent) concentration—often, by convention, the logarithmic transformation of a molar concentration is applied.[3,4] However, there is no reason why other endpoints cannot be analysed in this way where some form of potency is to be assessed and a direct relationship may be derivable, *e.g.* irritation, reproductive capacity, *etc.* The slope *a* of the relationship in eqn (18.1) reflects the relative response of the two species, *i.e.* a slope of unity indicates a comparable response. Relative response may be affected by the test protocols, *i.e.* a longer test may allow hydrophobic chemicals to achieve equilibrium and provide a more accurate measurement of effect. The constant *c* of the relationship provides a relative estimate of species sensitivity—a species is deemed to be more sensitive if a lower concentration of the chemical produces a similar toxic response.

The examples below are directed towards acute aquatic effects, but it should be noted that such direct correlations may, in certain circumstances, be applicable for predicting acute mammalian toxicity.[5]

Inevitably with regression analysis, a whole variety of statistical tests can be applied to assess goodness-of-fit (see Chapter 9). Most useful for assessing goodness-of-fit are the square of the correlation coefficient adjusted for degrees of freedom (r^2), the standard error of the estimate (s) and Fisher's statistic (F). One obviously hopes for a good statistical fit between the toxicity values within the limits of experimental variability. However, outliers to a strong relationship can provide important information about inter-species differences including metabolism, specific toxic mechanisms, reaching equilibrium, *etc.*[6] Identifying chemicals that do not fit the inter-species relationships is, in many ways, as important as it helps to define the domain where the relationship is applicable. The outliers also assist in the definition and utilisation of chemical categories for biological extrapolation—as described in more detail below.

The development of inter-species relationships has not been, nor should be, restricted to linear regression analysis. It should be remembered that regression analysis assumes error only in the dependent variable, whereas error will be present in both variables. Other types of regression analysis have been applied (*e.g.* Model II regression analysis[7] and an errors-in-variables model[8]) but these have not gained popularity. A method accounting for error in both variables is, however, described in Section 18.6 and more fully by Asfaw *et al.*[9] In addition, ranking techniques can be applied; these approaches are seldom utilised but may have some applicability within categories of chemicals.

18.2.2 Quantitative Structure–Activity–Activity Relationships (QSAARs)

Quantitative structure–activity–activity relationships (QSAARs) have been utilised occasionally and there is a growing interest in their application for more ambitious extrapolations (*i.e.* those that go a greater distance between taxa or for mammalian endpoints). QSAARs attempt to improve inter-species relationships (QAARs) by the inclusion of physico-chemical properties and/or structural descriptors that are commonly applied in QSAR analysis. These have the general form:

$$C_1 = aC_2 + b \, \text{Prop}_n \ldots + c \tag{18.2}$$

where:

Prop_n is a physico-chemical property or structural descriptor
b is the regression coefficient.

As many descriptors can be included as required, although in practice as small a number as possible is preferred (preferably a single additional descriptor). The theory is that 'poor' inter-species relationships may be improved by the inclusion of terms that account for the inter-species variability or differences. Thus, the logarithm of the octanol–water partition coefficient (log P) may be used if it is able to add in more information in terms of distribution or uptake; parameters relating to metabolism could be considered, *etc.* Developers of QSAARs should be cautioned not to place too much reliance on the physico-chemical properties and/or structural descriptors as these may bias the equation and effectively negate the influence and input of the surrogate toxicity parameter—hence the equation will effectively turn into a QSAR.

18.3 Prediction Models

Related to, but not the same as, inter-species relationships of toxicity are 'prediction models' (PMs). The PM is an unambiguous algorithm for converting *in vitro* data into predictions of pharmacotoxicological endpoints in animals or humans.[10,11] The concept of PMs is particularly advanced in the area of the development of an alternative test method for the replacement of an animal test. The alternative itself is a combination of a test system and a prediction model. In this context they tend to have a regulatory implication for the validation of *in vitro* alternatives and may be similar in form to QAARs; they may be based on regression analysis (for quantitative endpoints) or discriminant analysis (for qualitative endpoints) or even simple statements (if . . . then) or their combination into a decision making process.[10] PMs tend to be written into Test Guidelines and become a formal statement for the relationship between two tests.

18.4 Examples of Extrapolation of Toxicity Between Species: Acute Aquatic Toxicity

This section considers the possibilities of using inter-species relationships of toxicity and QSAARs and demonstrates how toxicity can be extrapolated within and beyond taxa to provide evidence for the identification of a hazard associated with a chemical. Aquatic toxicity has been chosen to illustrate this effect, but it could also relate to mammalian endpoints.

18.4.1   Inter-Species Relationships of Toxicity: Within Taxa Extrapolation of Aquatic Toxicity

The success of inter-species relationships of toxicity relies on the intrinsic ability of the surrogate species to mimic the effect of a chemical on the species to be replaced. It is inevitable—and well established[12]—that the closer the species, the greater the likelihood of success. To illustrate this phenomenon, one can consider the acute toxicity of miscellaneous pesticides to the bluegill and rainbow trout. LeBlanc reported the toxicity of 13 pesticides[12] of varying chemical structure and mode of toxic action to these two fish. Analysis of the acute lethal toxicity (LD_{50}) reveals the following inter-species relationship:

$$\text{Trout } LD_{50} = 0.97 \text{ Bluegill } LD_{50} - 0.11$$
$$n = 13, \quad r^2 = 0.93 \tag{18.3}$$

where n is the number of observations.

The data are reported by LeBlanc and the relationship between the two sets of toxicity values[12]—as suggested by the statistical fit—is clearly very strong (albeit for a relatively small data set). Analysis of the slope (approximately unity) and intercept (approximately zero), respectively, show that the toxicity data of these two fish species correspond well and that they are equally sensitive. Interestingly, in view of later discussion, is that there is likely to be a mixture of (very specific) mechanisms in the pesticides.

Similar results have been obtained when extrapolating the toxicity of fish species within mechanisms of action, *i.e.* following the formation of a category or group of chemicals. Raevsky and co-workers demonstrated good fish–fish correlations for the toxicity of non-polar narcotics[13] and polar narcotics.[14] Whilst these observations are relevant, it is of greater significance that extrapolations can be made across mechanisms and modes as reported by LeBlanc.[12]

In terms of replacing whole animal tests, extrapolation within taxa is of limited value and will only be productive if there is a significant welfare advantage in using one test over another (but useful for combining data from different regions where different species are preferred—if equivalent, this approach can expand data sets for modelling). Despite this restriction, within taxa relationships of toxicity could be extremely useful in the framework of an Integrated Testing Strategy (ITS) to assist in the utilisation of 'non-standard' toxicity data. For instance, using fish acute toxicity as an example, historical data to 'non-standard' species, which may not be performed according to Organisation for Economic Co-operation and Development (OECD) Guidelines or Good Laboratory Practice (GLP) could be extrapolated to standard fish toxicity data with a high degree of confidence. Currently this is an area where some relationships are being formalised for aquatic effects (see Section 18.6), but less so for other endpoints. Despite this, such formalisations of relationships could provide a valuable source of information from historical data for a greater number of endpoints. Importantly, if the species are closely related (*i.e.* within taxa), extrapolations should be possible without the restrictions of mechanism/modes of toxic action.

18.4.2 Inter-Species Relationships of Toxicity: Extrapolation of Aquatic Toxicity Between Trophic Levels: General Models Across Mechanisms and Modes of Action

A greater challenge, albeit with the improved possibility of more useful inter-species models for the replacement of higher organisms, is the investigation of the relationships between lower and higher organisms. Again, aquatic toxicity is used to illustrate this process, although it is applicable, in theory, to any toxicity endpoint.[5]

To illustrate the possibilities for inter-species extrapolation, the results and findings of Kahn *et al.* have been analysed.[15] These authors took, as the higher species, toxicity data to the fathead minnow (summarised by Russom *et al.*[16]) and as the replacement lower species, toxicity data to the ciliated protozoan *Tetrahymena pyriformis* taken from various publications from the same laboratory.[17–23] The results from these two tests represent high quality and reliable data sets with the data measured in the same laboratories for each test (see Chapter 4). It should, however, also be noted that the *T. pyriformis* assay is not a regulatory accepted endpoint.

Kahn *et al.*[15] found a total of 364 compounds with toxicity data to both the fathead minnow (LC_{50}) and *T. pyriformis* (IGC_{50}); the compounds and their toxicity data are listed in Table 18.1. A simple regression analysis between the molar concentrations of the toxicities to the two species gives the following QAAR:

$$\log(1/LC_{50}) = 1.001 \ \log(1/IGC_{50}) + 0.557$$
$$n = 364, \ r^2 = 0.754, \ s = 0.642, \ F = 1109 \tag{18.4}$$

The relationship between the two sets of toxicity is shown graphically in Figure 18.1. There is a strong trend of increasing toxicity between both species but considerable scatter about the plot. The slope (of unity) suggests the two sets of potency values are concordant; the intercept suggests that the fish assay is more sensitive than the protozoan by about 0.5 log units. The increased sensitivity of the fish is to be expected for a number of reasons including:

- the greater complexity of the organism (though it may have a greater defensive metabolism capacity);
- the length of the test, *i.e.* 96 hours *vs.* 40 hours; and
- the protocol—the *T. pyriformis* assay uses an organic growth medium which itself may sorb the chemicals and reduce the level of toxicant.

There are at least two possibilities for improving the statistical quality (*i.e.* goodness-of-fit) of eqn (18.4). The first is the identification and removal of outliers (some comments are made regarding this in the next section). The

Table 18.1 Toxicity data to the fathead minnow (LC$_{50}$), *Tetrahymena pyriformis* (IGC$_{50}$) for the compounds considered in eqn (18.4), extra descriptors included in eqn (18.5), ordered according to mechanism of action. The non-polar narcotic compounds are considered in eqn (18.6). Data matrix adapted from Kahn *et al.*[15]

Name	CAS	log (1/LC$_{50}$) mmol l^{-1} P. pinephales	log (1/IGC$_{50}$) mmol l^{-1} T. pyriformis	P$_C^{avg}$	#Nrel	HACA2
Compounds assumed to be acting by a non-polar narcotic mechanism of action						
methanol	67561	−2.96	−2.67	0.987	0.000	1.174
dimethylsulfoxide	67685	−2.64	−2.49	0.785	0.000	0.000
ethanol	64175	−2.50	−1.99	0.985	0.000	1.010
2-propanol	67630	−2.16	−1.88	0.983	0.000	0.979
acetone	67641	−2.09	−2.20	1.078	0.000	0.875
2-methyl-2-propanol	75650	−1.94	−1.79	0.982	0.000	0.926
1-propanol	71238	−1.88	−1.75	0.984	0.000	1.065
2-butanone	78933	−1.65	−1.75	1.054	0.000	0.793
1-butanol	71363	−1.37	−1.43	0.984	0.000	1.027
2-methyl-1-propanol	78831	−1.29	−1.37	0.984	0.000	0.894
3-pentanone	96220	−1.25	−1.46	1.040	0.000	0.702
N,N-diethylethanolamine	100378	−1.18	−1.50	0.979	0.044	1.063
2-pentanone	107879	−1.16	−1.22	1.040	0.000	0.804
3-methyl-1-pentyn-3-ol	77758	−1.09	−1.32	1.106	0.000	0.923
3-methyl-2-butanone	563804	−1.00	−1.17	1.040	0.000	0.809
cyclohexanol	108930	−0.85	−0.77	0.980	0.000	0.954
cyclohexanone	108941	−0.80	−1.23	1.031	0.000	0.846
1-pentanol	71410	−0.73	−1.03	0.984	0.000	1.046
4-methyl-2-pentanone (methyl isobutyl ketone)	108101	−0.72	−1.21	1.030	0.000	0.826
methyl acetate	79209	−0.68	−1.60	1.063	0.000	0.000
2-hexanone	591786	−0.63	−1.34	1.030	0.000	0.792
2,4-dimethyl-3-pentanol	600362	−0.15	−0.71	0.981	0.000	0.846
5-methyl-2-hexanone	110123	−0.14	−0.65	1.022	0.000	0.797

acetophenone	98862	−0.13	−0.46	1.021	0.000	0.833
2-heptanone	110430	−0.06	−0.49	1.023	0.000	0.804
1-hexanol	111273	0.02	−0.38	0.983	0.000	1.027
3,3-dimethyl-2-butanone	75978	0.06	−1.44	1.029	0.000	0.789
dimethyl phthalate	131113	0.21	−0.44	1.039	0.000	0.000
propyl acetate	109604	0.23	−1.24	1.033	0.000	0.000
hexylamine	111262	0.25	−0.22	0.984	0.046	0.506
N,N-dimethylaniline	121697	0.28	0.28	0.978	0.050	0.250
benzyl-*tert*-butanol	103059	0.39	−0.07	0.978	0.000	0.812
toluene	108883	0.43	−0.50	0.979	0.000	0.000
1-heptanol	111706	0.53	0.11	0.983	0.000	1.027
2-octanone	111137	0.55	−0.15	1.018	0.000	0.792
benzene	71432	0.65	−0.12	0.980	0.000	0.000
2-ethyl-1-hexanol	104767	0.66	0.17	0.981	0.000	1.080
5-nonanone	502567	0.66	0.07	1.014	0.000	0.673
N-propyl sulfide	111477	0.74	0.00	0.982	0.000	0.351
α,α,α-4-tetrafluoro-*m*-toluidine	2357473	0.77	0.77	0.969	0.059	0.504
chlorobenzene	108907	0.82	−0.13	0.978	0.000	0.000
monobromobenzene	108861	0.94	0.08	0.976	0.000	0.000
N,N-diethylaniline	91667	0.96	0.67	0.977	0.039	0.187
2-nonanone	821556	0.97	0.66	1.014	0.000	0.833
1-octanol	111875	0.98	0.58	0.983	0.000	1.027
p-xylene	106423	1.08	0.12	0.979	0.000	0.000
1,2-dichlorobenzene	95501	1.19	0.53	0.976	0.000	0.000
isopropylbenzene	98828	1.28	0.69	0.980	0.000	0.000
naphthalene	91203	1.32	−0.12	1.099	0.000	0.000
1-nonanol	143088	1.40	0.86	0.983	0.000	0.000
2-decanone	693549	1.51	0.58	1.011	0.000	1.027
hexyl acetate	142927	1.52	−0.04	1.015	0.000	0.792
1-bromohexane	111251	1.68	0.94	0.982	0.000	0.000
pentyl ether	693652	1.70	1.09	0.982	0.000	0.000
3,4-dichlorotoluene	95750	1.74	1.07	0.976	0.000	0.000
1,2,4-trichlorobenzene	120821	1.78	1.08	0.974	0.000	0.000

Table 18.1 (*continued*)

Name	CAS	$\log (1/LC_{50})$ $mmol\,l^{-1}$ P. pimephales	$\log (1/IGC_{50})$ $mmol\,l^{-1}$ T. pyriformis	P_C^{avg}	#Nrel	HACA2
1-decanol	112301	1.82	1.34	0.983	0.000	1.027
butyl benzene	104518	1.83	1.25	0.980	0.000	0.000
biphenyl	92524	1.90	1.05	0.975	0.000	0.000
amylbenzene	538681	1.94	1.79	0.981	0.000	0.000
2-undecanone	112129	2.06	1.53	1.008	0.000	0.815
1-bromoheptane	629049	2.09	1.49	0.982	0.000	0.000
2-dodecanone	6175491	2.19	1.67	1.006	0.000	0.821
1-undecanol	112425	2.22	1.96	0.983	0.000	1.027
1-dodecanol	112538	2.27	2.16	0.983	0.000	1.027
1-bromooctane	111831	2.36	1.87	0.982	0.000	0.000
1-tridecanol	112709	2.59	2.37	0.983	0.000	1.027
2-tridecanone	593088	2.74	2.12	1.004	0.000	0.821
7-tridecanone (dihexyl ketone)	462180	2.79	1.52	1.004	0.000	0.672
Other mechanisms of toxic action (unspecified)						
diethanolamine	111422	−2.65	−1.03	0.979	0.056	2.309
2-methyl-2,4-pentanediol	107415	−1.96	−1.96	0.979	0.000	1.660
3,3-dimethylglutaric acid	4839467	−1.94	−0.66	1.062	0.000	1.655
triethanolamine	102716	−1.90	−1.75	0.975	0.040	3.150
ethyl carbamate (urethane)	51796	−1.77	−1.65	1.067	0.077	0.355
propionic acid, sodium salt	137406	−1.70	−1.45	1.077	0.000	0.879
trimethyl phosphate	512561	−1.70	−1.82	0.959	0.000	0.000
acetonitrile	75058	−1.60	−2.28	1.364	0.167	0.508
2-methyl-3-butyn-2-ol	115195	−1.59	−1.49	1.137	0.000	0.904
1-amino-2-propanol	78966	−1.53	−0.93	0.981	0.071	1.401
2-aminoethanol	141435	−1.53	−1.01	0.982	0.091	1.509
propionitrile	107120	−1.44	−1.97	1.222	0.111	0.504

Compound	CAS					
1-methylpiperazine	109013	−1.36	−0.96	0.981	0.105	0.589
1,4-dicyanobutane	111693	−1.25	−1.54	1.233	0.125	0.979
1-(2-aminoethyl)piperazine	140318	−1.23	−0.79	0.979	0.125	0.907
2-(ethylamino)ethanol	110736	−1.22	−0.77	0.982	0.059	1.284
1,3-diaminopropane	109762	−1.21	−0.70	0.984	0.133	1.027
1,4-bis(3-aminopropyl)piperazine	7209383	−1.19	−0.66	0.979	0.105	1.475
1,2-diaminopropane	78900	−1.13	−0.56	0.983	0.133	0.958
diethylacetamide	685916	−1.11	−1.54	1.022	0.048	0.806
1-benzylpyridinium-3-sulfonate	69723940	−0.99	−1.50	1.016	0.036	0.088
2-butanone oxime	96297	−0.99	−1.07	1.058	0.067	0.753
2-picoline	109068	−0.98	−1.01	0.978	0.071	0.466
5-chloro-2-pyridinol	4214793	−0.94	−0.75	1.040	0.083	1.265
3-chloro-1-propanol	627305	−0.93	−1.40	0.980	0.000	0.889
acetoxime	127060	−0.88	−1.25	1.082	0.083	0.745
3-acetamidophenol (3-hydroxyacetanilide)	621421	−0.87	0.16	1.015	0.050	1.888
2-cyanopyridine	100709	−0.84	−0.79	1.104	0.167	0.911
2-methoxyethylamine	109853	−0.84	−1.79	0.980	0.071	0.493
benzamide	55210	−0.74	−0.91	1.029	0.063	1.150
4-acetamidophenol (4-hydroxyacetanilide)	103902	−0.73	−0.82	1.017	0.050	1.905
propylamine	107108	−0.72	−0.71	0.986	0.077	0.542
2-chloro-3-pyridinol	6636788	−0.68	−0.04	0.975	0.083	1.331
4-picoline	108894	−0.64	−0.88	0.978	0.071	0.493
2,4,5-trimethyloxazole	20662844	−0.61	−1.01	1.030	0.059	0.387
1,6-dicyanohexane	629403	−0.59	−0.77	1.162	0.091	1.015
2-ethylpyridine	100710	−0.59	−0.87	0.979	0.059	0.417
(+/−)-secbutylamine	13952846	−0.58	−0.67	0.985	0.063	0.474
Cis-3-hexen-1-ol	928961	−0.58	−0.81	1.036	0.000	1.090
butylamine	109739	−0.56	−0.57	0.985	0.063	0.506
2-methylimidazole	693981	−0.54	−0.91	1.066	0.167	0.812
benzoic acid, sodium salt	532321	−0.53	−0.92	1.026	0.000	0.818

Table 18.1 (*continued*)

Name	CAS	log (1/LC$_{50}$) mmol l^{-1} P. pimephales	log (1/IGC$_{50}$) mmol l^{-1} T. pyriformis	P_C^{avg}	#Nrel	HACA2
1,2-dimethylpropylamine	598743	−0.51	−0.71	0.983	0.053	0.432
pyrrole	109977	−0.50	−1.09	1.085	0.100	0.317
anthranilamide	88686	−0.46	−0.87	1.084	0.111	1.527
(*tert*)-butyl acetate	540885	−0.45	−1.49	1.023	0.000	0.000
3-aminoacetophenone	99036	−0.45	−0.82	1.018	0.053	1.302
hexanoic acid	142621	−0.44	−0.21	1.029	0.000	0.878
2-bromo-3-pyridinol	6602320	−0.43	0.00	0.974	0.083	1.366
allyl cyanide	109751	−0.43	−1.48	1.297	0.100	0.493
trans-3-hexen-1-ol	928972	−0.43	−0.78	1.036	0.000	1.090
ethyl acetate	141786	−0.42	−1.30	1.044	0.000	0.000
1-chloro-2-propanol	127004	−0.41	−1.49	0.980	0.000	0.953
methyl methacrylate	80626	−0.41	−1.22	1.107	0.000	0.000
amylamine	110587	−0.31	−0.48	0.985	0.053	0.533
2,2,2-trichloroethanol	115208	−0.30	−0.46	0.973	0.000	0.907
2,6-pyridinedicarboxylic acid	499832	−0.28	0.12	1.072	0.059	1.947
2,2-dichloroacetamide	683727	−0.27	−0.98	1.126	0.111	1.060
6-chloro-2-picoline	18368633	−0.26	−0.48	0.976	0.071	0.318
cyclohexanone oxime	100641	−0.26	−0.80	1.033	0.053	0.783
2-hydroxyethyl methacrylate	868779	−0.24	−1.08	1.080	0.000	1.063
6-chloro-2-pyridinol	16879020	−0.22	−0.29	1.041	0.083	1.181
3-picoline	108996	−0.19	−0.99	0.978	0.071	0.528
acrylamide	79061	−0.19	−0.79	1.240	0.100	1.191
4-toluidine	106490	−0.17	−0.05	0.978	0.059	0.527
diethyl benzylphosphonate	1080326	−0.17	−0.43	0.970	0.000	0.000
4-acetylpyridine	1122549	−0.14	−0.87	1.022	0.063	1.306
2,4-pentanedione (acetylacetone)	123546	−0.13	−0.27	1.101	0.000	1.596

2-acetyl-1-methylpyrrole	932161	−0.11	−0.69	1.076	0.056	0.863
acetaldoxime	107299	−0.11	−0.89	1.136	0.111	0.906
pyridine	110861	−0.10	−1.32	0.978	0.091	0.494
3-hydroxy-2-nitropyridine	15128822	−0.08	0.87	1.032	0.143	1.228
2-nitrophenol	88755	−0.06	0.67	1.100	0.067	0.927
aniline	62533	−0.05	−0.23	0.978	0.071	0.527
3-(3-pyridyl)-1-propanol	2859678	−0.04	−0.84	0.977	0.048	1.398
3-cyano-4,6-dimethyl-2-hydroxypyridine	769288	−0.03	−0.70	1.166	0.105	1.589
2-dimethylaminopyridine	5683330	−0.02	−0.55	1.018	0.105	0.600
2-amino-5-bromopyridine	1072975	−0.01	0.49	1.040	0.154	0.891
2-amino-4-chloro-6-methylpyrimidine	5600215	0.01	−0.47	1.175	0.200	0.999
nitrobenzene	98953	0.01	0.14	0.967	0.071	0.000
benzylamine	100469	0.02	−0.24	0.978	0.059	0.527
N-methylaniline	100618	0.03	0.06	0.978	0.059	0.376
4-nitroaniline	100016	0.04	1.88	1.183	0.125	0.506
4-methoxyphenol	150765	0.05	−0.14	0.975	0.000	0.979
1,2-bis(4-pyridyl)ethane	4916578	0.09	−0.03	0.976	0.077	0.996
diethyl succinate	123251	0.09	−0.85	1.045	0.000	0.000
4-nitrobenzamide	619807	0.10	0.18	1.020	0.111	1.145
2-phenyl-3-butyn-2-ol	127662	0.11	−0.18	1.059	0.000	0.852
salicylamide (2-hydroxybenzamide)	65452	0.13	−0.24	1.147	0.059	2.052
5-ethyl-2-methylpyridine	104905	0.17	−0.18	0.979	0.050	0.445
6-methyl-5-hepten-2-one	110930	0.17	−0.45	1.061	0.000	0.803
2-chloroethanol	107073	0.18	−1.42	0.979	0.000	1.037
2-butyn-1,4-diol	110656	0.21	−1.88	1.184	0.000	2.138
3-methoxyphenol	150196	0.22	−0.33	0.974	0.000	0.956
4-ethylaniline	589162	0.22	0.05	0.979	0.050	0.493
quinoline	91225	0.22	0.09	1.104	0.059	0.456
4-benzoylpyridine	14548460	0.25	−0.09	1.004	0.044	1.284

Table 18.1 (continued)

Name	CAS	log (1/LC$_{50}$) mmol l^{-1} P. pimephales	log (1/IGC$_{50}$) mmol l^{-1} T. pyriformis	P_C^{avg}	#Nrel	HACA2
vanillin (3-methoxy-4-hydroxybenzaldehyde)	121335	0.26	−0.03	1.059	0.000	1.792
3-ethoxy-4-hydroxybenzaldehyde	121324	0.28	0.02	1.048	0.000	1.835
3-butyn-1-ol	927742	0.29	−1.84	1.190	0.000	0.985
3-methylphenol	108394	0.29	−0.06	0.977	0.000	1.028
2,4-dichlorobenzamide	2447792	0.30	−0.36	1.027	0.063	1.055
resorscinol	108463	0.30	−0.65	0.973	0.000	2.000
benzothiazole	95169	0.33	−0.26	1.080	0.071	0.751
2-tolualdehyde	529204	0.36	−0.01	1.019	0.000	0.838
N-ethylbenzylamine	14321278	0.37	0.22	0.980	0.044	0.272
(+,−)-4-pentyn-2-ol	2117115	0.38	−1.63	1.137	0.000	0.964
2-hydroxy-4-methoxyacetophenone	552410	0.38	0.55	1.136	0.000	1.683
1,5-hexadien-3-ol	924414	0.41	0.25	1.105	0.000	0.876
2-tolunitrile	529191	0.42	−0.24	1.077	0.063	0.462
phenol	108952	0.46	−0.21	0.977	0.000	0.984
2,3-dibromopropanol	96139	0.49	−0.49	0.972	0.000	0.974
4-nitrophenol	100027	0.49	1.42	1.097	0.067	0.958
4-dimethylaminobenzaldehyde	100107	0.51	0.23	1.009	0.046	1.141
1-hexen-3-ol	4798441	0.52	−0.81	1.037	0.000	0.990
isopropyl methacrylate	4655349	0.53	−0.88	1.069	0.000	0.000
methyl 4-cyanobenzoate	1129357	0.54	−0.06	1.099	0.053	0.474
N,N-dimethylbenzylamine	103833	0.55	0.12	0.979	0.044	0.194
methyl 2,4-dihydroxybenzoate	2150472	0.56	0.61	1.162	0.000	1.901
4-bromoaniline	106401	0.56	1.01	0.974	0.071	0.507
1-bromobutane	109659	0.57	−0.18	0.982	0.000	0.000

carbon tetrachloride	56235	0.57	−0.02	0.971	0.000	0.000
5-bromovanillin	2973764	0.59	0.62	1.108	0.000	1.630
2,4,5-trimethoxybenzaldehyde	4460860	0.60	−0.10	1.068	0.000	0.822
2-chloro-4-methylaniline	615656	0.60	0.18	0.976	0.059	0.505
5-hydroxy-2-nitrobenzaldehyde	42454068	0.60	0.33	1.139	0.059	1.758
4-chloroaniline	106478	0.61	1.35	0.976	0.071	0.506
4-amino-2-nitrophenol	119346	0.63	0.88	1.099	0.118	1.482
2,6-dinitrophenol	573568	0.67	0.54	1.087	0.118	0.749
ethyl 4-aminobenzoate	94097	0.67	0.70	1.010	0.044	0.516
2,3,4,5,6-pentafluoroaniline	771608	0.69	0.20	1.190	0.071	0.431
butyraldehyde	123728	0.69	−0.38	1.055	0.000	0.857
1-fluoro-4-nitrobenzene	350469	0.70	0.25	0.964	0.071	0.000
triphenylphosphine oxide	791286	0.71	0.77	0.958	0.000	0.000
heptylamine	111682	0.72	0.21	0.984	0.040	0.524
1-chloro-2-nitrobenzene	88733	0.73	0.63	0.967	0.071	0.000
2-amino-5-chlorobenzonitrile	5922601	0.73	0.44	1.235	0.133	0.937
2-methylvaleraldehyde	123159	0.73	−0.47	1.031	0.000	0.804
3-nitrotoluene	99081	0.73	0.05	0.969	0.059	0.000
4-ethoxybenzaldehyde	10031820	0.73	0.07	1.010	0.000	0.836
2-ethoxyethyl methacrylate	2370630	0.76	−0.78	1.052	0.000	0.000
hexanal (hexylaldehyde)	66251	0.76	−0.17	1.031	0.000	0.846
di-3-butyn-2-ol	2028639	0.78	−0.40	1.190	0.000	0.995
α,α,α-4-tetrafluoro-2-toluidine	393395	0.78	−0.02	1.019	0.059	0.445
3-pyridinecarboxaldehyde	500221	0.81	−0.15	1.028	0.077	1.343
butyl acetate	123864	0.81	−0.49	1.025	0.000	0.000
4-methylphenol	106445	0.82	−0.18	0.977	0.000	0.984
valeraldehyde	110623	0.82	−0.02	1.041	0.000	0.846
2-butyn-1-ol	764012	0.84	−0.87	1.187	0.000	1.069
4-allyl-2-methoxyphenol(eugenol)	97530	0.84	0.42	1.012	0.000	0.937
diethyl phthalate	84662	0.84	0.23	1.028	0.000	0.000
4-ethoxy-2-nitroaniline	616864	0.85	0.76	1.057	0.087	0.431
2,4-dimethylphenol	105679	0.87	0.07	0.977	0.000	0.803

Table 18.1 (*continued*)

Name	CAS	log (1/LC_{50}) mmol l^{-1} P. pimephales	log (1/IGC_{50}) mmol l^{-1} T. pyriformis	P_C^{avg}	#Nrel	HACA2
2,4-dinitrotoluene	121142	0.87	0.87	0.958	0.105	0.000
methyl 4-nitrobenzoate	619501	0.88	0.40	1.009	0.050	0.000
methyl 4-chloro-2-nitrobenzoate	42087809	0.89	0.82	1.009	0.050	0.000
2-methylphenol	95487	0.89	−0.29	0.977	0.000	0.908
1-chloro-3-nitrobenzene	121733	0.92	0.73	0.966	0.071	0.000
2,4-dimethoxybenzaldehyde	613456	0.92	−0.06	1.130	0.000	0.822
2,3-benzofuran	271896	0.93	−0.11	1.130	0.000	0.000
2-chloro-4-nitroaniline	121879	0.93	0.75	1.182	0.125	0.451
2-methylbutyraldehyde	96173	0.94	−0.31	1.041	0.000	0.816
3,4-dimethylphenol	95658	0.94	0.12	0.977	0.000	1.002
2-allylphenol	1745819	0.95	0.33	1.020	0.000	1.000
4-phenylpyridine	939231	0.98	0.66	0.974	0.048	0.491
α-decalactone	706149	0.98	0.49	1.008	0.000	0.000
2-chloro-6-methylbenzonitrile	6575093	1.00	0.46	1.076	0.063	0.428
2,4,6-trimethylphenol	527606	1.02	0.28	0.977	0.000	0.834
2,4-dihydroxybenzaldehyde	95012	1.02	0.73	1.217	0.000	2.666
2-nitrobenzaldehyde	552896	1.02	0.17	1.020	0.063	0.830
benzaldehyde	100527	1.03	−0.20	1.027	0.000	0.839
diethyl malonate	105533	1.04	−1.00	1.056	0.000	0.000
2-chlorophenol	95578	1.05	0.18	0.975	0.000	0.983
diethyl adipate	141286	1.05	−0.13	1.032	0.000	0.000
2,6-diisopropylaniline	24544045	1.06	0.78	0.979	0.031	0.454
4-ethylphenol	123079	1.07	0.21	0.978	0.000	0.993
catechol	120809	1.08	0.75	0.975	0.000	1.922
2,4-dinitroaniline	97029	1.09	0.53	1.174	0.167	0.412
4-propylphenol	645567	1.09	0.64	0.978	0.000	1.002

benzophenone	119619	1.09	0.87	1.004	0.000	0.758
butyl butyrate	109217	1.09	0.52	1.015	0.000	0.000
4-bromophenyl 3-pyridyl ketone	14548459	1.11	0.82	1.002	0.044	1.238
ethyl benzoate	93890	1.14	−0.01	1.012	0.000	0.000
3-methylindole	83341	1.17	0.37	1.111	0.053	0.335
4-butylaniline	104132	1.17	1.07	0.979	0.039	0.519
methyl 2,5-dichlorobenzoate	2905693	1.17	0.81	1.015	0.000	0.000
4-nitrobenzaldehyde	555168	1.18	0.20	1.018	0.063	0.856
methyl 4-chlorobenzoate	1126461	1.19	0.42	1.015	0.000	0.000
2′,4′-dichloroacetophenone	2234164	1.21	0.62	1.019	0.000	0.809
ethyl caproate (ethyl hexanoate)	123660	1.21	0.06	1.014	0.000	0.000
2,3,6-trimethylphenol	2416946	1.22	0.28	0.977	0.000	0.855
2,4-dinitrophenol	51285	1.23	1.08	1.168	0.118	0.937
2,4-dichlorophenol	120832	1.32	1.04	0.973	0.000	0.976
4-chlorophenol	106489	1.32	0.55	0.975	0.000	0.959
3,4-dichloroaniline	95761	1.33	1.37	0.974	0.071	0.515
3,5-dibromo-4-hydroxybenzonitrile	1689845	1.34	1.16	1.155	0.071	1.271
3-benzyloxyaniline	1484260	1.34	0.88	1.002	0.036	0.504
2-chloroaniline	95512	1.35	−0.25	0.976	0.071	0.497
4-isopropyl benzaldehyde	122032	1.35	0.67	1.011	0.000	0.849
4-nitro-3-(trifluoromethyl)-phenol	88302	1.36	1.65	1.069	0.056	0.997
2-hydroxyethyl acrylate	818611	1.38	0.69	1.100	0.000	1.089
octylamine	111864	1.40	0.35	0.984	0.036	0.506
3-bromothiophene	872311	1.42	−0.04	1.078	0.000	0.369
4-chloro-3-methyl phenol	59507	1.42	0.80	0.975	0.000	0.959
isovaleraldehyde	590863	1.42	−0.33	1.040	0.000	0.811
nonanenitrile (octyl cyanide)	2243278	1.42	0.62	1.056	0.037	0.513
2-phenylphenol	90437	1.44	1.09	0.976	0.000	0.979
4-(*tert*)-butylphenol	98544	1.46	0.91	0.978	0.000	0.984

Table 18.1 (continued)

Name	CAS	$\log(1/LC_{50})$ mmol l^{-1} P. pimephales	$\log(1/IGC_{50})$ mmol l^{-1} T. pyriformis	P_C^{avg}	#Nrel	HACA2
4-dimethylamino-cinnamaldehyde	6203185	1.47	0.52	1.035	0.039	1.195
1-naphthol	90153	1.49	0.75	1.098	0.000	0.884
dimethyl nitroterephthalate	5292455	1.56	0.43	1.033	0.039	0.000
2-propyn-1-ol (propargyl alcohol)	107197	1.58	−1.07	1.293	0.000	1.060
benzyl methacrylate	2495376	1.58	0.65	1.041	0.000	0.000
4-phenoxyphenol	831823	1.58	1.36	0.974	0.000	0.996
2-hydroxypropyl acrylate	999611	1.59	0.65	1.079	0.000	0.943
ethyl acrylate	140885	1.60	0.52	1.106	0.000	0.000
2,4,6-trichlorophenol	88062	1.61	1.41	1.035	0.000	0.799
N-butyl sulfide	544401	1.61	1.04	0.982	0.000	0.369
p-phenoxybenzaldehyde	67367	1.63	1.26	1.000	0.000	0.831
salicylaldoxime	94677	1.63	0.25	1.149	0.059	1.828
diethyl benzylmalonate	607818	1.66	0.71	1.020	0.000	0.000
2-chloro-5-nitrobenzaldehyde	6361213	1.68	0.53	1.015	0.063	0.825
phenyl 4-aminosalicylate	133119	1.68	1.41	1.091	0.036	1.456
2,4,5-tribromoimidazole (nominal)	2034222	1.70	1.43	1.087	0.222	0.484
2,4,6-tribromophenol	118796	1.70	2.03	0.969	0.000	0.757
dibutyl succinate	141037	1.71	0.51	1.025	0.000	0.000
2,3,4-trichloroaniline	634673	1.73	1.35	1.108	0.071	0.477
salicylaldehyde (2-hydroxybenzaldehyde)	90028	1.73	0.42	1.145	0.000	1.756
2,5-dinitrophenol	329715	1.74	0.95	1.088	0.118	0.931
chloroacetonitrile	107142	1.75	0.85	1.361	0.167	0.442
isobutyl acrylate	106638	1.79	0.29	1.069	0.000	0.000

1-heptyn-3-ol	7383199	1.80	−0.27	1.086	0.000	0.992
4-*tert*-pentylphenol	80466	1.80	1.23	0.977	0.000	1.011
flavone	525826	1.80	1.41	1.023	0.000	0.768
2-vanillin(3-methoxysalicylaldehyde)	148538	1.80	0.38	1.014	0.000	1.648
4-chlorobenzaldehyde	104881	1.81	0.40	1.025	0.000	0.857
4-hexyloxyaniline (nominal conc.)	39905572	1.81	1.38	0.978	0.030	0.515
dicumarol	66762	1.82	1.70	1.092	0.000	1.564
nonylamine	112209	1.82	1.70	0.984	0.032	0.515
4,6-dimethoxy-2-hydroxybenzaldehyde	708769	1.83	0.62	1.127	0.000	1.692
dibutyl adipate	105997	1.85	0.79	1.017	0.000	0.000
4-nitrophenyl phenyl ether	620882	1.91	1.58	1.031	0.040	0.000
4-chlorocatechol	2138229	1.96	1.06	0.973	0.000	1.756
2-fluorobenzaldehyde	446526	1.96	0.08	1.024	0.000	0.808
diethyl sebacate	110407	1.98	1.35	1.018	0.000	0.000
2,4-dichlorobenzaldehyde	874420	1.99	1.04	1.024	0.000	0.829
cyclohexyl acrylate	3066715	2.02	0.76	1.049	0.000	0.000
2′,3′,4′-trichloroacetophenone	13608872	2.05	1.34	1.018	0.000	0.793
dibenzofuran	132649	2.05	1.42	0.973	0.000	0.000
4,6-dinitro-2-cresol (4,6-dinitro-2-methylphenol)	534521	2.06	1.72	1.130	0.100	0.785
malononitrile	109773	2.07	0.22	1.616	0.286	0.966
allyl methacrylate	96059	2.11	−0.68	1.130	0.000	0.000
hexyl acrylate	2499958	2.14	0.71	1.049	0.000	0.000
2-decyn-1-ol	4117140	2.16	0.99	1.051	0.000	1.086
1-benzoylacetone	93914	2.17	0.57	1.044	0.000	1.505
decylamine	2016571	2.18	2.06	0.983	0.029	0.506
5-bromosalicylaldehyde	1761611	2.19	1.11	1.142	0.000	1.694
4-phenylazophenol	1689823	2.23	1.66	0.974	0.080	1.348
pentaflurobenzaldehyde	653372	2.25	0.82	1.146	0.000	0.750
2-propen-1-ol (allyl alcohol)	107186	2.26	−1.92	1.099	0.000	1.069

Table 18.1 (*continued*)

Name	CAS	$\log (1/LC_{50})$ mmol l^{-1} P. pimephales	$\log (1/IGC_{50})$ mmol l^{-1} T. pyriformis	P_C^{avg}	#Nrel	HACA2
tetrachlorocatechol	1198556	2.29	1.70	1.191	0.000	1.618
5-chlorosalicylaldehyde	635938	2.31	1.01	1.144	0.000	1.709
diethyl chloromalonate	14064109	2.31	0.63	1.056	0.000	0.000
carbazole	86748	2.33	0.91	1.039	0.046	0.329
2,4,5-trichlorophenol	95954	2.34	2.10	1.035	0.000	0.980
2,3,4,6-tetrachlorophenol	58902	2.35	2.18	1.190	0.000	0.789
2,3,5,6-tetrachlorophenol	935955	2.35	2.22	1.191	0.000	0.838
1,4-dinitrobenzene	100254	2.37	1.30	0.957	0.125	0.000
di-*n*-butylorthophthalate	84742	2.44	1.60	1.015	0.000	0.000
diisobutyl phthalate	84695	2.49	1.44	1.015	0.000	0.000
3,5-dibromosalicylaldehyde	90595	2.52	1.64	1.077	0.000	1.687
dibromoacetonitrile	3252435	2.56	2.40	1.354	0.167	0.397
dibutyl fumarate	105759	2.56	1.49	1.054	0.000	0.000
triphenyl phosphate	115866	2.57	1.81	0.974	0.000	0.000
2,4,6-triiodophenol	609234	2.59	2.67	0.966	0.000	0.789
N-undecyl cyanide	2437254	2.62	1.90	1.037	0.028	0.513
4-*tert*-butyl-2,6-dinitrophenol	4097498	2.65	1.80	1.035	0.069	0.696
pentachloropyridine	2176627	2.73	1.68	1.040	0.091	0.188

2,3,4,5-tetrachlorophenol	4901513	2.75	2.71	1.190	0.000	0.901
2,6-di(*tert*)butyl-4-methylphenol	128370	2.78	1.80	0.978	0.000	0.480
undecylamine	7307553	2.91	2.33	0.983	0.027	0.515
2,3,5,6-tetrachloroaniline	3481207	2.93	1.76	1.192	0.071	0.452
2,2′-methylenebis(4-chlorophenol)	97234	2.94	3.09	0.973	0.000	1.568
pentachlorophenol	87865	3.04	2.05	1.191	0.000	0.814
allyl isothiocyanate	57067	3.06	2.06	1.236	0.091	0.301
1,3,5-trichloro-2,4-dinitrobenzene	6284839	3.09	2.19	1.169	0.125	0.000
2-methyl-1,4-naphthoquinone	58275	3.19	1.54	1.079	0.000	1.467
4-nonylphenol	104405	3.20	2.47	0.980	0.000	0.984
4-octylaniline	16245797	3.23	2.43	0.980	0.026	0.527
acrolein (2-propenal)	107028	3.52	1.87	1.239	0.000	0.920
1,3-dichloro-4,6-dinitrobenzene	3698837	3.72	2.57	1.168	0.125	0.000
1,5-dichloro-2,3-dinitrobenzene	28689089	3.72	2.60	1.016	0.125	0.000
pentabromophenol	608719	3.72	2.66	1.185	0.000	0.820
2,2′-methylene bis(3,4,6-trichlorophenol)	70304	4.29	3.04	1.095	0.000	1.261
N-vinylcarbazole	1484135	4.78	2.24	1.120	0.039	0.173

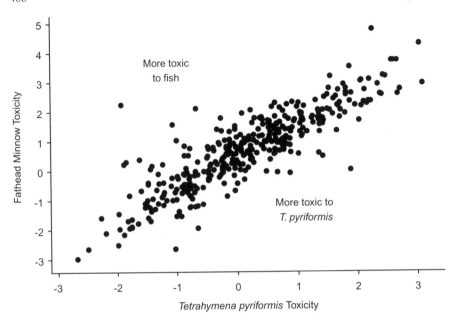

Figure 18.1   Plot of toxicities to the fathead minnow and *T. pyriformis* for 364 organic chemicals, as described by eqn (18.4). Data from Kahn *et al.*[15]

second is the inclusion of further descriptors to form a QSAAR. Kahn *et al.*[15] reported a significant improvement to eqn (18.4) by the inclusion of three further descriptors calculated using the CODESSA software:

$$\log(1/\text{LC}_{50}) = 0.964 \, \log(1/\text{IGC}_{50}) + 3.55 \, P_C^{avg} - 5.88 \, \#N_{rel}$$
$$- 0.341 \, HACA2 - 2.69 \qquad (18.5)$$
$$n = 364, \; R^2 = 0.824, \; s = 0.545, \; F = 421$$

where:

P_C^{avg} is the average bond order of a carbon atom
$\#N_{rel}$ is the relative number of nitrogen atoms
$HACA2$ is the area-weighted surface charge of the hydrogen-bonding acceptor atoms in the molecule.

The precise meaning of the descriptors included in eqn (18.5) is difficult to determine with regard as to why they should improve the inter-species relationship. They may, however, be responsible for improving the prediction of toxicity for the reactive compounds which are identified as outliers in the next section. Whether or not eqn (18.5) provides any significant improvement over a QSAR without the need for *T. pyriformis* data is difficult to ascertain.

18.4.3 Inter-Species Relationships of Toxicity: Extrapolation of Aquatic Toxicity Between Trophic Levels: Improvement of Models Through Consideration of Mechanism of Action and Category Formation

Analysis of eqn (18.4) to identify outliers indicates that compounds whose toxicity to fish is consistently under-predicted by *T. pyriformis* have specific mechanisms of toxicity. Typically such compounds are small and intrinsically reactive *via* various electrophilic mechanisms (see Chapter 13). For instance, the toxicity to fish of allyl and propargyl alcohols is significantly under-predicted by *T. pyriformis*. These are pro-electrophiles and undergo oxidation *via* alcohol dehydrogenase to an unsaturated aldehyde or ketone, which can act *in vivo* as a powerful electrophile. The presence of two unsaturated carbon–carbon double bonds permits second Michael-type addition and potential cross-linking within a single target bio-macromolecule.[24,25] It is likely that this metabolic route is absent in the protozoan.

The analysis of outliers from eqn (18.4) provides an understanding of how mechanism of action will assist in developing inter-species models. Compounds with specific mechanisms of action, or which may be metabolised to specifically acting compounds, are likely to be more toxic to fish than predicted by eqn (18.4). Thus, if these compounds can be identified before the analysis, they could be prioritised for *in vivo* testing; alternatively, caution could be placed on the predicted value.

As well as the identification of reactive compounds, identifying non-reactive compounds will also assist in the development of robust inter-species relationships.[26] To illustrate this principle, compounds likely to act by the non-polar narcosis mechanism of action were identified from within the data set created by Kahn *et al.*;[15] these compounds are noted in Table 18.1. This is, in effect, a category formed on a mechanistic basis and could be achieved using an appropriate profiler such as those in the OECD QSAR Application Toolbox (see Chapter 16), although for the purposes of this exercise, the grouping was performed manually. Compounds were chosen that have a high probability of being non-polar narcotic according to Ellison *et al.*;[27] this should not be considered a definitive list but does include compounds with a high likelihood of being non-polar narcotic, *i.e.* aliphatic alcohols and ketones, aromatic compounds with alkyl and/or halogen derivatives, *etc.* In total, 69 compounds were identified as belonging to the non-polar narcosis mechanism of action. For this chemical grouping (*i.e.* category), the following inter-species relationship is formed:

$$\log(1/\mathrm{LC}_{50}) = 1.10 \ \log(1/\mathrm{IGC}_{50}) + 0.488$$
$$n = 69, \ r^2 = 0.920, \ s = 0.41, \ F = 784$$

$$(18.6)$$

The inter-species relationship of toxicity for the non-polar narcotics category is shown in Figure 18.2. Equation 18.6 for the non-polar narcotics shows a very high statistical fit, considerably greater than for all compounds (and hence

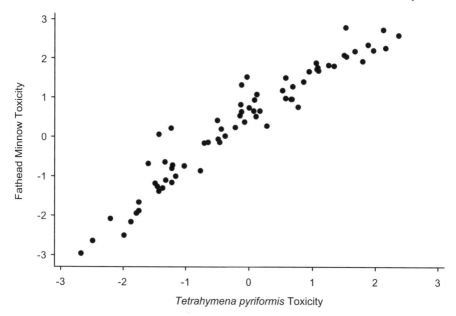

Figure 18.2   Plot of toxicities to the fathead minnow and *T. pyriformis* for the 69 organic chemicals in the non-polar narcosis category, as described by eqn (18.6). Data from Kahn *et al.*[15]

many different mechanisms) together. There is a slight increase in slope and decrease in intercept, but it is not possible to say if this is of true statistical significance.

The use of methods to form categories can therefore be of great assistance in developing and utilising inter-species relationship across trophic levels. As noted elsewhere (Chapters 16 and 17), a number of tools can be used to form categories that will be of assistance in the use of QAARs. With regard to aquatic toxicity, the following may be of use:

- The Verhaar[28] rules as well as later amendments.[29] These rules are implemented in a number of pieces of freely available software, *e.g.* the OECD (Q)SAR Application Toolbox and Toxtree developed by the European Commission Joint Research Centre (JRC). It should be noted that the OECD Toolbox has a number of methods for categorising for aquatic toxicity mechanism, *e.g.* the OASIS assignment mechanisms
- The Russom rules for predicting toxicity.[16]

In addition, whilst not directly related to identifying non-polar narcotics *per se*, evidence from the following may be useful to form categories:

- The assignment of compounds to a class of compound associated with non-polar narcosis by the US Environmental Protection Agency (US

EPA) ECOSAR software—implemented in the OECD (Q)SAR Application Toolbox.

- The absence of structural alerts associated with protein reactivity, *e.g.* from the protein binding profiler in the OECD (Q)SAR Application Toolbox or the rules of Enoch *et al.*[24]
- The absence of metabolic routes unique to the higher species. This may be more difficult to ascertain without toxicity data for structural analogues. However, some common routes of toxification in fish (for example) are known and, if these are not likely, then one may assume their absence.

The user of these approaches should not discount other evidence that may assist in grouping chemicals—this is described more fully in the next section. Such evidence may be from (non-animal) tests and could give confidence that the compound is a non-polar narcotic, for instance:

- evidence that toxic potency is well predicted by a non-polar narcosis QSAR for other species (hence no reactivity or specific mechanisms are present);
- the lack of reactivity in an *in chemico* assay (see Chapter 13); and
- the absence of toxicity promoted by endpoints that are commonly associated with protein reactivity (*e.g.* skin and respiratory sensitisation)

The grouping of compounds together to form a mechanistic category has also been shown to be relevant for non-narcotic mechanisms. For instance, Dimitrov *et al.*[30] showed good correlations for the toxicity to fathead minnow and *T. pyriformis* of aldehydes, which are likely to act as Schiff bases. Rationalisation of inter-species relationships has been taken in another direction by Kramer *et al.*[31] They demonstrated a good relationship between basal cytotoxicity (IC_{50}) and acute fathead minnow toxicity data for 82 industrial organic chemicals. As with eqn (18.4) and eqn (18.6), the cytotoxicity data were less sensitive than fish acute toxicity data, although the difference was greater (an order of magnitude). The mode of action of the chemical was found to be a significant reason for the general variation in the inter-species (regression) relationship. These results supported the notion that:

- the bioavailability of hydrophobic (high log P) and volatile (high Henry's Law constant) chemicals is significantly lower in *in vitro* assays than in the fish bioassay; and
- multiple cell types and endpoints should be included to mimic the modes of action possible in the whole organism.

Thus, consideration of the limitations of the test assays should be borne in mind for extrapolations.

Extending the consideration of physico-chemical properties also assists the understanding of inter-species relationships between what appear to be very different species. For some time, investigators have considered the relationship of acute toxicity between fish and mammals,[32,33] although such an attempt is fundamentally flawed unless the exposure route and conditions are taken into account. To explain this statement, Veith and co-workers demonstrated that only for mammalian inhalation studies is a steady-state achieved and this may form the basis of extrapolations between fish and mammalian acute toxicity.[34,35] As a result, there should theoretically be a lower correlation between fish acute toxicity and mammalian acute toxicity from the non-inhalation routes of administration.[34,35]

18.5 Endpoint-to-Endpoint Read-Down of Toxicity: Extrapolation of Toxicity Between Effects

The most ambitious methods for extrapolation of toxicity utilise evidence from one toxic effect to provide information relating to the possibility of another effect. At the outset extreme caution should be exercised; this statement should not be construed that test results from one assay could act as a substitute for another unrelated assay. However, within a mechanistic framework and correctly formed category, information may be inferred which, together with other knowledge, may allow for a weight of evidence approach to be attempted.

The formation of categories and application of other lines of evidence can assist in the identification of compounds with a number of toxic hazards. For instance, a number of toxic effects are a result of reactivity with biological macromolecules; examples include skin and respiratory sensitisation, excess acute aquatic toxicity, mutagenicity, liver and other organ toxicity, *etc.* Such toxicities may be initiated by a reactive chemical through events such as the formation of a covalent bond with, for example, a protein or DNA or production of a free radical (see Chapter 13). However, it is acknowledged that there is a spectrum of reactivity which means that not all reactive compounds will show the same toxicity effect.

In the previous section, approaches are outlined for the identification of compounds acting by non-polar narcosis (*i.e.* non-reactive toxicity). Conversely, there are a number of methods to identify compounds with reactive toxicity including:

- mechanistic organic chemistry;[24,25]
- the observation of acute toxicity above baseline
- the measurement of reactivity; and
- the utilisation of so-called *in chemico* assays.[36]

The latter approach, the use of *in chemico* assays, is described in Chapter 13. It relies on the formation of covalent bonds between an electrophilic compound and a nucleophilic reactant such as a glutathione. Interest in *in chemico* approaches has grown with the appreciation that reactive toxicity has long been a challenge to model (quantitatively) by *in silico* techniques.[37]

The key to using information within a mechanistic category across toxic effects is to appreciate that it is the molecular initiating events that are in common (see Chapter 14). If commonality in an initiating event can be established, then possibilities can be explored for using this information. At the time of writing this will appear to be a speculative concept to many and will require further effort for this information to be widely accepted. Perhaps the best method to achieve acceptance is through the use of case studies.

18.5.1 Predicting Reactive Toxicity Across Endpoints: Example(s) with Michael-type Acceptors

Considering the topic of protein reactivity to form mechanistically based categories, six domains of reactivity have been defined—albeit with a particular emphasis on skin sensitisation (these are described more fully in Chapter 13 and Figure 13.3 in particular). One of the best characterised electrophilic mechanisms, with regard to toxicity, is Michael addition.[38] Electro(nucleo)philic interactions of Michael acceptors include the addition of an SH group to the β-carbon atom (β-C atom) of a carbon–carbon double or carbon–carbon triple bond. The important molecular substructure—a polarised α,β-unsaturated configuration—results in a relatively diffuse and polarisable electron density of the olefinic π-bond, making Michael acceptors among the softest electrophiles.[39]

A variety of methods can be applied to identify compounds likely to act as Michael-type acceptors, *e.g.* OECD (Q)SAR Application Toolbox, Enoch *et al.*'s SMARTS strings, *etc.*[24] Should a chemical be identified as being 'reactive' and able to be placed in a known chemistry domain, then this opens up the possibility of extrapolating across endpoints. For instance, there could be strong evidence for a compound's skin sensitisation potential to be assessed from knowledge of its chemistry and excess toxicity above baseline in an *in vitro* assay. It must be stressed that this is not to suggest that skin sensitisation can be predicted from cytotoxicity. When compounds stray towards the edge of a domain (*e.g.* if they contain a fragment for which little or nothing is known it may modulate reactivity), *in chemico* information could provide further lines of evidence to support a decision.[37]

With the increased use of category formation, it is important to address how the compounds and their toxicity—in mechanistically formed categories—can be extrapolated. To form the category, a number of lines of evidence can be put forward including the presence of structural alerts, excess toxicity above non-polar narcosis and experimentally determined *in chemico* data. The outliers from eqn (18.1) (*i.e.* allyl and propargyl alcohols which can act as Michael-type acceptors) have been demonstrated to show toxicity above non-polar narcosis in *T. pyriformis*.[40] Such compounds would also be expected to be skin sensitisers due to their intrinsic reactivity, although few published skin sensitisation data exist to confirm or confound this theory.

Thus, allows forming categories of molecules on the basis of mechanistic understanding allows not only for read-across within an endpoint, but also between endpoints. This point needs to be addressed carefully and on a case-by-case basis. It is really only providing further information to support a hypothesis and should not, in itself, be viewed as a direct replacement. However, such conclusions may be found by thoughtful evaluation of data compiled for an ITS—something which is not currently performed and which may give added value and confidence to a compilation of data.

18.6 US Environmental Protection Agency Web-Based Inter-species Correlation Estimation (Web-ICE) Application

In order to make the process of applying extrapolations between the toxicity values of different species easier, an excellent tool is available from the US EPA, *i.e.* the Web-ICE application.[41–43] This is available free from http://nsdi.epa.gov/ceampubl/fchain/webice/.

The Interspecies Correlation Estimation Application (ICE) estimates the acute toxicity (LC_{50}/LD_{50}) of a chemical to a target species, genus, or family from the known toxicity of the chemical to a surrogate species with test data. ICE models are least square regressions of the relationship between surrogate and target species. They are based on a database of acute toxicity values including:

- median lethal water concentrations for aquatic species (LC_{50}; $\mu g\,L^{-1}$); and
- median lethal oral doses for wildlife species (LD_{50}; $mg\,kg^{-1}$ bodyweight).

In addition to direct toxicity estimation from a surrogate to target species, Web-ICE contains a Species Sensitivity Distribution (SSD) module that estimates the toxicity of all predicted species available for a common surrogate. Acute toxicity values generated by Web-ICE are expressed as a logistic cumulative probability distribution function in the SSD module to estimate an associated Hazardous Concentration (HC) or Hazardous Dose (HD).[42]

For aquatic species (fish and invertebrates), the database of acute toxicity used in the development of ICE models included 5501 EC/LC_{50} values for 180 species and 1,266 chemicals. The database was compiled from US EPA and public domain sources including the US EPA ECOTOX database (http://cfpub.epa.gov/ecotox/), other EPA sources, Mayer and Ellersieck[44] and the open literature (for list of references, see Raimondo *et al*).[45,46] Data used in model development were selected to adhere to the standard acute toxicity test condition requirements of the American Society for Testing and Materials (ASTM) (including earlier editions)[47] and the US EPA Office of Prevention, Pesticides, and Toxic Substances.[48] Data were standardised for test conditions and organism life stage to reduce variability.

The wildlife (birds and mammals) database comprised 4,329 acute, single oral dose LD_{50} values ($mg\,kg^{-1}$ bodyweight) for 156 species and 951 chemicals.

The data were collected from the open literature[49–53] and from data sets compiled by governmental agencies of the United States (US EPA) and Canada (Environment Canada).[54,55] Data were standardised by using only data for adult animals and data for chemicals of technical grade or formulations with >90% active ingredient. Open-ended toxicity values (*i.e.* >100 mg kg^{-1} or <100 mg kg^{-1}) and duplicate records among multiple sources were not included in model development. When data were reported as a range (*i.e.* 100–200 mg kg^{-1})[49] or data were collected from multiple sources for a species and chemical, the geometric mean of the values was used.

Models were developed using least squares methodology in which both variables are independent and subject to measurement error.[9] For species level models developed from aquatic and wildlife databases, an algorithm was written to pair every species with every other species by common chemical. Three or more common chemicals per pair were required for inclusion in the analysis. For each species pair, a linear model was used to calculate the regression equation stated in eqn (18.1).

Genus (aquatic only) and family level models were similarly developed by pairing each surrogate species with each genus or family by common chemical. Predicted genera and families required unique toxicity values for two or more species within the taxon. Toxicity values for the surrogate species were removed in cases where it was compared to its own genus or family.

ICE models were only developed between two aquatic taxa or two wildlife taxa; there are no models to predict toxicity to aquatic taxa from a wildlife species, or *vice versa*. Only models that had a significant relationship (p-value <0.05) are included in Web-ICE.

The following summarises the number of significant models developed from the aquatic and wildlife databases for different taxonomic levels:

   i) Aquatic species: 780 models comparing 77 species to 77 species
  ii) Aquatic genera: 289 models comparing 62 species to 28 genera
 iii) Aquatic family: 374 models comparing 69 species to 27 families
  iv) Wildlife species: 560 models comparing 49 species to 49 species
   v) Wildlife family: 292 models comparing 49 species to 16 families.

The uncertainty of each model was assessed using leave-one-out cross-validation. As suggested in Section 18.4, there was a strong relationship between taxonomic distance and cross-validation success rate, with uncertainty increasing with larger taxonomic distance.[56]

18.7 Recent Developments

The cellseus project is providing useful information on extrapolations between species.[57] In particular much work has been performed using fish embryo tests as a replacement for the acute fish test.[58]

18.8 Conclusions

It is possible to extrapolate toxicity data between species and endpoints. There are a variety of approaches to achieve this, but it would appear that the most accurate extrapolations are when the mode and/or mechanism of action are taken into account through the use of grouping or category formation. The US EPA has provided an extensive and freely available tool (Web-ICE) to support the extrapolation of acute toxicity. The theory can also be applied, with caution, to extrapolation between endpoints. There are possibilities to use these types of relationships within ITS, thus facilitating and encouraging the use of non-standard data to supplement the information gathering process.

18.9 Acknowledgments

This project was sponsored by the Department for Environment, Food and Rural Affairs (Defra) through the Sustainable Arable Link Programme. This work also was supported by the EU 6th Framework Integrated Project OSIRIS (www.osiris-reach.eu; contract no. GOCE-ET-2007-037017).

References

1. T. G. Vermeire, A. J. Baars, J. G. M. Bessems, B. J. Blaauboer, W. Slob and J. J. A. Muller, in *Risk Assessment of Chemicals. An Introduction*, ed. C. J. van Leeuwen and T. G. Vermeire, Springer, Dordrecht, The Netherlands, 2007, pp. 227–280.
2. T. P. Traas and C. J. van Leeuwen, in *Risk Assessment of Chemicals. An Introduction*, ed. C. J. van Leeuwen and T. G. Vermeire, Springer, Dordrecht, The Netherlands, 2007, pp. 281–356.
3. M. T. D. Cronin and J. C. Dearden, *Quant. Struct.-Act. Relat.*, 1995, **14**, 117.
4. M. T. D. Cronin, T. I. Netzeva, J. C. Dearden, R. Edwards and A. D. P. Worgan, *Chem. Res. Toxicol.*, 2004, **17**, 545.
5. M. Sjöström, A. Kolman, C. Clemedson and R. Clothier, *Toxicol. In vitro*, 2008, **22**, 1405.
6. J. S. Jaworska, R. S. Hunter and T. W. Schultz, *Arch. Environ. Contam. Toxicol.*, 1995, **29**, 86.
7. S. K. Janardan, C. S. Olson and D. J. Schaeffer, *Ecotoxicol. Environ. Saf.*, 1984, **8**, 531.
8. G. W. Suter II and A. E. Rosen, *Environ. Sci. Technol.*, 1988, **22**, 548.
9. A. Asfaw, M. R. Ellersieck and F. L. Mayer, *Interspecies Correlation Estimations (ICE) for Acute Toxicity to Aquatic Organisms and Wildlife. II. User Manual and Software*, US Environmental Protection Agency, National Health and Environmental Effects Research Laboratory, Gulf Ecology Division, Gulf Breeze, FL, 2003. EPA/600/R-03/106.

10. G. Archer, M. Balls, L. H. Bruner, R. D. Curren, J. H. Fentem, H.-G. Holzhütter, M. Liebsch, D. P. Lovell and J. A. Southee, *Altern. Lab. Anim.*, 1997, **25**, 505.

11. A. P. Worth and M. Balls, *Altern. Lab. Anim.*, 2001, **29**, 135.

12. G. A. LeBlanc, *Environ. Toxicol. Chem.*, 1984, **3**, 47.

13. O. A. Raevsky, V. Yu Grigor'ev, E. E. Weber and J. C. Dearden, *QSAR Comb. Sci.*, 2008, **27**, 1274.

14. O. A. Raevsky, V. Yu Grigor'ev, J. C. Dearden and E. E. Weber, *QSAR Comb. Sci.*, 2009, **28**, 163.

15. I. Kahn, U. Maran, E. Benfenati, T. I. Netzeva, T. W. Schultz and M. T. D. Cronin, *Altern. Lab. Anim.*, 2007, **35**, 15.

16. C. L. Russom, S. P. Bradbury, S. J. Broderius, D. E. Hammermeister and R. A. Drummond, *Environ. Toxicol. Chem.*, 1997, **16**, 948.

17. J. R. Seward and T. W. Schultz, *SAR QSAR Environ. Res.*, 1999, **10**, 557.

18. T. W. Schultz, *Chem. Res. Toxicol.*, 1999, **12**, 1262.

19. M. T. D. Cronin, N. Manga, J. R. Seward, G. D. Sinks and T. W. Schultz, *Chem. Res. Toxicol.*, 2001, **14**, 1498.

20. J. R. Seward, M. T. D. Cronin and T. W. Schultz, *SAR QSAR Environ. Res.*, 2001, **11**, 489.

21. G. D. Sinks and T. W. Schultz, *Environ. Toxicol. Chem.*, 2001, **20**, 917.

22. M. T. D. Cronin, A. O. Aptula, J. C Duffy, T. I. Netzeva, P. H. Rowe, I. V Valkova and T.W. Schultz, *Chemosphere*, 2002, **49**, 1201.

23. T. W. Schultz, M. T. D. Cronin, T. I. Netzeva and A. O. Aptula, *Chem. Res. Toxicol.*, 2002, **15**, 1602.

24. S. J. Enoch, J. C. Madden and M. T. D. Cronin, *SAR QSAR Environ. Res.*, 2008, **19**, 555.

25. S. J. Enoch, M. T. D. Cronin, T. W. Schultz and J. C. Madden, *Chem. Res. Toxicol.*, 2008, **21**, 513.

26. M. T. D. Cronin, J. C. Dearden and A. J. Dobbs, *Sci. Total Environ.*, 1991, **109/110**, 431.

27. C. M. Ellison, M. T. D. Cronin, J. C. Madden and T. W. Schultz, *SAR QSAR Environ. Res.*, 2008, **19**, 751.

28. H. J. M. Verhaar, C. J. van Leeuwen and J. L. M. Hermens, *Chemosphere*, 1992, **25**, 471.

29. S. J. Enoch, M. Hewitt, M. T. D. Cronin, S. Azam and J. C. Madden, *Chemosphere*, 2008, **73**, 243.

30. S. Dimitrov, Y. Koleva, T. W. Schultz, J. D. Walker and O. Mekenyan, *Environ. Toxicol. Chem.*, 2004, **23**, 463.

31. N. I. Kramer, J. L. M. Hermens and K. Schirmer, *Toxicol. In vitro*, 2009, **23**, 1372.

32. M. T. D. Cronin and J. C. Dearden, in *QSAR: Quantitative Structure-Activity Relationships in Drug Design*, ed. J. L. Fauchere, Liss, New York, 1989, pp. 407–410.

33. D. Delistraty, B. Taylor and R. Anderson, *Ecotox. Environ. Saf.*, 1998, **39**, 195.

34. D. Mackay, J. A. Arnot, E. P. Petkova, K. B. Wallace, D. J. Call, L. T. Brooke and G. D. Veith, *SAR QSAR Environ. Res.*, 2009, **20**, 393.

35. G. D. Veith, E. P. Petkova and K. B. Wallace, *SAR QSAR Environ. Res.*, 2009, **20**, 567.

36. A. O. Aptula, G. Patlewicz, D. W. Roberts and T. W. Schultz, *Toxicol. In Vitro*, 2006, **20**, 239.

37. M. T. D. Cronin, F. Bajot, S. J. Enoch, J. C. Madden, D. W. Roberts and J. Schwöbel, *Altern. Lab. Anim.*, 2009, **49**, 513.

38. T. W. Schultz, J. W. Yarbrough, R. S. Hunter and A. O. Aptula, *Chem. Res. Toxicol.*, 2007, **20**, 1359.

39. A. Jacobs, *Understanding Organic Reaction Mechanisms,* Cambridge University Press, Cambridge, 1997.

40. T. W. Schultz, T. I. Netzeva, D. W. Roberts and M. T. D. Cronin, *Chem. Res. Toxicol.*, 2005, **18**, 330.

41. J. Awkerman, S. Raimondo and M. G. Barron, *Environ. Sci. Technol.*, 2008, **42**, 3447.

42. S. D. Dyer, D. J. Versteeg, S. E. Belanger, J. G. Chaney and F. L. Mayer, *Environ. Sci. Technol.*, 2006, **40**, 3102.

43. S. D. Dyer, D. J. Versteeg, S. E. Belanger, J. G. Chaney, S. Raimondo and M. G. Barron, *Environ. Sci. Technol.*, 2008, **42**, 3076.

44. F. L. Mayer and M. R. Ellersieck, *Manual of Acute Toxicity: Interpretation and Database for 410 Chemicals and 66 Species of Freshwater Animals,* US Fish and Wildlife Service, Washington DC, 1986, Resource Publication 160.

45. S. Raimondo, D. N. Vivian, C. Delos and M. G. Barron, *Environ. Toxicol. Chem.*, 2008, **27**, 2599.

46. S. Raimondo, D. N. Vivian and M. G. Barron, *Ecotoxicology*, 2009, **18**, 918.

47. American Society for Testing and Materials, *Standard Guide for Conducting Acute Toxicity Tests with Fishes, Macroinvertebrates, and Amphibians*, ASTM, Philadelphia, PA, 2007, E 729-96(2007).

48. US Environmental Protection Agency, *Ecological Effects Test Guidelines. OPPTS 850.1075 Fish Acute Toxicity Test, Freshwater and Marine*, US EPA, Washington DC, 1996, EPA 712-C-96-118.

49. R. H. Hudson, R. K. Tucker and M. A. Haegele, *Handbook of Toxicity of Pesticides to Wildlife,* US Fish and Wildlife Service, Washington DC, 1984Resource Publication 153.

50. E. W. Shafer Jr and W. A. Bowles Jr, *Arch. Environ. Contam. Toxicol.*, 1985, **14**, 111.

51. E. W. Shafer Jr. and W. A. Bowles Jr, *Toxicity, Repellency or Phototoxicity of 979 Chemicals to Birds, Mammals and Plants,* US Department of Agriculture, Fort Collins, CO, 2004, Research Report No. 04-01.

52. E. W. Shafer Jr, W. A. Bowles Jr and J. Hurlbut, *Arch. Environ. Contam. Toxicol.*, 1983, **12**, 355.

53. G. J. Smith, *Pesticide Use and Toxicology in Relation to Wildlife: Organophosphorus and Carbamate Compounds,* US Department of the Interior, Washington DC, 1987, Resource Publication 170.

54. A. Baril, B. Jobin, P. Mineau and B. T. Collins, *A Consideration of Inter-Species Variability in the Use of the Median Lethal Dose (LD50) in Avian Risk Assessment,* Canada Wildlife Service, Ottawa, 1994, Technical Report No. 216.
55. P. Mineau, A. Baril, B. T. Collins, J. Duffe, G. Joerman and R. Luttik, *Rev. Environ. Contam. Toxicol.,* 2001, **170**, 13.
56. S. Raimondo, P. Mineau and M. G. Barron, *Environ. Sci. Technol.,* 2007, **41**, 5888.
57. K. Schirmer, K. Tanneberger, N. I. Kramer, D. Völker, S. Scholz, C. Hofner, L. E. J. Lee, N. C. Bols and J. L. M. Hermeus, *Aquat. Toxicol.,* 2008, **90**, 128.
58. E. Lammer, G. J. Carr, K. Wendler, J. M. Rawlings, S. E. Belanger and T. Braunbeck, *Comp. Biochem. Physiol. C Toxicol. Pharmocol.,* 2009, **149**, 196.

CHAPTER 19

Expert Systems for Toxicity Prediction

J. C. DEARDEN

School of Pharmacy and Chemistry, Liverpool John Moores University, Byrom Street, Liverpool L3 3AF, UK

'It is the mark of an instructed mind to rest easy with the degree of precision which the nature of the subject permits, and not to seek an exactness where only an approximation of the truth is possible.'

Aristotle (384–322 BC)

19.1 Introduction

It is now widely acknowledged and accepted that, for reasons of time, cost and animal usage, prediction methods are valid and essential tools in the armoury of those seeking to assess the toxicity of chemicals.[1,2] Such methods of course require toxicity data on which to base their predictions, and there is now such a plethora of those data (albeit still with huge gaps) that systematic approaches are needed in order to classify and use them efficiently and effectively. This is the role of the expert system, which has been defined as:

'any formalised system, not necessarily computer-based, which enables a user to obtain rational predictions about the toxicity of chemicals. All expert systems for the prediction of chemical toxicity are built upon experimental data representing

Issues in Toxicology No.7
In Silico Toxicology: Principles and Applications
Edited by Mark T. D. Cronin and Judith C. Madden
© The Royal Society of Chemistry 2010
Published by the Royal Society of Chemistry, www.rsc.org

one or more toxic manifestations of chemicals in biological systems (the data-base) and/or rules derived from such data (the rulebase)'.[3]

Expert systems can thus hold and use vast amounts of information and, in general, are rapid in operation. They can also be used by non-experts, although it must be added that expertise is often necessary for interpretation of their output.

Broadly speaking, expert systems are of two types.[4,5] One is based on the use of induced rules such as quantitative structure–activity relationships (QSARs) and may be called an automated rule induction system or a correlative system; generally such systems afford little or no mechanistic interpretation. The other is based on rules derived from toxicological knowledge, which are likely to have a strong mechanistic basis, and are known as knowledge-based systems. A simple example of such a rule is 'a primary aromatic amine is probably carcinogenic'. Some expert systems contain both types of rules.

Any model is only as good as the experimental data on which it is based. However, to be broadly useful, a model needs to cover a wide region of chemical and toxicological space. This raises two problems. First, although many toxicity data are available, there is still a dearth in many areas. Fortunately, efforts are now under way to make more data available in an accessible form such as ToxML (Toxicology XML standard), DSSTox (Distributed Structure-Searchable Toxicity Database Network) and ACToR (Aggregated Computational Toxicology Resource)[6]—see Chapter 3 for more details. Secondly, the greater the region of chemical and toxicological space covered, the more likely it is that chemicals will exert their toxicological effect(s) by different mechanisms, which means that there is a greater likelihood that predictions will be faulty (less accurate). This is especially so for complex endpoints such as carcinogenicity,[7] and explains why at present predictions from *in silico* models for some endpoints are not very reliable. It should be noted that most, if not all, commercial expert systems are continually being updated, *e.g.* Derek for Windows.[8]

Depending on the level of similarity of a query compound to one or more compounds in the training set used to develop an expert system, one can have a high or low level of confidence in a predicted toxicity value. It is therefore important to have an indication of whether or not the query compound lies within what is termed the 'applicability domain' (AD)[9,10] or the 'optimum prediction space' (OPS).[11] A number of toxicity prediction software programs now incorporate this facility. The concept of 'applicability domain' is discussed in more detail in Chapter 12.

The predictive ability of a knowledge-based expert system can be assessed in terms of:

- its sensitivity (the ratio of correctly predicted toxic compounds to the total number of toxic compounds tested);
- its specificity (the ratio of correctly predicted non-toxic compounds to the total number of non-toxic compounds tested); and
- its concordance (the ratio of all correctly predicted compounds to the total number of compounds tested).

In the regulatory environment, high sensitivity is to be preferred to reduce the risk of false negative predictions.[7] However, most expert systems currently have only moderate sensitivity.[12] One reason for this is that, because a given type of toxicity probably covers only a small region of toxicological space, it is easier to predict correctly that a chemical will be non-toxic than that it will be toxic.

In 2002, a meeting of QSAR experts from industry, regulatory authorities and academia was held in Setúbal, Portugal, to formulate a set of guidelines for the validation of QSARs and QSPRs (quantitative structure–property relationships)—in particular for regulatory purposes. These guidelines, which were later adapted and adopted by the Organisation for Economic Co-operation and Development (OECD),[13] are that a valid QSAR or QSPR should have:

(1) A defined endpoint
(2) An unambiguous algorithm
(3) A defined domain of applicability
(4) Appropriate measures of goodness of fit, robustness and predictivity
(5) A mechanistic interpretation, if possible.

These guidelines are now known as the OECD Principles for the Validation of (Q)SARs (see Chapter 11), although they are intended to apply also to QSPRs and presumably to expert systems as well. The OECD has provided a checklist to provide guidance on their interpretation.[14] Matthews and Contrera[15] have presented a useful critique of the application of the OECD Principles to prediction of human health effects.

It must be remembered that experimental data contain errors (discussed further in Chapter 4). Some common sources of error are:[16–18]

- different testing protocols in different laboratories;
- inaccurate data transcription; and
- incorrect chemical structures.

Cronin[19] has stated that the most fundamental concerns (about toxicity prediction) address the quality and relevance of the data. There is no easy way to check for these problems. In an investigation of the toxicity data landscape, Judson *et al.*[20] stated that they did not check on the correctness of the data from the original source. In effect, one generally has to take data on trust.

Annex XI[21] of the European Union's REACH (Regulation, Evaluation, Authorisation and restriction of CHemicals) Regulation states that:

'Results obtained from valid (Q)SARs may indicate the presence or absence of a certain dangerous property. Results of (Q)SARs may be used instead of testing when the following conditions are met:

- *results are derived from a (Q)SAR model whose scientific validity has been established;*

- *results are adequate for the purposes of classification and labelling and risk assessment;*
- *adequate and reliable documentation of the method is provided.'*

Annex XI, or indeed any other REACH documentation, gives no indication as to whether results from expert systems may be used instead of testing. However, it seems reasonable to assume that they can be used provided the above conditions are met. Bassan and Worth[22] included expert systems in their review of the computational tools that can be used in the prediction of toxicity for regulatory needs.

There is a significant literature on expert systems, including a number of reviews.[3–5,12,23–25]

19.2 Available Expert Systems

The expert systems discussed below are listed in Table 19.1. It is hoped that the list is comprehensive, but I apologise for any inadvertent omissions.

19.2.1 TOPKAT

TOPKAT (TOxicity Prediction by Komputer Assisted Technology) is a QSAR-based system that was developed by Health Designs Inc. of Rochester, NY, in the 1980s.[26] It is now owned by Accelrys[27] and is incorporated in its Discovery Studio suite as DS-TOPKAT. Its Optimum Prediction Space (OPS) technology gives an indication of whether or not a query compound falls within the applicability domain of the appropriate training set. It is also possible to carry out a similarity search to check the performance of DS-TOPKAT in predicting the toxicity of a chemical that is structurally similar to a query compound.

Discovery Studio 2.5 contains two versions of DS-TOPKAT. The earlier version uses structural, electronic, topological and electrotopological molecular descriptors; the new version uses molecular fingerprints in addition, and uses Bayesian for classification models and partial least squares (PLS) for regression models. Users can incorporate in-house data to improve the models.

The endpoints currently available in DS-TOPKAT are:

- rodent carcinogenicity;
- weight-of-evidence rodent carcinogenicity;
- rat oral LD_{50};
- rat chronic LOAEL (lowest observed adverse effect level);
- rat MTD (maximum tolerated dose);
- rat inhalation toxicity LC_{50};
- Ames mutagenicity;
- developmental toxicity potential;
- skin sensitisation;
- fathead minnow LC_{50};

Table 19.1   Toxicity prediction software.

Software	Operator	Endpoint(s) predicted	Website
TOPKAT	Accelrys	numerous	www.accelrys.com
Derek (DfW)	Lhasa	numerous	www.lhasalimited.org
MCASE (MC4PC)	Multicase	numerous	www.multicase.com
PASS	Academic	numerous	www.chem.ac.ru/Chemistry/ Soft/PASS.en.html
FDA QSAR	Leadscope	numerous	www.leadscope.com
HazardExpert	CompuDrug	several	www.compudrug.com
ToxBoxes/Tox Suite	ACDLabs (incorporating Pharma Algorithms)	several	www.acdlabs.com
ADMET Predictor	SimulationsPlus	several	www.simulations-plus.com
q-TOX	Quantum Pharmaceuticals	several	http://q-pharm.com
LAZAR	in-silico toxicology	several	http://lazar.in-silico.de
TerraQSAR	Terrabase	several	www.terrabase-inc.com
TIMES	Academic	several	www.oasis-lmc.org/ ?section = software& swid = 4
CAESAR	Consortium	several	www.caesar-project.eu/ software
T.E.S.T.	US EPA	several	www.epa.gov/nrmrl/std/ cppb/qsar/
ECOSAR	US EPA	aquatic toxicity	www.epa.gov/oppt/new-chems/tools/21ecosar.htm
CSGenoTox	ChemSilico	mutagenicity	www.chemsilico.com
OncoLogic	LogiChem	carcinogenicity	www.epa.gov/oppt/new-chems/tools/oncologic.htm
QikProp	Schrödinger	hERG	www.schrodinger.com
StarDrop	Optibrium	hERG	www.optibrium.com
ADMEWORKS Predictor	FQS Poland	carcinogenicity mutagenicity	www.fqs.pl
COMPACT	Academic	several	None
BfR Decision Support system	Regulatory	several	www.bfr.bund.de
Insilicofirst	Consortium	numerous	www.insilicofirst.com

- *Daphnia magna* EC_{50};
- rabbit skin irritancy; and
- eye irritancy.

An early study[28] obtained sensitivities, specificities and concordances of around 98% for prediction of carcinogenicity and mutagenicity, whilst Gombar *et al.*[29] reported sensitivities from 86 to 89% and specificities from 86 to 97% for prediction of developmental toxicity potential. In an examination of skin irritation prediction with TOPKAT, Mombelli[30] reported a sensitivity of 71% (with seven compounds outside the OPS) and a specificity of 63% (with two compounds outside the OPS). Patlewicz *et al.*[31] found TOPKAT to predict skin sensitisation with a sensitivity of 78%, a specificity of 46% and a concordance of 71%. Venkatapathy *et al.*[32] found that TOPKAT predicted LOAEL values for five chemical classes quite well, with the percentage of chemicals predicted within a factor of 3 in the LOAEL ranging from 88 to 100%. Mombelli[30] found TOPKAT to comply reasonably well with the OECD Principles for the Validation of (Q)SARs.

19.2.2 Derek (Derek for Windows)

Derek for Windows (DfW now marked as Derek Nexus)[33] is a knowledge-based expert system developed by Lhasa Limited[34] from the DEREK (Deductive Estimation of Risk from Existing Knowledge) system. The latter originated from Schering Agrochemicals and was aimed at capturing the toxicological knowledge of retiring scientist Derek Sanderson.[35]

The DfW knowledge base comprises alerts, reasoning rules and examples—all of which may contribute to the toxicity predictions. The alerts are supported by evidence and a list of references from which this was derived. The reasoning rules describe relationships between factors such as individual alerts, toxicological endpoints, species, physico-chemical properties and toxicological data, and allow a level of confidence to be allocated with each prediction.[24] Levels of confidence are expressed by the range 'open', 'impossible', 'improbable', 'doubted', 'equivocal', 'plausible', 'probable' and 'certain',[25] and take account of whether a query compound falls within the applicability domain of the relevant model.[9]

Coverage is most extensive for endpoints such as carcinogenicity, mutagenicity, chromosome damage and skin sensitisation. However, DfW additionally includes many other endpoints, namely α-2-γ-globulin nephropathy, anaphylaxis, bladder urothelial hyperplasia, cardiotoxicity, cerebral oedema, chloracne, cholinesterase inhibition, cumulative effect on white cell count and immunology, cyanide-type effects, developmental toxicity, genotoxicity, hepatotoxicity, hERG channel inhibition, high acute toxicity, irritation of the skin, gastrointestinal tract, respiratory tract, and eye, lachrymation, methaemoglobinaemia, nephrotoxicity, neurotoxicity, occupational asthma, ocular toxicity, oestrogenicity, peroxisome proliferation, phospholipidosis, photoallergenicity, photocarcinogenicity, photogenotoxicity, phototoxicity, pulmonary toxicity, respiratory sensitisation, teratogenicity, testicular toxicity, thyroid toxicity, and uncoupling of oxidative phosphorylation. It also incorporates, through adaptors to external programs, factors such as log *P* (octanol–water partition coefficient) and skin permeability; links to other software for the calculation of other properties can be created as required. There is usually one update release of DfW per year.

Lhasa Limited is a not-for-profit organisation and is membership-based, whereby users who license its software become members and have voting rights for electing the board of trustees, which manages the company. Members are encouraged to contribute data and knowledge to the continuing improvement of the software. Currently there are over 150 member organisations, most of which are commercial.

There have been numerous published evaluations of the performance of DfW; Marchant et al.[33] listed 15 such studies published between 2003 and 2007 covering carcinogenicity, genotoxicity, mutagenicity, organ toxicity, reproductive toxicity, skin irritation and skin sensitisation, but point out that the results of any evaluation need to be interpreted with caution, for a variety of reasons. One such study, concerning skin sensitisation, was that of Patlewicz et al.[31] They used a set of 210 chemicals, but found that about 35 of those were in the DfW training set. Of the remaining 175 chemicals, 130 were correctly predicted and 45 were incorrectly predicted (concordance 74.3%). A prediction was considered positive if an alert was identified with a confidence level of 'plausible' or higher. DfW was found to be good at identifying potential sensitisers (sensitivity 82%), but poor at identifying non-sensitisers (specificity 50%). Patlewicz et al.[31] commented that that was perhaps not surprising because DfW has been developed for the precautionary purpose of identifying potential sensitisers. However, this result appears to be at variance with the comments of Muster et al.,[12] who reported that expert systems generally yielded specificities of about 80%, but gave sensitivities of around 50%. This latter is perhaps to be expected, since non-toxic compounds usually occupy a much larger region of chemical space than do toxic compounds, so that it is easier to predict lack of toxicity than to predict toxicity.[23]

An investigation of the prediction of genotoxicity and rodent carcinogenicity of 60 pesticides by DfW was carried out by Crettaz and Benigni.[36] For genotoxicity, they found a sensitivity of 37%, a specificity of 71% and a concordance of 60%; for carcinogenicity they reported a sensitivity of 69%, a specificity of 47% and a concordance of 53%.

Hulzebos and Posthumus[37] found that the accuracy of prediction of DfW was 60% for skin sensitisation, and 75% for genotoxicity and carcinogenicity, for a limited number of chemicals. Irritation and reproductive toxicity were predicted poorly.

DfW has been found to comply well with the OECD Principles for the Validation of (Q)SARs.[30,38]

19.2.3   MCASE (MC4PC)

The CASE (Computer Automated Structure Evaluation) technology was developed by Professor Gilles Klopman at Case Western University, Cleveland, Ohio, in the 1980s.[39] It comprises several different programs—MCASE, ToxAlert and CASETOX. Some years ago the US Food and Drug Administration (FDA) collaborated with CASE to modify and improve the system.[40] The latest version of MCASE is called MC4PC and runs on a PC; an overview has been given by Klopman et al.[41]

MC4PC uses automated machine learning to create and detect structural alerts. Activities are characterised on a linear CASE units scale from 10 to 99, and are interpreted thus:

10–19	Inactive
20–29	Marginal
30–39	Active
40–49	Very active
50–99	Extremely active.

The algorithm tabulates, for each molecule of a training set, all fragments of a molecule between two and ten interconnected groups comprising heavy atoms. It then identifies those fragments that are associated with the relevant activity and which are termed biophores. Some continuous properties (*e.g.* log *P*, aqueous solubility, HOMO/LUMO energies, partial charges and graph indices) are also used. The fragments and other properties are then incorporated into a QSAR of the form:

$$CASE\,units = constant + a(fragment\,1) + b(fragment\,2)$$
$$+ c(fragment\,3) + ... \qquad (19.1)$$

No *a priori* mechanistic knowledge is used in the CASE approach, but the fragments associated with an activity are often found to have close links with known mechanisms.[40,42]

MC4PC gives, for each query compound, a percentage probability that the compound does possess the given activity, and also an indication of whether or not the compound lies within the applicability domain of the model. Users can incorporate additional compounds into the training sets, or create their own training sets. Software updates are released at intervals.

The number of toxicity endpoints covered by MC4PC is very large—probably the largest number available in any expert system. The endpoints include acute toxicity in various mammals, adverse effects in humans, carcinogenicity, cytotoxicity, developmental toxicity, teratogenicity, ecotoxicity, enzyme inhibition, genetic toxicity, skin irritation, eye irritation and allergic reactions. The modules for human cardiac disorders, for example, include conduction disorders, coronary artery disorders, electrocardiogram disorders, heart failure disorders, myocardial disorders, rate rhythm disorders, cardiac palpitations, QT-interval prolongation, torsade, tachycardia, and myocardial infarction.

There are very few independent published studies of the performance of MC4PC, but Contrera *et al.*[43] found that MC4PC predicted carcinogenicity for a set of 1,540 compounds with 61% sensitivity, 71% specificity and 66% concordance. It has been shown[42] that MC4PC satisfies the OECD Principles for the Validation of (Q)SARs.

19.2.4   PASS

Professor V. V. Poroikov and his co-workers at the Institute of Biomedical Chemistry of the Russian Academy of Medical Sciences in Moscow developed PASS (Prediction of biological Activity Spectra for Substances) for the prediction of over 1,500 biological activity endpoints, including a number of toxicity endpoints.[44,45] PASS uses a novel method called Multilevel Neighbourhoods of Atoms to select molecular sub-structures that are related to given biological activities.

PASS uses training sets currently totalling over 70,000 compounds. It yields semi-quantitative predictions of activity as probabilities that a query compound will exhibit a specified activity (Pa) or will not do so (Pi). For example, levamisole, a known anthelmintic and immunomodulator, has a PASS-predicted probability of 93.6% to be active as an anthelmintic, of 85.4% to be active as an immunomodulator, of 0.5% to be inactive as an anthelmintic, and of 0.8% to be inactive as an immunomodulator.[46] The reliability of a prediction is considered to be high if Pa $> 70\%$. However, lower probabilities may still be worth investigating, as they offer the chance to discover a new chemical entity.[46]

PASS has been shown to be effective in predicting adverse effects of drugs. In a study of the top 200 drugs on the market, PASS was able to predict 83% of their known adverse and toxic effects.[47]

An investigation into the use of PASS to predict the rodent carcinogenicity of 412 chemicals from the National Toxicological Program (NTP) and 1,190 chemicals from the Carcinogenic Potency Database (CPDB)[45] found that the mean accuracy of prediction calculated by leave-one-out cross-validation was 73–80% for the NTP compounds, and was 90% for the CPDB compounds. Prediction accuracy was lower when calculated by leave-20%-out cross-validation (63% for the CPDB compounds). A further study of rodent carcinogenicity prediction of 293 compounds by PASS[48] found a sensitivity of 81%, a specificity of 74% and a concordance of 76%.

19.2.5   FDA QSAR

A collaboration between Leadscope Inc.[49] and the US FDA has led to the development of a wide range of toxicity prediction modules. These are:

- rodent carcinogenicity suite (six rat and mouse models);
- non-human genetic toxicity suite (three groups: gene mutation (12 models), DNA damage (three models) and clastogenicity (five models));
- non-human reproductive toxicity suite (six male and three female models);
- non-human developmental toxicity suite (seven models, all but one covering three species—rat, mouse and rabbit);
- non-human neurotoxicity suite (four models);
- non-human miscellaneous toxicity endpoints suite (three groups—*in vitro* genetic toxicity; five models); *in vitro* sister chromatid exchange (two models); mouse lymphoma (one model);

- *in vivo* and *in vitro* carcinogenicity (three models);
- human adverse hepatobiliary effects suite (five models); and
- adverse urinary tract effects suite (six rodent models).

In all models, an indication of whether a query compound falls within the applicability domain of the model is defined by nearest neighbour analysis. All models were constructed at the FDA and FDA/ICSAS publications document the weight-of-evidence methods and data sources employed.

The models use predefined structural features stored in a template library[50] and some calculated properties such as molecular weight, number of hydrogen bonds and number of rotatable bonds, which are correlated with toxicity data. Users can incorporate new structural features and/or numerical properties, and can also incorporate new compound sets into the software. Roberts *et al.*[51] have given details of the Leadscope software.

19.2.6 HazardExpert/ToxAlert

HazardExpert, developed by CompuDrug,[52,53] was one of the first knowledge-based expert systems available, having first appeared in 1987. It operates by a combination of rule-based toxicological knowledge, QSAR models and fuzzy logic; the last enables HazardExpert to simulate different exposure conditions. It gives toxicity predictions for several species. ToxAlert, which is based on HazardExpert, flags compounds in a screening library or other compound collections for hazards associated with specific toxicophores. It is based on an improved version of the knowledge base implemented in HazardExpert, and as well as giving overall toxicity categorisation, provides probability percentages for various toxicity endpoints. It provides an indication of whether or not a query compound falls within its applicability domain.[30]

HazardExpert's predicted endpoints include carcinogenicity, mutagenicity, teratogenicity, membrane irritation, skin sensitisation, immunotoxicity and neurotoxicity. Its chemical database and rule base are accessible to, and can be modified by, the user. Through a link to its sister program, Metabol-Expert, it is able to make toxicity predictions for metabolites. An early assessment of HazardExpert's ability to predict carcinogenicity[54] found, for a test set of 80 chemicals, a sensitivity of 36% and a specificity of 81%. Mombelli[30] found HazardExpert 3.1 to comply reasonably well with the OECD Principles for the Validation of (Q)SARs, but observed that it predicted skin irritation poorly.

19.2.7 ToxBoxes/Tox Suite

Pharma Algorithms[55] developed a suite of programs, ToxBoxes, for predicting toxicity. In 2009, the company merged with ACDLabs[56] and the latter now offers the same suite under the name of ACD/Tox Suite. ACDLabs is now the sole distributor of the software.

The toxicity modules currently available are: Ames mutagenicity, hERG potassium channel inhibition, organ-specific toxicity in various species, rat LD_{50}, mouse LD_{50}, aquatic toxicity, skin and eye irritation, and oestrogen receptor binding affinity. The hERG and mutagenicity modules are trainable, and a reliability index is provided for each prediction as an indication of the applicability of the model to the query compound. The models are built using a mechanistic approach involving the construction of molecular fragmental parameters; this allows the identification of reactive functional groups, and enables the construction of two-dimensional (2-D) pharmacophores that can be converted into three-dimensional (3-D) models.[57]

The mutagenicity module is reported to give about 95% concordance,[58] the hERG potassium channel inhibition module gives >85% concordance,[59] the rodent LD_{50} modules have RMSEs of <0.5 log unit,[60] and the rabbit eye and skin irritation models have concordances of 77.6% and 73.6% respectively.[61]

19.2.8 ADMET Predictor

ADMET Predictor is a QSAR-based system developed by SimulationsPlus for the prediction of ADMET properties (absorption, distribution, metabolism, elimination, toxicity).[62] Its current toxicity modules include oestrogen receptor toxicity, maximum recommended therapeutic dose, fathead minnow LC_{50}, rat and mouse carcinogenicity (as TD_{50}, a quantitative measure), mutagenicity, hERG-encoded K^+ channel affinity, *Tetrahymena pyriformis* $pIGC_{50}$, and human liver adverse effects of drugs, and these can be customised. The SimulationsPlus website[62] gives details of the performance of several of these modules. For example, for hERG-encoded K^+ channel affinity, the mean absolute error (MAE) of prediction for an 82-compound training set was 0.44, and for a 14-compound test set it was 0.45. For mutagenicity, ten Ames models yielded concordances ranging from 78% to 96%.

There do not as yet appear to be any peer-reviewed publications examining the performance of any ADMET Predictor toxicity modules.

19.2.9 q-TOX

Quantum Pharmaceuticals[63] has developed a novel approach to the prediction of toxicities. Following the work of Fliri *et al.*,[64,65] who showed that experimental values of molecular activities against a large group of proteins can be used for the prediction of a wide range of biological effects, the company's co-founders (Peter Fedichev and Andre Vinnik)[63] developed a toxicity prediction system, q-TOX, using publicly available toxicological knowledge to establish relationships between calculated protein binding constants and various toxicity endpoints. It was found that in order to obtain many toxicity endpoints it was sufficient to obtain IC_{50} values for several dozen proteins and interpret them.

The endpoints currently available in q-TOX are:

- hERG binding;
- human maximum recommended daily dose (MRDD);
- human maximum recommended therapeutic dose;
- human TD_{50}; and
- LD_{50} values for rat, mouse, rabbit and dog.

In addition, the software will give qualitative indications of a wide range of potential side effects including gastrointestinal, hepatic, renal, cardiac, neural, pulmonary and behavioural.

The Quantum Pharmaceuticals website[63] gives a root mean square deviation of 1.18 log unit for the prediction of hERG binding, whilst that for MRDD is reported as about 1 log unit. Due to the novel approach used in q-TOX the usual considerations of applicability domain probably do not apply. It is possible for the software to be customised for a specific user. A disadvantage of the software appears to be the length of time required to perform the calculations, which is reported[63] to be 5–10 hours per molecule.

19.2.10 LAZAR

LAZAR (LAZy structure-Activity Relationships) was developed by Professor Christoph Helma of the University of Freiburg, Germany, and is marketed by *in silico* technology.[66] It uses a modified K-nearest-neighbour algorithm to make predictions for a query compound by searching a database with chemical structures and experimental data for compounds that are similar to the query compound.[67,68] The software was developed on the predicate that inaccurate predictions are frequently not the result of poor algorithms, but of insufficient or inaccurate experimental data. LAZAR provides the rationales for its predictions, and supplies a confidence index indicating whether or not a query compound lies within the applicability domain of the training database.

A number of endpoints are available including mutagenicity, rat, mouse and hamster carcinogenicity (12 endpoints), human liver toxicity (six endpoints), FDA maximum recommended daily dose, and fathead minnow LC_{50}. Helma[68] reported that LAZAR predicted *Salmonella* mutagenicity with 85% concordance and rodent carcinogenicity with 86% concordance. To date there are no peer-reviewed comparative studies of the performance of LAZAR.

19.2.11 TerraQSAR

Several of TerraBases's QSAR prediction modules relate to toxicity.[69] They are:

- TerraQSAR – *Daphnia* (*Daphnia magna* 48-hour LC_{50});
- TerraQSAR – OMAR (mouse and rat oral LD_{50});
- TerraQSAR – RMIV (rat and mouse intravenous LD_{50});
- TerraQSAR – FHM (fathead minnow LC_{50});

- TerraQSAR – E2-RBA (oestrogen receptor binding affinity); and
- TerraQSAR – SKIN (skin irritation).

In addition, according to the TerraBase website,[69] TerraQSAR – HIV-1, a module for the prediction of human immunodeficiency virus EC_{50}/IC_{50}, is in preparation.

The computation of each toxicity endpoint uses molecular fragments.[70] A probabilistic neural networks methodology is used in the models to account for non-rectilinear correlations.[71,72] Details of training set performance are given for some of the modules.[69] For example, for the TerraQSAR – *Daphnia* module, the leave-one-out cross-validation RMSE was 0.18 log unit, whilst that for TerraQSAR – RMIV was 0.12 log unit. In a comparative assessment of several QSAR approaches to the prediction of fathead minnow toxicity, Pavan *et al.*[73] found that TerraQSAR – FHM gave the best predictions. TerraQSAR meets the requirements of the OECD Principles for the Validation of (Q)SARs.[74]

19.2.12 TIMES

The TIMES (TIssue MEtabolism Simulator) platform[75] is a hybrid expert system developed at Bourgas University, Bulgaria, using funding and data from a consortium comprising industry and regulators. It was devised for the prediction of skin sensitisation,[76] but now also includes mutagenicity and ER/AR (estrogen receptor/androgen receptor) binding affinities.[75]

The software is unique in that it takes account of metabolism, so that the toxicity not only of a query compound, but also of likely metabolites, can be predicted.[77] It is based on the identification of structural alerts in a training set using the COREPA software.[78–80] These, together with physico-chemical factors such as steric and electronic properties, solubility and lipophilicity allow for the predictive classification of a query compound as sensitising or non-sensitising.

Roberts *et al.*[81] carried out a validation assessment of TIMES-SS (skin sensitisation module), using an external test set of 16 sensitisers and 24 non-sensitisers. An overall concordance of 83% was found, with a sensitivity of 75% and a specificity of 87.5%. For the TIMES mutagenicity module, Mekenyan *et al.*[77] found an overall concordance of 94%, with a sensitivity of 82% and a specificity of 94%. No validation figures are as yet available for the performance of the ER/AR module. Patlewicz *et al.*[82] have described the four-stage applicability domain that TIMES-SS utilises, and have shown that TIMES-SS is scientifically valid in accordance with the OECD Principles for the Validation of (Q)SARs.[13]

19.2.13 CAESAR

Models from the CAESAR (Computer Assisted Evaluation of industrial chemical Substances According to Regulations) project[83] have recently been developed by a consortium led by the Istituto di Ricerche Farmacologische Mario Negri in Milan and funded by the European Commission. They are free to use from the project website and offer predictions for four toxicity endpoints—skin sensitisation, mutagenicity, carcinogenicity and developmental toxicity. The methods used vary; for example, the skin sensitisation module uses a QSAR model based on Dragon descriptors, whilst the developmental toxicity module uses methods based on random forest and adaptive fuzzy partition. No information is yet available on the performance of CAESAR (but see Section 19.5).

19.2.14 T.E.S.T.

The US Environmental Protection Agency (US EPA) has recently made available for free download from its website[84] a toxicity prediction program called Toxicity Estimation Software Tool (T.E.S.T.). Its current endpoints (version 4.0) include 96-hour fathead minnow LC_{50}, 48-hour *Tetrahymena pyriformis* IGC_{50}, rat oral LD_{50}, bioconcentration factor and reproductive toxicity.

T.E.S.T. allows the estimation of toxicity using several different QSAR methodologies, including an hierarchical method, the FDA method,[85] multiple linear regression, a group contribution method, a nearest neighbour method and a consensus method that takes the average of all the above model predictions. A random forest method is also used but only for reproductive toxicity.

For each endpoint, test set predictions showed that the consensus method gave the best results. For example, in the fathead minnow module, individual root mean squared error (RMSE) values ranged from 0.821 to 0.990, whilst the consensus method had RMSE = 0.785.

The T.E.S.T. software appears to conform with most of the OECD Principles for the Validation of (Q)SARs. There do not appear as yet to be any peer-reviewed assessments of the performance of any of the modules within T.E.S.T.

19.2.15 ECOSAR

ECOSAR (ECOlogical Structure-Activity Relationships) is an expert system that estimates the aquatic toxicity of chemicals. Developed by the US EPA, it can be downloaded free of charge from its website.[86] ECOSAR uses QSARs to estimate both acute and chronic toxicities of chemicals to aquatic organisms such as fish, aquatic invertebrates such as daphnids, and aquatic plants such as green algae.

ECOSAR contains a library of class-based QSARs, overlaid with a decision tree based on expert rules for selecting the appropriate QSAR for a query compound. Currently, ECOSAR version 1.00 can identify over 120 chemical classes and allows access to over 440 QSARs. The QSARs are based on the octanol–water partition coefficient (log P, log K_{ow}). Input is the SMILES notation of the query compound (or its CAS number) and its log P value. If the latter is not available, ECOSAR calculates it from sub-routine KOWWIN.

Hulzebos and Posthumus[37] found that ECOSAR yielded a 67% overall accuracy of prediction, averaged across a number of chemical classes. Reuschenbach et al.[87] recently published an assessment of the performance of a slightly earlier version of ECOSAR, version 0.99g. Using defined toxicity classes (very toxic, toxic, harmful, not harmful) corresponding to given ranges of LC_{50} or EC_{50} values, they found correct predictions of toxic effects of 69% for fish, 64% for *Daphnia*, and 60% for algae for a very large test set of almost 1,100 compounds. Quantitative correlations of observed *versus* predicted LC_{50} or EC_{50} values were found generally to be very poor; for example, for a fish toxicity test set of 425 compounds, the r^2 value was only 0.31. Reuschenbach et al.[87] commented that, for some chemical classes, only a very limited training set was used to develop the QSAR model and that expert knowledge should be used in order to decide whether or not ECOSAR can be used with a specific substance for prediction of a specific endpoint. No peer-reviewed assessments of the performance of the latest version (1.00) of ECOSAR have been published to date. ECOSAR has been found[38] to comply reasonably well with the OECD Principles for the Validation of (Q)SARs.[13]

Future developments of ECOSAR will include:

- increasing the number of data points in training sets;
- refining data sets and regression analysis for neutral organics and chemical classes with excess toxicity, and investigating outliers identified during statistical analysis;
- investigating predictive methods for inorganics and organometallics; and
- incorporating a limited number of QSARs for polymer evaluation.

19.2.16   CSGenoTox

ChemSilico[88] is a company predicated largely on the development of software for the prediction of physico-chemical properties such as aqueous solubility and octanol–water partition coefficient (log P). However, it recently introduced its first toxicity prediction module, CSGenoTox. The name is something of a misnomer for it predicts Ames mutagenicity and not genotoxicity *per se*. It is a QSAR-based method and uses three types of descriptor developed by Kier and Hall:

- molecular connectivity indices;[89]
- electrotopological state indices;[90] and
- kappa shape indices.[91]

An artificial neural network approach is used to take account of non-rectilinear dependencies in the correlations. The software does not appear to have a facility for indicating whether or not a query compound is within its applicability domain.

A training set of 2963 compounds was used in the model development, and a 338-compound external validation set was used to assess the model accuracy. ChemSilico claim[88] that for the external validation set, sensitivity was 82%, specificity was 87%, and overall concordance was 84%.

19.2.17 OncoLogic

As its name implies, OncoLogic is an expert system for the prediction of carcinogenicity only. It was developed jointly by the US EPA and an expert systems developer, LogiChem Inc., and can be downloaded free of charge from the US EPA website.[92] OncoLogic is a rule-based system requiring the input of a considerable amount of information about a query compound, including some physico-chemical properties, chemical stability, route of exposure, bioactivation and detoxification, genotoxicity, and other supportive data.[92,93] It comprises four independent sub-systems for estimating the carcinogenicity of organic compounds, polymers, metals and metal-containing compounds, and fibres. Each sub-system involves firstly a fixed, hierarchical decision-tree procedure, followed by the evaluation of carcinogenicity based on the structural and other information provided by the user. For example, the metals sub-system can take into account such factors as exposure, usage, oxidation state, physical state, solubility and dissociation products.

The largest sub-system is that for organics, and contains many thousands of discrete rules derived from expert knowledge. There are separate modules for over 50 chemical classes. The user has to define the class of the query compound and OncoLogic cannot deal with query compounds outside its defined chemical classes. Neither is it possible for users to modify OncoLogic's rulebase. The output from OncoLogic is a semi-quantitative prediction of carcinogenic potential—namely low, marginal, low-moderate, moderate, moderate-high and high, together with the underlying scientific rationale.[92] For validation purposes, OncoLogic, together with other expert systems, was used to predict the carcinogenic potential of 26 chemicals in a NTP[94] prospective predictive exercise. OncoLogic was found to be one of the best performers.[95]

19.2.18 QikProp

QikProp is an expert system developed and marketed by Schrödinger Inc.[96] Most of the properties that it predicts are physico-chemical, but it offers one toxicity endpoint—namely hERG potassium ion channel blockade, which is related to prolongation of the QT interval of the cardiac action potential. This is a serious problem and several drugs (*e.g.* sertindole and cisapride) possessing this type of toxicity have been withdrawn from the market.

QikProp's hERG module is unusual in that it is based on a homology model of the receptor site derived by Farid *et al.*[97] from induced fit docking of 12 known hERG blockers. It was suggested that hERG blockers bind to the intracellular potassium conduction cavity of hERG *via* a mechanism involving extensive hydrophobic and ring stacking interactions with combinations of Tyr652 and Phe656 side-chains from at least two monomers. There are no reports available of the performance of the QikProp hERG module in predicting potential hERG blockers.

19.2.19   StarDrop

Optibrium is a new company formed to market and develop the ADME prediction suite known as StarDrop;[98] the suite was previously owned by Bio-Focus, and prior to that by Inpharmatica, where it was known as Admensa. Most of the properties predicted by StarDrop are physico-chemical, but also include some absorption and binding properties. It has one toxicity module—namely hERG potassium channel blockade in mammalian cells.

The hERG prediction model uses 158 descriptors, including a number of physico-chemical properties and counts of different atomic and/or functional groups and fragments. The model was trained on 135 compounds, with a 17-compound validation set and a 16-compound test set. For the combined validation and test sets, the model achieved an R^2 value of 0.72 and a RMSE of 0.64 log unit. A standard deviation of prediction is given for a query compound, with a large value indicating that the compound is outside the prediction space of the model, so that the prediction should be treated with caution. It is possible to incorporate a user's in-house data into the module and also to run a user's own hERG models within StarDrop. There are to date no peer-reviewed assessments of the performance of StardDrop's hERG model.

19.2.20   ADMEWORKS Predictor

Like some other ADMET prediction software, ADMEWORKS Predictor[99] covers mainly ADME endpoints, but also includes two toxicity endpoints—carcinogenicity and mutagenicity. Its models are based largely on physico-chemical, topological, geometrical and electronic descriptors derived from the ADAPT software developed by Jurs[100] and from MOPAC,[101] although some fragment count methods are also used. It is possible to build customised prediction models by integration with ADMEWORKS ModelBuilder.

No information appears to be available as to the performance of the ADMEWORKS Predictor toxicity modules, and it is not known whether the software gives an indication of the confidence level of predictions for a query compound.

19.2.21   COMPACT

COMPACT (Computer-Optimised Parametric Analysis of Chemical Toxicity) is a non-commercial toxicity prediction program developed at the University of Surrey, UK.[102,103] It uses only two molecular descriptors, namely the difference (ΔE) between E_{HOMO} and E_{LUMO} (energies of the highest occupied and lowest unoccupied molecular orbitals) respectively, and molecular area/depth^2 (a/d^2), to predict specificity for cytochrome P450 enzymes, particularly those involved in carcinogenicity and mutagenicity. For example, a 'COMPACT ratio' (defined as (a/d^2)×ΔE) of >0.25 is indicative of P4501 specificity, and hence of carcinogenicity.

In a study of 80 NTP chemicals, Brown *et al.*[54] found that COMPACT predicted carcinogenicity with a sensitivity of 71% and a specificity of 67%. In another COMPACT study of the mutagenicity of 100 NTP chemicals, Lewis *et al.*[103] found a concordance of 63%.

COMPACT is not available commercially, but is included here for completeness and for comparative purposes.

19.2.22 BfR Decision Support System

The BfR decision support system was developed by the German Federal Institute for Health Protection of Consumers and Veterinary Medicine (BgVV)[104,105]—now the Bundesinstitut für Risikobewertung (BfR).[106] It uses a decision tree approach, and its current endpoints are skin and eye irritation and corrosion. Zinke and Gerner[104] reported a concordance of 63% for severely damaging effects in all endpoints, and of 40% for minor damaging effects. This compares with a concordance of 74% for local irritation/corrosion using MCASE.[107]

19.2.23 Insilicofirst

A very new collaboration, Insilicofirst, between Lhasa Limited, Leadscope, Multicase and Molecular Networks GmbH offers toxicity assessment through a single common user interface.[108] Currently a user can obtain Insilicofirst toxicity predictions and data from each of the consortium's expert systems (*quod vide*) for query compounds, although an integrated prediction system is under consideration. It may be noted that Molecular Networks GmbH offers prediction of many properties, in particular metabolic and degradation reactions.

19.3 Independent Comparative Assessment of Software Performance

It has been pointed out[109] that most of the publications by software developers claim good predictions, often with better than 90% concordance. However, valid comparisons of performance need to be carried out independently of software developers using a given test set. Of course, no test set gives perfect coverage of chemical or biological space, and different test sets will almost always give different results.

Probably the earliest comparative assessment of toxicity prediction software was that by the US National Toxicology Program,[94] whereby model developers

Table 19.2   Results of first and second NTP carcinogenicity prediction tests.

System	1st test % concordance	2nd test % concordance
Oncologic	–	67
TOPKAT	58	–
DEREK	59	38
CASE	49	18
COMPACT	54	44
Human experts	75	–

were invited to submit their predictions of the carcinogenicity of 40 untested chemicals, which were then subjected to rodent carcinogenicity testing. Eleven model developers submitted predictions, including some who had developed commercial software.[110] A second NTP test utilised 26 chemicals.[95] The results from the commercial software (plus those from COMPACT and from human experts) are shown in Table 19.2.

As can be seen, the results were not encouraging, *ex facie*, with the software performing little better (or in some cases worse) than random selection. However, it has been pointed out[111] that the substances that the NTP tested were not a random selection from the chemical universe but included a majority of suspect substances. Hence the goal was not to separate carcinogens from non-carcinogens (which is what the expert systems were designed to do), but rather to separate actual carcinogens from possible carcinogens. Benigni[112] has analysed the results of the NTP and other external validation tests in detail.

Contrera *et al.*[43] compared rodent carcinogenicity predictions from MC4PC and MDL-QSAR.[113] (MDL-QSAR is not an expert system, but contains the programs, algorithms and statistical tools to develop QSARs.) They used a training set of 1,540 compounds and predictive performance was assessed by a leave-many-out 10% cross-validation procedure. MDL-QSAR was slightly better than MC4PC; sensitivities were 63% and 61%, specificities were 75% and 71%, and concordances were 69% and 66%, respectively.

Matthews *et al.*[114] extended this study by including another expert system, Leadscope PDM, and another model-building system, BioEpisteme,[115] and by increasing the size of the training set to 1,572 compounds. The study required high specificities and models were adjusted to obtain specificities of >80%, which had the effect of reducing sensitivities. Results for MC4PC, MDL-QSAR, BioEpisteme and Leadscope PDM were sensitivities of 39.8%, 46.1%, 45.9% and 47.5%, and specificities of 86.5%, 80.5%, 80.6% and 89.7%, respectively.

Prediction of Ames mutagenicity of over 400 compounds by DEREK and TOPKAT was determined.[116] For 409 compounds predicted by DEREK, sensitivity was 46%, specificity was 69% and concordance was 64.5%; for 399 compounds predicted by TOPKAT, sensitivity was 62%, specificity was 81% and concordance was 73%. Interestingly, these TOPKAT results were based on 303 compounds within the TOPKAT applicability domain (AD) and 96 compounds outside the AD, and were slightly better than those based only on compounds within the AD.

DEREK, TOPKAT and MCASE Ames mutagenicity predictions were compared by Snyder *et al.*[117] DEREK gave a sensitivity of 52%, a specificity of 75% and a concordance of 74% for 372 compounds; for TOPKAT the corresponding figures were 43%, 85% and 82% respectively for 339 compounds, whilst for MCASE they were 48%, 93% and 90% respectively for 357 compounds. These two studies demonstrate the different results that can be obtained with different test sets.

Lewis *et al.*[118] compared the performance of COMPACT and HazardExpert in predicting human carcinogenicity of 14 compounds; COMPACT gave 71% concordance and HazardExpert gave 57% concordance. Mombelli[30] evaluated the performance of DEREK, HazardExpert and TOPKAT in predicting the skin irritation ability of 116 compounds. Sensitivities were found to be DEREK 8%, HazardExpert 21%, and TOPKAT 60%, and specificities were found to be DEREK undetermined, HazardExpert 0%, and TOPKAT 60%. It should be noted that the version of DEREK used by Mombelli was marketed by LHASA Llc, Newton, MA, and not by Lhasa Limited, Leeds, UK.[34]

19.4 Consensus Modelling

Improved predictivity can often be achieved by combining the predictions from two or more expert systems and/or QSARs;[119,120] this is most probably because individual prediction errors are to some extent averaged out.[121] A consensus prediction can be simply the arithmetic mean of all predictions, or it can involve using a leverage-weighted mean, in which the most predictive model has the greatest contribution.[120] For classification endpoints, using consensus-positive predictions from any two or more programs can yield improved predictivity.[114]

Consensus modelling has been shown to give improved predictivity for soil–water partitioning,[122] oestrogen receptor binding[123] and genotoxicity.[121] However, Hewitt *et al.*[120] examined the use of consensus modelling of QSAR predictions for a number of endpoints, including toxicity to fathead minnow and to *Tetrahymena pyriformis*, and found no consistent significant benefits in terms of statistical fit and predictivity in the consensus models compared with single regression models.

There have been few studies to date of consensus modelling of toxicity predictions from expert systems. Lewis *et al.*[118] found that a combination of COMPACT and HazardExpert was able correctly to predict carcinogenicity of 14 known human carcinogens by utilising the ability of COMPACT to predict P450-mediated activity and the ability of HazardExpert to predict direct-acting alkylating agents by their structural alerts.

Contrera *et al.*[43] found that, by merging MC4PC and MDL-QSAR carcinogenicity predictions for 1,540 compounds, they were able to improve sensitivity from 61% and 63% respectively to 67%, specificity from 71% and 75% to 84%, and concordance from 66% and 69% to 76%. Matthews *et al.*[114] examined five carcinogenicity prediction systems (MC4PC, MDL-QSAR, BioEpisteme, Leadscope and DfW) using 1,572 compounds and found that accepting only consensus-positive predictions from any two programs yielded the best results, with sensitivity of 57.5% and specificity of 85.3%. Lagunin *et al.*[48] found that rodent carcinogenicity prediction of 293 compounds by PASS[46] and a QSAR method, CISOC-PSCT,[124] was somewhat improved by consensus modelling. Sensitivity, specificity and concordance were 81%, 74% and 76% respectively for PASS, and 36%, 89% and 77% respectively for CISOC-PSCT, whilst the consensus model yielded 69%, 86% and 83% respectively.

As mentioned in Section 19.2.14, the US EPA's T.E.S.T. program[84] uses five different approaches to toxicity prediction, together with a consensus prediction that is the arithmetic average of the five individual predictions. The validation checks reported[84] showed that for all endpoints, the consensus predictions were better than the individual predictions.

It is to be hoped that future consensus work will further improve the toxicity predictivity of expert systems.

19.5 Software Performance with *Tetrahymena pyriformis* Test Set

The *Tetrahymena pyriformis* toxicity data for the 350-compound test set used in this study were taken from Enoch *et al.*[125] and Ellison *et al.*[126]

Two expert systems, ADMET Predictor from SimulationsPlus[62] and T.E.S.T. from the US EPA[84] have a *Tetrahymena pyriformis* toxicity prediction module. SimulationsPlus kindly ran the test set used in this study through its module and obtained a reasonably good correlation of observed *vs.* predicted IGC_{50} values:

$$\log 1/IGC_{50}(\text{observed}) = 1.04 \log 1/IGC_{50}(\text{predicted}) - 0.021 \qquad (19.2)$$

$$n = 350 \, r^2 = 0.701 \, s = 0.433 \, F = 816.9$$

Figure 19.1 shows the plot of observed vs. predicted $\log 1/IGC_{50}$ values from ADMET Predictor.

The consensus predictions from T.E.S.T. were somewhat better:

$$\log 1/IGC_{50}(\text{observed}) = 1.06 \log 1/IGC_{50}(\text{predicted}) - 0.023 \qquad (19.3)$$

$$n = 349 \, r^2 = 0.751 \, s = 0.395 \, F = 1048.5$$

T.E.S.T. did not make any predictions for one compound (2,3-benzofuran).

Figure 19.2 shows the plot of observed *vs.* predicted $\log 1/IGC_{50}$ values from T.E.S.T.

In both sets of predictions, five compounds had predicted $\log 1/IGC_{50}$ values that were too low by more than 1 log unit; these were 4-amino-2-cresol, 2-aminophenol, methoxyhydroquinone, methylhydroquinone and 3-methyl-4-nitrophenol. Again in both sets of predictions, four compounds had predicted $\log 1/IGC_{50}$ values that were too high by more than 1 log unit; these were 4-

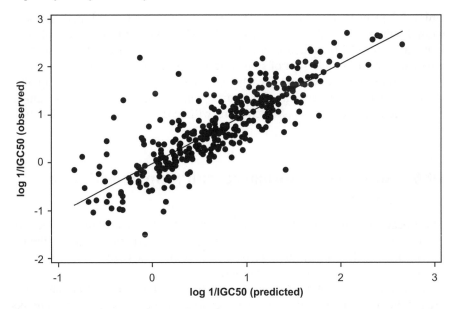

Figure 19.1 Observed *Tetrahymena pyriformis* toxicities *vs.* those predicted by ADMET Predictor.

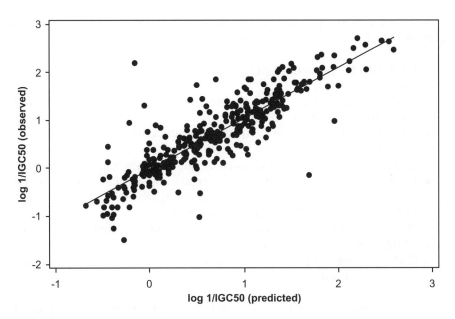

Figure 19.2 Observed *Tetrahymena pyriformis* toxicities *vs.* those predicted by T.E.S.T.

amino-2,3-xylenol (in ADMET Predictor only), 2.4,6-tris(dimethylamino-methyl)phenol (in T.E.S.T. only), 4-hydroxybenzoic acid, 4-hydroxyphenyl-acetic acid and 2,4,6-trinitrophenol. One can postulate reasons for these large errors such as the conversion of some phenols to the corresponding more toxic semiquinone and ionisation of carboxylic acids leading to reduced uptake. However, with the exception of the carboxylic acids, there are many similar compounds in the test set studied here with considerably lower prediction errors.

19.6 Software Performance with Skin Sensitisation Test Set

The Local Lymph Node Assay (LLNA) skin sensitisation data for the 210 compound test set studied here were taken from Gerberick *et al.*[127] Binary classification values (1 or 0) were based on compounds with EC3 values of $\geq 10\%$ being taken to be non-sensitisers.

Predictions were obtained from seven expert systems. In all but one case, predictions were made using the latest versions of the expert systems. Thanks are due to the developers/marketers of DS-TOPKAT, DfW, MC4PC and PASS for kindly running the compounds through their latest skin sensitisation modules (DS-TOPKAT 2.5, DfW 11, MC4PC 2.0.0.219, and PASS 2009.2). The HazardExpert predictions were obtained with version 3.1, as access to the latest version was not possible. One molecule, streptomycin, was too large to be handled by most of the expert systems.

Assessment of the performance of the expert systems is based on binary classification (1 or 0) values. The results are given in Table 19.3.

- DS-TOPKAT predictions were 'strong' (classed as 1), 'weak' and 'none' (both classed as 0).
- DfW predictions were 'certain', 'probable', 'plausible' (all classed as 1), and 'nothing to report' (classed as 0).
- MC4PC predictions were' 'active', 'possibly active' (both classed as 1), 'probably inactive' and 'inactive' (both classed as 0). MC4PC reported no prediction for 11 compounds, as being out of its applicability domain.
- PASS gives the probability that a compound is active (Pa) and the probability that it is inactive (Pi). Compounds with Pa > Pi were classed as 1, and those with Pi > Pa were classed as 0.
- HazardExpert reported 'class 3' (classed as 1) and 'no alert' (classed as 0).
- TIMES-SS reported activity as 1 or 0.
- CAESAR (version 1.0.0.3) reported activity as 'active' (classed as 1) and 'inactive' (classed as 0).

From a regulatory viewpoint, high sensitivity (proportion of true positive predictions) is desirable,[7] since it means a low proportion of false negative predictions. On that basis alone, CAESAR and DfW predictions are preferable.

Table 19.3 Software prediction of skin sensitisation.

Software	Sensitivity	Specificity	Concordance
DS-TOPKAT	60.8%	60.7%	60.8%
DfW[a]	93.1%	33.6%	62.5%
MC4PC[b]	65.3%	48.5%	56.6%
PASS	40.7%	60.7%	51.0%
HazardExpert	11.8%	96.3%	55.2%
TIMES-SS	63.9%	47.6%	55.4%
CAESAR	95.1%	30.6%	61.9%
Consensus 1[c]	75.8%	53.0%	64.1%
Consensus 2[d]	75.8%	52.0%	63.6%

[a]Predictions are for mouse. Predictions for human are: 93.1%, 36.5% and 63.9% respectively.
[b]Predictions do not include 11 out-of-domain compounds.
[c]Consensus 1 used two or more identical predictions from TOPKAT, MC4PC and TIMES-SS.
[d]Consensus 2 used four or more identical predictions from all seven expert systems.

On the other hand, they both have very low specificity, which means a high proportion of false positive predictions and could result in rejection of true non-sensitisers.

DS-TOPKAT gave the most consistent results in that sensitivity, specificity and concordance values were all very similar. However, at about 60%, none of them was very different from random (*i.e.* 50%), so DS-TOPKAT cannot be regarded as a particularly good predictor of skin sensitisers or non-sensitisers.

On the assumption that good sensitivity together with reasonable specificity is a realistic current goal, then from the results in Table 19.3, DS-TOPKAT, MC4PC and TIMES-SS are moderately good, although not satisfactory. These three expert systems were then examined to see whether a consensus approach could improve their performance, since consensus modelling has been found (see Section 19.3) to improve predictivity.

Following the approach used by Matthews *et al.*,[114] in which a majority prediction was used, two different criteria were used:

- two or more identical predictions from DS-TOPKAT, MC4PC and TIMES-SS; and
- four or more identical predictions from all seven expert systems.

It can be seen from Table 19.3 that both methods gave very similar results—considerably better than those of individual expert systems—on the assumption that high sensitivity and reasonable specificity are realistic goals in the regulatory environment.[7] It is therefore recommended that, if possible, a consensus method is used for expert system prediction of skin sensitisation.

19.7 Conclusions

There are now a considerable number of expert systems available for the prediction of a range of toxicity endpoints. All depend on the quality and availability of experimental data; these are to some extent still lacking and are, in some cases, of dubious accuracy. Hence a prime ongoing need for *in silico* prediction is more and better experimental data.

There has been a rapid increase in recent years in the number of expert systems available for toxicity prediction. Many of these meet the OECD Principles for the Validation of (Q)SARs,[13] but some do not—or at least it is not clear, from the sometimes limited information on company websites, whether or not they do so. It is recommended that all developers and marketers of toxicity prediction software publish this information. Most expert systems give information as to the confidence that one can have in a toxicity prediction for a query compound. It is recommended that all expert systems give such information.

Faced with the need to make a toxicity prediction, how should one decide which expert system to use? There are a number of considerations to be taken into account. First, of course, is the selection of one or more expert systems that offer the requisite endpoints. Then comes the key aspect of cost. A few of the expert systems considered here (*e.g.*, TIMES, CAESAR, ECOSAR and Oncologic) are freely available. The price of others varies greatly, with the most expensive costing many thousands of dollars. However, it is often possible to purchase or license specific modules, to pay the software company to run one's compound(s) through their appropriate toxicity module(s), or even to obtain one-off predictions free of charge. If details of putative mechanisms of action and appropriate literature references are required, then the choice is limited. DfW undoubtedly offers the most information in this respect.

For REACH work, it is important that the expert system used accords with the OECD Principles for the Validation of (Q)SARs (see Section 19.1). This has been indicated (where known) for each of the expert systems discussed in Section 19.2.

The accuracy of expert systems for toxicity prediction is generally not high. This is undoubtedly due in part to poor and insufficient data on which to develop the software, but is subject also to the modelling approach adopted. For these reasons, using a consensus of predictions from two or more expert systems has been found to yield better results. It is recommended that this approach be adopted wherever possible, though it is recognised that this would be more expensive than using a single expert system.

19.8   Acknowledgments

I am grateful to my colleagues Dr Steve Enoch, Dr Mark Hewitt and Mrs Claire Ellison for their computational contributions.

References

1. V. K. Gombar, B. E. Mattioni, C. Zwickl and J. T. Deahl, in *Computational Toxicology: Risk Assessment for Pharmaceutical and Environmental Chemicals*, ed. S. Ekins, Wiley, Hoboken, NJ, 2007, pp. 183–215.

2. D. E. Johnson, A. D. Rodgers and S. Sudarsanam, in *Computational Toxicology: Risk Assessment for Pharmaceutical and Environmental Chemicals*, ed. S. Ekins, Wiley, Hoboken, NJ, 2007, pp. 725–749.
3. J. C. Dearden, M. D. Barratt, R. Benigni, D. W. Bristol, R. D. Combes, M. T. D. Cronin, P. N. Judson, M. P. Payne, A. M. Richard, M. Tichý, A. P. Worth and J. J. Yourick, *Altern. Lab. Anim.*, 1997, **25**, 223.
4. A. M. Richard, *Mutat. Res.*, 1994, **305**, 73.
5. R. D. Combes and P. N. Judson, *Pestic. Sci.*, 1995, **45**, 179.
6. A. M. Richard, C. Yang and R. S. Judson, *Toxicol. Mech. Meth.*, 2008, **18**, 103.
7. B. Simon-Hettich, A. Rothfuss and T. Steger-Hartmann, *Toxicology*, 2006, **224**, 156.
8. K. Langton, G. Y. Patlewicz, A. Long, C. A. Marchant and D. A. Basketter, *Contact Derm.*, 2006, **55**, 342.
9. T. I. Netzeva, A. P. Worth, T. Aldenberg, R. Benigni, M. T. D. Cronin, P. Gramatica, J. S. Jaworska, S. Kahn, G. Klopman, C. A. Marchant, G. Myatt, N. Nikolova-Jeliaskova, G. Y. Patlewicz, R. Perkins, D. W. Roberts, T. W. Schultz, D. W. Stanton, J. J. M. van de Sandt, W. Tong, G. Veith and C. Yang, *Alt. Lab. Anim.*, 2005, **33**, 155.
10. J. Jaworska, N. Nikolova-Jeliaskova and T. Aldenberg, *Alt. Lab. Anim.*, 2005, **33**, 445.
11. V. K. Gombar, *SAR QSAR Environ. Res.*, 1999, **10**, 371.
12. W. Muster, A. Breidenbach, H. Fischer, S. Kirchner, L. Müller and A. Pähler, *Drug Discov. Today*, 2008, **13**, 303.
13. Organisation for Economic Co-operation and Development, *OECD Principles for the Validation, for Regulatory Purposes, of (Quantitative) Structure–Activity Relationship Models*, OECD, Paris, 2004, www.oecd.org/dataoecd/33/37/37849783.pdf [accessed March 2010].
14. Organisation for Economic Co-operation and Development, *The Report from the Expert Group on (Quantitative) Structure–Activity Relationships [(Q)SARs] on the Principles for the Validation of (Q)SARs*, OECD, Paris, 2004, OECD Environment Health and Safety Publications Series on Testing and Assessment No. 49, ENV/JM/MONO(2004)24, www.olis.oecd.org/olis/2004doc.nsf/LinkTo/NT00009192/$FILE/JT00176183.PDF [accessed March 2010].
15. E. J. Matthews and J. F. Contrera, *Expert Opin. Drug Metab. Toxicol.*, 2007, **3**, 125.
16. J. C. Dearden, M. T. D. Cronin and K. L. E. Kaiser, *SAR QSAR Environ. Res.*, 2009, **20**, 241.
17. E. Gottmann, S. Kramer, B. Pfahringer and C. Helma, *Environ. Health Perspect.*, 2001, **109**, 509.
18. D. Young, T. Martin, R. Venkatapathy and P. Harten, *QSAR Comb. Sci.*, 2008, **27**, 1337.
19. M. T. D. Cronin, *ALTEX*, 2006, **23**, Special Issue, 365.
20. R. Judson, A. Richard, D. J. Dix, K. Houck, M. Martin, R. Kavlock, V. Dellarco, T. Henry, T. Holderman, P. Sayre, S. Tan, T. Carpenter and E. Smith, *Environ. Health Perspect.*, 2009, **117**, 685.

21. http://ec.europa.eu/environment/chemicals/reach/reviews_en.htm [accessed March 2010].
22. A. Bassan and A. P. Worth, in *Computational Toxicology: Risk Assessment for Pharmaceutical and Environmental Chemicals*, ed. S. Ekins, Wiley, Hoboken, NJ, 2007, pp. 751–775.
23. J. C. Dearden, in *Computer Applications in Pharmaceutical Research and Development*, ed. S. Ekins, Wiley, Hoboken, NJ, 2006, pp. 469–494.
24. C. A. Marchant, in *Virtual ADMET Assessment in Target Selection and Maturation*, ed. B. Testa and L. Turski, IOS Press, Amsterdam, 2006, pp. 237–248.
25. P. N. Judson, in *Computational Toxicology: Risk Assessment for Pharmaceutical and Environmental Chemicals*, ed. S. Ekins, Wiley, Hoboken, NJ, 2007, pp. 521–543.
26. K. Enslein, *Toxicol. Ind. Health*, 1988, **4**, 479.
27. www.accelrys.com
28. K. Enslein, V. K. Gombar and B. W. Blake, *Mutat. Res.*, 1994, **305**, 47.
29. V. K. Gombar, K. Enslein and B. W. Blake, *Chemosphere*, 1995, **31**, 2499.
30. E. Mombelli, *Alt. Lab. Anim.*, 2008, **36**, 15.
31. G. Patlewicz, A. O. Aptula, E. Uriarte, D. W. Roberts, P. S. Kern, G. F. Gerberick, I. Kimber, R. J. Dearman, C. A. Ryan and D. A. Basketter, *SAR QSAR Environ. Res.*, 2007, **18**, 515.
32. R. Venkatapathy, C. J. Moudgal and R. M. Bruce, *J. Chem. Inf. Comput. Sci.*, 2004, **44**, 1623.
33. C. A. Marchant, K. A. Briggs and A. Long, *Toxicol. Mech. Meth.*, 2008, **18**, 177.
34. www.lhasalimited.org
35. D. M. Sanderson and C. G. Earnshaw, *Hum. Exp. Toxicol.*, 1991, **10**, 261.
36. P. Crettaz and R. Benigni, *J. Chem. Inf. Model.*, 2005, **45**, 1864.
37. E. M. Hulzebos and R. Posthumus, *SAR QSAR Environ. Res.*, 2003, **14**, 285.
38. E. Hulzebos, D. Sijm, T. Traas, R. Posthumus and L. Maslankiewicz, *SAR QSAR Environ. Res.*, 2005, **16**, 385.
39. G. Klopman, *J. Am. Chem. Soc.*, 1984, **106**, 7315.
40. E. J. Matthews, R. D. Benz and J. F. Contrera, *J. Mol. Graphics Model.*, 2000, **18**, 605.
41. G. Klopman, J. Ivanov, R. Saiakhov and S. Chakravarti, in *Predictive Toxicology*, ed. C. Helma, Taylor & Francis, Boca Raton, FL, 2005, pp. 423–457.
42. R. D. Saiakhov and G. Klopman, *Toxicol. Mech. Meth.*, 2008, **18**, 159.
43. J. F. Contrera, N. L. Kruhlak, E. J. Matthews and R. D. Benz, *Regul. Toxicol. Pharmacol.*, 2007, **49**, 172.
44. V. Poroikov and D. Filimonov, in *Predictive Toxicology*, ed. C. Helma, Taylor & Francis, Boca Raton, FL, 2005, pp. 459–478.
45. A. A. Lagunin, J. C. Dearden, D. A. Filimonov and V. V. Poroikov, *Mutat. Res.*, 2005, **586**, 138.

46. www.chem.ac.ru/Chemistry/Soft/PASS.en.html [accessed March 2010].
47. V. Poroikov, D. Akimov, E. Shabelnikova and D. Filimonov, *SAR QSAR Environ. Res.*, 2001, **12**, 327.
48. A. Lagunin, D. Filimonov, A. Zakharov, W. Xie, Y. Huang, F. Zhu, T. Shen, J. Yao and V. Poroikov, *QSAR Comb. Sci.*, 2009, **28**, 806.
49. www.leadscope.com
50. C. Yang, K. Cross, G. J. Myatt, P. E. Blower and J. F. Rathman, *J. Med. Chem.*, 2004, **47**, 5984.
51. G. Roberts, G. J. Myatt, W. P. Johnson, K. P. Cross and P. E. Blower, *J. Chem. Inf. Comput. Sci.*, 2000, **40**, 1302.
52. www.compudrug.com
53. M. P. Smithing and F. Darvas, in *Food Safety Assessment*, ed. J. W. Finley, S. F. Robinson and D. J. Armstrong, American Chemical Society, Washington DC, 1992, ACS Symposium Series 484, pp. 191–200.
54. S. J. Brown, A. A. Raja and D. F. V. Lewis, *Alt. Lab. Anim.*, 1994, **22**, 482.
55. Much of the content of the Pharma Algorithms website (www.ap-algorithms.com) has moved to the ACDLabs website or is no longer available..
56. www.acdlabs.com
57. P. Japertas, R. Didziapetris and A. Petrauskas, *Mini-Revs. Med. Chem.*, 2003, **3**, 797.
58. R. Didziapetris, K. Lanevskij and P. Japertas, *Chem. Res. Toxicol.*, 2008, **21**, 2450.
59. L. Juska, R. Didziapetris and P. Japertas, *Toxicol. Lett.*, 2008, **180S**, S153.
60. P. Japertas, R. Didziapetris and A. Petrauskas, *Chem. Res. Toxicol.*, 2007, **20**, 1998.
61. P. J. Jurgutis, R. Didziapetris and P. Japertas, *Chem. Res. Toxicol.*, 2007, **20**, 2010.
62. www.simulations-plus.com
63. http://q-pharm.com
64. A. F. Fliri, W. T. Loging, P. F. Thadeio and R. A. Volkmann, *Proc. Nat. Acad. Sci. U. S. A.*, 2005, **102**, 261.
65. A. Fliri, W. Loging, P. Thadeio and R. Volkmann, *J. Med. Chem.*, 2005, **48**, 6918.
66. http://lazar.in-silico.de
67. C. Helma, in *Predictive Toxicology*, ed. C. Helma, Taylor & Francis, Boca Raton, FL, 2005, pp. 479–499.
68. C. Helma, *Mol. Divers.*, 2006, **10**, 147.
69. www.terrabase-inc.com
70. K. L. E. Kaiser, S. P. Niculescu and T. W. Schultz, *SAR QSAR Environ. Res.*, 2002, **13**, 57.
71. K. L. E. Kaiser and S. P. Niculescu, *Chemosphere*, 1999, **38**, 3237.
72. K. L. E. Kaiser and S. P. Niculescu, *Water Qual. Res. J. Canada*, 2001, **36**, 619.
73. M. Pavan, A. P. Worth and T. I. Netzeva, Comparative Assessment of QSAR Models for Aquatic Toxicity, European Commission Joint

Research Centre, Ispra, Italy, 2005, EUR 21750, http://ecb.jrc.ec.europa.eu/qsar/information-sources/ [accessed March 2010].

74. K. L. E. Kaiser, personal communication to J. C. Dearden, 31 July 2009.
75. www.oasis-lmc.org/?section = software&swid = 4
76. S. D. Dimitrov, L. K. Low, G. Y. Patlewicz, P. S. Kern, G. D. Dimitrova, M. H. I. Comber, R. D. Phillips, J. Niemela, P. T. Baily and O. G. Mekenyan, *Int. J. Toxicol.*, 2005, **24**, 189.
77. O. G. Mekenyan, S. D. Dimitrov, T. S. Pavlov and G. D. Veith, *Curr. Pharm. Des.*, 2004, **10**, 1273.
78. O. G. Mekenyan, J. M. Ivanov, S. H. Karabunarliev, S. P. Bradbury, G. T. Ankley and W. Karcher, *Environ. Sci. Technol.*, 1997, **31**, 3702.
79. O. G. Mekenyan, N. Nikolova, S. Karabunarliev, S. Bradbury, G. Ankley and B. Hansen, *Quant. Struct.-Act. Relat.*, 1999, **18**, 139.
80. S. Bradbury, V. Kamenska, P. Schmieder, G. Ankley and O. Mekenyan, *Toxicol. Sci.*, 2000, **58**, 253.
81. D. W. Roberts, G. Patlewicz, S. D. Dimitrov, L. K. Low, A. O. Aptula, P. S. Kern, G. D. Dimitrova, M. I. H. Comber, R. D. Phillips, J. Niemelä, C. Madsen, E. B. Wedebye, P. T. Bailey and O. G. Mekenyan, *Chem. Res. Toxicol.*, 2007, **20**, 1321.
82. G. Patlewicz, S. D. Dimitrov, L. K. Low, P. S. Kern, G. D. Dimitrova, M. I. H. Comber, A. O. Aptula, R. D. Phillips, J. Niemelä, C. Madsen, E. B. Wedebye, D. W. Roberts, P. T. Bailey and O. G. Mekenyan, *Regul. Toxicol. Pharmacol.*, 2007, **48**, 225.
83. www.caesar-project.eu/software
84. www.epa.gov/nrmrl/std/cppb/qsar
85. J. F. Contrera, E. J. Matthews and R. D. Benz, *Regul. Toxicol. Pharmacol.*, 2003, **38**, 243.
86. www.epa.gov/oppt/newchems/tools/21ecosar.htm
87. P. Reuschenbach, M. Silvano, M. Dammann, D. Warnecke and T. Knacker, *Chemosphere*, 2008, **71**, 1986.
88. www.chemsilico.com
89. L. B. Kier and L. H. Hall, *Molecular Connectivity in Structure-Activity Analysis*, Wiley, New York, 1986.
90. L. H. Hall and L. B. Kier, *Molecular Structure Description: The Electrotopological State,* Academic Press, New York, 1999.
91. L. B. Kier and L. H. Hall, in *Topological Indices and Related Descriptors in QSAR and QSPR*, ed. J. Devillers and A. T. Balaban, Gordon & Breach, Reading, UK, 1999, pp. 491–562.
92. www.epa.gov/oppt/newchems/tools/oncologic.htm
93. Y.-T. Woo and D. Y. Lai, in *Predictive Toxicology*, ed. C. Helma, Taylor & Francis, Boca Raton, FL, 2005, pp. 385–413.
94. http://ntp.niehs.nih.gov
95. R. Benigni and R. Zito, *Mutat. Res.*, 2004, **566**, 49.
96. www.schrodinger.com
97. R. Farid, T. Day, R. A. Friesner and R. A. Pearlstein, *Bioorg. Med. Chem.*, 2006, **14**, 3160.

98. www.optibrium.com/stardrop
99. www.fqs.pl
100. http://research.chem.psu.edu/pcjgroup/adapt.html
101. www.openmopac.net
102. D. V. Parke, C. Ioannides and D. F. V. Lewis, *Alt. Lab. Anim.*, 1990, **18**, 91.
103. D. F. V. Lewis, C. Ioannides and D. V. Parke, *Mutat. Res.*, 1993, **291**, 61.
104. S. Zinke and I. Gerner, *Alt. Lab. Anim.*, 2000, **28**, 609.
105. I. Gerner, S. Zinke, G. Graetschel and E. Schlede, *Alt. Lab. Anim.*, 2000, **28**, 665.
106. www.bfr.bund.de
107. H. S. Rosenkranz, Y. P. Zhang and G. Klopman, *Alt. Lab. Anim.*, 1998, **26**, 779.
108. www.insilicofirst.com
109. E. Benfenati and G. Gini, *Toxicology*, 1997, **119**, 213.
110. R. Benigni, *Mutat. Res.*, 1997, **387**, 35.
111. A. M. Richard and R. Benigni, *SAR QSAR Environ. Res.*, 2002, **13**, 1.
112. R. Benigni, *Drug Discov. Today: Technologies*, 2004, **1**, 457.
113. www.mdli.com
114. E. J. Matthews, N. L. Kruhlak, R. D. Benz and J. F. Contrera, *Toxicol. Mech. Meth.*, 2008, **18**, 189.
115. www.prousresearch.com
116. N. F. Cariello, J. D. Wilson, B. H. Britt, D. J. Wedd, B. Burlinson and V. Gombar, *Mutagenesis*, 2002, **17**, 321.
117. R. D. Snyder, G. S. Pearl, G. Mandakas, W. N. Choy, F. Goodsaid and L. Y. Rosenblum, *Environ. Mol. Mutagen.*, 2004, **43**, 143.
118. D. F. V. Lewis, M. G. Bird and M. N. Jacobs, *Hum. Exp. Toxicol.*, 2002, **21**, 115.
119. M. Ganguly, N. Brown, A. Schuffenhauer, P. Ertl, V. J. Gillet and P. A. Greenidge, *J. Chem. Inf. Model.*, 2006, **46**, 2110.
120. M. Hewitt, M. T. D. Cronin, J. C. Madden, P. H. Rowe, C. Johnson, A. Obi and S. J. Enoch, *J. Chem. Inf. Model.*, 2007, **47**, 1460.
121. J. R. Votano, M. Parham, L. H. Hall, L. B. Kier, S. Oloff, A. Tropsha, Q. Xie and W. Tong, *Mutagenesis*, 2004, **19**, 365.
122. P. Gramatica, E. Giani and E. Papa, *J. Mol. Graph. Model.*, 2007, **25**, 755.
123. W. Tong, Q. Xie, H. Hong, L. Shi, H. Fang and R. Perkins, *Environ. Health Perspect.*, 2004, **112**, 1249.
124. Q. Liao, J. H. Yao, F. Li, S. G. Yuan, J. P. Doucet, A. Panaye and B. T. Fan, *SAR QSAR Environ. Res.*, 2004, **15**, 217.
125. S. J. Enoch, M. T. D. Cronin, T. W. Schultz and J. C. Madden, *Chemosphere*, 2008, **71**, 1225.
126. C. M. Ellison, M. T. D. Cronin, J. C. Madden and T. W. Schultz, *SAR QSAR Environ. Res.*, 2008, **19**, 751.
127. G. F. Gerberick, C. A. Ryan, P. S. Kern, H. Schlatter, R. J. Dearman, I. Kimber, G. Y. Patlewicz and D. A. Basketter, *Dermatitis*, 2005, **16**, 157.

CHAPTER 20

Exposure Modelling for Risk Assessment

J. MARQUART

TNO Quality of Life, Food and Chemical Risk Analysis, Utrechtseweg 48, PO Box 360, 3700 AJ Zeist, The Netherlands

20.1 Introduction: Hazard, Exposure, Risk

The risk of a substance is determined by its hazard and the relevant exposure of the target organism to the substance. 'Hazard' is defined as follows by the International Programme on Chemical Safety (IPCS) as: '*An Inherent property of an agent or situation having the potential to cause adverse effects when an organism, system, or (sub)population is exposed to that agent*'.[1] The hazard of a substance has multiple facets. There can, for example, be different relevant 'endpoints' (an 'endpoint' is, for example, a measureable toxic effect such as acute toxicity or carcinogenicity) and different relevant timeframes. Hazards may also be qualitatively and quantitatively different depending on the route of exposure.

The hazardous properties of a substance may be inherent, but they will only lead to actual (unwanted or adverse) effects when there is exposure to the substance, as indicated by the following definition. 'Exposure' is defined as the: '*Concentration or amount of a particular agent that reaches a target organism, system, or (sub)population in a specific frequency for a defined duration*' or as '*Contact between an agent and a target. Contact takes place at an exposure surface over an exposure period*'.[1,2] Just like hazard, exposure also has several facets. In fact, the facets are largely the same, *e.g.* differences in routes

Issues in Toxicology No.7
In Silico Toxicology: Principles and Applications
Edited by Mark T. D. Cronin and Judith C. Madden
© The Royal Society of Chemistry 2010
Published by the Royal Society of Chemistry, www.rsc.org

and timeframes. There are also differences in 'type of exposure'. Exposure *via* inhalation to a liquid substance may, for example, only be to vapours of the substance or also to aerosols (finely dispersed droplets in air). Exposure *via* inhalation to a solid substance can be to different particle size distributions.

'Risk' is defined by IPCS as: '*The probability of an adverse effect in an organism, system, or (sub)population caused under specified circumstances by exposure to an agent.*'[1] This probability depends on the hazard of the substance and on the exposure in the relevant situation.

The scope of regulatory-driven or voluntary programmes studying hazards, exposure and risks is generally to prevent or minimise the occurrence of adverse effects. All steps making up a total programme of prevention or minimisation of occurrence of adverse effects are called 'risk analysis'.[1] Figure 20.1 shows the steps in risk analysis as defined by IPCS. For specific definitions, see ref. 1.

'Exposure assessment' is defined as: '*The process of estimating or measuring the magnitude, frequency, and duration of exposure to an agent, along with the number and characteristics of the population exposed. Ideally, it describes the sources, pathways, routes, and the uncertainties in the assessment*'.[1] This process is an important element in risk assessment. Where hazard characterisation often leads to some kind of threshold (*i.e.* a dose or exposure concentration of an agent below which a stated effect is not observed or expected to occur),[1] there is a need for exposure assessment to lead to quantifiable levels that can be compared to such thresholds. In many risk assessment processes, the number and characteristics of the population exposed are not specifically assessed. A conclusion of an unacceptable risk is reached if the estimated exposure level is above the threshold or if the probability of an exposure level above a threshold is higher than that considered to be acceptable.

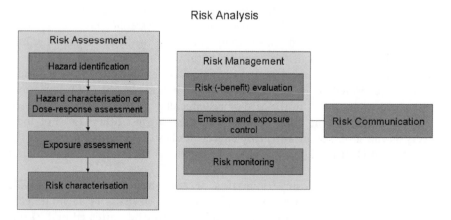

Figure 20.1 Steps in the IPCS risk analysis process.[1]

20.2 Types of Exposure Estimates used in Risk Characterisation in REACH

To enable a proper quantitative risk characterisation, a threshold and an exposure estimate need to be available. In many situations exposure can occur *via* different routes:

- oral (*e.g.* ingestion of chemicals in food or drink);
- inhalation (of concentrations in air); or
- dermal (contact of chemicals with the skin).

Therefore, separate thresholds are often calculated per route. In addition, because the effects of a chemical not only depend on the intensity of exposure but also on the duration, there are often separate thresholds for different timeframes. In occupational health, for example, the occupational exposure limits generally have a defined duration of eight hours (a work shift), but for acutely acting substances, short-term exposure limits (generally over 15 minutes) are also often derived.[3] For consumers, the Derived No Effect Levels (DNELs) for chronic exposures are generally derived as values over 24 hours because consumers can be exposed for 24 hours to chemicals in their living environment.

In the European Union's REACH legislation, DNELs are the thresholds used in risk characterisation.[4] DNELs will usually be derived for workers and for the general population (including both consumers and people exposed *via* the environment), whereas DNELs for the general population will be lower because the general population contains more sensitive groups (*e.g.* small children, elderly and people with specific illnesses). Those types of DNELs considered relevant in REACH are presented in Figure 20.2.[5]

DNELs for repeated exposure (chronic) and systemic effects will, in most cases, have to be derived. For workers, the oral route is not considered relevant, but for the general population it is. DNELs for acute exposure may also be needed when the substance has to be classified for acute effects and peak exposure is possible. DNELs for local exposure will be relatively uncommon. For inhalation a differentiation between local and systemic effects is not very useful because both types of effect occur as a result of the same exposure and the DNEL is expressed in the same units ($mg\,m^{-3}$). For dermal exposure, a differentiation between local and systemic effects is logical because systemic effects are expected to be expressed in $mg\,kg^{-1}$ bodyweight (bw) day^{-1} and local effects in $mg\,cm^{-2}$. In the case of systemic effects, the DNEL is supposed to be an external value, although the unit may suggest otherwise. For local effects on the skin, specifically acute effects, studies do generally not allow the derivation of DNELs. Thus a DNEL for acute local effects on the skin will not be derived very often.

In principle, the exposure estimation process should be able to provide conclusions for exposure estimates for all relevant DNELs. To facilitate easy comparison, the exposure estimates should preferably be expressed in the same units as the DNELs. However, even without a DNEL, a quantitative exposure

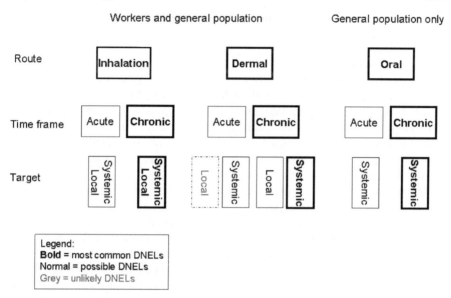

Figure 20.2 Types of DNELs used in REACH.

estimate may be useful to assess the risks of effects. For endpoints that generally do not have an established, quantitative DNEL (*e.g.* irritation, corrosion, sensitisation), a qualitative risk characterisation is expected. The exposure estimation for the qualitative risk characterisation should also include a characterisation of the intensity of exposure that can be expressed quantitatively.

With respect to the most common DNELs that will be found in risk characterisation in REACH, exposure estimation methods should preferably be able to provide conclusions relating to:

- long-term (averaged) exposure by inhalation (in $mg\,m^{-3}$);
- long-term (averaged) exposure *via* the skin (in $mg\,kg^{-1}\,bw\,day^{-1}$ and in $mg\,cm^{-2}$); and
- long-term (averaged) exposure *via* the oral route for the general population only (in $mg\,kg^{-1}\,bw\,day^{-1}$).

Exposure estimates over shorter periods are helpful as well—especially for inhalation and oral exposure—because thresholds for short-term dermal exposure are less common.

For workers, the average time of long-term exposure estimates should be equal to that of the full shift DNELs that are derived (eight hours), while for the general population, the average time of 24 hours is common. No specific guidance on short-term average times is available, but short-term DNELs for workers are often derived with average times of 15 minutes, while for the general population, relevant 'short-term exposure' can range from minutes to hours.

20.3 Methods for Exposure Assessment in Regulatory Processes

Exposure assessment can be performed in many ways. In workplaces, measurement of personal exposure *via* inhalation is often possible. There are also methods for sampling contamination from the skin, but these are less commonly used. Performing measurements for the exposure of consumers or the general population is less common. Exposure estimation in regulatory processes based on measurements is therefore generally restricted to inhalation exposure for workers and, even for that type of exposure, measurements for most substances are not available.

The use of measured exposure levels in regulatory processes is not straightforward. Many measured exposure levels are archived with very limited contextual information, making their interpretation very difficult. For REACH it is, for example, expected that the 'safe use' is described by assessing the operational conditions and risk management measures that lead to exposure levels below the DNELs. However, in many measurement programmes, the description of conditions and risk management measures is minimal and many data sets do not allow, for instance, differentiation between situations based on the percentage of the measured substance in products used. Only in a small number of cases is a list of measured values available, with some free text indicating the jobs of the workers involved. This kind of information can seldom be used for regulatory risk characterisation processes.[6] Some guidelines for ensuring good quality of measured data are presented by Tielemans *et al.*[7,8] and Money and Margary.[9]

In occupational hygiene and epidemiological studies, observational techniques are also used to provide semi-quantitative estimates of exposure. But due to the large effort needed to observe (many) situations and because the resulting parameters are usually not quantitative exposure levels, these methods are not very useful for regulatory risk characterisation processes.

Finally, there are several types of quantitative exposure models that estimate exposure levels based on inputs for a number of determinants of exposure. Such exposure models can be very simple, *e.g.* the German easy-to-use workplace control scheme[10] based on COSHH Essentials (www.coshh-essentials.org.uk; developed by the UK's Health and Safety Executive), or very complicated, *e.g.* Consexpo[11] or ART.[12] These models form the further focus of this chapter.

20.4 Types of Human Exposure Models

Basically, there are two types of human exposure models:

- **Mathematical models** that calculate processes of emission, distribution, transmission and removal of substances in specific areas; and
- **Statistical models** that base estimations on the relationships between exposure levels that occur and determinants or modifiers of these exposure levels.

A more recent development is the use of Bayesian techniques to combine model outcomes (either from physico-chemical models or from statistical models) with additional information into new combined results. In this case, the results from the models are treated as the 'prior estimate' and the other information (often actual measurement results) are used to modify this into the 'posterior estimate'.[13,14]

20.4.1 Mathematical Models

The mathematical models as intended here are models that try to model actual physical and chemical processes leading to emission of substances from a source, dispersion through an area and finally exposure of workers. Models that estimate evaporation of volatile liquids as 'source strength' and the (average) concentration in an area resulting from processes of dispersion and removal from the area are quite common. Examples of such models have been presented by authors such as Nicas[15] and the US Environmental Protection Agency (US EPA) (*e.g.* IT Environmental Programs,[16] GEOMET Technologies[17]). The American Industrial Hygiene Association (AIHA) has published a book summarising many aspects of mathematical models.[18] ConsExpo is a consumer exposure model that includes elements of these models in part of its equations.[11] These models often actually consist of two parts:

- modelling generation rate (the amount of vapour generated per unit of time); and
- modelling dispersion, dilution and removal rates.

In general, evaporation is calculated based on relatively simple assumptions regarding the actual processes in a liquid product and on the boundary of such a product. It is, for example, generally assumed that liquid mixtures are ideal mixtures, *i.e.* that the partial vapour pressure of substances is equal to their molar percentage in the mixture. However, there are also options to calculate evaporation more scientifically, based on the exact composition of the mixture, *e.g.* using the SUBFAC method.[19] ConsExpo has the option to choose between instant emission of all the substance, emission of all the substance over the duration of use and evaporation calculated from activity coefficients and composition of the mixture.

Emission of dusts from handling solids cannot be estimated easily on the basis of physical laws. Therefore, to use the physico-chemical models with solids, a generation rate will have to be estimated in another way, *e.g.* by estimating a loss of product due to the action. When a generation rate has been estimated, the physico-chemical models calculate concentrations in rooms or parts of rooms by accounting for mixing of contamination in air, removal by ventilation, addition from outside sources and removal by other so-called 'sinks'. Most models used in regulatory areas focus only on mixing and removal by ventilation. The basic assumption is that there are no relevant outside sources and no other removal mechanisms than ventilation.

Models for mixing and ventilation can be very complex, *e.g.* modelling transfer of air from one part of the room to another, based on knowledge or estimation of air flows. However, in general, the most used models of this type assume that a room is generally one box which is instantaneously homogeneously mixed. In some cases there is a factor to correct for non-perfect mixing, or a room can be divided into two separate boxes—one around the source and the other the remainder of the room.

Some models link several rooms (boxes) together, taking account of the flows in and out of adjacent rooms. An example is the 'multichamber exposure model' (MCEM) developed by the US EPA.[20]

A basic equation used for estimating a concentration in a room at a certain point in time is the 'well mixed room' equation (20.1) that assumes that the air in a room is well mixed:

$$V * \frac{dC_{Aroom}}{dt} = QC_{Ain} + G_A - K_{sink}C_{Aroom} - QC_{Aroom} \tag{20.1}$$

where:

C_{Aroom} = uniform room concentration of substance A (mg m^{-3})
t = time (min)
Q = airflow (m^3 min^{-1})
$C_{A\ in}$ = incoming air concentration of substance A (mg m^{-3})
G_A = mass generation rate of substance A (mg min^{-1})
K_{sink} = sink rate (m^3 min^{-1})
V = volume of air in the room (m^3)

20.4.2 Statistical Models

In this chapter, all models that estimate exposure levels based on inputs and known or assumed effects of modifiers that are not calculated based on basic physical and chemical laws are considered to be statistical models. However, a further sub-division can be made within this group. There are models in which the relation between inputs and exposure levels is based on knowledge of exposure levels modified by expert judgement, without any formal statistical analyses of the relations between input values and exposure levels. These models generally can be pictured as a flow diagram, guiding the user through the options and finally resulting in an exposure level for each possible combination of options. These exposure levels may be derived by experts from, for example, an exposure database. An example of such a model is EASE, which was used extensively in the past in regulatory risk assessments in Europe.[21] This model has been replaced by other models in recent years and its use is discontinued. One of the replacements is the European Centre for Ecotoxicology and Toxicology of Chemicals (ECETOC) TRA worker,[22,23] which was developed to be an improved and more user-friendly version of EASE.

Other models base their relationships between exposure levels and input values of parameters in the model on actual statistical analyses of sets of exposure levels and inputs. This can be done by direct analyses of the relation between exposure levels and inputs *via* multiple regression or mixed modelling as, for example, has been achieved for RISKOFDERM.[24] It can also be performed by first constructing an 'exposure score' based on the combination of inputs used and then 'calibrating' the exposure scores by an analysis of the relation between exposure scores and exposure levels. This has, for example, been done for Stoffenmanager[14,25] and ART.[12]

20.5 Detailed Description of the Models

A number of mathematical models that are used in regulatory exposure assessments and can be found on the internet are discussed briefly below. A number of statistical models are then described.

20.5.1 ConsExpo

ConsExpo was created by RIVM in The Netherlands[11] and is actually a suite of models in a common shell. It is the tool for modelling consumer exposure to substances used in consumer products (non-food) and is often used in EU regulatory risk assessments.

The model includes modelling of inhalation exposure as well as dermal exposure. For inhalation exposure, the tool uses generation rate modelling for volatiles or more simple estimates of emission of all relevant substance instantaneously or evenly over the duration of use. No specific generation rate estimation is available for use of solid products. There is a method to estimate values for generated aerosols in spray applications. The modelling of dispersion is *via* a box model, assuming instantaneous mixing through the whole room.

The tool has several options for dermal exposure:

- instant application;
- constant rate application;
- rubbing off from surfaces;
- migration (leaching); and
- diffusion (from articles to skin).

For the use of chemical products such as paint, no source strength estimation is included. The contamination on the outside of the skin needs to be entered manually. The model includes a quantitative structure–activity relationship (QSAR) for dermal absorption and can estimate dermal uptake as the final result.

For oral exposure, the tool calculates intake based on:

- direct oral consumption (*e.g.* swallowing a piece of eraser);
- constant ingestion rate;

- migration (*e.g.* sucking on a baby's dummy); and
- migration from packaging material (*via* food or drink).

RIVM has gathered default values (*e.g.* for product use rates, room sizes, *etc.*) for several parameters in a number of equations for typical consumer product uses. These defaults are compiled into fact sheets, e.g. relating to children's toys. There is also a general fact sheet with, for example, anthropometric data and details on housings. These fact sheets are also available in a database that can be used from ConsExpo.

The outputs of ConsExpo are estimates of event-average concentrations in the air or dermal or oral uptake. An event in this case is a one-time of application of a product or use of an article, including any follow-up emissions due to evaporation after application and exposure due to people staying in rooms where concentrations are still not sufficiently reduced due to evaporation. From the basic event-based exposure, longer term averages can also be calculated. Distributed results from a Monte Carlo simulation and graphs can be part of the output. A sensitivity analysis can also be performed.

Exposure estimates from ConsExpo are:

- air concentration (external, $mg\,m^{-3}$);
- internal inhalation dose ($mg\,kg^{-1}$ bw);
- external dermal exposure as dermal load ($mg\,cm^{-2}$ of exposed skin) or external dose ($mg\,kg^{-1}$ bw);
- internal dermal dose ($mg\,kg^{-1}$ bw); and
- external and internal oral exposure ($mg\,kg^{-1}$ bw).

In addition to point estimates, output distributions of these estimates can be derived when probabilistic options have been chosen (*e.g.* inputs as distributions instead of as single values).

ConsExpo version 4.1 (or older versions) and the fact sheets can be downloaded from the RIVM website (www.rivm.nl/en/healthanddisease/productsafety/ConsExpo.jsp).

20.5.2 ECETOC TRA Consumer

The ECETOC Targeted Risk Assessment tool (ECETOC TRA) includes a consumer exposure assessment element based on ConsExpo. It uses conservative defaults for many inputs and the more simple equations from ConsExpo to derive a so-called Tier 1 assessment.

The model requires input on the type of consumer product that is the source of the exposure. Products and articles are organised in Product Categories and Article Categories which conform to the use descriptor system used in REACH.[26] A number of categories have subcategories. For instance, the Product Category 'Adhesives, sealants' has four subcategories—'Glues, hobby use', 'Glues DIY-use (carpet glue, tile glue, wood parquet glue)', 'Glue from spray' and 'Sealants'. Users choose either one or more Product or Article Categories, or one or more subcategories. The Product or Article Category results are calculated from the worst-case subcategories.

The model calculates conservative estimates based on conservative defaults included in the program. However, for some inputs (*e.g.* product ingredient fraction, skin contact area and amount of product used per application), users can override the defaults by entering more specific values.

The model calculates the following results:

- room average air concentrations (called 'inhalation exposure' by the model) in $mg\,m^{-3}$ averaged over the exposure duration;
- external inhalation exposure ($mg\,kg^{-1}\,bw\,day^{-1}$);
- external dermal exposure ($mg\,kg^{-1}\,bw\,day^{-1}$); and
- external oral exposure ($mg\,kg^{-1}\,bw\,day^{-1}$)

The results do not account for absorption percentages or uptake rates. They can be used as worst-case values for uptake by implicitly assuming that the absorption percentages are 100% for all routes.

ECETOC TRA (version 2.0) can be downloaded from the ECOTEC website (www.ecetoc.org/tra) together with a user guide and supporting documents.

20.5.3   Wall Paint Exposure Model

The Wall Paint Exposure Model (WPEM), created by GEOMET Technologies for the US EPA, is designed to assess the inhalation exposure of consumers and/or professionals to (volatile) substances due to painting of walls and ceilings in a building, including the evaporation afterwards. The model is not used regularly in regulatory processes in the EU. It estimates generation rates based on (semi-)experimental models resulting from statistical analyses of small chamber experiments on emission of substances from either alkyd or latex paints. The reducing effect of a few sinks (carpet and wallboard) and re-emission from sinks are also taken into account. The model then calculates one or two zone air concentrations *via* mass balance models. Based on the calculated concentrations, the model can also calculate the lifetime average daily dose by taking into account factors such as the number of painting events and activity patterns (time, location and breathing rate).

The model calculates the following results:

- highest instantaneous concentration a person is exposed to;
- highest 15-minute average exposure;
- highest eight-hour average exposure;
- average daily exposure (over the number of years of exposure); and
- lifetime average daily exposure.

All exposure values are external (average) room concentration values, presented in both $mg\,m^{-3}$ and parts per million (ppm).

Based on these data the model also calculates:

- single event dose (mg);
- acute potential dose rate ($mg\,kg^{-1}\,bw\,day^{-1}$), which is the dose over one day;
- average daily dose ($mg\,kg^{-1}\,bw\,day^{-1}$), which is the dose averaged over the number of years of exposure; and
- lifetime average daily dose ($mg\,kg^{-1}\,bw\,day^{-1}$).

All these dose values are external values that account for breathing rate but not for absorption percentage or uptake rate.

WPEM can be downloaded from the US EPA website (www.epa.gov/oppt/exposure/pubs/wpemdl.htm).

20.5.4   Multi-chamber Concentration and Exposure Model

The Multi-chamber Concentration and Exposure Model (MCCEM), created by GEOMET Technologies for US EPA, models the dispersion of contamination and reduction due to sinks through different rooms of a building from emissions generated in the building.[20] As far as I know this model has not yet been used in any EU regulatory risk assessment. The model itself does not estimate emissions or sinks. Emission rates and sink rates (if considered relevant) must be estimated by the user by another means and entered into the model. The model estimates air concentrations in as many as four zones for a given building using a mass balance approach. The model contains a database with data on zone or area volumes, interzonal air flows, and whole-house exchange rates. The model estimates average and peak indoor air concentrations, with averaging durations between an hour and one year, as well as seasonal or annual exposure profiles using a long-term model. All estimated values are external values that do not take into account absorption percentage or uptake rate of the chemical.

MCCEM can be downloaded from the US EPA website (www.epa.gov/oppt/exposure/pubs/mccemdl.htm).

20.5.5   ECETOC TRA Worker

Whereas the ECETOC TRA consumer model is a mathematical model, the ECETOC TRA worker tool (whose second version is specifically designed to be a Tier 1 tool for use for REACH purposes), is a statistical model based on the older EASE model.[21,22] It estimates exposures by leading the user through a number of choices and providing basic exposure levels for each type of task. These basic exposure levels are based on EASE but have been modified. The modifications were partly made to link more directly to the Process Categories

(PROC) used in REACH, partly to improve user-friendliness and partly to improve the basic estimates based on experience and judgement by industry experts. The original EASE estimates were reported to be based on an analysis of measured data in relation to the options of that model,[21] though this analysis has never been published.

In ECETOC TRA worker, a basic estimate is made for each PROC included in the model. Separate basic estimates have been derived for industrial and professional use situations. These basic estimates can be modified to take account of:

- the use of local exhaust ventilation;
- working outside;
- duration of exposure within a full shift;
- percentage of substance in a product used; and
- the use of respiratory protective equipment (RPE).

The model estimates inhalation exposure to vapours or dust and dermal exposure. Not all modifiers are used for all types of exposures. Except for the use of local exhaust ventilation, all modifiers have fixed values. The modifier value for local exhaust ventilation depends on the PROC.

The output of ECETOC TRA worker consists of the following results:

- initial inhalation exposure not accounting for, for example, short duration of exposure or for effect of RMM (ppm for liquids and $mg\,m^{-3}$ for solids)—this is the value that would be estimated if a pure substance would be used for more than four hours per day indoors without local exhaust ventilation (LEV) or personal protective equipment (PPE);
- full shift inhalation exposure accounting for all exposure modifiers (ppm or $mg\,m^{-3}$);
- estimated exposed skin area (cm^2);
- full shift dermal exposure levels ($mg\,kg^{-1}\,bw\,day^{-1}$); and
- total exposure (inhalation + dermal, $mg\,kg^{-1}\,bw\,day^{-1}$),

Inhalation exposure is estimated in ppm for vapours and $mg\,m^{-3}$ for solids, but the values in ppm are also recalculated to $mg\,m^{-3}$ based on the molar weight of the substance. This value is also used to add to the dermal value for calculation of the total exposure. A default body weight of 70 kg for workers is used. All exposure estimates are external values, not taking into account absorption percentages or uptake rates.

In addition to the separate worker exposure tool of ECETOC TRA, there is also an integrated tool that has worker, consumer and environmental exposure parts. However, at the time of writing, there were still some differences between the separate worker tool and the integrated tool due to errors during the development of the integrated tool.

ECETOC TRA is the most used Tier 1 model for exposure estimation under REACH. ECETOC TRA workers (version 2) is available as a spreadsheet (with visual basic programming) that can be downloaded from the ECOTEC website (www.ecetoc.org/tra) together with a user guide and supporting documents.

20.5.6 EMKG-EXPO-TOOL

The EMKG-EXPO-TOOL is a tool developed in Germany by BAuA based on the COSHH Essentials control banding tool (see Section 20.3). It uses just a few exposure determinants:

- volume used;
- vapour pressure or dustiness
- whether exposure is only limited to $<15\,\mathrm{min\,day^{-1}}$; and
- whether the application is on more than $1\,\mathrm{m^2}$ (liquids only).

The tool can be used to estimate rough Tier 1 exposure levels very quickly. The estimates are based on exposure levels that are taken from COSHH Essentials. These values can be derived from the documentation for COSHH Essentials but cannot be found in COSHH Essentials itself. The input parameters allow only very limited differentiation because they are all in a small number of categories. Due to its very simple nature, the EMKG-EXPO-TOOL is not used very often in regulatory risk assessment processes.

The results presented by the EMKG-EXPO-TOOL are full shift inhalation exposure estimates for solids ($\mathrm{mg\,m^{-3}}$) or liquids (ppm). The results are concentrations of substances in the air (around the worker) not accounting for respiration volumes, absorption percentage or uptake rate.

The EMKG-EXPO-TOOL can be downloaded as a spreadsheet with visual basic programming from the REACH helpdesk (www.reach-clp-helpdesk.de/en/Exposure/Exposure.html). There is no supporting documentation. The first sheet of the workbook contains some supporting information on, for example, limitations of the tool.

20.5.7 Stoffenmanager

Stoffenmanager, created by a Dutch consortium of TNO, ArboUnie and BECO, can be used as a Tier 1 + tool for regulatory exposure assessments. Originally, Stoffenmanager was developed as a risk ranking and risk management tool for small and medium sized enterprises (SMEs). The original version contained a qualitative exposure ranking system followed by a risk banding system. The qualitative tool also includes separate modules for purposes such as storage of chemicals and chemical registries. The dermal part of Stoffenmanager is only qualitative and is not discussed further.[27]

The exposure ranking of the inhalation section was based on exposure scores calculated according to separate scores for each determinant. In a later version, a calibration was performed to link the final exposure scores to actual exposure levels. This was achieved by statistically analysing the relation between exposure scores and a large number of actual exposure measurements for which the relevant scores could be assigned.[12] Afterwards the estimated exposure levels were validated using a new smaller set of exposure levels. The validation showed some room for improvement of the quantitative model and the complete set of measured data was used to perform an improved calibration.[25]

The current version of Stoffenmanager (4.0) estimates inhalation exposure levels to vapours and dusts. The relation between Stoffenmanager scores calculated by the model and exposure levels was determined in the last calibration separately for four different types of substances/situations:

- (very) low volatile substances (including solids used in liquid mixtures);
- volatile substances;
- solid products; and
- dusts emitted from wood or stone materials by comminuting activities.

Stoffenmanager provides the following results:

- task-based inhalation exposure levels in $mg\,m^{-3}$; and
- full shift average inhalation exposure levels in $mg\,m^{-3}$.

The estimates are external values that do not account for respiratory volume, absorption percentage or uptake rate.

The basic task-based value provided is the 90th percentile of the distribution of outputs fitting to the inputs. However, a set of further percentiles (50th, 75th and 95th), as well as a graphical presentation of the resulting distribution, can also be derived for the task-based estimate.

The 90th percentile value is used for calculating a full shift average (if required) based on one or more activities with a total duration of up to 480 minutes. It is assumed that the exposure is 0 for the duration between the total of the entered activities and the full shift duration of 480 minutes.

Stoffenmanager is a freely available web tool with password protection to ensure confidentiality of data. It can be found on a dedicated website (www.stoffenmanager.nl). At the time of writing, the daily average option (full shift average) could only be found by logging in for 'Quantitative exposure assessment'.

20.5.8 RISKOFDERM

RISKOFDERM is a model for estimating potential dermal exposure (*i.e.* not accounting for any personal protective equipment) resulting from the use of chemicals. It was built by a group of institutes—TNO in the Netherlands, UK Health and Safety Executive (HSE) and the Finnish Institute of Occupational Health (FIOH)—in a larger project with a total of 15 partners that participated in gathering relevant information for the development of the model. A control banding tool for dermal exposure was also developed during the project.

RISKOFDERM is based on a conceptual model of dermal exposure which was used to guide the clustering of activities in so-called Dermal Exposure Operation Units (DEO Units) and to guide the gathering of relevant information (*e.g.* on determinants of exposure). A relatively large number of potential dermal exposure measurements were performed and a large number of parameters considered to be potential determinants were gathered. Statistical analyses were undertaken to derive equations linking the inputs to potential determinants to measured exposure levels and these were used to derive model equations. The part of the variation in exposure levels not explained by the model equations was used to construct variation estimates used in calculating output distributions.

The types of activities (clustered in DEO Units) for which exposure estimates can be made are:

- handling of contaminated surfaces (more specifically: product transfer such as weighing, dumping);
- wiping;
- application of products with a hand-held tool (*e.g.* a brush);
- application of products by spraying;
- immersion of objects into products; and
- mechanical treatment of objects.

Estimated exposure levels are for products (which can be a single substance or a chemical mixture such as a paint). However, exposure levels for components of products can be calculated outside of the model by assuming a linear relation with the fraction of the component in the product:

$$\text{Component exposure} = \text{Fraction} \times \text{Product exposure} \qquad (20.2)$$

Separate estimates are made for potential dermal exposure to the hands (with a fixed area of $820\,\text{cm}^2$) and to the remainder of the body (fixed area of $18{,}720\,\text{cm}^2$) for those situations where sufficient measured data are available to develop meaningful equations.

The results presented by **RISKOFDERM** are:

- potential dermal exposure rates of the product for hands and remainder of the body ($\mu\text{L}\,\text{min}^{-1}$ for liquid products and $\text{mg}\,\text{min}^{-1}$ for solids)—median as well as a distribution of percentiles; and
- potential dermal exposure loading of the product for hands and remainder of the body (μL for liquids and mg for solids).

The RISKOFDERM tool is available in two versions.

A deterministic version in a spreadsheet can be downloaded free from the TNO website (www.tno.nl) *via* the following route: [Markets > Chemistry >

Products > Toxicology Centre – leader in intelligent testing > Human Exposure Assessment and Risk Management] or by searching the website for 'RISKOFDERM'. A guidance document for this spreadsheet tool can also be downloaded. This version allows only single values of input parameters to be used.

A beta-version of a probabilistic webtool of RISKOFDERM is available (after registration) *via* the Health & Safety Laboratory website (http://xnet.hsl.gov.uk/riskofderm/). In this version, (simple) distributions of input parameters can be entered, allowing a probabilistic analysis of potential dermal exposure depending on variable input values. This version also includes a two-dimensional probabilistic method to account for model uncertainty. At the time of writing the web version has not been published officially as it is awaiting formal quality assurance. In addition, the web version may be extended in the future with more options for different probability distributions. The website contains background on the model and on the probabilistic calculations performed.

RISKOFDERM is the Tier 2 model for dermal exposure described in the guidance documents published by the European Chemicals Agency (EChA) on information requirements and chemical safety assessment.[28]

20.5.9 Advanced REACH Tool (ART)

The Advanced REACH Tool (ART) has been developed specifically to be a Tier 2 inhalation exposure assessment tool for use in REACH. It can also be used for other situations where worker inhalation exposure needs to be assessed. The tool has two parts: 1) a statistical model and 2) the possibility to update the estimates made by the model with information from measurements using Bayesian methodology.

The basis of the model part of the tool is the same conceptual source–receptor model that is underneath the exposure estimations in Stoffenmanager. However, the model has been fully rebuilt and is much more refined than in Stoffenmanager.

The basis of the model consists of determining the influence of modifiers on all steps from emission to actual exposure in the form of scores. These scores are multiplied to reach a final score. A calibration exercise with approximately 2,700 good quality measurements has linked exposure scores to real exposure levels. Mixed effects models were used to evaluate the association between model scores and measurements. The calibration of the model showed that (with 90% confidence) the true geometric mean for an exposure situation is within a factor 4–5 of the predicted geometric mean. Separate calibrations were conducted for exposure to dust, mist (liquid aerosols), vapours and fumes.

For liquids, the model requires input of the activity coefficient of the substance in the mixture. The default value used is 1. The user is referred to the UNIFAC estimation method for more specific estimations.[29]

The model estimates can be combined in ART with results from measurements. In the final version of ART, the intention is to have an in-built database with exposure measurements and related context.

Exposure data from the database can be used to update the outcome. In this process the weight of exposure data will depend on the analogy of the situation in which the exposure data were gathered with the situation under assessment. The user will be able to select data for this process from the database. It will also be possible to enter data that are available to the user. This can already be done in ART version 1.0. In ART version 1.0 the data entered are, by default, assumed to be highly similar to the exposure situation under assessment and therefore have a high weight in the updating process of the estimates.

The model integrated in ART has a high resolution compared with other models due to the many options that can be chosen for, for example, local exhaust ventilation. The scores for all options have been based as much as possible on the available data. However, the number of options for which available data are sufficient to base its effect on is very limited, so expert judgment has been used to interpolate and sometimes extrapolate. ART is relatively unique in presenting both an output distribution as well as confidence intervals around the percentiles estimated.

The use of measured data can, on the one hand, change the estimates of the output distribution. On the other hand, the use of measured data is very important in lowering the confidence intervals around the estimates. The effect on the confidence intervals depends on the type of data. Data from more dataset, more facilities and more workers have more effect. Large datasets also have more effect than small datasets.

The results that ART presents are:

- full shift exposure; and
- long-term exposure.

ART calculates an overall distribution for full shift exposures by combining the between company, between worker and within worker variance components. The percentiles calculated do not necessarily apply to any specific individual in the population.

The calculation of the distribution of workers' long-term average (mean) exposure uses separate estimates of between-company, between-worker and within-worker variation. Percentiles from this distribution correspond directly to specific individuals within the population exposure distribution, *e.g.* 10% of workers have a mean exposure in excess of the 90th percentile. From a scientific perspective, the distribution of long-term average exposure is the most appropriate one when considering chronic health endpoints or comparing with chronic toxicity data.

For each type of exposure, the 50th, 75th, 90th, 95th and 99th percentiles (all in $mg\,m^3$) are presented in the beta-version and the 80th, 90th, 95th and 99th percentile confidence limits are calculated.

20.6 Advantages and Disadvantages of the Models

The main advantage of mathematical models is the fact that they are independent of measured data for deriving the relationships they use to estimate exposure levels from determinants of exposure. They are based on scientific knowledge and equations from physics or chemistry. However, the actual processes leading to exposure are very complex. The relations used to estimate, for example, concentrations in a room are generally a very substantial simplification. For instance, a room is considered to be well-mixed, which makes the position of the worker compared to the source less relevant. In real life, many rooms are not well-mixed and the position of the worker is very relevant for exposure level. In addition, several processes such as adsorption of substances to surfaces, agglomeration of solid particles, re-suspension of solids by airflows, *etc.* are generally not taken into account in the mathematical models. A particular issue is that it is very difficult to model is evaporation of substances from complex mixtures. Models that are too simple may lead to very inaccurate or biased exposure estimates.

Whereas inhalation exposure processes are relatively well understood and modelled, skin exposure processes are very complex. Contamination occurs due to direct contact with a product, deposition from the air and contact with contaminated surfaces. In addition, the skin may be decontaminated by contact with cleaner surfaces. Therefore, mathematical models are not generally used in estimating skin exposure.

Statistical models tend not to follow the physical or chemical laws of nature exactly, but relate exposure levels to determinants based on combinations of determinants and exposure levels found in real data. In this way, they can take account of all processes involved even if they are not well-understood. Since they are based on real exposure levels, they can for instance take account of the relative position of the worker and source. Statistical models can also provide information on the uncertainty of the modelling, which can be very useful for drawing relevant conclusions.

The main disadvantage of the statistical models is that they can never be used with confidence outside the scope of the data on which they were built. If certain processes were not part of the data set, the model cannot estimate exposure levels accurately. This disadvantage has been partly remediated by filling gaps in knowledge and data through expert judgement. In some cases (*e.g.* ECETOC TRA worker) there is no real statistical analysis underneath the assumed relations between determinants and exposure estimates. The use of expert judgement makes it possible to estimate a wide range of situations based on relatively limited data. However, the certainty and accuracy of the estimates are very unclear when no external validation has been performed.

20.6.1 ConsExpo

ConsExpo is a mathematical model. The models in the full tool range from very simple to more complex. Although the full tool has not been validated some components have been. Its wide range and mathematical basis means it has a large scope.

For consumer exposure, the fact that there are not many other more useful models is the reason why ConsExpo—in spite of its uncertainties and lack of validation—is the tool most used for consumer exposure assessment in European statutory risk assessments. The existence of a number of fact sheets with default values makes it easy to use in many situations. However, the default values in the fact sheets are also not very well-validated.

20.6.2   ECETOC TRA Consumer

The consumer part of ECETOC TRA is very simple to use and therefore very attractive for a quick risk assessment. But again it is based on defaults that are not very well validated. When situations do not fit the (often rather conservative) defaults, another model or method is needed to estimate exposures. ECETOC TRA consumer is only useful for Tier 1 assessments.

20.6.3   Wall Paint Exposure Model

The Wall Paint Exposure Model is a mathematical model. Its strong point is that it is dedicated to a specific exposure situation and built for that situation only. Like ConsExpo and ECETOC TRA consumer, it does not take account of the position of the exposed person in the room. It has not generally been used in Europe mainly because of its rather specific scope and because the situations within its scope can be estimated reasonably well by ConsExpo.

20.6.4   Multi-chamber Concentration and Exposure Model

The Multi-chamber Concentration and Exposure Model has a substantial disadvantage in that it does not model emission of substance from a source, which in many situations, may be the most difficult part of the full exposure estimation. In most regulatory risk assessments, the relative worst-case situations are assessed. The worst case is generally the emission of a substance from a source in a room where a person also is located. Therefore, MCCEM appears not very useful for regulatory risk assessments such as the ones performed under the REACH legislation.

20.6.5   ECETOC TRA Worker

ECETOC TRA worker is more complex in its structure than its consumer counterpart. It is a statistical model that is intended to be conservative. Because it is still relatively simple and widely supported by the chemical industry in Europe, it is a very useful tool under the REACH legislation. However, a major disadvantage is the lack of knowledge about the validity of the model, which is related to the fact that the way in which exposure values have been chosen to fit to exposure situations is not very transparent.

20.6.6 EMKG-EXPO-TOOL

The EMKG-Expo-Tool is a very simple tool. It has a small number of determinants such that it is, on the one hand, extremely simple to use, but, on the other hand, (logically) it cannot discriminate very well between different situations. Therefore, it is not very useful for anything more than a very rough indication of possible exposure levels. It is also unknown how valid its estimates are.

20.6.7 Stoffenmanager

Stoffenmanager is useful when exposure situations are somewhat more complex than the situations covered by, for example, ECETOC TRA worker. It can take account of sources that are further away and the influence of room sizes. Through these parameters the options to account for, for example, more or less automated processes are better than for ECETOC TRA. Its validity is relatively well-known because an earlier version was validated against independent measured data and the latest version is based on the full set of calibration and validation data used so far.

20.6.8 RISKOFDERM

The major advantage of RISKOFDERM is that it is based directly on measured data and therefore is of known validity within its scope. Another advantage is that information on the uncertainty of the model for different situations is provided. A number of determinants not in for example, ECETOC TRA, are also taken into account (*e.g.* use of rate and direction of application). A major disadvantage is that many dermal exposure situations were not studied during its development and are therefore out of scope.

20.6.9 Advanced REACH Tool (ART)

The Advanced REACH Tool is the most recent addition to the set of tools for estimating exposure levels in a regulatory setting, though it is only relevant for inhalation exposure. It is the tool with the most wide-ranging set of operational conditions and risk management measures as determinants of exposure in its scope. It can, for instance, differentiate between different types of local exhaust ventilation. Its treatment of far field sources is more complete than that of Stoffenmanager. A major advantage is the option to use the statistical model estimates made with ART in combination with actual measured data in a Bayesian process. A disadvantage is that the model is relatively complex and therefore not suited for a Tier 1 assessment. The model estimates are based on a calibration with real measured data, but have not yet been validated against another set of independent data.

20.7 Conclusions

The range of models that can be used to estimate exposures in a regulatory risk assessment process ranges from very simple (*e.g.* EMKG-EXPO-TOOL) to complex (*e.g.* some parts of ConsExpo and ART with Bayesian updates). Only limited options are available for consumer exposure assessment, all in the forms of mathematical models. ECETOC TRA consumer may be used in a Tier 1 assessment with ConsExpo as a further tier tool. None of the consumer models has been properly validated over its full range.

For worker exposure assessment there is a greater choice of models, largely due to the fact that worker exposure has been studied more widely. The ECETOC TRA worker tool can be used as a Tier 1 model, although one has to take into account that the validity of its estimates is not very well-known. For a further Tier model, the Advanced REACH Tool appears at present the most logical choice for inhalation exposure. It has the same conceptual basis as Stoffenmanager, is also calibrated against measured data, and has more options. For dermal exposure, RISKOFDERM is the most logical choice for further Tier exposure assessment. It is calibrated against measured data, but its scope is not as wide as that of ART.

In general, modelling of dermal exposure (for consumers as well as workers) is far less advanced than modelling of inhalation exposure.

References

1. International Programme on Chemical Safety, *IPCS Risk Assessment Terminology. Part 1: IPCS/OECD Key Generic Terms used in Chemical Hazard/Risk Assessment. Part 2: IPCS Glossary of Key Exposure Assessment Terminology*, World Health Organization, Geneva, 2004.
2. V. Zartarian, T. Bahadori and T. McKone, *J. Expos. Anal. Environ. Epidemiol.*, 2005, **15**, 1.
3. Scientific Committee Group on Occupational Exposure Limits, *Methodology for the Derivation of Occupational Exposure Limits: Key Documentation*, SCOEL, Employment and Social Affairs, Luxembourg, 1998.
4. Corrigendum to Regulation (EC) No 1907/2006 of the European Parliament and of the Council of 18 December 2006 concerning the Registration, Evaluation, Authorisation and Restriction of Chemicals (REACH), establishing a European Chemicals Agency, amending Directive 1999/45/EC and repealing Council Regulation (EEC) No 793/93 and Commission Regulation (EC) No. 1488/94 as well as Council Directive 76/769/EEC and Commission Directives 91/155/EEC, 93/67/EEC, 93/105/EC and 2000/21/EC (Official Journal of the European Union L136 of 30 December 2006), *Off. J. Eur. Union*, 2007, 2007, **L136**, 3–280.

5. European Chemicals Agency, *Guidance on Information Requirements and Chemical Safety Assessment. Chapter R.8: Characterisation of dose [concentration]-response for human health*, EChA, Helsinki, 2008, http://guidance.echa.europa.eu/docs/guidance_document/information_requirements_en.htm [accessed March 2010].
6. H. Marquart, C. Northage and C. Money, *J. Expo. Sci. Environ. Epidemiol.*, 2007, **17**, S16.
7. E. Tielemans, Y. Christopher, H. Marquart, M. Groenwold and J. van Hemmen, *Ann. Occup. Hyg.*, 2002, **46**, 559.
8. E. Tielemans, H. Marquart, J. de Cock, M. Groenewold and J. van Hemmen, *Ann. Occup. Hyg.*, 2002, **46**, 287.
9. C. D. Money and S. A. Margary, *Ann. Occup. Hyg.*, 2002, **46**, 279.
10. Bundesanstalt für Arbeitsschutz und Arbeitsmedizin [Federal Institute for Occupational Safety and Health], *Easy to Use Workplace Control Scheme for Hazardous Substances: A practical guide for the application of the German Hazardous Substances Ordinance by small and medium-sized enterprises working with hazardous substances without workplace limit values*, BAuA, Dortmund, 2006, www.baua.de/nn_18306/en/Topics-from-A-to-Z/Hazardous-Substances/workplace-control-scheme.pdf [accessed March 2010].
11. J. E. Delmaar, M. V. D. Z. Park and J. G. M. van Engelen, *ConsExpo 4.0 Consumer Exposure and Uptake Models Program Manual*, RIVM, Bilthoven, The Netherlands, 2005, RIVM Report 320104004/2005.
12. E. Tielemans, T. Schneider, H. Goede, M. Tischer, N. D. Warren, P. Ritchie, H. Kromhout, M. van Tongeren, J. J. van Hemmen and J. Cherrie, *Ann. Occup. Hyg.*, 2008, **52**, 577.
13. P. Hewett, P. Logan, J. Mulhausen, G. Ramachandran and S. Banerjee, *J. Occup. Environ. Hyg.*, 2006, **3**, 568.
14. E. Tielemans, N. D. Warren, T. Schneider, M. Tischer, P. Ritchie, H. Goede, H. Kromhout, J. J. van Hemmen and J. Cherrie, *J Expo. Sci. Environ. Epidemiol.*, 2007, **17**, 72.
15. M. Nicas, *Chem. Health Saf.*, 2003, **Jan–Feb**, 14.
16. IT Environmental Programs, Inc. *Preparation of Engineering Assessments, Volume I: CEB Engineering Manual*, US Environmental Protection Agency Office of Toxic Substances, Washington DC, 1991, Contract No. 68-D8-0112.
17. GEOMET Technologies, 2001. *Wall Paint Exposure Model (WPEM): Version 3.2. User's Guide,* US EPA Office of Pollution Prevention and Toxics and National Paint and Coatings Association, Washington DC, 2001, www.epa.gov/oppt/exposure/pubs/wpemman.pdf [accessed March 2010].
18. C. Keil, *Mathematical Models for Estimating Occupational Exposure to Chemicals,* American Industrial Hygiene Association Publications, Fairfax, VA, 2000.
19. E. Olsen, I. Olsen, E. Wallström and D. Rasmussen, *Ann. Occup. Hyg.*, 1992, **36**, 637.

20. M. D. Koontz and N. L. Nagda, *Indoor Air*, 1991, **1**, 593.

21. J. Tickner, J. Friar, K. S. Creely, J. W. Cherrie, D. E. Pryde and J. Kingston, *Ann. Occup. Hyg.*, 2005, **49**, 103.

22. European Centre for Ecotoxicology and Toxicology of Chemicals, *Targeted Risk Assessment*, ECETOC, Brussels, 2004, Technical Report No. 93.

23. European Centre for Ecotoxicology and Toxicology of Chemicals, *Addendum to ECETOC Targeted Risk Assessment (Technical Report No. 93)*, ECETOC, Brussels, 2009, Technical Report No. 107.

24. N. D. Warren, J. Marquart, Y. Christopher, J. Laitinen and J. J. van Hemmen, *Ann. Occup. Hyg.*, 2006, **50**, 491.

25. J. Schinkel, W. Fransman, H. Heussen, H. Kromhout, H. Marquart and E. Tielemans, *Occup. Environ. Med.*, 2010, **67**, 125.

26. European Chemicals Agency, *Guidance on Information Requirements and Chemical Safety Assessment. Chapter R.12: Use descriptor system.* EChA, Helsinki, 2010, http://guidance.echa.europa.eu/docs/guidance_document/ information_requirements_en.htm [accessed March 2010].

27. H. Marquart, H. Heussen, M. Le Feber, D. Noy, E. Tielemans, J. Schinkel, J. West and D. van der Schaaf, *Ann. Occup. Hyg.*, 2008, **52**, 429.

28. European Chemicals Agency, *Guidance on Information Requirements and Chemical Safety Assessment. Chapter R.14: Occupational Exposure Estimation.* EChA, Helsinki, 2008, http://guidance.echa.europa.eu/docs/ guidance_document/information_requirements_en.htm [accessed March 2010].

29. A. A. Fredenslund, R. Jones and J. M. Prausnitz, *Am. Inst. Chem. Eng. J*, 1975, **21**, 1086.

CHAPTER 21

Toxicokinetic Considerations in Predicting Toxicity

J. C. MADDEN

School of Pharmacy and Chemistry, Liverpool John Moores University, Byrom Street, Liverpool L3 3AF, UK

21.1 Introduction

Toxicokinetics is the study of how the concentration of a toxicant in the body as a whole, and within individual compartments thereof, varies over time. Internal exposure in combination with external exposure and the compound's inherent toxicity together determine the true extent of the toxicity that may be elicited by a xenobiotic. As such, an understanding of the toxicokinetics of a compound is as important as knowledge of its intrinsic toxicity when predicting the overall response *in vivo*. Historically, most of the work on studying the time course of a xenobiotic in the body was related to pharmacological studies of drugs, giving rise to the term pharmacokinetics. Whilst the same parameters are considered in terms of determining the time course of a toxicant in the body, the term toxicokinetic is used to distinguish the time course of a compound with unwanted effects. Biokinetic modelling is a generic term sometimes used to encompass the study of the time course of any compound with either desirable or undesirable effects.[1]

Clearly, toxicity to an organism is not dependent on intrinsic toxicity alone, either of the parent or of a transformation product thereof. It will only be capable of causing that effect if it reaches the appropriate site of action in

Issues in Toxicology No.7
In Silico Toxicology: Principles and Applications
Edited by Mark T. D. Cronin and Judith C. Madden
© The Royal Society of Chemistry 2010
Published by the Royal Society of Chemistry, www.rsc.org

sufficient concentration to elicit a response. The likelihood of a compound reaching any given site depends on both the external and internal exposure.

External exposure relates to the likelihood of the compound reaching an organism, *i.e.* the concentration present in the environment to which the organism is exposed. Humans may be exposed to toxicants from different sources, *e.g.* those present in the air, those in direct contact with the skin and those ingested *via* food, water or unintentionally (concepts of external exposure are discussed in Chapter 20).

The aim of this chapter is to discuss the role of internal exposure, *i.e.* uptake and distribution to sites of action within an organism and ultimate elimination from the body. The toxicokinetic properties of a compound will determine the extent to which it is taken up into the body, the sites to which it will distribute, and the length of time it will persist at a given site or in the body as a whole. These factors are crucial in determining the *in vivo* response to a potential toxicant, as they may have a significant modulating effect on overall toxicity. This is an area where the importance of *in silico* tools for prediction is being increasingly recognised. This chapter provides a definition of key toxicokinetic properties relating to absorption, distribution, metabolism and elimination (ADME properties), as well as providing information on *in silico* tools to predict these properties. The chapter also introduces the concept of physiologically-based toxicokinetic modelling (PBTK) and the increasing role of these models in overall risk assessment.

21.2   Internal Exposure

As toxicants are predominantly delivered to their site of action *via* the blood stream, internal exposure is frequently characterised by determining the concentration of the toxicant in plasma (or blood) as a function of time. A typical plasma concentration–time profile for a toxicant is shown in Figure 21.1.

The overall internal exposure is represented by the shaded area in Figure 21.1; this is referred to as the area under the concentration–time curve, *i.e.* the *AUC*. The maximum concentration that a compound attains (in plasma) is referred to as C_{max} and the time to reach this maximal concentration is referred to as T_{max}. Whilst the *AUC* provides an indication of the amount present in plasma over time, many factors contribute to this concentration–time profile. These relate to the absorption, distribution, metabolism and elimination (ADME) properties of the compound. As these factors control the amount of toxicant in the body over time, ultimately they determine the overall response observed in the organism. Table 21.1 provides a brief definition of key ADME properties and these are discussed briefly below.

21.2.1   Absorption

The first step in internal exposure is absorption from the external environment into the internal system. Common routes are:

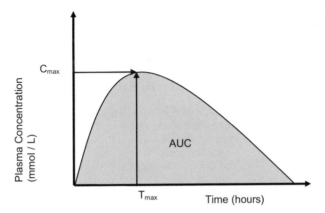

Figure 21.1 Plasma concentration–time curve for a typical toxicant.

 (i) uptake *via* the inhalation route—for atmospheric pollutants and compounds present in industrial settings, particularly volatile organic compounds
 (ii) percutaneous absorption—for compounds in direct contact with the skin such as cleaning agents, cosmetics or personal care products
 (iii) absorption across the gastrointestinal tract—for compounds ingested intentionally or unintentionally including *via* the food chain or water supply.

Absorption may be recorded as the percentage of a compound which crosses a given barrier; for example, percentage human intestinal absorption (*%HIA*) refers to the relative amount of compound taken up *via* the gastrointestinal tract. Absorption *via* the skin is usually recorded as a flux measurement (cm hour^{-1}), or uptake may be expressed as a ratio of the concentrations present in the external and internal environments (*e.g.* lung : air partition coefficients). Compounds are commonly absorbed *via* the transcellular route, *i.e.* passive or carrier-mediated diffusion through lipoidal cell membranes. Certain small, hydrophilic molecules can be absorbed *via* the paracellular route, *i.e.* passing between cells. Active transport processes, requiring energy to transport compounds, are selective processes by which some compounds may be taken up into cells or removed from them. Absorption alone does not determine compound entry into the systemic circulation; local or first pass metabolism may reduce the amount available systemically. Influx and efflux transporters, identified in many tissues, also modulate the amount entering (or leaving) individual tissues. The role of metabolism and influx/efflux transporters are considered separately (*vide infra*).

21.2.2 Distribution

Following successful absorption into the body, the next step to consider is the process of distribution. In a mammalian model, compounds are distributed to

Table 21.1 Definition of key ADME properties.

Properties	Definition
Absorption	**The processes by which compounds are taken up into the body via gastrointestinal tract, inhalational or dermal routes, etc.**
% Abs	Percentage of available compound that is absorbed across a barrier.
% HIA	Percentage that is absorbed across the human gastro-intestinal tract.
Skin permeability (K_p)	Permeability of a solute through skin, determined by flux measurements.
Distribution	**The factors affecting the extent to which a compound will distribute into different regions of the body.**
% PPB; f_b; f_u	Percentage of compound bound to plasma proteins; fraction bound to plasma proteins; fraction unbound (*i.e.* free fraction).
V_d; V_{du}	Apparent volume of distribution, *i.e.* the hypothetical volume into which a drug distributes; V_{du} is the volume of distribution for the unbound fraction of drug.
Tissue : blood PCs	Tissue : blood partition coefficients (the ratio of concentrations between tissue and blood).
BBB partitioning	Blood–brain barrier partitioning, frequently expressed as ratio of concentrations between brain and blood (serum/plasma) or expressed categorically to indicate likely ($+$) or not likely ($-$) to enter brain.
Milk : plasma PC (m : p)	Ratio of concentration between breast milk and plasma.
CI; TI	Clearance index; transfer index used to express placental transfer of compounds usually as a ratio using antipyrine as a marker.
Transporter affinity (K_i)	Relates to the affinity of compounds for a wide range of transporters, (*e.g.* P-gp/MDR1, OATP, *etc.*).
Metabolism	**The process by which xenobiotics are transformed (usually into a compound that can be more readily excreted). Of significance is the nature of the metabolite, the enzyme responsible for the catalysis of the process and the rate at which it occurs.**
% metabolism	The overall percentage of the compound that is metabolised.
K_m	Binding affinity for metabolic enzymes.
Elimination/Excretion	**Removal of compound from the system.**
Clearance	The volume of blood from which a compound is completely removed in a given time (units are typically mL min^{-1} or mL min^{-1} kg^{-1}).
Cl_{tot}; Cl_h; Cl_r	Total clearance by all routes; clearance by hepatic route (*i.e.* metabolism); clearance by renal route (*i.e.* urinary excretion).
% exc	The percentage of compound excreted unchanged in urine.
Composite parameters	**Parameters that are the result of more than one ADME property**
F	*Bioavailability*: the fraction of a given dose that enters the systemic circulation, determined by both absorption characteristics and avoidance of local (or first pass) metabolism.
$t_{1/2}$	*Half-life*, *i.e.* the time taken for the concentration of a compound in the body to fall by half. Dependent on the volume of distribution and clearance rate.

Table 21.1 (*Continued*)

Properties	Definition
AUC	Area under blood or plasma concentration–time curve; indicates overall exposure and is influenced by all ADME parameters.
C_{max}	Maximum concentration reached (usually measured in blood or plasma).
T_{max}	Time to reach maximum concentration.

their site of action *via* the blood stream. Compounds will partition between blood and tissues depending on the relative affinity of the compound for each compartment, relative solubility in each phase and ability to cross membranes. A key factor in determining overall distribution of a compound is its relative affinity to bind to tissue or plasma proteins. Compounds with a high percentage of plasma protein binding (% *PPB*) are more likely to remain in blood and be less widely distributed. Plasma proteins include human serum albumin (HSA) which possesses several binding sites, predominantly binding lipophilic and acidic compounds. Another serum protein, α-1 acid glycoprotein, has a higher affinity for neutral or basic compounds than acidic species. Compounds with a higher affinity for tissue binding will distribute more widely, consequently, tissue : blood partition coefficients (*i.e.* the ratio of the concentration of compound in tissue compared to blood) provide a useful indication of the relative partitioning between blood and individual tissues. Different tissues contain different relative concentrations of water, lipids, phospholipids and proteins; hence xenobiotics may have higher affinity for one tissue type than another. Information relating to the solubility of compounds in different tissue constituents has been used to develop tissue : blood partition coefficients.[2,3]

Partitioning into certain tissues is associated with specific concerns and this has led to more research being carried out in these areas. Entry into the central nervous system can lead to significant toxic effects, hence blood–brain barrier (BBB) partitioning has been extensively studied.[4–9] Passage between cells (*i.e.* transcellular transport) is restricted at this barrier because junctions between cells are much tighter in this region. Additionally efflux transporters are heavily expressed, providing a protective mechanism. Partitioning across the placenta (reported as a placental clearance or transfer index) is also of concern due to the potential to elicit toxicity in the developing foetus.[10,11] Similarly milk : plasma partitioning is a concern for neonates, particularly as milk may be the sole food source.[12,13] Partitioning into adipose tissue may result in long-term storage of lipophilic compounds, prolonging their toxic effect or leading to accumulative effects.

Globally the overall extent of distribution of a compound in the body is given by the volume of distribution (*Vd*), *i.e.* a hypothetical volume into which a compound distributes as shown in eqn (21.1):

$$Vd = \frac{Dose}{Co} \tag{21.1}$$

where:

$$Co = \text{concentration in blood or plasma}$$
$$Dose = \text{amount of compound administered.}$$

If a compound is highly distributed to tissues, then the concentration present in blood is low and the volume of distribution is high. Conversely a drug which predominantly remains in blood will have a low volume of distribution. This is important as it is only compounds in blood which are presented to the eliminating organs; hence compounds with a high volume of distribution will remain in the body for longer. Overall distribution is a concern as the more widely a compound distributes, the more sites are exposed to potential toxic effects. Distribution is significantly influenced by the activity of transporter processes.

21.2.3 Role of Transporters

Recently there has been a paradigm shift in understanding the mechanisms *via* which compounds are transported both into and out of cells, affecting uptake or removal from the different tissues or organs of the body. It is now recognised that, rather than simple passage through membranes, influx and efflux transporters play a significant role in the transport of many compounds. The most widely reported transporter is P-glycoprotein (P-gp). This transporter was first identified because of its high expression in tumours, rendering anti-cancer agents ineffective as drugs were actively transported out of the cell;[14] hence it is referred to as multidrug resistance protein (MDR1). P-gp is a member of one of the superfamilies of transporters—the ATP binding cassette (ABC) transporters. Over 40 members of this family have already been identified[15] and there is the potential to discover more.

The Na^+/glucose co-transporter was the first of the solute carrier (SLC) superfamily of transporters to be identified. SLC transporters are more often associated with influx than efflux processes acting as symporters or antiporters without requiring an energy source. SLC transporters are highly abundant, aiding the transport of many exogenous and endogenous compounds. It has been proposed that the network of efflux transporters may have developed as an evolutionary protective mechanism.[16] Transporters frequently act in concert with metabolising enzymes, removing compounds from the cell and re-presenting them to the metabolising enzymes. The high expression of transporters at the blood–brain barrier provides support to the theory of the protective role of efflux transporters. Transporters such as members of the organic anion transporter family (OATPs), large neutral amino acid transporters (LATs), novel organic cation transporters (OCTNs) and several others have been identified in this region. These transporters are associated not only with cell membranes but also contribute to the distribution of compounds within the cellular structures, *e.g.* mitochondria possess their own transporters. A great deal of research is currently being undertaken in this area with transporter proteins being identified in all organs and tissues of the body. Lee *et al.*[15] give an excellent overview of the current state of knowledge concerning transporters including classifications, distribution and activity.

Simple models for absorption or distribution of compounds, based on hydrophobicity or hydrogen-bonding alone and not taking into account transporter action may be considered an over-simplification and will result in incomplete models. Indeed, substrates for transporters have been shown to be outliers in predictive models for uptake. As transporters play such a key role in determining the distribution of toxicants within organisms, their ability to modulate toxic activity of chemicals should not be underestimated. As knowledge concerning the role of transporters increases, improvements can be made in developing *in silico* models for absorption and distribution.

21.2.4 Metabolism

Of all the toxicokinetic properties, the ability of a xenobiotic to undergo metabolism has the most significant modulating effect on toxic potential. Metabolism may render a toxic compound inactive or may convert an innocuous compound into a highly toxic substance within the organism. The influence of metabolism on predicting overall toxicity has been the subject of intense investigation with metabolic effects frequently being cited as being responsible for significant outliers in models for toxicity. Metabolising enzymes have been found in all tissues including skin, gut, blood-brain barrier, placenta and kidneys, although the liver is the main site of metabolism in mammals.[17] Compounds entering the body *via* the gastrointestinal tract are subject to local and 'first pass' metabolism. Following absorption from the gut, compounds enter the hepatic portal vein and travel directly to the liver. Here they may be metabolised prior to entering the general circulation. As an evolutionary protective mechanism this frequently involves a detoxification process, though compounds may be converted to highly reactive or toxic species.

Phase 0 metabolism refers to the ability of the compound to enter the site of metabolism and may be influenced by transporter mechanisms, discussed above. Phase I metabolism refers to the functionalisation or exposure of a functional group on the molecule. It involves reactions such as oxidation of nitrogen or sulfur groups, aliphatic or aromatic hydroxylation, de-amination, de-alkylation, dehalogenation, *etc.* These reactions are generally catalysed by the cytochrome P450 superfamily of enzymes. These enzymes are distributed widely across species with over 50 being expressed in humans. (Updates on nomenclature and characterisation are available from http://drnelson.uthsc.edu/cytochromeP450.html). They have been well-studied, predominantly in relation to their ability to metabolise drugs. Together CYP2C9, CYP3A4 and CYP2D6 have been shown to be responsible for the metabolism of 90% of all drugs. Although isoforms CYP1A1, CYP1A2, CYP2A6, CYP2B1 and CYP2E1 have been shown to have little role in drug metabolism, they are known to catalyse the activation of pro-carcinogenic environmental pollutants into carcinogenic species,[18] rendering them of particular interest to toxicity studies.

Phase II metabolism refers to the conjugation of either a parent compound, or its phase I metabolite, with a polar group. These processes, catalysed *via* transferase enzymes, include conjugation with reduced glutathione, glucuronidation, sulfation or acetylation. They act to increase the water solubility of the compound, enabling it to be readily excreted *via* the kidneys or into bile.

Phase III metabolism refers to the exit of the metabolite from the cell which again demonstrates the concerted interactions between transporters and enzymes.[19] Metabolites may be excreted into bile and from there re-enter the gastrointestinal system. They may then be removed from the body *via* excretion in faeces or may be re-absorbed into the blood. Polar metabolites are more readily excreted in urine than lipophilic parent compounds.

In toxicity predictions, there are three key factors to consider in terms of metabolism:

(i) The nature of the metabolite is the most important factor—specifically whether it is more or less toxic than the parent.
(ii) The extent to which it is formed, *i.e.* is it a major or minor metabolite?
(iii) The rate at which the compound is metabolised. Slow metabolism of hydrophobic compounds may result in bioaccumulation and prolonged toxic effects. Conversely, rapid metabolism of a toxicant may reduce its overall potential to induce damage. For efficiently metabolised compounds, rate of presentation to the liver may be the rate limiting step in detoxification. This can lead to differential toxicities being observed in sub-categories of populations (*e.g.* those with reduced hepatic function or blood flow), a subject discussed in more detail in Section 21.4.2.

When considering the results of toxicity assays (*e.g.* the Ames mutagenicity assay), it is essential to ascertain whether or not the reaction was performed in the presence or absence of metabolising enzymes, as this can significantly affect the results. Typically in such studies the enzymatic systems are included allowing the effects of metabolism to be accounted for. Comparative studies, where enzymes are included in one system and excluded from another, enable metabolic effects to be differentiated. The enzyme systems commonly incorporated are obtained by centrifuging liver homogenate. Centrifugation at 9,000*g* allows the 'S9' fraction to be collected which comprises both cytosolic and endoplasmic reticulum bound enzymes. Centrifuging the S9 fraction allows the microsomes to be separated; these are the endoplasmic reticulum bound enzymes and exclude cytosolic enzymes. In a toxicity study, the S9 fraction or the microsomal fraction may be used or no metabolising enzymes may be added to the system.

Whilst the focus of this chapter is consideration of metabolism associated with human or mammalian species, metabolism by environmental species is also extremely important. For example, plants may produce metabolites of pesticides; microbes within soil may convert active environmental pollutants to inactive ones and *vice versa*, and there are numerous other examples.

21.2.5 Elimination/Excretion

Elimination is the process by which compounds are ultimately removed from the body. Although it is possible for compounds to be eliminated *via* exhaled air, sweat and faeces, renal excretion of polar molecules (parent or metabolite) is the key route of elimination for most compounds. Renal excretion involves glomerular filtration of unbound drug and active tubular secretion (involving transporters such as P-gp), balanced with passive tubular reabsorption of unionised weak acids and bases.

Removal of compounds from the body is expressed as clearance rates and refers to the hypothetical volume of blood that would be completely cleared of compound per unit time (typically recorded as mL min^{-1}, mL hr^{-1} or mL min^{-1} kg^{-1}). Total clearance (Cl_{tot}) relates to clearance *via* all routes; alternatively, clearance *via* different routes can be considered individually, *e.g.* hepatic clearance (Cl_h) and renal clearance Cl_r.

21.2.6 Composite Parameters

Key individual ADME parameters are discussed above. However the composite parameters, *i.e.* bioavailability (F) and half-life ($t_{1/2}$), are arguably the most important ADME parameters.

Bioavailability refers to the fraction of unchanged compound that reaches the systemic circulation by a given route compared with that following an intravenous injection. It is typically calculated using eqn (21.2):

$$F = \frac{Dose \times AUC_{oral}}{Dose \times AUC_{iv}} \tag{21.2}$$

where AUC_{oral} and AUC_{iv} are the areas under the plasma concentration–time curve following oral and intravenous administration respectively. *Dose* is the amount of compound administered by the given route.

To be bioavailable, a compound must be absorbed from the site of administration, avoiding metabolism (local or first pass) and removal *via* efflux transporters to enter the systemic circulation. Bioavailability modulates toxic effect as it determines the internal exposure of the compound. Allied with this parameter is the *half-life* of the compound in the body which determines the duration of effect or potential for bioaccumulation and possible chronic responses. *Half-life* ($t_{\frac{1}{2}}$) is influenced by both volume of distribution (Vd) and clearance (Cl) as shown in eqn (21.3).

$$t^{1}/_{2} = \frac{0.693\,Vd}{Cl} \tag{21.3}$$

A larger volume of distribution is reflected by lower concentrations circulating in the blood; hence less is presented to the organs of elimination and the compound persists for longer in the body. Storage of lipophilic compounds (*e.g.* in adipose) can prolong half-life. Intuitively it would be expected that a higher clearance rate would result in a shorter half-life as blood is rapidly cleared of the compound.

21.3 Predicting ADME Parameters

As the overall toxicity elicited by a compound may be determined as much by its ADME properties as its intrinsic toxicity, prediction of ADME properties is essential in assessing the overall risk posed by a substance. Use of *in silico* tools to predict and optimise pharmacokinetic properties of drugs led to major improvements in drug development between 1991 and 2001 as the role of pharmacokinetics in modulating biological activity was recognised.[20] Similarly models for predicting toxicity are improved by consideration of the role of toxicokinetics which determines the concentration at the site of action. Whilst discussion of *in silico* tools in this book has predominantly referred to toxicity prediction, many of the tools are equally applicable to the prediction of ADME properties. For example, the workflow presented in Chapter 2 showing the steps involved in making an *in silico* prediction could be applied equally to the prediction of ADME properties, bearing in mind the same caveats and limitations.

Public availability of large, high quality data sets for ADME properties is limited, although increasing awareness of the importance of such properties is leading to more data being published. Examples of available datasets that may be useful for ADME model development are given in Table 21.2 and the references contained therein.

Much of the available ADME data and models relate to pharmaceutical compounds. This presents two difficulties. Firstly, the available data are skewed towards those compounds which are pharmacokinetically viable. Secondly, the chemical space covered by such models may not be representative of the chemical space of environmentally relevant chemicals. Indeed, the lack of ADME information for non-pharmaceuticals presents a real challenge in overall risk assessment for industrial chemicals.

Another important consideration in the modelling of ADME properties is that, whereas toxicity can be related to specific interactions, ADME characteristics are often associated with more 'global' properties of the molecule such as the influence of the logarithm of the octanol–water partition coefficient (log P) of the compound on its ability to permeate membranes. This has led to the generation of simplified 'rules-of-thumb', albeit that these are generally based on pharmaceutical compounds such as Lipinski's 'rule of fives'.[44] This suggests that poor oral absorption is associated with:

- molecular weight > 500;
- log P > 5;
- number of hydrogen bond donors > 5; and
- number of hydrogen bond acceptors > 10.

Table 21.2 Available data sets for ADME properties.[a]

Properties	Information available	Reference
Human intestinal absorption	Data for 648 chemicals	22
Human oral bioavailability	Data for 805 compounds	23
Human oral bioavailability	Data for 302 drugs	24
Human oral bioavailability/ human intestinal absorption	Oral bioavailability, fraction absorbed, fraction escaping gut wall elimination; fraction escaping hepatic elimination data for 309 compounds	25
Skin permeability	K_p data for 124 compounds	26
Skin permeability	K_p data for 101 chemicals	27
Protein binding data	Percentage bound to human plasma protein for 1,008 compounds	28
Volume of distribution	Data for 199 drugs in humans	29
Tissue : air partitioning	Data for 131 compounds partitioning into human blood, fat, brain, liver, muscle and kidney (incomplete data for certain tissues)	30
Tissue : blood partitioning	Data for 46 compounds partitioning into kidney, brain, muscle, lung, liver, heart and fat (incomplete data for certain tissues)	31
Air : brain partitioning	Human and rat air : brain partition coefficients for 81 compounds	32
Blood : brain partitioning	Blood/plasma/serum : brain partitioning data for 207 drugs in rat	33
Blood : brain partitioning	Log blood–brain barrier partitioning values for 328 compounds	7
Blood-brain barrier penetration	Binary data for 415 compounds (classified as **BBB** penetrating or non-penetrating)	8
Blood-brain barrier penetration	Binary data for 1,593 compounds (classified as **BBB** crossing or non-crossing)	9
Placental transfer	Placental clearance index values for 86 compounds and transfer index values for 58 compounds	10
Clearance	Data for total clearance in human for 503 compounds	34
Metabolic pathways	Catalogue of all known bioactivation pathways of functional groups or structural motifs commonly used in drug design using 464 reference sources.	35
CYP metabolism	List of 147 drugs with the CYP isoform predominantly	36

Table 21.2   (*Continued*)

Properties	Information available	Reference
	responsible for their metabolism (CYP3A4, CYP2D6 and CYP2C9)	
Clearance; plasma protein binding; volume of distribution	Total clearance, renal clearance, plasma protein binding and volume of distribution data for 62 drugs in humans	37
Milk : plasma partitioning	Concentration ratio data for 179 drugs and pollutants	12
Transporter data	117 substrates and 142 inhibitors of P-gp; 54 substrates and 21 inhibitors of MRP2; 41 substrates and 38 inhibitors of BCRP	38
Transporter data	Expression in 28 tissues of the ten most common ABC and SLC transporter mRNAs	15
	Ratio of concentration in tissue for 16 drugs in MDR wildtype *versus* knockout mice;	
	Tissue : plasma ratio of 13 drugs in wildtype *versus* knockout mice	
PgP data substrates and non-substrates	Binary classification of 203 compounds as P-gp substrates ($+$) or non-substrates ($-$)	39
% urinary excretion; % plasma protein binding; clearance; volume of distribution; half-life, time to peak concentration, peak concentration	A compilation of ADME data for approximately 320 drugs (incomplete data for some drugs)	40
Half-life; therapeutic, toxic and fatal blood concentrations	Data for over 500 drugs (incomplete data for some drugs)	41
Volume of distribution; % plasma protein binding; % HSA binding (from HPLC retention data)	Data for 179 drugs (incomplete data for percentage plasma protein binding for some drugs)	42
Toxicogenomics microarray data	Literature references for data on 36 compounds	43

[a]Updated and adapted from Madden.[21]

Similarly, useful 'rules-of-thumb' for solubility, permeability, bioavailability, volume of distribution, plasma protein binding, entry and binding within the central nervous system, P-gp efflux, hERG and CYP450 inhibition were published by Gleeson.[45] These rules indicated how global molecular properties such as molecular weight and log *P* could influence the given ADME and, hence, toxicity properties. In addition to rules-of-thumb, many other *in silico* models have been developed to predict ADME properties. Table 21.3 and the

Table 21.3   Examples of models/reviews for the *in silico* prediction of ADME properties.

Property	References for models or reviews
Human intestinal absorption	44–48
Bioavailability	24, 48, 49
Skin permeability (K_p)	26, 27, 50
% Binding to plasma proteins	28, 51
Volume of distribution	28, 52–54
Tissue : blood partition coefficients	30, 32
Blood brain barrier partitioning	4–9
Placental transfer	10
Milk : plasma ratio	12,13
% Urinary excretion of unchanged compound	36
Clearance	34
Metabolism	55–58
Half-life	58,59
General reviews of *in silico* ADME prediction	60–63

references therein provide a few examples of *in silico* models and reviews for the prediction of ADME properties; however many more models have been published and a review of all existing ADME models is beyond the scope of this chapter.

There has also been a rapid expansion in the number of freely available and commercial software to predict ADME properties. A summary of the software available and a brief indication of the software's capabilities are given in Table 21.4 and the references therein.

As discussed above, metabolism is of particular relevance to toxicity prediction because of the potential to generate toxic metabolites from innocuous parent compounds. For this reason methods to predict metabolism have generated much interest in overall risk assessment. The work of Kalgutkar *et al.*[35] is useful as it presents known bioactivation pathways for organic compounds. Examples of software to predict metabolism are included in Table 21.4. Many of these models are based on knowledge of general reaction chemistry or specific, known metabolic pathways relevant to the functional groups identified within the molecule.

Use of the Organisation for Economic Co-operation and Development (OECD) Quantitative Structure–Activity Relationship (QSAR) Application Toolbox as a tool to identify or predict metabolites warrants specific attention here (see Chapter 16 for more details). This is a freely available piece of software[64] into which a compound of interest is entered. A search is made of the software's databases to determine if there are known skin or liver metabolites; if not these can be predicted using the skin and liver metabolism simulators within the program. The advantage of this approach is that once the metabolites have been identified, prediction of their toxicity can be carried out using the same methodology as for the parent compounds. This provides a more accurate indication of a compound's true toxicity potential. Within the

Table 21.4 Software for the prediction of ADME properties and selected commercial databases.[a]

Software provider	Software package	Predicted ADME properties	Website
Accelrys	Discovery Studio ADMET	Absorption, BBB penetration, plasma protein binding, CYP2D6 binding Physico-chemical properties/toxicity	www.accelrys.com
Bioinformatics and Molecular Design Research Centre (BMDRC)	PreADME	Physico-chemical properties Permeation through MDCK and Caco-2 cells BBB permeation Human intestinal absorption Skin permeability Plasma protein binding	www.bmdrc.org/04_product/01_preadme.asp
BioRad	Know-it-All	Bioavailability, BBB permeability, half-life, absorption, plasma protein binding, volume of distribution, rule of five violations, metabolism Physico-chemical properties	www.bio-rad.com
Chemistry Software Store	SLIPPER	Physico-chemical properties, absorption	www.timtec.net/software/slipper/introduction.htm
ChemSilico	CSBBB CSHIA CSPB Other modules	BBB partitioning Human intestinal absorption Plasma protein binding Physico-chemical properties/toxicities	www.chemsilico.com
CompuDrug	MetabolExpert MexAlert Rule of 5 Other modules	Metabolic fate of compounds Likelihood of first pass metabolism Calculates rule of five parameters Physico-chemical properties/toxicities	www.compudrug.com

Laboratory of Mathematical Chemistry, University Professor Assen Zlatarov, Bourgas, Bulgaria	TIMES	Metabolic pathways	http://oasis-lmc.org/?section = software&swid = 4
OECD	OECD (Q)SAR Application Toolbox	Skin and liver metabolic simulators (based on TIMES)	
Genego	MetaDrug	Platform for the prediction of drug metabolism and toxicity	www.genego.com/metadrug.php
Lhasa	Meteor	Metabolic fate of compounds	www.lhasalimited.org
Molecular Discovery	Metasite	Metabolic transformations	www.moldiscovery.com/index.php
	Volsurf +	Absorption, solubility, protein binding, volume of distribution, metabolic stability, BBB permeability	
MultiCASE	META/METAPC	Metabolic transformations	www.multicase.com/products/products.htm
	MCASE ADME module	Oral bioavailability, protein binding, urinary excretion, extent of metabolism, volume of distribution	
Molecular Networks	MetaboGen	Predicts metabolites and ranks according to likelihood of occurrence	www.molecular-networks.com/products/metabogen
ACD Labs	ADME Boxes	P-gp substrate specificity Absorption Bioavailability Plasma protein binding Volume of distribution Physico-chemical properties	www.acdlabs.com/products/pc_admet

Table 21.4 (*continued*)

Software provider	Software package	Predicted ADME properties	Website
QuantumLead	q-ADME	Half-life, absorption, Caco-2 permeability, volume of distribution, humans serum albumin binding Physico-chemical properties/toxicity	www.q-lead.com/adme_pk
Schrödinger	Qik Prop	Caco-2 and MDCK cell permeability, **BBB** permeation, serum albumin binding Physico-chemical properties	www.schrodinger.com
Simulations Plus	ADMET Predictor	Intestinal permeability, absorption, **BBB** permeation, volume of distribution, plasma protein binding	www.simulations-plus.com
	GastroPlus	Physico-chemical properties Physiological models for different species; dosage form effects One, two and three compartment models Complete **PBPK** models	
Provider	**Package**	**Commercial databases**	**Website**
Sunset Molecular	Wombat-PK	Database containing $>6,500$ clinical pharmacokinetic measurements for 1125 compounds (bioavailability, percentage excretion, percentage plasma protein bound, clearance, volume of distribution, half-life, **BBB** permeation, metabolising enzymes)	www.sunsetmolecular.com
University of Washington	Metabolism and transport drug-interaction database	Database for enzyme and transporter interactions	www.druginteractioninfo.org

[a]Updated and adapted from Madden.[21]

Toolbox parents and metabolites can both be profiled according to known toxicity mechanisms (*e.g.* binding to oestrogen receptors). From this, a category of similar compounds can be generated which enables the toxicity of the parent or metabolite of interest to be predicted. More information on these category formation and read-across approaches is given in Chapters 15 and 17 and an example of the role of the metabolic simulator is given in Chapter 24 (see Section 24.5.5).

There are many *in silico* models available for the prediction of fundamental ADME properties. These may be useful in determining the modulating effects that toxicokinetic properties have on the ability of a potential toxicant to elicit a toxic effect. This provides important supplementary information that can be incorporated into integrated testing strategies or weight-of-evidence approaches. Information from such models may also be useful in prioritising testing of compounds of concern. Figure 21.2 gives a diagrammatic representation of how ADME information obtained from *in silico* studies can be combined with *in vitro* and *in vivo* information to aid predictions of overall risk of toxicity that may be posed by a compound. Weight-of-evidence approaches and integrated testing strategies are discussed in Chapters 22 and 23, respectively.

The discussion above has described how ADME properties influence the overall plasma concentration-time curve (shown in Figure 21.1) and how these global effects can be modelled. However, much more information can be gained by considering the concentration–time curve within specific organs or tissues. Physiologically-based toxicokinetic models (PBTK) models provide a much more accurate representation of the compound's distribution in the body over time and, as such, have gained wider interest of late as they allow for organism- or population-specific physiological information to be considered in the overall modelling process.

21.4   Physiologically-Based Toxicokinetic Modelling

A traditional dose–response curve for a given toxicological effect can be represented as shown in Figure 21.3.

One problem with this traditional dose–response plot is that toxicity is correlated with an administered dose, whereas the concentration at the site of action may be very different. This creates an issue in that one compound may appear to be less toxic than another (*e.g.* when comparing LD_{50} values). However, the relative concentrations reaching the active site may, in fact, be responsible for the differences in observed toxicity. In other cases, the parent itself may not be responsible for any toxic effect and the dose–response relationship may be due to metabolite effects. This creates considerable problems when creating accurate models to predict toxicity.

As stated previously, a compound can only elicit a toxic response if it reaches the appropriate site of action in sufficient concentration for sufficient time. So far in this chapter ADME properties have been considered in general terms. In this section, the concept of using PBTK modelling to predict a toxicant's

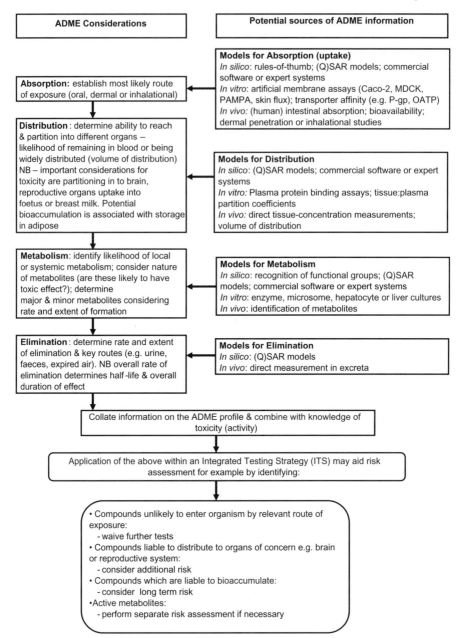

Figure 21.2 Diagrammatic representation of how ADME information can be incorporated into overall risk assessment.

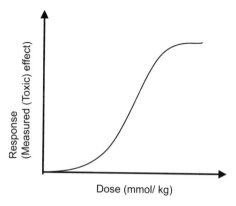

Figure 21.3 A typical dose–response curve.

concentration over time, within specific organs or tissues, is discussed. This subject is analogous to the physiologically-based pharmacokinetic modelling (PBPK) of therapeutic compounds.

The toxicokinetic profile of a compound in any given organism can be much better understood if the physiology of the organism is taken into consideration during modelling. The aim of PBTK is to provide a fully integrated view of how a toxicant will be dealt with by the organism, taking into account physico-chemical properties of the compound in addition to organism-specific differences in physiology.

Figure 21.1 represented the internal exposure of a toxicant in relation to the concentration in plasma over time. Conceptually, it is possible to determine concentration–time curves for any particular organ or tissue, although this is much more difficult to achieve experimentally and computationally. Modelling the exposure of an individual organ allows the toxicity elicited at that site to be determined much more accurately. As a simplified example, a compound which has a higher affinity for liver tissue than for blood may elicit toxicity to liver in excess of that predicted by circulating, systemic levels. Hence, models that are based on dose administered rather than focussing on the concentration in the relevant biophase may provide misleading results.

PBTK models use differential equations, solved simultaneously, to provide information on how the concentration in each compartment (*e.g.* an individual organ) of the system changes over time. PBTK properties integrate both compound-specific data and species (or even subject)-specific information. Compound-specific information relates to physico-chemical properties such as log *P*, *pK$_a$* or *solubility*; these values can be measured or predicted using *in silico* techniques (more details are provided in Chapter 5). The accuracy of their determination can have a significant effect on the accuracy of the PBTK prediction. Species-dependent information includes physiological factors such as organ volumes or blood flow rates.

Information on differential effects elicited in sub-populations, within a species, can be used to generate subject-specific models. For example, in population-based pharmacokinetic modelling consideration can be given to subjects with disease states that impair renal or hepatic function. These impaired functions may lead to higher doses being present at the active site for affected subjects within a population. PBTK models enable such factors to be modelled. ADME properties are generally a combination of both compound-specific and species-specific information. For example, metabolic clearance of a compound is affected not only by its ability to reach the site of metabolism (this may be directly influenced by the compound's log P), but also by the presence of the metabolising enzyme (expression and relative abundance of enzymes varies considerably between species and potentially between subjects within the same species).

21.4.1 Generating PBTK Models

Consider a representation of a mammalian system, *i.e.* one in which compounds are transported to different sites *via* the blood stream, as shown in Figure 21.4. In this representation, or model, eight compartments (excluding blood) are shown for simplicity. A PBTK model may include every organ of the body as an individual compartment or may 'lump' certain organs together. For example, rapidly perfused organs such as brain and lung may be combined. Similarly, poorly perfused organs such as adipose and bone may be combined.

Once a compound is present in the systemic blood circulation it is delivered to each organ *via* the arterial blood (the liver also receives blood from the gastrointestinal tract *via* the hepatic portal vein). The concentration in blood entering the organ can be defined as C_{in}. The concentration in blood leaving the organ is defined as C_{out}. Although, following equilibration, C_{in} may equal C_{out}

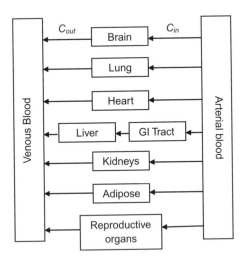

Figure 21.4 Diagrammatic representation of a physiologically-based toxicokinetic model.

for certain organs, for others (*i.e.* those with metabolic or excretory capacity) C_{out} will be lower. Differential equations can be used to solve for the concentration in tissue (C_{tissue}) over time, thus giving a full concentration–time profile within each organ. To physically measure concentrations in each tissue over time is highly expensive in terms of both time and animal usage. Consequently, it is preferable to develop models which predict this time course. Factors that affect concentrations in organs are those introduced earlier, *i.e.* the compound-specific and species-specific properties, blood flow to and from the organ, eliminating capacity of the organ (*e.g.* renal excretion or hepatic clearance), extent of plasma protein binding, relative affinity of the compound for organ *versus* blood (*i.e.* tissue : blood partition coefficients), *etc.* Input of these parameters into a model can provide an estimate of the overall plasma concentration–time curve from which relevant toxicokinetic parameters (*e.g.* half-life or total clearance) can be estimated.

21.4.2 Using PBTK Models

What is helpful in fully determining the concentration–time profile in an individual organ is that the information can then be used to determine the concentration–response curve within the organ (*i.e.* how the concentration in the individual organ correlates with toxic response). As concentration in individual organs may be very different from the concentration in plasma, use of organ concentrations provide a more accurate prediction of toxic response.

Another advantage of PBTK models is that they allow for better extrapolation between species, *i.e.* *in vivo–in vivo* correlations. Parameters used in the models can be scaled between different species using known physiological parameters.[65] *In vitro* data can also be readily incorporated and used to improve models, *e.g.* metabolic clearance measured in human hepatocytes can be used to replace values from rat models. Not only is it possible to improve predictions between species, it is also possible to improve predictions for subpopulations within a species. For example, altered organ blood flow rates could be input to represent reduced cardiac output in elderly subjects. Lower rates of hepatic clearance can be used to predict organ concentrations in neonates or those with hepatic impairment whose metabolic capacity may be lower. Similarly input parameters may be modified to take account of differences due to other disease states, sex, age or genetic variation. A compilation of relevant physiological parameters in healthy and health-impaired elderly patients has been compiled by Thompson *et al.*[66] Overall PBTK models provide a much more accurate representation of the behaviour of a xenobiotic within an organism with the flexibility to extrapolate between species and populations. The development of PBTK models and how the information gained from these may be used is shown in Figure 21.5. Thompson *et al.*[67] provide a review of approaches for using PBPK models in risk assessment, in particular their use in simulating internal exposure to environmental toxicants (as opposed to reliance

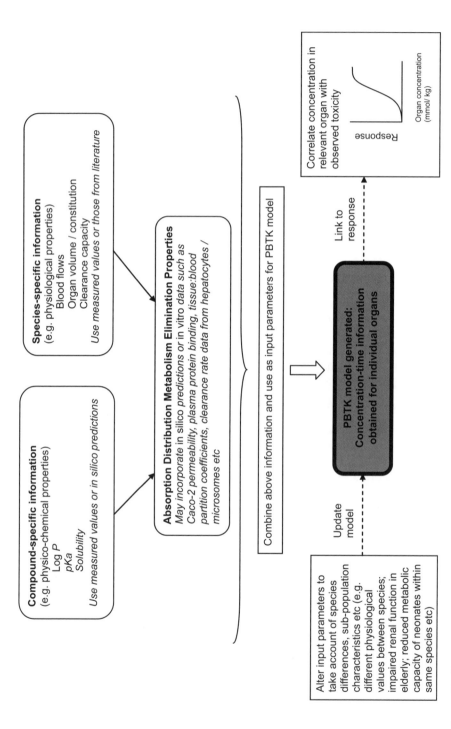

Figure 21.5 Development of PBTK models and use of information from these models.

on external concentration) for which little measured data are available. A useful overview of toxicokinetic considerations in toxicity testing for risk assessment is given by Vermeire *et al.*[68]

An increasing number of programs are available to perform the calculations associated with generating fully-integrated PBTK models. Selected examples of such software are given in Table 21.5.

Table 21.5 Example software used in PBPK/PBTK modelling.

Software and References	Capabilities
Cloe® PK www.cyprotex.com/ cloepredict/	Simulates concentration–time course in blood and major tissues and organs. Predicts renal excretion, hepatic metabolism and absorption. Integrates physico-chemical and experimental ADME screening data.
MeGen http://xnet.hsl.gov.uk/ megen/	Comprises database of species-specific anatomical and physiological parameters and an open-source PBPK model equation generator (based on information supplied by user).
WinNonLin® www.pharsight.com/ products/prod_home.php	Software for pharmacokinetic, pharmacodynamic and non-compartmental analysis. Calculates AUC, C_{max}, *etc.* for single/multiple dose or steady state conditions. Library of built-in pharmacokinetic (PK), pharmacodynamic (PD) and PK/PD models. Capacity for users to integrate own models and import/export to other statistical software.
ascIX (math model software) www.acslx.com	Provides a pharmacokinetic toolkit for PK/PBPK/ PD analysis. Predicts drug dispersion and residual levels. Tools can be used independently or combined with user's models.
NONMEM® www.iconplc. com/technology/products/ nonmem/	(Previously property of University of California, now distributed *via* ICON Development Solutions). NONMEM itself is a general program for model fitting. PREDPP is a package of subroutines that computes predictions for population pharmacokinetic/pharmacodynamic data analysis.
Symcyp www.simcyp.com	A population-based ADME simulator that enables modelling and simulation of drug absorption, distribution, metabolism and excretion (ADME) in virtual populations (incorporating inter-subject physiological variability). Predicts drug–drug interactions and pharmacokinetic outcomes in clinical populations. Contains databases for human physiological, genetic and epidemiological information that can be integrated with *in vitro* data to predict *in vivo* outcomes.

21.5 Conclusions and Outlook

When making an overall prediction of the potential toxicity of a compound there are many factors to consider in addition to the inherent toxicity of the compound. This chapter has introduced concepts relating to ADME/ toxicokinetic considerations. A greater understanding of the factors affecting the internal exposure will lead to more accurate models being developed for toxicity. Models that predict bioavailability and metabolism in particular have already been recognised as key components in toxicity prediction. As greater understanding is acquired of the physiological processes involved (*e.g.* metabolic interactions between different compounds and the role of transporters), the more reliable predictions will become. There is scope for much improvement in these models, in particular, current predictive models tend to be biased towards pharmaceutical compounds and there is a need to develop models based on compounds from other areas of chemical space, such as industrial chemicals or cosmetic products. Chapter 9 discussed the role of three-dimensional models in predicting toxicity and how knowledge acquired in the pharmaceutical arena could be used to solve problems in toxicology. The same is also true of toxicokinetics; there is a great deal of expertise in pharmacokinetic modelling that can be usefully applied here. Knowledge of toxicokinetics can then be incorporated into weight-of-evidence approaches and integrated testing strategies for the *in silico* prediction of toxicity.

21.6 Acknowledgements

The funding of the European Union 6th Framework Programme OSIRIS Integrated Project (GOCE-037017-OSIRIS) is gratefully acknowledged.

References

1. M. B. d'Yvoire, P. Prieto, B. J. Blaauboer, F. Y. Bois, A. Boobis, C. Brochot, S. Coecke, A. Freidig, U. Gundert-Remy, T. Hartung, M. Jacobs, T. Lave, D. E. Leahy, H. Lennernas, G. D. Loizou, B. Meek, C. Pease, M. Rowland, M. Spendiff, J. Yang and M. Zeilmaker, *Altern. Lab. Anim.*, 2007, **35**, 661.
2. P. Poulin and K. Krishnan, *Toxicol. Appl. Pharmacol.*, 1996, **136**, 126.
3. S. Haddad, P. Poulin and K. Krishnan, *Chemosphere*, 2000, **40**, 839.
4. S. Kortagere, D. Chekmarev, W. J. Welsh and S. Ekins, *Pharm. Res.*, 2008, **25**, 1836.
5. J. Shen, Y. P. Du, Z. Y. Zhao, G. X. Liu and Y. Tang, *QSAR Comb. Sci.*, 2008, **27**, 704.
6. U. Norinder and M. Haberleinn, *Adv. Drug Deliv. Rev.*, 2002, **54**, 291.
7. D. A. Konovalov, D. Coomans, E. Deconinck and Y. V. Heyden, *J. Chem. Inf. Model.*, 2007, **47**, 1648.

8. H. Li, C. W. Yap, C. Y. Ung, Y. Xue, Z. W. Cao and Y. Z. Chen, *J. Chem. Inf. Model.*, 2005, **45**, 1376.
9. Y. H. Zhao, M. H. Abraham, A. Ibrahim, P. V. Fish, S. Cole, M. L. Lewis, M. J. deGroot and D. P. Reynolds, *J. Chem. Inf. Model*, 2007, **47**, 170.
10. M. Hewitt, J. C. Madden, P. H. Rowe and M. T. D. Cronin, *SAR QSAR Environ. Res.*, 2007, **18**, 57.
11. M. Hewitt, C. M. Ellison, S. J. Enoch, J. C. Madden and M. T. D. Cronin, *Reprod. Toxicol.* 2010, **29**, *in press.*.
12. M. H. Abraham, J. Gil-Lostes and M. Fatemi, *Eur. J. Med. Chem.*, 2009, **44**, 2452.
13. S. Agatonovic, L. H. Ling, S. Y. Tham and R. G. Alany, *J. Pharm. Biomed. Anal.*, 2002, **29**, 103.
14. R. L. Juliano and V. Ling, *Biochim. Biophys. Acta*, 1976, **55**, 152.
15. E. J. D. Lee, C. B. Lean and L. M. G. Limenta, *Expert Opin. Drug Metab. Toxicol.*, 2009, **5**, 1369.
16. B. Sarkadi, L. Homolya, G. Szakacs and A. Varadi, *Physiol. Rev.*, 2006, **86**, 1179.
17. D. A. Williams, in *Foye's Principles of Medicinal Chemistry*, ed. W. O. Foye, T. L. Lemke and D. A. Williams, Wolters Kluwer Health, Lippincott, Williams & Wilkins, Baltimore, MD, 2007, pp. 253–326.
18. G. R. Wilkinson, in *Goodman and Gilman's: The Pharmacological Basis of Therapeutics*, 10th Edition ed. J. G. Hardman and L. E. Limbird, McGraw Hill, New York, 2001, pp. 3–29.
19. G. Szakacs, A. Varadi, C. Ozvegy-Laczka and B. Sarkadi, *Drug Discov. Today*, 2008, **13**, 379.
20. E. H. Kerns and L. Di, *Drug-Like Properties: Concepts, Structure Design and Methods,* Elsevier, Burlington, VT, 2008.
21. J. C. Madden, in *Recent Advances In QSAR Studies*, ed. T. Puzyn, J. Leszczynski, M. T. D. Cronin, Springer, London, 2010, pp. 283–304.
22. T. Hou T, J. Wang, W. Zhang and X. Xu, *J. Chem. Inf. Model.*, 2007, **47**, 208.
23. http://modem.ucsd.edu/adme/databases/databases_bioavailability.htm [accessed March 2010].
24. T. L. Moda, C. A. Monanari and A. D. Andricopulo, *Bioorg. Med. Chem.*, 2007, **15**, 7738.
25. M. V. S. Varma, R. S. Obach, C. Rotter, H. R. Miller, G. Chang, S. J. Steyn, A. El-Kattan and M. D. Troutman, *J. Med. Chem.*, 2010, **53**, 1098.
26. G. Lian and L. Chen, *J Pharm. Sci.*, 2008, **97**, 584.
27. S. C. Basak, D. Mills and M. M. Mumtaz, *SAR QSAR Environ. Res.*, 2007, **18**, 45.
28. J. R. Votano, M. Parham, M. L. Hall, L. H. Hall, L. B. Kier, S. Oloff and A. Tropsha, *J. Med. Chem.*, 2006, **49**, 7169.
29. M. P. Gleeson, N. J. Waters, S. W. Paine and A. Ma Davis, *J. Med. Chem.*, 2006, **49**, 1953.
30. S. C. Basak, D. Mills and B. D. Gute, *QSAR SAR Environ. Res.*, 2006, **17**, 515.

31. H. Zhang, *J Chem. Inf. Model.*, 2005, **45**, 121.
32. M. H. Abraham, A. Ibrahim and W. E. Acree Jr, *Eur. J. Med. Chem.*, 2006, **41**, 494.
33. M. H. Abraham, A. Ibrahim, Y. Zhao and W. E. Acree Jr, *J Pharm. Sci.*, 2006, **95**, 2091.
34. C. W. Yap, Z. R. Li and Y. Z. Chen, *J. Mol. Graph. Model.*, 2006, **24**, 383.
35. A. S. Kalgutkar, I. Gardner, R. S. Obach, C. L. Schaffer, E. Callegari, K. R. Henne, A. E. Mutlib, D. K. Dalvie, J. S. Lee, Y. Nakai, J. P. O'Donnell, J. Boer and S. P. Harriman, *Curr. Drug Metab.*, 2005, **6**, 161.
36. N. Manga, J. C. Duffy, P. H. Rowe and M. T. D. Cronin, *SAR QSAR Environ. Res.*, 2005, **16**, 43.
37. J. V. Turner, D. J. Maddalena and D. J. Cutler, *Int. J. Pharmacol.*, 2004, **270**, 209.
38. M. Takano, R. Yumoto and T. Murakami, *Pharmacol. Ther.*, 2006, **109**, 137.
39. M. A. Cabrera, I. Gonzalez I, C. Fernandez, C. Navarro and M. Bermejo, *J. Pharm. Sci.*, 2006, **95**, 589.
40. K. E. Thummel and G. G. Shen, in *Goodman and Gilman's The Pharmacological Basis of Therapeutics*, 10th edn., J. Hardman and E. Limbird, McGraw-Hill, New York, 2001, pp. 1917–2023.
41. M. Schultz and A. Schmoldt, *Pharmazie*, 1997, **52**, 895.
42. F. Hollósy, K. Valkó, A. Hersey, S. Nunhuck, G. Keri and C. Bevan, *J. Med. Chem.*, 2006, **49**, 6958.
43. S. Ekins, *J. Pharmacol. Toxicol. Meth.*, 2006, **53**, 38.
44. C. A. Lipinski, F. Lombardo, B. W. Dominy and P. J. Feeney, *Adv. Drug Deliv. Rev.*, 1997, **23**, 3.
45. M. P. Gleeson, *J. Med. Chem.*, 2008, **51**, 817.
46. M. Lobell, M. Hendrix, B. Hinzen and J. Keldenrich, *Chem. Med. Chem.*, 2006, **1**, 1229.
47. J. Keldenich, *Chem. Biodiversity*, 2009, **6**, 2000.
48. T. Hou and J. Wang, *Exp. Op. Drug Metab. Toxicol.*, 2008, **4**, 759.
49. D. F. Veber, S. R. Johnson, H. Y. Cheng, B. R. Smith and K. D. Kopple, *J. Med. Chem.*, 2002, **45**, 2615.
50. P. H. Lee, R. Conradi and V. Shanmugasundaram, *Bioorg. Med. Chem. Lett.*, 2010, **20**, 69.
51. G. Colmenarejo, *Med. Res. Rev.*, 2003, **23**, 275.
52. T. Ghafourian, M. Barzegar-Jalali, S. Dastmalchi, T. Khavari-Khorasani, N. Hakimiha and A. Nokhodchi, *Int. J. Pharm.*, **319**, 82.
53. F. Lombardo, R. S. Obach, M. Y. Shalaeva and F. Gao, *J. Med. Chem.*, 2004, **47**, 1242.
54. X. F. Sui, J. Sun, X. Wu, H. Y. Li, J. F. Liu and Z. G. He, *Curr. Drug Metab.*, 2008, **9**, 574.
55. H. Sun and D. O. Scott, *Chem. Biol. Drug Des.*, 75, **1**, 3.
56. J. C. Madden and M. T. D. Cronin, *Expert Opin. Drug Metab. Toxicol.*, 2006, **2**, 545.
57. M. P. Payne, in *Predicting Chemical Toxicity and Fate*, ed. M. T. D. Cronin and D. J. Livingstone, CRC Press, Boca Raton, FL, 2004, pp. 205–227.

58. C. Quinones, J. Caceres, M. Stud and A. Martinez, *Quant. Struct.-Act. Relat.*, 2000, **19**, 448.
59. C. Quinones-Torrelo, S. Sagrado, R. M. Villaneuva-Camanas and M. J. Medina-Hernandez, *J. Chromatogr. B*, 2001, **761**, 13.
60. J. Gola, O. Obrezanova, E. Champness and M. Segall M, *QSAR Comb. Sci.*, 2006, **25**, 1172.
61. P. S. Kharkar, *Curr. Top. Med. Chem.*, 2010, **10**, 116.
62. S. Ekins, C. L. Waller and P. W. Swann, *J. Pharmacol. Toxicol. Meth.*, 2000, **44**, 251.
63. J. C. Dearden, *Expert Opin. Drug Metab. Toxicol.*, 2007, **3**, 635.
64. www.oecd.org/document/23/ 0,3343,en_2649_34379_33957015_1_1_1_1,00.html.
65. S. Thomas, *Altern. Lab. Anim.*, 2009, **37**, 497.
66. C. M. Thompson, D. O. Johns, B. Sonawane, H. A. Barton, D. Hattis, R. Tardif and K. Krishnan, *J. Toxicol. Environ. Health B*, 2009, **12**, 1.
67. C. M. Thompson, B. Sonawane, H. A. Barton, R. S. DeWoskin, J. C. Lipscomb, P. Schlosser, W. A. Chiu and K. Krishnan, *J. Toxicol. Environ. Health B*, 2008, **11**, 519.
68. T. G. Vermeire, A. J. Baars, J. G. M. Bessems, B. J. Blaauboer, W. Slob and J. J. A. Muller in *Risk Assessment of Chemicals: An Introduction*, second edition, ed. C. J. van Leeuwen and T. G. Vermeire, Springer, The Netherlands, 2007, pp. 227–280.

CHAPTER 22

Multiple Test In Silico Weight-of-Evidence for Toxicological Endpoints

T. ALDENBERG[a] AND J. S. JAWORSKA[b]

[a] RIVM, Antonie van Leeuwenhoeklaan 9, Bilthoven NL-3721MA, The Netherlands; [b] Procter and Gamble, Temselaan 100, 1853 Strombeek-Bever, Brussels, Belgium

22.1 Introduction

Addressing a chemical's potential to adversely affect human health or the environment will be increasingly more challenging due to new legislation to reduce and/or eliminate testing on animals *in vivo*.[1] As a result, *in vitro* and *in silico* methods will play a more prominent role in chemical risk assessment in the future. These tests have a limited potential to replace *in vivo* tests, so multiple test combinations (*i.e.* batteries) of *in vitro* and/or *in silico* tests will be necessary to address an *in vivo* toxicological endpoint.[2]

It is therefore essential to develop a conceptual framework for integration of test data coming from different sources to allow for integrated and reliable endpoint assessment, generally indicated as Integrated Testing Strategies (ITS)—see Chapter 23. Such a decision–analytic framework will be beneficial by yielding a more comprehensive basis to guide decisions, and increasing the capability to combine and reuse existing data.

Until now the predominant form of data integration is assessment of weight-of-evidence (WoE), mainly by combining and qualitatively integrating different

Issues in Toxicology No.7
In Silico Toxicology: Principles and Applications
Edited by Mark T. D. Cronin and Judith C. Madden
© The Royal Society of Chemistry 2010
Published by the Royal Society of Chemistry, www.rsc.org

lines of evidence through expert judgment.[3–8] However, qualitative WoE approaches are considered not sufficient.[9] In order to improve the transparency, consistency and objectivity of the assessments, the need for more formal approaches to data integration was recognised in 2008.[10] Similar objectives are pursued in evidence-based toxicology[11] and in the quantitative approach to statistical WoE on the basis of likelihood ratios and odds ratios.[7]

Expert systems in toxicological risk prediction in use by governmental regulatory agencies may exploit different knowledge representation paradigms and modelling approaches.[12,13] Conceptual requirements for a multi-test decision framework, based on integrating multiple evidence and a decision-theoretic setting, have recently been formulated.[14] They identify three main conceptual requirements. A multi-test decision framework should:

- be probabilistic, in order to quantify uncertainties and dependencies;
- be consistent by allowing reasoning in both causal and predictive directions; and
- support a cyclic hypothesis and data-driven approach, where the hypotheses can be updated when new data arrive.

Such a framework would allow for evidence maximisation and reduction of uncertainty.

The formal ITS development framework that potentially meets these requirements can be founded on probabilistic systems called Bayesian networks.[15–22] We call them Bayesian Network Integrated Testing Strategies, or BNITS.[14] In this chapter we:

- explain the Bayesian methodology behind small multiple test networks, addressing the problem of conditional dependence between tests;
- quantify WoE given different—possibly conflicting—test results;
- assess the uncertainty in the network models through logistic regression and model selection;
- relate quantitative WoE uncertainty to parameter uncertainty in *in silico* Bayesian networks; and
- demonstrate how small Bayesian network models can be evaluated in a spreadsheet-like format.

To demonstrate *in silico* estimation in multiple test Bayesian networks, we have applied the methodology to *in vitro* genotoxicity test data.[23] However, we use these data only for illustrative purposes, having no intention to comment on the value of these tests nor to criticise or amend in any way how to perform genotoxicity assessment.

22.2 Two-Test Training Dataset

The two-test example is adapted from a review of rodent carcinogenicity data (Table 22.1).[23] The group of hazardous chemicals—according to the gold

Table 22.1 Example data for 243 hazardous compounds (G^+) and 105 non-hazardous chemicals (G^-) subjected to tests A and B. A positive test result is denoted as 1, a non-positive ('negative') result as 0.

A	B	G^+	G^-	sum
1	1	123	20	143
1	0	12	7	19
0	1	74	44	118
0	0	34	34	68
	sum	243	105	348

standard, rodent carcinogenicity—is denoted as G^+. The group of non-hazardous compounds is indicated as G^-. We consider initially only the first two *in vitro* tests tabulated, which we call A and B. A positive test result is denoted as 1, a non-positive test result, also indicated as 'negative', as 0. (Equivocal, or otherwise compromised, test results were assigned to these discrete values: data labelled 'E', '+R', '+**', and '^', were taken as positive. Data labelled 'TC' in the original paper were removed). Supposing that this is the data we have, the question is how to classify a (new) chemical as (non-) hazardous on the basis of either one, or a joint, test result.

22.3 One-Test Bayesian Inference

Here we employ the standard Bayesian statistical treatment of (medical) diagnostic tests.[24–27]

Bayes' Theorem states that the probability of a (new) chemical to be hazardous, given a positive result for some test T, denoted as $\Pr(G = G^+|T = 1)$, is proportional to the prior probability of the chemical to be hazardous, $\Pr(G = G^+)$, times the likelihood of the test being positive for a hazardous compound, denoted as $\Pr(T = 1|G = G^+)$:

$$\Pr(G = G^+|T = 1) = c_1 \cdot P(G = G^+) \cdot \Pr(T = 1|G = G^+) \qquad (22.1)$$

The vertical bar, |, is used to express the conditionally 'given that'. Here, c_1 is a constant of proportionality.

G is a probabilistic variable with possible values of G^+ (hazardous) or G^- (non-hazardous). Probabilistic variable T models the test result with possible values of 1 (positive) and 0 (non-positive). We simplify notation by using only values:

$$\Pr(G^+|T_1) = c_1 \cdot \Pr(G^+) \cdot \Pr(T_1|G^+) \qquad (22.2)$$

Analogously, the probability that the chemical is non-hazardous (G^-), also given a *positive* test, is given by (using the same shorthand notation from now on):

$$\Pr(G^-|T_1) = c_1 \cdot \Pr(G^-) \cdot \Pr(T_1|G^-) \qquad (22.3)$$

It is not immediately obvious that the constants of proportionality in eqn (22.2) and eqn (22.3) are the same, but this follows from probability theory (ref. 19, p. 9).

The two prior probability numbers sum to one, representing the discrete probability distribution $(\Pr(G^+), \Pr(G^-))$. In probability, this distribution is called a Bernoulli distribution.

The likelihood term in the right-hand side of eqn (22.2) is the sensitivity of the test, also called the true positive fraction:[27]

$$S_e = \Pr(T_1|G^+) \tag{22.4}$$

The likelihood term in eqn (22.3) is the false positive fraction, or one minus the specificity of the test:

$$1 - S_p = \Pr(T_1|G^-) \tag{22.5}$$

This is the chance that a positive test occurs for a non-hazardous chemical.

'Likelihood' is the statistical term for the probability of an actual or possible result (*i.e.* 'datum') given the 'parameter' hazardous or non-hazardous. Taking the datum as fixed, its chance of occurring depends on the true value of the parameter. Likelihoods typically do *not* sum to one, which can be understood from the fact that sensitivity and specificity of a test need not, necessarily, be equal to each other.

The terms $(\Pr(G^+|T_1), \Pr(G^-|T_1))$ comprise the posterior distribution of the compound belonging to one of the groups, given a positive test result. These do sum to one, and also constitute a Bernoulli distribution. It follows that the constant of proportionality in eqn (22.2) and eqn (22.3), c_1, is given by:

$$c_1 = \frac{1}{\Pr(G^+) \cdot \Pr(T_1|G^+) + \Pr(G^-) \cdot \Pr(T_1|G^-)} \tag{22.6}$$

In case of a non-positive test result, T_0, eqn (22.2), eqn (22.3) and eqn (22.6) become:

$$\Pr(G^+|T_0) = c_0 \cdot \Pr(G^+) \cdot \Pr(T_0|G^+) \tag{22.7}$$

$$\Pr(G^-|T_0) = c_0 \cdot \Pr(G^-) \cdot \Pr(T_0|G^-) \tag{22.8}$$

$$c_0 = \frac{1}{\Pr(G^+) \cdot \Pr(T_0|G^+) + \Pr(G^-) \cdot \Pr(T_0|G^-)} \tag{22.9}$$

For a non-positive test, the likelihoods in eqn (22.7) and eqn (22.8) are:

$$1 - S_e = \Pr(T_0|G^+) \tag{22.10}$$

$$S_p = \Pr(T_0|G^-) \tag{22.11}$$

22.3.1 One-Test Bayesian Network

This one-test Bayesian model can be depicted as a simple Bayesian network involving two variables, as shown in Figure 22.1.

The ovals signify the probabilistic variables G and T, and are labelled accordingly. The arrow points into the causal direction.[20,21,28] The hazard of the compound causes a reliable test to be positive, not the other way around. A non-hazardous substance is likely to yield a negative (non-positive) test. These numbers are specified by the conditional probabilities $\Pr(T|G)$ in the four combinations, as summarised in Table 22.2, in relation to the diagnostic test performance indicators of sensitivity and specificity.

In Bayesian network jargon, these conditional probabilities constitute the Conditional Probability Table (CPT) for the arrow of the network.[16,17] G is the *parent* node of T; T is the *child* node.

The conditional probabilities in Table 22.2 sum to one, vertically. The statistical concept of likelihood (*i.e.* the probability of a fixed test result, *e.g.* $T=1$, in relation to the chemical class) is read-off horizontally for each test result. In this way, the CPT has two interpretations—one vertically as the specification of the causal dependence of T on G, and one horizontally as the statistical likelihood of a fixed test result as a function of the chemical class.

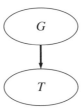

Figure 22.1   One-test Bayesian network of a chemical to be classified in a class G (hazardous or non-hazardous) on the basis of a binary test T.

Table 22.2   Conditional probabilities $\Pr(T|G)$ for specifying the causality on the arrow in the Bayesian network of Figure 22.1.

T	G^+	G^-
1	S_e	$1 - S_p$
0	$1 - S_e$	S_p
sum	1	1

Equations (22.2), (22.3), (22.6), (22.7), (22.8) and (22.9) may be combined into the general form in which Bayes' Theorem is usually stated:

$$Pr(G|T) = \frac{Pr(G) \cdot Pr(T|G)}{\sum_{G} Pr(G) \cdot Pr(T|G)} \qquad (22.12)$$

It is important to realise that, for the simplest case of a binary classification on the basis of a binary test, eqn (22.12) stands for four identities in total, when the variables are given values for each combination.

An important point to understand is that, although the flow of (presumed) causality is from G to T, the inference that Bayes' Theorem allows goes in the opposite direction, *i.e.* from evidence on T to probable values of G. This is a potential source of confusion as researchers want to reason from data to hypothesis, and sometimes draw arrows in the inference direction.

22.3.2 One-test Odds and Weight-of-Evidence

A convenient shorthand version of Bayes' Theorem follows, when dividing eqn (22.2) by eqn (22.3), and eqn (22.7) by eqn (22.8), respectively (ref. 20, p. 34):

$$\begin{cases} \dfrac{Pr(G^+|T_1)}{Pr(G^-|T_1)} = \dfrac{Pr(G^+)}{Pr(G^-)} \cdot \dfrac{Pr(T_1|G^+)}{Pr(T_1|G^-)} \\[3mm] \dfrac{Pr(G^+|T_0)}{Pr(G^-|T_0)} = \dfrac{Pr(G^+)}{Pr(G^-)} \cdot \dfrac{Pr(T_0|G^+)}{Pr(T_0|G^-)} \end{cases} \qquad (22.13)$$

Each term is expressed as 'odds', *i.e.* ratios of probabilities. This halves the number of equations, as one may always back-calculate probabilities from odds and vice versa. Thus if:

$$O^+ = \frac{P^+}{P^-}, \text{ and } P^+ + P^- = 1 \qquad (22.14)$$

then we have

$$(P^+, P^-) = \left(\frac{O^+}{1 + O^+}, \frac{1}{1 + O^+} \right) \qquad (22.15)$$

Usually, only one of eqn (22.13) applies, since the test comes out either positive, or non-positive. The prevailing result is called the *evidence*.

The standard assumption made in Bayesian analysis is that evidence is 'not uncertain', as it is given in relation to the unambiguous outcome in a certain case. Some software systems allow for uncertain (soft) evidence, but this awaits advances in fields such as probabilistic logic and imprecise probability to come to fruition.

Note that the pre-test or prior odds $\Pr(G^+)/\Pr(G^-)$ (ref. 27, p. 18) are the same in both eqns (22.13), as they do not depend on the test result. On many occasions when we are indifferent to the nature of the (new) chemical, the prior is assumed to be a (discrete) uniform distribution. This 'fifty–fifty' prior is given by $(0.5, 0.5)$, and the prior odds are 1 to 1. Using the uniform prior to model indifference goes back as far as Laplace. (Philosophical problems with Laplace' Principle of Insufficient Reason relate to the continuous uniform distribution as a model of indifference.)

In modern probability theory, the uniform prior is also the reference prior for a discrete distribution.[29,30] The uniform prior objectivises the analysis and— extremely important—if additional information on the chemical is available such as chemical fragments, it can be included in the Bayesian network explicitly to account for them.

In the case of the uniform reference prior, eqn (22.13) simplifies to:

$$\begin{cases} \dfrac{\Pr(G^+|T_1)}{\Pr(G^-|T_1)} = \dfrac{\Pr(T_1|G^+)}{\Pr(T_1|G^-)} \\[3mm] \dfrac{\Pr(G^+|T_0)}{\Pr(G^-|T_0)} = \dfrac{\Pr(T_0|G^+)}{\Pr(T_0|G^-)} \end{cases} \tag{22.16}$$

The post-test or posterior odds, *i.e.* $\Pr(G^+|T_1)/\Pr(G^-|T_1)$ and $\Pr(G^+|T_0)/\Pr(G^-|T_0)$, respectively, do depend on the test result—both with a non-uniform and a uniform prior.

The so-called diagnostic likelihood ratio (ref. 27, p. 17) for a positive test result is:

$$\mathrm{DLR}_1 = \frac{\Pr(T_1|G^+)}{\Pr(T_1|G^-)} = \frac{S_e}{1 - S_p} \tag{22.17}$$

and, for a non-positive test result, the diagnostic likelihood ratio is:

$$\mathrm{DLR}_0 = \frac{\Pr(T_0|G^+)}{\Pr(T_0|G^-)} = \frac{1 - S_e}{S_p} \tag{22.18}$$

We gain further insight when taking natural logarithms in eqn (22.13):

$$\begin{cases} \ln\left(\dfrac{\Pr(G^+|T_1)}{\Pr(G^-|T_1)}\right) = \ln\left(\dfrac{\Pr(G^+)}{\Pr(G^-)}\right) + \ln\left(\dfrac{\Pr(T_1|G^+)}{\Pr(T_1|G^-)}\right) \\[3mm] \ln\left(\dfrac{\Pr(G^+|T_0)}{\Pr(G^-|T_0)}\right) = \ln\left(\dfrac{\Pr(G^+)}{\Pr(G^-)}\right) + \ln\left(\dfrac{\Pr(T_0|G^+)}{\Pr(T_0|G^-)}\right) \end{cases} \tag{22.19}$$

For each test result, the posterior (post-test) log odds are the arithmetical sum of the prior (pre-test) log odds and the log diagnostic likelihood ratio. This permits the posterior log odds to be decomposed in an additive manner into prior log odds and test diagnostic log odds.

Log odds can be used as measures of statistical weight-of-evidence (WoE).[7,14,31–33] Good reports that weighting evidence in this way dates back to cryptanalytic work by Alan Turing during World War II at Bletchley Park in Buckinghamshire in the UK. Turing proposed a unit of WoE—similar to the decibel in acoustics—called deciban (db), which we adopt as the definition of quantitative WoE:

$$\text{WoE} = 10 \cdot \log_{10}(odds) \approx 4.343 \cdot \ln(odds) \qquad (22.20)$$

(The names 'ban' and 'deciban' referred to stationery from the town of Banbury in Oxfordshire on which their computations were performed.) As $10 \cdot \log_{10}(5/4) = 0.9691$, an event estimated with odds 5 to 4 approximately carries 1 db of evidence. This was considered as a convenient unit value from the point of view of human perception.

It follows that eqn (22.19) becomes, after multiplying by $10/\ln(10) = 4.343$:

$$\begin{cases} \text{WoE}_1^{\text{post}-\text{test}} = \text{WoE}^{\text{pre}-\text{test}} + \text{WoE}_1^{\text{test}} \\ \text{WoE}_0^{\text{post}-\text{test}} = \text{WoE}^{\text{pre}-\text{test}} + \text{WoE}_0^{\text{test}} \end{cases} \qquad (22.21)$$

each expressed in db. Note that this quantitative measure of WoE depends on the test result.

In case of the uniform (reference) prior, the pre-test WoE is zero, since the prior odds equal 1, meaning that no weight is added to the diagnostic (test) WoE whatever the test result. With the fifty–fifty prior, the post-test WoE for a particular test result equals the diagnostic (test) WoE for that test result.

22.3.3 One-Test Data Weight-of-Evidence Estimate

Table 22.3 shows two illustrative examples. The data for each of the tests, *A* and *B*, were derived from Table 22.1 by summing the number of chemicals over the other test.

Table 22.3 Two example one-test weight-of-evidence calculations (data from Table 22.1).

Test *A*	Data G^+	G^-	sum	Conditional probabilities G^+	G^-	DLR	WoE	Posterior probabilities G^+	G^-	sum
1	135	27	162	**0.556**	0.257	2.160	3.35	0.684	0.316	1.000
0	108	78	186	0.444	**0.743**	0.598	− 2.23	0.374	0.626	1.000
sum	243	105	348	1.000	1.000					
B	G^+	G^-	sum	G^+	G^-	DLR	WoE	G^+	G^-	sum
1	197	64	261	**0.811**	0.610	1.330	1.24	0.571	0.429	1.000
0	46	41	87	0.189	**0.390**	0.485	− 3.14	0.327	0.673	1.000
sum	243	105	348	1.000	1.000					

The conditional probabilities are estimated as fractions calculated from the raw data. (Later in the chapter we introduce a more sophisticated way of model fitting.) Sensitivity and specificity are printed in Table 22.3 in **bold**. The diagnostic likelihood ratios (DLR) are calculated from the conditional probabilities on the same line at some fixed test result (1 or 0). So, for a positive test A, $DLR_1 = 0.556/0.257 = 2.160$, and $DLR_0 = 0.444/0.743 = 0.598$, and so on. (We present rounded results of un-rounded computer results.) These DLRs are converted to WoE values (expressed in db) by applying eqn (22.20) to the DLRs.

Since we assume the uniform reference prior, we can calculate the posterior probabilities directly from the likelihood by normalising the likelihood values. Thus, for a positive test A, the likelihood (0.556, 0.257) is taken; these numbers sum to 0.813, so the posterior probabilities become $(0.556/0.813, 0.257/0.813) = (0.684, 0.316)$, which sums to 1. Alternatively, one could use the DLR odds directly through eqn (22.15), so $2.160/(1 + 2.160) = 0.684$, and $1/(1 + 2.160) = 0.316$, which results in the same posterior probabilities.

These examples illustrate some important issues. The sensitivity of test A is not very good (0.556), while the specificity is better (0.743). The complement of the specificity is low (0.257), which compensates for the meagre 0.556, as 0.556 is divided by 0.257 to get a $DLR_1 = 2.160$ for test A. This leads to a reasonable WoE_1 of 3.35 db in the case of a positive test A. Apparently, a bleak sensitivity can be counterbalanced by a good specificity and still lead to a reasonable WoE for a positive test result.

This obviously works the other way around. Although the specificity of test A is fairly good, the complement of the sensitivity (0.444) weakens it, leading to a $WoE_0 = -2.23$ db—in fact not as good as $WoE_1 = +3.35$. (A negative WoE indicates a situation of the DLR odds against the chemical being hazardous.)

Test B is even more dramatic in that the specificity of test B (0.390) is below 0.5. However, the sensitivity (0.811) is quite good. The false positive fraction, *i.e.* complement (0.610) of the specificity (0.390), counteracts this good sensitivity, leading to a poor $WoE_1 = 1.24$ db of a positive result for test B. Now the benefit of Turing's db measure becomes evident, as 1 db is considered insignificant in human reasoning; we must conclude that test B, despite its great sensitivity, has a barely significant WoE for a positive test result. The surprising effect, in turn, is that the dismal specificity of test B (0.390) is corrected by the low complement (0.189) of the sensitivity (0.811). The value of 0.485 for the diagnostic likelihood ratio of a non-positive test result leads to a workable $WoE_0 = -3.14$ db against the compound being hazardous.

We observe here that the absolute values of the test coefficients, sensitivity and specificity, are not the full story when assessing test performance. The diagnostic ratios—and logarithmic WoEs for that matter—depend on both coefficients.

22.4 Two-Test Battery Bayesian Inference

Bayes' Theorem, as expressed in eqn (22.12), was motivated from the point of view of a two-group classification (hazardous *vs.* non-hazardous compound) on the basis of the outcome of a single binary (laboratory) test. However, Bayes' Theorem is very general and can be applied to multiple tests (*i.e.* test batteries) without much change in notation.

Suppose we have two tests, S and T. We consider a battery test of S and T, with four possible outcomes: $(S = 1, T = 1)$, $(S = 1, T = 0)$, $(S = 0, T = 1)$, and $(S = 0, T = 0)$. Taken as two-component Boolean vectors, they can be conceived as 'single' battery test results. Similar to the T_1 and T_0 shorthand notation (see above), we write (S_1, T_1) or (S_1, T_0), and so on.

Bayes' Theorem for the two-test battery then becomes very similar to eqn (22.12):[25,34]

$$\Pr(G|S, T) = \frac{\Pr(G) \cdot \Pr(S, T|G)}{\sum\limits_{G} \Pr(G) \cdot \Pr(S, T|G)} \tag{22.22}$$

Note that we leave out the battery parentheses inside probability statements. The only difference is that the test (battery) has four values instead of two. However, this means that almost all of our previous equations for one-test transfer to the two-test battery are virtually unmodified, including diagnostic likelihood ratios and WoE.

The generic eqn (22.22) breaks down into eight identities:

$$\begin{cases}
\Pr(G^+|S_1,T_1) = c_{11} \cdot \Pr(G^+) \cdot \Pr(S_1,T_1|G^+) \\
\Pr(G^-|S_1,T_1) = c_{11} \cdot \Pr(G^-) \cdot \Pr(S_1,T_1|G^-) \\
\Pr(G^+|S_1,T_0) = c_{10} \cdot \Pr(G^+) \cdot \Pr(S_1,T_0|G^+) \\
\Pr(G^-|S_1,T_0) = c_{10} \cdot \Pr(G^-) \cdot \Pr(S_1,T_0|G^-) \\
\Pr(G^+|S_0,T_1) = c_{01} \cdot \Pr(G^+) \cdot \Pr(S_0,T_1|G^+) \\
\Pr(G^-|S_0,T_1) = c_{01} \cdot \Pr(G^-) \cdot \Pr(S_0,T_1|G^-) \\
\Pr(G^+|S_0,T_0) = c_{00} \cdot \Pr(G^+) \cdot \Pr(S_0,T_0|G^+) \\
\Pr(G^-|S_0,T_0) = c_{00} \cdot \Pr(G^-) \cdot \Pr(S_0,T_0|G^-)
\end{cases} \tag{22.23}$$

The conditional probabilities $\Pr(S_1,T_1|G^+)$, $\Pr(S_1,T_0|G^+)$, *etc.* for the hazardous group and $\Pr(S_1,T_1|G^-)$, $\Pr(S_1,T_0|G^-)$, *etc.* for the non-hazardous group are the battery equivalents of sensitivity and specificity, respectively. As we have more combinations, the naming requires further attention (see the next section on conditional dependence and correlation). Let us call them 'joint Cooper statistics'.

For each group of chemicals, hazardous and non-hazardous, the conditional probabilities must sum to 1, respectively, since the four battery results exhaust all the possibilities. Therefore, the Conditional Probability Table (CPT) of the battery network is as shown in Table 22.4.

Table 22.4   Two-test battery Conditional Probability Table (CPT) for the Bayesian network of Figure 22.2.

S,T	G^+	G^-
(1,1)	$p_{11} = \Pr(S_1, T_1 \mid G^+)$	$q_{11} = \Pr(S_1, T_1 \mid G^-)$
(1,0)	$p_{10} = \Pr(S_1, T_0 \mid G^+)$	$q_{10} = \Pr(S_1, T_0 \mid G^-)$
(0,1)	$p_{01} = \Pr(S_0, T_1 \mid G^+)$	$q_{01} = \Pr(S_0, T_1 \mid G^-)$
(0,0)	$p_{00} = \Pr(S_0, T_0 \mid G^+)$	$q_{00} = \Pr(S_0, T_0 \mid G^-)$
sum	1	1

When reading Table 22.4 in the horizontal direction, we obtain the likelihood for each battery result. Thus $(\Pr(S_1, T_1 \mid G^+), \Pr(S_1, T_1 \mid G^-))$ is the likelihood of battery result (1,1), $(\Pr(S_1, T_0 \mid G^+), \Pr(S_1, T_0 \mid G^-))$ is the likelihood of battery result (1,0), and so on.

22.4.1   Two-Test Battery Bayesian Network

By emphasising the similarity to the single-test network in Figure 22.1, and the analogy between the CPTs in Tables 22.2 and 22.4, it follows that we have created a 'single' battery Bayesian network, as depicted in Figure 22.2.

Figure 22.2 for a two-test battery is equivalent to the Bayesian network with separate variables for each test.[14] This follows from probability theory, which states that the joint probability of two variables (tests) equals the product of the probability of the first variable (test) times the probability of the second variable (test) given the first test. The latter Bayesian network is more flexible, but the definition of WoE and the model fitting, as developed in Section 22.5, is more complicated. In this chapter, we place emphasis on the fact that joint multiple test WoE is almost as simple as one-test WoE.

22.4.2   Two-Test Odds and Weight-of-Evidence

Analogous to eqn (22.13), we build four identities, expressed as odds:

$$\begin{cases} \dfrac{\Pr(G^+ \mid S_1, T_1)}{\Pr(G^- \mid S_1, T_1)} = \dfrac{\Pr(G^+)}{\Pr(G^-)} \cdot \dfrac{\Pr(S_1, T_1 \mid G^+)}{\Pr(S_1, T_1 \mid G^-)} \\[2mm] \dfrac{\Pr(G^+ \mid S_1, T_0)}{\Pr(G^- \mid S_1, T_0)} = \dfrac{\Pr(G^+)}{\Pr(G^-)} \cdot \dfrac{\Pr(S_1, T_0 \mid G^+)}{\Pr(S_1, T_0 \mid G^-)} \\[2mm] \dfrac{\Pr(G^+ \mid S_0, T_1)}{\Pr(G^- \mid S_0, T_1)} = \dfrac{\Pr(G^+)}{\Pr(G^-)} \cdot \dfrac{\Pr(S_0, T_1 \mid G^+)}{\Pr(S_0, T_1 \mid G^-)} \\[2mm] \dfrac{\Pr(G^+ \mid S_0, T_0)}{\Pr(G^- \mid S_0, T_0)} = \dfrac{\Pr(G^+)}{\Pr(G^-)} \cdot \dfrac{\Pr(S_0, T_0 \mid G^+)}{\Pr(S_0, T_0 \mid G^-)} \end{cases} \qquad (22.24)$$

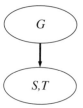

Figure 22.2 Two-test battery Bayesian network of a chemical to be classified in a class *G* (hazardous or non-hazardous) on the basis of two binary tests, *S* and *T*.

Thus, we recognise the pre-battery (prior) odds, $\mathrm{Pr}(G^+)/\mathrm{Pr}(G^-)$, and four diagnostic likelihood ratios, one for each battery result:

$$\begin{cases} \mathrm{DLR}_{11} = \dfrac{\mathrm{Pr}(S_1,T_1|G^+)}{\mathrm{Pr}(S_1,T_1|G^-)} \\[2mm] \mathrm{DLR}_{10} = \dfrac{\mathrm{Pr}(S_1,T_0|G^+)}{\mathrm{Pr}(S_1,T_0|G^-)} \\[2mm] \mathrm{DLR}_{01} = \dfrac{\mathrm{Pr}(S_0,T_1|G^+)}{\mathrm{Pr}(S_0,T_1|G^-)} \\[2mm] \mathrm{DLR}_{00} = \dfrac{\mathrm{Pr}(S_0,T_0|G^+)}{\mathrm{Pr}(S_0,T_0|G^-)} \end{cases} \tag{22.25}$$

Analogous to equations (22.21), we have four identities linking pre-battery, battery, and post-battery WoE:

$$\begin{cases} \mathrm{WoE}_{11}^{\text{post-batt}} = \mathrm{WoE}^{\text{pre-batt}} + \mathrm{WoE}_{11}^{\text{batt}} \\ \mathrm{WoE}_{10}^{\text{post-batt}} = \mathrm{WoE}^{\text{pre-batt}} + \mathrm{WoE}_{10}^{\text{batt}} \\ \mathrm{WoE}_{01}^{\text{post-batt}} = \mathrm{WoE}^{\text{pre-batt}} + \mathrm{WoE}_{01}^{\text{batt}} \\ \mathrm{WoE}_{00}^{\text{post-batt}} = \mathrm{WoE}^{\text{pre-batt}} + \mathrm{WoE}_{00}^{\text{batt}} \end{cases} \tag{22.26}$$

Again, we will assume a uniform reference prior, leading to nil pre-battery WoE. As stated earlier, the more tests, models or structure information included in the network, the more we feel justified to adhere to an indifferent prior, as we believe that the nodes in the network must do their diagnostic work. Obviously, in situations where prior information is available, or needed, the term is there to be used in the above equations.

22.4.3 Two-Test Data Weight-of-Evidence Estimate

Given the similarity between the single test and single battery network, the WoE tabulation becomes almost automatic. In Table 22.5, we estimate the WoE for each test combination by using empirical fractions calculated from Table 22.1 as estimates of the CPT. The appropriate DLRs and WoEs are calculated by row. This can be easily implemented in a spreadsheet.

Table 22.5 Two-test weight-of-evidence calculation for joint test results (data from Table 22.1).

Battery A,B	Data			Conditional probabilities				Posterior probabilities		
	G^+	G^-	sum	G^+	G^-	DLR	WoE	G^+	G^-	sum
(1,1)	123	20	143	0.506	0.190	2.657	4.24	0.727	0.273	1.000
(1,0)	12	7	19	0.049	0.067	0.741	− 1.30	0.426	0.574	1.000
(0,1)	74	44	118	0.305	0.419	0.727	− 1.39	0.421	0.579	1.000
(0,0)	34	34	68	0.140	0.324	0.432	− 3.64	0.302	0.698	1.000
sum	243	105	348	1.000	1.000					

Table 22.5 is a concise form of Table 5 in ref. 14, which was derived from the two-test layout in ref. 35 and ref. 36. The results are obviously identical.

We observe that, for an equivocally positive battery result (1,1), we have 4.24 db WoE—greater than the WoEs for each test individually in Table 22.4. A double non-positive result (0,0) has WoE = − 3.64 db. The conflicting battery results (1,0) and (0,1) have a negative WoE of − 1.30, and − 1.39, respectively, *i.e.* slightly against the chemical being hazardous. The absolute value approaches the Turing deciban unit of odds 5 to 4, so these WoEs seem relatively minor. It appears that combining a result in the second test that conflicts with the one in the first test may largely annihilate the WoE built up in the first test.

22.4.4 Two-Test Conditional Dependence and Conditional Correlation

The two-test CPT in Tables 22.4 and 22.5 lists the conditional probabilities of joint test results in a battery for each class of chemicals (hazardous and non-hazardous). The sensitivity and specificity coefficients of the respective single tests can be related to the conditional probabilities of two-test battery results. With the notation in Table 22.4, we have for the single test sensitivities:[35]

$$\begin{cases} Se_1 = p_{11} + p_{10} \\ Se_2 = p_{11} + p_{01} \end{cases} \tag{22.27}$$

And the individual test specificities are:

$$\begin{cases} Sp_1 = q_{01} + q_{00} \\ Sp_2 = q_{10} + q_{00} \end{cases} \tag{22.28}$$

In the Vacek two-test layout,[14,35,36] the single test performances emerge as marginal entries. (This can be understood if we look back at how we estimated these coefficients from the raw data in Tables 22.1 and 22.3.)

Independence of the two tests in the hazardous group would imply:

$$p_{11} = \Pr(S_1, T_1 | G^+) = \Pr(S_1 | G^+) \cdot \Pr(T_1 | G^+) = Se_1 \cdot Se_2 \qquad (22.29a)$$

This is called conditional independence (ref. 16, p. 11). If conditional independence holds for p_{11}, then it automatically holds for the other entries in the hazardous group:

$$p_{10} = Se_1 \cdot (1 - Se_2) \qquad (22.29b)$$

$$p_{01} = (1 - Se_1) \cdot Se_2, \qquad (22.29c)$$

$$p_{00} = (1 - Se_1) \cdot (1 - Se_2) \qquad (22.29d)$$

Conditional independence in the non-hazardous group would be similarly expressed as:

$$q_{00} = \Pr(S_0, T_0 | G^-) = \Pr(S_0 | G^-) \cdot \Pr(T_0 | G^-) = Sp_1 \cdot Sp_2. \qquad (22.30)$$

Again, this would imply independence in the other entries.

In theory, two tests may be conditionally dependent in one group, but not in the other.

The differences $p_{11} - Se_1 \cdot Se_2$ and $q_{00} - Sp_1 \cdot Sp_2$ are the conditional covariances, modelling departure(s) from conditional independence.[36,37] This leads naturally to conditional correlations for each group, respectively:[38–40]

$$\rho_{G^+} = \frac{p_{11} - Se_1 \cdot Se_2}{\sqrt{Se_1 \cdot (1 - Se_1) \cdot Se_2 \cdot (1 - Se_2)}} \qquad (22.31)$$

$$\rho_{G^-} = \frac{q_{00} - Sp_1 \cdot Sp_2}{\sqrt{Sp_1 \cdot (1 - Sp_1) \cdot Sp_2 \cdot (1 - Sp_2)}} \qquad (22.32)$$

If the tests are conditionally independent for a chemical class, then the conditional correlation is zero. On the contrary, complete conditional dependence arises in either of the chemicals groups when there is zero probability of getting conflicting test results in that group. For example, if in the hazardous group $p_{11} = Se_1 = Se_2$ so that $p_{10} = 0$ and $p_{01} = 0$, then $\rho_{G^+} = 1$. Similar absolute correlation may occur in the non-hazardous group. Essentially, this battery reduces to a single test.

To show the conditional dependence, we rearrange Table 22.5 in the Vacek two-test layout (Table 22.6), and calculate the conditional correlation.

In this setup, the single test performance indicators, Se_1, Se_2, Sp_1 and Sp_2, as given by eqn (22.27) and eqn (22.28), are easily visualised as marginal sums (in **bold**). The lack of independence in both groups follows from $p_{11} = 0.506$ *not* being equal to $Se_1 = 0.556$ times $Se_2 = 0.881$, and $q_{00} = 0.324$ *not* equalling the product of $Sp_1 = 0.743$ and $Sp_2 = 0.390$.

Table 22.6   Rearrangement of Table 22.5 to make one-test sensitivity and specificity coefficients (boldface) visible and calculate conditional correlations according to eqn (22.31) and eqn (22.32).

$A\backslash B$	\multicolumn{3}{c}{G^+}			\multicolumn{3}{c}{G^-}		
	1	0	*sum*	1	0	*sum*
1	123	12	135	20	7	27
0	74	34	108	44	34	78
sum	197	46	243	64	41	105

$A\backslash B$	1	0	*sum*	1	0	*sum*
1	0.506	0.049	**0.556**	0.190	0.067	0.257
0	0.305	0.140	0.444	0.419	0.324	**0.743**
sum	**0.811**	0.189	1.000	0.610	**0.390**	1.000

Conditional correlation G^+	Conditional correlation G^-
0.287	**0.158**

The conditional correlations, eqn (22.31) and eqn (22.32) for the hazardous and non-hazardous group are estimated to be 0.287 and 0.158, respectively. As complete conditional correlation would yield a value of 1, these correlations seem relatively low. This is a good thing since tests that correlate high carry less information than if they were independent.

22.4.5   Two-Test Conditional Independence Example

In order to see what a conditionally independent data set would look like, we have artificially constructed one by replacing the conditional probabilities or likelihoods in Tables 22.5 and 22.6 by the respective products of the marginal sensitivity and specificity coefficients, and then multiplying by the number of chemicals in each group. This leads to the conditionally independent data set in Table 22.7.

Note that for battery results (1,1) and (1,0), the total number of chemicals is $125.9 + 36.1 = 162$, the same as the $143 + 19 = 162$ compounds in Table 22.5. This is true for the other summations as well. In this way, the sensitivity and specificity coefficients for the individual tests are the same, but now we have independence, as $Se_1 = 0.556$ times $Se_2 = 0.881$ is *equal* to the conditional probability 0.450 of battery result (1,1) in hazardous group G^+, instead of 0.506 in Table 22.5 and 22.6, and so on for the other joint sensitivity and specificity coefficients.

The WoEs resulting from independence are 4.58 and -5.38, respectively, for the non-conflicting battery results, (1,1) and (0,0), compared with 4.24 and

Table 22.7 Artificially constructed data set with the same one-test sensitivity and specificity coefficients as in Table 22.6, but with conditional independence in both chemical classes: conditional correlations are nil.

Battery	Artificial independent data			Conditional probabilities				Posterior probabilities		
A,B	G^+	G^-	sum	G^+	G^-	DLR	WoE	G^+	G^-	sum
(1,1)	109.4	16.5	125.9	0.450	0.157	2.874	4.58	0.742	0.258	1.000
(1,0)	25.6	10.5	36.1	0.105	0.100	1.047	0.20	0.512	0.488	1.000
(0,1)	87.6	47.5	135.1	0.360	0.453	0.796	−0.99	0.443	0.557	1.000
(0,0)	20.4	30.5	50.9	0.084	0.290	0.290	−5.38	0.225	0.775	1.000
sum	243.0	105.0	348.0	1.000	1.000					

– 3.64 in Table 22.5. In particular, the double non-positive, (0,0), has grown in magnitude. The WoEs for the conflicting battery results, (1,0) and (0,1), change from – 1.30 and – 1.39 in Table 22.5 to 0.20 and – 0.99, hence expressing more neutral WoEs. Further analysis could reveal whether these effects can be generalised and related to the conditional correlation coefficients.

22.5   Two-Test Data Fitting and Model Selection

By comparing Tables 22.5 and 22.7, we gain a feeling for the sensitivity of the quantitative WoE values, estimated from the observed number of chemicals at each battery result. The WoEs relate directly to the posterior probabilities for a (new) chemical to belong to the hazardous or non-hazardous class, given a battery result.

We have employed straightforward point estimates of the conditional probabilities calculated from the empirical fractions. However, we know that the binomial parameter is uncertain when we estimates it from smaller numbers. Therefore, how much confidence do we have in the WoE estimates and related posterior probabilities?

Modelling data tables such as Table 22.1 is the domain of categorical data analysis. The posterior probabilities of a chemical to belong to either of the two classes (groups) can be modelled by logistic regression (ref. 41, p. 95). For the theoretical background, we refer to the monographs by Agresti, especially the AIDS and AZT example in the original handbook (ref. 42, p. 182) and in the introductory volume (ref. 43, p. 110). An overview is also given in Chapter 10.

Let us reconsider the data from Table 22.1 in Table 22.8. Instead of estimating the conditional probabilities directly by dividing 123 by 243, 20 by 105, and so on, we first try to model the data by studying the fractions π_{AB} and $1 - \pi_{AB}$ horizontally as a function of the test results of A and B, given the number of each battery result as fixed.

The fractions are assumed to be binomial for each fixed number of battery results and the four binomials are assumed to be independent. Thus, we now can take into account the fact that 12 out of 19 has the problem of being a small sample, while 123 out of 143 may lead to a more reliable estimate. Raw fractions do not take that into account.

Table 22.8   Number of hazardous chemicals as fraction of the number of compounds for each joint test result combination: 123/143 = 0.860, 12/19 = 0.632, *etc.* (data from Table 22.1).

A	B	G^+	G^-	sum	π_{AB}	$1 - \pi_{AB}$	sum
1	1	123	20	143	0.860	0.140	1.000
1	0	12	7	19	0.632	0.368	1.000
0	1	74	44	118	0.627	0.373	1.000
0	0	34	34	68	0.500	0.500	1.000
	sum	243	105	348			

22.5.1   Binomial Multiple Logistic Regression

We assume the binomial parameters π_{AB} to be a linear logistic function of A and B, for example:

$$\text{logit}(\pi_{AB}) = \ln\left(\frac{\pi_{AB}}{1 - \pi_{AB}}\right) = \beta_0 + \beta_1 \cdot A + \beta_2 \cdot B \qquad (22.33)$$

Or written out in full:

$$\begin{cases} \pi_{AB} = \dfrac{\exp(\beta_0 + \beta_1 \cdot A + \beta_2 \cdot B)}{1 + \exp(\beta_0 + \beta_1 \cdot A + \beta_2 \cdot B)} \\[4mm] 1 - \pi_{AB} = \dfrac{1}{1 + \exp(\beta_0 + \beta_1 \cdot A + \beta_2 \cdot B)} \end{cases} \qquad (22.34)$$

This polynomial model yields the possibility to undertake model selection when we try different polynomials, all being sub-models of a full, so-called saturated model. One can say that the sub-models are nested. The models considered are listed in Table 22.9.

  There are two advantages in fitting these binomial logistic regression models. First, for a given model, we can assess the parameter significance and associated confidence limits. This yields information on the consistency of the contributions of a test in the battery. Secondly, one can compare alternative models and apply model selection criteria to select a most reasonable model.

22.5.2   Logistic Regression Results and the Most Reasonable Model

Here we are interested in modelling the numbers of chemicals for each battery result directly, while calculating WoE from the fitted cell numbers later on. The results of fitting the five logistic regression models of Table 22.9 to the data in Table 22.8 are given in Table 22.10.

Table 22.9   Five binomial logistic regression models as a function of the test results. All are sub-models of the saturated model, **M12Sat**, which fits the data in Table 22.8 exactly given the fixed number of chemicals per battery test result.

Model	Equation
M0	$\text{logit}\,(\pi_{AB}) = \beta_0$
M1	$\text{logit}\,(\pi_{AB}) = \beta_0 + \beta_1 \cdot A$
M2	$\text{logit}\,(\pi_{AB}) = \beta_0 + \beta_2 \cdot B$
M12	$\text{logit}\,(\pi_{AB}) = \beta_0 + \beta_1 \cdot A + \beta_2 \cdot B$
M12Sat	$\text{logit}\,(\pi_{AB}) = \beta_0 + \beta_1 \cdot A + \beta_2 \cdot B + \beta_{12} \cdot A \cdot B$

Table 22.10 Regression results for the five binomial logistic models defined in Table 22.9, fitted to the data in Table 22.9, fitted to the data in Table 22.8. **M12Sat** is the saturated model and fits exactly, but all predictors are insignificant. **M12** is the most attractive model (lowest BIC and highly significant predictors). See the text for a further explanation.

Model	Predictors	Estimate	Standard error	Lower confidence limit	Upper confidence limit	p-value	*	nPar	Deviance	G^2	Degrees of freedom	Significance	AIC^a	BIC^b
M0	1	**0.839**	0.117	0.61	1.07	0.000		1	426.2	35.28	3	0.000	428.2	432.0
M1	1	**0.325**	0.149	0.03	0.62	0.029		2	399.0	8.08	2	0.018	403.0	410.7
	A	**1.284**	0.258	0.78	1.79	0.000	***							
M2	1	**0.115**	0.215	−0.31	0.54	0.592		2	411.1	20.19	2	0.000	415.1	422.8
	B	**1.009**	0.259	0.50	1.52	0.000	***							
M12	1	**−0.113**	0.225	−0.55	0.33	0.616		3	392.4	1.48	1	0.225	398.4	409.9
	A	**1.119**	0.267	0.60	1.64	0.000	***							
	B	**0.703**	0.272	0.17	1.24	0.010	**							
M12Sat	1	**0.000**	0.243	−0.48	0.48	1.000	—	4	390.9	0.00	0	—	398.9	414.3
	A	**0.539**	0.534	−0.51	1.59	0.313	—							
	B	**0.520**	0.308	−0.08	1.12	0.092	.							
	A · B	**0.758**	0.616	−0.45	1.96	0.219	—							

[a] Akaike Information Criterion.
[b] Bayesian Information Criterion.

Next to the *Model* column is the *Predictors* column. *A* and *B* stand for the respective predictors (values 0 and 1) and '1' indicates the constant in the regression, β_0. The estimated regression coefficients (the β's in Table 22.9) are given under *Estimate*. Then follow the standard errors of the coefficients, as well as lower and upper confidence limits. If the lower and upper confidence limits have different signs, then the coefficient does not differ significantly from zero. This is also indicated by the *p*-value and the associate significance codes borrowed from the statistical package R:[46]

- '***' means $0 < p < 0.001$
- '**' means $0.001 < p < 0.01$;
- '*' means $0.01 < p < 0.05$;
- '.' means $0.05 < p < 0.1$;
- '–' means $0.1 < p < 1$.

The significance of the constants is not given, as it is unimportant, but we are interested in the significance of the coefficients of the test variables.

Table 22.10 shows that the coefficients of the saturated model **M12Sat** are all insignificant. This is interesting as this model fits the data exactly. Apparently, by working with the empirical fractions in earlier sections, we have implicitly used a model that is insignificant in all its coefficients and may exhibit large prediction uncertainty.

The books by Agresti explain how the deviance column, which is -2 times the log likelihood of the model fit, is a measure of how well the model fits.[42,43] The saturated model fits best, the single constant **M0** fits worst. The G^2 column gives the differences between deviances with respect to the saturated model. These have a χ^2 distribution, with degrees of freedom equal to the reduction in number of parameters with respect to the saturated model.

The linear **M12** model has a significance of 0.225, which indicates that this model can equally well describe the data if, in fact, the saturated model holds. The other models, **M0**, **M1** and **M2** do not come even close and are unlikely to have generated the data.

In model selection, the AIC (Akaike Information Criterion) and BIC (Bayesian, or Schwarz, Information Criterion) are measures that penalise the deviance (*i.e.* measure of fit) for the number of parameters. Minimising AIC or BIC reduces the danger of overfitting. The respective theoretical derivations are quite complicated.[47] However, the final algorithms are very simple. The AIC subtracts twice the number of parameters from the deviance, while BIC subtract the natural log of the number of data records.

The subject of model selection is still under development. We think that the AIC is too mild and has a tendency to overfit. It is known that the AIC can favour too complex models and for 'infinite' sample sizes may stick to the wrong model (lack of consistency). We observe here that **M12** and **M12Sat** have practically the same AIC, while the significance of the parameters differs dramatically. This may imply that **M12Sat** will give quite unreliable, perhaps unusable, predictions.

The BIC measure of model selection favours relatively small models. Newer information-theoretic developments in model selection based on MDL (Minimum Description Length) are basically equivalent to BIC. BIC ranks **M12** first, then **M1**, followed by **M12Sat**, **M2**, and finally the test-independent constant **M0**. We conclude that, in this nested model set, **M12** makes the most sense.

We can use model **M12** in a variety of ways. The simplest is to use it to estimate the *expected* cell numbers. The equation is:

$$\text{logit}(\hat{\pi}_{AB}) = -0.113 + 1.119 \cdot A + 0.703 \cdot B \qquad (22.35)$$

or:

$$\hat{\pi}_{AB} = \frac{\exp(-0.113 + 1.119 \cdot A + 0.703 \cdot B)}{1 + \exp(-0.113 + 1.119 \cdot A + 0.703 \cdot B)} \qquad (22.36)$$

This is equivalent to using a regression line for a single prediction, neglecting confidence or prediction limits. A future extension would be to simulate the uncertainty of the coefficients and obtain the predictive distributions of cell numbers and WoE.

Table 22.11 compares the raw data and WoE measures based on it with the expected numbers on the basis of eqn (22.36) of model **M12**.

The WoE estimate based on **M12** yields 3.78 db for the double positive battery result instead of 4.24 on the basis of un-modelled fractions. The WoE estimate of the double non-positive battery result increases in absolute value from -3.64 to -4.13 db. The WoEs in the case of conflicting battery results,

Table 22.11   Data and predicted values by linear logistic regression model **M12**, the most attractive model (Table 22.10). **M12** fits the one-test data (Table 22.3) exactly for both A and B. Conditional correlations for G^+ and G^- are 0.246 and 0.244.

		Data			Conditional probabilities			
A	B	G^+	G^-	sum	G^+	G^-	DLR	WoE
1	1	123	20	143	0.506	0.190	2.657	4.24
1	0	12	7	19	0.049	0.067	0.741	-1.30
0	1	74	44	118	0.305	0.419	0.727	-1.39
0	0	34	34	68	0.140	0.324	0.432	-3.64
	sum	243	105	348	1.000	1.000		
		Model M12			Conditional probabilities			
A	B	G^+	G^-	$\hat{\pi}_{AB}$	G^+	G^-	DLR	WoE
1	1	121.1	21.9	**0.847**	0.498	0.209	2.387	3.78
1	0	13.9	5.1	**0.732**	0.057	0.048	1.182	0.73
0	1	75.9	42.1	**0.643**	0.312	0.401	0.779	-1.08
0	0	32.1	35.9	**0.472**	0.132	0.342	0.386	-4.13
	sum	243.0	105.0		1.000	1.000		

(1,0) and (0,1), are reduced somewhat and end up at roughly the deciban unit value of 0.73 and -1.08, respectively.

A nice feature of the model is that the one-test cell numbers as shown in Table 22.3 are predicted exactly. This means that the single-test sensitivities and specificities (Table 22.6) are exact too. This cannot be accidental. We conjecture that a two-test linear logistic model fitted to two-test data will always fit the one-test data without error.

At first sight, one would not expect this relatively simple linear model to exhibit conditional correlation. But it does: the conditional correlations are 0.246 and 0.244 for G^+ and G^-. Compared to 0.287 and 0.158 in Table 22.6 they are more even, but this seems difficult to generalise.

22.5.3 Weight-of-Evidence Relates to the Logistic Parameters Linearly

There are two more advantages to having the cell numbers modelled in this way. The first is that we obtain a very simple explicit expression for the WoE at each battery result, as shown below.

Looking at Table 22.11, we denote the number of cases of each battery result as n_{AB}, i.e. 143, 19, 118 and 68 in the upper left sub-table. The column totals, 243 and 105 are n_{G+} and n_{G-}, respectively. A likelihood row, under *Conditional probabilities* becomes $n_{AB} \cdot \hat{\pi}_{AB}/n_{G+}$ and $n_{AB} \cdot (1 - \hat{\pi}_{AB})/n_{G-}$. The DLR (Diagnostic Likelihood Ratio) on substitution with model **M12** can be expressed as:

$$\text{DLR}_{AB} = \left[\frac{\hat{\pi}_{AB}}{1 - \hat{\pi}_{AB}} \right] \cdot \frac{n_{G-}}{n_{G+}} = \left[\exp\left(\hat{\beta}_0 + \hat{\beta}_1 \cdot A + \hat{\beta}_2 \cdot B \right) \right] \cdot \frac{n_{G-}}{n_{G+}} \quad (22.37)$$

Taking the natural logarithm and multiplying by 4.343, we get:

$$\text{WoE}_{AB} = 4.343 \cdot \left(\hat{\beta}_0 + \hat{\beta}_1 \cdot A + \hat{\beta}_2 \cdot B - \ln(n_{G+}/n_{G-}) \right) \quad (22.38)$$

As the A and B values are 0 and 1 in all combinations, it turns out that the WoEs are simple linear sums of the appropriate regression coefficients. The reader may like to check that the last WoE column in the lower right sub-table of Table 22.11 can be calculated directly from the regression coefficients in eqn (22.35) and the number of chemicals in each group of the training set.

22.5.4 Outlook: Weight-of-Evidence Uncertainty

The second advantage of the modelled cell numbers is that the uncertainty of the regression coefficients translates directly to the WoE estimates. Hence, these are uncertain too. By simulation of the posterior for the binomial logistic regression model, we can calculate confidence (credibility) limits of the *in silico* WoE, given a joint test (battery) result.

An immediate consequence is that, for model **M12**, the uncertainty of WoE_{11} is expected to be greater than that of WoE_{10}, as the former is calculated from three parameters (β_0, β_1 and β_2) while the latter is calculated from two (β_0 and β_1). WoE_{00} depends only on the uncertainty of the intercept, β_0.

Similar reasoning applies to the comparison of independent estimates of each test separately *versus* the joint fit **M12** above. When the tests are fitted independently, the double positive WoE will depend on four parameters (two for each test), while the joint fit double positive (1,1) depends only on three parameters. The uncertainty of the latter may be smaller despite the fact that the individual training sets may be larger, which leads to smaller parameter uncertainty. This is an interesting programme for further research.

22.6   Conclusions

Prediction of a toxicological property (endpoint) of a chemical on the basis of laboratory test information is very similar to a medical diagnosis problem. The diagnostic accuracy of tests is important, as the possible occurrence of false positives and false negatives may lead to unreliable predictions. By combining results of multiple tests, one tries to increase the amount of information on the chemical and toxicological properties and to reduce uncertainty in the predictions.

Bayesian statistics is the premier methodology to estimate the probability of a chemical or toxicological property to be present, given information obtained from a set of tests—be it laboratory results, related substances or organisms/cells, or otherwise summarised knowledge. For medical tests, so-called Diagnostic Likelihood Ratios (DLRs) quantify the information contained in the test with regard to the odds of the toxicological property present.

The main concept of this chapter is to consider the logarithm of DLRs as quantitative measures of test performance. In statistics, this concept is actually quite old and dates back to cryptanalytic work by Alan Turing during World War II.

The advantage of logarithms of diagnostic likelihood ratios and pre-test or prior information is that contributions of each add up to form the log odds of the posterior prediction of the property at concern being present. These terms can be interpreted as 'weights', and this model for weighting strength of evidence by addition and subtraction benefits human and machine-based assessment experience.

The medical diagnosis model for the prediction of toxicological endpoints also transfers to combining multiple test results. The natural inference engine is called Bayesian networks and has been recently proposed as the main framework for building Integrated Testing Strategies.[14]

As in every branch of statistics, the step to multivariate assessment is beset with difficulties. In this chapter, we extend the single test Bayesian diagnosis model to a single-layer-multiple-test network when joint measurements are available for each test considered in relation to a gold standard. Veterinary statisticians have been modelling multiple test diagnosis systems for a number of years and have identified the possible effects of correlation between tests. In Bayesian network parlance, this is called 'conditional dependence' and measures of conditional correlation between two tests have been developed. General conditional dependence between tests is handled by the Bayesian network definition through conditional probability tables, the entries of which are defined on each arc (arrow) of the Bayesian network.

Under the proviso of the availability of joint test results in the presence of a gold standard, the diagnostic likelihood ratios and the quantitative measure of weight-of-evidence (WoE) based on them can be generalised from single test estimation to multiple test systems, almost without modification. We have demonstrated spreadsheet-like calculations that do not require other sophisticated software. The example can be easily extended from two-test data to joint multi-test data.

Turing developed a unit of WoE, analogous to the decibel in acoustics, to quantify the strength of evidence that can be readily applied to single and multiple test results. Thus, we obtain measures of WoE quantifying test results that are in agreement, as well as results that are in conflict. We have seen how an amount of WoE obtained for one test can be annihilated by the result of another test.

Although these calculations can be based directly on tables of cross-classified test results over the groups of chemicals evaluated (*e.g.* hazardous *vs.* non-hazardous), we have also addressed the question of the small-sample character of such data. Cross-classified data can be modelled with advanced regression analysis such as binomial logistic regression. These techniques are also applicable to evaluate the contribution of each test to the 'diagnosis' and to select the best model(s). This permits final judgement on the strength of tests to function in Integrated Test Strategies, although the selection of training data should also be taken into account. If heterogeneity lurks in the training set, and chemical or test-related information could be added in the regressions, one could re-evaluate the strength and relative contribution of each test and act accordingly, *e.g.* by dividing into subgroups.

A major result in our regression analysis is the direct linear relationship between quantitative *in silico* WoE, thus defined, and the regression coefficients in the categorical data analysis. This would allow further assessment of the uncertainty of WoE and the evaluation of optimal testing strategies on the basis of information content. A decision-analytic framework admits addition of utility-based calculations including costs, animal welfare and economic values.

Acknowledgements

The funding of the European Union Integrated Research Project **OSIRIS**: Optimized Strategies for Risk Assessment of Industrial Chemicals through Integration of Non-Test and Test Information, GOCE-CT-2007-037017, is gratefully acknowledged.

References

1. European Chemicals Agency, *Guidance on Information Requirements and Chemical Safety Assessment. Part A: Introduction to the Guidance Document (Guidance for the implementation of REACH)*, ECHA, Helsinki, 2008,

http://guidance.echa.europa.eu/docs/guidance_document/information_
requirements_en.htm [accessed March 2010].

2. S. Hoffmann and T. Hartung, *Hum. Exp. Toxicol.*, 2006, **25**, 497.
3. G. R. Batley, G. A. Burton, P. M. Chapman and V. E. Forbes, *Hum. Ecol. Risk Assess.*, 2002, **8**, 1517.
4. G. A. Burton, G. E. Batley, P. M. Chapman, V. E. Forbes, E. P. Smith, T. B. Reynoldson, C. E. Schlekat, P. J. Den Besten, A. J. Bailer, A. S. Green and R. L. Dwyer, *Hum. Ecol. Risk Assess.*, 2002, **8**, 1675.
5. P. M. Chapman, B. G. McDonald and G. S. Lawrence, *Hum. Ecol. Risk Assess.*, 2002, **8**, 1489.
6. T. B. Reynoldson, E. P. Smith and A. J. Bailer, *Hum. Ecol. Risk Assess.*, 2002, **8**, 1613.
7. E. P. Smith, I. Lipkovich and K. Ye, *Hum. Ecol. Risk Assess.*, 2002, **8**, 1585.
8. V. Thybaud, M. Aardema, J. Clements, K. Dearfield, S. Galloway, M. Hayashi, D. Jacobson-Kram, D. Kirkland, J. T. MacGregor, D. Marzin, W. Ohyama, M. Schuler, H. Suzuki and E. Zeiger, *Mutat. Res.*, 2007, **627**, 41.
9. E. Silbergeld, *Toxicol. Lett.*, 2008, **180**(Suppl. 1), S18.
10. Organisation for Economic Co-operation and Development, *Workshop on Integrated Approaches to Testing and Assessment*, OECD, Paris, 2008, OECD Environment Health and Safety Series on Testing and Assessment No. 88, ENV/JM/MONO(2008)10.
11. P. S. Guzelian, M. S. Victoroff, N. C. Halmes, R. C. James and C. P. Guzelian, *Hum. Exp. Toxicol.*, 2005, **24**, 161.
12. M. T. D. Cronin, in *Predicting Chemical Toxicity and Fate, ed. M. T. D. Cronin and D. J. Livingstone,* CRC Press, Boca Raton, FL, 2004, pp. 413–427.
13. S. Parsons and P. McBurney, in *Predictive Toxicology, ed. C. Helma,* Taylor & Francis, Boca Raton, FL, 2005, pp. 135–175.
14. J. S. Jaworska, S. Gabbert and T. Aldenberg, *Regul. Toxicol. Pharmacol.*, 2010, **57**, 157.
15. E. Castillo, J. M. Gutiérrez and A. S. Hadi, *Expert Systems and Probabilistic Network Models,* Springer-Verlag, New York, 1997.
16. F. V. Jensen and T. D. Nielsen, *Bayesian Networks and Decision Graphs,* Springer, New York, 2007.
17. U. B. Kjaerulff and A. L. Madsen, *Bayesian Networks and Influence Diagrams. A Guide to Construction and Analysis,* Springer, New York, 2008.
18. R. E. Neapolitan, *Probabilistic Reasoning in Expert Systems. Theory and Algorithms,* Wiley, New York, 1990.
19. R. E. Neapolitan, *Learning Bayesian Networks,* Pearson/ Prentice Hall, Upper Saddle River, NJ, 2004.
20. J. Pearl, *Probabilistic Reasoning in Intelligent Systems: Networks of Plausible Inference,* Morgan Kaufmann, San Mateo, CA, 1988.
21. J. Pearl, *Causality. Models, Reasoning, and Inference,* Cambridge University Press, Cambridge, 2000.
22. R. G. Cowell, A. P. Dawid, S. L. Lauritzen and D. J. Spiegelhalter, *Probabilistic Networks and Expert Systems. Exact Computational Methods for Bayesian Networks,* Springer, New York, 2007.

23. D. J. Kirkland, M. Aardema, L. Henderson and L. Müller, *Mutat. Res.*, 2005, **584**, 1.
24. M. J. Campbell and D. Machin, *Medical Statistics. A Commonsense Approach*, John Wiley & Sons, Chichester, 1993.
25. I. A. Gardner, *Aust. Vet. J.*, 2002, **80**, 758.
26. R. M. McDowell and J. S. Jaworska, *SAR QSAR Environ. Res.*, 2002, **13**, 111.
27. M. S. Pepe, *The Statistical Evaluation of Medical Tests for Classification and Prediction*, Oxford University Press, Oxford, 2003.
28. J. Pearl, *Stat. Surv.*, 2009, **3**, 96.
29. J. M. Bernardo, in *Handbook of Statistics, ed. D. K. Dey and C. R. Rao*, Elsevier, 2005, **Vol. 25**, pp. 17–19.
30. J. M. Bernardo and A. F. M. Smith, *Bayesian Theory*, Wiley, Chichester, 2000.
31. I. J. Good, *Biometrika*, 1979, **66**, 393.
32. I. J. Good, in *Bayesian Statistics 2, ed. J. M. Bernardo, M. H. DeGroot, D. V. Lindley and A. F. M. Smith*, Elsevier, Amsterdam, 1985, pp. 249–270.
33. I. J. Good, in *Encyclopedia of Statistical Sciences, ed. S. Kotz, N. L. Johnson and C. B. Read*, Wiley, New York, 1988, **vol. 8**, pp. 651–656.
34. B. S. Kim and B. H. Margolin, *Environ. Health Perspect. Suppl.*, 1994, **102**, 127.
35. I. A. Gardner, H. S. Stryhn, P. Lind and M. T. Collins, *Prev. Vet. Med.*, 2000, **45**, 107.
36. P. M. Vacek, *Biometrics*, 1985, **41**, 959.
37. L. A. Thibodeau, *Biometrics*, 1981, **37**, 801.
38. A. J. Branscum, I. A. Gardner and W. O. Johnson, *Prev. Vet. Med.*, 2005, **68**, 145.
39. N. Dendukuri and L. Joseph, *Biometrics*, 2001, **57**, 158.
40. M. P. Georgiadis, W. O. Johnson, I. A. Gardner and R. Singh, *Appl. Stat.*, 2003, **52**, 63.
41. T. Hastie, R. Tibshirani and J. Friedman, *The Elements of Statistical Learning. Data Mining, Inference, and Prediction*, Springer, New York, 2001.
42. A. Agresti, *Categorical Data Analysis*, Wiley-Interscience, New York, 2002.
43. A. Agresti, *An Introduction to Categorical Data Analysis*, Wiley-Interscience, New York, 2007.
44. M. S. Pepe, W. Leisenring and C. Rutter, in *Handbook of Statistics 18: Bioenvironmental and Public Health Statistics, ed. P. K. Sen and C. R. Rao*, North-Holland, Amsterdam, 2000, pp. 397–422.
45. W. Leisenring and M. S. Pepe, *Biometrics*, 1998, **54**, 444.
46. R Development Core Team, *R: A Language and Environment for Statistical Computing. Reference Index*, R Foundation for Statistical Computing (www.r-project.org), Vienna, 2009, version 2.10.1, http://cran.r-project.org/doc/manuals/refman.pdf [accessed March 2010].
47. K. P. Burnham and D. R. Anderson, *Model Selection and Multimodel Inference. A Practical Information-Theoretic Approach*, Springer, New York, 2002.

CHAPTER 23

Integrated Testing Strategies and the Prediction of Toxic Hazard

M. BALLS

Fund for the Replacement of Animals in Medical Experiments (FRAME), Russell & Burch House, 96–98 North Sherwood Street, Nottingham NG1 4EE, UK

23.1 Introduction

Over the past two or three decades, there has been mounting pressure to reduce reliance on laboratory animal testing for predicting hazards and assessing risks in relation to industrial chemicals and chemical products of various kinds including pharmaceuticals and agrochemicals, household and personal care products. Initially, this pressure was largely the result of ethical and animal welfare concerns, but the emphasis is now increasingly on scientific and economic considerations. It is increasingly recognised, and admitted, that animal tests cannot be a scientifically credible basis for predicting hazard in humans because of species differences of various kinds, so they cannot provide information which can reliably be used in risk assessment and risk management in humans. In addition, they cost too much and can take too long to perform. These problems have been brought sharply into focus by both the introduction of the EU REACH regulation, which requires the evaluation and registration of tens of thousands of new and existing chemicals, and the mounting crisis in the pharmaceutical industry caused by an ongoing decline in the introduction

Issues in Toxicology No.7
In Silico Toxicology: Principles and Applications
Edited by Mark T. D. Cronin and Judith C. Madden
© The Royal Society of Chemistry 2010
Published by the Royal Society of Chemistry, www.rsc.org

of new products and the increasing rate of withdrawal of drugs which have been placed on the market as a result of unacceptable adverse effects not detected during preclinical testing.

In response to this situation, a huge amount of effort is being invested in the development of non-animal test procedures which make use of existing data and bioinformatics, novel *in chemico*, *in silico* and *in vitro* approaches and, where possible, ethical studies in humans. The result is an increasing complexity and variety of methods based on mechanisms of action and modes of action at the molecular, cellular and tissue/organ/system levels, which are suited to answering highly specific questions rather than to providing information on toxic effects of all kinds, as in the one-model-suits-all systems once thought to be provided by laboratory rats, mice, dogs or macaques.

This means that the available tests will have to be used intelligently and selectively in combination as batteries and/or in tiered hierarchical schemes in what have come to be known as Integrated Testing Strategies (ITS). Before an ITS can be considered acceptable for use, especially in compliance with regulatory requirements, it must be established that it is relevant and reliable for a particular purpose (*e.g.* the classification and labelling of chemicals in terms of their toxicity, or for identifying particular hazards such as carcinogenicity, eye irritation, hepatotoxicity, nephrotoxicity, neurotoxicity, reproductive toxicity, skin irritation or skin sensitisation). This, in turn, requires that principles for the development, evaluation, acceptance and use of ITS need to be established and agreed by all stakeholders. This chapter is concerned with some of the considerations which will need to be taken into account as these principles are elaborated and agreed.

23.2   Essential Nature and Uses of ITS

ITS are hierarchical in nature and are used in a stepwise fashion progressing from the evaluation of pre-existing data and a consideration of the physico-chemical properties of a test item, *via in silico* modelling to the use of *in vitro* methods followed—sometimes and where necessary—by *in vivo* tests in animals and in human volunteers. In the case of pharmaceuticals (and some other products such as personal care products), the marketing and use of the product at the population level should be looked on as the final part of the ITS, conducted *via* post-marketing surveillance and reporting.

Individual tests can be duplicative or confirmatory (*i.e.* the result of one test can be used with a comparable result from another test to strengthen the conclusion reached about a particular toxic hazard) or they can be additive or complementary (*i.e.* they can provide different kinds of information which, taken together, can support a particular conclusion, perhaps as part of a weight-of-evidence approach). Sometimes, when other considerations have to be taken into account such as the patent life of a new chemical entity, tests may take place in parallel rather than in sequence. Here, the time factor would affect the affordability of this kind of application of the ITS.

The stepwise approach usually represents a decision-tree scheme, *i.e.* at the conclusion of each step, decisions can be made about whether or not further testing is advisable and, if so, what tests should or should not be performed. This will depend on the specific purpose of the study being performed, which may be aimed, for example, at the classification and labelling of chemicals, or deciding whether or not a chemical might be a skin sensitiser or a carcinogen.

The data provided by different tests employed in an ITS are unlikely to be of equal value, so decisions need to be taken about the relative weighting of the tests—especially where the prediction from one test may conflict with that from another test (see discussion in Chapters 4 and 22).

Bearing in mind the importance of cost, throughput and manageability, as well as common sense, the use of the variety of ITS which are available also requires a strategy, especially where a company is evaluating a series of new chemicals which may have beneficial and marketable prospects. For example, given that hepatotoxicity, nephrotoxicity and neurotoxicity are three important types of acute toxic effect and if an ITS was available for each of them, a positive result in the hepatotoxicity ITS would usually make it unnecessary to perform the nephrotoxicity ITS or the neurotoxicity ITS since a decision about the development of the test item could be taken on the basis of the result from the hepatotoxicity ITS alone. In another situation, if an ITS predicts that a chemical will be a strong skin allergen in humans and a decision about its development and use is taken on that basis, then the application of ITS for systemic toxic hazard is unlikely to be necessary. In other words, the use of Integrated Testing Strategies themselves requires a strategy, *i.e.* a selective, stepwise and decision-tree approach.

23.3 Historical Development of the Concept of ITS

One of the first integrated approaches to the application of *in vitro* test systems to the hazard assessment of chemicals was the Integrated Toxicity Testing Scheme (ECITTS) developed in 1991 by the European Group for Alternatives in Toxicity Testing (ERGATT) and the Swedish Board for Laboratory Animals (CFN).[1] This arose from the Organisation for Economic Co-operation and Development (OECD) High Production Volume (HPV) Chemicals Programme, which began in 1987 and which was the forerunner of the US HPV programme and the EU REACH system. The developers of ECITTS recognised the general role of basal cytotoxicity and took into account the main forms of target organ/system toxicity and the importance of biokinetics (Figure 23.1). They also went so far as to identify a set of 80 calibration chemicals for use in establishing the proposed scheme, which was designed to provide information on all the main types of toxic hazard conventionally evaluated in animal studies.

Another important early step was taken at the European Centre for the Validation of Alternative Methods (ECVAM) Workshop 16 in 1994,[2] when a three-stage scheme for acute toxicity testing was proposed (Figure 23.2). This

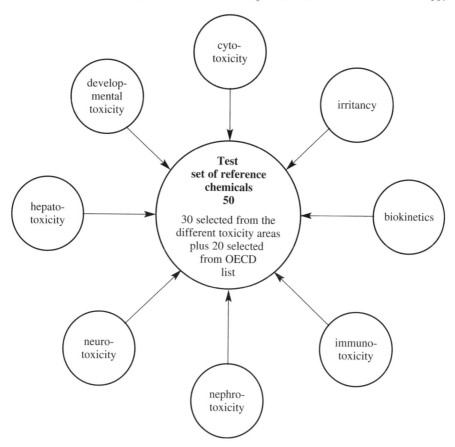

Figure 23.1 The ERGATT/CFN Integrated Toxicity Testing Scheme (ECITTS), adapted with permission from Walum *et al.*[1]

took into account basal cytotoxicity, biotransformation, selective cytotoxicity and cell-specific function toxicity (where the toxicant affects processes which may not be critical for the affected cells themselves, but which are critical for the organisms as a whole, *e.g.* suppression of the release of a hormone).

These approaches were followed up in a number of ECVAM-sponsored workshops and task force meetings and ECVAM-funded studies. For example, the ECVAM Integrated Testing Strategies Task Force built on ECITTS (Figures 23.3 and 23.4) to propose a generic scheme for chemical toxicity (Figure 23.5) and a specific scheme for neurotoxicity (Figure 23.6), now with the addition of *in silico* methods, in between the evaluation of existing data and the use of *in vitro* tests.[3]

Later, an international in-depth discussion took place at the Interagency Coordinating Committee on the Validation of Alternative Methods (ICC-VAM) International Workshop on *In vitro* Methods for Assessing Acute Systemic Toxicity at Arlington, VA, USA, in 2000. This discussion led to a variety

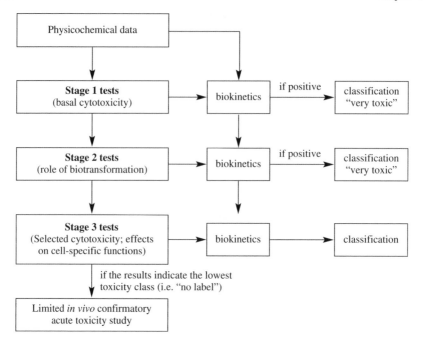

Figure 23.2 A proposed scheme for the classification and labelling of chemicals according to their potential acute toxicities, adapted with permission from Seibert *et al.*[2]

Type of building block	Details
Experimental	Collection of *in vitro* data for: 1. Biokinetic parameters, including biotransformation. 2. Cytotoxicity and neurotoxicity data (in different *in vitro* systems).
Modelling	1. Incorporation of *in vitro* data in a physiologically based biokinetic model. 2. Determination of target tissue concentration. 3. Modelling of toxicological response. 4. Prediction of systemic toxicity.
Validation	1. Validation of the model against *in vivo* kinetics. 2. Validation of the model against *in vivo* toxicity

Figure 23.3 The building blocks of ECITTS, reproduced with permission from Blaauboer *et al.*[3]

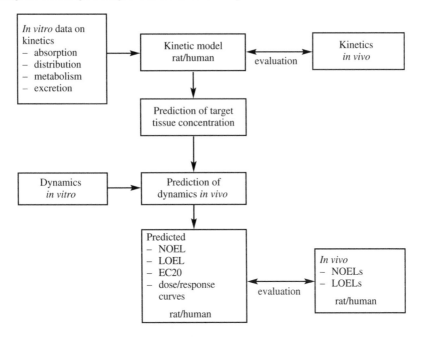

EC20 = concentration of test chemical which causes 20% of the maximum observed response; NOEL = no observed effect level; LOEL = lowest observed effect level.

Figure 23.4   Interrelationships between the building blocks of ECITTS, reproduced with permission from Blaauboer *et al.*[3]

of suggestions on the integrated use of *in chemico*, *in silico* and *in vitro* methods.[4] For example, a breakout group which considered *in vitro* methods for organ-specific toxicity proposed a stepwise scheme which was similar to the earlier ERGATT/CFN and ECVAM schemes, but which also emphasised the importance of the effects of toxicants on the barrier functions of epithelial cells (Figure 23.7).

Other breakout groups considered *in vitro* screening methods for assessing acute toxicity, *in vitro* methods for toxicokinetics (ADME) and the selection of reference chemicals for validation of *in vitro* acute toxicity methods.

One of the first publications on the use of ITS in relation to the REACH system was that of Combes *et al.*,[5] who proposed an overall strategy for the use of existing data and *in silico* and *in vitro* methods in tier-testing schemes (Figure 23.8), taking into account the prediction of human exposure, compound prioritisation, identifying the missing data on toxic hazard needed to improve risk assessment and the obtaining of such missing information, as well as the threshold of regulatory concern concept, read-across, reverse risk assessment and predicting metabolism.

This was followed by a number of other initiatives. For example, the QSAR and Modelling Research Group in the School of Pharmacy and Chemistry,

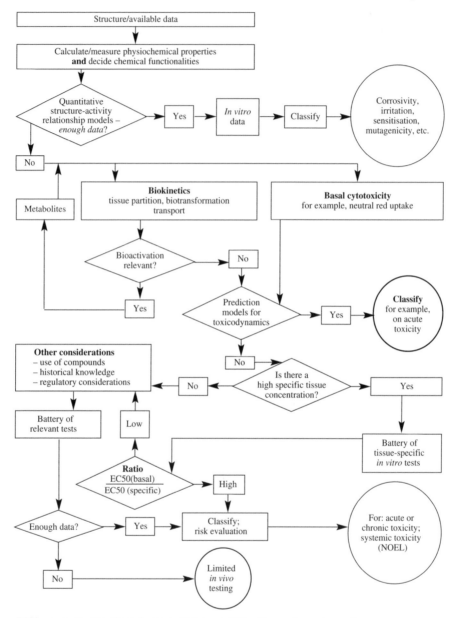

EC50 = concentration of test chemical which causes 50% of the maximum observed response;
NOEL = no observed effect level.

Figure 23.5   A generic scheme for evaluating chemical toxicity, reproduced with permission from Blaauboer *et al.*[3]

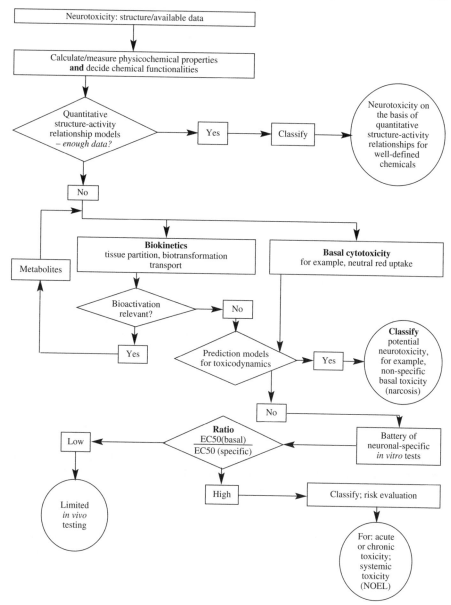

EC50 = concentration of test chemical which causes 50% of the maximum observed response; NOEL = no observed effect level.

Figure 23.6 A specific scheme for evaluating neurotoxicity, reproduced with permission from Blaauboer *et al.*[3]

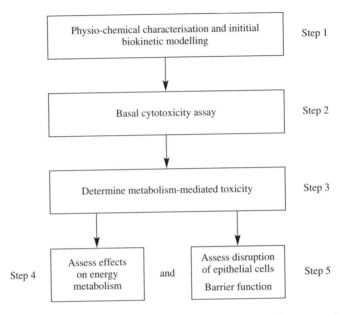

Figure 23.7 A proposed scheme for assessing acute toxicity using non-animal methods, adapted with permission from the ICCVAM workshop report.[4]

Liverpool John Moores University (LJMU), and the Fund for the Replacement of Animals in Medical Experiments (FRAME) undertook a project funded by the UK Department for Environment, Food and Rural Affairs (Defra) which aimed to assess the status of alternatives and to develop ITS for their use in relation to the REACH system.[6] A series of ITS were published in various issues of *ATLA – Alternatives to Laboratory Animals* (later republished together in an ATLA supplement[7]). They dealt with ITS in general and proposed decision-tree strategies with respect to the REACH regulation for environmental toxicity, mutagenicity and carcinogenicity, skin corrosion and irritation, skin sensitisation, acute systemic toxicity and toxicokinetics, eye irritation, developmental and reproductive toxicity (including endocrine disruption), and repeat dose toxicity. The scheme for developmental and reproductive toxicity is shown as an example (Figure 23.9).[8]

23.4 ITS and Risk Assessment

It has long been recognised that:

- the use of hazard predictions derived from non-animal tests and testing schemes requires a different paradigm for risk assessment; and
- the aim of such tests and schemes should be to provide information of direct relevance to human risk rather than to provide conclusions which would otherwise have been obtained in laboratory animal studies.

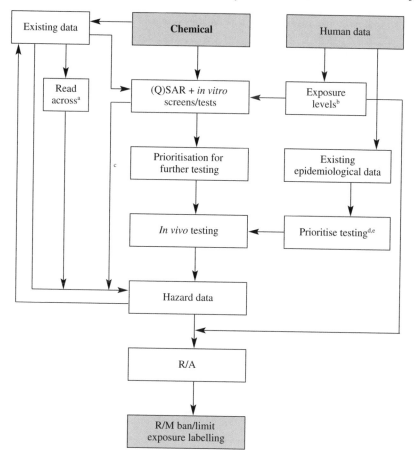

[a]*As performed by the Health and Safety Executive;* [b]*if unavailable, generate before* in vivo *testing;* [c]*in a few cases, where* in vitro *tests have been accepted as replacements;* [d]*focus on subchronic/chronic, if acute human data are available;* [e]*Prioritisation of testing: low priority and high priority are relative phrases, meaning that chemicals of high priority need to be subjected to risk assessment before chemicals of low priority. Other information, including uses and benefits, availability of substitutes, relative toxic potencies and levels of anticipated human exposures, are all factors that should be taken into account when assigning further priorities within these two broad categories.*

Figure 23.8   A scheme for the integrated testing of chemicals, reproduced with permission from Combes *et al.*[5]

A detailed discussion of risk assessment is beyond the scope of this chapter, but the need for new approaches to risk assessment has been discussed in a number of important reports and comments, of which some examples are given.[5,9–12] Combes *et al.*[5] produced an overall decision-tree scheme (Figure 23.10), while Combes *et al.*[9] first considered the use of animal data in relation to the threshold dose concept (*i.e.* the level of exposure below which no toxic effect is

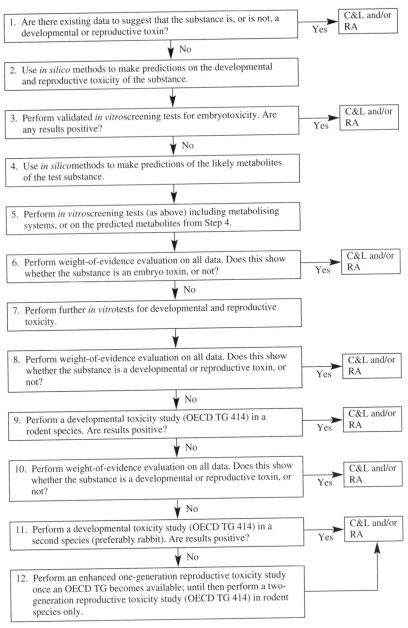

1. Are there existing data to suggest that the substance is, or is not, a developmental or reproductive toxin? — Yes → C&L and/or RA

No ↓

2. Use *in silico* methods to make predictions on the developmental and reproductive toxicity of the substance.

↓

3. Perform validated *in vitro* screening tests for embryotoxicity. Are any results positive? — Yes → C&L and/or RA

No ↓

4. Use *in silico* methods to make predictions of the likely metabolites of the test substance.

↓

5. Perform *in vitro* screening tests (as above) including metabolising systems, or on the predicted metabolites from Step 4.

↓

6. Perform weight-of-evidence evaluation on all data. Does this show whether the substance is an embryo toxin, or not? — Yes → C&L and/or RA

No ↓

7. Perform further *in vitro* tests for developmental and reproductive toxicity.

↓

8. Perform weight-of-evidence evaluation on all data. Does this show whether the substance is a developmental or reproductive toxin, or not? — Yes → C&L and/or RA

No ↓

9. Perform a developmental toxicity study (OECD TG 414) in a rodent species. Are results positive? — Yes → C&L and/or RA

No ↓

10. Perform weight-of-evidence evaluation on all data. Does this show whether the substance is a developmental or reproductive toxin, or not? — Yes → C&L and/or RA

No ↓

11. Perform a developmental toxicity study (OECD TG 414) in a second species (preferably rabbit). Are results positive? — Yes → C&L and/or RA

No ↓

12. Perform an enhanced one-generation reproductive toxicity study once an OECD TG becomes available; until then perform a two-generation reproductive toxicity study (OECD TG 414) in rodent species only.

C&L = classification and labelling; RA = risk assessment.

Figure 23.9 An ITS for developmental and reproductive toxicity, reproduced with permission from Grindon *et al.*[8]

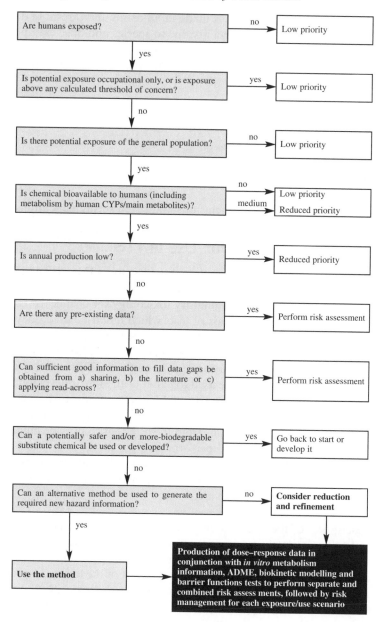

Figure 23.10　An overall decision-tree scheme for chemicals testing, reproduced with permission from Combes *et al.*[5]

expected to occur) and assumptions made in adjusting the threshold dose and comparing it with exposure, uncertainty due to species and dose extrapolation, and the arbitrary nature of adjustment factors.

These authors then turned to the use of data provided by the new technologies including:

- toxicogenomics;
- extrapolation from *in vitro* to *in vivo*;
- adjustment factors for *in vitro* data;
- biokinetic modelling; and
- applying *in vitro* data in quantitative risk assessment.

The report concluded with 13 conclusions and seven recommendations.[9]

23.5   *In vitro* Methods for Use in ITS

In common with many other suggestions for ITS, the LJMU/FRAME/Defra schemes outline the possible structures of ITS appropriate for the main kinds of questions that must be asked in regulatory toxicology for chemicals. However, while they discuss some of the kinds of tests which might be used, they do not indicate precisely what test or tests should be used at each stage of the ITS. Developments in relation to *in silico* approaches are discussed in Chapters 16, 17 and 19, so this section will focus on *in vitro* methods. This is highly relevant since *in silico* and *in vitro* should be used together both in their development and in their application, not only as parts of ITS appropriate to particular circumstances but also because predictions from one approach can be used to complement or guide the conclusions from the other, or to provide guidance on which tests should be selected and performed.

Tests for use in an ITS should be developed according to acceptable criteria which are not dissimilar for *in silico* and *in vitro* methods. In the case of *in vitro* tests, the following should be provided:

- a description of the test including, where possible, its mechanistic basis;
- a statement on the purpose of the test in relation to a specific toxic hazard;
- a defined endpoint and a defined endpoint assay procedure;
- a clear indication of the expression of the result of the test and a prediction model (an unambiguous algorithm) to link it to the toxic effect of interest;
- an indication of appropriate measures for indicating the predictivity of the test;
- an indication of the defined domain of applicability of the test;
- an indication of how the test should (or should not) be used in relation to other tests; and
- evidence concerning the status of the test in the validation process.

The concept of the applicability domain is well-established for *in silico* procedures (see Chapter 12), but has only recently come to be considered in a formal way in relation to *in vitro* tests. Previously, while test developers had been encouraged to indicate the kinds of chemicals or physical forms of chemicals for which their methods were appropriate, any limitations in the breadth of their applicability were often seen as a kind of weakness in comparison with what were believed to be all-purpose animal test procedures.

An enormous amount of effort is currently being invested in the development of many kinds of human-oriented preparations which reflect progress being made at the molecular, cell and tissue/organ/system levels.[13] These include:

- exploitation of the 'omics' (genomics, proteomics, metabolonomics);[5]
- the production of recombinant cell lines containing the key enzymes involved in human drug metabolism;[14]
- the development of dynamic (perfused), integrated organotypic multi-system preparations of various kinds such as the HurelTM microfluidic biochip with integrated tissue/organ compartments[15] and the integrated discrete multiple organ co-culture (IdMOC®) system;[16] and
- reconstituted human tissues (*e.g.* skin, liver lobule and lymph node).[17,18]

The goal is the development of a 'human on a chip'.[13]

At the moment, it is not clear how these kinds of systems will be made available for use in ITS in terms of their manageability, throughput, cost and commercial exploitation. This is one of the great challenges which must be met if the promise of ITS is to be realised.

23.6 Evaluation and Application of ITS

The pressure to develop ITS arises from various sources including:

- the European Commission because of the need to limit the scale of additional animal testing to comply with the REACH regulation,[19,20] and the 7th Amendment to the EU Cosmetics Directive, which calls for a ban on the animal testing of cosmetic ingredients;[21,22]
- the US Food and Drug Administration which has produced its *Critical Path Initiative* and other proposals for tackling the crisis resulting from the failure to market new products and the withdrawal of drugs already on the market because of safety considerations;[23] and
- the US National Academy of Sciences (NAS) which has produced an impressive report on *Toxicity Testing in the 21st Century: A Vision and A Strategy*.[24]

These pressures all have in common the recognition that animal models can no longer be relied upon to provide a satisfactory basis for predicting human hazards or risks, and that more effort should be put into the development, evaluation, acceptance and use of what FRAME would describe as replacement alternative methods and strategies, particularly when they are of direct relevance to humans. The NAS report[24] stated that:

'. . . *a suite of new tests typically will be needed to replace an individual* in vivo *test, given that apical findings can trigger multiple mechanisms . . . although it is current practice to validate a single test . . . the new paradigm would routinely entail validation of test batteries and would use multivariate comparisons.*'

This is all very encouraging but how will the emerging ITS be evaluated, validated and accepted? Surely, this must begin within the commercial industries that produce the chemicals and products of various kinds which are subject to regulations designed to protect workers and consumers, and the environment in general. However, given the likelihood of an increasing number and complexity of the *in vitro* methods which could be used in ITS, it is unlikely that many, or even any, individual company could maintain the breadth of skills and experience, as well as the minimal frequency of performance essential to maintain the high quality of performance on which the application of any ITS would inevitably depend. This will open up opportunities for contract testing laboratories, each of which could offer a selection of the available test procedures. The company with the chemical or product to be tested would design the specific ITS which was appropriate for the particular circumstances, then commission the contractors to do the testing and return the results for stepwise analysis and decision-making by the client company.

If this scenario comes about, it will be the producing companies which first evaluate the ITS and select the schemes and tests which are appropriate for their products in the light of in-house considerations and subsequent regulatory requirements. They are also likely to be involved in the development of the schemes and the tests, as in the case of Unilever, for example, which has established an impressive and sophisticated research programme for this purpose.[8] Within the last five or so years, the company has established more than 100 multi-faceted activities with external partners which have focused primarily on skin allergy, carcinogenicity and genotoxicity, inhalation toxicity and repeated dose systemic toxicity.

The question of the validation of ITS was discussed by Combes[25] in the light of a comment by Hartung[26] that '. . . *at this moment we lack most of the tools to compose and to validate . . . testing strategies. They certainly require more substances to be tested, both to determine the proper composition of the tests and to validate the strategy. The main question . . . is whether we have to and can validate individual tests separately or only the overall strategy.*'

Combes felt that neither the statement in the NAS report[24] nor Hartung's[26] represented the then current position, and proposed the following criteria for validating test batteries and ITS:

(1) Each of the component methods must previously have been individually validated in a study that satisfies the internationally-accepted criteria for validation.

(2) A clear purpose and applicability domain for the test battery or ITS should be identified.

(3) The purpose and applicability domain should reflect the purposes and applicability domains for all the component tests in the test battery or ITS.

(4) The test battery or ITS should be validated against a set of chemicals which cover the whole of the applicability domain, so as to allow validation for its intended purpose to be properly assessed.

(5) The set of chemicals used for validation should comprise a subset of each of the sets of chemicals that were used in the validation of all the component tests in the test battery or ITS, and which are representative of the applicability domains of the individual methods.

(6) The set of chemicals should be chosen so as to allow the performance of a tiered decision-tree.

ITS should be assessed from toxicity predictions based on data obtained at each decision point (*e.g.* chemicals with well-defined hazard labelling and classification, and chemicals with well-defined No Observable Adverse Effect Levels for quantitative risk assessment).

If the requirements for validation are too demanding, they may set impractical hurdles which are too high to be cleared. Nevertheless, it is clearly essential that tests—and the schemes which they use—are independently evaluated for their reliability and relevance for their stated purposes before they are considered acceptable for use in compliance with regulatory requirements. There are a number of serious dangers here:

(1) The demands of the validation proposed could be weakened in the face of the high costs—in terms of human and financial resources and animal lives—when the politicians and regulators come face-to-face with the implication of the regulations they have put in place. This particularly applies to the REACH legislation where various ways of waiving the need for additional testing are being introduced and there is talk of the acceptance of 'suitable' tests (*i.e.* tests that satisfy the criteria for test development but which have not been independently validated). The words used among the 248 pages of Regulation No. 1907/2006 of the European Parliament and of the Council of 18 December 2006[27] were as follows:

'*Results obtained from suitable* in vitro *methods may indicate the presence of a certain dangerous property or may be important in relation to a mechanistic understanding, which may be important for the assessment. In this context, "suitable" means sufficiently well developed according to internationally agreed test development criteria (e.g. the ECVAM criteria for entry into the pre-validation process).*'

But how can a test method be accepted as 'suitable' for use in compliance with the REACH system if it has not been independently shown to be relevant and reliable for its stated purpose (*i.e.* to have been validated)?

(2) As in the past, the regulatory authorities may be unwilling to permit the replacement of the conventional animal procedures with which they feel comfortable, especially when no procedures are in place for the removal of guidelines from their test guideline programmes, as FRAME discovered found when it proposed amendments to the OECD Human Health Effects Test Guidelines.[28,29] Hence, whatever may be said at one level about the need for alternative tests and ITS in relation to the REACH system, the regulatory authorities appear to remain entrenched in their demands for results according to the current animal test guidelines.

(3) One of the biggest problems in validating tests and evaluating ITS is obtaining appropriate sets of reference chemicals. Animal test data are not suitable for use in evaluating the performance of *in silico* and *in vitro* tests aimed at predicting toxic hazard in humans, since their ability to predict effects in humans is limited and usually unknown. For these reasons, it can be argued that the currently accepted tests cannot be used to identify human carcinogens or human reproductive toxins. In addition, animal test data can also be so highly variable as to be an inadequate basis for any kind of comparison, as in the case of the Draize rabbit eye irritation test.[30] Against this background, it is obvious why one of the most successful studies conducted to date has been the international validation study on the 3T3 neutral red uptake phototoxicity test, where independently selected human data were used as the basis for *in vivo/in vitro* comparisons.[31]

23.7   Securing the Regulatory Acceptance of ITS

A number of points are made below with respect to how best to secure the acceptance of ITS by regulatory bodies.

(1) If a rigid validation process were applied too widely, the whole situation could become so complicated as to be unmanageable and it would be impossible to achieve the acceptance of ITS for any purpose. The first essential step is to recognise the consequences of failing to develop satisfactory procedures for the evaluation and acceptance of ITS, and the second step is to overcome the problems which can be foreseen. One suggestion is as follows: There should be a rigorous evaluation of what candidate tests or ITS could be considered worth the effort—after objective comparison with their rivals and without undue influence from those with vested interests. This should primarily be the responsibility of those who need to use the ITS (*i.e.* the companies producing the materials to be tested) in collaboration with those capable of developing the tests and the schemes.

(2) Validation does not always require practical multi-laboratory studies as there are circumstances where a weight-of-evidence approach is acceptable and preferable.[32] Such an approach would be advisable in the case of ITS. Rather than having regulatory bodies prescribe precisely what needs to be done in all possible circumstances (which leads to thousands of pages of prescription and guidance, as in the case of the REACH

system,[33] in addition to the hundreds of pages in the Regulation itself[27]), it should be left to the producing companies to submit their proposals. These should have clearly outlined strategies and strong scientific justification, backed by documented experience and in consultation and/or collaboration with a validation authority such as ECVAM, on what would be sufficient, in particular circumstances, given the nature of the chemical or product, its projected use and the likely human exposure.

(3) This would mean a move away from the generalised 'one-system-fits-all' and 'one-guideline fits-all' checklist approach currently favoured by those responsible for the regulation of chemicals and certain kinds of chemical products toward the specific case-by-case justification which has long applied to pharmaceuticals.

(4) The company seeking to conform with the regulatory requirements in order to get a chemical or products registered and authorised would be expected to produce a dossier in which the following would be considered:
   • the nature and projected use of the test item, with an evaluation of likely human exposure;
   • an outline of the ITS employed;
   • selection of the tests to be used (from validated and accepted sets), their applicability domains, the relationships between them, their relative weighting, the order of their performance, the result provided, its expression and its relationship to the toxic hazard of concern;
   • the decision points to be used and the decisions to be reached; and
   • how the outcome would be expressed in terms of the probability of the reliability of its predictions in particular circumstances.

The dossier could also reflect the specific experience of those performing the tests or applying the ITS, especially when tests of equal scientific merit were available.

(5) The type of overall decision to be made is a crucial aspect of the application of any ITS in relation to in-house and regulatory requirements. Hierarchical sets of decision levels could be foreseen such as:
   (a) an in-house decision whether or not to further develop a chemical or product;
   (b) labelling a chemical or product as likely lead to a particular kind of toxicity under certain circumstances;
   (c) concluding that a chemical or product would not represent a toxic hazard in terms of specific kinds of toxicity, exposure scenarios and exposure levels.

Level (a) is already being used by companies, for example, by using *in silico* methods alone; *in silico* methods may also be appropriate for level (b)—at least in some circumstances. However, the goal must be to establish the use of *in silico* and *in vitro* methods in intelligent ITS so that decisions at level (c) could be achieved without the need for data from animal tests (which, by the way, given the problem of species differences, might unduly complicate and frustrate the use of sophisticated systems focused on human toxicology and pathophysiology).

Approaches for assigning MoA and/or MeoA	Examples	Applications
Chemical and/or structural features	Structural alerts	Structural alerts for mutagenicity, carcinogenicity, skin sensitisation, etc. Profilers in OECD (Q)SAR Application Toolbox, Toxmatch, etc.
	Fragment/residual analysis	Identification of structural fragments associated with excess toxicity above narcosis for ecotoxicity, as well as structural features associated with narcosis
	Physicochemical property profiles	Partitioning, molecular size, polarity, hydrogen bonding, etc. associated with nine MoAs
(Q)SARs incorporating physicochemical properties and structural (geometrical) features	Direct prediction of mechanism from a (Q)SAR	Predictions from a (Q)SAR of MeoA for acute aquatic toxicity
	3-D-QSAR/CoMFA models for activity related to binding to receptors	Identification of compounds able to bind to certain receptors e.g. the oestrogen receptor
	Structural (geometrical) properties	Identification of compounds able to bind to particular receptors, e.g. planarity for activity related to ArH receptor
Molecular, chemical and/or biological responses, such as receptor binding and mutagenicity — possibly making use of gene up-regulation and down-regulation results from techniques such as toxicogenomics	Observations from standard toxicity tests	Biochemical etc. responses indicating mechanisms of action, e.g. Fish Acute Toxicity Syndromes
	Observations from other toxicity tests, e.g. mutagenicity tests	DNA binding mechanisms
	In vitro tests, including receptor binding etc. and appropriate use and application of batteries of such tests	Identification of modes and mechanisms of acute aquatic toxicity
	Peptide reactivity tests for electrophilicity	Glutathione depletion tests to provide evidence of skin sensitisation
	Tests for nucleophilicity	Prediction of reactivity (nucleophilicity)
	Tests for oxygen stress	Evidence of free-radical formation or oxidative stress
	Molecular techniques (e.g. microarrays and application of toxicogenomics)	Direct -omic evidence and fingerprints of mechanisms of action
Integrative/apical biological response such as lethality, etc.	Acute/chronic ratio in *in vitro* toxicity	Acute/chronic ratio in ecotoxicity may guide whether the chemical acts by a narcotic mechanism
Physicochemical properties and bioavailability	Properties	e.g. volatility relating to inhalation effects, solubility relating to aquatic toxicity
	Ability to be bioavailable	i.e. abiotic and biotic stability, transport and distribution properties

Figure 23.11 Methods to establish modes and mechanisms of toxic action, reproduced with permission from Vonk *et al.*[34]

(6) The only way of solving many of the current problems, which will become available in the medium term (not, let us hope, only in the longer term) is to have sufficient understanding of the mechanisms of action and modes of action involved in humans so that they can be effectively represented in both tests and ITS. This was the focus of an important workshop held in 2008 as part of the EU OSIRIS Integrated Project, which included a summary of methods for establishing modes and mechanisms of toxic action, in what emerged as a kind of ITS (Figure 23.11).[34] This also provides a good example of the challenging variety of methods that are becoming available. The aim must be to escape from alerts and correlations by coming to know much more about what really goes on when chemicals meet sub-cellular molecules and structures, cells and tissues.

(7) Finally, it must be emphasised that no test or ITS should be seen as carved in stone (a criticism which can be levelled at the current animal test procedures). Procedures of any kind become more or less acceptable and useful as experience is gained with them. They will inevitably be overtaken at some point as new procedures come on stream based on deeper levels of understanding and better methods of analysis. That, above all, is the main reason why it is the specific dossier—produced by those responsible for the manufacture and/or marketing of the chemical or product, who are ultimately responsible for ensuring that it is acceptably safe for use—on which the full exploitation of modern science and escape from flawed animal models will depend.

Meanwhile, many experts and stakeholders will wrestle with the REACH regulation and the guidance offered by the European Commission and the European Chemicals Agency in the hope that there will be opportunities for rational ways forward. The Institute for Health and Consumer Protection at the European Commission's Joint Research Centre will make new contributions on alternative methods and ITS,[35] and the discussion will go on, as exemplified by a thoughtful RAINBOW Workshop report, *Integrating* in vivo, in vitro *and* in silico *methods for chemical assessment: problems and prospects*, due to be published in *ATLA –Alternatives to Laboratory Animals* in 2010.[36]

References

1. E. Walum, M. Balls, V. Bianchi, B. Blaauboer, G. Bolcsfoldi, A. Guillouzo, G. A. Moore, L. Odland, C. Reinhardt and H. Spielmann, *Altern. Lab. Animal*, 1992, **20**, 406.

2. H. Seibert, M. Balls, J. H. Fentem, V. Bianchi, R. H. Clothier, P. J. Dierickx, B. Ekwall, M. J. Garle, M. J. Gómez-Lechón, L. Gribaldo, M. Gülden, M. Liebsch, E. Rasmussen, R. Roguet, R. Shrivastava and E. Walum, *Altern. Lab. Anim.*, 1996, **24**, 499.

3. B. J. Blaauboer, M. D. Barratt and J. B. Houston, *Altern. Lab. Anim.*, 1999, **27**, 229.

4. ICCVAM, *Report of the International Workshop on In vitro Methods for Assessing Acute Systemic Toxicity*, National Institutes of Health, Washington DC, 2001, NIH Publication No. 01-4499.

5. R. Combes, M. Barratt and M. Balls, *Altern. Lab. Anim.*, 2003, **31**, 7.

6. C. Grindon, R. Combes, M. T. D. Cronin, D. W. Roberts and J. F. Garrod, *Altern. Lab. Anim.*, 2006, **34**, 407.

7. *Altern. Lab. Anim.*, 2008, 36 Suppl. 1, 1–147.

8. C. Grindon, R. Combes, M. T. D. Cronin, D. W. Roberts and J. F. Garrod, *Altern. Lab. Anim.*, 2008, **36**, 65.

9. R. Combes, M. Balls, P. Illing, N. Bhogal, J. Dales, G. Duvé, V. Feron, C. Grindon, M. Gülden, G. Loizou, R. Priston and C. Westmoreland, *Altern. Lab. Anim.*, 2006, **34**, 621.

10. Health Council of The Netherlands, *Toxicity Testing: A More Efficient Approach*, The Health Council of The Netherlands (Gezondheidsraad), The Hague, 2001, U 2260/ES/mj/442-B3.

11. B. J. Blaauboer and M. E. Andersen, *Arch. Toxicol.*, 2007, **81**, 385.

12. B. J. Blaauboer, *Toxicol. Lett.*, 2008, **180**, 81.

13. P. Carmichael, M. Davies, M. Dent, J. Fentem, S. Fletcher, N. Gilmour, C. MacKay, G. Maxwell, L. Merolla, C. Pease, F. Reynolds and C. Westmoreland, *Altern. Lab. Anim.*, 2009, **37**, 595.

14. J. Doehmer, *Altern. Lab. Anim.*, 2009, **37**(Suppl. 1), 29.

15. G. T. Baxter, *Altern. Lab. Anim.*, 2009, **37**(Suppl. 1), 11.

16. A. Li, *Altern. Lab. Anim.*, 2009, **37**, 377.

17. C. Giese, C. Demmler, R. Ammer, S. Hartmann, A. Lubitz, L. Miller, T. Mueller and U. Marx, *Artif. Organs*, 2006, **30**, 803.

18. U. Marx, in *In vitro Drug Testing – Breakthroughs and Trends in Cell Culture Technologies*, ed. U. Marx and V. Sandig, 2007, Wiley-VCH, Weinheim, p. 271.

19. European Chemical Agency, *Guidance in a Nutshell. Chemical Safety Assessment*, EChA, Helsinki, Finland, 2009, http://guidance.echa.europa.eu/docs/guidance_document/nutshell_guidance_csa_en.pdf [accessed March 2010].

20. C. Grindon and R. Combes, *Altern. Lab. Anim.*, 2008, **36**(Suppl. 1), 1.

21. Directive 2003/15/EC of the European Parliament and of the Council of 27 February 2003 amending Council Directive 76/768/EEC on the approximation of the laws of the Member States relating to cosmetic products, *Off. J. Eur. Union*, 2003, **L66**, 26–35.

22. COLIPA, *Working Together to Replace Animal Testing*, COLIPA, Brussels, 2009, www.thefactsabout.co.uk/files/10320091546293744_Colipa_BrochureJaber-Def-5.pdf [accessed March 2010].

23. US Food and Drug Administration, *Challenge and Opportunity on the Critical Path to New Medicinal Products*, US Department of Health and Human Services, Food and Drug Administration, Silver Spring, MD, 2004.

24. National Research Council, *Toxicity Testing in the 21st Century: A Vision and A Strategy*, National Academy Press, Washington DC, 2007.

25. R. Combes, *Altern. Lab. Anim.*, 2007, **35**, 375.

26. T. Hartung, *ALTEX*, 2007, **24**, 67.

27. Corrigendum to Regulation (EC) No 1907/2006 of the European Parliament and of the Council of 18 December 2006 concerning the Registration, Evaluation, Authorisation and Restriction of Chemicals (REACH), establishing a European Chemicals Agency, amending Directive 1999/45/EC and repealing Council Regulation (EEC) No 793/93 and Commission Regulation (EC) No 1488/94 as well as Council Directive 76/769/EEC and Commission Directives 91/155/EEC, 93/67/EEC, 93/105/EC and 2000/21/EC (Official Journal of the European Union L396 of 30 December 2006), *Off. J. Eur. Union*, 2007, **L136**, 3–280.

28. R. D. Combes, I. Gaunt and M. Balls, *Altern. Lab. Anim.*, 2004, **32**, 168.

29. M. Balls and R. D. Combes, *Altern. Lab. Anim.*, 2006, **34**, 105.

30. M. Balls, P. A. Botham, L. H. Bruner and H. Spielmann, *Toxicol. In vitro*, 1995, **9**, 871.

31. H. Spielmann, M. Balls, J. Dupuis, W. J. W. Pape, G. Pechovitch, O. de Silva, H. G. Holzhütter, R. Clothier, P. Desolle, F. Gerberick, M. Liebsch, W. W. Lovell, T. Maurer, U. Pfannenbecker, J. M. Potthast, M. Csato, D. Sladowski, W. Steiling and P. Brantom, *Toxicol. In vitro*, 1998, **12**, 305.

32. M. Balls, P. Amcoff, S. Bremer, S. Casati, S. Coecke, R. Clothier, R. Combes, R. Corvi, R. Curren, C. Eskes, J. Fentem, L. Gribaldo, M. Halder, T. Hartung, S. Hoffmann, L. Schechtman, L. Scott, H. Spielmann, W. Stokes, R. Tice, D. Wagner and V. Zuang, *Altern. Lab. Anim.*, 2006, **34**, 608.

33. European Chemical Agency, *Guidance on Information Requirements and Chemical Safety Assessment*, EChA, Helsinki, Finland, 2009, http://guidance.echa.europa.eu/docs/guidance_document/information_requirements_en.htm [accessed March 2010].

34. J. A. Vonk, R. Benigni, M. Hewitt, M. Nendza, H. Segner, D. van de Meent and M. T. D. Cronin, *Altern. Lab. Anim.*, 2009, **37**, 557.

35. Anon., 'EPAA Annual Conference 2008',European Commission Joint Research Centre News & Events [online], 3 November 2008, http://ec.europa.eu/dgs/jrc/index.cfm?id = 1410&obj_id = 2110&dt_code = EVN &lang = en [accessed March 2010].

36. E. Benfenati, G. Gini, S. Hoffmann and R. Luttik, *Altern. Lab. Anim.*, 2010, **38**, 153.

CHAPTER 24

Using In Silico *Toxicity Predictions: Case Studies for Skin Sensitisation*

M. T. D. CRONIN AND J. C. MADDEN

School of Pharmacy and Chemistry, Liverpool John Moores University, Byrom Street, Liverpool L3 3AF, UK

24.1 Introduction

The intention of Chapters 1–23 is to lead the reader through the methods used to make *in silico* predictions of toxicity. The chapters variously describe sources of toxicity data, methods to calculate descriptors, statistical approaches, techniques to group molecules into categories and a number of commercial and freely available software packages to assist in estimating toxicity. What is lacking with these chapters, and often in the literature and guidance on *in silico* techniques, is an illustration of how the predictions may be used to assess the toxicity of a compound. The aim of this chapter, therefore, is to show what information could be obtained from *in silico* methods and how this can be interpreted in an informal weight-of-evidence approach.

The approach taken in this chapter is to assess two chemicals for skin sensitisation using a variety of *in silico* methods. Skin sensitisation was chosen as the endpoint for a number of reasons. It is of considerable relevance in terms of the REACH regulation and the Cosmetics Directive. There is considerable

Issues in Toxicology No.7
In Silico Toxicology: Principles and Applications
Edited by Mark T. D. Cronin and Judith C. Madden
© The Royal Society of Chemistry 2010
Published by the Royal Society of Chemistry, www.rsc.org

mechanistic understanding of this endpoint.[1] The predictive methodologies are supported (indirectly for the two chosen compounds) by *in chemico* reactivity data.[2,3] Lastly, and most importantly for this chapter, there are a number of *in silico* predictive methods for skin sensitisation based on a variety of methods [structure–activity relationships (SAR), quantitative structure–activity relationships (QSAR), chemical grouping, *etc.*][4] which are supported by experimental data that have not been included in the training sets for these models.[5,6]

24.2 Forming a Consensus: Integrating Predictions

For successful use of *in silico* techniques, it must be remembered that a prediction is merely a further piece of information regarding a compound. An individual predicted value should not be thought of as a direct replacement for an *in vivo* test unless there is great confidence in that prediction. The concept of forming a consensus from several different prediction methods is not new.[7–14] It should be noted that use of the term 'consensus' in this chapter indicates the application of predictions from different techniques. Other workers have used the term 'consensus-QSAR' to indicate creating an ensemble of different models using the same technique, on the same data set, by selecting different descriptors.[7–9]

There are some interesting factors regarding using different methods to make predictions. For a start (as illustrated in this study), the models are often based on different training sets. This will often mean that the models contain different information, cover different areas of chemical space and hence will be associated with individual applicability domains (see Chapter 12 for a further discussion of definition of a domain). The last point may mean that in a consensus approach, the compound may be in the domain of some models and out of the domain of others. Thus flexibility may be required if formal methods (*e.g.* Bayesian approaches see Chapter 22) are used to combine together predictions.[15]

There are a number of reports of using different approaches to form a consensus from QSAR models. As noted above, a compilation of *in silico* predictions can be made from the same, or similar, modelling techniques. Typically these are regression-based QSARs where different groups of descriptors may have been selected from a large pool of variables by a genetic algorithm.[7–10] More relevant to this chapter are attempts to compile and form a consensus of *in silico* predictions from the different techniques (typically expert systems).[11–13] This approach may have significant advantages for predicting the hazard associated with human health effects since the different methods may capture various mechanistic information. A number of approaches to weight or average the predictions have been proposed for toxicity prediction.[14] Lastly, it should not be forgotten that the Integrated Testing Strategies (ITS) introduced in Chapter 23 are a more sophisticated form of consensus model that integrate *in silico* predictions with *in vitro* and *in chemico* data.

24.3.   Choice of Chemicals for Analysis

Two chemicals were chosen for analysis in this study:

- 4-amino-2-nitrophenol (CAS Number 119-34-6); and
- 1,14-tetradecanediol (CAS Number 19812-64-7).

The chemical structures of the two test compounds are shown in Figure 24.1.

These chemicals were chosen for a number of reasons, some of which are illustrative and would not necessarily occur in a real predictive study (*i.e.* availability of test data). Key among the choice of these compounds to illustrate the methods is that *in vivo* skin sensitisation data are available from the literature—namely Roberts *et al.* reported Local Lymph Node Assay (LLNA) data for these compounds.[6] The LLNA data were taken from a study that reported the predictions for 40 compounds for which LLNA data were measured in order to assess the predictive performance of the TIMES-SS expert system.[5,6] Ellison *et al.* have also utilised the whole of this test set to develop a weight-of-evidence approach; from this we have considered only two compounds, albeit in greater detail.[5] The two test compounds were not in the original training sets of the model, making the predictions more realistic of a real-life scenario.

The LLNA data show that 4-amino-2-nitrophenol is a skin sensitiser whilst 1,14-tetradecanediol was a non-sensitiser. In terms of their chemistry, both are simple organic compounds but from different chemical classes. From a review of the data reported below, it would appear that 4-amino-2-nitrophenol is more commonly used industrially (possibly as a hair dye precursor). It is important to note that the data set reported by Roberts *et al.* was selected rationally to challenge the TIMES-SS,[6] so structural space was the consideration rather than choosing industrially important chemicals.

24.4.   *In Silico* Prediction Methods and *In Chemico* Data

24.4.1   *In Silico* Tools

In an Integrated Testing Strategy (see Chapter 23), it is usual to search for existing data before moving onto further methods. In this study, existing data relating to skin sensitisation were sought from:

- OECD (Q)SAR Application Toolbox (see Chapter 16 for further information);
- OECD eChemPortal (see Chapter 3 for further information); and
- US Environmental Protection Agency (US EPA) ACToR toxicity data resources (see Chapter 3 for further information)

Following the search of existing data, predictions were made (or retrieved from the literature) from a number of (Q)SARs, categories and expert systems. It is

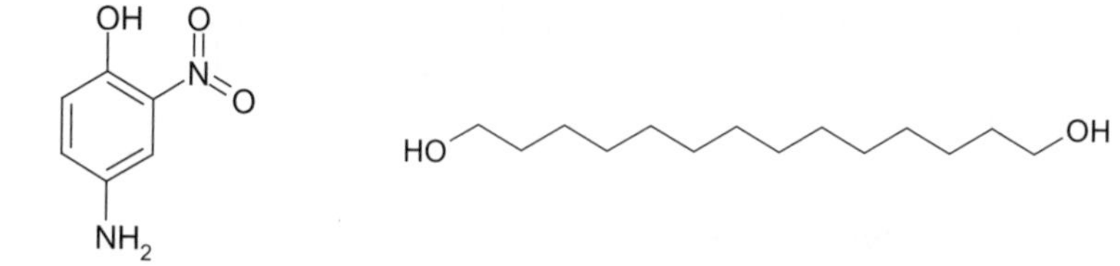

4-amino-2-nitrophenol
119-34-6

1,14-tetradecanediol
19812-64-7

Figure 24.1   Names, CAS numbers and structures of the two compounds considered in the exercise described in this chapter.

Table 24.1   Summary of the (Q)SAR, category formation and expert systems approaches used to predict the skin sensitisation potential of the test compounds, along with an indication of where to find further information.

In Silico *method*	*Brief description*	*Sources of further information*
SMARTS strings for reactivity	Identification of fragments associated with known mechanisms of electrophilic reactivity.	Chapter 7 Chapter 13 Enoch *et al.*[16]
TIMES-SS	A hydrid QSAR system combining rules for protein reactivity with prediction of potential skin metabolites. Please note that the predictions reported here were taken from Roberts *et al.*[6]	Chapter 19, Roberts *et al.*[6]
Derek for Windows (now renamed Derek Nexus)	Identification of structural alerts known to be associated with skin sensitisation. Please note that the predictions reported here were taken from Ellison *et al.*[5]	Chapter 19
OECD (Q)SAR Application Toolbox	Profiles chemicals (in this study on the basis of protein reactivity) in an attempt to form categories of molecules.	Chapter 16
CAESAR model for skin sensitisation	Performs a QSAR prediction on the basis of the 'Adaptive Fuzzy Partition (AFP)' technique using eight 2-D descriptors from the Dragon software. (AFP is a form of discriminant analysis that predicts so called 'fuzzy' group membership.)	Chapter 10, Chapter 19

not the purpose of this chapter to describe these systems in detail, so reference is made to other information sources. The identity and brief description of the models utilised and the reference sources are given in Table 24.1.[16]

As noted in Chapter 19, there are many other methods to predict toxicity; other expert systems that could have been utilised include TOPKAT and

M-CASE (see Chapter 19; Section 19.2). These systems were not used here due to commercial considerations, but there is no reason why they (and any other suitable approaches) could not be utilised elsewhere.

24.4.2   *In Chemico* Reactivity Data

Another approach to consider was not *in silico*, but the use of physico-chemical reactivity data. *In chemico* data are determined experimentally and relate the intrinsic reactivity of a compound to one or more amino acids or proteins. In terms of soft nucleophiles, the most frequently applied assays have utilised glutathione (GSH), cysteine and a mix of amino acids and peptides. Two assays have been described by Gerberick and co-workers[2,17,18] and Schultz and co-workers[19–22] for a mixture of three synthetic peptides or GSH respectively. A further suite of *in chemico* assays has been developed by Natsch and co-workers.[23–25] The different approaches were reviewed recently by Cronin *et al.*[3] It must be stressed that *in chemico* information is not a direct one-to-one replacement for skin sensitisation testing. It merely provides evidence that a compound is able to bind to proteins. Therefore, any knowledge gained from this approach should be treated in the same manner as for the SMARTS string, or protein binding profiler in the OECD (Q)SAR Application Toolbox.

Until recently, *in chemico* reactivity data were spread across the scientific literature going back as far as the 1930s. A compilation of *in chemico* data was published in 2010 by Schwöbel *et al.*[26] In addition, further information is available from the website for a recent project in this area.[27] Both these sources were searched for suitable data, not only for the test compounds themselves but for suitable structural and mechanistic analogues.

24.5   Existing Data, *In Silico* Predictions and *In Chemico* Data for 4-Amino-2-Nitrophenol

The existing data and *in silico* predictions for skin sensitisation for 4-amino-2-nitrophenol are summarised in Table 24.2. Individual predictions are discussed and commented upon below.

24.5.1   Existing Toxicological Data and Information Relating to the Skin Sensitisation Potential of 4-Amino-2-Nitrophenol

No skin sensitisation data for 4-amino-2-nitrophenol were found in the OECD (Q)SAR Application Toolbox. However, it should be noted that the compilation of data in the OECD Toolbox (version 1.1) was created before the publication of Roberts *et al.* and newer versions will be updated.[6]

Table 24.2   Summary of existing data and *in silico* predictions for 4-amino-2-nitrophenol.

In Silico *method*	*Prediction*
Existing data and information	No skin sensitisation test data
SMARTS strings for reactivity	Identified as a pro-Michael acceptor.
TIMES-SS	Positive (reported by Roberts *et al.*[6])
Derek for Windows	Two alerts identified. This led to the conclusion of this compounds being a plausible human skin sensitisation (reported by Ellison *et al.*[5]).
OECD (Q)SAR Application Toolbox	No protein binding – a category has not been formed.
CAESAR model for skin sensitisation	Active

The OECD eChemPortal provided access to the National Library of Medicine's Hazardous Substance Data Bank (HSDB), which contained a report of dermal exposure to 4-amino-2-nitrophenol but no mention of contact allergy. No further information was retrieved from ACToR.

24.5.2   SMARTS Strings for Reactivity

The compound is identified as belonging to the mechanism of action related to being a Michael acceptor. This is defined in more detail in the original publication[16] as well as in Chapters 7 and 13; the structural characteristics are shown in Figure 13.3. The fact it is identified as a 'pro-Michael acceptor' indicates that it requires transformation to the Michael acceptor. This may be due to abiotic transformation or metabolic transformation to the corresponding quinone—see the discussion below regarding the OECD Toolbox (Section 24.5.5). If it is accepted that this compound could be transformed to the quinone, then this alert is appropriate and it is highly probable that the compound (or more precisely its metabolite) will have the capability to bind to proteins. It should be remembered that the ability to bind to a protein is not a direct prediction of skin sensitisation. However, this can be thought of as the molecular initiating event (MIE) as defined in Chapter 14 (particularly in reference to Sections 14.4 and 14.8). Therefore it indicates a strong likelihood that this compound could be a skin sensitiser.

24.5.3   TIMES-SS

Roberts *et al.* reported that the TIMES-SS software predicted 4-amino-2-nitrophenol to be a skin sensitser.[6] This is stated as being due to its ability to act as pro-Michael acceptor. This is useful supporting evidence to the findings from the SMARTS strings.

24.5.4   Derek for Windows

The compound fired two alerts in Derek for Windows for skin sensitisation.[5] These were:

- aromatic primary or secondary amine; and
- phenol or precursor.

The supporting data in Derek for Windows (not shown) appear to indicate that either alert could be feasible and may lead to skin sensitisation. It is important to note that, as with the SMARTS strings, the presence of an alert does not automatically mean the compound is a skin sensitiser. However, there appears to be little confounding evidence from this system and the compound is not shown to have a skin permeability profile that may preclude sensitisation.

It is interesting to compare the information of SMARTS and Derek for Windows alerts. The SMARTS alerts provide information at a mechanistic level (in terms of organic chemistry). This is related to general phenomena associated with protein binding, so could be related to a number of endpoints such as respiratory sensitisation, mutagenicity and elevated acute aquatic toxicity.[28] It forms an excellent method to profile the compound and develop a category, if required. The Derek for Windows alerts are endpoint-specific and based on a single chemical fragment. This relates them more to mechanistic biology. There is no particular disadvantage to either approach for predicting skin sensitisation, but the user of these methods should be aware of the intrinsic differences.

24.5.5   OECD (Q)SAR Application Toolbox (OECD Toolbox)

The compound was entered into the OECD Toolbox (Ver 1.1). It was profiled using the protein binding profiler (which contains similar mechanistic chemistry to the SMARTS strings and TIMES-SS); 4-amino-2-nitrophenol did not contain any alerts for protein binding. Subsequently, the skin metabolism simulator was run on the molecule. This identified two potential metabolites, which are shown in Figure 24.2. The metabolites were profiled using the protein binding profiler. Both

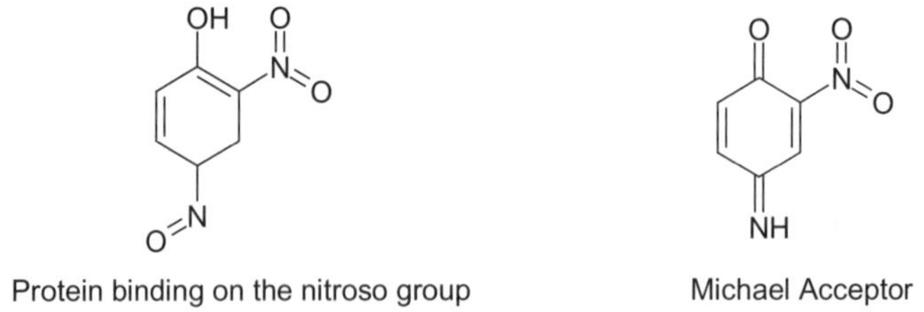

Protein binding on the nitroso group   Michael Acceptor

Figure 24.2   The two metabolites of 4-amino-2-nitrophenol predicted by the OECD (Q)SAR Application Toolbox and the possible mechanisms of protein binding.

metabolites were associated with the Michael acceptor domain, with the structural fragments responsible for nucleophilic substitution noted in Figure 24.2.

The fact that metabolites were found to be reactive has several important implications. First, it demonstrates the importance of considering metabolism with regard to toxicity prediction (see the discussion in Sections 12.3 and 21.2.4). The second is how to utilise the information regarding metabolites. In this context it is straightforward as there are only two predicted metabolites. For more structurally complex molecules the number of metabolites can be far greater—thus the model user must make a decision as to how far this should be taken. Lastly, this confirms that the information provided by the SMARTS strings and TIMES-SS implicitly includes metabolic information and that it is important for this compound. The reader should note that there is some duplication of information between the OECD Toolbox and TIMES-SS.

24.5.6   CAESAR Model for Skin Sensitisation

The CAESAR model for skin sensitisation predicted 4-amino-2-nitrophenol as being a skin sensitiser. The basis of this prediction is on a QSAR for 'fuzzy classes', *i.e.* where there is no clear boundary. The prediction provides 'class indices' which may be interpreted in terms of them being the likelihood of class membership (see Chapter 10, Section 10.2.3.3). The indices were: active = 0.902; inactive = 0.098. Thus, there is a greater than 90% likelihood of 4-amino-2-nitrophenol being a skin sensitiser. There is (at the time of performing this prediction) no formal description of the CAESAR model other than stating the statistical method and descriptors utilised. However, in addition to the prediction, the CAESAR output does provide the six nearest neighbours, as determined by k-Nearest Neighbours (see Chapter 10; Section 10.4) using the molecular descriptor set. The six nearest neighbours determined by the CAESAR model, together with their similarity to 4-amino-2-nitrophenol, are shown in Figure 24.3.

Whilst the nearest neighbours to 4-amino-2-nitrophenol shown in Figure 24.3 were not selected on a mechanistic basis, they give considerable confidence to the prediction. They indicate that the model is able to perform similarity searching intuitively. All six similar compounds are defined as skin sensitisers. A skin sensitiser in the CAESAR model is defined as producing a significant effect (*i.e.* EC3) in the Local Lymph Node Assay. Non-sensitisers elicit no sensitisation response as defined by Gerberick *et al.*[29] Some compounds very relevant chemically to 4-amino-2-nitrophenol are identified as its nearest neighbours. For instance, structures **1**, **2**, **3** and **6** in Figure 24.3 have amino and/or hydroxy groups in a 1,2 or 1,4 configuration. These compounds fit into the pro-Michael acceptor mechanistic domain defined by Roberts *et al.*[6] and Enoch *et al.*[16] In addition, structures **1** and **2** in Figure 24.3 also include a nitro substituent. There is no mechanistic reason why the nitro group should reduce skin sensitisation potential and it is useful confirmatory information that the nitro-substituted compounds are skin sensitisers. Thus, whilst detailed documentation

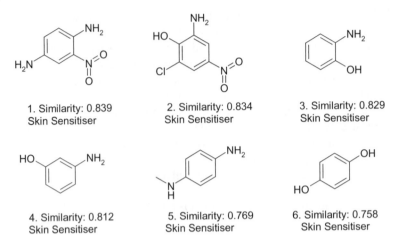

1. Similarity: 0.839
Skin Sensitiser

2. Similarity: 0.834
Skin Sensitiser

3. Similarity: 0.829
Skin Sensitiser

4. Similarity: 0.812
Skin Sensitiser

5. Similarity: 0.769
Skin Sensitiser

6. Similarity: 0.758
Skin Sensitiser

Figure 24.3   The six nearest neighbours to 4-amino-2-nitrophenol determined by the CAESAR model for skin sensitisation together with their similarity to 4-amino-2-nitrophenol and skin sensitisation activity.

is lacking regarding the CAESAR model, there is good confirmatory evidence to support the prediction that 4-amino-2-nitrophenol is a skin sensitiser.

24.5.7   *In Chemico* Data

The compilation of *in chemico* data published by Schwöbel *et al.*[26] was searched in addition to as well as further information available electronically.[27] A search of these sources revealed no available *in chemico* reactivity data for 4-amino-2-nitrophenol. However, in order to utilise this approach, *in chemico* data for compounds sharing the same chemical mechanism (pro-Michael acceptor) and structural class (*i.e.* aminophenols) were sought. A useful data set representing the *in chemico* reactivity of pro-Michael acceptors to GSH (as per the Schultz assay)[19–22] was found.[30] Within these results, GSH reactivity data for four compounds are helpful in confirming the theories regarding reactivity; these data are summarised in Figure 24.4. Aniline and 3-aminophenol were found to be non-reactive to GSH; 2-aminophenol and 4-aminophenol were found to be reactive to GSH. This is in accordance with the suggested possible metabolic routes outlined in the discussion above and confirms that 3-aminophenol is not reactive—probably due to the fact it is not able to form the corresponding *ortho*- or *para*-quinone. As before, there is no evidence to suggest that the presence of the nitro group as a substituent will decrease reactivity. These *in chemico* data were measured in the absence of a metabolic system, so it is assumed that abiotic transformation (oxidation in air) was responsible for the formation of the quinone.

    The evidence from *in chemico* reactivity, therefore, whilst not for 4-amino-2-nitrophenol itself, does confirm the reactivity ability of compounds that fall

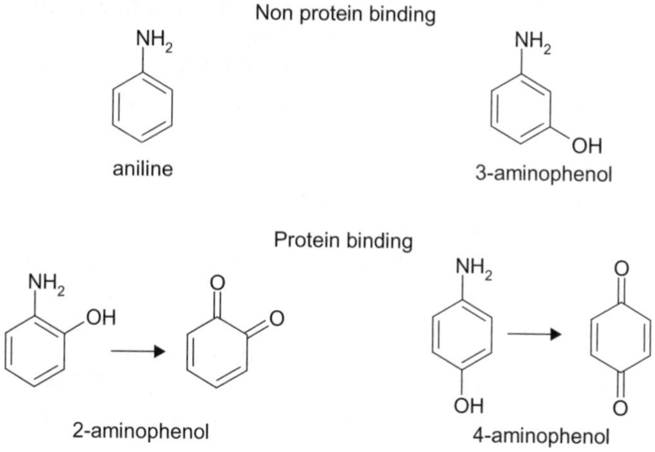

Figure 24.4 *In chemico* reactivity of aniline and aminophenols: 2- and 4-aminophenol are thought to be reactive via abiotic transformation to the corresponding quinone as shown. The data have been retrieved from a freely available source.[30]

into the pro-Michael acceptor domain and specifically that they have the capability to act as soft electrophiles, which is important for protein binding. As stated above, this should not be used as direct evidence of skin sensitisation, but it does support the concept that this molecule is capable of the molecular initiating event for skin sensitisation.

24.5.8   Other Supporting Information

Further evidence to support a hypothesis can be drawn from many sources. For instance, Chapter 18 discusses the concept of endpoint–endpoint extrapolation as a means to confirm a mechanism. The 4-aminophenols have very commonly been associated with elevated acute aquatic toxicity.[31,32] This suggests that covalent protein binding is occurring in other species. They were reported by Dupuis and Benezra to be capable of oxidation to the quinone[33]and there is likely to be much other supporting information for their ability to be protein reactive.

24.5.9   Conclusion from *In Silico* and *In Chemico* Information

There is overwhelming evidence to consider 4-amino-2-nitrophenol as a skin sensitiser. This is from the range of (Q)SAR predictions that are positive, but more importantly from the mechanistic information on reactivity that is supported by the *in chemico* evidence. The metabolism of the compound is important and does need to be considered either implicitly with the model, or explicitly *via* a metabolic simulator. Unfortunately no test, exposure or clinical data are available to support these conclusions.

With regard to the use of QSAR, formal QSAR Model Reporting Formats (Chapter 11; Section 11.7) are not presented here but could be created. Ellison *et al.*[5] have stated that 4-amino-2-nitrophenol is within the applicability domain of all models. There is evidence for this assumption regarding the applicability domain from the supporting information provided by Derek for Windows and also the CAESAR model.

The above discussion illustrates the types of evidence that may be drawn together to enable a skin sensitiser to be identified. In this case there is a test result for the LLNA which confirms that this compound is a skin sensitiser.[6] In addition to making predictions for compounds with unknown activities, this type of analysis will enable the predictive performance of models and their integration—as well as the process of forming a consensus—to be developed.

24.6   *In Silico* Predictions and *In Chemico* Data for 1,14-Tetradecanediol

The existing data and *in silico* predictions for skin sensitisation for 1,14-tetradecanediol are summarised in Table 24.3. Individual predictions are discussed and commented upon below.

24.6.1   Existing Toxicological Data and Information Relating to the Skin Sensitisation Potential of 1,14-Tetradecanediol

No skin sensitisation data for 1,14-tetradecanediol were found in the OECD (Q)SAR Application Toolbox. It should be noted that the compilation of data in the OECD Toolbox (version 1.1) was created before the publication of Roberts *et al.*[6] and newer versions will be updated.

Table 24.3   Summary of existing data and *in silico* predictions for 1,14-tetradecanediol.

In Silico *method*	*Prediction*
Existing data and information	No skin sensitisation test data
	No other information of, for example, occupational exposure
SMARTS strings for reactivity	No fragments associated with an electrophilic mechanism found.
TIMES-SS	Negative (reported by Roberts *et al.*[6])
Derek for Windows	Nothing to report (reported by Ellison *et al.*[5])
OECD (Q)SAR Application Toolbox	No protein binding—a category has not been formed.
CAESAR model for skin sensitisation	Active

The OECD eChemPortal provided no records for 1,14-tetradecanediol when searched for by both CAS number and chemical name. The compound was found within ACToR, but no information relevant to skin sensitisation was found.

24.6.2 SMARTS Strings for Reactivity

No fragments relevant to electrophilic reactivity associated with protein binding were observed in 1,14-tetradecanediol. This does not mean in itself that a compound is non-sensitiser since metabolism or other mechanisms (*e.g.* free radical formation) may occur. However, this would appear to be unlikely given the discussion below; therefore the lack of a structural fragment for electrophilicity for 1,14-tetradecanediol indicates a strong likelihood that this compound could be a non sensitiser.

24.6.3 TIMES-SS

Roberts *et al.* reported that the TIMES-SS software predicted 1,14-tetradecanediol to be a non-sensitser.[6] This is due to the absence of electrophilic features related to protein binding and because it is not metabolised to a reactive compound. This is useful supporting evidence to the findings from the SMARTS strings.

24.6.4 Derek for Windows

The compound did not fire any alerts for skin sensitisation in Derek for Windows. The actual output from Derek for Windows is 'Nothing to Report', which is a more accurate reflection of the prediction. The outcome of 'Nothing to Report' (and therefore an open prediction) has previously been understood as a negative prediction.[34–37] While this is strictly incorrect, Ellison *et al.* reported that skin sensitisation is a well-studied endpoint and therefore, as with the Toolbox, the majority of mechanisms are likely to be known.[5] The knowledge within Derek for Windows is likely to have a good coverage of the endpoint and hence treating 'Nothing to Report' as a sign of inactivity has been considered appropriate for this study. Although this finding has been taken in this study to indicate the compound is a non-sensitiser, readers are cautioned against using the 'Nothing to Report' outcome from Derek for Windows as a lack of activity for other contexts.

24.6.5 OECD (Q)SAR Application Toolbox

The compound was entered into the OECD Toolbox. It was profiled using the protein binding profiler (which contains similar mechanistic chemistry to the SMARTS strings and TIMES-SS); 1,14-tetradecanediol did not contain any alerts for protein binding. Subsequently, the skin metabolism simulator was run on the compound. This identified no metabolites. A large number of potential metabolites were identified by the other metabolic simulators (gastrointestinal tract, liver and microbial), many of which were profiled as being capable of protein reactivity. However, metabolism by non-skin routes was not considered to be relevant for skin sensitisation.

As with the SMARTS strings, absence of protein binding in the skin does not automatically imply a non-sensitiser, but in combination with other evidence of the lack of other mechanisms or effects, it can be considered to be strong evidence.

24.6.6 CAESAR Model for Skin Sensitisation

The CAESAR model for skin sensitisation predicted 1,14-tetradecanediol as being a skin sensitiser. The 'class indices' for the prediction were: active = 1.00; inactive = 0.000. Thus, according to the CAESAR model prediction, there is a great deal of certainty (*i.e.* 100% likelihood) of 1,14-tetradecanediol being a skin sensitiser. As before, it is useful to consider the six nearest neighbours determined by the CAESAR model. These, together with their similarity to 1,14-tetradecanediol, are shown in Figure 24.5.

The nearest neighbours to 1,14-tetradecanediol are all in a gross structural sense, very similar, *i.e.* they all comprise long alkyl chains, without branching, with a functional group (typically a bromine atom) on the terminal carbon(s). However, none of the compounds are an aliphatic alcohol on their own. It is highly probable that all the compounds selected by CAESAR are electrophilic.

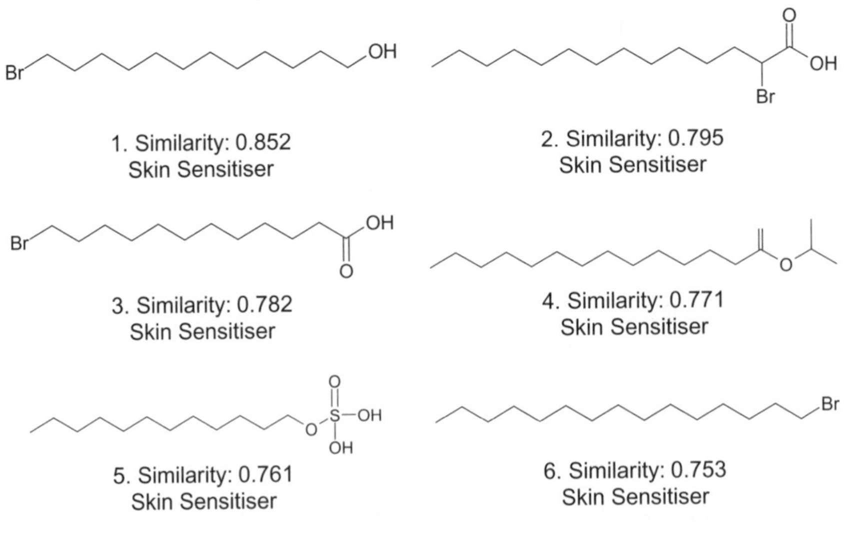

1. Similarity: 0.852
Skin Sensitiser

2. Similarity: 0.795
Skin Sensitiser

3. Similarity: 0.782
Skin Sensitiser

4. Similarity: 0.771
Skin Sensitiser

5. Similarity: 0.761
Skin Sensitiser

6. Similarity: 0.753
Skin Sensitiser

Figure 24.5   The six nearest neighbours to 1,14-tetradecanediol determined by the CAESAR model for skin sensitisation together with their similarity to 1,14-tetradecanediol and skin sensitisation activity.

Profiling in the OECD Toolbox and reference to Roberts' mechanistic scheme in Figure 13.3 suggests that the nearest neighbours are capable of nucleophilic substitution, particularly by the S_N2 mechanism, where bromine or OSO_2 will act as the leaving group. The skin sensitisation potential of the bromoalkanes is well-established in the literature. One compound (structure **4** in Figure 24.5) has no obvious leaving group and its reason for being a skin sensitiser cannot be explained in terms of protein binding.

Therefore, it seems likely that the CAESAR model is making its prediction naively in terms of mechanistic chemistry. Five of the six similar compounds in structural space have a mechanistic rationale for being a skin sensitiser. This mechanistic rationale (*i.e.* nucleophilic substitution) is lacking in 1,14-tetradecanediol. Therefore, the prediction from the CAESAR model must be treated with some caution and only low confidence can be assigned to it. This example demonstrates the need to resort to mechanistic information in toxicity prediction—not only to build models, but for their interpretation and validation.

24.6.7 *In Chemico* Data

No *in chemico* reactivity data were obtained for 1,14-tetradecanediol from Schwöbel *et al.*[26] or elsewhere. However, as this compound belongs to the class of saturated aliphatic alcohols, further data were searched for these compounds. Reactivity data for two aliphatic alcohols were obtained—namely 1-butanol and iso-propanol.[17–18,24,25] Whilst these are structurally distinct from 1,14-tetradecanediol, they do fall within the same category or grouping (*i.e.* saturated aliphatic alcohols). Figure 24.6 summarises the *in chemico* reactivity data for these compounds. In all assays, these two saturated aliphatic alcohols are non-reactive. Interestingly, *in chemico* data for one further aliphatic alcohol were found. This is a halogen-substituted alcohol (2-iodoethanol). There is evidence of reactivity for this compound, even though the reference is over 75 years old![39] This again ties in with a mechanistic interpretation, iodine will act as a leaving group hence enabling the S_N2 mechanism.

Thus, even without direct reactivity data for 1,14-tetradecanediol, *in chemico* information (including historical information) serves to confirm the hypothesis

Compound	*In Chemico* Test Results	Reference
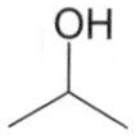 OH iso-propanol	Non reactive to glutathione as well as various amino acids and peptides	Gerberick et al;[18] Natsch et al[25]
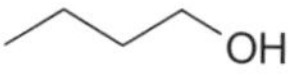 OH 1-butanol	Non reactive to glutathione as well as various amino acids and peptides	Gerberick GF et al;[17,18] Natsch et al[24,25]
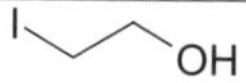 I OH 2-iodoethanol	Reactive (slow) with glutathione and cysteine	Goddard DR et al[39]

Figure 24.6 *In chemico* reactivity of iso-propanol, 1-butanol and 2-iodoethanol.

that a saturated aliphatic alcohol will not be reactive to proteins, hence giving further weight to the conclusion that this compound is a non-sensitiser.

24.6.8   Other Supporting Information

Saturated aliphatic alcohols are unreactive in many toxicological endpoints. For instance, they are well-recognised as being non-polar narcotics in acute aquatic toxicity studies.[40] This would be supportive of them not being capable of protein binding. Anecdotally, there is good evidence that saturated aliphatic alcohols are not skin sensitisers.[41]

24.6.9   Conclusion from *In Silico* and *In Chemico* Information

With the exception of the prediction from the CAESAR model, all the evidence suggests that 1,14-tetradecanediol is a non-sensitiser. The CAESAR prediction is put in doubt by the comprehensive evidence that 1,14-tetradecanediol is not capable of reaction with proteins, which comes both from mechanistic organic chemistry and *in chemico* data. There appear to be no obvious skin metabolites. Unfortunately no test, exposure or clinical data are available to support these conclusions. With regard to the use of QSAR, Ellison *et al.* have stated that 1,14-tetradecanediol is within the applicability domain of all models considered (albeit with the issue of mechanism in the CAESAR model).[5] As is already known, this information is valuable to rationalise the negative test result reported by Roberts *et al.*[6]

24.7   Conclusions and Recommendations

This exercise is for illustration only, but it serves to demonstrate how to interpret and use *in silico* and *in chemico* test results. The findings illustrate how, by using a number of predictions, greater certainty and confidence can be placed on an assessment; they also provide a useful rationalisation of predictions. The user can adapt these approaches to predict any endpoint and to include any further methods, models or information (*e.g. in vitro* data), or omit any that have been applied here.

What is missing from this chapter is a formal process to integrate the test results—termed 'an informal weight-of-evidence' in the Introduction (Section 24.1). There is a need more for research into methods to formally develop a consensus and integrate test results. This may be in terms of statistical methods such as Bayesian analysis (Chapter 22) or simpler approaches as presented by Ellison *et al.*[5] and Hewitt *et al.*[42]

24.8   Acknowledgements

This project was sponsored by the Department for Environment, Food and Rural Affairs (Defra) through the Sustainable Arable Link Programme. The funding of the EU FP6 InSilicoTox Marie Curie Project (MTKD-CT-2006-42328) is gratefully acknowledged. The advice of Dr Johannes Schwöbel, Liverpool John Moores University, is gratefully appreciated in the understanding of the *in chemico* data.

References

1. R. J. Vandebriel and H. van Loveren, *Crit. Rev. Toxicol.*, 2010, **40**, 389–404.

2. F. Gerberick, M. Aleksic, D. Basketter, S. Casati, A.-T. Karlberg, P. Kern, I. Kimber, J.-P. Lepoittevin, A. Natsch, J. M. Ovigne, R. Costanza, H. Sakaguchi and T. W. Schultz, *Altern. Lab. Anim.*, 2008, **36**, 215.

3. M. T. D. Cronin, F. Bajot, S. J. Enoch, J. C. Madden, D. W. Roberts and J. Schwöbel, *Altern. Lab. Anim.*, 2009, **49**, 513.

4. G. Patlewicz, A. O. Aptula, D. W. Roberts and E. Uriarte, *QSAR Comb. Sci.*, 2008, **27**, 60.

5. C. M. Ellison, J. C. Madden, P. Judson and M. T. D. Cronin, *Mol. Informat.*, 2010, **29**, 97.

6. D. W. Roberts, G. Patlewicz, S. D. Dimitrov, L. K. Low, A. O. Aptula, P. S. Kern, G. D. Dimitrova, M. I. H. Comber, R. D. Phillips, J. Niemelä, C. Madsen, E. B. Wedebye, P. T. Bailey and O. G. Mekenyan, *Chem. Res. Toxicol.*, 2007, **20**, 1321.

7. P. Gramatica, P. Pilutti and E. Papa, *J. Chem. Inf. Comput. Sci.*, 2004, **44**, 1794.

8. P. Gramatica, E. Giani and E. Papa, *J. Mol. Graphics Model.*, 2006, **25**, 755.

9. H. Zhu, A. Tropsha, D. Fourches, A. Varnek, E. Papa, P. Gramatica, T. Oberg, P. Dao, A. Cherkasov and I. V. Tetko, *J. Chem. Inf. Model.*, 2008, **48**, 766.

10. M. Hewitt, M. T. D. Cronin, J. C. Madden, P. H. Rowe, C. Johnson, A. Obi and S. J. Enoch, *J. Chem. Inf. Model.*, 2007, **47**, 1460.

11. E. J. Matthews, N. L. Kruhlak, R. D. Benz and J. F. Contrera, *Toxicol. Mech. Meth.*, 2008, **18**, 189.

12. T. Abshear, G. M. Banik, M. L. D'Souza, K. Nedwed and C. Peng, *SAR QSAR Environ. Res.*, 2006, **17**, 311.

13. V. E. Kuz'min, E. N. Muratov, A. G. Artemenko, E. V. Varlamova, L. Gorb, J. Wang and J. Leszczynski, *QSAR Comb. Sci.*, 2009, **28**, 664.

14. J. R. Votano, M. Parham, L. H. Hall, L. B. Kier, S. Oloff, A. Tropsha, Q. Xie and W. Tong, *Mutagenesis*, 2004, **19**, 365.

15. J. Jaworska, A. Harol, P. Kern and G. F. Gerberick, *Toxicol. Lett.*, 2009, **189**, S239.

16. S. J. Enoch, J. C. Madden and M. T. D. Cronin, *SAR QSAR Environ. Res.*, 2008, **19**, 555.

17. G. F. Gerberick, J. D. Vassallo, R. E. Bailey, J. G. Chaney, S. W. Morrall and J.-P. Lepoittevin, *Toxicol. Sci.*, 2004, **81**, 332.

18. G. F. Gerberick, J. D. Vassallo, L. M. Foertsch, B. B. Price, J. G . Chaney and J.-P. Lepoittevin, *Toxicol. Sci.*, 2007, **97**, 417.

19. D. W. Roberts, T. W. Schultz, E. M. Wolf and A. O. Aptula, *Chem. Res. Toxicol.*, 2010, **23**, 228.

20. T. W. Schultz, J. W. Yarbrough and M. Woldemeskel, *Cell Biol. Toxicol.*, 2005, **21**, 181.

21. J. W. Yarbrough and T. W. Schultz, *Chem. Res. Toxicol.*, 2007, **20**, 558.

22. T. W. Schultz, J. W. Yarbrough, R. S. Hunter and A. O. Aptula, *Chem. Res. Toxicol.*, 2007, **20**, 1359.

23. D. W. Roberts and A. Natsch, *Chem. Res. Toxicol.*, 2009, **22**, 592.

24. A. Natsch and H. Gfeller, *Toxicol. Sci.*, 2008, **106**, 464.

25. A. Natsch, R. Emter and G. Ellis, *Toxicol. Sci.*, 2009, **107**, 106.

26. J. A. H. Schwöbel, Y. K. Koleva, F. Bajot, S. J. Enoch, M. Hewitt, J. C. Madden, D. W. Roberts, T. W. Schultz and M. T. D. Cronin, *Chem. Rev.*, 2010, in press.

27. www.inchemicotox.org

28. Y. K. Koleva, J. C. Madden and M. T. D. Cronin, *Chem. Res. Toxicol.*, 2008, **21**, 2300.

29. G. F. Gerberick, C. A. Ryan, P. S. Kern, H. Schlatter, R. Dearman, I. Kimber, G. Patlewicz and D. Basketter, *Dermatitis*, 2005, **16**, 157.

30. www.inchemicotox.org/Results/Year_1_in_chemico_test_results.pdf [accessed March 2010].

31. A. O. Aptula, D. W. Roberts, M. T. D. Cronin and T. W. Schultz, *Chem. Res. Toxicol.*, 2005, **18**, 844.

32. M. T. D. Cronin, A. O. Aptula, J. C Duffy, T. I. Netzeva, P. H. Rowe, I. V. Valkova and T. W. Schultz, *Chemosphere*, 2002, **49**, 1201.

33. G. Dupuis and C. Benezra, *Allergic Contact Dermatitis to Simple Chemicals: A Molecular Approach,* Dekker, New York, 1982.

34. E. Hulzebos, D. Sijm, T. Traas, R. Posthumus and L. Maslankiewicz, *SAR QSAR Environ. Res.*, 2005, **16**, 385.

35. L. Maslankiewicz, E. Hulzebos, T. G. Vermeire, J. J. A. Muller and A. H. Piersma, *Can Chemical Structure Predict Reproductive Toxicity?* RIVM, Bilthoven, The Netherlands, 2005, RIVM report 601200005/2005, http://rivm.openrepository.com/rivm/bitstream/10029/7374/1/601200005.pdf [accessed March 2010].

36. G. Patlewicz, A. O. Aptula, E. Uriarte, D. W. Roberts, P. S. Kern, G. F. Gerberick, I. Kimber, R. J. Dearman, C. A. Ryan and D. A. Basketter, *SAR QSAR Environ. Res.*, 2007, **18**, 515.

37. G. M. Pearl, S. Livingstone-Carr and S. K. Durham, *Curr. Top. Med. Chem.*, 2001, **1**, 247.

38. D. A. Basketter, D. W. Roberts, M. T. D. Cronin and E. W. Scholes, *Contact Derm.*, 1992, **27**, 137.

39. D. Goddard and M. Schubert, *Biochem. J.*, 1935, **29**, 1009.

40. C. M. Ellison, M. T. D. Cronin, J. C. Madden and T. W. Schultz, *SAR QSAR Environ. Res.*, 2008, **19**, 751.
41. M. T. D. Cronin and D. A. Basketter, in *Trends in QSAR and Molecular Modelling 92*, ed. C.G. Wermuth, Escom, Leiden, 1993, pp. 297–298.
42. M. Hewitt, C. M. Ellison, S. J. Enoch, J. C. Madden and M. T. D. Cronin, *Reprod. Toxicol.*, 2010, 29, in press.

APPENDIX 1

Tetrahymena Pyriformis *Toxicity Data*

This data set represents 350 chemicals with published toxicity data to *Tetrahymena pyriformis*. The *T. pyriformis* toxicity assay is described in detail in Chapter 4, Section 4.5.3.2. The original toxicity data reported here are taken from the literature. Compounds 1–100 are non-polar narcotics and were selected from ref. 1 to give a good representation of chemical groups. The remaining compounds (101–350) were selected from ref. 2 to be representative of a number of different, mechanisms of action.

These data are provided in good faith with the intention that the reader may be able to reproduce some of the statistical analyses in Chapters 9 and 12. They are not intended to be definitive QSAR data sets and should not be used as such. The data sets are available electronically from www.rsc.org/Publishing/eBooks/index.asp

References

1. C. M. Ellison, M. T. D Cronin, J. C. Madden and T. W. Schultz, *SAR QSAR Environ. Res.*, 2008, **19**, 751.
2. S. J. Enoch, M. T. D. Cronin, T. W. Schultz and J. C. Madden, *Chemosphere*, 2008, **71**, 1225.

Issues in Toxicology No.7
In Silico Toxicology: Principles and Applications
Edited by Mark T. D. Cronin and Judith C. Madden
© The Royal Society of Chemistry 2010
Published by the Royal Society of Chemistry, www.rsc.org

Appendix 1.1 Chemical information for *Tetrahymena pyriformis* toxicity data used in Chapters 9 and 12.

ID	Name	SMILES	CAS	Mechanism[a]	Group[b]	Chapter 12 Training to test sets split
1	4-chloroanisole	Clc(ccc1OC)cc1	623-12-1	NPN	Anisole	Training
2	4-methylanisole	COc1ccc(C)cc1	104-93-8	NPN	Anisole	Training
3	3,5-dichloroanisole	COc1cc(Cl)cc(Cl)c1	33719-74-3	NPN	Anisole	Training
4	2-chloroanisole	COc1c(Cl)cccc1	766-51-8	NPN	Anisole	Test
5	3-chloroanisole	COc1cc(Cl)ccc1	2845-89-8	NPN	Anisole	Test
6	4-allyl anisole	C=CCc1ccc(OC)cc1	140-67-0	NPN	Anisole	Test
7	2,3,4-trichloroanisole	COc1c(Cl)c(Cl)c(Cl)cc1	54135-80-7	NPN	Anisole	Training
8	allylbenzene	c1ccccc1CC=C	300-57-2	NPN	Benzene derivatives	Training
9	4-xylene	c1(C)cc(C)cc1	106-42-3	NPN	Benzene derivatives	Training
10	cumene (isopropylbenzene)	CC(C)c1ccccc1	98-82-8	NPN	Benzene derivatives	Training
11	1,2,4-trimethylbenzene (pseudocumene)	Cc1c(C)cc(C)cc1	95-63-5	NPN	Benzene derivatives	Training
12	*n*-butylbenzene	CCCCc1ccccc1	104-51-8	NPN	Benzene derivatives	Test
13	*n*-amylbenzene	CCCCCc1ccccc1	538-68-1	NPN	Benzene derivatives	Training
14	bromobenzene	Brc(ccc1)cc1	108-86-1	NPN	Benzene derivatives	Training
15	1-bromo-4-ethylbenzene	Brc(ccc1CC)cc1	1585-07-5	NPN	Benzene derivatives	Training
16	4-bromo-1-chlorobenzene	c1(Cl)ccc(Br)cc1	106-39-8	NPN	Benzene derivatives	Training
17	4-fluorobromobenzene	Brc(ccc1F)cc1	406-00-4	NPN	Benzene derivatives	Training
18	1,4-dibromobenzene	c1(Br)ccc(Br)cc1	106-37-6	NPN	Benzene derivatives	Training
19	(2-bromoethyl)benzene (phenethyl bromide)	c1ccccc1CCBr	103-63-9	NPN	Benzene derivatives	Training
20	1-bromo-3-phenylpropane ((3-bromopropyl)benzene)	BrCCCc1ccccc1	637-59-2	NPN	Benzene derivatives	Training
21	1-chloro-3-phenylpropane	ClCCCc1ccccc1	104-52-9	NPN	Benzene derivatives	Training
22	(2-chloroethyl)benzene (phenethyl chloride)	CC(Cl)c1ccccc1	622-24-2	NPN	Benzene derivatives	Training
23	chlorobenzene	Clc1ccccc1	108-90-7	NPN	Benzene derivatives	Training
24	1,2,4-trichlorobenzene	Clc1c(Cl)cc(Cl)cc1	120-82-1	NPN	Benzene derivatives	Training
25	1,2-dichlorobenzene (o-dichlorobenzene)	Clc1c(Cl)cccc1	95-50-1	NPN	Benzene derivatives	Training
26	1,4-dichlorobenzene	Clc1c(Cl)ccc1	106-46-7	NPN	Benzene derivatives	Training

Appendix 1.1 *(continued)*

ID	Name	SMILES	CAS	Mechanism[a]	Group[b]	Chapter 12 Training to test sets split
27	1-bromo-2,3-dichlorobenzene	Brc1c(Cl)c(Cl)ccc1	56961-77-4	NPN	Benzene derivatives	Test
28	1-bromo-2,6-dichlorobenzene	Brc1c(Cl)cccc1Cl	19393-92-1	NPN	Benzene derivatives	Training
29	1,2,3-trichlorobenzene	Clc1c(Cl)c(Cl)ccc1	87-61-6	NPN	Benzene derivatives	Test
30	1,3-dichlorobenzene	Clc1cc(Cl)ccc1	541-73-1	NPN	Benzene derivatives	Test
31	1,3,5-trichlorobenzene	Clc1cc(Cl)cc(Cl)c1	108-70-3	NPN	Benzene derivatives	Training
32	biphenyl	c(ccc1c(cccc2)cc2)cc1	92-52-4	NPN	Bi-phenyl	Training
33	4-phenyltoluene (4-methyl biphenyl)	c1ccccc1c2ccc(C)cc2	644-08-6	NPN	Bi-phenyl	Training
34	4-ethylbiphenyl	c1ccccc1c2ccc(CC)cc2	5707-44-8	NPN	Bi-phenyl	Training
35	3,3-diphenyl-1-propanol	c1ccccc1C(c2ccccc2)CCO	20017-67-8	NPN	Di-phenyl	Training
36	1,1-diphenyl-2-propanol	c1ccccc1C(c1ccccc1)C(O)C	29338-49-6	NPN	Di-phenyl	Training
37	2,2-diphenylethanol	c1ccccc1C(c2ccccc2)CO	1883-32-5	NPN	Di-phenyl	Training
38	(+/−)-1,2-diphenyl-2-propanol	c1ccccc1CC(C)(c2ccccc2)O	5342-87-0	NPN	Di-phenyl	Training
39	phenyl acetate	c(cc1OC(=O)C)cc1	122-79-2	NPN	Ester	Training
40	methyl phenoxyacetate	c(cc1OCC(=O)OC)cc1	2065-23-8	NPN	Ester	Training
41	2,3-benzofuran	O(C=C1)c2c(C1)c1c2	271-89-6	NPN	Furan	Training
42	naphthalene	c(ccc1cc2)cc1cc2	91-20-3	NPN	Naphthalene	Training
43	1-methylnaphthalene	c(ccc1c(c2)C)cc1cc2	90-12-0	NPN	Naphthalene	Training
44	1-methoxynaphthalene	c1cc2ccc(OC)c2cc1	2216-69-5	NPN	Naphthalene	Training
45	1-ethoxynaphthalene	c1cc2ccc(OCC)c2cc1	5328-01-8	NPN	Naphthalene	Training
46	1-phenylnaphthalene	c(ccc1c(c2)-c(ccc3)cc3)cc1cc2	605-02-7	NPN	Naphthalene	Training
47	2-methylnaphthalene	c1cc2cc(C)ccc2cc1	91-57-6	NPN	Naphthalene	Training
48	1,4-dimethylnaphthalene	c1cc2c(C)ccc(C)c2c1	571-58-4	NPN	Naphthalene	Test
49	1,2-dimethylnaphthalene	c1cc2cc(C)c(C)c2c1	573-98-8	NPN	Naphthalene	Training
50	2-vinylnaphthalene	c1cc2ccccc2cc1C=C	827-54-3	NPN	Naphthalene	Training
51	styrene (vinyl benzene)	c1ccccc1C=C	100-42-5	NPN	Naphthalene	Training
52	1-naphthalenemethanol	c1cc2ccc(CO)c2cc1	4780-79-4	NPN	Naphthalene	Training
53	1-naphthaleneethanol	c1cc2ccc(CCO)c2cc1	773-99-9	NPN	Naphthalene	Test
54	2-naphthalene methanol	c1cc2cc(CO)cc2cc1	1592-38-7	NPN	Naphthalene	Training
55	2-naphthaleneethanol	c1cc2cc(CCO)cc2cc1	1485-07-0	NPN	Naphthalene	Training
56	1-fluoronaphthalene	Fc(cc1)c(ccc2)c1c2	321-38-0	NPN	Naphthalene	Training
57	1-chloronaphthalene	Clc(ccc1)c(ccc2)c1c2	90-12-1	NPN	Naphthalene	Training

#	Name	SMILES	CAS		Class	Set
58	1-bromonaphthalene	Brc(ccc1)pc(ccc2)c1c2	90-11-9	NPN	Naphthalene	Training
59	1-iodonaphthalene	Ic(ccc1)pc(ccc2)c1c2	90-14-2	NPN	Naphthalene	Training
60	phenethyl alcohol	c1ccccc1CCO	60-12-8	NPN	Phenethyl alcohol	Training
61	(sec)phenethyl alcohol	c1ccccc1C(C)O	98-85-1	NPN	Phenethyl alcohol	Training
62	2-phenyl-1-propanol (b-methylphenethyl alcohol)	CC(c1ccccc1)CO	1123-85-9	NPN	Phenethyl alcohol	Training
63	(+ −)-1-phenyl-2-pentanol (a-propylphenethyl alcohol)	CCCC(O)Cc1ccccc1	705-73-7	NPN	Phenethyl alcohol	Training
64	b-ethylphenethyl alcohol (2-phenyl-1-butanol)	c1ccccc1C(CC)CO	89104-46-1	NPN	Phenethyl alcohol	Training
65	(+ −)-a-ethylphenethyl alcohol (1-phenyl-2-butanol)	c1ccccc1CC(CC)O	120055-09-6	NPN	Phenethyl alcohol	Training
66	4-acetylbiphenyl	c(ccc1c(ccc2C(=O)C)cc2)cc1	92-91-4	NPN	Phenone	Training
67	4-bromobenzophenone	c1ccccc1C(=O)c2ccc(Br)cc2	90-90-4	NPN	Phenone	Training
68	benzophenone	c(ccc1C(=O)-c(ccc2)cc2)cc1	119-61-9	NPN	Phenone	Training
69	4-chlorobenzophenone	Clc(ccc1C(=O)c(ccc2)cc2)cc1	134-85-0	NPN	Phenone	Training
70	3-chlorobenzophenone	c1c(Cl)ccccc1C(=O)c2ccccc2	1016-78-0	NPN	Phenone	Test
71	2-methylbenzophenone	c1(C)ccccc1C(=O)c2ccccc2	131-58-8	NPN	Phenone	Training
72	2-fluorobenzophenone	c1ccccc1C(F)c1C(=O)c2ccccc2	342-24-5	NPN	Phenone	Training
73	acetophenone	c1ccccc1C(=O)C	98-86-2	NPN	Phenone	Training
74	propiophenone	c1ccccc1C(=O)CC	93-55-0	NPN	Phenone	Training
75	butyrophenone	c1ccccc1C(=O)CCC	495-40-9	NPN	Phenone	Training
76	octanophenone	c1ccccc1C(=O)CCCCCCC	1674-67-9	NPN	Phenone	Training
77	hexanophenone	c1ccccc1C(=O)CCCCC	942-92-7	NPN	Phenone	Training
78	valerophenone	c1ccccc1C(=O)CCCC	1009-14-9	NPN	Phenone	Training
79	heptanophenone	c1ccccc1C(=O)CCCCCC	1671-75-6	NPN	Phenone	Training
80	1-phenyl-2-butanone	c1ccccc1CC(=O)CC	1007-32-5	NPN	Phenone	Test
81	4'-methyocyacetophenone (4-acetylanisole)	CC(=O)c1ccc(OC)cc1	100-06-1	NPN	Phenone	Test
82	4-bromotoluene	Brc(ccc1C)cc1	106-38-7	NPN	Toluene	Training
83	benzyl chloride (a-chlorotoluene)	ClCc(ccc1)cc1	100-44-7	NPN	Toluene	Test
84	α,α-dichlorotoluene (benzal chloride)	c1ccccc1C(Cl)Cl	98-87-3	NPN	Toluene	Training
85	benzyl bromide (a-bromotoluene)	BrCc(ccc1)cc1	100-39-0	NPN	Toluene	Training
86	3,4-dichlorotoluene	Cc1cc(Cl)c(Cl)cc1	95-75-0	NPN	Toluene	Training
87	benzal bromide (á,á-dibromotoluene)	BrCc(ccc1)cc1	100-39-0	NPN	Toluene	Training

     Appendix 1

Appendix 1.1 (*continued*)

ID	Name	SMILES	Mechanism[a]	Group[b]	Chapter 12 Training to test sets split
88	5-phenyl-1-pentanol	c1ccccc1CCCCCO	NPN	Aromatic alcohols (other)	Test
89	3-phenyl-1-propanol	c1ccccc1CCCO	NPN	Aromatic alcohols (other)	Test
90	4-phenyl-1-butanol	c1ccccc1CCCCO	NPN	Aromatic alcohols (other)	Training
91	(S + −)-1-phenyl-1-butanol	c1ccccc1C(O)CCC	NPN	Aromatic alcohols (other)	Training
92	(R + −)-1-phenyl-1-butanol	c1ccccc1C(CCC)O	NPN	Aromatic alcohols (other)	Training
93	(+ −)-2-phenyl-2-butanol	c1ccccc1C(O)(C)CC	NPN	Aromatic alcohols (other)	Training
94	(+ −)1-phenyl-1-propanol	c1ccccc1C(O)CC	NPN	Aromatic alcohols (other)	Training
95	1-phenyl-2-propanol	c1ccccc1CC(O)C	NPN	Aromatic alcohols (other)	Test
96	a,a-dimethylbenzenepropanol	c1ccccc1CCC(C)(C)O	NPN	Aromatic alcohols (other)	Test
97	6-phenyl-1-hexanol	c1ccccc1CCCCCCO	NPN	Aromatic alcohols (other)	Training
98	3-phenyl-1-butanol	CC(c1ccccc1)CCO	NPN	Aromatic alcohols (other)	Training
99	2-methyl-1-phenyl-2-propanol	CC(O)(C)Cc1ccccc1	NPN	Aromatic alcohols (other)	Test
100	1-indanol	C1Cc2ccccc2C1O	NPN	Aromatic alcohols (other)	Training
101	1,3,5-trihydroxybenzene	Oc1cc(O)cc(O)c1	PN	Phenol	Training
102	2-hydroxybenzylalcohol (salicylalcohol)	c(cc1CO)cc1O	PN	Phenol	Training
103	resorscinol	Oc1cc(O)ccc1	PN	Phenol	Training
104	4-(4-hydroxyphenyl)-2-butanone	CC(=O)CC1ccc(O)cc1	PN	Phenol	Training

105	3-methoxyphenol	c(cc(O)c1)cc1OC	150-19-6	PN	Phenol	Training
106	ethyl-4-hydroxy-3-methoxyphenylacetate	CCOC(=O)Cc1cc(OC)c(O)cc1	60563-13-5	PN	Phenol	Training
107	4-methoxyphenol	c(cc(O)c1)OC)c(c1)O	150-76-5	PN	Phenol	Training
108	3-cyanophenol	c(cc(O)c1)cc1C#N	873-62-1	PN	Phenol	Training
109	4-ethoxyphenol	c(cc(c1)OCC)c(c1)O	622-62-8	PN	Phenol	Training
110	4-hydroxypropiophenone	Oc(ccc1C(=O)CC)cc1	70-70-2	PN	Phenol	Training
111	3-hydroxybenzaldehyde	c(cc(O)c1)cc1C=O	100-83-4	PN	Phenol	Training
112	4-chlororesorcinol	Oc1cc(O)c(Cl)cc1	95-88-5	PN	Phenol	Training
113	2-fluorophenol	Fc(ccc1)c(c1)O	367-12-4	PN	Phenol	Test
114	4-hydroxybenzaldehyde	c(cc(c1)C=O)c(c1)O	123-08-0	PN	Phenol	Training
115	2-allylphenol	c(ccc1CC=C)cc1O	1745-81-9	PN	Phenol	Test
116	3-fluorophenol	Fc(ccc1)cc1O	372-20-3	PN	Phenol	Training
117	4-isopropylphenol	Oc(ccc1C(C)C)cc1	99-89-8	PN	Phenol	Training
118	2-hydroxy-4-methoxyacetophenone	CC(=O)c1c(O)cc(OC)cc1	552-41-0	PN	Phenol	Training
119	3-methyl-2-nitrophenol	Oc1c(N(=O)=O)c(C)ccc1	4920-77-8	PN	Phenol	Training
120	4-propylphenol	Oc(ccc1CCC)cc1	645-56-7	PN	Phenol	Training
121	2-hydroxy-4,5-dimethylacetophenone	CC(=O)c1c(O)cc(C)c(C)c1	36436-65-4	PN	Phenol	Training
122	2-methyl-3-nitrophenol	Oc1c(C)c(N(=O)=O)ccc1	5460-31-1	PN	Phenol	Training
123	3-chlorophenol	Clc(ccc1)cc1O	108-43-0	PN	Phenol	Test
124	4,6-dichlororesorcinol	Oc1cc(O)c(Cl)cc1(Cl)	137-19-9	PN	Phenol	Training
125	4-benzyloxyphenol	c1cccc1COc2ccc(O)cc2	103-16-2	PN	Phenol	Test
126	3-iodophenol	Ic(ccc1)cc1O	626-02-8	PN	Phenol	Training
127	4-bromo-2,6-dimethylphenol	Oc1c(Br)cc1C	2374-05-−2	PN	Phenol	Training
128	2,3-dichlorophenol	Clc(ccc1)c(Cl)c1O	576-24-9	PN	Phenol	Test
129	5-pentylresorcinol	Oc1cc(O)cc(CCCCC)c1	500-66-3	PN	Phenol	Training
130	4-phenylphenol	Oc1cc(c2cccc2)cc1	92-69-3	PN	Phenol	Training
131	benzyl-4-hydroxybenzoate	Oc1ccc(C(=O)OCc2ccccc2)cc1	94-18-8	PN	Phenol	Training
132	4-hexyloxyphenol	Oc(ccc1OCCCCCC)cc1	18979-55-0	PN	Phenol	Training
133	4-hexylresorcinol	Oc1c(O)c(CCCCCC)cc1	136-77-6	PN	Phenol	Test
134	2,4,5-trichlorophenol	Clc(cc(Cl)c1O)c(Cl)c1	95-95-4	PN	Phenol	Training
135	2-ethylhexyl-4'-hydroxybenzoate	CCCC(CC)COC(=O)c1ccc(O)cc1	5153-25-3	PN	Phenol	Test
136	4-hydroxyphenylaceticacid	c1(CC(=O)O)ccc(O)cc1	156-38-7	PN	Phenol	Test
137	3-hydroxybenzylalcohol	c(cc(O)c1)cc1CO	620-24-6	PN	Phenol	Training
138	4-hydroxybenzoicacid(4-carboxylphenol)	Oc(ccc1C(=O)O)cc1	99-96-7	PN	Phenol	Training

Appendix 1.1   (*continued*)

ID	Name	SMILES	CAS	Mechanism[a]	Group[b]	Chapter 12 Training to test sets split
139	3-hydroxy-4-methoxybenzylalcohol	Oc1c(OC)ccc(CO)c1	4383-06-6	PN	Phenol	Training
140	4-hydroxy-3-methoxybenzylamineHCL	NCc1cc(OC)c(O)cc1	7149-10-2	PN	Phenol	Test
141	4-hydroxyphenethylalcohol	Oc(ccc1CCO)cc1	501-94-0	PN	Phenol	Training
142	3-hydroxybenzoicacid(3-carboxylphenol)	c(cc(O)c1)cc1C(=O)O	99-06-9	PN	Phenol	Test
143	4-hydroxybenzamide	Oc1ccc(C(=O)N)cc1	619-57-8	PN	Phenol	Training
144	4-hydroxy-3-methoxybenzylalcohol	OCc1ca(OC)c(O)cc1	498-00-0	PN	Phenol	Training
145	2,6-dimethoxyphenol	Oc1c(OC)c1O)cc1OC	91-10-1	PN	Phenol	Training
146	2,4,6-tris(dimethylaminomethyl)phenol	Oc1c(CN(C)C)cc(CN(C)C)cc1CN(C)C	90-72-2	PN	Phenol	Training
147	salicylicacid	c(ccc1C(=O)O)cc1O	69-72-7	PN	Phenol	Test
148	2-methoxyphenol(guaiacol)	c1c(OC)c(O)ccc1	90-05-1	PN	Phenol	Training
149	5-methylresorcinol	Oc1c(O)cc(C)c1	504-15-4	PN	Phenol	Training
150	4-hydroxybenzylcyanide(4-methylcyanophenol)	Oc(ccc1CC#N)cc1	14191-95-8	PN	Phenol	Training
151	3-hydroxyacetophenone	c(cc(O)c1)cc1C(=O)C	121-71-1	PN	Phenol	Training
152	2-ethoxyphenol	c1c(OCC)c(O)ccc1	94-71-3	PN	Phenol	Test
153	4-hydroxyacetophenone(4-acetylphenol)	Oc(ccc1C(=O)C)cc1	99-93-4	PN	Phenol	Training
154	3-ethoxy-4-methoxyphenol	c1c(OC)c(OCC)cc(O)c1	65383-57-5	PN	Phenol	Training
155	o-cresol(2-methylphenol)	c(ccc1C)cc1O	95-48-7	PN	Phenol	Training
156	salicylamide(2-hydroxybenzamide)	c(ccc1C(=O)N)cc1O	65-45-2	PN	Phenol	Training
157	phenol	c(ccc1O)cc1	108-95-2	PN	Phenol	Training
158	p-cresol(4-methylphenol)	c1cc(C)ccc1O	106-44-5	PN	Phenol	Training
159	4-hydroxy-3-methoxyphenethylalcohol	OCCc1cc(OC)c(O)cc1	2380-78-1	PN	Phenol	Training
160	3-acetamidophenol(3-hydroxyacetanilide)	c(ccc1NC(=O)C)c(O)c1	621-42-1	PN	Phenol	Test
161	isovanillin(3-hydroxy-4-methoxybenzaldehyde)	Oc(cc(c1)C=O)c(c1)OC	621-59-0	PN	Phenol	Training

162	4-hydroxy-3-methoxyacetophenone (Acetovanillone)	CC(=O)c1cc(OC)c(O)cc1	498-02-2	PN	Phenol	Training
163	3,5-dimethoxyphenol	Oc(cc(c1)OC)cc1OC	500-99-2	PN	Phenol	Test
164	2-hydroxyethylsalicylate	c1(C(=O)OCCO)c(O)cccc1	87-28-5	PN	Phenol	Training
165	m-cresol(3-methylphenol)	c(cc(c1)O)cc1C	108-39-4	PN	Phenol	Training
166	methyl-3-hydroxybenzoate	c(ccc1C(=O)OC)c(O)c1	19438-10-9	PN	Phenol	Training
167	vanillin(3-methoxy-4-hydroxybenzaldehyde)	Oc(ccc1C=O)c(c1)OC	121-33-5	PN	Phenol	Training
168	4-hydroxy-3-methoxybenzonitrile	N#Cc1cc(OC)c(O)cc1	4421-08-3	PN	Phenol	Test
169	3-ethoxy-4-hydroxybenzaldehyde	c1c(C=O)cc(OCC)c(O)c1	121-32-4	PN	Phenol	Test
170	4-fluorophenol	Fc(cc1O)cc1	371-41-5	PN	Phenol	Training
171	2-cyanophenol	c(ccc1#N)cc1O	611-20-1	PN	Phenol	Training
172	5-fluoro-2-hydroxyacetophenone	CC(=O)c1c(O)ccc(F)c1	394-32-1	PN	Phenol	Training
173	2,4-dimethylphenol	c(cc(O)c1C)c(c1)C	105-67-9	PN	Phenol	Training
174	2-hydroxyacetophenone	CC(=O)c1c(O)cccc1	582-24-1	PN	Phenol	Training
175	2,5-dimethylphenol	c(cc(c1O)c(c1)C)C	95-87-4	PN	Phenol	Test
176	methyl-4-hydroxybenzoate	Oc(ccc1C(=O)OC)cc1	99-76-3	PN	Phenol	Training
177	3,5-dimethylphenol	c(c(cc1O)c(c1)C)C	108-68-9	PN	Phenol	Test
178	4'-hydroxypropiophenone	Oc1cc(C(=O)CC)ccc1	70-70-2	PN	Phenol	Training
179	2,3-dimethylphenol	c(cc(O)c1C)c1C	526-75-0	PN	Phenol	Training
180	3,4-dimethylphenol	c(cc(c1C)c(O)c1	95-65-8	PN	Phenol	Test
181	2-ethylphenol	c(ccc1CC)cc1O	90-00-6	PN	Phenol	Test
182	syringaldehyde	C(=O)c1cc(OC)c(O)c(OC)c1	134-96-3	PN	Phenol	Training
183	salicylhydrazide	c(ccc1C(=O)NN)cc1O	936-02-7	PN	Phenol	Training
184	2-chlorophenol	Clc(ccc1)c(c1)O	95-57-8	PN	Phenol	Training
185	4-hydroxy-2-methylacetophenone	CC(=O)c1c(C)cc(O)cc1	875-59-2	PN	Phenol	Training
186	4-ethylphenol	c(cc1CC)c(c1)O	123-07-9	PN	Phenol	Training
187	3-ethylphenol	c(cc(O)cc1)cc1CC	620-17-7	PN	Phenol	Training
188	salicylaldoxime	c(ccc1C=NO)cc1O	94-67-7	PN	Phenol	Training
189	2,3,6-trimethylphenol	Oc1c(C)c(C)ccc1(C)	2416-94-6	PN	Phenol	Training
190	2,4,6-trimethylphenol	Oc1c(C)cc(C)cc1(C)	527-60-6	PN	Phenol	Training
191	2-hydroxy-5-methylacetophenone	CC(=O)c1c(O)ccc(C)c1	1450-72-2	PN	Phenol	Training
192	2-bromophenol	Brc(ccc1)c(c1)O	95-56-7	PN	Phenol	Training
193	5-bromo-2-hydroxybenzylalcohol	c1c(O)c(CO)cc(Br)c1	2316-64-5	PN	Phenol	Training
194	2,3,5-trimethylphenol	Oc(cc1C)c(c1C)C	697-82-5	PN	Phenol	Training
195	o-vanillin(3-methoxysalicylaldehyde)	c(cc(OC)c1O)cc1C=O	148-53-8	PN	Phenol	Test

Appendix 1.1 (*continued*)

ID	Name	SMILES	CAS	Mechanism[a]	Group[b]	Chapter 12 Training to test sets split
196	salicylhydroxamicacid	c1c(C(=O)NO)c(O)ccc1	89-73-6	PN	Phenol	Training
197	2-chloro-5-methylphenol	Oc1c(Cl)ccc(C)c1	615-74-7	PN	Phenol	Test
198	4-allyl-2-methoxyphenol(eugenol)	Oc1c(OC)cc(CC=C)cc1	97-53-0	PN	Phenol	Training
199	salicylaldehyde(2-hydroxybenzaldehyde)	c(ccc1C=O)cc1O	90-02-8	PN	Phenol	Training
200	2,6-difluorophenol	c1cc(F)c(O)c(F)c1	28177-48-2	PN	Phenol	Training
201	ethyl-3-hydroxybenzoate	CCOC(=O)c1cccc(O)c1	7781-98-8	PN	Phenol	Training
202	4-cyanophenol	c(cc1)C#N)c(c1)O	767-00-0	PN	Phenol	Training
203	4-propyloxyphenol	Oc1ccc(OCCC)cc1	18979-50-5	PN	Phenol	Training
204	4-chlorophenol	Clc(ccc1O)cc1	106-48-9	PN	Phenol	Training
205	ethyl-4-hydroxybenzoate	Oc(ccc1C(=O)OCC)cc1	120-47-8	PN	Phenol	Training
206	5-methyl-2-nitrophenol	Oc1c(N(=O)=O)ccc(C)c1	700-38-9	PN	Phenol	Training
207	2-bromo-4-methylphenol	c1c(C)cc(Br)c(O)c1	6627-55-0	PN	Phenol	Test
208	2,4-difluorophenol	c1c(F)cc(F)c(O)c1	367-27-1	PN	Phenol	Training
209	3-isopropylphenol	c(cc(O)c1)cc1C(C)C	618-45-1	PN	Phenol	Test
210	5-bromovanillin	Brc(cc1)C=O)c(O)c1OC	2973-76-4	PN	Phenol	Training
211	a,a-trifluoro-p-cresol	c1c(O)ccc1C(F)(F)F	402-45-9	PN	Phenol	Training
212	methyl-4-methoxysalicylate	COc1cc(O)c(C(=O)OC)cc1	5446-06--6	PN	Phenol	Training
213	4-bromophenol	Brc(ccc1O)cc1	106-41-2	PN	Phenol	Test
214	2-chloro-4,5-dimethylphenol	c1(C)c(C)cc(Cl)c(O)c1	1124-04-5	PN	Phenol	Training
215	4-butoxyphenol	Oc(ccc1OCCCC)cc1	122-94-1	PN	Phenol	Training
216	4-chloro-2-methylphenol	Clc(ccc1O)c1C	1570-64-5	PN	Phenol	Test
217	3-*tert*-butylphenol	c(ccc1C(C)(C)C)c(O)c1	585-34-2	PN	Phenol	Training
218	2,6-dichlorophenol	Clc(ccc1)c(O)c1Cl	87-65-0	PN	Phenol	Training
219	2-methoxy-4-propenylphenol	Oc1ccc(C=C)cc1OC	97-54-1	PN	Phenol	Training
220	3-chloro-5-methoxyphenol	c1(O)cc(Cl)cc(OC)c1	65262-96-6	PN	Phenol	Test
221	4-chloro-3-methylphenol	Clc(ccc1O)c(c1)C	35421-08-0	PN	Phenol	Training
222	2-isopropylphenol	c(ccc1C(C)C)cc1O	88-69-7	PN	Phenol	Test
223	2,6-dichloro-4-fluorophenol	c1c(Cl)c(O)c(Cl)cc1(F)	392-71-2	PN	Phenol	Training
224	4-iodophenol	Ic(ccc1O)cc1	540-38-5	PN	Phenol	Training
225	2,2'-biphenol	c1cc(O)c(c2c(O)cccc2)cc1	1806-29-7	PN	Phenol	Training

226	4-(*tert*)butylphenol	Oc(ccc1C(C)(C)C)cc1	98-54-4	PN	Phenol	Test
227	3,4,5-trimethylphenol	Oc(cc(c1C)C)cc1C	527-54-8	PN	Phenol	Training
228	2,2',4,4'-tetrahydroxybenzophenone	Oc1cc(O)c(C(=O)c2c(O)cc(O)cc2)cc1	131-55-5	PN	Phenol	Training
229	4-sec-butylphenol	Oc(ccc1C(C)CC)cc1	99-71-8	PN	Phenol	Test
230	3-hydroxydiphenylamine	Oc1cc(Nc2ccccc2)ccc1	101-18-8	PN	Phenol	Training
231	4-hydroxybenzophenone	O=C(c1ccccc1)(c2ccc(O)cc2)	1137-42-4	PN	Phenol	Training
232	2,4-dichlorophenol	Clc(ccc1O)cc1Cl	120-83-2	PN	Phenol	Training
233	2,4,6-tribromoresorcinol	Oc1c(Br)c(O)c(Br)cc1(Br)	2437-49-2	PN	Phenol	Training
234	benzyl-4-hydroxyphenyl ketone	c1ccc(CC(=O)c2ccc(O)cc2)cc1	2491-32-9	PN	Phenol	Test
235	4-chloro-3-ethylphenol	c1c(Cl)c(CC)cc(O)c1	14143-32-9	PN	Phenol	Test
236	2-phenylphenol	Oc1c(c2ccccc2)cccc1	90-43-7	PN	Phenol	Training
237	2,5-dichlorophenol	Clc(ccc1Cl)cc1O	583-78-8	PN	Phenol	Training
238	3-chloro-4-fluorophenol	c1c(F)c(Cl)cc(O)c1	2613-23-2	PN	Phenol	Test
239	3-bromophenol	Brc(ccc1)cc1O	591-20-8	PN	Phenol	Training
240	6-*tert*-butyl-2,4-dimethylphenol	Oc(cc(cc1O)c(c1)C)(C)(C)C	1879-09-0	PN	Phenol	Training
241	4-chloro-3,5-dimethylphenol	Clc(cc1O)c(c1)C	88-04-0	PN	Phenol	Training
242	2-hydroxybenzophenone	c1cc(O)c(C(=O)c2ccccc2)cc1	117-99-7	PN	Phenol	Training
243	4-*tert*-pentylphenol	Oc(ccc1C(C)(C)CC)cc1	80-46-6	PN	Phenol	Training
244	4-bromo-3,5-dimethylphenol	c1(O)c(Br)c(C)cc(O)c1	7463-51-6	PN	Phenol	Training
245	4-bromo-6-chloro-o-cresol	c1(Br)cc(Cl)c(O)c(C)c1	7530-27-0	PN	Phenol	Training
246	*p*-cyclopentylphenol	C2CCCC2c1ccc(O)cc1	1518-83-8	PN	Phenol	Training
247	2-*tert*-butylphenol	c(ccc1C)(C)(C)cc1O	88-18-6	PN	Phenol	Training
248	2-(*tert*)butyl-4-methylphenol	c(cc(O)c1C(C)(C)C)(C)c(c1)C	2409-55-4	PN	Phenol	Test
249	2-hydroxydiphenylmethane	Oc1ccccc1Cc2ccccc2	28994-41-4	PN	Phenol	Training
250	butyl-4-hydroxybenzoate	Oc1ccc(C(=O)OCCCC)cc1	94-26-8	PN	Phenol	Training
251	3-phenylphenol	Oc1cc(c2ccccc2)ccc1	580-51-8	PN	Phenol	Training
252	n-pentyloxyphenol	Oc1ccc(OCCCCC)cc1	18979-53-8	PN	Phenol	Training
253	2,4-dibromophenol	Brc(cc(Br)c1O)cc1	615-58-7	PN	Phenol	Test
254	2,4,6-trichlorophenol	Clc(cc(Cl)c1O)cc1Cl	88-06-2	PN	Phenol	Training
255	2-hydroxy-4-methoxybenzophenone	c1ccccc1C(=O)c2c(O)cc(OC)cc2	131-57-7	PN	Phenol	Training
256	isoamyl-4-hydroxybenzoate	Oc1ccc(C(=O)OCCC(C)C)cc1	6521-30-8	PN	Phenol	Training
257	3,5-dichlorosalicylaldehyde	C(=O)c1c(O)c(Cl)cc(Cl)c1	90-60-8	PN	Phenol	Training
258	4-cyclohexylphenol	Oc1ccc(C2CCCCC2)cc1	1131-60-8	PN	Phenol	Training
259	3,5-dichlorophenol	Clc(cc(Cl)c1)cc1O	591-35-5	PN	Phenol	Training
260	3,5-di-*tert*-butylphenol	c1c(C(C)(C)C)cc(C(C)(C)C)cc1(O)	1138-52-9	PN	Phenol	Training

Appendix 1.1 (*continued*)

ID	Name	SMILES	CAS	Mechanism[a]	Group[b]	Chapter 12 Training to test sets split
261	3,5-dibromosalicylaldehyde	c1C(=O)cc(Br)cc(Br)c1(O)	90-59-5	PN	Phenol	Training
262	3,4-dichlorophenol	Clc(ccc1O)c(Cl)c1	95-77-2	PN	Phenol	Test
263	4-bromo-2,6-dichlorophenol	c1(Br)cc(Cl)c(O)c(Cl)c1	3217-15-0	PN	Phenol	Training
264	2,6-di(*tert*)butyl-4-methylphenol(BTH)	Oc(c(cc1O)C(C)(C)C)c(c1)C(C)(C)C	128-37-0	PN	Phenol	Training
265	4-chloro-2-isopropyl-5-methylphenol	Clc(cc(c1O)C(C)C)c(c1)C	89-68-9	PN	Phenol	Training
266	2,4,6-tribromophenol	Brc(cc(Br)c1O)cc1Br	118-79-6	PN	Phenol	Training
267	4-heptyloxyphenol	Oc(ccc1OCCCCCCC)cc1	13037-86-0	PN	Phenol	Test
268	4-(*tert*)octylphenol	Oc(ccc1C(C)(C)CC(C)(C)C)cc1	140-66-9	PN	Phenol	Test
269	4-(4-bromophenyl)phenol	Oc1ccc(c2ccc(Br)cc2)cc1	29558-77-8	PN	Phenol	Training
270	3,5-diiodosalicylaldehyde	C(=O)c1c(O)c(I)cc(I)c1	2631-77-8	PN	Phenol	Training
271	2,3,5-trichlorophenol	Oc1c(Cl)c(Cl)cc(Cl)c1	933-78-8	PN	Phenol	Training
272	4-nonylphenol	Oc(ccc1CCCCCCCCC)cc1	104-40-5	PN	Phenol	Training
273	nonyl-4-hydroxybenzoate	O1cc(C(=O)OCCCCCCCCC)cc1	38713-56-3	PN	Phenol	Training
274	2,4-diaminophenol2HCl	c1c(O)c(N)cc(N)c1	95-86-3	PN	Phenol	Training
275	5-amino-2-methoxyphenol	c1(OC)c(O)cc(N)cc1	1687-53-2	PE	Phenol	Test
276	6-amino-2,4-dimethylphenol	Oc1c(C)cc(C)c(O)c1(N)	41458-65-5	PE	Phenol	Training
277	trimethylhydroquinone	Oc1c(C)c(C)c(O)c(C)c1	700-13-0	PE	Phenol	Training
278	methylhydroquinone	Oc1c(C)cc(O)cc1	95-71-6	PE	Phenol	Training
279	4-acetamidophenol(4-hydroxyacetanilide)	Oc(ccc1NC(=O)C)cc1	103-90-2	PE	Phenol	Training
280	3-aminophenol(3-hydroxyaniline)	c(cc(c1O)cc1N	591-27-5	PE	Phenol	Training
281	4-aminophenol(4-hydroxyaniline)	c(cc(c1)N)c(c1)O	123-30-8	PE	Phenol	Training
282	3-methylcatechol	Oc1c(O)c(C)ccc1	488-17-5	PE	Phenol	Training
283	2-amino-4-(*tert*)butylphenol	c1c(O)c(N)cc(C(C)(C)C)c1	1199-46-8	PE	Phenol	Training
284	4-methylcatechol	Oc1c(O)cc(C)cc1	452-86-8	PE	Phenol	Training
285	1,2,4-trihydroxybenzene	Oc1c(O)ccc(O)c1	533-73-3	PE	Phenol	Training
286	hydroquinone	Oc1ccc(O)cc1	123-31-9	PE	Phenol	Test
287	catechol	Oc1c(O)cccc1	120-80-9	PE	Phenol	Training
288	5-chloro-2-hydroxyaniline(2-amino-4-chlorophenol)	Clc(ccc1O)cc1N	95-85-2	PE	Phenol	Training
289	1,2,3-trihydroxybenzene	Oc1c(O)c(O)ccc1	87-66-1	PE	Phenol	Test

No.	Name	SMILES	CAS	Source	Class	Set
290	2-aminophenol(2-hydroxyaniline)	c(ccc1N)cc1O	95-55-6	PE	Phenol	Test
291	4-chlorocatechol	Oc1c(O)cc(Cl)cc1	2138-22-9	PE	Phenol	Training
292	chlorohydroquinone	Oc1c(Cl)cc(O)cc1	615-67-8	PE	Phenol	Training
293	4-amino-2-cresol	Oc1c(C)cc(N)cc1	2835-96-3	PE	Phenol	Training
294	2,3-dimethylhydroquinone	Oc1c(C)c(C)c(O)cc1	608-43-5	PE	Phenol	Training
295	4-amino-2,3-dimethylphenolHCL	c1(O)c(C)c(C)c(N)cc1	3096-69-3	PE	Phenol	Training
296	bromohydroquinone	Oc1c(Br)cc(O)cc1	583-69-7	PE	Phenol	Test
297	tetrachlorocatechol	Oc1c(O)c(Cl)c(Cl)c(Cl)c1(Cl)	1198-55-6	PE	Phenol	Training
298	phenylhydroquinone	Oc1c(c2ccccc2)cc(O)cc1	1079-21-6	PE	Phenol	Training
299	3,5-di(*tert*)butylcatechol	Oc1c(O)c(C(C)(C)C)cc(C(C)(C)C)c1	1020-31-1	PE	Phenol	Training
300	methoxyhydroquinone	Oc1c(OC)cc(O)cc1	824-46-4	PE	Phenol	Training
301	tetrafluorohydroquinone	Oc1c(F)c(F)c(O)c(F)c1F	771-63-1	PRC	Phenol	Training
302	tetrabromocatechol	Oc1c(O)c(Br)c(Br)c(Br)c1O	488-47-1	PRC	Phenol	Training
303	tetramethylhydroquinone	Oc1c(C)c(C)c(O)c(C)c1C	527-18-4	PRC	Phenol	Training
304	tetrachlorohydroquinone	Oc1c(Cl)c(Cl)c(O)c(Cl)c1Cl	87-87-6	PRC	Phenol	Training
305	2,3-dinitrophenol	c1cc(N(=O)=O)c(N(=O)=O)c(O)c1	66-56-8	WARU	Phenol	Training
306	2,3,5,6-tetrafluorophenol	c1c(F)c(F)c(F)c(O)c(F)c1(F)	769-39-1	WARU	Phenol	Test
307	2,6-diiodo-4-nitrophenol	Ic(cc(1)N(=O)=O)c(O)c1I	305-85-1	WARU	Phenol	Training
308	3,4,5,6-tetrabromo-o-cresol	Br(c(Br)c(Br)c1C)c(Br)c1O	576-55-6	WARU	Phenol	Training
309	2,4,6-trinitrophenol	Oc(c(cc1N(=O)=O)N(=O)=O)c(c1)N(=O)=O	29663-11-4	WARU	Phenol	Training
310	3,4-dinitrophenol	c1c(N(=O)=O)c(N(=O)=O)cc(O)c1	577-71-9	WARU	Phenol	Training
311	2,6-dinitrophenol	c1cc(N(=O)=O)c(O)c(N(=O)=O)c1	573-56-8	WARU	Phenol	Training
312	2,6-dichloro-4-nitrophenol	Clc(cc(c1)N(=O)=O)c(O)c1Cl	618-80-4	WARU	Phenol	Training
313	2,5-dinitrophenol	c1c(N(=O)=O)c(O)cc(N(=O)=O)c1	329-71-5	WARU	Phenol	Training
314	2,4-dinitrophenol	Oc(cc1N(=O)=O)c(c1)N(=O)=O	51-28-5	WARU	Phenol	Training
315	2,6-dinitro-*p*-cresol	Oc(cc1C)N(=O)=O)c(c1)N(=O)=O	609-93-8	WARU	Phenol	Training
316	4-bromo-2-fluoro-6-nitrophenol	Oc1c(F)cc(Br)cc1N(=O)=O	320-76-3	WARU	Phenol	Training
317	pentafluorophenol	Fc(c(F)c(F)c1O)c(F)c1F	771-61-9	WARU	Phenol	Test
318	4,6-dinitro-o-cresol(4,6-dinitro-2-methylphenol)	Oc(cc1N(=O)=O)=O)O)c(c1)N(=O)=O	534-52-1	WARU	Phenol	Training
319	2,4-dichloro-6-nitrophenol	c1(O)c(Cl)cc(Cl)cc1(N(=O)=O)	609-89-2	WARU	Phenol	Training
320	pentachlorophenol	Clc(c(Cl)c(Cl)c1O)c(Cl)c1Cl	87-86-5	WARU	Phenol	Test
321	2,3,5,6-tetrachlorophenol	Clc(cc(Cl)c(Cl)c1O)	935-95-5	WARU	Phenol	Test

Appendix 1.1 (*continued*)

ID	Name	SMILES	CAS	Mechanism[a]	Group[b]	Training to test sets split
322	pentabromophenol	Brc(c(Br)c(Br)c1O)c(Br)c1Br	608-71-9	WARU	Phenol	Training
323	2,3,4,5-tetrachlorophenol	Clc(c(Cl)c(Cl)c1O)c(Cl)c1	4901-51-3	WARU	Phenol	Training
324	3-nitrophenol	c(cc(O)c1)cc1N(=O)=O	554-84-7	SE	Phenol	Test
325	2-nitrophenol	c(ccc1N(=O)=O)cc1O	88-75-5	SE	Phenol	Training
326	3-fluoro-4-nitrophenol	Oc1cc(F)c(N(=O)=O)cc1O	394-41-2	SE	Phenol	Training
327	2,6-dibromo-4-nitrophenol	Brc(cc(c1)N(=O)=O)c(O)c1Br	99-28-5	SE	Phenol	Test
328	4-nitro-3-(trifluoromethyl)-phenol	Oc1cc(C(F)(F)F)c(N(=O)=O)cc1	88-30-2	SE	Phenol	Training
329	3-hydroxy-4-nitrobenzaldehyde	C(=O)c1cc(O)c(N(=O)=O)cc1	704-13-2	SE	Phenol	Training
330	5-hydroxy-2-nitrobenzaldehyde	c1c(N(=O)=O)c(C=O)cc(O)c1	42454-06-8	SE	Phenol	Training
331	2-amino-4-nitrophenol	Oc(ccc1N(=O)=O)c(c1)N	99-57-0	SE	Phenol	Training
332	4-methyl-2-nitrophenol	Oc1c(N(=O)=O)cc(C)cc1	119-33-5	SE	Phenol	Training
333	4-hydroxy-3-nitrobenzaldehyde	C(=O)c1cc(N(=O)=O)c(O)cc1	3011-34-5	SE	Phenol	Test
334	4-nitrosophenol	c(cc(c1)N=O)c(c1)O	104-91-6	SE	Phenol	Training
335	2-nitroresorcinol	Oc1c(N(=O)=O)c(O)ccc1	601-89-8	SE	Phenol	Test
336	4-methyl-3-nitrophenol	c1c(O)cc(N(=O)=O)c(C)c1	2042-14-0	SE	Phenol	Training
337	2-chloromethyl-4-nitrophenol	c1(O)c(CCl)cc(N(=O)=O)cc1	2973-19-5	SE	Phenol	Training
338	2-bromo-2'-hydroxy-5'-nitroacetanilide	BrCC(=O)Nc1c(O)ccc(N(=O)=O)c1	3947-58-8	SE	Phenol	Training
339	4-amino-2-nitrophenol	c(cc(O)c1N(=O)=O)c(c1)N	119-34-6	SE	Phenol	Training
340	2-fluoro-4-nitrophenol	Oc1c(F)cc(N(=O)=O)cc1	403-19-0	SE	Phenol	Training
341	5-fluoro-2-nitrophenol	c1(O)c(N(=O)=O)ccc(F)c1	446-36-6	SE	Phenol	Training
342	4-nitrocatechol	Oc1c(O)cc(N(=O)=O)cc1	3316-09-4	SE	Phenol	Training
343	2-amino-4-chloro-5-nitrophenol	Clc(cc(c1O)N)c(c1)N(=O)=O	6358-07-2	SE	Phenol	Training
344	4-fluoro-2-nitrophenol	Oc1c(N(=O)=O)cc(F)cc1	394-33-2	SE	Phenol	Training
345	4-nitrophenol	c(cc(c1)N(=O)=O)c(c1)O	100-02-7	SE	Phenol	Training
346	2-chloro-4-nitrophenol	Clc(cc(c1)N(=O)=O)c(O)c1	619-08-9	SE	Phenol	Test
347	4-chloro-6-nitro-*m*-cresol	Clc(cc(c1O)N(=O)=O)c(c1)C	7147-89-9	SE	Phenol	Training
348	3-methyl-4-nitrophenol	Oc(ccc1N(=O)=O)c(c1)C	2581-34-2	SE	Phenol	Training
349	4-bromo-2-nitrophenol	Oc1c(N(=O)=O)cc(Br)cc1	7693-52-9	SE	Phenol	Test
350	4-chloro-2-nitrophenol	Clc(ccc1O)cc1N(=O)=O	89-64-5	SE	Phenol	Training

[a]Mechanism of action: non-polar narcosis (NPN); polar narcosis (PN); pro-electrophile (PE); pro-redox cycler (PRC); weak acid respiratory uncouple (WARU); soft electrophile (SE).
[b]Arbitarily assigned chemical class.

Appendix 1.2   Toxicity data to *Tetrahymena pyriformis* and descriptors utilised in Chapters 9 and 12.

ID^a	log $(IGC_{50})^{-1b}$	log P	Solubility^c	$E_{LUMO}{}^d$	Molecular mass^e	Surface area^f	Volume^g
1	0.60	2.72	− 2.62	0.1867	142.6	151.6	95.0
2	− 0.12	2.62	− 2.37	0.5185	122.2	157.8	92.0
3	0.93	3.36	− 3.74	− 0.1936	177.0	162.5	110.7
4	0.22	2.72	− 2.53	0.1485	142.6	151.9	94.9
5	0.29	2.72	− 2.79	0.1208	142.6	151.3	95.0
6	0.64	3.47	− 3.24	0.4767	148.2	190.0	115.9
7	1.64	4.01	− 3.94	− 0.3721	211.5	173.9	125.5
8	0.19	3.39	− 2.82	0.4744	118.2	166.0	95.5
9	0.12	3.09	− 2.67	0.4874	106.2	150.8	84.8
10	0.69	3.45	− 3.21	0.5416	120.2	167.4	98.1
11	0.78	3.63	− 3.18	0.5052	120.2	168.9	97.1
12	1.25	4.01	− 3.92	0.5212	134.2	189.9	112.4
13	1.79	4.50	− 4.47	0.5198	148.3	209.6	125.4
14	0.08	2.88	− 2.90	0.0593	157.0	128.0	79.6
15	0.67	3.92	− 3.90	0.0530	185.1	167.1	105.8
16	0.56	3.53	− 3.62	− 0.3050	191.4	142.1	95.4
17	0.11	3.08	− 3.11	− 0.3027	175.0	139.2	86.9
18	0.68	3.77	− 4.16	− 0.3936	235.9	146.1	99.6
19	0.42	3.37	− 3.19	0.2252	185.1	168.0	106.4
20	0.69	3.87	− 3.83	0.3447	199.1	189.1	120.5
21	0.50	3.78	− 3.36	0.3621	154.6	186.0	115.8
22	0.17	3.21	− 2.97	0.1494	140.6	161.2	102.8
23	− 0.13	2.64	− 2.45	0.1542	112.6	125.9	75.0
24	1.16	3.93	− 3.96	− 0.4691	181.4	147.8	106.3
25	0.53	3.28	− 3.20	− 0.1423	147.0	138.8	90.4
26	0.53	3.28	− 3.20	− 0.1423	147.0	138.8	90.4
27	1.30	4.17	− 4.41	− 0.4361	225.9	152.0	110.3
28	1.34	4.17	− 4.41	− 0.4495	225.9	152.9	110.2
29	1.21	3.93	− 3.98	− 0.3645	181.4	148.5	105.8
30	0.56	3.28	− 3.29	− 0.1579	147.0	135.3	90.7
31	0.87	3.93	− 4.10	− 0.4023	181.4	145.9	106.5
32	1.05	3.76	− 3.73	− 0.0684	154.2	189.9	118.4
33	1.66	4.30	− 4.38	− 0.0678	168.2	205.9	131.4
34	1.97	4.80	− 4.63	− 0.0664	182.3	210.7	144.9
35	1.04	3.18	− 2.95	0.3480	212.3	230.4	179.2
36	0.75	3.10	− 2.89	0.2030	212.3	224.1	178.7
37	0.65	2.69	− 2.43	0.3632	198.3	213.9	154.9
38	0.80	3.65	− 3.36	0.3271	212.3	221.0	179.8
39	− 0.23	1.59	− 1.47	0.2614	136.2	159.1	98.1
40	− 0.21	1.61	− 1.62	0.2953	166.2	190.3	115.4
41	− 0.11	2.54	− 2.34	− 0.0635	118.1	137.6	79.3
42	0.33	3.17	− 2.96	− 0.2650	128.2	159.6	90.2
43	0.74	3.72	− 3.54	− 0.2668	142.2	175.9	102.9
44	0.80	3.25	− 3.30	− 0.1793	158.2	184.5	110.4
45	1.25	3.74	− 3.65	− 0.1530	172.2	203.8	123.4
46	2.04	4.93	− 4.90	− 0.3966	204.3	204.4	151.8
47	1.08	3.72	− 3.54	− 0.2340	142.2	178.4	103.1
48	1.43	4.26	− 4.07	− 0.2709	156.2	189.7	115.6
49	1.31	4.26	− 4.02	− 0.2422	156.2	192.3	115.7

Appendix 1.2 *(continued)*

ID^a	log $(IGC_{50})^{-1b}$	log P	Solubility^c	E_{LUMO}^d	Molecular mass^e	Surface area^f	Volume^g
50	1.24	4.07	− 3.80	− 0.4076	154.2	188.2	110.8
51	− 0.02	2.89	− 2.48	0.0186	104.2	144.2	78.3
52	0.13	2.25	− 1.87	− 0.3048	158.2	182.0	109.9
53	0.27	2.74	− 2.29	− 0.3479	172.2	198.0	124.2
54	0.39	2.25	− 1.73	− 0.3743	158.2	180.6	111.2
55	0.49	2.74	− 2.29	− 0.3140	172.2	201.6	123.8
56	0.39	3.37	− 3.15	− 0.5074	146.2	167.0	97.7
57	0.81	3.81	− 3.80	− 0.5188	162.6	172.0	106.0
58	1.86	4.06	− 4.18	− 0.5810	207.1	175.6	110.6
59	2.17	4.34	− 4.76	− 0.5868	254.1	180.4	116.7
60	− 0.68	1.57	− 0.75	0.4058	122.2	157.9	93.2
61	− 0.82	1.49	− 0.80	0.4974	122.2	155.2	92.6
62	− 0.40	1.98	− 1.38	0.4380	136.2	172.0	105.6
63	0.16	2.97	− 2.42	0.5261	164.3	209.6	136.5
64	− 0.11	2.48	− 1.90	0.3121	150.2	190.4	119.9
65	− 0.16	2.48	− 1.90	0.5270	150.2	189.6	127.4
66	0.92	3.44	− 3.57	− 0.6410	196.3	212.6	147.2
67	1.26	4.04	− 4.55	− 0.7434	261.1	217.9	158.8
68	0.87	3.15	− 3.25	− 0.4740	182.2	205.9	137.4
69	1.50	3.79	− 4.02	− 0.6911	216.7	216.5	153.9
70	1.55	3.79	− 4.02	− 0.6499	216.7	215.7	154.5
71	1.37	3.69	− 3.79	− 0.4340	196.3	208.7	151.5
72	0.96	3.35	− 3.52	− 0.5662	200.2	211.1	145.9
73	− 0.46	1.67	− 1.43	− 0.3616	120.2	152.1	87.8
74	− 0.07	2.16	− 2.05	− 0.3441	134.2	169.4	100.7
75	0.21	2.66	− 2.65	− 0.3433	148.2	187.7	113.8
76	1.89	4.62	− 4.64	− 0.3446	204.3	249.6	165.6
77	1.19	3.64	− 3.59	− 0.3436	176.3	213.0	139.6
78	0.56	3.15	− 3.07	− 0.3429	162.2	205.3	126.7
79	1.56	3.64	− 3.59	− 0.3436	176.3	213.0	139.6
80	− 0.21	1.96	− 1.96	0.4379	148.2	190.0	117.2
81	− 0.13	1.75	− 1.78	− 0.3074	150.2	176.9	107.6
82	0.47	3.43	− 3.37	0.0439	171.0	147.9	92.3
83	0.06	2.79	− 2.09	0.0748	126.6	147.9	90.2
84	0.11	2.97	− 2.92	− 0.1602	161.0	160.7	106.6
85	0.64	2.88	− 2.94	− 0.0502	171.0	150.8	94.8
86	1.07	3.83	− 3.75	− 0.1376	161.0	155.1	103.4
87	0.64	2.88	− 2.94	− 0.0502	171.0	150.8	94.8
88	0.53	3.04	− 2.49	0.4957	164.3	216.2	133.6
89	− 0.27	2.06	− 1.29	0.4752	136.2	179.9	106.8
90	0.12	2.55	− 1.80	0.4727	150.2	196.5	119.7
91	− 0.09	2.48	− 1.90	0.3823	150.2	187.7	119.4
92	− 0.01	2.48	− 1.90	0.3819	150.2	187.6	129.7
93	0.06	2.44	− 1.87	0.4077	150.2	180.2	119.8
94	− 0.43	1.98	− 1.38	0.3811	136.2	174.2	105.6
95	− 0.62	1.98	− 1.37	0.5686	136.2	176.0	105.7
96	− 0.07	2.93	− 2.39	0.4810	164.3	204.3	139.9
97	0.87	3.53	− 3.01	0.4947	178.3	236.6	146.5
98	0.01	2.48	− 1.90	0.4969	150.2	183.1	130.0
99	− 0.41	2.44	− 1.87	0.5319	150.2	184.5	121.6

Appendix 1.2   (*continued*)

ID^a	log $(IGC_{50})^{-1b}$	log P	Solubility^c	$E_{LUMO}{}^d$	Molecular mass^e	Surface area^f	Volume^g
100	− 0.58	1.93	− 1.32	0.3052	134.2	161.0	99.2
101	− 1.26	0.55	0.32	0.2469	126.1	134.7	79.4
102	− 0.95	0.60	0.36	0.4016	124.2	147.9	85.5
103	− 0.65	1.03	− 0.11	0.3206	110.1	128.6	72.6
104	− 0.50	1.48	− 1.09	0.3181	164.2	199.1	124.3
105	− 0.33	1.59	− 0.67	0.4138	124.2	146.6	85.9
106	− 0.23	1.92	− 2.37	0.1983	210.2	227.2	151.6
107	− 0.14	1.59	− 0.88	0.3129	124.2	147.2	85.7
108	− 0.06	1.61	− 1.21	− 0.5144	119.1	134.0	81.6
109	0.01	2.08	− 1.18	0.3369	138.2	165.8	98.9
110	0.05	1.68	− 1.45	− 0.3646	150.2	175.3	108.4
111	0.09	1.23	− 0.62	− 0.5467	122.1	138.7	81.7
112	0.13	1.68	− 1.21	− 0.0077	144.6	139.5	88.3
113	0.19	1.71	− 0.90	0.0131	112.1	128.1	73.1
114	0.27	1.23	− 0.67	− 0.4492	122.1	141.7	81.6
115	0.33	2.91	− 2.08	0.3485	134.2	168.5	102.1
116	0.38	1.71	− 1.09	0.0247	112.1	128.4	73.0
117	0.47	2.97	− 2.09	0.4457	136.2	175.9	104.8
118	0.55	2.05	− 1.53	− 0.4441	166.2	181.7	114.0
119	0.61	2.46	− 2.28	− 0.8763	153.1	159.4	100.6
120	0.64	3.04	− 2.35	0.4322	136.2	178.4	105.2
121	0.71	3.06	− 2.44	− 0.4734	164.2	192.0	119.2
122	0.78	2.46	− 1.84	− 1.0891	153.1	155.4	100.0
123	0.87	2.16	− 1.70	0.0187	128.6	132.7	81.5
124	0.97	2.32	− 1.91	− 0.2402	179.0	150.8	104.1
125	1.04	3.30	− 2.90	0.2822	200.2	216.6	145.7
126	1.12	2.68	− 2.73	− 0.0699	220.0	146.0	92.3
127	1.17	3.50	− 3.08	0.0851	201.1	171.9	111.6
128	1.28	2.80	− 2.24	− 0.2624	163.0	142.1	96.9
129	1.31	3.54	− 2.96	0.3444	180.3	222.9	138.8
130	1.39	3.28	− 2.60	− 0.0835	170.2	194.5	125.3
131	1.55	3.70	− 3.33	− 0.4186	228.3	235.0	159.1
132	1.64	4.05	− 3.50	0.3385	194.3	234.9	150.6
133	1.80	4.04	− 2.99	0.3266	194.3	229.1	151.7
134	2.10	3.45	− 3.24	− 0.5555	197.4	158.4	113.0
135	2.51	5.36	− 5.03	− 0.3666	250.4	282.8	209.6
136	− 1.50	0.95	0.02	0.1398	152.2	167.7	104.2
137	− 1.04	0.60	0.56	0.2008	124.2	144.5	87.1
138	− 1.02	1.39	− 0.98	− 0.4816	138.1	148.6	88.3
139	− 0.99	0.42	− 0.18	0.3337	154.2	173.6	105.8
140	− 0.97	0.41	0.34	0.3586	153.2	178.4	108.6
141	− 0.83	1.09	− 0.05	0.3339	138.2	164.8	99.7
142	− 0.81	1.39	− 0.91	− 0.5740	138.1	147.8	88.5
143	− 0.78	0.26	0.10	− 0.2448	137.1	152.2	89.4
144	− 0.70	0.42	− 0.18	0.4088	154.2	176.3	105.8
145	− 0.60	1.16	− 1.31	0.3881	154.2	173.3	105.9
146	− 0.52	0.77	− 0.21	0.4174	265.4	306.1	215.4
147	− 0.51	2.24	− 1.56	− 0.4571	138.1	147.9	88.2
148	− 0.51	1.34	− 1.24	0.3915	124.2	148.3	86.0
149	− 0.39	1.58	− 0.88	0.3006	124.2	145.4	85.5

Appendix 1.2   (*continued*)

ID^a	$\log (IGC_{50})^{-1b}$	$\log P$	$Solubility^c$	$E_{LUMO}{}^d$	$Molecular\ mass^e$	$Surface\ area^f$	$Volume^g$
150	− 0.38	1.08	− 0.78	0.0635	133.2	155.4	96.9
151	− 0.38	1.19	− 0.80	− 0.4631	136.2	155.4	94.6
152	− 0.36	1.83	− 1.65	0.3822	138.2	166.6	99.0
153	− 0.30	1.19	− 0.77	− 0.3787	136.2	160.6	94.6
154	− 0.30	1.65	− 1.26	0.3181	168.2	194.0	119.0
155	− 0.30	2.06	− 1.08	0.3961	108.2	140.3	78.5
156	− 0.24	1.03	− 0.72	− 0.1870	137.1	156.1	92.0
157	− 0.21	1.51	− 0.56	0.3971	94.1	122.7	65.8
158	− 0.18	2.06	− 1.07	0.4279	108.2	139.2	78.5
159	− 0.18	0.91	− 0.32	0.3242	168.2	195.6	120.0
160	− 0.16	0.27	− 0.93	0.2052	151.2	173.2	103.0
161	− 0.14	1.05	− 1.14	− 0.4883	152.2	165.2	101.8
162	− 0.12	1.02	− 1.28	− 0.4037	166.2	185.8	114.5
163	− 0.09	1.67	− 1.15	0.4154	154.2	171.9	106.0
164	− 0.08	1.63	− 0.83	− 0.4755	182.2	196.2	122.4
165	− 0.06	2.06	− 1.09	0.3766	108.2	138.0	78.9
166	− 0.05	2.00	− 1.35	− 0.4844	152.2	171.3	101.6
167	− 0.03	1.05	− 1.35	− 0.4878	152.2	167.2	101.8
168	− 0.03	1.44	− 1.78	− 0.4489	149.2	162.4	101.4
169	0.02	1.55	− 1.76	− 0.4654	166.2	184.4	114.5
170	0.02	1.71	− 0.95	0.0591	112.1	128.0	73.0
171	0.03	1.61	− 1.13	− 0.5094	119.1	136.8	81.6
172	0.04	2.17	− 1.60	− 0.7851	154.1	159.8	101.3
173	0.07	2.61	− 1.48	0.4033	122.2	157.7	91.2
174	0.08	1.97	− 1.26	− 0.5172	136.2	154.2	93.8
175	0.08	2.61	− 1.50	0.3453	122.2	158.0	91.1
176	0.08	2.00	− 1.41	− 0.3971	152.2	171.8	101.5
177	0.11	2.61	− 1.52	0.3829	122.2	157.5	91.8
178	0.12	1.68	− 1.16	− 0.4424	150.2	177.2	107.5
179	0.12	2.61	− 1.63	0.3861	122.2	154.8	91.2
180	0.12	2.61	− 1.42	0.4362	122.2	157.0	91.2
181	0.16	2.55	− 1.62	0.3859	122.2	154.3	92.2
182	0.17	0.88	− 1.28	− 0.4912	182.2	191.7	121.2
183	0.18	0.67	− 0.24	− 0.4429	152.2	162.1	101.9
184	0.18	2.16	− 1.40	0.0294	128.6	134.1	81.7
185	0.19	1.74	− 1.20	− 0.2875	150.2	174.2	107.9
186	0.21	2.55	− 1.72	0.4352	122.2	158.7	92.2
187	0.23	2.55	− 1.56	0.4027	122.2	157.1	92.5
188	0.25	1.39	− 0.81	− 0.3141	137.1	155.4	89.9
189	0.28	3.15	− 1.90	0.3828	136.2	174.8	103.8
190	0.28	3.15	− 1.95	0.4328	136.2	174.4	104.2
191	0.31	2.52	− 1.87	− 0.4826	150.2	171.9	106.7
192	0.33	2.40	− 1.89	− 0.0495	173.0	135.7	86.2
193	0.34	1.49	− 1.06	0.0406	203.0	163.4	105.9
194	0.36	3.15	− 2.31	0.3596	136.2	174.5	104.2
195	0.38	1.83	− 1.48	− 0.4538	152.2	166.1	101.8
196	0.38	0.91	− 0.52	− 0.4214	153.1	163.4	99.6
197	0.39	2.70	− 2.14	0.0576	142.6	150.6	94.3
198	0.42	2.73	− 2.34	0.3829	164.2	199.6	122.6
199	0.42	2.01	− 1.06	− 0.4340	122.1	143.1	81.9

Appendix 1.2 (*continued*)

ID^a	log $(IGC_{50})^{-1b}$	log P	Solubility^c	$E_{LUMO}{}^d$	Molecular mass^e	Surface area^f	Volume^g
200	0.47	1.91	− 1.25	− 0.3208	130.1	134.1	80.5
201	0.48	2.49	− 1.94	− 0.4527	166.2	187.5	114.7
202	0.52	1.61	− 1.12	− 0.4124	119.1	135.9	81.4
203	0.52	2.58	− 1.72	0.3384	152.2	182.0	111.8
204	0.55	2.16	− 1.60	0.0945	128.6	133.6	81.5
205	0.57	2.49	− 1.94	− 0.3672	166.2	189.6	114.6
206	0.59	2.46	− 2.29	− 0.9910	153.1	161.6	99.4
207	0.60	2.95	− 2.51	− 0.0120	187.0	154.4	99.0
208	0.60	1.91	− 1.21	− 0.3180	130.1	133.5	80.4
209	0.61	2.97	− 2.15	0.3983	136.2	174.3	104.7
210	0.62	1.94	− 2.55	− 0.7302	231.0	179.9	121.6
211	0.62	2.48	− 2.21	− 0.3481	162.1	157.4	100.6
212	0.62	2.68	− 2.24	− 0.4271	182.2	193.7	121.1
213	0.68	2.40	− 2.10	0.0204	173.0	138.5	86.0
214	0.69	3.25	− 2.54	0.0529	156.6	168.5	107.0
215	0.70	3.07	− 2.31	0.3388	166.2	198.6	124.7
216	0.70	2.70	− 2.04	0.1228	142.6	152.6	94.4
217	0.73	3.42	− 2.54	0.4311	150.2	185.0	117.5
218	0.74	2.80	− 2.16	− 0.2583	163.0	144.3	97.4
219	0.75	2.65	− 3.00	0.0054	164.2	198.6	119.5
220	0.76	2.24	− 1.69	0.0598	158.6	158.1	101.7
221	0.80	2.70	− 2.31	0.0930	142.6	150.3	94.2
222	0.80	2.97	− 2.08	0.4078	136.2	168.7	105.2
223	0.80	3.00	− 2.51	− 0.5677	181.0	150.1	104.4
224	0.85	2.68	− 2.71	0.0234	220.0	144.6	92.0
225	0.88	2.80	− 2.37	− 0.2386	186.2	199.0	134.0
226	0.91	3.42	− 2.54	0.4631	150.2	184.1	117.4
227	0.93	3.15	− 2.31	0.4297	136.2	172.6	104.1
228	0.96	2.78	− 2.79	− 0.7536	246.2	219.5	163.7
229	0.98	3.46	− 2.35	0.4451	150.2	189.6	118.4
230	1.01	2.46	− 2.65	0.1037	185.2	204.5	136.4
231	1.02	2.67	− 2.69	− 0.4842	198.2	213.8	144.8
232	1.04	2.80	− 2.42	− 0.2450	163.0	143.1	97.3
233	1.06	3.70	− 4.88	− 0.6545	346.8	172.2	133.2
234	1.07	2.90	− 2.65	− 0.3373	212.3	215.5	157.4
235	1.08	3.20	− 2.49	0.1403	156.6	167.4	107.9
236	1.09	3.28	− 2.50	− 0.1199	170.2	189.7	127.6
237	1.13	2.80	− 2.42	− 0.3252	163.0	142.9	97.5
238	1.13	2.36	− 1.70	− 0.2921	146.5	137.4	88.9
239	1.15	2.40	− 2.13	− 0.0743	173.0	136.5	86.2
240	1.16	4.52	− 3.78	0.4548	178.3	209.7	142.6
241	1.20	3.25	− 2.56	0.1467	156.6	168.7	106.9
242	1.23	3.44	− 3.07	− 0.6302	198.2	213.0	144.2
243	1.23	3.91	− 3.16	0.4611	164.3	197.5	141.9
244	1.27	3.50	− 3.08	0.1088	201.1	172.1	111.2
245	1.28	3.59	− 3.31	− 0.2327	221.5	166.6	114.5
246	1.29	3.84	− 3.08	0.4322	162.2	198.3	125.1
247	1.30	3.42	− 2.54	0.4360	150.2	186.7	117.3
248	1.30	3.97	− 3.21	0.4772	164.3	199.4	141.1
249	1.31	3.54	− 2.98	0.2436	184.2	210.9	141.2

Appendix 1.2 (*continued*)

ID^a	log $(IGC_{50})^{-1b}$	log P	Solubility^c	$E_{LUMO}{}^d$	Molecular mass^e	Surface area^f	Volume^g
250	1.33	3.47	− 3.09	− 0.3671	194.2	222.2	140.4
251	1.35	3.28	− 2.62	− 0.1625	170.2	193.4	125.6
252	1.36	3.56	− 2.93	0.3386	180.3	215.4	137.6
253	1.40	3.29	− 3.21	− 0.3480	251.9	152.5	106.3
254	1.41	3.45	− 3.21	− 0.5018	197.4	157.4	113.1
255	1.42	3.52	− 3.52	− 0.5743	228.3	219.9	165.8
256	1.48	3.89	− 3.46	− 0.3629	208.3	239.8	154.0
257	1.55	3.30	− 2.83	− 0.8936	191.0	163.0	113.3
258	1.56	4.33	− 3.51	0.4425	176.3	207.4	148.0
259	1.57	2.80	− 2.90	− 0.2853	163.0	143.6	97.4
260	1.64	5.33	− 4.68	0.4704	206.4	244.8	168.8
261	1.64	3.79	− 3.90	− 0.9239	279.9	170.5	122.2
262	1.75	2.80	− 2.65	− 0.2363	163.0	146.3	97.1
263	1.78	3.69	− 3.54	− 0.5145	241.9	162.0	117.6
264	1.80	5.03	− 4.58	0.3832	220.4	250.6	183.0
265	1.85	4.16	− 3.32	0.1136	184.7	196.7	133.5
266	2.03	4.18	− 4.56	− 0.6207	330.8	170.3	126.7
267	2.03	4.54	− 4.02	0.3384	208.3	254.1	163.6
268	2.10	5.28	− 4.63	0.4738	206.4	235.3	181.1
269	2.31	4.17	− 4.00	− 0.3980	249.1	214.3	145.5
270	2.34	4.34	− 5.05	− 0.9012	373.9	182.9	134.1
271	2.37	3.45	− 3.34	− 0.5781	197.4	154.6	112.6
272	2.47	5.99	− 5.15	0.4287	220.4	297.0	185.4
273	2.63	5.93	− 5.61	− 0.3677	264.4	312.4	205.3
274	0.13	− 0.87	0.64	0.5440	124.2	139.2	83.6
275	0.45	0.07	− 0.28	0.3582	139.2	158.5	96.4
276	0.89	1.69	− 1.65	0.4422	137.2	169.1	100.0
277	1.34	2.67	− 1.18	0.2160	152.2	176.6	112.2
278	1.86	1.58	− 0.31	0.2401	124.2	148.8	85.6
279	− 0.82	0.27	− 0.70	0.2535	151.2	174.2	103.3
280	− 0.52	0.24	− 0.18	0.5414	109.1	130.9	73.6
281	− 0.08	0.24	− 0.03	0.4392	109.1	132.4	74.9
282	0.28	1.58	− 0.88	0.2694	124.2	144.6	85.3
283	0.37	2.51	− 2.55	0.5218	165.3	192.2	125.7
284	0.37	1.58	− 0.70	0.3011	124.2	147.0	85.5
285	0.44	0.55	− 0.01	0.1099	126.1	133.3	79.3
286	0.47	1.03	0.07	0.2332	110.1	128.8	72.5
287	0.75	1.03	− 0.18	0.2967	110.1	128.8	72.6
288	0.78	1.24	− 1.80	0.1723	143.6	142.8	90.7
289	0.85	0.97	− 0.37	0.2694	126.1	135.4	79.4
290	0.94	0.60	− 0.53	0.4730	109.1	127.0	74.5
291	1.06	1.68	− 1.11	− 0.1712	144.6	139.0	89.8
292	1.26	1.68	− 0.87	− 0.1106	144.6	138.4	88.3
293	1.31	0.79	− 0.78	0.4348	123.2	151.5	86.6
294	1.41	2.13	− 0.79	0.2085	138.2	158.8	99.7
295	1.44	1.34	− 1.35	0.4164	137.2	161.9	99.2
296	1.68	1.92	− 1.64	− 0.1858	189.0	143.8	92.9
297	1.70	3.61	− 4.09	− 0.8307	247.9	172.2	134.3
298	2.01	2.80	− 2.37	− 0.1708	186.2	200.3	132.4

Appendix 1.2 (*continued*)

ID^a	$\begin{array}{c}log\\(IGC_{50})^{-1b}\end{array}$	$log\ P$	$Solubility^c$	$E_{LUMO}{}^d$	$\begin{array}{c}Molecular\\mass^e\end{array}$	$\begin{array}{c}Surface\\area^f\end{array}$	$Volume^g$
299	2.11	4.85	− 4.39	0.3071	222.4	246.5	188.3
300	2.20	0.86	− 0.63	0.2236	140.1	154.6	92.4
301	1.84	1.83	− 1.52	− 1.1222	182.1	152.4	101.9
302	0.98	4.59	− 5.65	− 0.9871	425.7	189.9	151.7
303	1.28	3.22	− 2.59	0.2127	166.2	196.2	125.3
304	2.11	3.61	− 3.51	− 0.9275	247.9	173.7	134.3
305	0.46	1.73	− 2.02	− 1.7988	184.1	163.4	112.5
306	1.17	2.31	− 1.81	− 0.9940	166.1	144.6	94.9
307	1.71	4.24	− 5.09	− 1.4214	390.9	188.4	138.9
308	2.57	5.62	− 6.51	− 0.8823	423.7	199.2	157.6
309	− 0.16	1.54	− 2.01	− 2.5342	229.1	185.2	126.7
310	0.27	1.73	− 1.44	− 1.8500	184.1	163.1	111.7
311	0.54	1.73	− 1.71	− 1.9519	184.1	162.4	106.1
312	0.63	3.20	− 2.65	− 1.4406	208.0	168.0	117.7
313	0.95	1.73	− 2.04	− 2.2615	184.1	164.4	106.7
314	1.08	1.73	− 1.97	-1.8876	184.1	163.2	106.7
315	1.23	2.27	-2.59	-1.8941	198.1	180.5	119.0
316	1.62	3.00	-3.48	-1.6506	236.0	163.4	113.7
317	1.64	2.51	-2.72	-1.2963	184.1	150.3	102.5
318	1.72	2.27	-2.47	-1.8288	198.1	183.7	119.3
319	1.75	3.20	-3.45	-1.5791	208.0	163.8	117.3
320	2.05	4.74	-4.94	-0.9774	266.3	176.3	142.8
321	2.22	4.09	-3.63	-0.8167	231.9	166.9	128.2
322	2.66	5.96	-7.27	-1.1935	488.6	198.4	164.7
323	2.71	4.09	-3.91	-0.7516	231.9	164.0	127.8
324	0.51	1.91	-1.35	-1.1594	139.1	142.9	86.6
325	0.67	1.91	-1.75	-1.0144	139.1	144.0	86.1
326	0.94	2.11	-1.57	-1.2847	157.1	147.6	93.3
327	1.36	3.69	-3.83	-1.4526	296.9	175.9	127.1
328	1.65	2.87	-2.58	-1.5847	207.1	172.8	121.6
329	0.27	1.63	-1.81	-1.7547	167.1	160.8	101.9
330	0.33	1.63	-1.23	-1.4867	167.1	161.8	104.3
331	0.48	0.99	-1.40	-0.9763	154.1	153.2	95.5
332	0.57	2.46	-2.34	-0.9668	153.1	160.4	99.2
333	0.61	1.63	-1.69	-1.4569	167.1	160.1	101.9
334	0.65	1.38	-0.62	-0.7985	123.1	134.6	79.9
335	0.66	1.43	-1.67	-1.3207	155.1	145.6	92.1
336	0.74	2.46	-1.60	-1.1106	153.1	159.5	99.0
337	0.75	2.71	-2.30	-1.1949	187.6	171.8	117.6
338	0.87	2.40	-3.26	-1.1130	275.1	214.9	145.2
339	0.88	0.64	-1.15	-1.1193	154.1	152.0	95.0
340	1.07	2.11	-1.57	-1.3338	157.1	149.5	93.7
341	1.13	2.11	-1.98	-1.2974	157.1	147.1	93.4
342	1.17	1.43	-1.17	-1.1675	155.1	150.4	93.3
343	1.17	1.64	-1.97	-0.9623	188.6	170.4	112.9
344	1.38	2.11	-2.15	-1.2936	157.1	148.0	93.4
345	1.42	1.91	-1.27	-1.0651	139.1	144.5	86.5
346	1.59	2.55	-2.07	-1.2637	173.6	155.8	102.1
347	1.64	3.10	-3.07	-1.3458	187.6	169.2	114.2

Appendix 1.2   (*continued*)

ID^a	$\log (IGC_{50})^{-1b}$	$\log P$	Solubility^c	$E_{LUMO}{}^d$	Molecular mass^e	Surface area^f	Volume^g
348	1.73	2.46	-1.86	-1.0068	153.1	159.3	98.8
349	1.87	2.80	-3.18	-1.2397	218.0	159.5	106.4
350	2.05	2.55	-2.57	-1.2299	173.6	153.2	102.0

^{a}ID numbers refer to Appendix 1.1.
^{b}Toxicity to *Tetrahymena pyriformis* in mmol L^{-1}.
^{c}log molar units.
^{d}Energy of the lowest unoccupied molecular orbital (eV).
^{e}Dalton.
^{f}$(\text{Å})^2$.
^{g}$(\text{Å})^3$.

APPENDIX 2
Skin Sensitisation (Local Lymph Node Assay) Data

This data set represents 210 chemicals with published skin sensitisation data assessed in the Local Lymph Node Assay (LLNA). The LLNA assay is described in detail in Chapter 4, Section 4.5.5. The original toxicity data reported here are taken from the literature and are the data (with some corrections) reported in ref. 1.

These data are provided in good faith with the intention that the reader may be able to reproduce some of the statistical analyses in Chapter 10. They are not intended to be definitive QSAR data sets and should not be used as such. The data sets are available electronically from www.rsc.org/Publishing/eBooks/index.asp and

References

1. G. F. Gerberick, C. A. Ryan, P. S. Kern, H. Schlatter, R. J. Dearman, I. Kimber, G. Y. Patlewicz and D. Basketter, *Dermatitis*, 2005, **16**, 157.

Issues in Toxicology No.7
In Silico Toxicology: Principles and Applications
Edited by Mark T. D. Cronin and Judith C. Madden
© The Royal Society of Chemistry 2010
Published by the Royal Society of Chemistry, www.rsc.org

Appendix 2.1 Chemical information for the skin sensitisation data set used in Chapter 10.

ID	Name	SMILES	CAS number
1	abietic acid	[H][C@]12CCC(=CC1=CC[C@@]3([H])[C@](C)(CCC[C@]23C)(O)=O)C(C)C	514-10-3
2	2-acetylcyclohexanone	O=C(C(C(=O)CCC1)C1)C	874-23-7
3	4-allylanisole	O(c(ccc(c1)CC=C)c1)C	140-67-0
4	2-amino-6-chloro 4-nitrophenol	O=[N+](c1cc(Cl)c(O)c(N)c1)[O−]	6358-09-4
5	2-(4-amino 2 nitrophenylamino)-ethanol (hc red #3)	[O−][N+](c1cc(N)ccc1NCCO)=O	2871-01-4
6	2-aminophenol	Oc(c(N)ccc1)c1	95-55-6
7	3-aminophenol	Oc(cccc1N)c1	591-27-5
8	alpha-amyl cinnamic aldehyde	CCCCC(=O)=C	122-40-7
9	2-(4-*tert*-amylcyclohexyl)acetaldehyde (qrm 2113)	CCC(C)(C)C1CCC(CC1)CC=O	620159-84-4
10	*trans*-anethol	COc1ccc(cc1)=C	104-46-1
11	aniline	Nc(cccc1)c1	62-53-3
12	benzaldehyde	O=Cc(cccc1)c1	100-52-7
13	1,2,4-benzenetricarboxylic anhydride (trimellitic anhydride)	O=C(OC(=O)c1ccc(C(=O)O)c2)c12	552-30-7
14	1,2-benzisothiazolin-3-one (proxel active)	O=C(NSc1cccc2)c12	2634-33-5
15	benzo[a]pyrene	c(c(cc1)ccc2)c2cc3)(c3cc(c4ccc5)c5)c14	50-32-8
16	benzocaine	O=C(OCc)c(ccc(N)c1)c1	94-09-7
17	*p*-benzoquinone	C1(=O)C=CC(=O)C=C1	106-51-4
18	benzyl benzoate	O=C(OCc(cccc1)c1)c(cccc2)c2	120-51-4
19	benzyl bromide	BrCc(cccc1)c1	100-39-0
20	benzylidene acetone (4-phenyl-3-buten-2-one)	CC(=O)=C	122-57-6
21	bis-1,3-(2′,5′-dimethylphenyl)-propane-1,3-dione	Cc1cc(c(c1)C(=O)CC(=O)c1cc(ccc1C)C)C	no CAS
22	bisphenol a-diglycidyl ether	O(C1COc(ccc(c2)C(c(ccc(OCC(O3)C3)c4)(C)C)c2)C1	1675-54-3
23	1-bromobutane	BrCCCC	109-65-9
24	1-bromodocosane	BrCCCCCCCCCCCCCCCCCCCCCC	6938-66-5

25	1-bromododecane	BrCCCCCCCCCCCC	143-15-7
26	12-bromododecanoic acid	OC(=O)CCCCCCCCCCCBr	73367-80-3
27	12-bromo-1-dodecanol	OCCCCCCCCCCCCBr	3344-77-2
28	1-bromoeicosane	BrCCCCCCCCCCCCCCCCCCCC	4276-49-7
29	1-bromoheptadecane	BrCCCCCCCCCCCCCCCCC	3508-00-7
30	1-bromohexadecane	BrCCCCCCCCCCCCCCCC	112-82-3
31	1-bromohexane	BrCCCCCC	111-25-1
32	3-bromomethyl-5,5-dimethyl-dihydro-2(3H)-furanone	CC1(C)CC(CBr)C(=O)O1	154750-20-6
33	1-bromononane	BrCCCCCCCCC	693-58-3
34	1-bromooctadecane	BrCCCCCCCCCCCCCCCCCC	112-89-0
35	1-bromopentadecane	BrCCCCCCCCCCCCCCC	629-72-1
36	1-bromotetradecane	BrCCCCCCCCCCCCCC	112-71-0
37	7-bromotetradecane	CCCCCCC(Br)CCCCCC	74036-97-8
38	2-bromotetradecanoic acid	O=C(O)C(Br)CCCCCCCCCCCC	10520-81-7
39	1-bromotridecane	BrCCCCCCCCCCCCC	765-09-3
40	1-bromoundecane	BrCCCCCCCCCCC	693-67-4
41	2,3-butanedione	O=C(C(=O)C)C	431-03-8
42	1-butanol	OCCCC	71-36-3
43	alpha-butyl cinnamic aldehyde	CCCC(=O)=C	7492-44-6
44	p-tert-butyl-a-ethyl hydrocinnamal (lilial)	O=CC(CC)Cc(ccc(c1)C(C)(C)C)C1	80-54-6
45	butyl glycidyl ether	O(C1COCCO)C1	2426-08-6
46	C4 azlactone	CCCCC1=NC(C)(C)C(=O)O1	176664-99-6
47	C6 azlactone	CCCCCCC1=NC(C)(C)C(=O)O1	176665-02-4
48	C9 azlactone	CCCCCCCCCC1=NC(C)(C)C(=O)O1	176665-04-6
49	C11 azlactone	CCCCCCCCCCCC1=NC(C)(C)C(=O)O1	176665-06-8
50	C15 azlactone	CCCCCCCCCCCCCCCC1=NC(C)(C)C(=O)O1	176665-09-1
51	C17 azlactone	CCCCCCCCCCCCCCCCCC1=NC(C)(C)C(=O)O1	176665-11-5
52	C19 azlactone	CCCCCCCCCCCCCCCCCCCC1=NC(C)(C)C(=O)O1	no CAS
53	camphorquinone	CC1(C)C2CC1(C)C(=O)C2=O	465-29-2
54	chlorobenzene	c(cccc1)(c1)Cl	108-90-7
55	1-chloro-2,4-dinitrobenzene	[O-][N+](c1cc([N+]([O-])=O)c(Cl)cc1)=O	97-00-7
56	1-chlorohexadecane	ClCCCCCCCCCCCCCCCC	4860-03-1

Appendix 2.1 (*continued*)

ID	Name	SMILES	CAS number
57	5-chloro 2 methyl 4 isothiazolin-3-one	S1N(C)C(=O)C=C1Cl	26172-55-4
58	1-chloromethylpyrene	ClCc1ccc2ccc3cccc4ccc1c2c34	1086-00-6
59	1-chlorononane	ClCCCCCCCCC	2473-01-0
60	1-chlorooctadecane	ClCCCCCCCCCCCCCCCCCC	3386-33-2
61	1-chlorotetradecane	ClCCCCCCCCCCCCCC	2425-54-9
62	cinnamic alcohol	OC=C	104-54-1
63	cinnamic aldehyde	O=C=C	104-55-2
64	citral	C=C/CC=C=O	5392-40-5
65	clotrimazole	Clc1ccccc1C(c2ccccc2)(c3ccccc3)n4ccnc4	23593-75-1
66	courmarin	c1cc2OC(=O)C=Cc2cc1	91-64-5
67	cyclamen aldehyde	O=CC(C)Cc(ccc(c1)C(C)C)c1	103-95-7
68	*trans*-2-decenal	CCCCCCC=C=O	3913-71-1
69	1,2-dibromo-2,4-dicyanobutane	C(#N)C(Br)(CCC(#N))CBr	35691-65-7
70	diethyl acetaldehyde	O=CC(CC)CC	97-96-1
71	diethylenetriamine	N(CCN)CCN	111-40-0
72	diethyl malate	CCOC(=O)=C/C(=O)OCC	141-05-9
73	1-(2',5'-diethylphenyl)butane-1,3-dione	CCc1ccc(c(c1)C(=O)CC(C)=O)CC	167998-76-7
74	diethylphthalate	O=C(OCC)c(c(ccc1)C(=O)OCC)c1	84-66-2
75	diethyl sulfate	O=S(=O)(OCC)OCC	64-67-5
76	3,4-dihydrocoumarin	O=C(Oc(c(ccc1)C2)c1)C2	119-84-6
77	dihydroeugenol	O(c(c(O)ccc1CCC)c1)C	2785-87-7
78	1,4-dihydroquinone	Oc(ccc(O)c1)c1	123-31-9
79	3-dimethylaminopropylamine	N(CCCN)(C)C	109-55-7
80	7,12-dimethylbenz[alpha]anthracene	c(c(c(c1)ccc2)c2)c(c(c3ccc4)c4)C)(c3C)c1	57-97-6
81	5,5-dimethyl-3-methylene-dihydro-2(3H)-furanone	CC1(C)CC(=C)C(=O)O1	29043-97-8
82	1-(2',5'-dimethylphenyl)butane-1,3-dione	CC(=O)CC(=O)c1cc(ccc1C)C	56290-55-2
83	dimethyl sulfate	O=S(=O)(OC)OC	77-78-1
84	dimethyl sulfoxide	O=S(C)C	67-68-5
85	dodecyl methane sulfonate	CCCCCCCCCCCCOS(C)(=O)=O	51323-71-8

			CAS Number
86	4-ethoxymethylene-2-phenyl-2-oxazolin-5-one (oxazolone)	CCO=C1/N=C(OC1=O)c2ccccc2	15646-46-5
87	3-ethoxy-1-(2',3',4',5'-tetra-methylphenyl)propane-1,3-dione	CCOC(=O)CC(=O)c1cc(c(c(c1C)C)C)C	170928-69-5
88	ethyl acrylate	O=C(OCC)C=C	140-88-5
89	ethyl benzoylacetate	O=C(OCC)CC(=O)c(cccc1)c1	94-02-0
90	ethylenediamine free base	NCCN	107-15-3
91	ethylene glycol dimethacrylate	O=C(OCCOC(=O)C(=C)C)C(=C)C	97-90-5
92	4-(N-ethyl-N-2-methansulphonamido-ethyl)-2-methyl-1,4-phenylenediamine CD3	O=S(=O)(NCCN(c(ccc(N)c1C)c1)CC)C	25646-71-3 CD3 92-09-1)
93	N-ethyl-N-nitrosourea	O=C(N(N=O)CC)N	759-73-9
94	ethyl vanillin	O=Cc(ccc(O)c1OCC)c1	121-32-4
95	eugenol	O(c(c(O)ccc1CC=C)c1)C	97-53-0
96	farnesal	C=C/CC=C=C=O	502-67-0
97	fluorescein-5-isothiocyanate	O=C(OC(c(c(Oc1cc(O)cc2)cc(O)c3)c3)(c12)c4ccc(N=C=S)c5)c45	3326-32-7
98	formaldehyde	O=C	50-00-0
99	furil	o1cccc1C(=O)C(=O)c2occc2	492-94-4
100	geraniol	C=C/CC=C	106-24-1
101	glutaraldehyde	O=CCCCC=O	111-30-8
102	glycerol	OCC(O)CO	56-81-5
103	glyoxal	O=CC=O	107-22-2
104	2,4-heptadienal	CC/C=C=C/C/C=O	5910-85-0
105	hexane	C(CCCC)C	110-54-3
106	trans-2-hexenal	O=CC=CCCC	6728-26-3
107	hexyl cinnamic aldehyde	CCCCCC(=O)=C	101-86-0
108	4-hydrobenzoic acid	O=C(O)c(ccc(O)c1)c1	99-96-7
109	hydroxycitronellal	O=CCC(CCCC(O)(C)C)C	107-75-5
110	2-hydroxyethyl acrylate	O=C(OCCO)C=C	818-61-1
111	3 and 4-(4-hydroxy-4-methylpentyl)-3-cyclo-hexene-1-carboxaldehyde (lyral)	O=CC(CC(=C1)CCCC(O)(C)C)C1	31906-04-4
112	2-hydroxypropyl methacrylate	O=C(OCC(O)C)C(=C)C	923-26-2

Appendix 2.1 *(continued)*

ID	Name	SMILES	CAS number
113	imidazolidinylurea	O=C(NCNC(=O)NC(NC(=O)N1CO)C1(=O))NC(NC(=O)N2CO)C2(=O))	39236-46-9
114	1-iodododecane	C(CCCCCCCCCCC)I	4292-19-7
115	1-iodohexadecane	C(CCCCCCCCCCCCCCC)I	544-77-4
116	1-iodohexane	C(CCCCC)I	638-45-9
117	1-iodononane	C(CCCCCCCC)I	4282-42-2
118	1-iodooctadecane	C(CCCCCCCCCCCCCCCCC)I	629-93-6
119	1-iodotetradecane	C(CCCCCCCCCCCCC)I	19218-94-1
120	isoeugenol	COc1cc(C=C	97-54-1
121	isononanoyl chloride	ClC(=O)CCCCCC(C)C	57077-36-8
122	isopropanol	OC(C)C	67-63-0
123	isopropyl eugenol	CC(C)Oc1cc(CC=C)ccc1O	51474-90-9
124	isopropyl isoeugenol	C=C	no CAS
125	isopropyl myristate	O=C(OC(C)C)CCCCCCCCCCCCC	110-27-0
126	kanamycin	O(C(CO)C(O)C(O)C1O)CN)C1OC(C(O)C(OC(OC(C(O)C2N)CO)C2O)C(N)C3)C3N	59-01-8
127	lactic acid	O=C(O)C(O)C	50-21-5
128	lauryl gallate (dodecyl gallate)	O=C(OCCCCCCCCCCCC)c(cc(O)c(O)c1O)c1	1166-52-5
129	R(+) limonene	C(=CCC(C(=O)C1)(C1)C	5989-27-5
130	linalool	OC(C=C)(CCC=C(C)C)C	78-70-6
131	2-mercaptobenzothiazole	N(c(cS1)ccc2)c2)=C1S	149-30-4
132	4'-methoxyacetophenone	O=C(c(ccc(OC)c1)c1)C	100-06-1
133	2-methoxy-4-methyl-phenol	O(c(c(O)ccc1C)c1)C	93-51-6
134	1-(p-methoxyphenol)-1-penten-3-one	CCC(=O)=C	104-27-8
135	4-(methylamino)phenol sulfate (metol)	Oc1cc(NC)cc1	55-55-0
136	2-methyl -4H,3,1-benzoxazin-4-one (product 2040)	O=C(OC(=Nc1cccc2)C)c12	525-76-8
137	alpha-methyl cinnamic aldehyde	C=O)=C/c1ccccc1	101-39-3
138	6-methylcoumarin	c1c(C)cc2C=CC(=O)Oc2c1	92-48-8
139	methyl dodecane sulfonate	CCCCCCCCCCCS(=O)(=O)OC	2374-65-4

No.	Name	SMILES	CAS
140	2,5-diamino-toluene	Nc(c(cc(N)c1)C)c1	95-70-5
141	3-methyleugenol	COc1c(O)ccc(CC=C)c1C	186743-26-0
142	5-methyleugenol	COc1cc(CC=C)c(C)cc1O	186743-25-9
143	6-methyleugenol	COc1cc(CC=C)cc(C)c1O	186743-24-8
144	methyl hexadecene sulfonate	CCCCCCCCCCCCCC=C/S(=O)(=O)OC	26452-48-2
145	methyl hexadecyl sulfonate	CCCCCCCCCCCCCCCCS(=O)(=O)OC	4230-15-3
146	5-methyl-2,3-hexanedione	O=C(C=O)OC(C)C	13706-86-0
147	p-methylhydrocinnamic aldehyde	O=CCc(ccc(c1)C)c1	5406-12-2
148	methyl 4-hydroxybenzoate (methylparaben)	O=C(OC)c(ccc(O)c1)c1	99-76-3
149	2-methyl-5-hydroxyethylaminophenol	Cc1ccc(cc1O)NCCO	55302-96-0
150	3-methylisoeugenol	COc1c(ccc(c1C)=C	186743-29-3
151	6-methylisoeugenol	COc1cc=C	13041-12-8
152	2-methyl-2H-isothiazol-3-one	S1N(C)C(=O)C=C1	2682-20-4
153	methyl methanesulfonate	O=S(=O)(OC)C	66-27-3
154	1-methyl-3-nitro-1-nitrosoguanidine	O=[N+](NC(N(C)N=O)=N)[O-]	70-25-7
155	N-methyl-N-nitrosourea	O=C(N(N=O)C)N	684-93-5
156	methyl 2-nonynoate	O=C(OC)C#CCCCCC	111-80-8
157	alpha-methylphenylacetaldehyde	O=CC(c(cccc1)c1)C	93-53-8
158	3-methyl-4-phenyl-1,2,5-thiadiazole-1,1-dioxide (MPT)	CC1=NS(=O)(=O)N=C1c1ccccc1	3775-21-1
159	methyl salicylate	O=C(OC)c(c(O)ccc1)c1	119-36-8
160	methyl 2-sulphophenyl octadecanoate	CCCCCCCCCCCCCCCCCC(O)C(=O)Oc1ccccc1S(O)(=O)=O	no CAS
161	2-methylundecanal	O=CC(CCCCCCCCC)C	110-41-8
162	1-naphthol	Oc(c(ccc1)cc2)c1)c2	90-15-3
163	4-nitrobenzyl bromide	[O-][N+](c1ccc(CBr)cc1)=O	100-11-8
164	2-nitro-p-phenylenediamine	[O-][N+](c1cc(N)ccc1N)=O	5307-14-2
165	nonanoyl chloride	O=C(CCCCCCCC)Cl	764-85-2
166	cis-6-nonenal	O=CCCCC=CCC	2277-19-2
167	octanoic acid	O=C(O)CCCCCCC	124-07-2
168	oleyl methane sulfonate	O=S(=O)(OCCCCCCCCC=CCCCCCCCC)C	35709-09-2
169	oxalic acid	O=C(O)C(=O)O	144-62-7
170	palmitoyl chloride	O=C(CCCCCCCCCCCCCCC)Cl	112-67-4

Appendix 2.1 (*continued*)

ID	Name	SMILES	CAS number
171	pationic 138C (sodium lauryl lactylate)	CCCCCCCCCCCC(OC(C(OC(C(O)=O)C)=O)C)=O	13557-75-0
172	penicillin G	O=C(NC(C(=O)N1C(C(=O)O)C(S2)(C)C)C12)Cc(cccc3)c3	61-33-6
173	pentachlorophenol	Oc(c(c(c1Cl)Cl)Cl)c1Cl	87-86-5
174	perillaldehyde	O=CC(=CCC(C(=C)C)C1)C1	2111-75-3
175	phenylacetaldehyde	O=CCc(cccc1)c1	122-78-1
176	phenyl benzoate	O=C(Oc(cccc1)c1)c(cccc2)c2	93-99-2
177	1,4-phenylenediamine	Nc(ccc(N)c1)c1	106-50-3
178	3-phenylenediamine	Nc(cccc1N)c1	108-45-2
179	1-phenyl-2-methylbutane-1,3-dione	c(ccc1C(=O)C(C(=O)C)C)cc1	6668-24-2
180	1-phenyloctane-1,3-dione	CCCCC(=O)CC(=O)c1ccccc1	no CAS
181	1-phenyl-1,2-propanedione	O=C(c(cccc1)c1)C(=O)C	579-07-7
182	potassium dichromate	O[Cr](=O)(O[Cr]((=O)(O)=O)=O	7778-50-9
183	beta-propriolactone	O=C(OC1)C1	57-57-8
184	propylene glycol	OCC(O)C	57-55-6
185	3-propylidenephthalide	CC=C1/OC(=O)c2ccccc12	17369-59-4
186	propylparaben	O=C(OCCC)c(ccc(O)c1)c1	94-13-3
187	pyridine	n(cccc1)c1	110-86-1
188	resorcinol	Oc(cccc1O)c1	108-46-3
189	saccharin	O=C(NS(=O)(=O)c1cccc2)c12	81-07-2
190	salicylic acid	O=C(O)c(c(O)ccc1)c1	69-72-7
191	sodium lauryl sulfate	CCCCCCCCCCCCOS(=O)(O)=O	151-21-3

No.	Name	SMILES	CAS
192	sodium 3,5,5-trimethylhexanoyloxybenzenesulfonate	CC(CC(C)(C)C)CC(Oc1c(S(O)(=O)=O)ccc1)=O	94612-91-6
193	streptomycin sulfate	Long SMILES – refer to ChemIDPlus	3810-74-0
194	sulfanilamide	O=S(=O)(N)c(ccc(N)c1)c1	63-74-1
195	sulfanilic acid	O=S(=O)(O)c(ccc(N)c1)c1	121-57-3
196	tartaric acid	O=C(O)C(O)C(O)C(=O)O	87-69-4
197	tetrachlorosalicylamilide	O=C(Nc(ccc(c1Cl)Cl)c1)c(c(O)c(cc2Cl)Cl)c2	1154-59-2
198	2,2,6-tetramethyl-heptane-3,5-dione	O=C(C)(O)CC(=O)C(C)(C)C	1118-71-4
199	1-(2′,3′4′5′-trimethylphenyl)-3-(4′-*tert*-butyl-phenyl)propane-1,3-dione	Cc1cc(c(c1C)C)C)C(=O)CC(=O)c1ccc(cc1)C(C)(C)C	55846-68-9
200	1-(2′,3′,4′,5′-tetramethylphenyl)butane-1,3-dione	CC(=O)CC(=O)c1cc(c(c1C)C)C)C	167998-73-4
201	tetramethylthiuram disulfide	N(C(=S)SSC(N(C)C)=S)(C)C	137-26-8
202	2,4,6-trichloro-1,3,5-triazine (cyanuric chloride)	n(c(nc(n1)Cl)Cl)c1Cl	108-77-0
203	4,4-trifluro-1-phenylbutane-1,3-dione	O=C(c(cccc1)c1)CC(=O)C(F)(F)F	326-06-7
204	1-(3′,4′,5′-trimethoxyphenyl)-4-dimethyl-pentane-1,3-dione	COc1cc(cc(c1OC)OC)C(=O)CC(=O)C(C)(C)C	135099-98-8
205	1,1,3-trimethyl-2-formylcyclohexa-2,1-diene (safranal)	O=CC(=C(C=CC1)C)C1(C)C	116-26-7
206	3,5,5-trimethylhexanoyl chloride	O=C(CC(C)(C)CC(C)C)Cl	36727-29-4
207	undec-10-enal	O=CCCCCCCCCC=C	112-45-8
208	vanillin	O=Cc(ccc(O)c1OC)c1	121-33-5
209	vinylidene dichloride	C(=C)(Cl)Cl	75-35-4
210	vinyl pyridine	n(ccc(c1)C=C)c1	100-43-6

Appendix 2.2   Skin sensitisation data and descriptors utilised in Chapter 10.

ID^a	$LLNA$ $EC_3^{b,c}$	$LLNA$ $Class^d$	$LLNA$ binary $class^e$	E_{LUMO}^f	Molecular $mass^g$	$Volume^h$
1	15	Weak	0	0.4457	302.5	250.3
2	NC	Non	0	0.5810	140.2	112.7
3	18	Weak	0	0.4858	148.2	115.9
4	2.2	Moderate	1	− 1.1670	188.6	111.7
5	2.2	Moderate	1	− 0.7235	197.2	132.2
6	0.4	Strong	1	0.4741	109.1	75.2
7	3.2	Moderate	1	0.5411	109.1	75.3
8	11	Weak	0	− 0.1710	202.3	170.8
9	37	Weak	0	0.9023	196.4	174.4
10	2.3	Moderate	1	0.0498	148.2	111.3
11	89	Weak	0	0.6394	93.1	68.5
12	NC	Non	0	− 0.4349	106.1	75.1
13	9.2	Moderate	1	− 1.9946	192.1	113.7
14	2.3	Moderate	1	− 0.4922	151.2	93.6
15	0.0009	Extreme	1	− 1.1107	252.3	164.6
16	NC	Non	0	− 0.0170	165.2	115.7
17	0.0099	Extreme	1	− 1.7351	108.1	71.6
18	17	Weak	0	− 0.3540	212.3	162.1
19	0.2	Strong	1	− 0.0500	171.0	94.8
20	3.7	Moderate	1	− 0.6314	146.2	108.0
21	NC	Non	0	− 0.3523	280.4	211.7
22	1.5	Moderate	1	0.3641	340.5	270.0
23	NC	Non	0	0.8284	137.0	80.5
24	8.3	Moderate	1	0.8251	389.6	314.0
25	18	Weak	0	0.8259	249.3	184.3
26	18	Weak	0	0.8021	279.3	194.1
27	6.9	Moderate	1	0.8193	265.3	191.3
28	6.1	Moderate	1	0.8248	361.5	287.7
29	4.8	Moderate	1	0.8251	319.4	248.9
30	2.3	Moderate	1	0.8257	305.4	236.1
31	10	Weak	0	0.8277	165.1	106.3
32	3.6	Moderate	1	0.4559	207.1	126.6
33	NC	Non	0	0.8266	207.2	145.2
34	15	Weak	0	0.8251	333.5	262.1
35	5.1	Moderate	1	0.8248	291.4	223.1
36	9.2	Moderate	1	0.8252	277.3	209.9
37	21	Weak	0	0.7743	277.3	213.2
38	3.4	Moderate	1	− 0.2067	307.3	223.4
39	10	Weak	0	0.8253	263.3	196.9
40	20	Weak	0	0.8261	235.2	171.3
41	11	Weak	0	− 0.0870	86.1	67.8
42	NC	Non	0	3.4238	74.1	66.7
43	11	Weak	0	− 0.1767	188.3	157.4
44	19	Weak	0	0.3120	204.3	167.3
45	31	Weak	0	2.3902	130.2	112.0
46	1.8	Moderate	1	0.2726	169.2	138.8
47	1.3	Moderate	1	0.2703	197.3	166.4
48	2.8	Moderate	1	0.2702	239.4	206.1
49	16	Weak	0	0.2699	267.5	232.5

Appendix 2.2 (*continued*)

ID^a	LLNA $EC_3^{b,c}$	LLNA $Class^d$	LLNA binary $class^e$	E_{LUMO}^f	Molecular $mass^g$	$Volume^h$
50	18	Weak	0	0.2696	323.6	285.7
51	19	Weak	0	0.2693	351.6	312.2
52	26	Weak	0	0.2696	379.7	339.0
53	10	Weak	0	− 0.2679	166.2	134.5
54	NC	Non	0	0.1547	112.6	74.9
55	0.05	Extreme	1	− 2.1178	202.6	115.8
56	9.1	Moderate	1	1.5059	260.9	231.3
57	0.009	Extreme	1	− 0.5539	149.6	91.4
58	0.005	Extreme	1	− 1.0794	250.7	164.4
59	NC	Non	0	1.5073	162.7	140.9
60	16	Weak	0	1.5060	289.0	257.4
61	20	Weak	0	1.5063	232.9	205.5
62	21	Weak	0	0.0165	134.2	98.5
63	3	Moderate	1	− 0.7757	132.2	94.3
64	13	Weak	0	− 0.1505	152.3	132.8
65	4.8	Moderate	1	− 0.1834	344.9	259.1
66	NC	Non	0	− 0.9305	146.1	94.1
67	22	Weak	0	0.2940	190.3	154.1
68	2.5	Moderate	1	− 0.1152	154.3	139.2
69	0.9	Strong	1	− 0.7039	265.9	135.0
70	76	Weak	0	0.9397	100.2	91.8
71	5.8	Moderate	1	2.8183	103.2	87.8
72	5.8	Moderate	1	− 0.4989	172.2	131.8
73	9.6	Moderate	1	− 0.0883	218.3	174.4
74	NC	Non	0	− 0.6600	222.3	162.3
75	3.3	Moderate	1	− 0.6542	154.2	106.7
76	5.6	Moderate	1	− 0.0424	148.2	100.6
77	6.8	Moderate	1	0.4249	166.2	125.3
78	0.11	Strong	1	0.2331	110.1	72.5
79	2.2	Moderate	1	2.9459	102.2	92.6
80	0.006	Extreme	1	− 0.8200	256.4	187.6
81	1.8	Moderate	1	− 0.0381	126.2	103.2
82	13	Weak	0	− 0.3598	190.3	146.2
83	0.19	Strong	1	− 0.7398	126.1	80.2
84	72	Weak	0	0.8056	78.1	58.1
85	8.8	Moderate	1	− 0.7027	264.5	215.0
86	0.003	Extreme	1	− 0.8901	217.2	146.4
87	33	Weak	0	− 0.1321	248.3	192.8
88	28	Weak	0	0.0415	100.1	76.5
89	NC	Non	0	− 0.5464	192.2	140.3
90	2.2	Moderate	1	3.2876	60.1	52.4
91	28	Weak	0	0.0146	198.2	152.4
92	0.6	Strong	1	− 0.5671	271.4	196.0
93	1.1	Moderate	1	0.0488	117.1	80.9
94	NC	Non	0	− 0.4526	166.2	114.7
95	13	Weak	0	0.3930	164.2	122.3
96	12	Weak	0	− 0.1567	220.4	197.0
97	0.14	Strong	1	− 1.1800	389.4	274.6
98	0.61	Strong	1	0.7932	30.0	23.8

Appendix 2.2 *(continued)*

ID^a	$LLNA$ $EC_3^{b,c}$	$LLNA$ $Class^d$	$LLNA$ binary $class^e$	E_{LUMO}^f	Molecular $mass^g$	$Volume^h$
99	NC	Non	0	− 0.6704	190.2	131.2
100	26	Weak	0	1.0713	154.3	136.7
101	0.1	Strong	1	0.6330	100.1	78.8
102	NC	Non	0	3.0270	92.1	68.2
103	1.4	Moderate	1	− 0.7622	58.0	40.2
104	4	Moderate	1	− 0.6011	110.2	88.9
105	NC	Non	0	3.7357	86.2	85.7
106	5.5	Moderate	1	− 0.1152	98.2	86.8
107	11	Weak	0	− 0.1798	216.3	183.9
108	NC	Non	0	− 0.4808	138.1	88.3
109	33	Weak	0	0.8862	172.3	150.0
110	1.4	Moderate	1	− 0.1019	116.1	84.8
111	17	Weak	0	0.8545	210.3	181.2
112	NC	Non	0	0.1032	144.2	117.3
113	24	Weak	0	− 0.2210	388.4	265.0
114	13	Weak	0	0.4611	296.3	190.2
115	19	Weak	0	0.4606	352.4	242.1
116	NC	Non	0	0.4631	212.1	112.5
117	24	Weak	0	0.4617	254.2	151.3
118	NC	Non ·	0	0.4605	380.5	267.9
119	14	Weak	0	0.4608	324.3	216.2
120	1.2	Moderate	1	− 0.0403	164.2	118.1
121	2.7	Moderate	1	0.2474	176.7	147.3
122	NC	Non	0	3.5754	60.1	54.4
123	NC	Non	0	0.2474	192.3	150.6
124	0.6	Strong	1	− 0.0170	192.3	148.2
125	44	Weak	0	1.2386	270.5	240.1
126	NC	Non	0	1.9745	484.6	351.2
127	NC	Non	0	0.8208	90.1	66.0
128	0.3	Strong	1	− 0.7443	338.5	258.6
129	69	Weak	0	1.1952	136.3	122.5
130	30	Weak	0	1.1861	154.3	136.5
131	1.7	Moderate	1	− 0.6653	167.2	104.5
132	NC	Non	0	− 0.3092	150.2	107.7
133	5.8	Moderate	1	0.4219	138.2	98.8
134	9.3	Moderate	1	− 0.5770	190.3	140.3
135	0.8	Strong	1	0.4668	123.2	87.8
136	0.7	Strong	1	− 0.8490	161.2	105.2
137	4.5	Moderate	1	− 0.5794	146.2	112.8
138	NC	Non	0	− 0.8922	160.2	106.9
139	0.39	Strong	1	− 0.7362	264.5	214.8
140	0.2	Strong	1	0.6081	122.2	88.9
141	32	Weak	0	0.2225	178.2	137.1
142	13	Weak	0	0.4076	178.2	135.6
143	17	Weak	0	0.3733	178.2	135.1
144	0.8	Strong	1	− 0.5866	318.6	262.0
145	NC	Non	0	− 0.7362	320.6	266.7
146	26	Weak	0	0.0088	128.2	107.4
147	14	Weak	0	0.2533	148.2	117.7

Appendix 2.2 (*continued*)

ID^a	LLNA $EC_3^{b,c}$	LLNA $Class^d$	LLNA binary $class^e$	E_{LUMO}^f	Molecular $mass^g$	$Volume^h$
148	NC	Non	0	− 0.3972	152.2	101.5
149	0.4	Strong	1	0.5847	167.2	121.3
150	3.6	Moderate	1	0.0378	178.2	135.4
151	1.6	Moderate	1	0.0354	178.2	132.7
152	1.9	Moderate	1	− 0.2086	115.2	75.7
153	8.1	Moderate	1	− 0.7303	110.1	73.0
154	0.03	Extreme	1	− 0.9594	147.1	90.0
155	0.05	Extreme	1	0.1592	103.1	67.5
156	2.5	Moderate	1	0.1729	168.3	139.2
157	6.3	Moderate	1	0.1891	134.2	103.3
158	1.4	Moderate	1	− 1.3738	208.2	137.4
159	NC	Non	0	− 0.4967	152.2	101.2
160	2	Moderate	1	− 0.9065	454.7	369.1
161	10	Weak	0	0.8996	184.4	170.1
162	1.3	Moderate	1	− 0.2474	144.2	97.0
163	0.05	Extreme	1	− 1.3526	216.0	114.2
164	0.4	Strong	1	− 0.6840	153.2	95.0
165	1.8	Moderate	1	0.2529	176.7	143.6
166	23	Weak	0	0.8601	140.2	125.1
167	NC	Non	0	1.0281	144.2	121.3
168	25	Weak	0	− 0.6975	346.6	292.0
169	15	Weak	0	− 0.6657	90.0	53.4
170	8.8	Moderate	1	0.2522	274.9	234.6
171	15	Weak	0	0.8139	344.5	283.6
172	30	Weak	0	− 0.0346	334.4	247.7
173	20	Weak	0	− 0.9773	266.3	142.8
174	8.1	Moderate	1	− 0.0823	150.2	127.3
175	3	Moderate	1	0.1601	120.2	90.7
176	20	Weak	0	− 0.4811	198.2	145.6
177	0.16	Strong	1	0.6410	108.2	76.8
178	0.49	Strong	1	0.7584	108.2	77.3
179	29	Weak	0	− 0.5217	176.2	134.2
180	11	Weak	0	− 0.4727	218.3	173.9
181	1.3	Moderate	1	− 0.6420	148.2	106.8
182	0.08	Extreme	1		218.0	95.5
183	0.15	Strong	1	0.9133	72.1	52.2
184	NC	Non	0	3.1698	76.1	61.7
185	3.7	Moderate	1	− 0.9202	174.2	121.1
186	NC	Non	0	− 0.3675	180.2	127.6
187	72	Weak	0	0.1381	79.1	57.4
188	NC	Non	0	0.3208	110.1	72.7
189	NC	Non	0	− 1.2894	183.2	107.5
190	NC	Non	0	− 0.5912	138.1	87.9
191	14	Weak	0	− 0.5797	266.4	209.5
192	6.4	Moderate	1	− 0.8322	314.4	239.5
193						
194	NC	Non	0	− 0.3260	172.2	107.3
195	NC	Non	0	− 0.5580	173.2	105.5
196	8.7	Moderate	1	0.3118	150.1	99.6

Appendix 2.2 *(continued)*

ID^a	LLNA $EC_3^{b,c}$	LLNA $Class^d$	LLNA binary $class^e$	E_{LUMO}^f	Molecular $mass^g$	$Volume^h$
197	0.04	Extreme	1	−0.9967	351.0	217.5
198	27	Weak	0	0.5165	184.3	159.7
199	NC	Non	0	−0.4223	336.5	272.2
200	8.3	Moderate	1	−0.1734	218.3	173.6
201	5.2	Moderate	1	−1.9473	240.4	171.1
202	0.09	Extreme	1	−1.3118	184.4	102.0
203	20	Weak	0	−0.7748	216.2	142.1
204	NC	Non	0	−0.4152	294.4	231.2
205	7.5	Moderate	1	−0.2999	150.2	125.9
206	2.7	Moderate	1	0.2874	176.7	148.3
207	6.8	Moderate	1	0.8605	168.3	150.3
208	NC	Non	0	−0.4781	152.2	101.9
209	NC	Non	0	0.3792	96.9	59.0
210	1.6	Moderate	1	−0.2666	105.2	76.6

[a]ID numbers refer to Appendix 2.1.
[b]The EC3 value is the percentage concentration (by weight) of a test chemical that stimulates the lymph nodes in excess of three times that observed for a vehicle treated group.
[c]NC indicates that sensitisation was not detected at any of the concentrations tested.
[d]Classification of skin sensitisation response.
[e]Arbitrary binary classification of skin sensitisation (non and weak = 0; moderate and strong = 1).
[f]Energy of the lowest unoccupied molecular orbital (eV).
[g]Dalton.
[h]$(Å)^3$.

Subject Index